FLIGHT MANUAL

USAF SERIES
F-84F-25 and Later Aircraft

PUBLISHED UNDER AUTHORITY OF THE SECRETARY OF THE AIR FORCE

Commanders are responsible for bringing this publication to the attention of all air force personnel cleared for operation of subject aircraft.

See Weekly Index, T.O. 0-1-1A for current status of Safety of Flight Supplements.

CHANGE NOTICE

LATEST CHANGED PAGES SUPERSEDE
THE SAME PAGES OF PREVIOUS DATE

Insert changed pages into basic publication. Destroy superseded pages.

5354 AF BAFB ALA 8/7/61 3700

15 FEBRUARY 1958
CHANGED 11 JULY 1961

INSERT LATEST CHANGED PAGES. DESTROY SUPERSEDED PAGES.

LIST OF EFFECTIVE PAGES

NOTE: The portion of the text affected by the changes is indicated by a vertical line in the outer margins of the page.

TOTAL NUMBER OF PAGES IN THIS PUBLICATION IS 520 CONSISTING OF THE FOLLOWING:

Page No.	Issue
*Title	11 Jul 1961
*A	11 Jul 1961
B	15 Mar 1961
C Blank	30 Dec 1960
i	Original
ii thru iiA	30 May 1960
iiB Blank	30 May 1960
iii	30 May 1960
iv	Original
1-1 thru 1-2	30 May 1960
1-3	Original
1-4	30 May 1960
1-5 thru 1-10	Original
1-11 thru 1-12	30 May 1960
1-13	Original
1-14	30 May 1960
1-15 thru 1-16	Original
1-16A	30 Jun 1958
1-16B Blank	30 Jun 1958
1-17	30 Jun 1958
1-18 thru 1-20	Original
1-21 thru 1-22	30 May 1960
1-23	Original
1-24	30 Jun 1958
1-25	Original
1-26 thru 1-29	30 May 1960
1-30	Original
1-31 thru 1-35	30 May 1960
1-36 thru 1-38	Original
1-39	30 May 1960
1-40 thru 1-42	Original
1-43 thru 1-45	30 May 1960
1-46 thru 1-48	Original
1-49	30 May 1960
1-50 thru 1-51	Original
1-52	30 Jun 1958
1-53	30 May 1960
1-54 thru 1-56	30 Jun 1958
1-57 thru 1-58	Original
1-59	30 Jun 1958
1-60	30 May 1960
1-61	Original
1-62 thru 1-64A	30 May 1960
1-64B Blank	30 May 1960
1-65	30 Jun 1958
1-66	Original
1-66A	30 May 1960
1-66B Blank	30 May 1960
1-67 thru 1-68	30 May 1960
1-69 thru 1-70	Original
1-71 thru 1-76	30 May 1960
2-1 thru 2-2	30 May 1960
2-3 thru 2-5	15 Mar 1961
2-6 thru 2-8	30 May 1960
2-10	15 Mar 1961
2-10A Added	15 Mar 1961
2-10B Blank	15 Mar 1961
2-11	30 May 1960
2-12	15 Mar 1961
2-13	30 May 1960
2-14	15 Mar 1961
2-15	30 May 1960
2-16	Original
2-17	30 May 1960
2-18	15 Mar 1961
2-19	30 May 1960
2-20 thru 2-21	15 Mar 1961
2-22 thru 2-24	30 May 1960
2-25 thru 2-38 Deleted	30 May 1960
3-1 thru 3-10E	30 May 1960
3-10F Blank	30 May 1960
3-11 thru 3-12	30 May 1960
3-13 thru 3-15	Original
3-16 thru 3-18	30 May 1960
3-19	Original
3-20	30 Dec 1960
3-21 thru 3-26	Original
3-27 thru 3-28	30 May 1960
3-29 thru 3-42 Deleted	30 May 1960
4-1	30 May 1960
4-2 thru 4-5	Original
4-6	15 Mar 1961
4-7	Original
4-8 thru 4-10A	30 May 1960
4-10B Blank	30 May 1960
4-11 thru 4-12B	30 May 1960
4-13 thru 4-18	30 May 1960
4-19 thru 4-20	Original
4-20A thru 4-20B	30 May 1960
4-21	30 May 1960
4-22	Original
4-23	15 Mar 1961
4-24	30 May 1960
4-25	15 Mar 1961
4-26 thru 4-26A	30 May 1960
4-26B Blank	30 May 1960
4-27 thru 4-28	Original
4-29	30 Jun 1958
4-30 thru 4-36	Original
4-37	30 Jun 1958
4-38	30 May 1960
4-38A	30 Jun 1958
4-38B Blank	30 Jun 1958
4-39	30 Jun 1958
4-40	Original
4-41	30 Jun 1958
4-42	15 Mar 1961
*5-1	11 Jul 1961
5-2 thru 5-4B	30 May 1960
5-5 thru 5-6	30 May 1960
5-7	Original
5-8	30 May 1960
5-9	30 May 1960
5-10	Original
6-1	Original
6-2 thru 6-4	30 May 1960
6-5	Original
6-6	30 Jun 1958
6-7 thru 6-8	Original
6-9	30 May 1960
6-10	Original
6-11	30 May 1960
6-12 Blank	30 May 1960
7-1	30 May 1960
7-2 thru 7-4	30 May 1960
7-5	15 Mar 1961
7-6	30 May 1960
7-7	Original
7-8 thru 7-8A	30 May 1960
7-8B Blank	30 May 1960
7-9	Original
7-10 thru 7-12	30 May 1960
7-13 thru 7-18 Deleted	30 May 1960
9-1	30 May 1960
9-2	Original

REPUBLIC F-84 THUNDERJET PILOT'S FLIGHT OPERATING INSTRUCTIONS

ISBN 978-1-4303-1044-0

ADDITIONAL COPIES OF THIS PUBLICATION MAY BE OBTAINE AS FOLLOWS: USAF

USAF ACTIVITIES. – In accordance with T.O. 00-5-2.

NAVY ACTIVITIES. – Use Publications and Forms Order Blank (NavWeps 140) and submit in accordance with instruction thereon. For listing of available material and details of distribution see Naval Aeronautics Publications Index NavWeps 00-500.

TABLE OF CONTENTS

IMPORTANT

In order that you will gain the maximum benefits from this manual it is imperative that you read this page and the following page carefully.

This Flight Manual is applicable to F-84F-25 and subsequent airplanes designed by Republic Aviation Corporation and manufactured by both Republic Aviation and General Motors Corporations. Unless otherwise specified, information is applicable to all airplanes. However, where information applies to specific airplanes, these airplanes will be identified by code letters which appear at the top right corner of the paragraph. The code letters assigned to aricraft series are as follows:

BLOCK DESIGNATION CODE

Code	*Model*	*Ser. Nos.*
A	F-84F-25RE	51-1621 thru 51-1760
	-25GK	51-9357 thru 51-9409
B	-30RE	51-1761 thru 51-1827
	-30GK	51-9410 thru 51-9454
	-30RE	52-6355 thru 52-6422
C	-35RE	51-17061 thru 51-17088
	-35GK	51-9455 thru 51-9503
	-35RE	52-6423 thru 52-6522
		52-7018 thru 52-7049
D	-40GK	51-9504 thru 51-9547
	-40RE	52-6523 thru 52-6642
		52-7050 thru 52-7089
	-41GK	52-8767 thru 52-8834
E	-45RE	52-6643 thru 52-6812
		52-7090 thru 52-7114
	-46RE	52-7115 thru 52-7126
	-46GK	52-8835 thru 52-8982
F	-50RE	52-6813 thru 52-6907
	-51RE	52-7127 thru 52-7191
	-51GK	52-8983 thru 52-9128
G	-55RE	52-6908 thru 52-7007
	-56RE	52-7192 thru 52-7228
		52-10510 thru 52-10538
		52-7008 thru 52-7017
H	-61RE	53-6532 thru 53-6715
J	-66RE	53-6716 thru 53-6835
K	-71RE	53-6836 thru 53-6955
	-76RE	53-6956 thru 53-7075
	-81RE	53-7076 thru 53-7230

NOTE: The suffix letters RE identify airplanes manufactured by Republic Aviation Corporation while the suffix letters GK identify airplanes manufactured by General Motors Corporation.

NOTE

This manual specifies product improvement changes by block numbers, i.e., "On F-84F-40RE and later aircraft, the speed brake switch has been incorporated in the throttle control." Many of these improvements are being effected on aircraft, prior to the production effective point of the modification by RETROFIT ACTION accomplished in accordance with Time Compliance Technical Orders. It is suggested that operating personnel become familiar with these technical orders, since the actual configuration of early aircraft may differ individually, depending on the status of Time Compliance Technical Orders compliance.

* *

SCOPE.

This manual contains all the information necessary for safe and efficient operation of the F-84F. These instructions do not teach basic flight principles, but are designed to provide you with a general knowledge of the aircraft, its flight characteristics, and specific normal and emergency operating procedures. Your flying experience is recognized, and elementary instructions have been avoided.

SOUND JUDGEMENT.

The instructions in this manual are designed to provide for the needs of a crew inexperienced in the operation of this aircraft. This book provides the best possible operating instructions under most circumstances, but it is a poor substitute for sound judgment. Multiple emergencies, adverse weather, terrain, etc., may require modification of the procedures contained herein.

PERMISSIBLE OPERATIONS.

The Flight Manual takes a "positive approach" and normally tells you only what you can do. Any unusual operation or configuration (such as asymmetrical loading) is prohibited unless specifically covered in the Flight Manual. Clearance must be obtained from ARDC before any questionable operation is attempted which is not specifically covered in the Flight Manual.

STANDARDIZATION.

Once you have learned to use one Flight Manual you will know how to use them all—closely guarded standardization assures that the scope and arrangement of all Flight Manuals are identical.

ARRANGEMENT.

The manual has been divided into 10 fairly independent sections, each with its own table of contents. The objective of this subdivision is to make it easy both to read the book straight through when it is first received and thereafter to use it as a reference manual. The independence of these sections also makes it possible for the user to rearrange the book to satisfy his personal taste and requirements. The first three sections cover the minimum information required to safely get the aircraft into the air and back down again. Before flying any new aircraft these three sections must be read thoroughly and fully understood. Section IV covers all equipment not essential to flight but which permits the aircraft to perform special functions. Sections V and VI are obvious. Section VII covers lengthy discussions on any technique or theory of operation which may be applicable to the particular aircraft in question. The experienced pilot will probably be aware of the information in this section but he should check it for any possible new information. The contents of the remaining sections are fairly obvious.

YOUR RESPONSIBILITY.

These Flight Manuals are constantly maintained current through an extremely active revision program. Frequent conferences with operating personnel and constant review of UR's, accident reports, flight test reports, etc., assure inclusion of the latest data in these manuals. In this regard, it is essential that you do your part! If you find anything you don't like about the book, let us know right away. We cannot correct an error whose existence is unknown to us.

PERSONAL COPIES, TABS AND BINDERS.

In accordance with the provisions of AFR 5-13, flight crew members are entitled to have personal copies of the Flight Manuals. Flexible, loose leaf tabs and binders have been provided to hold your personal copy of the Flight Manual. These good-looking, simulated-leather binders will make it much easier for you to revise your manual as well as to keep it in good shape. These tabs and binders are secured through your local materiel staff and contracting officers.

HOW TO GET COPIES.

If you want to be sure of getting your manuals on time, order them before you need them. Early ordering will assure that enough copies are printed to cover your requirements. Technical Order 00-5-2 explains how to order Flight Manuals so that you automatically will get all changes, revisions, and Safety of Flight Supplements. Basically, all you have to do is order the required quantities in the Publication Requirements Table(T.O. 0-3-1). Talk to your Senior Materiel Staff Officer—it is his job to fulfill your Technical Order requests. Make sure to establish some system that will rapidly get the manuals and Safety of Flight Supplements to the flight crews once they are received on the base.

SAFETY OF FLIGHT SUPPLEMENTS.

Safety of Flight Supplements are used to get information to you in a hurry. Safety of Flight Supplements use the same number as your Flight Manual, except for the addition of a suffix letter. Supplements covering loss of life will get to you in 48 hours; those concerning serious damage to equipment will make it in 10 days. You can determine the status of Safety of Flight Supplements by referring to the Weekly Index (T.O. 0-1-1A). This is the only way you can determine whether a supplement has been rescinded. The title page of the Flight Manual and title block of each Safety of Flight Supplement should also be checked to determine the effect that these publications may have on existing Safety of Flight Supplements. It is critically important that you remain constantly aware of the status of all supplements—you must comply with all existing supplements but there is no point in restricting the operation of your aircraft by complying with a supplement that has been replaced or rescinded. If you have ordered your Flight Manual on the Publications Requirements Table, you automatically will receive all supplements pertaining to your aircraft. Technical Order 00-5-1 covers some additional information regarding these supplements.

WARNINGS, CAUTIONS, AND NOTES.

For your information, the following definitions apply to the "Warnings," "Cautions," and "Notes" found throughout the manual:

WARNING—Operating procedures, practices, etc., which will result in personal injury or loss of life if not carefully followed.

CAUTION—Operating procedures, practices, etc., which if not strictly observed will result in damage to equipment.

NOTE—An operating procedure, condition, etc., which it is essential to emphasize.

This page intentionally left blank.

MB-8 FLIGHT COMPUTER.

The MB-8 Flight Computer for this airplane will be available in the near future. This computer is designed to provide pilots with compact cruise control data which will aid in preparation of flight plans, inflight operation, and emergency inflight planning and operation. The computer is a five disc, metal and plastic circular computer with a canvas carrying case. Three of the discs can be used with any airplane and are referred to as "standard discs." The remaining discs contain data only for this airplane and are described as "data discs." The standard discs and carrying case are carried in Class 05-A and are available through normal supply channels. The data discs are distributed automatically to all bases having this airplane. New or revised discs are issued each time the performance data in the Flight Manual is revised. The performance data in the computer and the manual is always kept current and consistent. If you have not yet received your computer, see your Base Operations Officer or T.O. 5F5-1-1. Reference should also be made to T.O. 5F5-1-1 and the appendix of this manual for information on the operation of the computer.

CHECK LIST.

The Flight Manual now contains only amplified check lists; the condensed check lists have been issued as a separate cardboard technical order. For the T.O. number and date of the check list applicable to this Manual see the back of the title page. Order your check list as you would any technical orders. Line items in the Flight Manual and applicable cardboard check lists are identical as pertains to arrangement and item number. The cardboard check lists are designed for use with binders having plastic envelopes into which the individual cards are placed. These binders are available through normal Air Force supply channels.

COMMENTS AND QUESTIONS.

Comments and questions regarding any phase of the Flight Manual program are invited and should be forwarded through your Command Headquarters to F-84 Weapons System, Brookley AFB, Mobile, Alabama, ATTN: MONW.

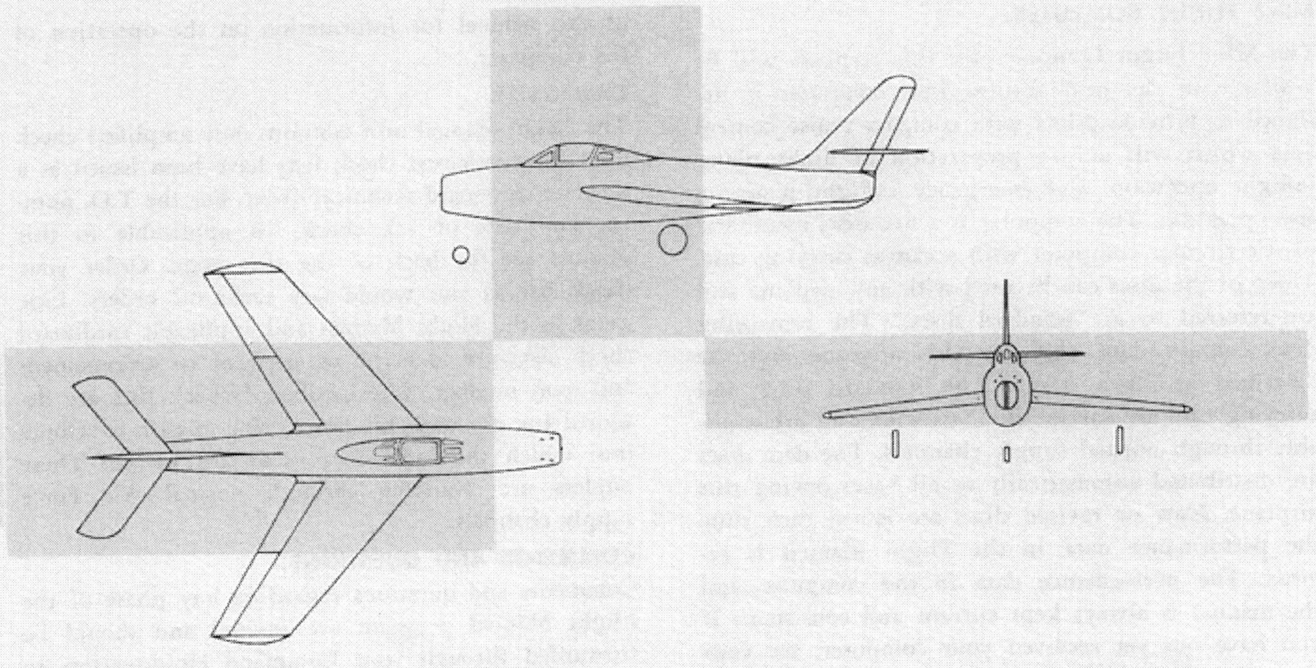

F-84F-25
& LATER AIRCRAFT

Figure 1—1

SECTION 1 DESCRIPTION

TABLE OF CONTENTS

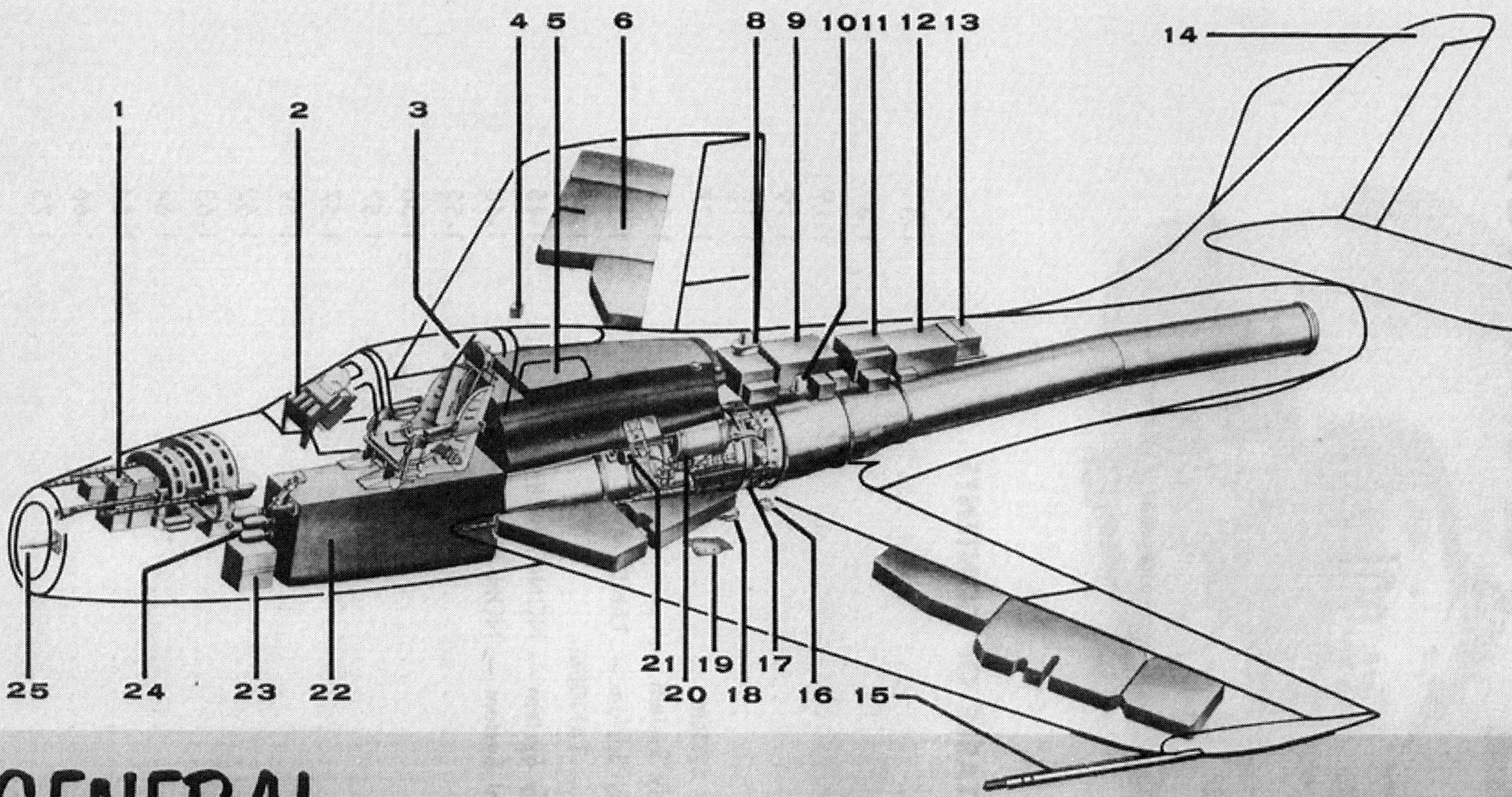

GENERAL ARRANGEMENT DIAGRAM

1 *Radar Equipment*
2 *Gun-Bomb-Rocket Sight*
3 *Ground Refueling Receptacle (C and LATER)*
4 *External Power Receptacle*
5 *Main Fuel Tank*
6 *Wing Fuel Tank*
7 *(Deleted)*
8 *Radio Compass Loop Antenna*
9 *Command Radio Equipment*
10 *Directional Indicator Equipment*
11 *Identification Radio Equipment*
12 *Auto-Pilot Equipment*
13 *APW-11 Radar Equipment*
14 *Command Radio Antenna*
15 *Pitot Tube (J and LATER)*
16 *Radar Antenna*
17 *Jet Engine*
18 *Radio Compass Antenna*
19 *Air Refueling Receptacle (Air and Ground A and B)*
20 *Oil Tank*
21 *Hydraulic Reservoir*
22 *Forward Fuel Tank*
23 *Batteries*
24 *Liquid oxygen converter*
25 *Pitot Tube (A thru H)*

AIRCRAFT.

The Republic Aviation Corporation F-84F-25 and later is a single place, swept-back fighter designed for flight in the subsonic and sonic speed ranges at high altitude. It is powered by a high thrust turbojet engine and lends itself readily to fighter-bomber and long range characteristics. It can carry exceptionally large external loads and yet retain its high speed performance and stability. Outstanding design features are air refueling, automatic pilot, pneumatic-type emergency landing gear system and hydraulic actuator controls. Pilot comfort is assured by providing greater quantities of air at comfortable temperatures and cabin pressurization at lower effective altitude. Visibility is improved by use of a double wall canopy and a dry air circulating system for anti-fogging purposes and a flat front windshield.

AIRCRAFT GROSS WEIGHT.

The gross weight of the aircraft averages from 18,500 LB for the clean aircraft to 25,380 LB for the clean aircraft plus four external tanks.

AIRCRAFT DIMENSIONS.

Approximate overall dimensions of the aircraft are as follows:

Length (Incl stabilizers)	43.3 FT
Wing Span	33.6 FT
Height (to top of fin)	
A thru C	14.5 FT
D and later	15.00 FT
Tread	20.4 FT
Ground Clearance	
Stabilator tip	8.4 FT
Wing tip	3.7 FT
Pylon tank	1.0 FT

ENGINE.

The aircraft is powered by an axial flow jet propulsion engine, AF Model J65-W-3, B-3; W-7 or B-7. The J65-W-3 and B-3 engines are rated at 7200 pounds thrust while the J65-W-7 and B-7 are rated at 7800 pounds thrust. Air enters an intake in the nose of the aircraft and compressed progressively through a 13 stage compressor section of the engine and discharged through a two stage turbine. An ignition and a fuel priming system is provided for starting the engine. The air inlet to the engine compressor is provided with a retractable screen to prevent foreign objects from entering the compressor during ground operation. An auxiliary air inlet door is provided on each side of the forward fuselage to allow more air to enter the annular duct when the need arises. These doors open automatically during high speed engine operation at sea level and close automatically when the demand for air is reduced. Supplementary thrust for take-off is supplied by externally hung ato units.

ENGINE FUEL CONTROL SYSTEM.

The engine fuel control system (figure 1—12) is designed to automatically maintain constant RPM during normal operation; however, during emergency operation, all compensating fuel circuits are by-passed, and the system then does not automatically maintain constant RPM. The engine fuel control system consists mainly of an engine-driven, booster pump, and a constant-output, dual element fuel pump and a regulator. The failure of either element of the dual fuel pump will not affect the operation of the remaining pump and sufficient fuel will be available from the single pump for 100 per cent RPM operation at sea level on a 37.8°C (100°F) day. The engine driven booster pump provides a head of pressure of approximately 50 PSI for the dual element fuel pump. The emergency fuel system by-passes the temperature and altitude compensating provisions in the fuel control and must be manually selected.

Note

Emergency operation is possible only when electrical power is available from the primary bus, since its selection is dependent on solenoid valves.

Main Fuel Control.

The fuel control is basically an engine speed governing control which by-passes main fuel pump output when necessary so as to maintain a fixed engine speed for a given throttle control position. The control prevents overspeeding beyond the maximum governed speed, preserves constant engine speed by compensating for changes in air density with changes in altitude, provides increased fuel during acceleration at such a rate as to increase the acceleration while avoiding surge and preventing excessive exhaust gas temperatures, limits fuel flow during engine deceleration in such a manner as to increase deceleration rate while avoiding flame-outs, and provides a means for selection of emergency operation whenever a failure occurs in the main fuel control system. As engine RPM tends to exceed prescribed limits the speed governor metering valve acts to restrict further fuel flow. A metering valve is also actuated by altitude and air temperature sensing bellows to provide correct fuel flow for these variables. A dual fuel pump by-pass valve which permits excessive fuel to return to the inlet side of the fuel pump is also provided. This by-pass valve operates from two sources. From one source, it protects against excessive compressor pressure rise by sensing inlet and outlet compressor pres-

MAIN DIFFERENCE TABLE

	F-84F-25 and later
Engine	*J65-W-3, J65-B-3, J-65-W-7 or J-65-B-7*
Starter	*COMBUSTION*
Fuel System	*Manual Fuel Tank Selector* *Air Refueling* *Single Point Ground Refueling C & LATER* *Emergency Fuel System*
Air Intake	*Fuselage Nose*
Wing and Empennage	*Swept Back Wing and Stabilator* *Modified and J & LATER — Drag Chute*
Flight Controls	*Conventional Aileron and Rudder With Hyd Boost* *Modified and K & LATER — One Piece Stabilator With Tandem Hydraulic Actuator* *Modified and E & LATER — Spoilers*
Trim	*Adjustable Spring Capsules*
Canopy	*Clamshell Jettisonable*
Gun Sight	*A-4 With APG-30*
Command Radio	*A thru G — AN/ARC-33* *H and Later — AN/ARC-34*
Inverters	*Unmodified A & B — 1500VA* *Modified and D & LATER — 2250VA*
Speed Brake	*Unmodified A thru C — Two Position* *Modified and D & LATER—Multi-position*
Eng Screens	*A thru G — Manual Control* *H & LATER — Automatic Control*
Ejection Seat	*A thru D — Right Hand Jettison Control, Manual Adjustment* *E & LATER — Left or Right Hand Jettison Control, Electric Height Adjustment*

Figure 1—3

INSTRUMENT PANEL A thru C

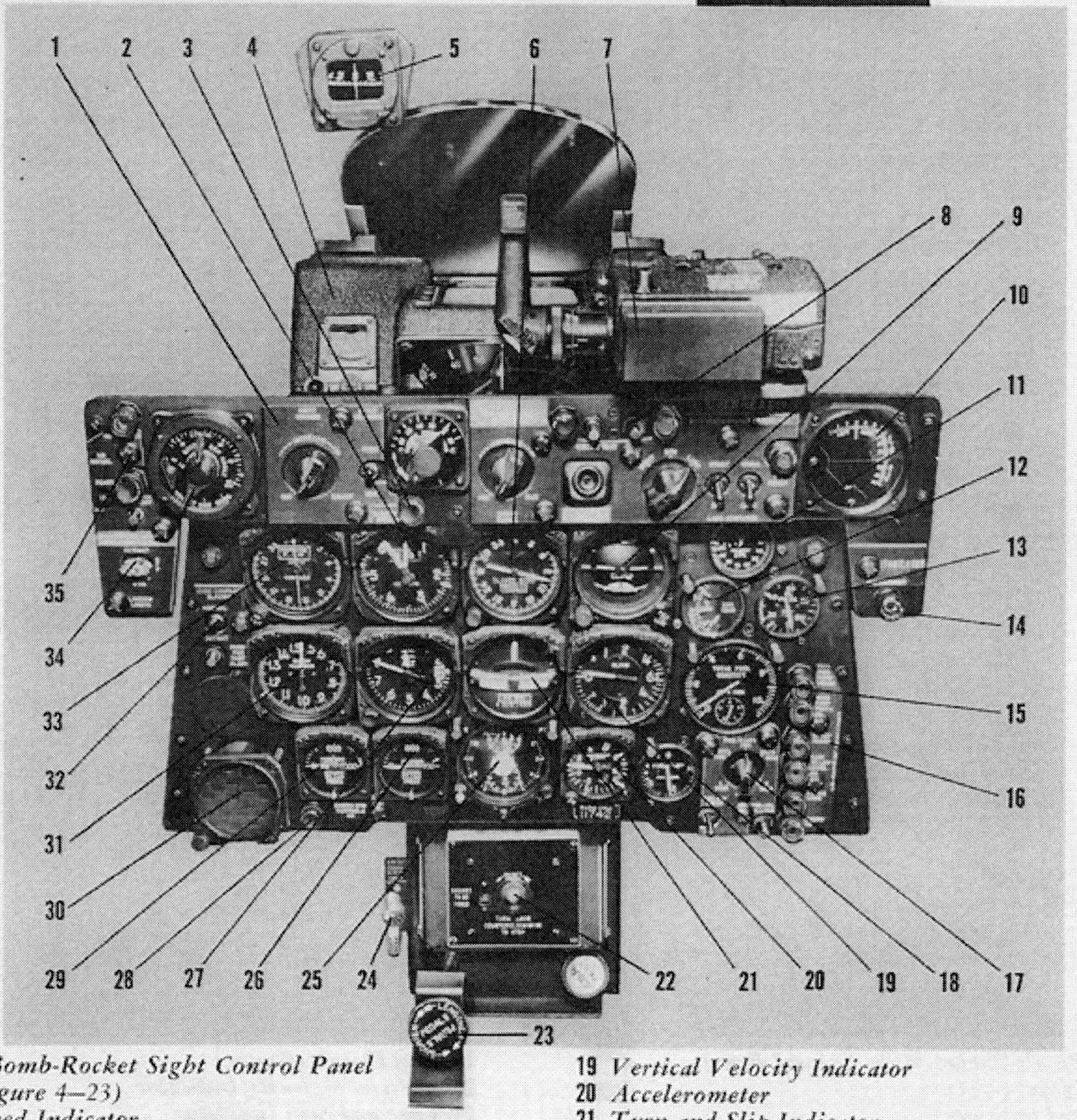

1 *Gun-Bomb-Rocket Sight Control Panel (See figure 4—23)*
2 *Airspeed Indicator*
3 *Canopy Open Indicator Light*
4 *Gun-Bomb-Rocket Sight*
5 *Magnetic Compass*
6 *Directional Indicator*
7 *Gun Camera*
8 *Engine Overheat and Fire Warning System (See figure 1—36)*
9 *Attitude Indicator*
10 *Dive and Roll Indicator*
11 *Engine Tachometer*
12 *Engine Exhaust Gas Temperature Indicator*
13 *Engine Oil Pressure Indicator*
14 *Stabilator Mechanical Advantage Indicator*
15 *Fuel Quantity Indicator (See figure 1—21)*
16 *Fuel System Warning Lights (See figure 1—20)*
17 *Fuel Quantity Selector Switches (See figure 1—21)*
18 *Fuel Flow Indicator*
19 *Vertical Velocity Indicator*
20 *Accelerometer*
21 *Turn and Slip Indicator*
22 *Rocket Release Control (See figure 4—29)*
23 *Rudder Pedal Adjusting Knob*
24 *Drag Chute Control (See figure 1—35)*
25 *Clock*
26 *Hydraulic Pressure Gage-Boost System (Non-Tandem)*
27 *Altimeter*
28 *Emergency Hydraulic Controls On Indicator Light (Non-Tandem)*
29 *Hydraulic Pressure Gage-Utility System (Non-Tandem) (Duplex Gage on Tandem)*
30 *Flight Command Indicator*
31 *Machmeter*
32 *Directional Indicator Controls (See figure 4—17)*
33 *Radio Compass Indicator*
34 *Landing Lights Switch*
35 *Air Refueling Controls (See figure 4—13)*

Figure 1—4

INSTRUMENT PANEL D thru G

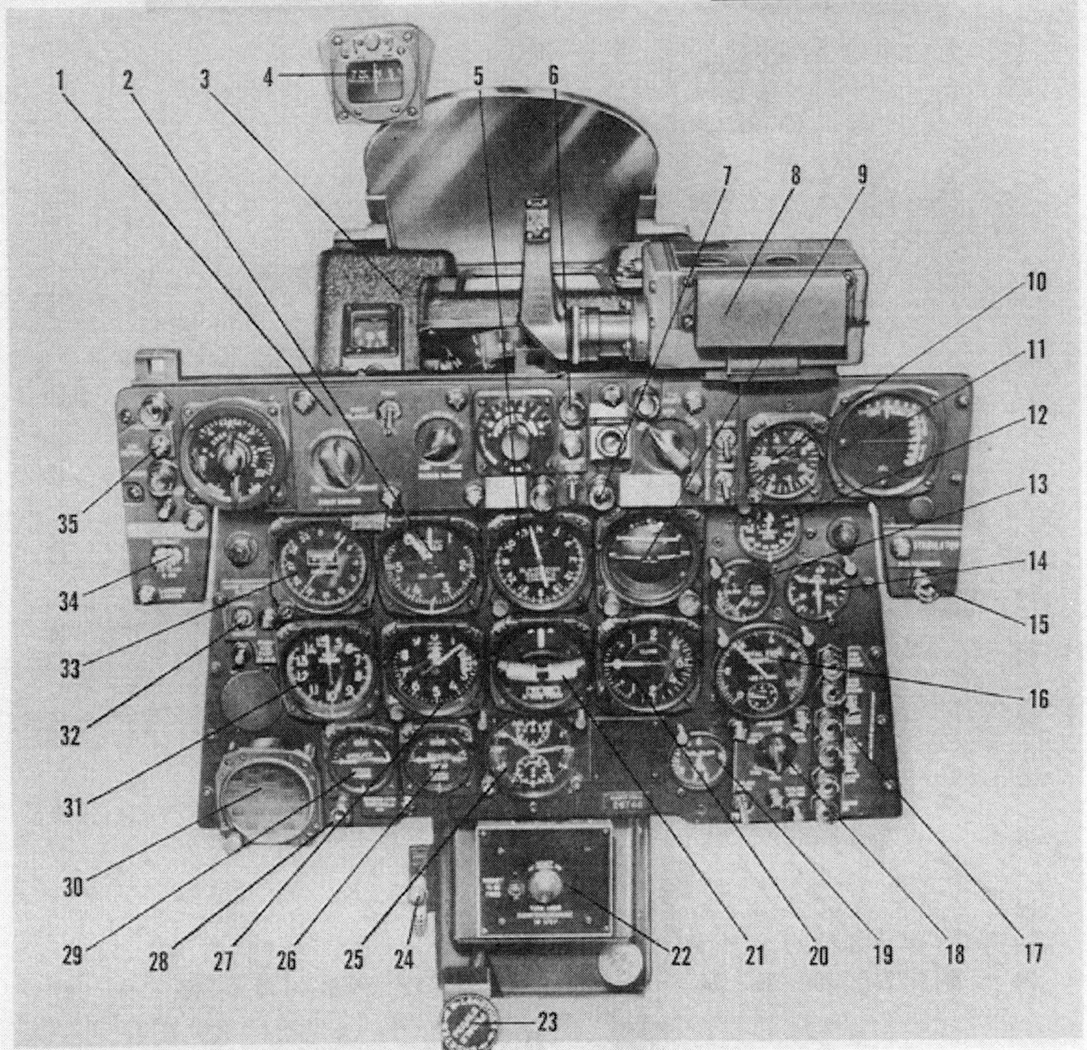

1 *Gun-Bomb-Rocket Sight Control Panel (See figure 4–24)*
2 *Airspeed Indicator*
3 *Gun-Bomb-Rocket Sight*
4 *Magnetic Compass*
5 *Directional Indicator*
6 *Canopy Open Indicator Light*
7 *Engine Overheat and Fire Warning System (See figure 1–36)*
8 *Gun Camera*
9 *Attitude Indicator*
10 *Accelerometer*
11 *Dive and Roll Indicator*
12 *Engine Tachometer*
13 *Engine Exhaust Gas Temperature Indicator*
14 *Engine Oil Pressure Indicator*
15 *Stabilator Mechanical Advantage Indicator*
16 *Fuel Quantity Indicator (See figure 1–21)*
17 *Fuel System Warning Lights (See figure 1–20)*
18 *Fuel Quantity Selector Switch(See figure 1–21)*
19 *Fuel Flow Indicator*
20 *Vertical Velocity Indicator*
21 *Turn and Slip Indicator*
22 *Rocket Release Control (See figure 4–29)*
23 *Rudder Pedal Adjusting Knob*
24 *Drag Chute Control (See figure 1–35)*
25 *Clock*
26 *Hydraulic Pressure Gage-Boost System (Non-Tandem)*
27 *Altimeter*
28 *Emergency Hydraulic Controls On Indicator Light (Non-Tandem)*
29 *Hydraulic Pressure Gage-Utility System (Non-Tandem) (Duplex Gage on Tandem)*
30 *Flight Command Indicator*
31 *Machmeter*
32 *Directional Indicator Controls (See figure 4–17)*
33 *Radio Compass Indicator*
34 *Landing Lights Switch*
35 *Air Refueling Controls (See figure 4–13)*

Figure 1–5

INSTRUMENT PANEL

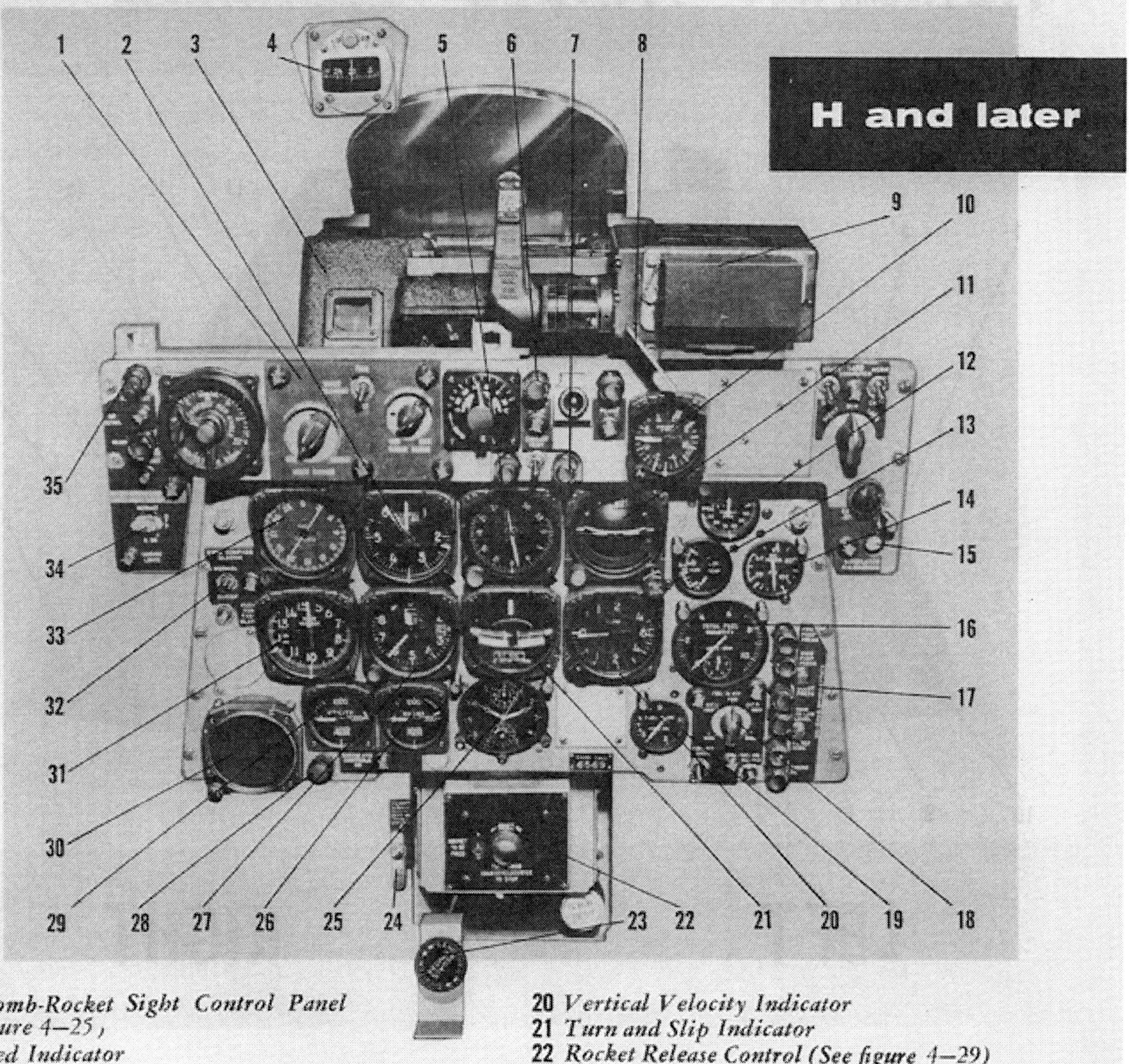

1 *Gun-Bomb-Rocket Sight Control Panel (See figure 4—25)*
2 *Airspeed Indicator*
3 *Gun-Bomb-Rocket Sight*
4 *Magnetic Compass*
5 *Directional Indicator*
6 *Canopy Open Indicator Light*
7 *Engine Overheat and Fire Warning System (See figure 1—36)*
8 *Gun-Bomb-Rocket Sight Mechanical Caging Lever*
9 *Gun Camera*
10 *Accelerometer*
11 *Attitude Indicator*
12 *Engine Tachometer*
13 *Engine Exhaust Gas Temperature Indicator*
14 *Engine Oil Pressure Indicator*
15 *Stabilator Mechanical Advantage Indicator*
16 *Fuel Quantity Indicator (See figure 1—21)*
17 *Fuel System Warning Lights (See figure 1—20)*
18 *Fuel Quantity Selector Switch (See figure 1—21)*
19 *Fuel Flow Indicator*
20 *Vertical Velocity Indicator*
21 *Turn and Slip Indicator*
22 *Rocket Release Control (See figure 4—29)*
23 *Rudder Pedal Adjusting Knob*
24 *Drag Chute Control (See figure 1—35)*
25 *Clock*
26 *Hydraulic Pressure Gage-Boost System (Non-Tandem)*
27 *Altimeter*
28 *Emergency Hydraulic Controls On Indicator Light (Non-Tandem)*
29 *Hydraulic Pressure Gage-Utility System (Non-Tandem) (Duplex Gage on Tandem)*
30 *Flight Command Indicator*
31 *Machmeter*
32 *Directional Indicator Controls (See figure 4—17)*
33 *Radio Compass Indicator*
34 *Landing Lights Switch*
35 *Air Refueling Controls (See figure 4—13)*

Figure 1—6

AUXILIARY PANELS TYPICAL

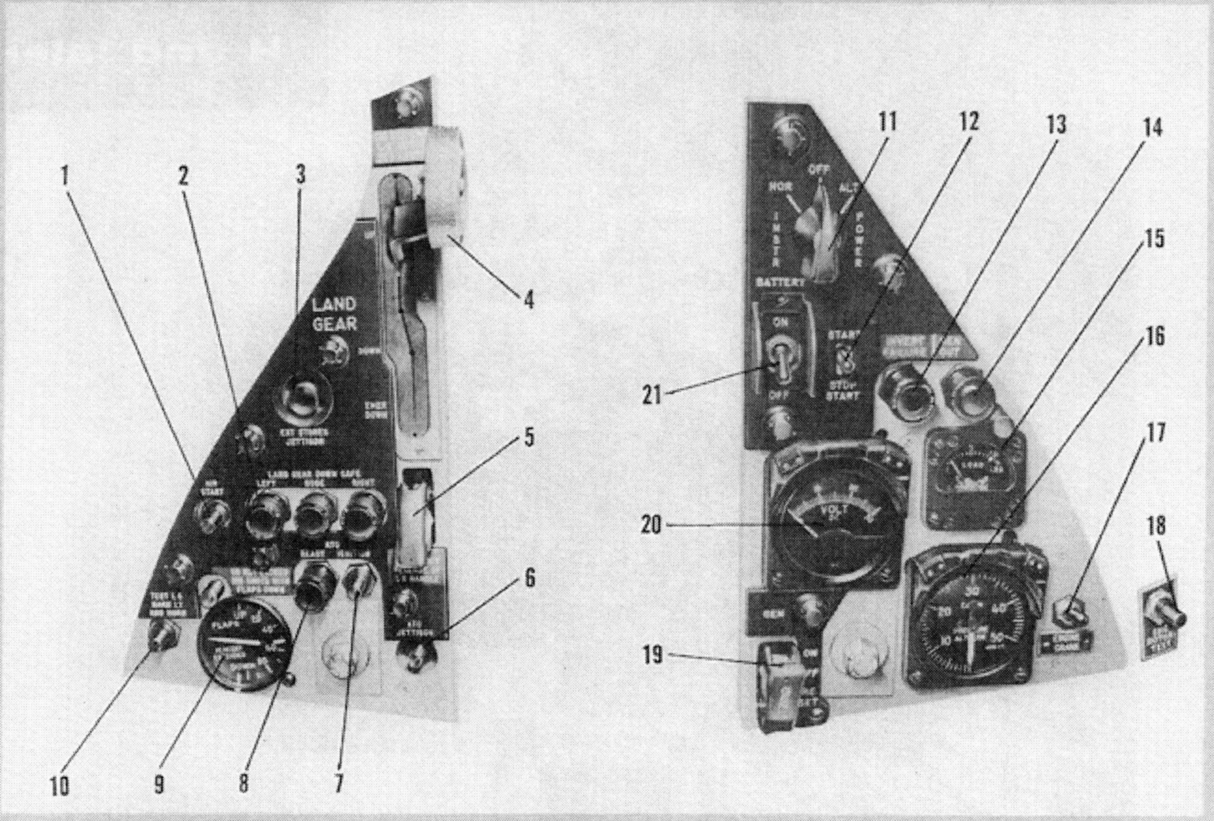

LEFT

RIGHT

1 *Engine Air Start Switch*
2 *Landing Gear Indicator Lights (Position Indicators on unmodified A thru G)*
3 *External Stores Jettison Switch*
4 *Landing Gear Selector Handle*
5 *Emergency Landing Gear Release Switch*
6 *ATO Jettison Switch*
7 *ATO Ignition Switch*
8 *ATO Ready Warning Light*
9 *Landing Flap Position Indicator*
10 *Landing Gear Warning Light and Horn Test Switch*

11 *Instrument Power Switch*
12 *Engine Starter Switch*
13 *Inverter Failure Warning Light*
14 *Generator-Out Indicator Light*
15 *Loadmeter*
16 *Cabin Altimeter*
17 *Engine Crank Switch*
18 *Engine Rotor-Test Switch*
19 *Generator Switch*
20 *Voltmeter*
21 *Battery Switch*

Figure 1—7

LEFT HAND CONSOLE

A thru G

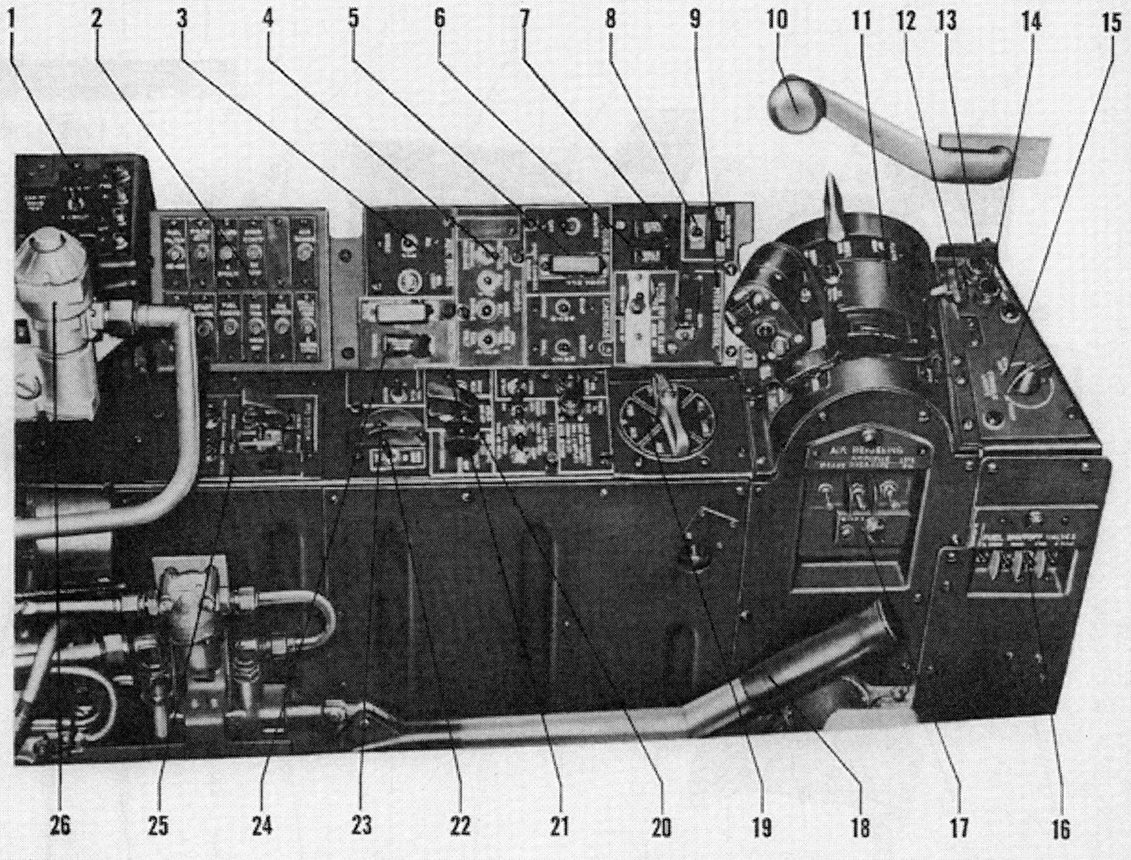

1 *Canopy Squib Test Switch and Indicator Lights (See figure 1—38)*
2 *Auxiliary Circuit Breaker Panel (See figure 1—24)*
3 *Auxiliary Bombs Control Panel (See figure 4—27)*
4 *Bomb Control Panel (See figure 4—26)*
5 *Guns and Camera Control Panel (See figure 4—21)*
6 *Inboard Pylon Jettison Switch*
7 *Outboard Pylon Jettison Switch*
8 *Rudder Trim Switch*
9 *Flight Control Panel (See figures 1—30 & 1—34)*
10 *Canopy Control Lever*
11 *Throttle Quadrant (See figure 1—15)*
12 *Throttle Friction Lock*
13 *Aileron Neutral Trim Indicator Light*
14 *Rudder Neutral Trim Indicator Light*
15 *Console Lights Rheostat*
16 *Fuel System Shut-off Valves (See figure 1—19)*
17 *Air Refueling Switches (See figure 4—13)*
18 *Hydraulic Hand Pump (Non-Tandem Aircraft)*
19 *Fuel System Control Panel (See figure 1—13)*
20 *Spoiler Shut-off Switch (Modified & D and Later)*
21 *Pitot Heater Switch*
22 *Pneumatic Compressor Switch*
23 *Ato Ready Switch*
24 *Rocket Control Panel (See figure 4—28)*
25 *Flight Control Hydraulic System Switches, Non-Tandem (located on Flight Control Panel on Tandem Aircraft)*
26 *Anti-g Valve*

Figure 1—8

LEFT HAND CONSOLE

H and LATER

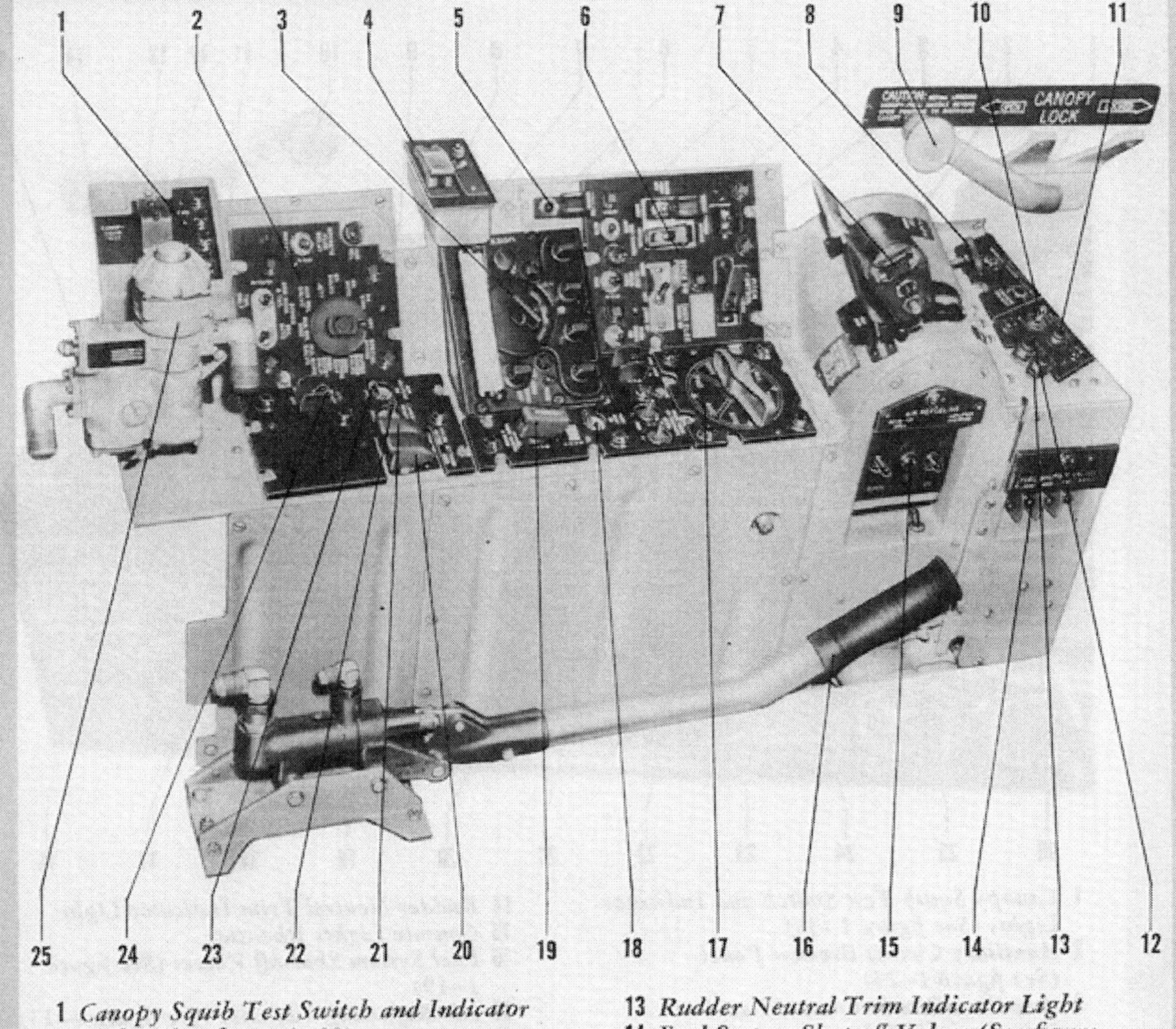

1 *Canopy Squib Test Switch and Indicator Lights (See figure 1–38)*
2 *Armament Control Panel*
3 *Command Radio Control Panel–AN/ARC-34 (See figure 4–6)*
4 *AN/ARC-34 Emergency Battery Switch and Press to Test (Modified)*
5 *Spoilers Shut-off Switch*
6 *Flight Controls Panel*
7 *Throttle Quadrant (See figure 1–15)*
8 *Throttle Friction Lock*
9 *Canopy Control Lever*
10 *Jettison Circuit Breaker Check Switch*
11 *Jettison Circuit Breaker Check Light*
12 *Aileron Neutral Trim Indicator Light*
13 *Rudder Neutral Trim Indicator Light*
14 *Fuel System Shut-off Valves (See figure 1–20)*
15 *Air Refueling Switches (See figure 4–13)*
16 *Hydraulic Hand Pump (Non-Tandem)*
17 *Fuel System Control Panel (See figure 1–13)*
18 *Pitot Heat Switch*
19 *Gun Arming Switch*
20 *Inboard Pylon Jettison Switch*
21 *Outboard Pylon Jettison Switch*
22 *Gun Camera Switch*
23 *Gun Heater Switch*
24 *Ato Ready Switch*
25 *Anti-g Valve*

Figure 1–9

RIGHT HAND CONSOLE

A thru G

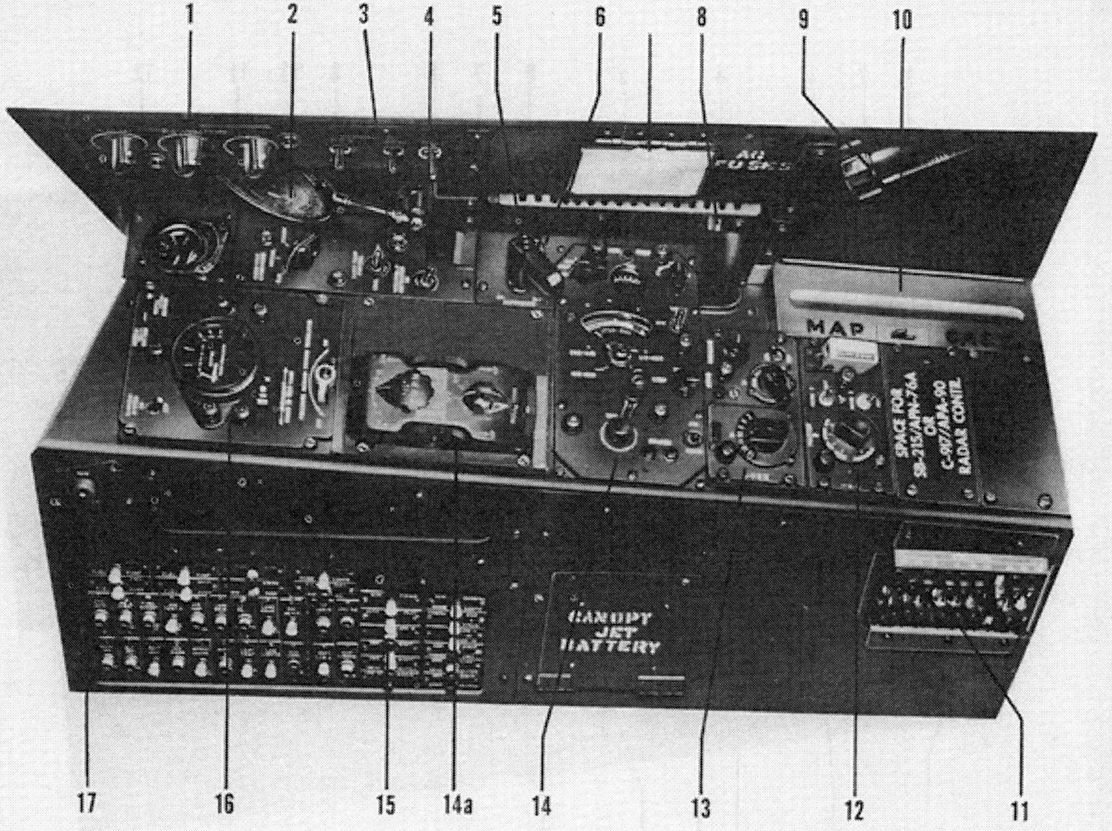

1 *Interior and Exterior Lights Control Panel (See figure 4—9)*
2 *Batori Computer*
3 *Cabin Temperature and Pressure Control (See figure 4—3)*
4 *Windshield Defroster Switch*
5 *Defroster Control*
6 *Canopy Dry Air Switch*
7 *A-C Fuse Panel*
8 *Side Air Outlet Shut-off*
9 *Cockpit Light*
10 *Map Case*
11 *Spare Fuses*
12 *Radar Control Panel AN/APX-6 (See figure 4—7)*
13 *Command Radio Control Panel AN/ARC-33 (See figure 4—5)*
14 *Radio Compass Control Panel AN/ARN-6 (See figure 4—8)*
14a *Autopilot Control Panel*
15 *Main Circuit Breaker Panel (See figure 1—24)*
16 *Oxygen Regulator (See figure 4—10)*
17 *Main Inverter Circuit Breaker*

Figure 1—10

RIGHT HAND CONSOLE

H and LATER

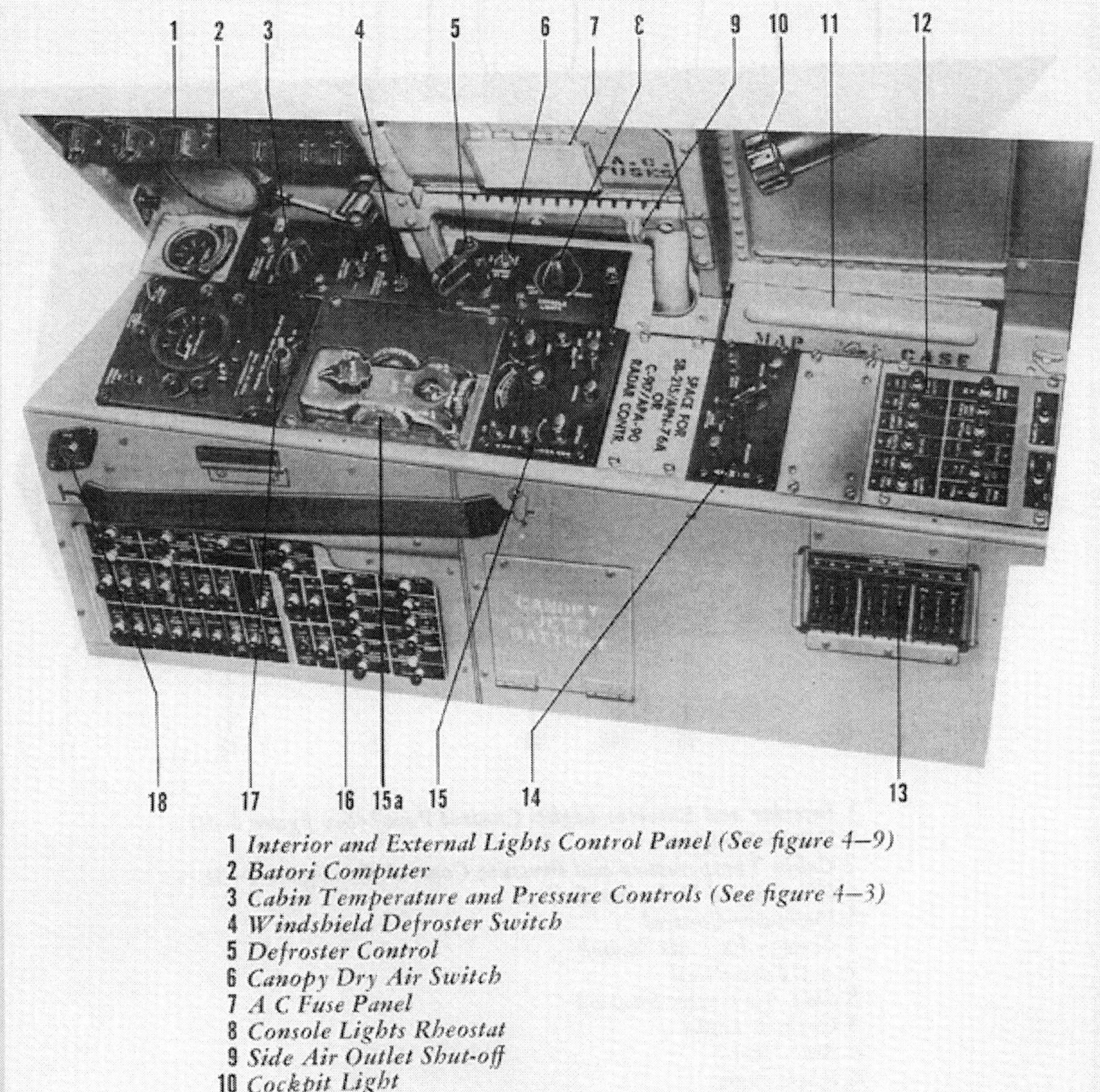

1 *Interior and External Lights Control Panel (See figure 4—9)*
2 *Batori Computer*
3 *Cabin Temperature and Pressure Controls (See figure 4—3)*
4 *Windshield Defroster Switch*
5 *Defroster Control*
6 *Canopy Dry Air Switch*
7 *A C Fuse Panel*
8 *Console Lights Rheostat*
9 *Side Air Outlet Shut-off*
10 *Cockpit Light*
11 *Map Case*
12 *Auxiliary Circuit Breaker Panel (See figure 1—24)*
13 *Spare Fuses*
14 *Radar Ground Support Control Panel AN/APW-11A*
15 *Radio Compass Control Panel — AN/ARN-6 (See figure 4—8)*
15a *Autopilot Control Panel*
16 *Main Circuit Breaker Panel (See figure 1—24)*
17 *Oxygen Regulator (See figure 4—10)*
18 *Main Inverter Circuit Breaker*

Figure 1—11

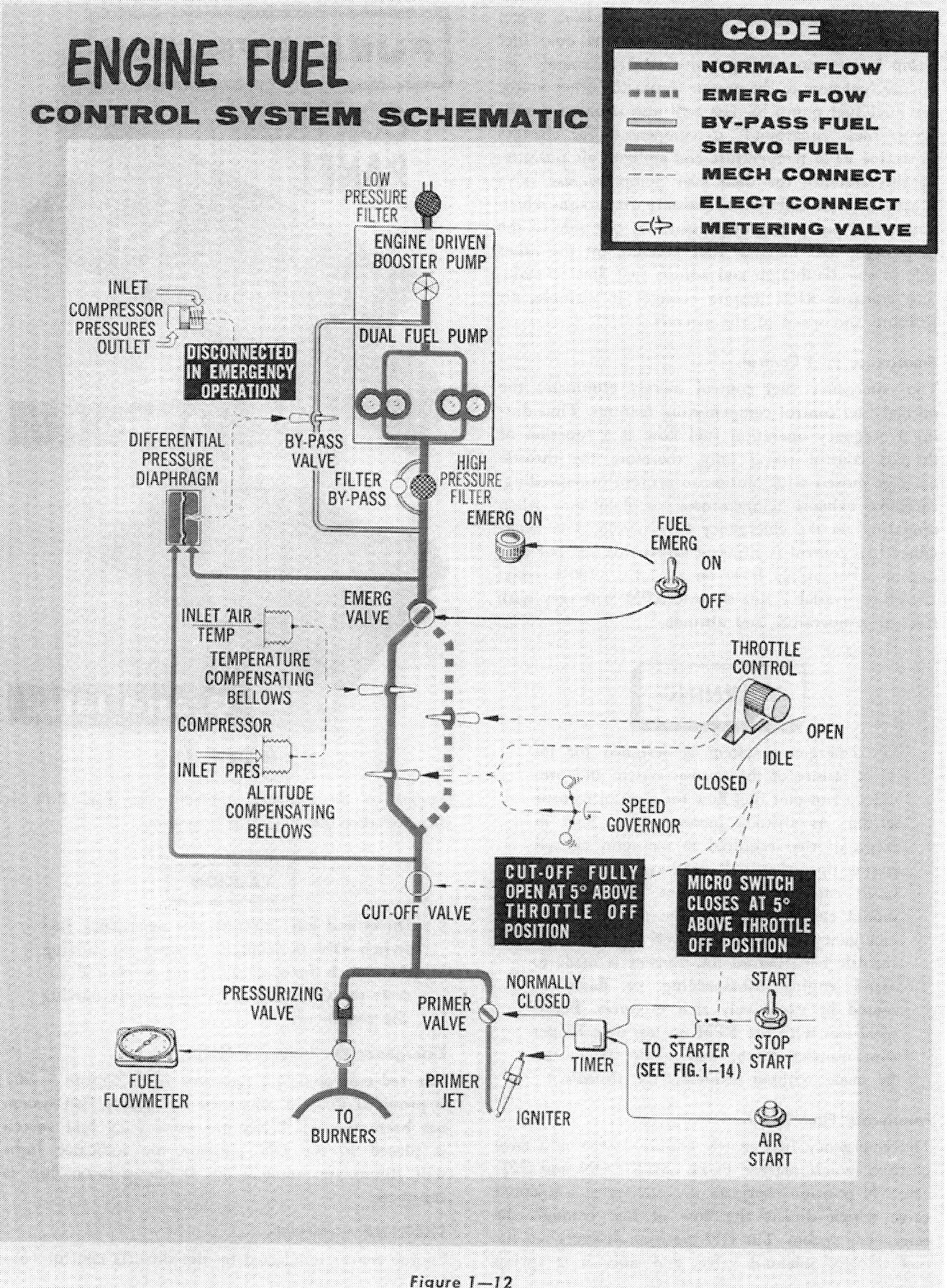
ENGINE FUEL
CONTROL SYSTEM SCHEMATIC
CODE
NORMAL FLOW
EMERG FLOW
BY-PASS FUEL
SERVO FUEL
MECH CONNECT
ELECT CONNECT
METERING VALVE
LOW PRESSURE FILTER
ENGINE DRIVEN BOOSTER PUMP
INLET
COMPRESSOR PRESSURES
OUTLET
DISCONNECTED IN EMERGENCY OPERATION
DUAL FUEL PUMP
DIFFERENTIAL PRESSURE DIAPHRAGM
BY-PASS VALVE
FILTER BY-PASS
HIGH PRESSURE FILTER
EMERG ON
FUEL EMERG
ON
OFF
EMERG VALVE
INLET AIR TEMP
TEMPERATURE COMPENSATING BELLOWS
COMPRESSOR
INLET PRES
ALTITUDE COMPENSATING BELLOWS
THROTTLE CONTROL
OPEN
IDLE
CLOSED
SPEED GOVERNOR
CUT-OFF VALVE
CUT-OFF FULLY OPEN AT 5° ABOVE THROTTLE OFF POSITION
MICRO SWITCH CLOSES AT 5° ABOVE THROTTLE OFF POSITION
PRESSURIZING VALVE
NORMALLY CLOSED
PRIMER VALVE
START
STOP
START
TIMER
TO STARTER (SEE FIG.1-14)
FUEL FLOWMETER
PRIMER JET
IGNITER
TO BURNERS
AIR START

Figure 1—12

sures which are balanced against a spring-load. When compressor pressure rise is excessive, the dual fuel pump by-pass opens to permit fuel "runaround," reducing fuel flow to the engine. From the other source the dual fuel pump by-pass will also open to permit excess fuel "runaround" to compensate for changes in engine RPM temperature and ambient air pressure. In this instance the dual fuel pump by-pass valve is actuated by a differential pressure diaphragm which senses fuel pump output pressure on one side of the diaphragm and metered fuel pressure on the other side of the diaphragm and adjusts fuel flow to maintain constant RPM despite changes in altitude, air pressure and speed of the aircraft.

Emergency Fuel Control.

The emergency fuel control merely eliminates the normal fuel control compensating features. Thus during emergency operation fuel flow is a function of throttle control travel only, therefore the throttle must be moved with caution to prevent overspeeding, excessive exhaust temperatures, or flame-out, when operating on the emergency fuel system. The emergency fuel control is adjusted to provide 100 per cent engine RPM at sea level on a 37.8°C (100°F) day; therefore, available full throttle RPM will vary with free air temperature and altitude.

WARNING

The emergency system is designed for in-flight failure of the normal system and provides a constant fuel flow for a given throttle setting. As altitude increases, fuel flow in excess of that required to maintain normal engine operation will produce engine overspeed conditions. Therefore, if the pilot should elect to go from the normal to the emergency system above 6000 feet, he must throttle back before the transfer is made to avoid engine overspeeding or flame-out, caused by excessively rich mixtures. Below 6000 feet with the RPM no less than 85 per cent transfer to the emergency system can be made without reducing the throttle.

Emergency Fuel Switch.

The emergency fuel switch (figure 1—13) is a two-position switch, marked FUEL EMERG ON, and OFF. The ON position energizes the fuel transfer solenoid valve which directs the flow of fuel through the emergency system. The OFF position de-energizes the fuel transfer solenoid valve, and since it is spring loaded to the normal position, the fuel flow is directed through the normal system.

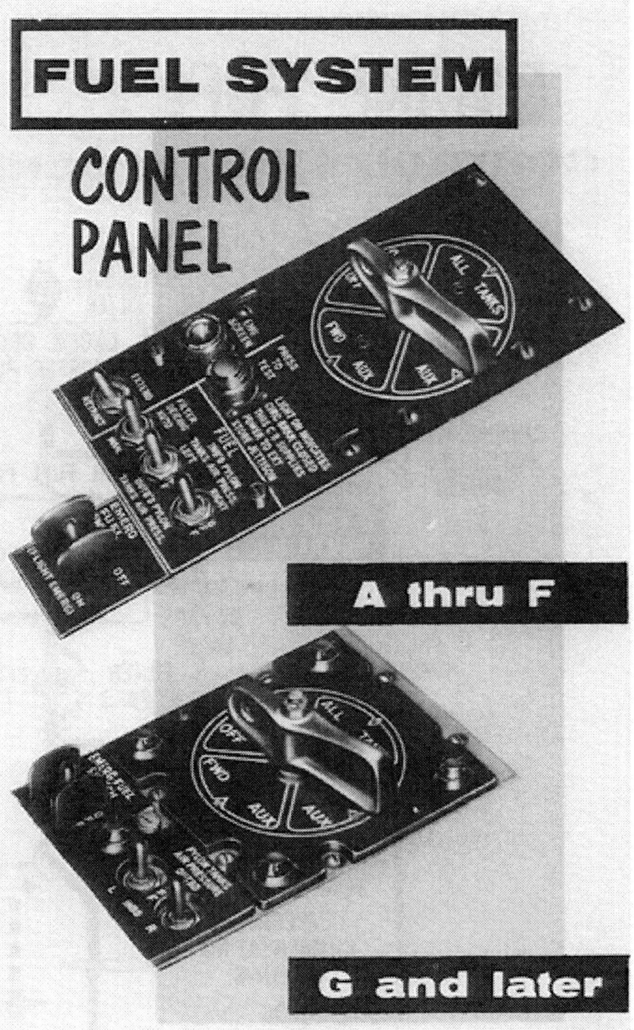

Figure 1—13

CAUTION

On G and later aircraft, the emergency fuel switch ON position is selected by moving the switch forward while on A thru F aircraft the ON position is selected by moving the switch aft.

Emergency-On Indicator Light.

The red emergency-on indicator light (figure 1—20) is provided to show when the emergency fuel system has been selected. When the emergency fuel switch is placed in the ON position the indicator light will illuminate immediately if the primary bus is energized.

THROTTLE CONTROL.

Engine power is selected by the throttle control (fig-

ENGINE STARTER SYSTEM SCHEMATIC

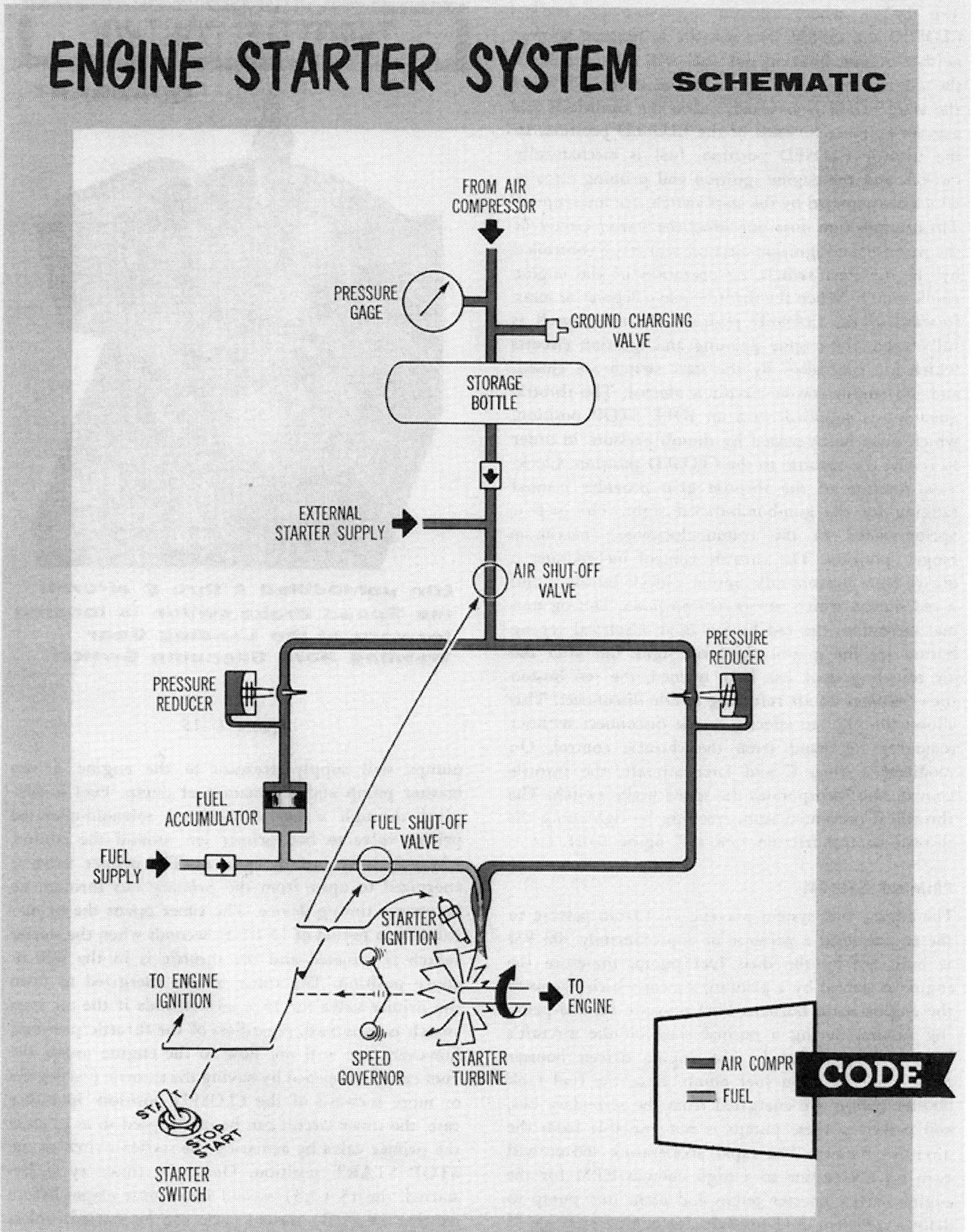

Figure 1—14

ure 1—15) whose extreme positions are marked CLOSED and OPEN. The throttle is designed so that neither engine ignition nor fuel will be supplied to the engine, nor will the starter be energized when the start switch is actuated, unless the throttle is five degrees or more forward of the CLOSED position. In the throttle CLOSED position, fuel is mechanically cut off, and the engine ignition and priming circuits, which are actuated by the start switch, are interrupted. Throttle position does not affect the starter circuit or the priming and ignition circuits, which are controlled by the air start switch, or operation of the engine crank switch. When the throttle is five degrees or more forward of the CLOSED position, the fuel cut-off is fully open, the engine priming and ignition circuits which are controlled by the start switch are closed, and the engine starter circuit is alerted. The throttle quadrant is supplied with an IDLE STOP position, which must be by-passed by thumb pressure in order to return the throttle to the CLOSED position. Clockwise rotation of the throttle grip provides manual ranging for the gun-bomb-rocket sight. The grip is spring-loaded to the counterclockwise (maximum range) position. The throttle control incorporates a microphone press-to-talk button (black button), and a red button which serves two purposes. During normal operation, the red button is an electrical caging button for the gun-bomb-rocket sight, but after the air refueling door has been opened, the red button then becomes an air refueling nozzle disconnect. This allows the pilot to effect a nozzle disconnect without removing his hand from the throttle control. On modified A thru C and later aircraft, the throttle control also incorporates the speed brake switch. The throttle is prevented from creeping by tightening the throttle control friction lock (12, figure 1—8).

THROTTLE CONTROL

QUADRANT

(On unmodified A thru C aircraft the Speed Brake switch is located forward of the Landing Gear Warning Horn Silencing Switch

Figure 1—15

PRIMING SYSTEM.

The engine fuel system prevents fuel from passing to the engine until a pressure of approximately 100 PSI is built up by the dual fuel pump, therefore the engine is started by a priming system which by-passes the engine main burners. Fuel pressure for the priming system, during a normal start on the aircraft's battery, is provided by the engine driven booster pump and the main fuel pump, since the fuel tank booster pumps are energized from the secondary bus, and power to these pumps is not available from the aircraft's battery. The rapid accelerating starter will turn up the engine to a high enough RPM for the engine driven booster pump and main fuel pump to deliver sufficient fuel pressure for engine starting. If a start is made using an external power source the secondary bus is energized and the fuel tank booster pumps will supply pressure to the engine driven booster pump and the main fuel pump. Fuel is supplied through a normally closed solenoid-operated primer valve to two primer jets around the engine, where it is ignited by igniters. The primer valve is energized to open from the primary bus through an automatic timing device. The timer opens the primer valve for a period of 15 (±3) seconds when the starter switch is actuated and the throttle is in the idle or above position. The timer is also energized to open the primer valve for 15 (±3) seconds if the air start switch is depressed, regardless of the throttle position; however, fuel will not flow to the engine unless the fuel cut-off is opened by having the throttle five degrees or more forward of the CLOSED position. In either case, the timer circuit can be interrupted so as to close the primer valve by actuating the starter switch to the STOP START position. Once the timer cycle has started, the 15 (±3) second limit must elapse before another 15 (±3) second cycle can be started, unless the timer circuit is interrupted by placing the starter switch in the STOP START position.

IGNITION.

The fuel-air mixture in the engine is ignited by an automatic ignition system which incorporates two igniters. Once the engine is started the ignition system is no longer used, as burning in the combustion chambers is continuous. Power for ignition is supplied by two high tension vibrators which derive their power from the primary bus. The system incorporates a timer device which supplies ignition continuously for 15 (±3) seconds; controls are also provided for interrupting the 15 (±3) second ignition cycle. The ignition system is operated by the start or air start switches which are discussed under starting system.

This page intentionally left blank.

STARTING SYSTEM.

Engine starting torque is provided by a rapid accelerating combustion type starter (figure 1–14) mounted on the front of the engine. The starter is a self contained unit and is capable of turning the engine up to approximately 24 per cent RPM in 3 to 3½ seconds. Torque is developed by a high speed turbine, driven through reduction gears. Fuel for the turbine is supplied from the airplane's fuel system, while the air supply is stored in a high pressure storage bottle which is automatically refilled by the pneumatic compressor. The capacity of the storage bottle is sufficient for one start. The fuel and air mixture is automatically fed to the turbine and ignited when the starter switch is actuated and the throttle is five degrees or more forward of the CLOSED position. A centrifugally operated switch cuts off the fuel and air supply when the turbine reaches its maximum RPM, while a pressure switch cuts the starter out if maximum speed is not reached, or if the centrifugal switch is inoperative. Electrical power for operating the starter is supplied from the primary bus. The engine ignition and primer systems operate concurrently with the starter, and are designed to function for a period of 15 seconds if the throttle is five degrees or more forward of the CLOSED position, and will not operate when the throttle is closed. Controls are also provided for starting the engine during air starts. A switch is provided so that the engine can be turned over slowly and its rotation checked without the possibility of the engine starting.

Note

When starting the engine on the aircraft's battery various warning lights will illuminate until the engine RPM reaches a speed that will cut in the generator and energize the secondary bus.

Starter Switch.

The starter switch (figure 1–7) is provided to actuate the engine starter and the automatic ignition and priming timer. It is a three position switch spring-loaded to the center position. The two extreme positions are START and STOP START with the center position unmarked. Held for a period of approximately one second in the START position ignites the combustion starter and energizes the engine ignition and priming timer unit, if the throttle is five degrees or more forward of the CLOSED position. The starter will operate until the engine RPM reaches approximately 24 per cent, or for approximately four seconds if maximum RPM is not reached. The ignition and priming timer will operate for 15 (±3) seconds only after the throttle is advanced approximately five degrees or more from the CLOSED position. The START position is not recommended for use during an air start unless low engine RPM and low altitude make it necessary, as damage to the starter will result. If the starter switch is momentarily held in the STOP START position, the combustion starter will be turned off, and the automatic engine ignition and priming timer will be de-energized immediately. For starting, electrical power is supplied from the primary bus which is energized by the airplane's battery, if the battery switch is in the ON position. Fuel tank booster pumps will be inoperative but the rapid acceleration of the engine, when starting, will be sufficient for the engine driven booster pump and the main pump to supply fuel for starting providing the starter fuel accumulator is full. A start can also be accomplished with electrical power supplied from an external power source. In the event of a false start, external power should be used for a second attempt in order to permit the fuel tank booster pumps to refill the starter fuel accumulator.

Air Start Switch.

The air start switch (figure 1–7) is provided to supply engine ignition and priming fuel during an air start, when the engine is windmilling and starting torque is not required. A ground check of the ignition system can also be made using the air start switch. In this case the throttle must be in the CLOSED position so that fuel will not flow to the engine primers. The switch is a push button type marked AIR START. When depressed momentarily, the automatic timing unit is energized from the primary bus to supply ignition to the engine and open the priming solenoid for a period of 15 (±3) seconds, regardless of the throttle position. However, priming fuel will not flow to the primers unless the throttle is advanced approximately five degrees or more from the CLOSED position. Ignition can be stopped, and the primer valve closed, if the timer circuit is interrupted by placing the starter switch in the STOP START position.

ENGINE CRANK SWITCH.

The engine crank switch (figure 1–7) by-passes the throttle control and is provided so that engine rotation and starter operation may be checked during ground inspection without starting the engine. The switch is a two position type with one position marked ENGINE CRANK and the other position unmarked. When held in the ENGINE CRANK position with the throttle CLOSED, power from the primary bus is supplied to start the combustion starter. Ignition or primer fuel will not be supplied to the engine in this instance. The combustion starter will stop when its maximum rpm is reached, or after four seconds of operation.

ENGINE ROTOR TEST SWITCH.

The engine rotor test switch (figure 1–7) is provided so that the engine rotor can be turned slowly to check its rotation. The switch is marked ENG ROTOR TEST and when depressed opens an electrically operated pneumatic valve which supplies compressed air to the combustion chamber of the starter. The compressed air operates the starter slowly. The engine will continue to rotate as long as the switch is depressed and the air supply is sufficient. The engine rotor test switch is powered from the primary bus.

ENGINE SCREEN.

The engine screens are provided to protect the engine compressor from foreign objects during ground operation. However, the screens are retractable during flight so as to prevent loss of thrust and minimize high exhaust gas temperatures caused by engine screen icing. Engine thrust is decreased approximately 4.4 per cent with the screens extended. The screens are located in the air intake duct just forward of the engine and are operated with hydraulic pressure. The hydraulic pressure passes through a solenoid valve that is powered from the primary bus. A light indicates operation and position of the screens. On H and later aircraft the engine screens have an automatic control feature, that may be selected by the pilot, to assure that the screens are extended when the landing gear is extended.

A thru G

Engine Screen Switch.

The engine screen switch (figure 1–13) positions a solenoid valve in the hydraulic system to operate the engine screens. The two switch positions are EXTEND and RETRACT. When the switch is placed in the EXTEND or the RETRACT position the engine screen will move to the respective position and the solenoid valve will automatically return to neutral. The switch is left in the selected position after the screens have cycled.

H and LATER

Engine Screen Switch.

The engine screen switch (figures 1–30 and 1–34) installed on H and later aircraft is a three position switch marked RETR, EXT, and AUTO. The switch positions a solenoid valve in the hydraulic system to operate the screens. When the switch is placed in the RETR or EXT positions the engine screens will move to the respective positions and the solenoid valve will automatically return to neutral. The switch is left in the selected position after the screen has cycled. In the AUTO position, engine screen operation is automatic and the screens will automatically extend and remain extended as long as the main landing gear is in the extended position. The engine screens will also retract automatically after a time delay of 3½ (±1) minutes when the landing gear reaches the up and locked position. The time delay feature is provided so that enough altitude is gained before the screens retract after take-off so that emergency procedures can be initiated in the event of engine failure caused by debris entering the engine compressor section. If manual selection is desired, the switch is placed in the desired position to retract or extend the screens.

Engine Screen Warning Light.

The engine screen warning light (figures 1–13, 1–30 and 1–34), will illuminate whenever the engine screen

Figure 1–16

switch is repositioned and will remain illuminated until both screens have completed their cycle to either the fully extended or retracted position if power is available from the primary bus.

ENGINE INSTRUMENTS.

The engine tachometer and the exhaust gas temperature indicator are self generated electrical instruments which do not require power from the airplane's electrical system. The fuel quantity indicator, the fuel flow indicator and the oil pressure indicator are operated from the AC power circuit and will operate from either the main or alternate inverters.

ASSIST TAKE-OFF SYSTEM.

Provision is made for the installation of four solid fuel ato units on an expendable adapter which is suspended from retractable hooks on the underside of the fuselage (figure 1–16). The units are used to supply additional thrust for high gross weight or shorter run take-offs. Each unit is rated at 1000 pounds thrust and once ignited, burns continuously for 14 seconds. After ignition takes place the units cannot be turned off or the adapter cannot be jettisoned while the units are producing thrust. Power to ignite the units is supplied from the primary bus. The adapter is jettisoned and the hooks on the fuselage are retracted by means of hydraulic pressure which is controlled by an electrically operated shut-off valve powered from the primary bus.

ATO READY SWITCH.

The ato ready switch (23, figure 1–8; 22, figure 1–9) is provided as a safety switch to prevent accidental firing of the ato units. It is a switch type circuit breaker with two positions marked OFF and ATO READY. The ATO READY position arms the ato ignition system and iluminates the ato ready warning light (figure 1–7) which is marked ATO READY.

ATO IGNITION SWITCH.

The ato ignition switch (figure 1–7) is a push button type switch marked ATO IGNITION. When depressed to the ATO IGNITION position a circuit is completed to the ignition posts of the ato units if the ato ready switch is in the ATO READY position.

ATO JETTISON SWITCH.

The ato unit adapter, suspended from retractable hooks under the fuselage, is hydraulically jettisoned after the ato units have burned out. The jettisoning system is controlled by the ato jettison switch (figure 1–7). The switch is a push button type jettison switch energizes the ato jettison solenoid shut-off valve to allow hydraulic pressure to release the adapter and retract the hooks into the fuselage.

OIL SYSTEM.

The engine oil system is entirely automatic and requires no attention from the pilot. Lubricating oil is contained in an oil tank which is mounted on the upper left side of the engine compressor housing. The oil tank capacity is 6.0 U. S. gallons of which 4.27 gallons is usable. Oil, under pressure, is delivered to the front bearings and accessory drives where it is scavenged and returned to the oil tank. Two metering pumps, in the oil pump, supply metered oil to the center and rear engine bearings where it is mixed with air and forms a mist. This oil is not scavenged but is vented overboard as a mist through a vent located on the underside of the fuselage. Oil consumption is 3 pounds (approximately 3.2 pints) per hour maximum. An AC operated oil pressure indicator on the main instrument panel (14, figures 1–5 and 1–6) indicates oil pressure if the main or alternate inverter is operating. The oil tank is designed to supply sufficient oil to engine during inverted flight for a period of approximately one minute. Oil grade and specifications are noted in the servicing diagram figure 1–41.

AIRCRAFT FUEL SYSTEM.

The aircraft fuel system (figure 1–17), is designed to provide automatic fuel transfer during normal operation without attention from the pilot. The airplane is basically equipped with four internal, self-sealing fuel tanks; a main tank installed behind the pilot, a forward tank under the cockpit floor and interconnected tanks in each wing. In addition, a jettisonable external tank can be carried on each of four pylons. Fuel from the external tanks is transferred to the main tank before wing or forward tank fuel is used. Normally tanks empty in the following order; external tanks, wing tanks, forward tank and the main tank. A restrictor is incorporated in the forward tank outlet to adjust fuel flow so as to automatically maintain the required cg location; the wing tanks are emptied at the same time or before the forward tank. Transfer of fuel from the external to the main tank is accomplished by means of air pressure which is manually controlled. Each internal tank incorporates a float which controls a shut-off valve in the transfer line to the tank. Fuel will flow to the tank until it is full then the float will automatically close the shut-off valve in the transfer line to that tank. These shut-off valves may be closed at any time to prevent fuel flow to any of the internal tanks. Internal tank fuel is pumped into the main tank by booster pumps which operate automatically according to the position of the

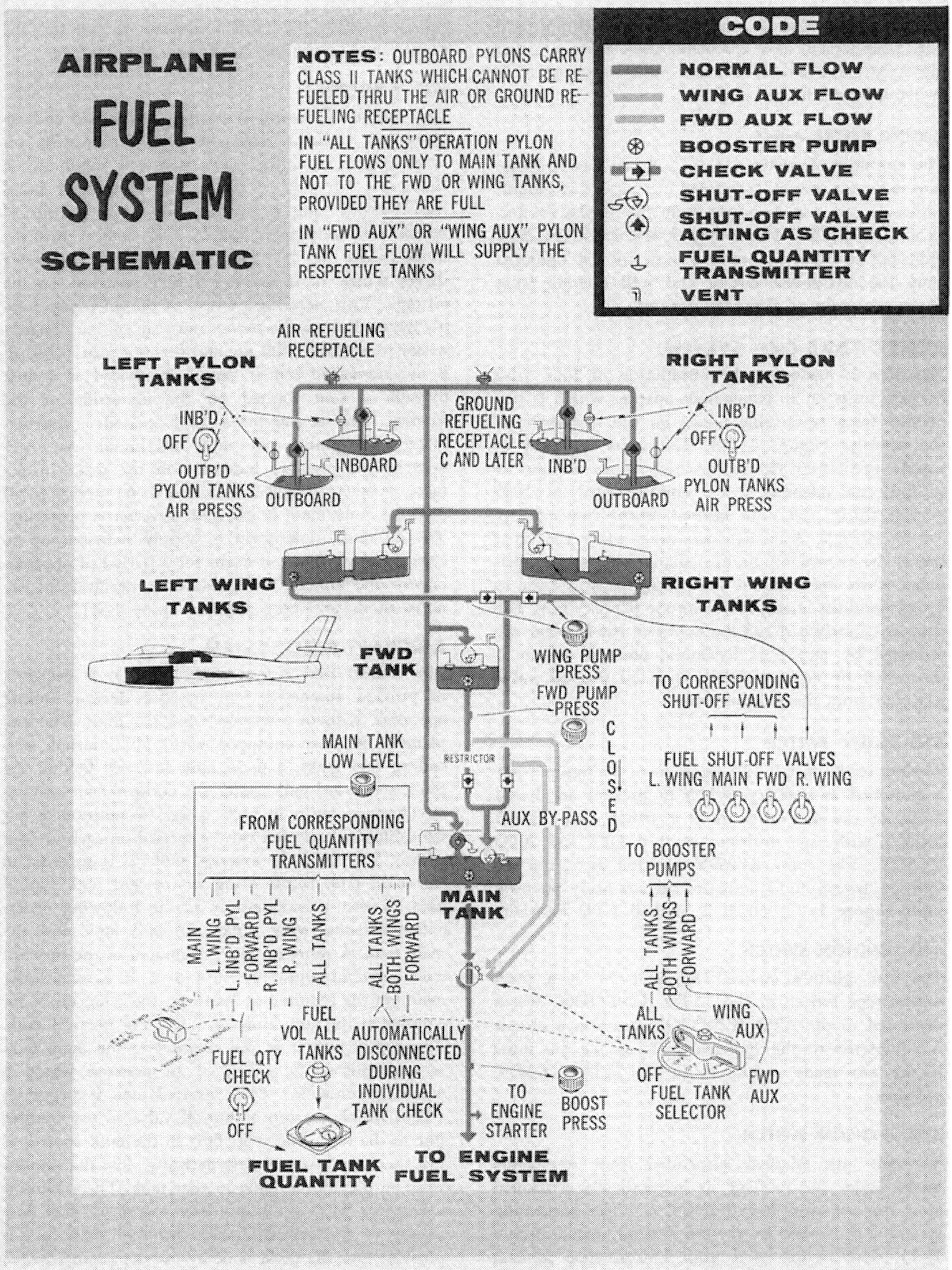
AIRPLANE
FUEL
SYSTEM
SCHEMATIC
NOTES: OUTBOARD PYLONS CARRY CLASS II TANKS WHICH CANNOT BE REFUELED THRU THE AIR OR GROUND REFUELING RECEPTACLE
IN "ALL TANKS" OPERATION PYLON FUEL FLOWS ONLY TO MAIN TANK AND NOT TO THE FWD OR WING TANKS, PROVIDED THEY ARE FULL
IN "FWD AUX" OR "WING AUX" PYLON TANK FUEL FLOW WILL SUPPLY THE RESPECTIVE TANKS
CODE
NORMAL FLOW
WING AUX FLOW
FWD AUX FLOW
BOOSTER PUMP
CHECK VALVE
SHUT-OFF VALVE
SHUT-OFF VALVE ACTING AS CHECK
FUEL QUANTITY TRANSMITTER
VENT
AIR REFUELING RECEPTACLE
LEFT PYLON TANKS
INB'D
OFF
OUTB'D
PYLON TANKS AIR PRESS
INBOARD
OUTBOARD
GROUND REFUELING RECEPTACLE C AND LATER
RIGHT PYLON TANKS
INB'D
OFF
OUTB'D
PYLON TANKS AIR PRESS
INB'D
OUTBOARD
LEFT WING TANKS
RIGHT WING TANKS
FWD TANK
WING PUMP PRESS
FWD PUMP PRESS
TO CORRESPONDING SHUT-OFF VALVES
MAIN TANK LOW LEVEL
RESTRICTOR
CLOSED
FUEL SHUT-OFF VALVES
L. WING MAIN FWD R. WING
AUX BY-PASS
FROM CORRESPONDING FUEL QUANTITY TRANSMITTERS
TO BOOSTER PUMPS
MAIN TANK
MAIN
L. WING
L. INB'D PYL
FORWARD
R. INB'D PYL
R. WING
ALL TANKS
ALL TANKS
BOTH WINGS
FORWARD
ALL TANKS
BOTH WINGS
FORWARD
FUEL VOL ALL TANKS
AUTOMATICALLY DISCONNECTED DURING INDIVIDUAL TANK CHECK
FUEL QTY CHECK
OFF
ALL TANKS
WING AUX
OFF
FWD AUX
FUEL TANK SELECTOR
TO ENGINE STARTER
BOOST PRESS
FUEL TANK QUANTITY
TO ENGINE FUEL SYSTEM

Figure 1—17

FUEL QUANTITY DATA

TANKS	USABLE FUEL GAL	USABLE FUEL POUNDS	FULLY SERVICED GAL	FULLY SERVICED POUNDS
Main	187.0	1215.5	187.4	1218.1
Fwd	164.0	1066.0	164.6	1070.3
L. Wing	90.0	585.0	92.0	598.0
R. Wing	109.0	708.5	111.8	726.7
L. Inboard Pylon—Type 1	222.0	1443.0	225.0	1462.5
R. Inboard Pylon—Type 1	222.0	1443.0	225.0	1462.5
R. Inboard Pylon—Type 1 Alt	450.0	2925.0	450.0	2925.0
L. Inboard Pylon—Type 1 Alt	450.0	2925.0	450.0	2925.0
L. Outboard Pylon—Type II	222.0	1443.0	225.0	1462.5
R. Outboard Pylon—Type II	222.0	1443.0	225.0	1462.5

USABLE FUEL TOTALS	GAL	POUNDS
Maximum Internal (Main, Forward, Wing)	550.0	3575.0
Internal plus Inboard Pylons	994.0	6461.0
Internal plus Inboard Pylons (Alt)	1450.0	9425.0
Internal plus Inboard and Outboard Pylons	1438.0	9347.0

NOTE Weight based on fuel at 6.5 pounds per gallon

Figure 1—18

fuel tank selector. The main tank is kept full until all external and internal fuel is used. A booster pump in the main tank supplies the engine fuel system. The pilot may vary the normal selector to allow a direct flow of fuel to the engine from either the internal wing or forward tanks. If the wing or forward tank is selected and the main tank shut-off valve is open, fuel will flow to the main tank until it is full then all fuel will flow from the selected tank to the engine. The system is provided with a flowmeter which indicates the rate of fuel flow to the engine. Also a fuel quantity indicator which indicates the amount of fuel remaining in the airplane except for unmodified Type II external tanks. Provisions for ground refueling of the airplane, from a single point, are made through an adapter installed in the air refueling receptacle on A and B airplanes. On C and later airplanes a ground refueling receptacle is installed on the underside of the right wing. The single point refueling equipment is designed to operate similar to the air refueling system. The refueling truck must be equipped with a single-point nozzle and must be capable of delivering fuel at 500 GPM at a pressure of 50 PSI. The airplane is provided with an air refueling system which is covered in Section IV.

FUEL SPECIFICATION AND GRADE.

Recommended fuel specification and grade are noted on the servicing diagram, figure 1—41.

ELECTRIC FUEL BOOSTER PUMPS.

An electric booster pump is provided in the main fuel tank to supply fuel to the engine fuel control system. Electric booster pumps in the wing and forward tanks normally transfer fuel from these tanks to the main tank but also may be used to supply the engine fuel control system directly by proper positioning of the fuel tank selector. All the booster pumps are electrically operated from the secondary bus and are controlled by a rotary switch mechanically connected to the fuel tank selector. At altitudes below 6,000 feet, full engine RPM may be maintained with a failed booster pump as the fuel can be recovered by direct suction of the engine driven booster pump. Satisfactory engine operation up to maximum range power

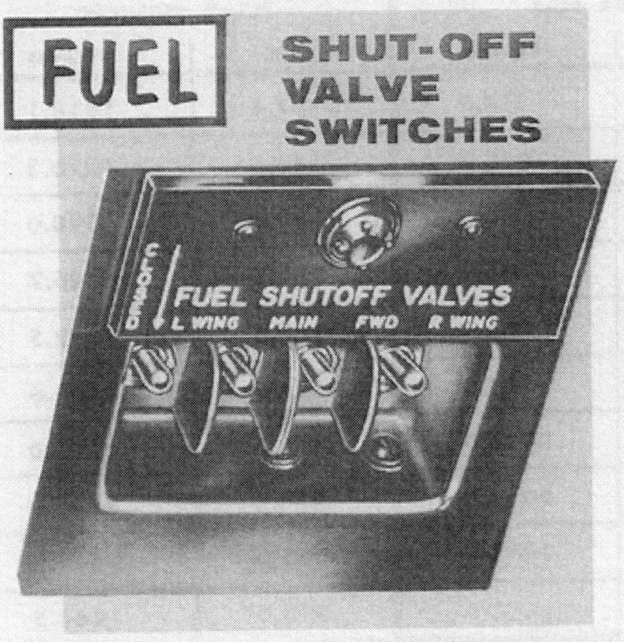

Figure 1—19

settings will result up to 20,000 feet under the most severe conditions when using JP-4 fuel. However, if booster pump failure is experienced, above 20,000 feet with JP-4 fuel and engine operation is satisfactory (no RPM drop or excessive RPM fluctuation) flight may be continued without reducing altitude. It must be remembered that prolonged flight at high altitudes, without booster pump operation, will shorten the engine fuel pump life due to the low inlet pressures. Fuel from the wing or forward tanks cannot be transferred to the main fuel tank without the aid of the booster pumps in the respective tanks, however, fuel may be fed directly to the engine from the forward tank but at slightly lower altitudes than would be available when feeding from the main tank. It is possible to feed in this manner from the wing tanks but operation under these conditions is not recommended since fuel in the tank with the failed pump cannot be recovered after the other tank is empty. There are no direct indicators to show when a booster pump is not operating, however, booster pump failure may be suspected as noted under fuel system indicators.

FUEL SHUT-OFF VALVE SWITCHES.

Shut-off switches (16, figure 1—8 and figure 1—19) are provided to close the fuel shut-off valves in the internal tanks so that fuel will not be transferred from the external to the internal or from the wing and forward to the main. The same fuel lines are used for transfer of fuel from the external to the internal tanks as is used to supply all tanks during air refueling. The fuel shut-off valves provide a means of controlling fuel flow to the individual internal tanks by closing a valve at the fuel line entrance to the tank. Each of the four fuel shut-off valves is controlled by a two-position toggle switch. The switches are marked L WING, MAIN, FWD and R WING with an arrow indicating the CLOSED position. During normal operation the switches are left in the up position, which allows fuel to transfer from the external to the internal tanks and also allows all tanks to be refueled from the refueling receptacle. By placing the fuel shut-off valve in the CLOSED position the respective fuel tank will not receive fuel by transfer or air refueling. When the main fuel shut-off switch is in the CLOSED position both valves in the main tank are closed and fuel from neither internal nor external tanks can be transferred. This system is provided to isolate each of the internal tanks, in the event of battle damage to the tank, or failure of the booster pump in the tank. The fuel shut-off valves are actuated by the primary bus.

FUEL TANK SELECTOR.

The fuel tank selector (figure 1—13) is a rotary control having four positions. The selector mechanically positions the fuel tank selector valve and indexes a rotary switch which opens and closes circuits to the internal tank booster pumps and to the fuel quantity indicator so that readings on the indicator will reflect the tank(s) being used. The four positions are as follows:

ALL TANKS.

The ALL TANKS (normal) position turns on all booster pumps and channels all internal fuel into the main tank. The fuel quantity indicator will read fuel in all tanks except outboard external tanks and unmodified Type II inboard external tanks. The main tank booster pump supplies fuel to the engine driven booster pump. As the fuel in the main tank drops to a predetermined level, a float in the tank opens a shut-off valve in the transfer line from external tanks and allows fuel from the external tanks to transfer to the main tank. When external tanks have emptied, fuel from the wing and forward tanks transfer to the main tank simultaneously. To maintain a favorable CG travel the flow is such that approximately 0 to 20 gallons will still be in the forward tank after the wing tanks have emptied. The remaining forward and main tank fuel is then used. The fuel quantity indicator will indicate fuel remaining in internal and Type I or modified Type II inboard external tanks.

FUEL SYSTEM WARNING LIGHTS

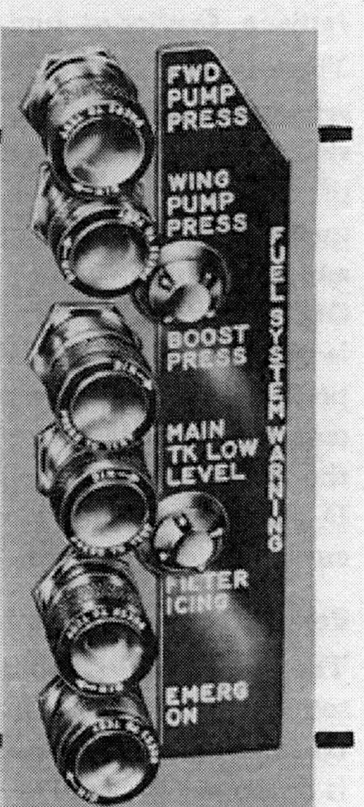

Figure 1—20

CAUTION

> Aircraft that do not have the fuel ejector incorporated in the main fuel tank, T.O. 1F-84-667, the following restrictions apply: With 700 pounds of fuel or less remaining in the main fuel tank, sustained uncoordinated or climbing turns, accelerations, and nose high attitudes can result in booster pump starvation and subsequent engine flame-out. Because of this, the following time limits should not be exceeded while performing any of the above described maneuvers when operating with the fuel tank selector in the ALL TANKS position.
>
> | MAIN TANK FUEL QUANTITY | TIME LIMIT |
> |---|---|
> | 700 pounds | 1.0 minute |
> | 400 pounds | 0.5 minute |

WING AUX.

The WING AUX position turns on the wing tank booster pumps; routes wing tank fuel directly to the engine driven booster pump and completes a circuit to the fuel quantity indicator so that the reading indicates fuel in the wing and Type I inboard external tanks only. No other booster pump will operate with the fuel tank selector in this position. When operating in WING AUX position with the fuel shut-off switches open fuel will feed to the engine and continue to transfer to the main tank until either the pylon tank float valve in the main tank closes if there is external fuel, or the internal tank float valve in the main tank closes if there is no external fuel. All remaining fuel in the external and wing tanks will then be available to the engine fuel system through the auxiliary feed lines.

CAUTION

> When operating with the fuel tank selector in the WING AUX position, avoid uncoordinated maneuvers, steep descents or rapid maneuvers, since these maneuvers may uncover the wing tank fuel outlets thereby causing a flame-out.

FWD AUX.

The FWD AUX position closes a cricuit to turn on the forward tank booster pump; indexes the selector valve so that forward tank fuel is fed directly to the engine driven booster pump of the engine fuel system and completes a circuit to the fuel quantity indicator so that the reading will indicate fuel in the forward and Type I or modified Type II inboard external tanks only. No other electric booster pump will operate with the fuel tank selector in this position. When operating in the FWD AUX position with the fuel shut-off switches open, fuel will feed to the engine and continue to transfer to the main tank until either the pylon tank float valve in the main tank closes if there is external fuel, or the internal tank float valve in the main tank closes if there is no external fuel. All remaining fuel in the external and forward tanks will then be available to the engine fuel system through the auxiliary feed lines.

CAUTION

> When operating with the fuel tank selector in the FWD AUX position with low fuel quantity remaining, avoid uncoordinated turns or nose down attitudes as these maneuvers may result in loss of fuel supply to the forward tank booster pump.

OFF.

The OFF position turns off the internal tank booster pumps and shuts off the fuel supply to the engine fuel system. The fuel quantity indicator will show the amount of fuel remaining in the internal tanks plus the fuel in Type I inboard external tanks.

PYLON TANKS AIR PRESSURE SWITCHES.

Fuel from the external tanks is transferred to the main tank by means of air pressure. The pylon tank air pressure switches (figure 1—13) control the flow of air for pressurizing the external tanks. On aircraft A thru G there are two switches; one marked LEFT and the other RIGHT. Each switch has three positions; INB'D PYLON TANKS AIR PRESS., OUTB'D PYLON TANKS AIR PRESS., and OFF. On H and later aircraft these two switches are marked L and R. Each switch has three positions; PYLON TANKS AIR PRESS. OUTBD, OFF and INBD. When in the INB'D or OUTB'D position on all aircraft, electrical power is directed to open the selected solenoid valve permiting air from the engine compressor to pressurize the respective tanks. In the OFF position the air pressure and vent ports are closed. The vent port is automatically opened during the air refueling cycle so as to permit fuel flow into the tanks. The air pressure port is spring-loaded to the closed position. Power from the primary bus keeps the port open, therefore, in the event of primary bus power failure the external fuel will not transfer.

CAUTION

The air refueling receiver switch must be in the CLOSED position in order to pressurize the pylon tanks.

PYLON TANK JETTISONING.

External Stores Jettison Switch.

The external stores jettison switch (figure 1—7) is a push button type twitch recessed in the panel to prevent accidental actuation. The switch is marked EXTERNAL STORES JETTISON, and when depressed will jettison the outboard and inboard pylon tanks together with the pylons on RE airplanes up to F-84F-45RE and all GK airplanes if the battery bus is energized. On F-84F-45RE and subsequent RE airplanes only the pylon tanks will be jettisoned and the pylons will be retained. Rockets will be jettisoned on all airplanes if the airplane is airborne and the primary bus is energized.

CAUTION

On airplanes up to F-84F-45RE and all GK airplanes the pylons will jettison together with any stores installed if the external stores jettison switch is depressed with the airplane in the static position. On F-84F-45RE and later RE airplanes only the stores will jettison if the external stores jettison switch is depressed with the airplane in the static position.

Jettison Outboard and Inboard Pylon Switches.

The inboard and outboard pylons can be jettisoned separately by actuating the respective pylon jettison switch (6, 7, figure 1—8; 20, 21, figure 1—9). Each of the two switches is guarded with a cover type guard. One guard is marked JETT INB'D PYLON and an arrow while the other is marked JETT OUTB'D PYLON with an arrow. When either switch is actuated in the direction of the arrow to the JETT position the respective pylons will be jettisoned simultaneously together with any stores carried on them if the primary bus is energized on modified A thru D and E and later aircraft or if the battery bus is energized on unmodified A thru D aircraft.

Bomb Release Switch.

The outboard or inboard pylon tanks can be jettisoned in the same manner as normally releasing the bombs by use of the bomb release switch. This switch is described in detail in Section IV.

WING PUMP PRESSURE WARNING LIGHT.

Wing booster pump pressure is indicated by an amber light (figure 1—20), marked WING PUMP PRESS. The light will illuminate if the pressure in the fuel line from either wing tank is below approximately nine psi and the primary bus is energized. The light-on condition indicates that one or both wing tanks are empty or the booster pump in one or both wing tanks has failed. In the event of a failed booster pump in one wing, fuel from that wing will not transfer to the main tank. However, the fuel can be utilized in WING AUX operation.

Figure 1—21

CAUTION

During normal operation, the wing pump pressure warning light will illuminate when one or both wing tanks are empty. However, in WING AUX operation, the engine is being fed directly from the wing tanks and at the first flicker or flash of the warning light the FWD AUX position on the fuel tank selector must be selected immediately (if fuel remains in the forward tank) to assure against a flame-out.

FORWARD PUMP PRESSURE WARNING LIGHT.

Forward tank booster pump pressure is indicated by an amber light (figure 1–20) marked FWD PUMP PRESS. The light will illuminate if the pressure in the fuel line from the forward tank is below approximately nine PSI and the primary bus is energized. The light-on condition indicates that the forward tank is empty or the booster pump has failed. In the event of a booster pump failure the forward tank fuel cannot be transferred to the main tank by gravity since the main tank is higher than the forward. However, the fuel can be utilized in FWD AUX operation.

MAIN TANK LOW LEVEL WARNING LIGHT.

The main tank low level warning light (figure 1–20) is a red light marked MAIN TK LOW LEVEL and will illuminate when the main tank fuel level is between 830 and 1120 pounds and will remain on until the main tank is refueled or the primary bus is deenergized.

BOOSTER PRESSURE WARNING LIGHT.

The booster pressure warning light (figure 1–20) is a red light marked BOOST. PRESS. and will illuminate if the primary bus is energized and the fuel pressure between the fuel selector valve and the engine driven booster pump is below approximately nine PSI.

Note

It is permissible for the booster pressure warning light to flicker when the emergency fuel system is switched ON or OFF and flicker or remain on momentarily during burst acceleration from any throttle setting. This is due to a momentary increase in fuel flow demanded by the engine above the output of the pump resulting in a decrease in pressure in the engine feed line causing the pressure switch to actuate the light.

FUEL QUANTITY INDICATOR.

The fuel quantity indicator (15, figure 1–4; 16, figure 1–5; 16, figure 1–6) is an AC powered electrical unit which operates from either the normal or alternate inverter. It reads the quantity of fuel remaining in any or all internal tanks in the airplane and Type 1 or modified Type II inboard external tanks in units of pounds. Fuel weight is calibrated by the combination of a float unit in each tank and a unit installed in the main tank which compensates for the density and temperature of fuel in the tanks. This combination assures a true indication of the weight of fuel in the airplane within a tolerance of approximately 4 per cent when operating with the fuel tank selector in the ALL TANKS positions; however, the indicator will read differently each time the airplane is serviced with fuel of a different density. A fuel volume switch is provided to change the electrical circuit so that full fuel volume is indicated on the same dial. The fuel volume reading is necessary in that it provides a means of indicating that the airplane is fully serviced regardless of fuel density. Three reference points are positioned on the dial to indicate full fuel for three different configurations.

WARNING

Aircraft that do not have the ejector incorporated in the main fuel tank, T.O. 1F-84-667, the following restrictions apply: With 700 LB of fuel or less remaining in the main tank, sustained uncoordinated or climbing turns, accelerations, and nose high attitudes can result in booster pump starvation and subsequent engine flame-out. A properly sealed main tank pump compartment will contain sufficient fuel to insure satisfactory engine operation, for the following periods, when performing any of the above maneuvers.

MAIN TANK FUEL QUANTITY	OPERATING TIME
700 LB	1.0 minute
400 LB	0.5 minute

Normal Reading.

Normally the reading on the large scale multiplied by 1000 plus the reading on the small scale multiplied by 100 indicates the fuel remaining in the airplane in pounds. With the fuel tank selector in the ALL TANKS or OFF position and the fuel quantity check switch OFF the fuel quantity indicator will read the total weight of internal and Type I or modified Type II inboard external fuel remaining in the airplane. With the fuel tank selector in the WING AUX or

FWD AUX position and the fuel quantity check switch OFF the quantity indicator will read the weight of fuel remaining in the respective tank plus the fuel remaining in the Class I inboard external tanks.

WARNING

Due to an inherent error in the system, on unmodified aircraft A thru E not incorporating T.O. 1F-84-267, the fuel quantity indicator reads approximately 25 per cent low when operating with the fuel tank selector in the WING AUX or FWD AUX position.

Fuel Volume Switch.

The fuel volume switch (17, figure 1–4; 18, figure 1–5; 18, figure 1–6) marked FUEL VOL ALL TKS is provided to change the fuel quantity indicator a-c circuit so that fuel volume instead of fuel weight is read on the fuel quantity indicator. This information is valuable when refueling on the ground or in the air as the fuel capacity of the airplane, in gallons, remains constant. There are three dots on the indicator, one at 4195 LB corresponds to full internal fuel only, the second dot at 7551 LB corresponds to full internal fuel plus two 230 GAL Type I inboard external tanks, the third dot at 10.761 LB corresponds to full internal fuel plus two 450 GAL Type I inboard external tanks. When the fuel volume switch is depressed with the fuel tank selector in the ALL TANKS position and the airplane fully serviced the pointer will stop at the dot corresponding to the airplane configuration. Validity of these readings would be affected if outboard external tanks or unmodified Type II inboard tanks are carried since these do not have fuel quantity transmitters.

Fuel Tank Quantity Selector Switch.

The fuel tank quantity selector switch (17, figure 1–4; 18, figure 1–5; 18, figure 1–6) is provided so that the fuel quantity in individual tanks can be determined. It is a rotary switch with the following positions; MAIN, L WING, EXT L INB'D, FWD, EXT R INB'D and R WING. When positioned to any one of these positions, the quantity of fuel in the respective tank will register on the fuel quantity indicator when the fuel quantity check switch is placed in the FUEL QTY CHECK position regardless of the position of the fuel tank selector. Unmodified Type II inboard external tanks will not register fuel remaining as these tanks are not equipped with fuel quantity transmitters.

Note

On A thru D aircraft not incorporating T.O. F-84-267, if the fuel quantity selector switch is positioned to the EXT R INB'D position without pylon tanks installed all internal fuel remaining will be indicated on the fuel quantity indicator when the fuel quantity check switch is positioned to the FUEL QTY CHECK position. If the quantity selector switch is positioned to EXT L INB'D position under the same conditions the quantity indicator will read zero.

CAUTION

On A and B aircraft not incorporating T.O. 1F-84-267 the fuel quantity remaining readings for individual tanks, obtained by use of the fuel tank quantity selector switch, will not add up to the total fuel remaining reading obtained with the fuel tank selector in the ALL TANKS position due to the inherent tolerances in the calibrating system. However, the total fuel quantity remaining will be correct to within approximately 4 per cent.

Fuel Quantity Check Switch.

The fuel quantity check switch (17, figure 1–4; 18, figure 1–5; 18, figure 1–6), is used when it is desired to ascertain the quantity of fuel in the individual tanks. The switch has two positions FUEL QTY CHECK and OFF. When placed in the FUEL QTY CHECK position the quantity of fuel remaining in the tank, indicated by the position of the fuel tank quantity selector switch, will register on the fuel quantity indicator. The OFF position will register fuel remaining in tanks as indicated by the fuel tank selector.

FUEL FLOW INDICATOR.

The fuel flow indicator (18, figure 1–4; 19, figure 1–5; 19, figure 1–6) registers the flow of fuel to the engine in units of pounds per hour. The indicator is AC powered from either the main or alternate inverter. It is operated by a fuel flow meter which is located in the fuel system between the pressurizing valve and the engine burners. All fuel flowing to the engine will be recorded whether from the normal or auxiliary systems. The fuel flow indicator is accurate to approximately one per cent on the high side to three per cent on the low side.

ELECTRICAL POWER SUPPLY SYSTEM.

The aircraft is equipped with both AC and DC electrical power systems (figures 1–22 and 1–23). All circuits are protected with either a circuit breaker or a fuse both of which are accessible from the cockpit. A 4.5 volt dry cell battery is installed as an emergency source of power for canopy jettisoning if the main electrical system fails. On H and later aircraft, a 26.26 volt, wet cell, silver zinc battery is provided for AN/ARC-34 radio.

CIRCUIT BREAKER PANELS.

Circuit breaker panels (figure 1–24) are provided to protect the various electrical circuits in the airplane. The circuit breakers are of the push button type and are pushed in to reset.

DC SYSTEM.

The 28-volt DC system is powered from a 400 ampere, engine-driven generator with a 24-volt battery as a standby source of DC power. The system also incorporates an external power receptacle for the accommodation of an external power cart. Electrical power is distributed through a three bus system consisting of: a battery bus, a primary bus, and a secondary bus. The battery bus services emergency equipment and remains energized regardless of the battery switch position or generator operation. The primary bus services equipment essential to flight and is energized by the battery, the generator, or the external power receptacle. The secondary bus services equipment not essential to flight and is energized by the generator or the external power receptacle. Therefore, in the event of generator failure in flight, all equipment not essential to flight will be automatically cut out since the secondary bus will cease to be energized, and battery power will be conserved for primary bus equipment, i.e., equipment essential to flight. Three circuit breakers, located in the battery compartment, are provided to deenergize circuits energized by the battery bus when ground personnel are working on the aircraft. The external power receptacle must be used for operational check of electrical equipment if the engine is not operating because the secondary bus cannot be energized by the aircraft's battery.

A THRU G

Battery Bus Circuit Breakers.

The battery bus is divided into three circuits each of which is protected by a circuit breaker. The three battery bus circuit breakers are installed in the battery compartment. On A THRU D aircraft they are marked: APX-6 DEST, PYLON JETT and CANOPY JETT. On E thru G aircraft they are marked: APX-6 DEST, EXT STORE JETT–BOMB ARM, and CANOPY JETT. When the circuit breakers are closed, all battery bus circuits except the canopy jettison circuit will be connected directly to the battery. The canopy jettison circuit is connected directly to the battery only when the battery switch is in the ON position. A press-to-test light is provided in the cockpit to indicate when the pylon jettison circuit breaker is closed on A THRU D aircraft. On E THRU G aircraft the test light is used to indicate that the external stores jettison-bomb arming circuit breaker is closed.

A THRU G

Circuit Breaker Test Light.

The circuit breaker test light on A THRU D aircraft is provided as an indicator to show that the pylon jettison and bomb arming circuit breaker, located in the battery compartment is closed. It is a green light marked: PRESS TO TEST-LIGHT ON INDICATES CIRC BRKR CLOSED–THIS CB SUPPLIES POWER TO BOMBS & PYLON JETTISON. On E THRU G aircraft the light is marked: PRESS TO TEST-LIGHT ON INDICATES CIRC BRKR CLOSED–THIS CB SUPPLIES POWER TO EXT STORE JETTISON. If the light illuminates when pressed, it indicates that the circuit breaker is closed and the circuit is energized. The light is powered by the battery bus.

H and LATER

Battery Bus Circuit Breakers.

The battery bus on H and later aircraft is divided as in earlier airplanes into three circuits, each of which is protected by a circuit breaker. The three circuit breakers are located in the battery compartment and are marked: RX JETT, EXT STORE JETT and CANOPY JETT. When the circuit breakers are closed, all battery bus circuits except the canopy jettison circuit will be connected directly to the battery. The canopy jettison circuit is connected directly to the battery only when the battery switch is in the ON position. A check switch and indicator light are provided in the cockpit to indicate whether or not the rocket jettison and external stores jettison circuit breakers are closed and the circuits energized. On modified and J and later aircraft a fourth circuit breaker marked DRAG CHUTE has been added to the circuit breaker panel, in the battery compartment and a press-to-test light is installed in the cockpit. The indicator light will illuminate when pressed if the drag chute circuit breaker is closed and the circuit is energized.

ELECTRICAL A thru G

POWER SUPPLY SYSTEM SCHEMATIC

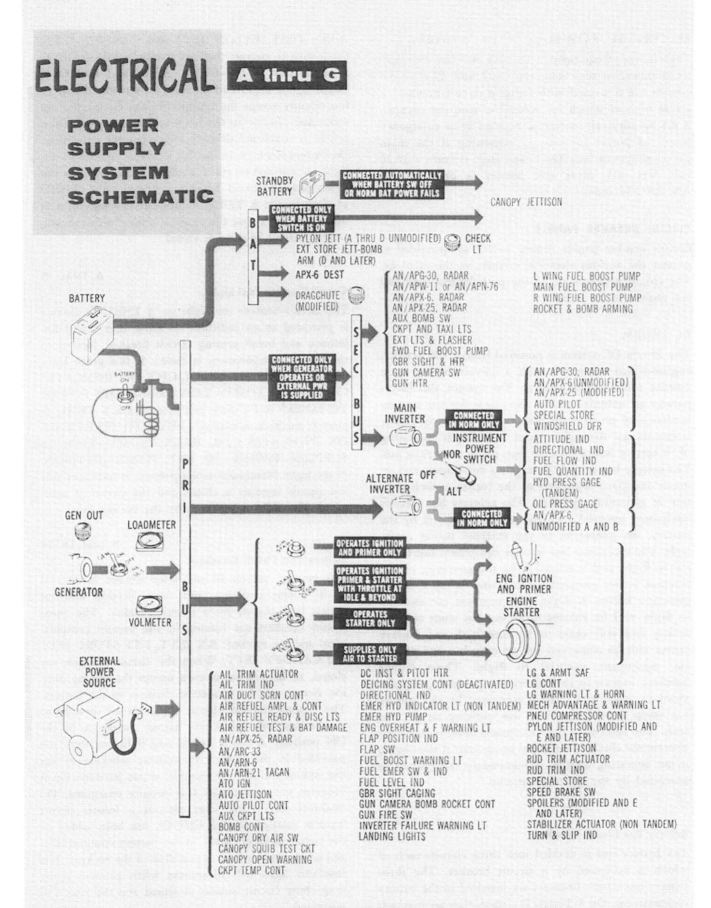

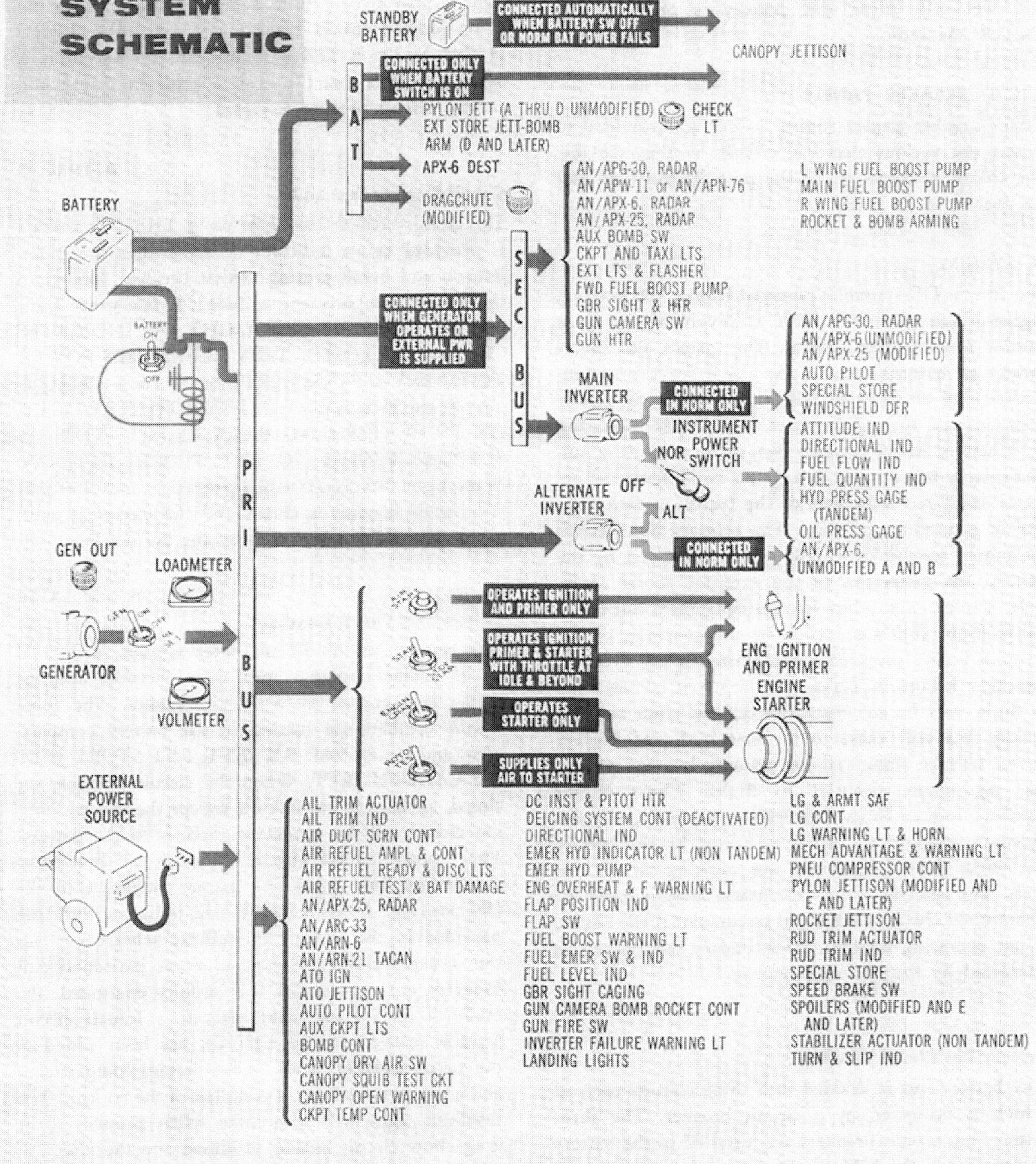

STANDBY BATTERY
CONNECTED AUTOMATICALLY WHEN BATTERY SW OFF OR NORM BAT POWER FAILS
CANOPY JETTISON
CONNECTED ONLY WHEN BATTERY SWITCH IS ON
BAT
PYLON JETT (A THRU D UNMODIFIED)
EXT STORE JETT-BOMB ARM (D AND LATER)
CHECK LT
APX-6 DEST
DRAGCHUTE (MODIFIED)
BATTERY
AN/APG-30, RADAR
AN/APW-11 or AN/APN-76
AN/APX-6, RADAR
AN/APX-25, RADAR
AUX BOMB SW
CKPT AND TAXI LTS
EXT LTS & FLASHER
FWD FUEL BOOST PUMP
GBR SIGHT & HTR
GUN CAMERA SW
GUN HTR
L WING FUEL BOOST PUMP
MAIN FUEL BOOST PUMP
R WING FUEL BOOST PUMP
ROCKET & BOMB ARMING
SEC BUS
CONNECTED ONLY WHEN GENERATOR OPERATES OR EXTERNAL PWR IS SUPPLIED
MAIN INVERTER
CONNECTED IN NORM ONLY
AN/APG-30, RADAR
AN/APX-6 (UNMODIFIED)
AN/APX-25 (MODIFIED)
AUTO PILOT
SPECIAL STORE
WINDSHIELD DFR
INSTRUMENT POWER SWITCH
NOR
OFF
ALT
ALTERNATE INVERTER
ATTITUDE IND
DIRECTIONAL IND
FUEL FLOW IND
FUEL QUANTITY IND
HYD PRESS GAGE (TANDEM)
OIL PRESS GAGE
AN/APX-6, UNMODIFIED A AND B
CONNECTED IN NORM ONLY
PRI BUS
GEN OUT
LOADMETER
GENERATOR
VOLMETER
EXTERNAL POWER SOURCE
OPERATES IGNITION AND PRIMER ONLY
OPERATES IGNITION PRIMER & STARTER WITH THROTTLE AT IDLE & BEYOND
OPERATES STARTER ONLY
SUPPLIES ONLY AIR TO STARTER
ENG IGNTION AND PRIMER
ENGINE STARTER
AIL TRIM ACTUATOR
AIL TRIM IND
AIR DUCT SCRN CONT
AIR REFUEL AMPL & CONT
AIR REFUEL READY & DISC LTS
AIR REFUEL TEST & BAT DAMAGE
AN/APX-25, RADAR
AN/ARC-33
AN/ARN-6
AN/ARN-21 TACAN
ATO IGN
ATO JETTISON
AUTO PILOT CONT
AUX CKPT LTS
BOMB CONT
CANOPY DRY AIR SW
CANOPY SQUIB TEST CKT
CANOPY OPEN WARNING
CKPT TEMP CONT
DC INST & PITOT HTR
DEICING SYSTEM CONT (DEACTIVATED)
DIRECTIONAL IND
EMER HYD INDICATOR LT (NON TANDEM)
EMER HYD PUMP
ENG OVERHEAT & F WARNING LT
FLAP POSITION IND
FLAP SW
FUEL BOOST WARNING LT
FUEL EMER SW & IND
FUEL LEVEL IND
GBR SIGHT CAGING
GUN CAMERA BOMB ROCKET CONT
GUN FIRE SW
INVERTER FAILURE WARNING LT
LANDING LIGHTS
LG & ARMT SAF
LG CONT
LG WARNING LT & HORN
MECH ADVANTAGE & WARNING LT
PNEU COMPRESSOR CONT
PYLON JETTISON (MODIFIED AND E AND LATER)
ROCKET JETTISON
RUD TRIM ACTUATOR
RUD TRIM IND
SPECIAL STORE
SPEED BRAKE SW
SPOILERS (MODIFIED AND E AND LATER)
STABILIZER ACTUATOR (NON TANDEM)
TURN & SLIP IND

Figure 1—22

ELECTRICAL H and LATER

POWER SUPPLY SYSTEM SCHEMATIC

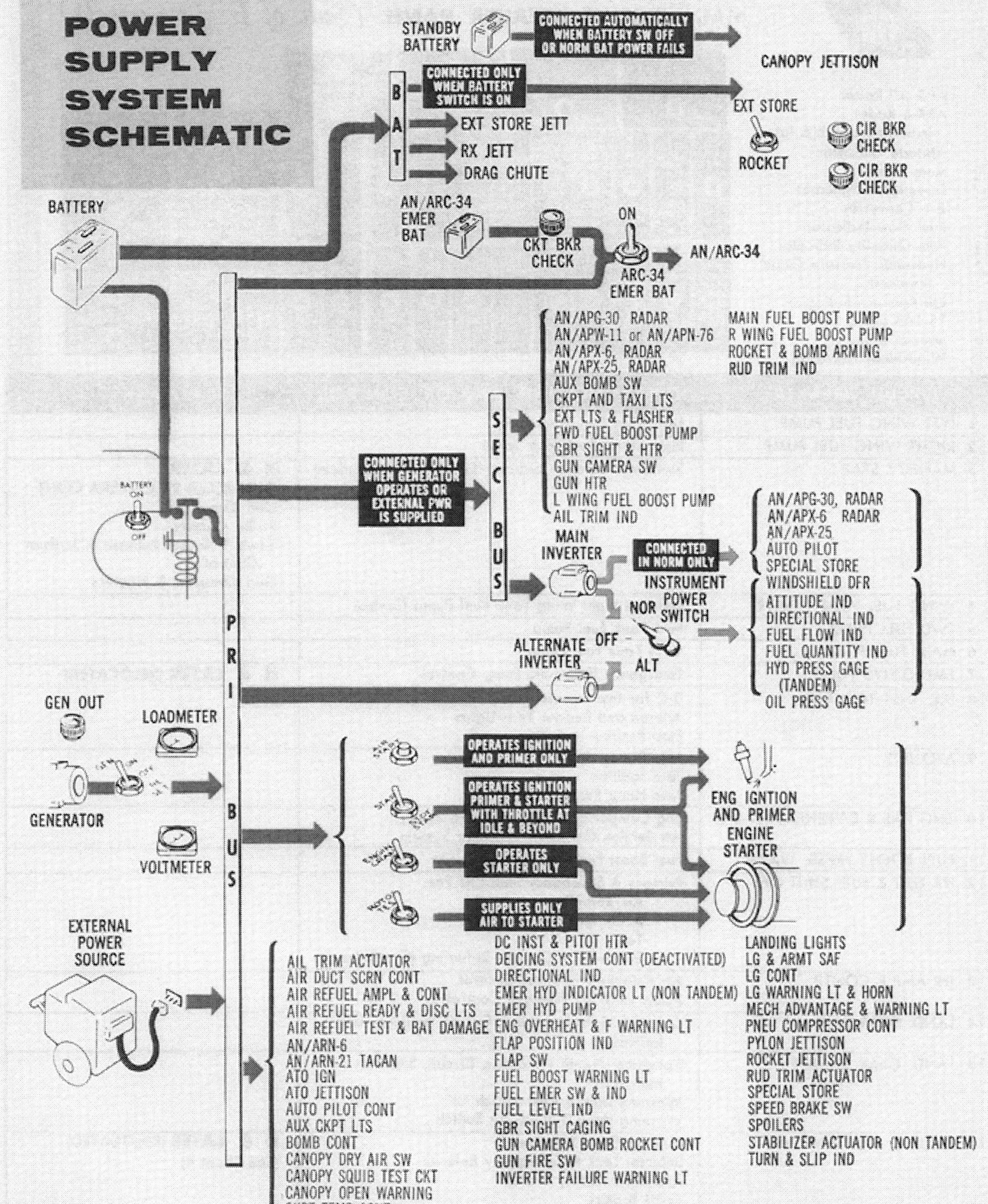

Figure 1—23

CIRCUIT BREAKER PANELS

MAIN CIRCUIT BREAKER PANEL *(Side of Right Console)*

APG-30 Radar
APX-6 Radar (modified and E & later)
Attitude Indicator
Auto Pilot
Directional Indicator
E-1 Converter
Fuel Flow Indicator
Fuel Quantity Indicator
Hydraulic Pressure Gage (tandem)
Oil Pressure Indicator
PS-103 Servo
Special Store
Windshield Defroster

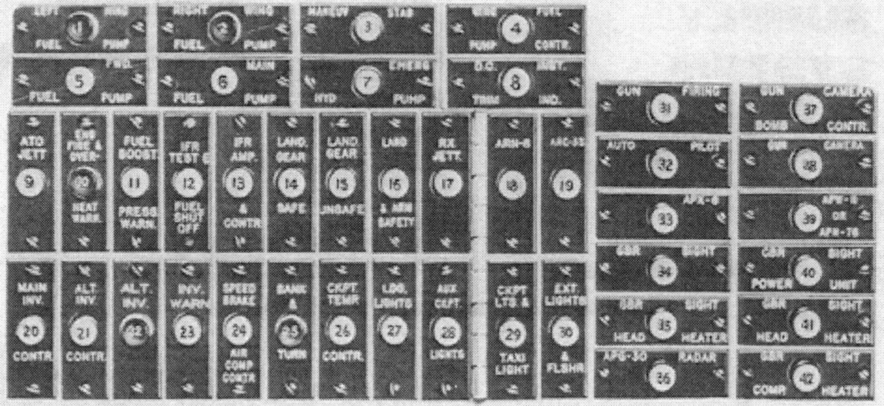

CIRCUIT BREAKER	OPERATING UNIT AND OR CONTROL	REMARKS
1 LEFT WING FUEL PUMP	Left Wing Fuel Tank Pump	
2 RIGHT WING FUEL PUMP	Right Wing Fuel Tank Pump	
3 MANEUV STAB	Stabilator Electric Actuator Control (Non-Tandem)	**H & LATER** GUN BOMB RX CAMERA CONT Gun Camera Strike Camera Bomb & Rocket Release & Jettison Control Gun Chargers & Heaters
4 WING FUEL PUMP CONTR	Left and Right Wing Tank Fuel Pump Control	
5 FWD FUEL PUMP	Fwd Tank Fuel Pump	
6 MAIN FUEL PUMP	Main Tank Fuel Pump	
7 EMERG HYD PUMP	Emergency Hydraulic Pump Control	**H & LATER** (RELOCATED)
8 D.C. INST TRIM IND	D-C for Instruments Aileron and Rudder Trim Lights Flap Position Indicator	
9 ATO JETT	Jato Ready Light Jato Ignition Jato Hook Retract Control	
10 ENG FIRE & OVERHEAT WARN.	Eng Compartment Fire Warning System Aft Section Overheat Warning System	
11 FUEL BOOST PRESS. WARN.	Fuel Boost Pressure Warning Light	
12 IFR TEST & FUEL SHUT OFF	Primary & Secondary Shut Off For: Air Refueling Battle Damage Test Single-Point Ground Refueling (C and Later)	
13 IFR AMP & CONTR	Air Refueling Receiver Control Pylon Tank Air Pressure Control	
14 LAND. GEAR SAFE	Complete Safe Indicating Circuit, Switches and Indicators	
15 LAND. GEAR UNSAFE	Complete Unsafe Indicating Circuit, Switches and Indicators Warning Light & Test Switch Warning Horn & Silencing Switch	
16 LAND. & ARM. SAFETY	Landing Gear Selector Selector Lock & Emergency Release Armament Safety Switch Boost Brakes	**H & LATER** (RELOCATED) (See Sheet 4)

Figure 1—24 (Sheet 1 of 4)

CIRCUIT BREAKER	OPERATING UNIT AND OR CONTROL	REMARKS
17 RX JETT	Rocket Jettison Solenoids (Rocket Jettison Squibs G and later)	**H & LATER** GUN FIRING Gun Firing Solenoids
18 ARN-6	Radio Compass	
19 ARC-33	Command Radio	**H & LATER** ARC-34 Command Radio
20 MAIN INV CONTR	Main Inverter Control (Selector Switch)	
21 ALT INV CONT	Alternate Inverter Control (Selector Switch) Auto Pilot Gyro Motors (E and Later)	
22 ALT INV	APX-6 Radar (unmodified A thru D) Attitude Indicator Directional Indicator Fuel Flow Indicator Fuel Quantity Indicator Hydraulic Pressure Indicator (Tandem) Oil Pressure Indicator	
23 INV WARN.	Inverter Failure Warning Light Oxygen Warning Light Oxygen Panel Light	
24 SPEED BRAKE AIR COMP CONTR	Speed Brake Control Pneumatic System Control	
25 BANK & TURN	Turn and Slip Indicator	
26 CKPT TEMP CONTR	Cabin Temperature Control Cabin Pressurization Canopy Seal Control Canopy Open Warning Light Canopy Anti-Fog (Dry Air) Windshield Defrost (Hot Air) Windshield Defrost Control (Electric)	**G & LATER** CKPT TEMP CONT SEAT ADJUST Same as A thru C but with seat elevation actuator and control added.
27 LDG LIGHTS	Landing Lights Landing Light Actuators	
28 AUX CKPT LIGHTS	Auxiliary Cockpit Lights C-4A Light	
29 CKPT LTS & TAXI LIGHT	Flight Placards Instrument Lights Taxi Light	
30 EXT LIGHTS FLSHR	Position Light Flasher Unit	**H & LATER** EXT LTS FLSHR & ARC-34 BAT HTR Position Lights & Flasher Unit ARC-34 Emerg Battery Heater
31 GUN FIRING	Gun Firing Solenoids	**H & LATER** AUTO PILOT Auto Pilot Servos
32 AUTO PILOT	Auto Pilot Servos	**H & LATER** AUX BOMBS Auxiliary Bomb Selector Auxiliary Bomb Release Light
33 APX-6	Identification Radar	
34 GBR SIGHT	Gun, Bomb & Rocket Sight	
35 GBR SIGHT HEAD HEATER	GBR Sight Head Heater	
36 APG-30 RADAR	Gun, Bomb & Rocket Laying Radar	
37 GUN CAMERA BOMB CONTR	Gun Firing, Bomb & Rocket Release & Jettison Control Gun Chargers & Heaters	**H & LATER** BOMBS & RX Bomb & Rocket Control
38 GUN CAMERA AILERON & RUDDER NEUTRAL TRIM INDICATOR (H AND LATER)	Gun Camera	**C THRU G** CAMERA GSAP & STRIKE Gun Camera Strike Camera
39 APW-11 or TACAN	Ground Support Radar or Airborne Navigational Radar	
40 GBR SIGHT POWER UNIT	GBR Sight Power Unit	
41 GBR SIGHT HEAD HEATER	GBR Sight Head Heater	
42 GBR SIGHT COMP HEATER	GBR Sight Computer Heater	

Figure 1—24 (Sheet 2 of 4)

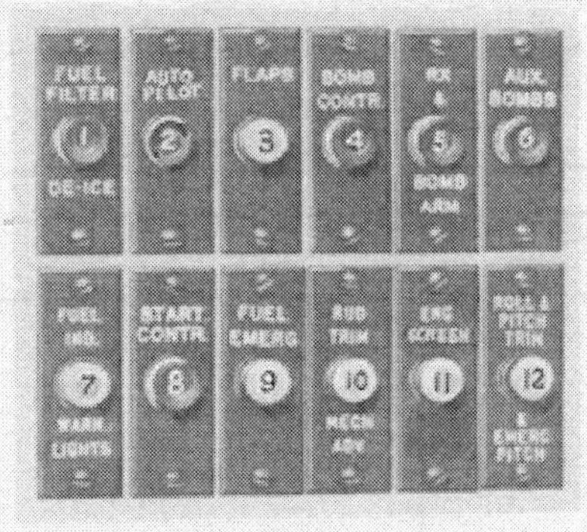

AUXILIARY CIRCUIT BREAKER PANEL

A thru G

(Location-Top of Left Console)

CIRCUIT BREAKER	OPERATING UNIT AND OR CONTROL	REMARKS
1 FUEL FILTER DE-ICE	Complete Fuel Filter De-Icing System	Deactivated
2 AUTO PILOT	Auto Pilot Gyro Motors (Some Aircraft)	**E THRU G** PYLON JETT Pylon Jettison Control & Pylon Squibs
3 FLAPS	Wing Flap Control (Spoiler Control on Modified Aircraft)	**E THRU G** FLAPS & SPOILER Wing Flap Control Spoiler Control
4 BOMB CONTR	Bomb Control	**E THRU G** RX & BOMB ARM Bomb Control & Rocket Arming
5 RX & BOMB ARM	**A THRU E** Rocket Bomb Arming	**G ONLY** EMERG HYD CONT Emergency Hydraulic Pump Control
6 AUX BOMBS	Auxiliary Bomb Selector and Indicator Light	
7 FUEL IND WARN. LIGHTS	Fwd Pump Pressure Warning Light Wing Pumps Pressure Warning Light Main Pump Pressure Warning Light Main Tank Low Level Warning Light	
8 START CONTR	Engine Start-Stop Start Engine Crank Rotor Test Air Start Inverter Failure Light	
9 FUEL EMERG	Fuel Control Emergency System and Indicator Light	
10 RUD TRIM MECH ADV	Rudder Trim Control Actuator & Indicator Mechanical Advantage Shifter & Control	
11 ENG SCREEN	Engine Screen Control and Warning Light	
12 ROLL & PITCH TRIM & EMERG PITCH	Aileron Trim Control & Actuator Pitch Trim Control & Actuator	**G ONLY** ROLL & PITCH TRIM Aileron Trim Control & Actuator Pitch Trim Control & Actuator

Figure 1—24 (Sheet 3 of 4)

AUXILIARY CIRCUIT BREAKER PANEL

H and LATER

(Location-Top of Right Console)

CIRCUIT BREAKER	OPERATING UNIT AND OR CONTROL	REMARKS
1 FUEL IND WARN LIGHTS	Fwd Tank Pressure Warning Light Wing Tanks Pump Pressure Warning Light Main Tank Pump Pressure Warning Light Main Tank Low Level Warning Light	
2 FUEL EMERG	Fuel Control Emergency System Control and Indicator Light	
3 FUEL FILTER DE-ICE	Complete Fuel Filter De-Icing System	**K AND LATER** (RELOCATED) (Deactivated)
4 START CONT	Engine Start-Stop Start Engine Crank Rotor Test Air Start Inverter Failure & Generator Out Press-to-Test Light	
5 ENG SCREEN	Engine Screen Control and Warning Light	
6 LAND GEAR & ARM SAFETY	Gear Selector Gear Selector Lock and Emergency Release Armament Safety Switch and Boost Brakes	
7 ROLL & PITCH TRIM	Aileron Trim Control & Actuator Pitch Trim Control & Actuator	
8 YAW TRIM MECH ADV	Rudder Trim Control Actuator and Indicator Light MA Shifter & Control	
9 FLAPS & SPOILER	Wing Flap Control Spoiler Control	
10 PYLON JETT	Pylon Jettison Selector, Control and Squibs	
11 EMER HYD CONT	Emergency Hydraulic Pump Control	
12 RX & BOMB ARM	Rocket & Bomb Arming	
13 EMER HYD PUMP	Emergency Hydraulic Pump	
14 MANEUV STAB	Stabilator Electric Actuator Control (non-tandem)	

Figure 1—24 (Sheet 4 of 4)

H and LATER

Jettison Circuit Breaker Check Switch.

A jettison circuit breaker check switch (10, figure 1–9) and indicator light installed in the cockpit is marked JETT CIRCUIT BREAKER CHECK and has three positions: EXT STORES, ROCKETS and an unmarked off position. It is used to determine whether or not the rocket jettison and external stores circuit breakers in the battery compartment are closed and whether the circuits are receiving power from the battery. When the switch is in the EXT STORES or ROCKETS position the check light will illuminate if the respective circuit breaker is closed and the circuit is energized by the battery.

H and LATER

Jettison Circuit Breaker Check Light.

A jettison circuit breaker check light (11, figure 1–11) installed in the cockpit is marked JETT CIRCUIT BREAKER CHECK. The check light will illuminate when the jettison circuit breaker check switch is in either the EXT STORES or ROCKET position and the respective jettisoning circuit breaker is closed and the circuit is energized by the battery.

MODIFIED and J and LATER

Drag Chute Circuit Breaker Check Light.

A press-to-test type drag chute circuit breaker check light (figure 1–35) is installed in the cockpit, adjacent to the drag chute control handle. When pressed the check light will illuminate if the drag chute circuit breaker in the battery compartment is closed and the drag chute control circuit is energized by the battery.

Battery Switch.

The battery switch (figure 1–7) is a two position on-off switch. The ON position connects the aircraft's battery to the primary bus through a relay which obtains its power directly from the battery. Approximately 18 volts are required to close the relay and it will remain closed down to approximately seven volts. Therefore, unless approximately 18 volts is available from the battery, the relay will not close. In the OFF position the battery is disconnected from the primary bus.

Generator Switch.

The generator switch (figure 1–7) positions are ON, OFF and RESET. The switch is guarded in the ON position. If the generator voltage is too high or too low, the generator is automatically disconnected from the electrical system. If the generator has cut-out due to over-voltages, the generator switch is positioned to RESET for a few seconds, to reset the generator field control relay, and then returned to ON. If the generator is not operating after positioning the generator switch to RESET and returning to the ON position, the OFF position is selected to disconnect the generator from the electrical system. The ON position connects the generator to the electrical system whenever the engine speed exceeds approximately 30 per cent which closes the reverse current relay.

Loadmeter.

The loadmeter (figure 1–7) is marked LOAD and indicates the load being drawn from the generator in per cent from zero to 100 per cent, with provisions for an additional 25 per cent reading to indicate over-load, and a minus 10 per cent reading to indicate discharge. The normal loadmeter reading is approximately 0.5 ($\pm$0.1).

Voltmeter.

The voltmeter (figure 1–7) indicates the voltage output of the generator.

Generator Out Indicator Light.

The generator out indicator light (figure 1–7) is a red light marked GEN OUT. Whenever the reverse current relay opens, the generator out indicator will illuminate, which shows that the generator is disconnected from the primary bus.

AC SYSTEM.

Unmodified A and B Airplanes.

The AC system is powered from two single phase 115-volt, 400 cycle, 1500 VA inverters; one is a main inverter and the other is an alternate inverter. The main inverter requires power from the primary and secondary buses and will therefore not be available if the secondary bus (or generator) fails. The alternate inverter is powered from the primary bus, and therefore, in event of generator failure, only AC powered instruments which are necessary for flight will operate, provided the instrument power switch is in the alternate position. A switch for selection of the inverters and a light for indication of failure of either inverter are located on right auxiliary instrument panel. With selector switch on normal, both inverters are in operation. If secondary bus fails it will be necessary to place selector switch to alternate position to energize instruments necessary for flight.

Modified Airplanes.

These aircraft incorporate a 2250 VA main inverter. Production changes resulted in only the main inverter operating with the selector switch in NOR position. On these aircraft all AC equipment is operated from the main inverter during normal operation. The alternate inverter is in operation only with selector switch in ALT position. The alternate inverter is used to energize necessary flight instruments only in the event of main inverter failure. The main inverter requires primary and secondary bus power while the alternate inverter requires only primary bus power.

Instrument Power Switch (Inverter Selection).

The instrument power switch (figure 1–7) has three positions, NOR, OFF and ALT. In the NOR position on unmodified A and B aircraft with 1500 VA inverters, both main and alternate inverters are operating if the secondary bus is energized. The main inverter supplies power to the instruments, the autopilot, the windshield defroster and the gun-bomb-rocket sight, while the alternate inverter supplies power to the identification radar. On modified airplanes, only the main inverter operates with selector switch in NOR position. On these airplanes, the main inverter supplies all AC power if the secondary bus is energized. The ALT position supplies power from the alternate inverter to the instruments if the primary bus is energized.

Inverter Failure Indicator Light.

Failure of the main or alternate inverter is indicated by illumination of the inverter failure indicator light (figure 1–7) which is marked INVERT FAILURE. If the instrument power switch is in the NOR position and the inverter failure indicator light illuminates, the main inverter is inoperative. (May be caused by inverter or generator failure.) Turning the instrument power switch to the ALT position will shift AC power output to the alternate inverter and the light should go out in a few seconds. If the inverter failure indicator light goes on with the instrument power switch in the ALT position, it indicates that the alternate inverter is not operating. The light is energized by the primary bus.

AC Fuse Box.

AN AC Fuse Box (7, figure 1–10) is provided on the right side of the cockpit with fuses to protect the AC circuits. A placard inside the fuse box door identifies each fuse and indicates the correct amperage. Spare fuses are provided in the spare fuse panel (11, figure 1–7) on the side of the right console. Fuses are provided for the following circuits:

Directional indicator (2 fuses)
Flowmeter
Oil Pressure Indicator
Instrument Power Converter
Fuel Level System
AN/APN-76 Radar **(Deleted)**
Auto Pilot
Windshield Defroster
AN/APG-30
AN/APX-6
Hydraulic Pressure Indicators (tandem aircraft)

PNEUMATIC POWER SUPPLY SYSTEM.

The pneumatic power supply system (figures 1–25 and 1–26) is provided to supply air pressure to recharge the engine starter air storage bottle and to provide an emergency means of extending the landing gear. Air is bled from the engine compressor, then pressurized by a hydraulically operated compressor and stored in two storage bottles. The system is fully automatic; the hydraulic pressure from the utility hydraulic system to the compressor motor is shut off when air pressure in the storage bottles reaches 3000 PSI. Manual shut-off is also provided. Modified and E and later aircraft not incorporating a priority system (figure 1–25) both bottles are charged simultaneously, in approximately 30 minutes. Unmodified A THRU C aircraft incorporating a priority system (figure 1–26) the storage bottle for emergency extension of the landing gear is charged first, taking approximately 3 minutes. At this time the engine starter valve is automatically opened and the compressor charges the engine starter air storage bottle, which requires about 25 minutes. When both bottles are charged the hydraulic shut-off valve is closed shutting off the compressor. The hydraulic shut-off valve is electrically controlled with primary bus power which is available only when the speed brakes are closed on A THRU G aircraft. However, on some A THRU G aircraft, this feature has been eliminated and the compressor switch alone controls the compressor. On H and later aircraft, the pneumatic compressor will be operative only when the landing gear is up and locked. This system is provided to cut out the hydraulic motor of the pneumatic compressor during periods when there is a greater than normal demand for pressure in the utility hydraulic system. The automatic cut-off switch in the left wheel well can be actuated on the ground to permit compressor operation with the landing gear extended.

CAUTION

The compressor should not be operated for more than 30 minutes continuously to prevent overheating and possible compressor explosion. Therefore, if the storage bottles take an excessive amount of time to refill, the compressor should be shut-off.

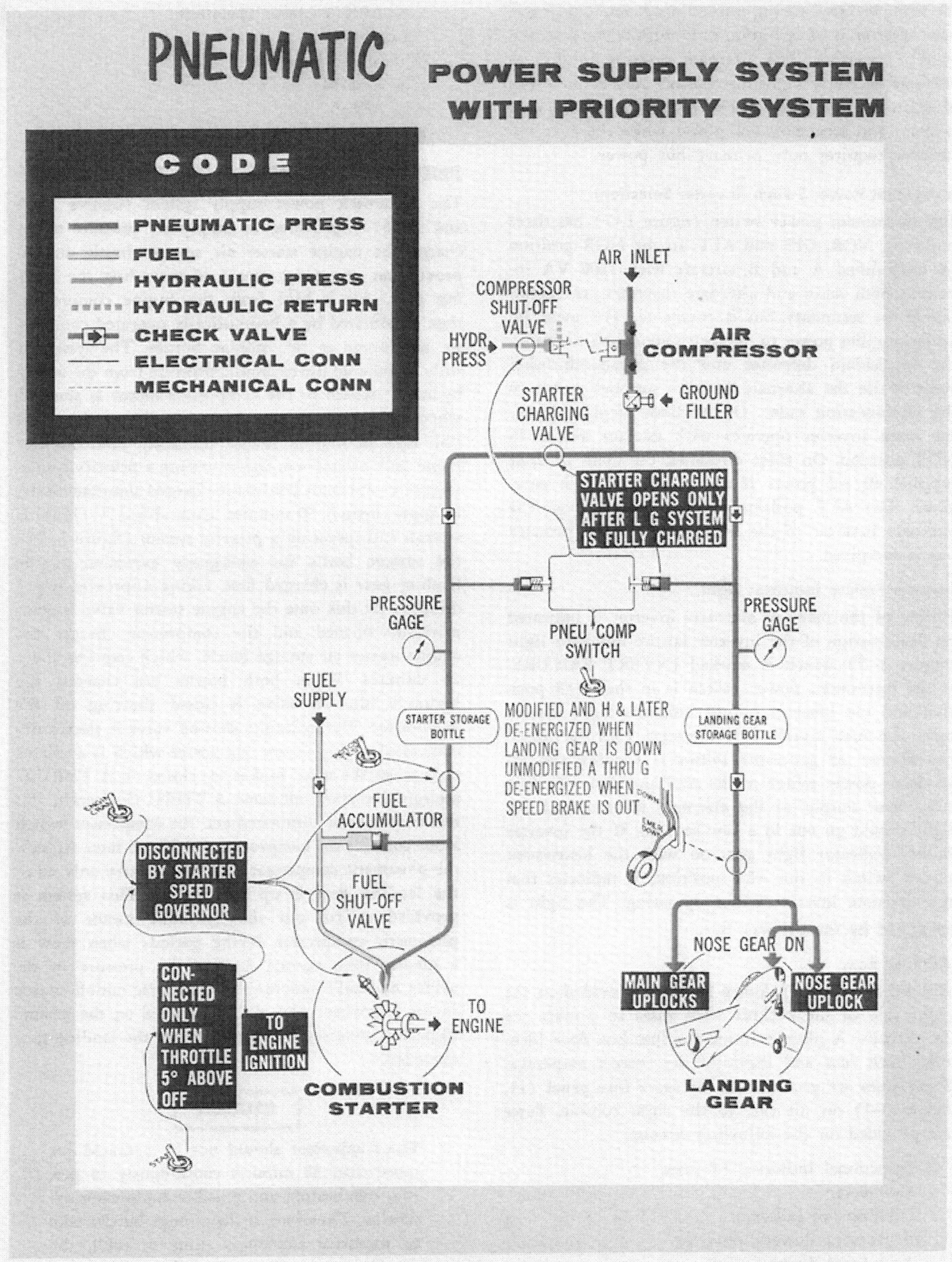
PNEUMATIC POWER SUPPLY SYSTEM WITH PRIORITY SYSTEM
CODE
PNEUMATIC PRESS
FUEL
HYDRAULIC PRESS
HYDRAULIC RETURN
CHECK VALVE
ELECTRICAL CONN
MECHANICAL CONN
AIR INLET
COMPRESSOR SHUT-OFF VALVE
HYDR PRESS
AIR COMPRESSOR
STARTER CHARGING VALVE
GROUND FILLER
STARTER CHARGING VALVE OPENS ONLY AFTER L G SYSTEM IS FULLY CHARGED
PRESSURE GAGE
PNEU COMP SWITCH
PRESSURE GAGE
FUEL SUPPLY
STARTER STORAGE BOTTLE
MODIFIED AND H & LATER DE-ENERGIZED WHEN LANDING GEAR IS DOWN
UNMODIFIED A THRU G DE-ENERGIZED WHEN SPEED BRAKE IS OUT
LANDING GEAR STORAGE BOTTLE
FUEL ACCUMULATOR
DISCONNECTED BY STARTER SPEED GOVERNOR
FUEL SHUT-OFF VALVE
NOSE GEAR DN
MAIN GEAR UPLOCKS
NOSE GEAR UPLOCK
CON-NECTED ONLY WHEN THROTTLE 5° ABOVE OFF
TO ENGINE IGNITION
TO ENGINE
COMBUSTION STARTER
LANDING GEAR

Figure 1—25

pressure and the other half marked POWER for the power system pressure. Each half section of the gage face is marked to show hydraulic pressure ranging from 0 to 2,000 PSI. Normal operating pressure, of the system, is approximately 1,500 PSI for both the utility and power system, depending on the amount of equipment operating. Emergency system pressure is indicated on the POWER half of the gage face. When the emergency pump is operating, the POWER indicator needle will show an emergency operating pressure of approximately 1,500 PSI in the power system. The hydraulic pressure gage is powered by the AC electrical system and will operate only when the instrument power switch is in the NOR or ALT position and the main or alternate inverter is operating.

TANDEM

FLIGHT CONTROL SYSTEMS.

The flight controls on tandem aircraft are identical to non-tandem aircraft except for the actuating cylinders and the hydraulic pressure supply. The hydraulic system has been redesigned to make the power and utility systems two completely separate and functionally independent systems. The emergency hydraulic system is a simple secondary source of pressure for the power system only. The aileron and stabilator actuators are tandem cylinders which allow surface actuation by either the power, utility or emergency hydraulic pressure. The spoilers and rudder are actuated only from the utility system pressure.

TANDEM

NORMAL SYSTEM.

During normal operation, hydraulic pressure from the engine driven utility pump is directed to the utility side of the aileron and stabilator tandem actuators, also to the spoiler and rudder actuators. An accumulator in the utility system stores a volume of pressurized hydraulic fluid to insure an adequate supply to the rudder and the utility side of the stabilator tandem actuator in the event of a greater-than-normal demand on the system is required. Hydraulic pressure from the engine driven power pump is directed to the power side of the aileron and stabilator tandem actuators, and to an accumulator. The accumulator is provided for the same purpose as the accumulator in the utility system. If the utility system pressure should fail, the ailerons and stabilator will actuate from the power system. The rudder will be controlled only through mechanical linkage and the spoilers will be inoperative as these surfaces are actuated by utility pressure only. If the power system pressure should fail, all surfaces will operate with utility system pressure. In the event of failure of both the utility and power hydraulic systems the emergency hydraulic system must be manually selected for control operation. Pressure gages are provided to indicate pressure in each of the systems.

TANDEM

Stabilator.

The stabilator is operated by a hydraulic actuator controlled by a mechanical linkage attached to the control stick. There is a direct mechanical linkage between the control stick and stabilator. However, due to the surface hinge moments, direct manual control is not possible without hydraulic power. Since the powered systems are irreversible (i.e., air loads on the surfaces are not felt on the controls) control feel is simulated by an artificial feel unit in each of the primary control systems. These units vary control feel in proportion to stick or rudder pedal deflection only and will not be affected by aircraft speed or altitude. The artificial feel unit is a spring capsule designed to give the pilot a sense of control feel by increasing the control forces as the controls are moved. The units are also used when trimming the airplane about the roll, yaw and pitch axes. This is accomplished with an actuator powered from the primary bus which repositions the spring capsule so that when the controls are moved to trim the airplane, the spring capsule is moved to a no-load position and the airplane will be trimmed about the axes with no load on the cockpit controls. A control stick damper (pitch only) is incorporated to restrict the speed at which the control stick can be moved to assist in preventing over control and porpoising. The damper is factory set and cannot be adjusted for individual pilot feel during flight. A mechanical advantage shifter is incorporated in the stabilator linkage to decrease the control stick sensitivity when the landing gear is retracted. When the landing gear is not up and locked, the stabilator moves through an arc of approximately 16½ degrees, leading edge down, (airplane nose up) to approximately 4½ degrees, leading edge up, (airplane nose down). This range of movement during high speed flight could cause over-controlling and porpoising. Therefore, when the landing gear reaches the up and locked position, a circuit is completed to an electric actuator which cycles the mechanical advantage shifter and reduces the control stick to control surface ratio from 1:1 to 1.8:1. The change in ratio reduces the travel of the stabilator to an arc of approximately 11 degrees leading edge down (airplane nose up) to four degrees leading edge up (airplane nose down). Control stick travel remains approximately the same. Therefore, stick travel per

HYDRAULIC POWER SUPPLY SYSTEM SCHEMATIC

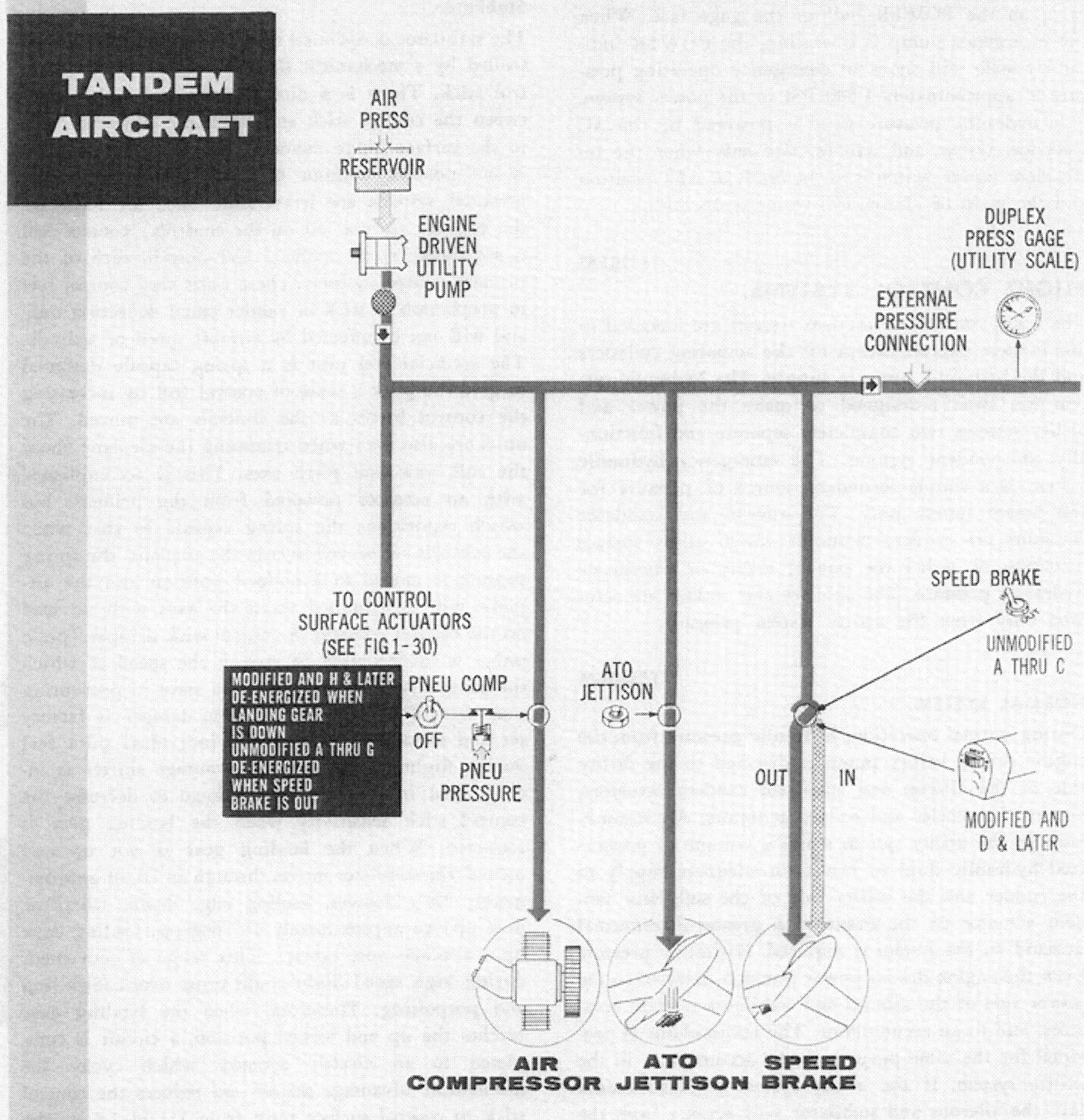

Figure 1—27 (Sheet 1 of 2)

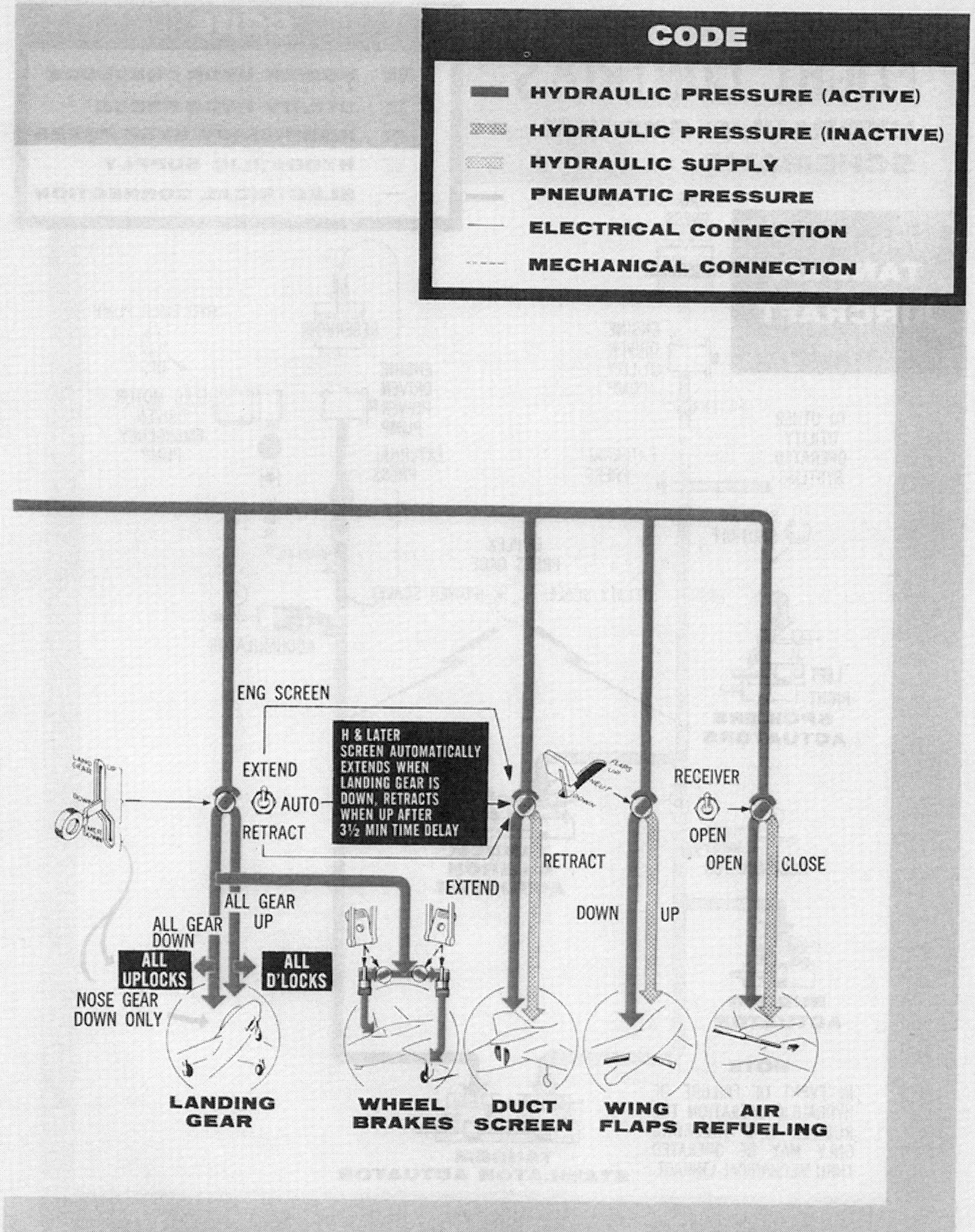

Figure 1—27 (Sheet 2 of 2)

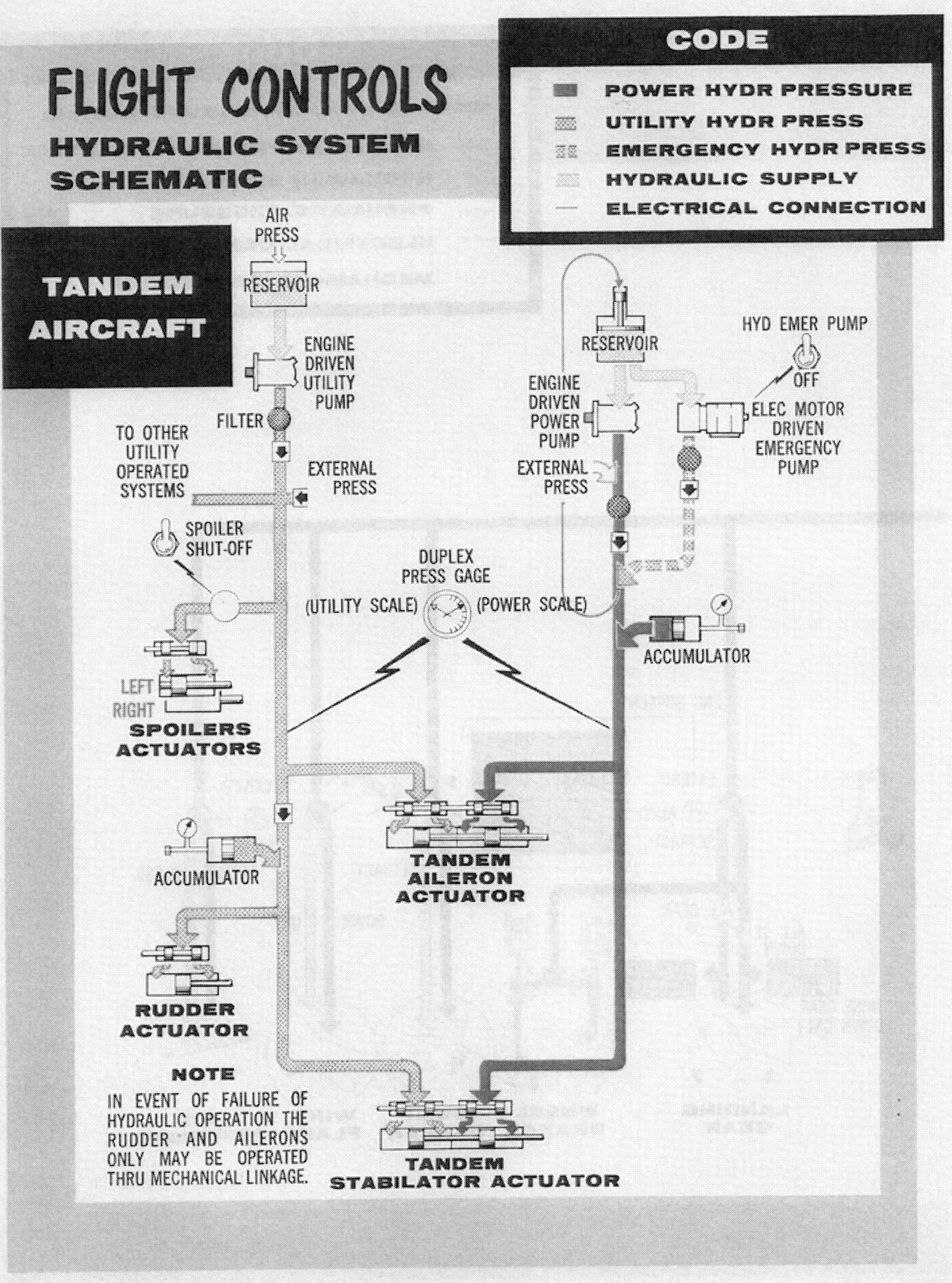
FLIGHT CONTROLS
HYDRAULIC SYSTEM SCHEMATIC
CODE
POWER HYDR PRESSURE
UTILITY HYDR PRESS
EMERGENCY HYDR PRESS
HYDRAULIC SUPPLY
ELECTRICAL CONNECTION
TANDEM AIRCRAFT
AIR PRESS
RESERVOIR
ENGINE DRIVEN UTILITY PUMP
FILTER
TO OTHER UTILITY OPERATED SYSTEMS
EXTERNAL PRESS
SPOILER SHUT-OFF
DUPLEX PRESS GAGE
(UTILITY SCALE)
(POWER SCALE)
LEFT
RIGHT
SPOILERS ACTUATORS
RESERVOIR
HYD EMER PUMP
OFF
ENGINE DRIVEN POWER PUMP
ELEC MOTOR DRIVEN EMERGENCY PUMP
EXTERNAL PRESS
ACCUMULATOR
TANDEM AILERON ACTUATOR
ACCUMULATOR
RUDDER ACTUATOR
NOTE
IN EVENT OF FAILURE OF HYDRAULIC OPERATION THE RUDDER AND AILERONS ONLY MAY BE OPERATED THRU MECHANICAL LINKAGE.
TANDEM STABILATOR ACTUATOR

Figure 1—28

degree of stabilator travel is increased to reduce control sensitivity by permitting finer control adjustment. The change in ratio is automatic and cannot be controlled by the pilot. When the right main landing gear outer door is up and locked, the stabilator control system begins to shift to the 1.8:1 ratio. Approximately eight seconds after the landing gear is up and locked, the mechanical advantage actuator has completed its cycle to change the stabilator controls into the 1.8:1 ratio. This requires a slight aft movement of the control stick which may feel to the pilot like a small nose down trim change. Conversely, when the landing gear is extended and the mechanical advantage ratio changes to 1:1, the opposite effect is noticeable. An indicator light is provided to indicate when the mechanical advantage actuator is shifting or is not in the proper ratio. A force of approximately 25 pounds is required to move the control stick to either extreme position while approximately three pounds are required to move it out of the neutral position. There are no controllable trim tabs provided.

TANDEM

Control Stick.

The control stick (figure 1–29) is conventional and incorporates a hand grip with the following controls: trim switch, bomb release switch, trigger, radar "out" switch, "roger" button for the radar beacon installation, and the autopilot release switch. These switches are discussed under the applicable systems.

TANDEM

Rudder Pedals.

The rudder pedals can be folded down for pilot comfort. Depressing the release lever on the inboard side of each pedal allows the foot pad of the pedal to fold toward the pilot. This allows the pilot to rest his heel on the base of the pedal and still maintain rudder control with his heel. Brakes cannot be applied with the pedals in the folded position. Brake application is accomplished by toe action in the normal manner when the pedals are in the normal position.

TANDEM

Rudder Pedal Adjusting Knob.

The rudder pedal adjusting knob (23, figure 1–4; 23, figure 1–5; 23, figure 1–6) adjusts both rudder pedals simultaneously to the desired fore and aft position.

TANDEM

Trim Switch.

Airplane trim about the roll and pitch axis is controlled by a five position trim switch (figure 1–29) which is spring-loaded to the center or off position. The lateral positions of the trim switch adjust the artificial feel unit in the aileron control system so as to change the neutral position of the spring capsule thereby trimming the aircraft about the roll axis. The fore and aft positions of the switch adjust the artificial feel unit in the stabilator control to a no-load position. Both trim circuits are energized from the primary bus if the pitch trim switch is in the NORM positon.

TANDEM

Rudder Trim Switch.

Airplane trim about the vertical axis is accomplished by adjusting the neutral position of the spring capsule in the rudder control system. The neutral position is selected by a three position YAW TRIM switch (figure 1–30) which is energized from the primary bus and spring-loaded to the off position. The other switch positions are marked L and R. The switch is held in the desired position until the airplane is trimmed about the vertical axis and then released.

TANDEM

Alternate Pitch Trim Switch.

The alternate pitch trim switch (figure 1–30) is provided as an override for the trim switch on the control stick. The switch is powered from the primary bus and has four positions. Two positions are marked STICK SW POWER: NORM and OFF. The other two positions are marked NOSE DN and NOSE UP. The NORM position supplies primary bus power to the control stick trim switch which controls the aileron and stabilator trim actuators. The OFF position disconnects primary bus power from the trim switch on the control stick so that the control stick trim switch can be isolated from the control system in the event of trim switch failure. In the OFF position both the

CONTROL STICK

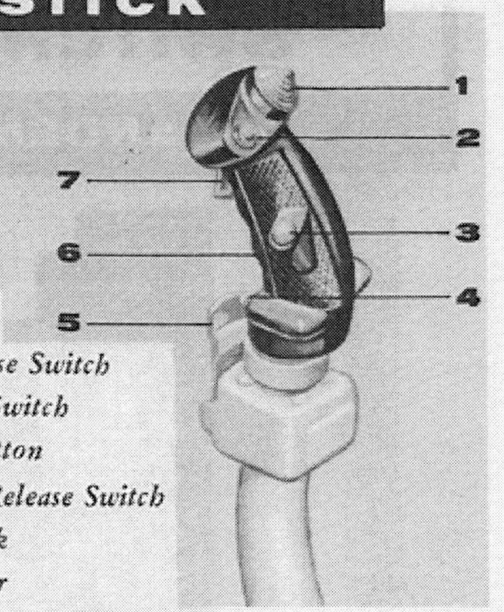

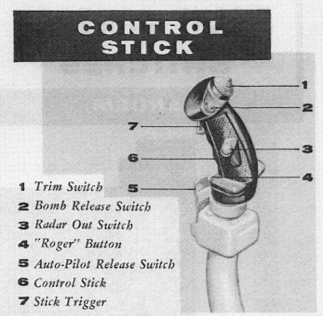

1 *Trim Switch*
2 *Bomb Release Switch*
3 *Radar Out Switch*
4 *"Roger" Button*
5 *Auto-Pilot Release Switch*
6 *Control Stick*
7 *Stick Trigger*

Figure 1—29

FLIGHT CONTROL SWITCHES

TANDEM

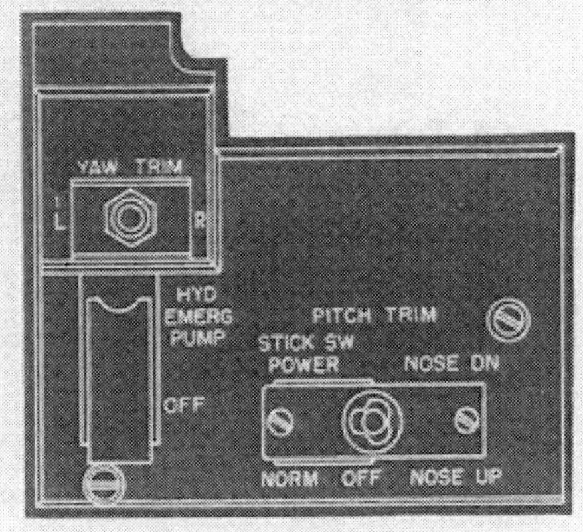

Modified A thru G

Modified H thru J and K

Figure 1—30

aileron and stabilator trim actuator circuits are inoperative. The NOSE DN and NOSE UP positions actuate the stabilator trim actuator to trim the aircraft on its pitch axis in the event of failure of the control stick trim switch.

TANDEM

Aileron and Rudder Neutral Trim Indicators.

The aileron and rudder neutral trim indicators (13, 14, figure 1–8; 12, 13, figure 1–9) are green lights that illuminate when the actuator in the respective artificial feel unit is in the neutral position regardless of the control-surface position. However, with hydraulic pressure available, the controls will seek the trimmed (no-load) position of the artificial-feel units. Therefore, if these units are in the neutral position, the control surfaces will be in the neutral position. The lights are energized from the secondary bus circuit. On K and later airplanes, the aileron neutral trim light will illuminate only if the trim switch on the control stick is depressed, and the actuator in the respective artificial feel unit is in the neutral position.

TANDEM

Mechanical Advantage Indicator Light.

The mechanical advantage indicator light (14, figure 1–4; 15, figure 1–5; 15, figure 1–6) is an amber light marked STABILATOR M-A WARNING. The light will illuminate whenever the mechanical advantage actuator is cycling from either extreme position to the other position and the primary bus is energized. The light will also illuminate if the mechanical advantage linkage is out of cycle i.e., the landing gear down and mechanical advantage not in the 1:1 ratio or if the landing gear is up and locked and the mechanical advantage is not in the 1.8:1 ratio.

TANDEM

EMERGENCY SYSTEM.

The emergency hydraulic system pressure is supplied by an electrically driven hydraulic pump which is directed to the power side of the aileron and stabilator tandem actuators. The emergency pump must be manually placed in operation through the emergency hydraulic pump switch and is used in the event of engine failure or similar malfunction which would render the utility and power hydraulic systems inoperative. The emergency pump receives its fluid supply from the power hydraulic system reservoir which is pressurized by the power or emergency hydraulic system pressure. Emergency pressure cannot be utilized to operate any hydraulic system powered by the utility system except the ailerons and stabilator.

TANDEM

Emergency Hydraulic Pump Switch.

The emergency hydraulic pump switch E (figure 1–30) is provided so that the emergency hydraulic system can be turned on since the emergency hydraulic system does not include an automatic cut-in feature. The switch has two positions marked HYD EMER PUMP and OFF and is guarded in the OFF position by a cover type guard. The HYD EMER PUMP position connects primary bus power to energize the emergency hydraulic system directly and supplies hydraulic pressure to the power side of the aileron and stabilator tandem actuators. The OFF position disconnects the emergency hydraulic pump from the electrical system. Emergency hydraulic pressure will be indicated on the POWER half of the dual hydraulic pressure gage. The battery life for emergency pump operation with the generator inoperative, is from approximately 15 minutes, provided no other equipment is operating, to approximately three minutes, depending on the amount of equipment operating.

NON-TANDEM

HYDRAULIC POWER SUPPLY SYSTEM.

Note

Unmodified or non-tandem aircraft may be readily identified by the presence of two individual hydraulic gages on the main instrument panel and the installation of a hydraulic hand pump on the cockpit floor near the left console.

The hydraulic power supply system (figure 1–31) is divided into three individual systems: a booster system which supplies pressure to the surface control actuators, a utility system which supplies all other hydraulic systems and an emergency system which supplies hydraulic pressure to the aileron and stabilator surface control actuators in the event of failure of the boost and utility systems. The boost and the utility systems are divided by check valves so that the boost system cannot supply pressure to the utility system. However, the utility system will supply hydraulic pressure not only to the utility system but also to the control system actuators if the demand arises. Hydraulic pressure in each of the boost and utility systems is provided by a variable delivery type engine driven pump. Each pump instantly and positively adjusts delivery to the demand of the respective hydraulic system. The pumps are designed so that fluid under pressure is supplied only when a system demands power. If power is not demanded the suction port of each piston is closed off and fluid under pressure is not supplied. When power is required the suction port is open. A zero-G type hydraulic reservoir has been installed in B and later aircraft to prevent damage to the hydraulic pump from cavitation when in flight under zero-G conditions. The pressure in the emergency hydraulic system is provided by an electrically driven pump. This system automatically takes over and supplies hydraulic pressure to the aileron and stabilizer actuators if the boost and the utility systems pressure drops to approximately 600 to 800 PSI. The aircraft hydraulic system is fully automatic and it is only necessary to move the selected control or switch to the desired position to obtain control of the selected system. A hydraulic hand pump is provided for ground operation of the utility system. A compressed air system is provided for emergency extension of the landing gear. Hydraulically operated systems are; surface control actuators, landing gear, landing flaps, air refueling system, speed brakes, booster wheel brakes, ato release, engine screen system and pneumatic compressor. These systems are described under applicable headings.

WARNING

Do not fly in a zero-G flight configuration for any extended period of time on A aircraft as hydraulic pressure may be lost due to hydraulic pump cavitation. If loss of pressure occurs, considerable time may elapse before hydraulic pressure is regained.

NON-TANDEM

HYDRAULIC HAND PUMP.

The hydraulic hand pump (18, figure 1–8; 16, figure 1–9) can be used to supply hydraulic pressure to the utility system for operating any hydraulic system except the flight control actuators, and the pneumatic compressor. The hand pump has a telescoping handle that is pulled out, rotated 90 degrees and moved aft about one-quarter inch before using. This provides a longer handle and clearance between the handle and console for greater ease in operation. To operate a system, place the selected control in the desired position, then operate the hand pump.

Note

The hand pump is basically used during ground maintenance. However, in an emergency, hydraulic systems powered by the utility pump, may be actuated by the hand pump. Refer to figure 1–32 for the number of strokes of the hand pump required to operate each system.

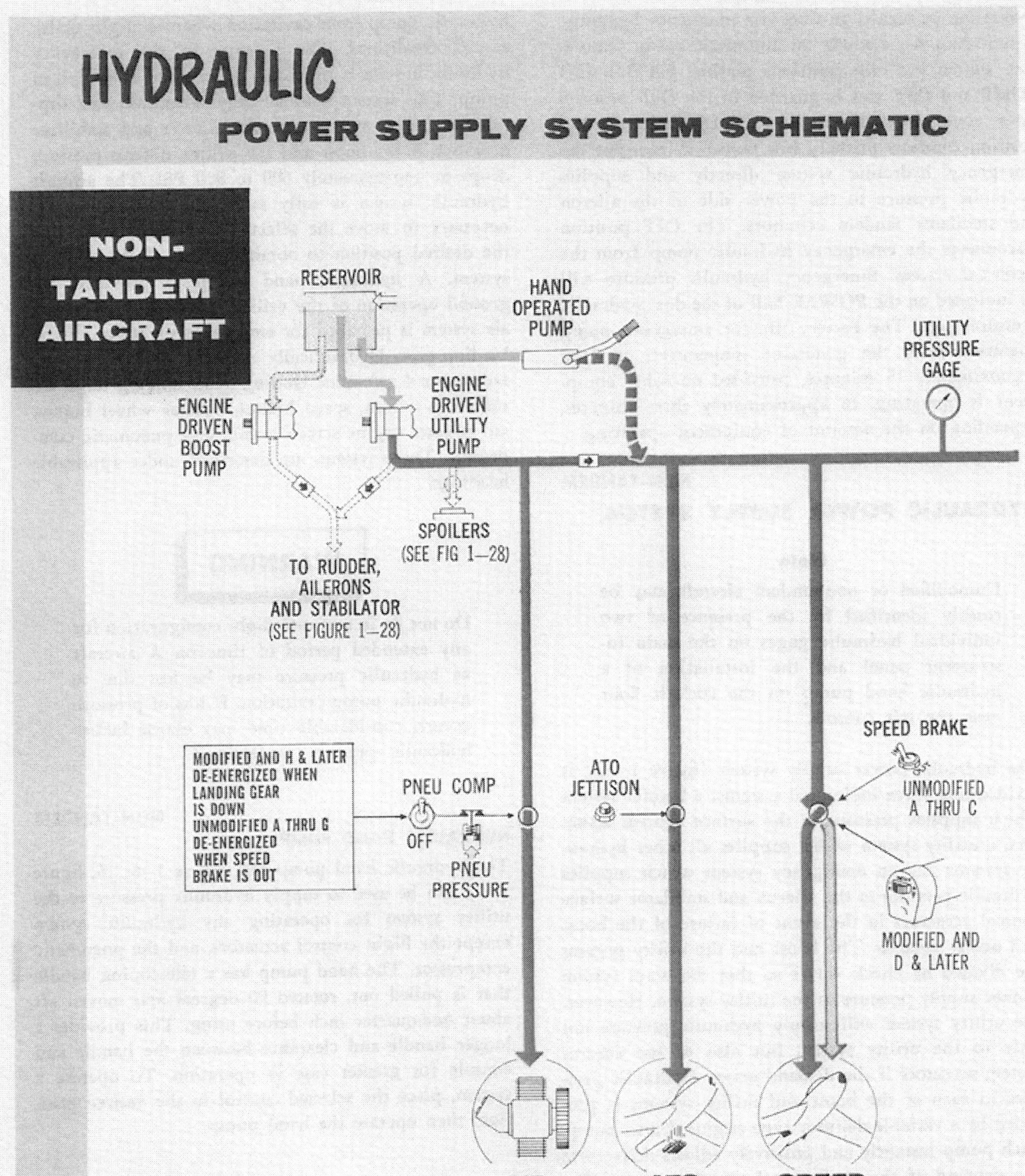
HYDRAULIC
POWER SUPPLY SYSTEM SCHEMATIC
NON-TANDEM AIRCRAFT
RESERVOIR
HAND OPERATED PUMP
UTILITY PRESSURE GAGE
ENGINE DRIVEN BOOST PUMP
ENGINE DRIVEN UTILITY PUMP
SPOILERS (SEE FIG 1—28)
TO RUDDER, AILERONS AND STABILATOR (SEE FIGURE 1—28)
MODIFIED AND H & LATER DE-ENERGIZED WHEN LANDING GEAR IS DOWN
UNMODIFIED A THRU G DE-ENERGIZED WHEN SPEED BRAKE IS OUT
PNEU COMP
OFF
PNEU PRESSURE
ATO JETTISON
SPEED BRAKE
UNMODIFIED A THRU C
MODIFIED AND D & LATER
AIR COMPRESSOR
ATO JETTISON
SPEED BRAKE

Figure 1—31 (Sheet 1 of 2)

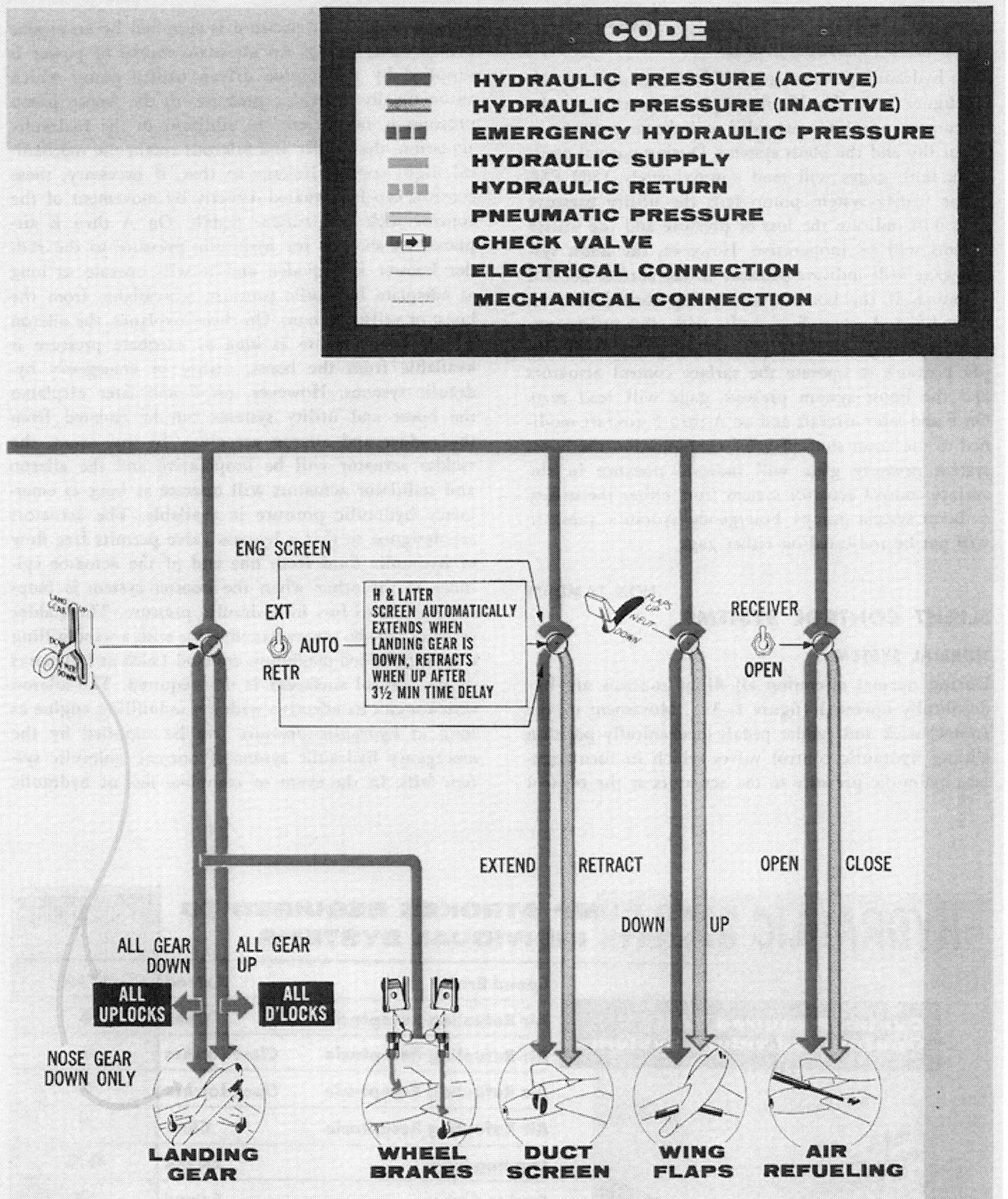

Figure 1—31 (Sheet 2 of 2)

NON-TANDEM

HYDRAULIC PRESSURE GAGES.

Two hydraulic pressure gages (26, 29, figure 1–4; 26, 29, figure 1–5; 26, 29, figure 1–6) located on the instrument panel are provided to indicate pressure in the utility and the boost systems. During normal operation both gages will read approximately 1500 PSI. If the utility system pump fails the utility pressure gage will indicate the loss of pressure and the utility systems will be inoperative. However, the boost system gage will indicate pressure in the surface control actuators. If the boost system pump should fail, on unmodified A thru E aircraft, with the utility system operating normally, the utility pump will supply pressure to operate the surface control actuators and the boost system pressure gage will read zero. On F and later aircraft and on A thru E aircraft modified to the boost shut-off valve configuration the boost system pressure gage will indicate pressure in the surface control actuator system from either the utility or boost system pumps. Emergency hydraulic pressure will not be indicated on either gage.

NON-TAMDEM

FLIGHT CONTROL SYSTEMS.

NORMAL SYSTEM.

During normal operation all flight controls are hydraulically operated (figure 1–33). Movement of the control stick and rudder pedals mechanically position sliding hydraulic control valves which in turn regulate hydraulic pressure to the actuators at the control surfaces. Hydraulic pressure is supplied by an engine driven boost pump. An alternate source of power is supplied by the engine driven utility pump which automatically provides pressure if the boost pump pressure is insufficient. In addition to the hydraulic actuation, the rudder and ailerons retain the mechanical flight control linkage so that, if necessary, these controls can be actuated directly by movement of the control stick and rudder pedals. On A thru E airplanes no shut-off for hydraulic pressure to the rudder booster is provided and it will operate as long as adequate hydraulic pressure is available from the boost or utility systems. On these airplanes, the aileron booster will operate as long as adequate pressure is available from the boost, utility or emergency hydraulic systems. However, on F and later airplanes the boost and utility systems can be isolated from the rudder and aileron actuators. In this event, the rudder actuator will be inoperative and the aileron and stabilator actuators will operate as long as emergency hydraulic pressure is available. The actuators are designed so that a by-pass valve permits free flow of hydraulic fluid from one end of the actuator cylinder to the other when the booster system is inoperative due to loss in hydraulic pressure. The rudder actuator will be somewhat effective with a windmilling engine provided maximum demand (such as full travel of the control surfaces) is not required. The aileron actuator can be effective with a windmilling engine as long as hydraulic pressure can be supplied by the emergency hydraulic system if normal hydraulic system fails. In the event of complete loss of hydraulic

HYDRAULIC HAND PUMP STROKES REQUIRED TO OPERATE INDIVIDUAL SYSTEMS

Cycles Required To Operate Individual Systems

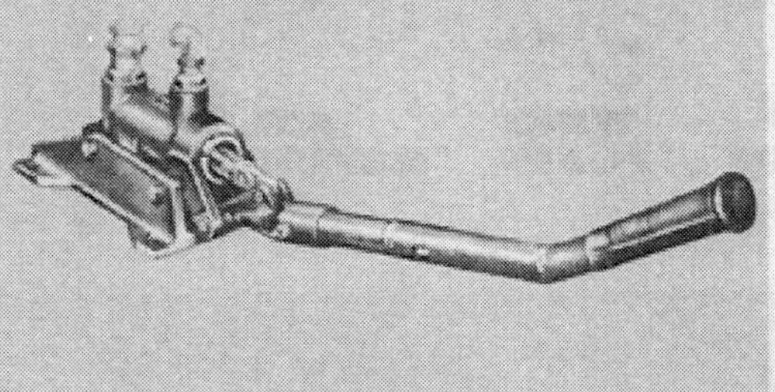

		APPROX. STROKES
Speed Brake	Extend	102
Air Refueling Receptacle	Open	4
Air Refueling Receptacle	Close Latches	4
Air Refueling Receptacle	Open Latches	4
Air Refueling Receptacle	Close	2
Landing Flaps	Extend	46
Engine Screen	Extend	8
Engine Screen	Retract	8
Landing Gear (Handle in down or emerg down — no pneumatic pressure)	Extend	80

Figure 1—32

pressure mechanical linkage to the aileron and rudder permit continued operation of these control surfaces. The conventional elevators and stabilizer have been combined to form a single surface known as the stabilator. The stabilator is designed to give the pilot longitudinal control of the airplane at all altitudes and speeds.

NON-TANDEM

Stabilator.

The stabilator is operated by a hydraulic actuator controlled by a mechanical linkage attached to the control stick. There is a direct mechanical linkage between the control stick and stabilator. However, due to the surface hinge moments, direct manual control is not possible without hydraulic power. Since the powered systems are irreversible (i.e., air loads on the surfaces are not felt on the controls) control feel is simulated by an artificial feel unit in each of the primary control systems. These units vary control feel in proportion to stick or rudder pedal deflection only and will not be affected by aircraft speed or altitude. The artificial feel unit is a spring capsule designed to give the pilot a sense of control feel by increasing the control forces as the controls are moved. The units are also used when trimming the airplane about the roll, yaw and pitch axes. This is accomplished with an actuator powered from the primary bus which repositions the spring capsule so that when the controls are moved to trim the airplane, the spring capsule is moved to a no-load position and the airplane will be trimmed about the axes with no load on the cockpit controls. A control stick damper (pitch only) is incorporated to restrict the speed at which the control stick can be moved to assist in preventing over-control and porpoising. The damper is factory set and cannot be adjusted for individual pilot feel during flight. A mechanical advantage shifter is incorporated in the stabilator linkage to decrease the control stick sensitivity when the landing gear is retracted. When the landing gear is not up and locked, the stabilator moves through an arc of approximately 16½ degrees leading edge down (airplane nose up) to approximately 4½ degrees leading edge up (airplane nose down). This range of movement during high speed flight could cause over-controlling and porpoising. Therefore, when the landing gear reaches the up and locked position, a circuit is completed to an electric actuator which cycles the mechanical advantage shifter and reduces the control stick to control surface ratio from 1:1 to 1.8:1. The change in ratio reduces the travel of the stabilator to an arc of approximately 11 degrees leading edge down (airplane nose up) to four degrees leading edge up (airplane nose down). Control stick travel remains approximately the same. Therefore, stick travel per degree of stabilator travel is increased to reduce control sensitivity by permitting finer control adjustment. The change in ratio is automatic and cannot be controlled by the pilot. When the right main landing gear outer door is up and locked, the stabilator control system begins to shift to the 1.8:1 ratio. Approximately eight seconds after the landing gear is up and locked, the mechanical advantage actuator has completed its cycle to change the stabilator controls into the 1.8:1 ratio. This requires a slight aft movement of the control stick which may feel to the pilot like a small nose down trim change. Conversely when the landing gear is extended and the mechanical advantage ratio changes to 1:1, the opposite effect is noticeable. An indicator light is provided to indicate when the mechanical advantage actuator is shifting or is not in the proper ratio. A force of approximately 25 pounds is required to move the control stick to either extreme position while approximately three pounds are required to move it out of the neutral position. There are no controllable trim tabs provided.

NON-TANDEM

Control Stick.

The control stick (figure 1–29) is conventional and incorporates a hand grip with the following controls: trim switch, bomb release switch, trigger, radar "out" switch; "roger" button for the radar beacon installation, and the autopilot release switch. These switches are discussed under the applicable systems.

NON-TANDEM

Rudder Pedals.

The rudder pedals can be folded down for pilot comfort. Depressing the release lever on the inboard side of each pedal allows the foot pad of the pedal to fold toward the pilot. This allows the pilot to rest his heel on the base of the pedal and still maintain rudder control with his heel. Brakes cannot be applied with the pedals in the folded position. Brake application is accomplished by toe action in the normal manner when the pedals are in the normal position.

NON-TANDEM

Rudder Pedal Adjusting Knob.

The rudder pedal adjusting knob (23, figure 1–4; 23, figure 1–5; 23, figure 1–6) adjust both rudder pedals simultaneously to the desired fore and aft position.

NON-TANDEM

Trim Switch.

Airplane trim about the roll and pitch axis is controlled by a five position trim switch (figure 1–29) which is spring-loaded to the center or off position. The lateral positions of the trim switch adjust the

FLIGHT CONTROLS
HYDRAULIC SYSTEM SCHEMATIC

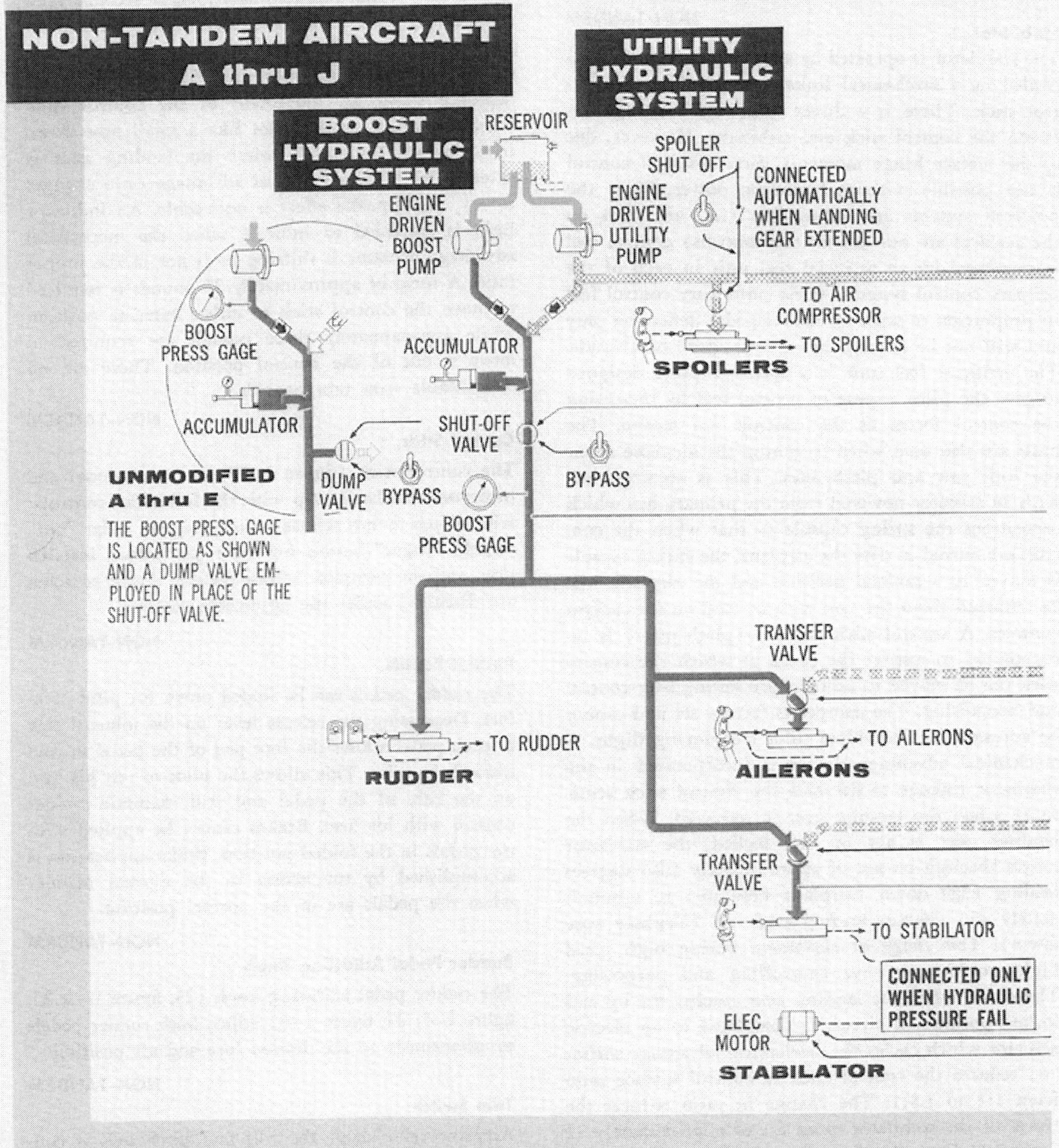
NON-TANDEM AIRCRAFT
A thru J
BOOST
HYDRAULIC
SYSTEM
UTILITY
HYDRAULIC
SYSTEM
RESERVOIR
ENGINE
DRIVEN
BOOST
PUMP
ENGINE
DRIVEN
UTILITY
PUMP
SPOILER
SHUT OFF
CONNECTED
AUTOMATICALLY
WHEN LANDING
GEAR EXTENDED
TO AIR
COMPRESSOR
TO SPOILERS
SPOILERS
BOOST
PRESS GAGE
ACCUMULATOR
ACCUMULATOR
SHUT-OFF
VALVE
BY-PASS
DUMP
VALVE
BYPASS
BOOST
PRESS GAGE
UNMODIFIED
A thru E
THE BOOST PRESS. GAGE
IS LOCATED AS SHOWN
AND A DUMP VALVE EM-
PLOYED IN PLACE OF THE
SHUT-OFF VALVE.
TRANSFER
VALVE
TO AILERONS
AILERONS
TO RUDDER
RUDDER
TRANSFER
VALVE
TO STABILATOR
CONNECTED ONLY
WHEN HYDRAULIC
PRESSURE FAIL
ELEC
MOTOR
STABILATOR

Figure 1—33 (Sheet 1 of 2)

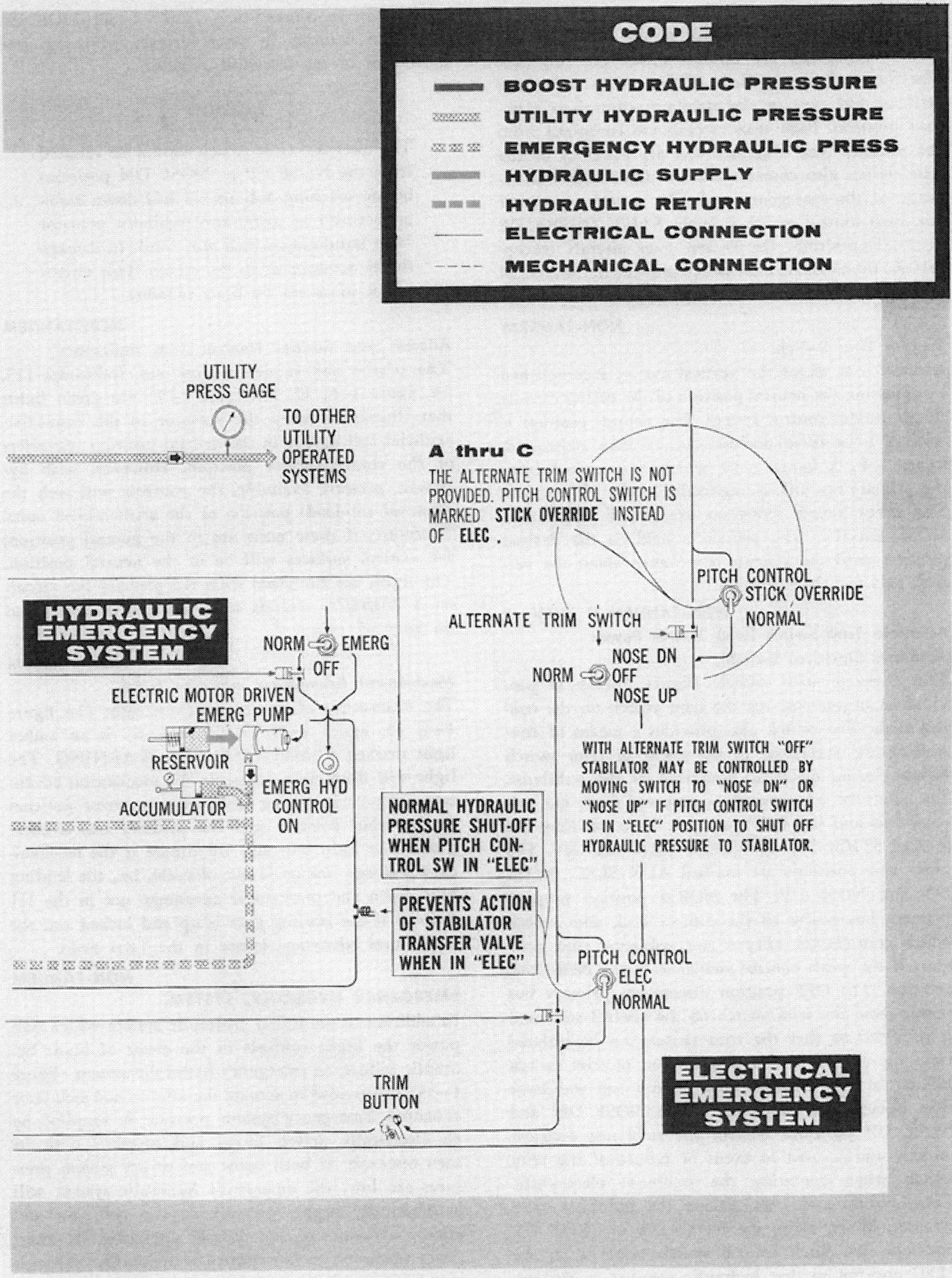
CODE
BOOST HYDRAULIC PRESSURE
UTILITY HYDRAULIC PRESSURE
EMERGENCY HYDRAULIC PRESS
HYDRAULIC SUPPLY
HYDRAULIC RETURN
ELECTRICAL CONNECTION
MECHANICAL CONNECTION
UTILITY PRESS GAGE
TO OTHER UTILITY OPERATED SYSTEMS
A thru C
THE ALTERNATE TRIM SWITCH IS NOT PROVIDED. PITCH CONTROL SWITCH IS MARKED STICK OVERRIDE INSTEAD OF ELEC .
PITCH CONTROL
STICK OVERRIDE
NORMAL
ALTERNATE TRIM SWITCH
HYDRAULIC EMERGENCY SYSTEM
NORM
EMERG
OFF
ELECTRIC MOTOR DRIVEN EMERG PUMP
RESERVOIR
ACCUMULATOR
EMERG HYD CONTROL ON
NOSE DN
OFF
NOSE UP
WITH ALTERNATE TRIM SWITCH "OFF" STABILATOR MAY BE CONTROLLED BY MOVING SWITCH TO "NOSE DN" OR "NOSE UP" IF PITCH CONTROL SWITCH IS IN "ELEC" POSITION TO SHUT OFF HYDRAULIC PRESSURE TO STABILATOR.
NORMAL HYDRAULIC PRESSURE SHUT-OFF WHEN PITCH CONTROL SW IN "ELEC"
PREVENTS ACTION OF STABILATOR TRANSFER VALVE WHEN IN "ELEC"
PITCH CONTROL
ELEC
NORMAL
ELECTRICAL EMERGENCY SYSTEM
TRIM BUTTON

Figure 1—33 (Sheet 2 of 2)

artificial feel unit in the aileron control system so as to change the neutral position of the spring capsule thereby trimming the aircraft about the roll axis. The fore and aft positions of the switch adjust the artificial feel unit in the stabilator control to a no-load position. Both trim circuits are energized from the primary bus. The fore and aft positions of the trim switch also controls the stabilator electrical actuator if the emergency hydraulic system fails or if the pitch control switch is in the STICK OVERRIDE or ELEC position. On D and later aircraft (except -41GK) the alternate trim switch must be in the NORM position in order for the trim switch to be operative.

NON-TANDEM

Rudder Trim Switch.

Aircraft trim about the vertical axis is accomplished by adjusting the neutral position of the spring capsule in the rudder control system. The neutral position is selected by a three position rudder trim switch (8, figure 1–8; 6, figure 1–9) which is energized from the primary bus and spring-loaded to the off position. The other switch positions are NOSE LEFT and NOSE RIGHT. The switch is held in the desired position until the aircraft is trimmed about the vertical axis and then released.

NON-TANDEM D THRU J*

Alternate Trim Switch (Stick Switch Power Auxiliary Electrical Switch).

The alternate trim switch (figure 1–34) is provided as an override for the trim switch on the control stick. The switch also provides a means of controlling the stabilator in the event of trim switch failure during electrical operation of the stabilator. The alternate trim switch is powered from the primary bus and has four positions. Two positions are marked STICK SW POWER: NORM and OFF. The other two positions are marked AUX ELEC: NOSE DN and NOSE UP. The NORM position supplies primary bus power to the control stick trim switch which controls the aileron and stabilator trim actuators if the pitch control switch is in the NORMAL position. The OFF position disconnects primary bus power from the trim switch on the control stick and is provided so that the trim switch can be isolated from the control system in the event of trim switch failure. In this position both aileron and stabilator trim circuits are inoperative. The NOSE DN and NOSE UP positions control the stabilator actuator directly and is used in event of failure of the trim switch when operating the stabilator electrically. These positions do not actuate the stabilator trim actuator. When using the NOSE DN or NOSE UP positions, the pitch control switch must be in the ELEC position so that hydraulic pressure to the stabilator will be dumped in A THRU E and -51GK aircraft and shut-off in later aircraft, allowing free movement of the electrical actuator.

CAUTION

The alternate trim switch should be released from the NOSE UP or NOSE DN positions before reaching full up or full down stabilator travel to avoid the stabilator actuator from bottoming which may result in damage to the actuator or to the clutch. This circuit is not protected by limit switches.

NON-TANDEM

Aileron and Rudder Neutral Trim Indicators.

The aileron and rudder neutral trim indicators (13, 14, figure 1–8; 12, 13, figure 1–9) are green lights that illuminate when the actuator in the respective artificial feel unit is in the neutral position regardless of the control-surface position. However, with hydraulic pressure available, the controls will seek the trimmed (no-load) position of the artificial-feel units. Therefore, if these units are in the nuetral position, the control surfaces will be in the neutral position. The lights are energized from the primary bus circuit on A THRU G aircraft and from the secondary bus on later aircraft.

NON-TANDEM

Mechanical Advantage Indicator Light.

The mechanical advantage indicator light (14, figure 1–4; 15, figure 1–5; 15, figure 1–6) is an amber light marked STABILATOR M-A WARNING. The light will illuminate whenever the mechanical advantage actuator is cycling from either extreme position to the other position and the primary bus is energized. The light will also illuminate if the mechanical advantage linkage is out of cycle, i.e., the landing gear down and mechanical advantage not in the 1:1 ratio or if the landing gear is up and locked and the mechanical advantage is not in the 1.8:1 ratio.

NON-TANDEM

EMERGENCY HYDRAULIC SYSTEM.

In addition to the utility hydraulic system which will power the flight controls in the event of boost hydraulic failure, an emergency hydraulic system (figure 1–33) is provided to actuate the aileron and stabilator actuators. Emergency system pressure is supplied by an electrically driven pump and supplied with its own reservoir. If both boost and utility system pressures are low, the emergency hydraulic system will automatically supply pressure to the stabilator and aileron actuators during normal operation. In emergency operation rudder control is maintained through the direct mechanical linkage only. The emergency

*Except -41GK aircraft.

FLIGHT CONTROL SWITCHES

NON-TANDEM

① A thru C and -41 GK

② D thru G except -41 GK

③ A thru G

④ H thru J

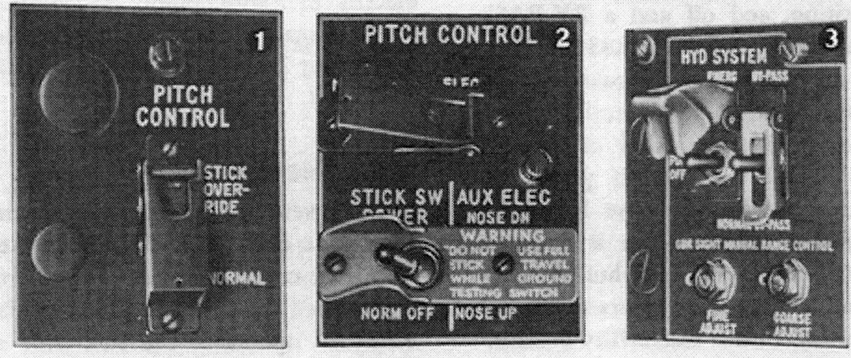

Figure 1—34.

hydraulic pump is energized by the primary bus and controlled from the cockpit for either automatic operation or manual selection. An indicator light in the cockpit illuminates when the emergency system is operating.

Note

In order to conserve pump life, the emergency pump should not be operated during ground operation of the hydraulic system, except for system checkout.

NON-TANDEM

Emergency Hydraulic Pump Switch.

The emergency hydraulic pump switch (25, figure 1—8; 6, figure 1—9) is provided so that the emergency hydraulic system can be cut-in directly in the event the automatic cut-in system is inoperative or to turn the emergency pump off; e.g. during ground maintenance when the engine is not running and normal hydraulic pressure is below 600 PSI. The switch has three positions EMERG, OFF and NORMAL, guarded in the NORMAL position by a cover type guard, and energized by the primary bus. The OFF position disconnects the emergency hydraulic pump from the electrical circuit and must be used during ground maintenance when the normal hydraulic pressure is below 600 PSI and the battery switch is ON. Otherwise, the emergency pump will operate continuously. The NORMAL position arms the hydraulic pressure switch so that when pressure in the boost and utility systems drops below 600 PSI the emergency hydraulic pump is energized, the aileron and stabilator hydraulic pressure lines are transferred to the emergency hydraulic system and the emergency hydraulic controls on indicator light will illuminate. The EMERG position by-passes the hydraulic pressure switch and turns on the emergency hydraulic pump, transfers the aileron and stabilator hydraulic pressure lines to the emergency hydraulic system and illuminates the emergency hydraulic controls on indicator directly. The EMERG position is used if the automatic cut-in system is inoperative. When operating in the EMERG position and pressure in the boost or utility system is regained, the emergency system will continue to operate as the EMERG position by-passes the pressure switch. The battery life for emergency pump duration when operating from the airplane's battery with the generator inoperative is from approximately 15 minutes, provided no other equipment is operating, to approximately three minutes depending on the amount of equipment operating.

Note

If primary bus power fails when operating in the NORMAL or EMERG positions, the spring-loaded aileron and stabilator transfer valves will return to the normal hydraulic system.

NON-TANDEM

Hydraulic By-Pass Switch.

The hydraulic by-pass switch (25, figure 1–8; 6, figure 1–9) is provided to relieve or shut-off hydraulic pressure in the boost and utility systems so that the emergency hydraulic system can be checked for operation prior to take-off. The switch has a momentary contact BY-PASS position, and off and a BY-PASS position. The momentary contact BY-PASS position can be used without moving the cover guard and is used when checking the emergency hydraulic system. This relieves the pilot of the possibility of leaving the switch in the BY-PASS position. If placed in either BY-PASS position, the by-pass valve is opened and boost and utility hydraulic pressure is dumped on A THRU E and -51GK aircraft and shut-off from the rudder, aileron and stabilator actuators on F and later aircraft except -51GK. In this case utility pressure will be maintained for utility systems operation. In the event the emergency hydraulic pump switch is in the NORMAL position and the pressure in the hydraulic line from the dump valve or shut-off valve to the actuators drops to approximately 600 PSI, a hydraulic pressure switch closes which turns the emergency hydraulic pump on, transfers the aileron and stabilator hydraulic pressure lines to the emergency system, and illuminates the emergency hydraulic controls on indicator, if the primary bus is energized. Pressure in the hydraulic line downstream of the shut-off valve on F and later airplanes is vented to return. The emergency hydraulic system accumulator is also pressurized so that the emergency hydraulic system will be available whenever the demand arises.

CAUTION

The by-pass switch should be used in flight only when it is desired to transfer from the normal hydraulic system to the emergency hydraulic system. The by-pass switch is not to be used in flight to transfer from the hydraulic to the electric actuator operation since hydraulic system operation is such that back pressures may develop and cause the actuator clutch to become disengaged from the electric motor.

NON-TANDEM

Emergency Hydraulic Controls On Indicator.

The emergency hydraulic controls on indicator (28, figure 1–4; 28, figure 1–5; 28, figure 1–6) is an amber light powered by the primary bus which will illuminate when the electrical circuit to the emergency hydraulic pump is energized. This condition occurs when the emergency hydraulic pump switch is in the NORMAL position and normal hydraulic pressure drops to approximately 600 PSI thereby actuating the hydraulic pressure switch or when the emergency hydraulic pump switch is placed in the EMERG position. Illumination of the emergency hydraulic controls on indicator light does not indicate that the electric hydraulic pump is operating or pressurizing the emergency system but only indicates that the emergency pump circuit is energized.

NON-TANDEM

EMERGENCY ELECTRIC SYSTEM.

In the event of failure of the emergency hydraulic system, an emergency electrical system is available for stabilator control. Rudder and aileron control will be maintained through the direct mechanical linkage. In electrical operation the stabilator movement is controlled by fore and aft movement of the trim switch on the control stick. With the pitch control switch in the NORMAL position and the primary bus energized when hydraulic pressure drops below 450 PSI, the stabilator automatically changes over to electrical operation and control transfers to the trim switch. Provisions are also made for manual transfer to electrical operation. During electrical operation of the stabilator, the control stick should be moved in the direction of stabilator travel with sufficient force to position the hydraulic control valve on the actuator. While not assisting in stabilator movement this action assures smooth operation of the electrical actuator by releasing any hydraulic pressure built up in the actuator cylinder and, at the same time applying any residual pressure available in the direction of stabilator travel.

NON-TANDEM

Pitch Control Switch.

The pitch control switch (9, figure 1–8; 6, figure 1–9) is provided as an emergency switch to transfer stabilator control from the hydraulic system to the electrical system in the event of failure of the hydraulic pressure switch. On A THRU C airplanes the switch has two positions: STICK OVERRIDE and NORMAL. On C and later airplanes the two positions are ELEC and NORMAL. The switch is guarded in the NORMAL position and energized from the primary bus. The guard on the pitch control switch is of the slotted type and is capable of holding the switch in either position. The NORMAL position allows stabilator control with the fore and aft positions of the control stick when hydraulic pressure in the normal or emergency hydraulic systems is adequate. If hydraulic pressure in the emergency system drops below approximately 450 PSI, actuation of a pressure switch changes the stabilator over to electrical operation and stabilator control transfers to the fore and aft positions of the trim button on the control stick. The STICK OVERRIDE or ELEC position by-passes the hydraulic pressure switch and electrically opens the by-pass valve which dumps hydraulic pressure in the boost and utility system on A thru E airplanes. On F and later airplanes the STICK OVERRIDE or ELEC position shuts off utility hydraulic system pressure to the rudder, aileron and stabilator actuators. Utility pressure is maintained for operation of the utility systems. The STICK OVERRIDE or ELEC position also shuts off emergency hydraulic pressure to the stabilator, disconnects the pitch trim feel device circuit and transfers stabilator control to the fore and aft positions of the trim button on the control stick. The aileron booster will be operated by the emergency hydraulic system if the emergency pump switch is in the NORMAL or EMERG positions and emergency hydraulic pressure is adequate.

WARNING

If the landing gear is up and locked the mechanical linkage from the control stick to the stabilator actuator or the actuator can be damaged if the trim switch is held in the fore and aft position until full travel of the stabilator is realized when operating the stabilator electrically. This occurs because the mechanical advantage linkage is in the 1.8:1 ratio and the limit of the stabilator electrical actuator travel is controlled by mechanical stops on the control stick which bottoms with the actuator at half travel. The actuator will continue to operate if the trim switch is held even after the control stick has reached its limit with resultant stretching of the control linkage. This condition is most prevalent in ground operation check out of the override system. In the event of runaway trim on D and later aircraft, the pitch control switch shall be in the ELEC position to insure proper control through the alternate trim switch.

WARNING

The STICK OVERRIDE or ELEC position must be used any time the electric actuator is used for stabilator operation except during electrical continuity ground check prior to take-off as outlined in Section II. The STICK OVERRIDE or ELEC position relieves hydraulic pressure in the system thereby preventing the actuator clutch from disengaging or the actuator from locking in one position.

LANDING FLAPS.

The landing flaps are partial span, plain, trailing edge type and extend from the aileron to the fuselage on each wing. The flaps are extended and retracted with hydraulic pressure and are controlled by an electrically operated selector valve powered from the primary bus. The flaps are locked in the up position by a mechanical lock in each of the actuating cylinders and are kept in the down position by hydraulic pressure. Intermediate positions of the flaps may be selected and the position will be shown on the flap position indicator. A restrictor is installed in the flap hydraulic system so that seven to 10 seconds are required to retract or extend the flaps. The flaps will not extend, or if extended, will retract if the air-speed is excessive. The flaps are not mechanically connected but are hydraulically synchronized so that they will extend simultaneously in flight. On F-84F-25RE up to -45RE the flap system is synchronized to within a maximum of 10 degrees differential during full travel and a differential of 10 degrees at the 50 per cent position during ground operation. On F-84F-45RE and -25 GK and later airplanes, this differential is a maximum of five degrees during full travel and a differential of six degrees at the 50 per cent position. However, with a cross-wind the differential may be greater. The flaps will synchronize due to air loads at a speed of approximately 80 knots. The flap position indicator may increase or decrease slightly when air loads are applied to the flaps and they become synchronized during the take-off run. In the event of hydraulic failure on non-tandem aircraft, the flaps can be lowered with, approximately, 46 strokes of the hydraulic hand pump if the landing gear has not been extended. If the landing gear has been extended by emergency procedure, the time and number of strokes required to lower the flaps will be greatly increased, as the down side of the main-landing-gear actuating cylinders must be pressurized before sufficient pressure is obtained to lower the flaps.

LANDING FLAP LEVER.

The landing flap lever (figure 1–15) is used for operation of the landing flaps. There are three positions, UP, NEUT and DOWN. The UP position electrically positions the hydraulic selector valve to direct hydraulic pressure to the up side of the flap actuators. When the flaps are fully up and the mechanical locks in the actuating cylinders are locked, the flap selector valve automatically returns to the neutral position which relieves the pressure on the up side of the actuating cylinder. In the NEUT position, hydraulic pressure in the down side of the flap actuating cylinder is trapped while pressure in the up side is relieved. This position is used for partial flap extension. When the flaps are lowered to the desired position the landing flap lever is returned to NEUT. The trapped hydraulic fluid maintains the flaps in the desired position. The DOWN position electrically positions the hydraulic selector valve to direct hydraulic pressure to the down side of the flap actuating cylinders where it is maintained until the lever is repositioned.

CAUTION

If the landing flap lever is in the DOWN position, on non-tandem A thru D aircraft, and the speed brakes are opened or closed, the flaps may momentarily blow up approximately 25 per cent due to a momentary drop in hydraulic pressure. This condition can be avoided by placing the flap lever in the NEUT position, after the flaps are fully extended.

LANDING FLAP POSITION INDICATOR.

Markings on the landing flap position indicator (figure 1–7) show the position of the landing flap in per cent with a range from zero to 100 per cent down if the primary bus is energized.

SPEED BRAKES.

A hydraulically operated speed brake is installed on each side of the aft fuselage. The speed brakes are designed as drag increasing devices for use during "let down" and maneuvers. A multi-position selector valve located in the aft fuselage utilizes hydraulic pressure to hold the doors opened or closed. The neutral position of the selector valve traps hydraulic fluid in the hydraulic lines to the speed brake actuating cylinders so that the speed brakes will remain in the open, closed, or any intermediate position. Speed brake operation is controlled by an electrical selector valve powered from the primary bus. In the event of failure of the primary bus, the selector valve will return to neutral position and the speed brakes will not open, or if open will close under air loads. Some unmodified aircraft have internal mechanical locks which lock the speed brakes in the closed(IN)position. On some unmodified A thru G aircraft the pneumatic compressor is inoperative if the speed brakes are not fully closed.

WARNING

On non-tandem aircraft a maintenance ground check is prescribed to detect aircraft possessing a momentary "stick-lock" condition. The ground check is as follows:
With engine at 80 to 100 per cent RPM, check boost, utility, and accumulator gages, which should indicate 1400 (±50) PSI); check main and emergency reservoirs for quantity (both tanks should read full). Cycle speed brakes a minimum of five complete cycles while operating stabilator. Stabilator motion should be smooth and steady. Pilot's of aircraft exhibiting unsatisfactory drops in pressure when check is conducted should not actuate speed brakes when a momentary "stick-lock" would be critical, i.e., close to the ground, in close formation, and etc. Pilots may disregard the warning when flying aircraft showing satisfactory stabilator operation during periodic check, if assured that prescribed periodic ground check has been conducted.

SPEED BRAKE SWITCH.

The speed brake switch is a three position sliding switch located on the throttle control (figure 1–15). The three positions are OUT, NEUT and IN. This switch, powered by the primary bus, controls the position of the hydraulic control valve. The OUT position indexes the control valve to apply pressure to open and hold the speed brake in the open position. The IN position indexes the control valve to close the speed brakes, and when the brakes are fully closed, hydraulic pressure is maintained to hold the brakes closed. The NEUT position indexes the control valve to trap hydraulic fluid in the hydraulic lines so that the speed brake will remain in any selected intermediate position. When the speed brake is closed during flight, the switch should be left in the IN position as hydraulic pressure may bleed off sufficiently to allow the speed brakes to open slightly if the NEUT position is selected. On unmodified A thru D aircraft the neutral position is not provided. On unmodified A thru C aircraft the speed brake switch is a two position toggle switch located immediately aft of the throttle friction lock.

MODIFIED and E and LATER

AILERON SPOILERS.

Spoilers are located on the upper surface of both wings, immediately forward of the flaps and are automatically actuated through the control stick lateral deflection for UP aileron only. The spoilers increase the roll performance (rate of roll) of the aircraft during high speed, low altitude flight. At the intermediate speeds the spoilers are less effective, but together with aileron displacement, rate of roll is more than adequate. At low speeds the spoilers are not effective, however, aileron effect is adequate for low speed flight. As the aileron moves up from zero to 10 degrees the corresponding spoiler moves from zero to approximately 45 degrees. Spoiler extension of approximately 45 degrees remains constant for the remainder of aileron travel. The spoilers are hydraulic actuated and controlled by a normally open shut-off valve powered from the primary bus. On unmodified A thru J aircraft the spoilers will only operate after the right main landing gear is up and locked. On modified and K and later aircraft the spoiler circuit is not affected by the position of the landing gear. A switch is also provided in the cockpit so that the spoilers can be turned off during flight. In the event of primary bus power failure, the spoilers cannot be turned off either by the switch in the cockpit or by extension of the landing gear.

MODIFIED and E and LATER

SPOILER SHUT-OFF SWITCH.

The spoiler shut-off switch (20, figure 1–8; 5, figure 1–9), is an on-off switch powered from the primary bus. The switch is marked SPOILER SHUT-OFF and is guarded in the on position. When in the on position, the spoilers will operate when the right main landing gear is up and locked on unmodified A thru J aircraft and on modified and K and later aircraft will operate regardless of the landing gear position. The spoilers can be made inoperative by placing the spoiler shut-off switch in the SPOILER SHUT-OFF position.

LANDING GEAR SYSTEM.

The retractable tricycle landing gear is hydraulically operated and electrically controlled by a landing gear selector valve powered by the primary bus. The valve is spring loaded to the neutral position and in the event of primary bus failure hydraulic pressure will not be available for landing gear operation or boost brakes. Each strut is enclosed by fairing doors that are flush with the contour of the wing and fuselage when the gear is retracted and remain open when the landing gear is extended. The nose wheel shock strut is mechanically shrunk as it is being retracted and automatically returns to its fully extended position when the gear is let down. Mechanical locks secure the three struts in the retracted or extended positions. Inadvertent retraction of the gear when the aircraft is on the ground is prevented by a solenoid lock which automatically prevents moving the landing gear selector handle. A switch is provided to override this safety system in emergencies. The main landing gear downlocks are spring-loaded to the locked position and are unlocked by hydraulic pressure when the struts are retracted. In an emergency the main gear uplocks are released by pneumatic pressure and the gears drop to the locked down position by gravity. During normal operation the nose gear is retracted and extended with hydraulic pressure. In an emergency the nose gear is unlocked and extended with pneumatic pressure. Ground safety locks are provided for maintenance purposes only. Landing gear indicator lights and a warning horn are provided to inform the pilot of the position of the landing gear struts. On A thru G aircraft an armament safety switch, incorporated on the main landing gear struts, prevents the guns from being fired as long as the weight of the aircraft keeps the shock struts compressed. On H and later aircraft the armament safety switch is incorporated in the landing gear selector handle. Also on the later aircraft a pneumatic compressor shut-off switch is actuated by the main landing gear uplock so as to cut the pneumatic compressor off when the landing gear is released from the up and locked position.

LANDING GEAR SELECTOR HANDLE.

The landing gear selector handle (figure 1–7) controls the extension and retraction of the landing gear struts for both the normal and emergency systems. The normal system control valves and indicators are energized from the primary bus. The valve is spring loaded to the neutral position and in the event of primary bus failure, hydraulic pressure will not be available for landing gear operation or booster brakes. The selector has a plastic wheel shaped knob which incorporates a red warning light. The selector handle has three positions; UP, DOWN and EMERG DOWN. The DOWN position electrically positions the hydraulic control valve to apply pressure to extend the landing gear and to energize the brake boosters. Hydraulic power is supplied to the brakes only when the landing gear selector handle is in the DOWN or EMERG DOWN position and primary bus power is available. An electrical solenoid prevents the landing gear selector from being moved from the DOWN position, with the shock struts compressed. This safety can be overridden if it is desirable to retract the gear before the weight is off the landing gear struts. The UP position electrically positions the hydraulic control valve so that hydraulic pressure releases the downlocks, retracts the landing gear and closes the fairing doors. Hydraulic pressure is maintained in the system as long as the selector handle is in the UP position. The EMERG DOWN position is used whenever the normal system fails to lower the gear. This position mechanically opens the pneumatic valve and compressed air unlocks all three gears and extends the nose gear. The main gears drop and lock by gravity. To position the landing gear selector handle from one position to another position, the handle must be pulled out toward the pilot first, then moved to the new position. When going from the DOWN to the EMERG DOWN position the handle must be pulled out toward the pilot first, then pushed inboard and down to the EMERG DOWN position. Once the EMERG DOWN position has been selected the handle cannot be returned to the DOWN position until the handle lock has been released by the ground crew and the system has been bled completely. An armament safety switch has been incorporated in the landing gear selector handle on H and later aircraft so that the armament system cannot be energized when the landing gear selector handle is in the DOWN or EMERG DOWN position. The landing gear selector on unmodified A thru G aircraft incorporates a NEUTRAL position which shuts off hydraulic pressure and relieves the hydraulic pressure in the landing gear system, so that in the event of leakage downstream of the selector valve, hydraulic fluid will not be lost. However, use of the NEUTRAL position is not recommended.

CAUTION

Once the landing gear is extended by the emergency procedure, the emergency landing gear release switch will not retract the gear. In the event that only the main gear can be locked down with the main system, the possibility exists of back pressure in the return hydraulic lines unlocking the main landing gear whenever the nose gear is lowered by means of the pneumatic system. However, within a short time, the pneumatic system should reactuate the downlock cylinder, relocking the main gears in the down position. Therefore, the landing gear position indicators must be rechecked to ascertain whether all gears are down and locked after the pneumatic system has been used to lower any gears.

Note

The landing gear selector handle should be left in the UP position during flight as the landing gear may unlock during high g maneuvers if the handle is not in the UP position.

EMERGENCY LANDING GEAR RELEASE SWITCH.

The emergency landing gear release switch (figure 1–7) is provided as an override switch for the landing gear selector handle solenoid lock. The switch is marked EMERGENCY RELEASE OF L G HANDLE and is protected by a cover type guard. When it is necessary to retract the landing gear, while the aircraft is on the ground, the cover guard is lifted and the emergency landing gear release switch is actuated. This energizes the landing gear selector handle solenoid lock which allows the selector handle to be moved to the UP position.

CAUTION

If the landing gear shock struts are overinflated, so as to be fully extended, it will be possible to move the landing gear selector to the UP position with the aircraft on the ground.

LANDING GEAR POSITION INDICATORS.

Green indicator lights, marked LEFT, NOSE and RIGHT (figure 1–7) illuminate when the respective landing gear is extended and in a safe, locked condition for landing. On unmodified A thru G aircraft three landing gear position indicators are installed and are marked LEFT, NOSE, and RIGHT. When the respective gear is locked down the outline of the wheel appears on the indicator. If the gear is in any position between locked down or up, alternate red and white stripes appear and when the gear is retracted the word UP appears on the indicator.

LANDING GEAR SELECTOR HANDLE LIGHT.

A red light, located in the plastic knob of the landing gear selector handle illuminates if the handle and gear positions are not compatible and the throttle is below minimum cruise.

Console Lights Rheostat.

On aircraft modified in accordance with T.O. 1F-84F-674 the landing gear selector handle light can be dimmed by placing the console lights rheostat (15, figure 1–8; 8, figure 1–11) between the DIM and BRIGHT positions. However, when the rheostat is in the OFF position the landing gear selector handle light will illuminate brightly.

LANDING-GEAR-WARNING HORN.

The landing-gear-warning horn, located on the aft wall of the cockpit, or the MA-1 warning signal, whichever is installed, will sound if the throttle is retarded below the 82-per-cent-RPM position when the landing gear is not down and locked.

Landing-Gear-Warning-Horn Silence Switch.

The landing-gear-warning-horn silence switch (figure 1–15) is a push-button-type switch marked LG WARN HORN SILENCE. The warning horn, or the MA-1 Warning Signal, whichever is installed, can be silenced by depressing the landing-gear-warning-horn silence switch momentarily. If the horn, or the warning signal, has been silenced, and the throttle is opened above the 82-per-cent-RPM position, then closed again, the horn, or signal, will start to sound again.

Landing-Gear-Warning-Light-and-Horn Test Switch.

The landing-gear-warning-light-and-horn test switch (figure 1–7) is the push-button-type switch marked TEST LG HANDLE LT AND HORN. If the test switch is pushed and held, the red light in the landing-gear-selector handle will illuminate, regardless of the position of the landing gear. This tests the electrical circuit and the lamp in the selector handle. The landing-gear-warning horn, or the MA-1 warning signal, will also sound if the throttle is positioned below the 82-per-cent-RPM position.

BRAKE SYSTEMS.

BRAKES — WHEEL.

Fluid for the brake control is provided by the utility hydraulic system. The system is designed so that foot pressure on the master cylinder, attached to each rudder pedal, is added to the hydraulic pressure of the brake system. System hydraulic pressure is available to the brakes only when the landing gear selector handle is in the DOWN or EMERG DOWN position and the primary bus is energized. The brakes will be effective with a minimum hydraulic pressure of approximately 550 PSI. If pressure is not available from the utility hydraulic system, the brakes will function through conventional action of the master brake cylinder when toe pressure is applied to the pedal. However, in this event approximately twice the toe pressure will be required for the brakes to be as effective as when hydraulic pressure is available. Parking brakes are not provided.

CAUTION

In the event of a primary bus failure, the spring loaded landing gear selector valve will move to the neutral position and boosted brakes will not be available.

BRAKING TECHNIQUE.

As is obvious, brakes themselves can merely stop the wheel from turning, but stopping the aircraft is dependent on the friction of the tires on the runway. For this purpose it is easiest to think in terms of coefficient of friction which is equal to the frictional force divided by the load on the wheel. It has been found that the optimum braking occurs with approximately a 15 to 20 per cent rolling skid; i.e., the wheel continues to rotate but has approximately 15 to 20 per cent slippage on the surface so that the rotational speed is 80 to 85 per cent of the speed which the wheel would have were it in free roll. As the amount of skid increases beyond this amount the coefficient of friction decreases rapidly so that with a 75 per cent skid the friction is of the order of 60 per cent of the optimum and, with a full skid, becomes even lower. There are two reasons for this loss in braking effectiveness with skidding; 1st, the immediate action is to scuff the rubber, tearing off little pieces which act almost like rollers under the tire; 2nd, the heat generated starts to melt the rubber and the molten rubber acts as a lubricant. NACA figures have shown that for an incipient skid with a load of approximately 10,000 pounds per wheel the coefficient of friction on dry concrete is as high as 0.8, whereas the coefficient is of the order of 0.5 or less with a 75 per cent skid. Therefore, if one wheel is locked during application of brakes there is a very definite tendency for the aircraft to turn away from that wheel and further application of brake pressure will offer no corrective action. Since the coefficient of friction goes down when the wheel begins to skid it is apparent that a wheel, once locked, will never free itself until brake pressure is reduced so that the braking effect on the wheel is less than the turning moment remaining with the reduced frictional force. It should also be noted that the actual turning moment which depends on the frictional force is the product of the coefficient of friction times the load on the wheel. Therefore braking pressure can result in locking the wheel more easily if the brakes are applied immediately after touchdown than if the same pressure is applied after the full weight of the aircraft is on the wheels, and a wheel once locked in this manner right after touchdown will not become unlocked as the load is increased as long as the brake pressure is maintained.

CAUTION

On aircraft equipped with Cagle Brakes that have not been modified in accordance with applicable Technical Order, some pumping of brakes may be required to obtain satisfactory braking action.

MODIFIED and J and LATER

DRAG CHUTE.

A 16-foot drag chute is installed in a compartment in the ventral fin to reduce landing roll, permitting the use of shorter runways and to serve as an added safety factor in the event of brake system failure or on wet, slippery runways. The drag chute and riser cable is packed in a canvas bag and stowed in an aluminum container which is in the ventral fin compartment. The compartment is equipped with doors, spring loaded to the open position, and an electrically actuated latch lock and unlock mechanism controlled by a pull-type tee handle. A shear pin has been provided as a safeguard and will break, releasing the drag chute, in the event that the drag chute is accidentally released at speeds in excess of 220 knots. A properly packed chute will deploy at speeds as low as 60 knots IAS in approximately four seconds. However, to insure consistent and proper operation, it is recommended that the chute be deployed above 75 knots IAS. To keep the chute blossomed after landing roll, approximately 70 per cent or sufficient power to attain an effective headwind of 10 knots is required. Experience has shown that 80 per cent power will cause scorching of the chute, when trailing directly behind the aircraft. Taxiing directly into strong headwinds should therefore be avoided. The drag chute should not be deployed before touchdown. Blossoming of the chute during landing flare-out will result in a faster rate of sink and a strong tendency to pitch nose down. Estimated data shows that chute deployment on a dry runway with 90-degree crosswinds in excess of 40 knots may cause the aircraft to leave the runway. Tests have been performed on dry runways in crosswinds up to 30 knots with excellent results. On icy runways, however, weathervaning, due to crosswinds as low as 10 knots, may result. Normally, the chute will be found to be of great value on icy runways. In the event of a go-around the drag chute must be jettisoned as the maximum airspeed the pilot can attain with the drag chute deployed is approximately 140 knots at 100 per cent RPM. The drag chute may be deployed or jettisoned if battery bus power is available.

CAUTION

The drag chute compartment is susceptible to condensation of moisture under certain atmospheric conditions. To minimize the possibility of the chute failing to deploy as a result of it being wet and freezing during flight a preflight inspection should be performed as outlined in Section II.

MODIFIED and J and LATER

Drag Chute Control Handle.

The drag chute control handle (figure 1–35) is a T-handle marked PULL TO DEPLOY: ROTATE 90°, PULL TO JETTISON. To deploy the drag chute, the handle is pulled aft approximately two inches, closing a microswitch which energizes the riser cable lock solenoid. The solenoid forces a plunger between the forward arms of a latch clamp locking the riser cable in the drag position. Movement of the plunger closes a second micro switch which energizes a solenoid in the left door spring capsule releasing the door latch and opening the door; the spring capsule on the right door is then free to force the right door to the open position. As the left door swings open, a cable attached to the door pulls the pilot chute rip pin, allowing the pilot chute and drag chute to be released in sequence. When the drag chute is released it extends approximately 35 feet behind the aircraft. On airplanes modified in accordance with T.O. 1F-84-728, a button is provided in the T-handle which prevents the pilot from returning the handle to the full forward position, inadvertently jettisoning the chute when it is in the deployed position. To jettison the chute when it is in the deployed position, the button in the handle is depressed, the T-handle is then rotated 90 degrees counterclockwise and is pulled aft a second time. This will electrically open the riser cable latch lock, releasing the riser cable and drag chute for jettisoning. The drag chute control circuit is powered by the battery bus.

MODIFIED and J and LATER

Drag Chute Power Test Indicator Light.

A press-to-test indicator light (figure 1–35) marked PRESS TO CHECK LT ON–CHUTE PWR ON, is located adjacent to the drag chute control handle.

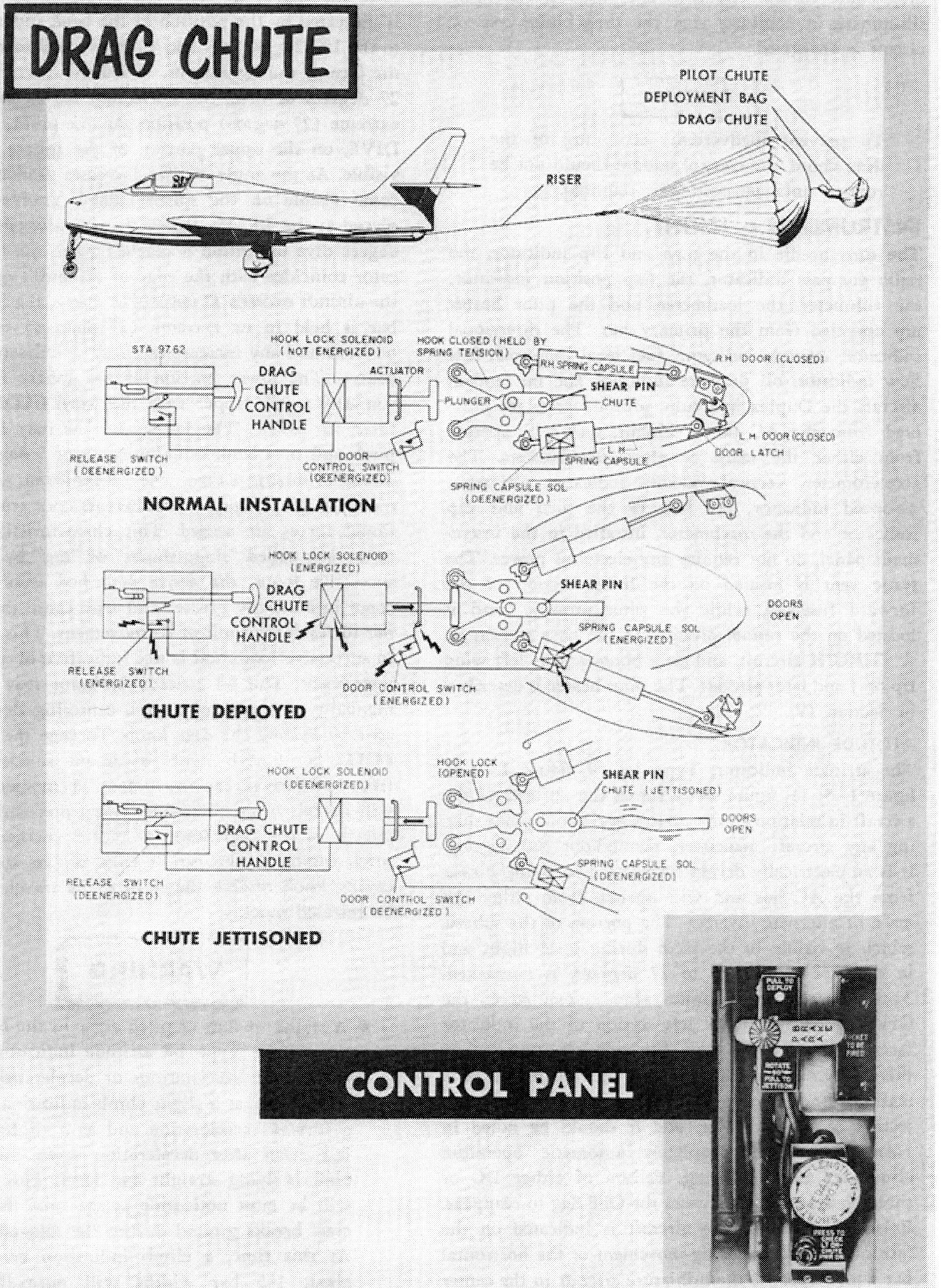
DRAG CHUTE
PILOT CHUTE
DEPLOYMENT BAG
DRAG CHUTE
RISER
STA 97.62
HOOK LOCK SOLENOID (NOT ENERGIZED)
HOOK CLOSED (HELD BY SPRING TENSION)
DRAG CHUTE CONTROL HANDLE
ACTUATOR
RH SPRING CAPSULE
R.H. DOOR (CLOSED)
SHEAR PIN
PLUNGER
CHUTE
L.H. SPRING CAPSULE
L.H. DOOR (CLOSED)
DOOR LATCH
RELEASE SWITCH (DEENERGIZED)
DOOR CONTROL SWITCH (DEENERGIZED)
SPRING CAPSULE SOL (DEENERGIZED)
NORMAL INSTALLATION
HOOK LOCK SOLENOID (ENERGIZED)
DRAG CHUTE CONTROL HANDLE
SHEAR PIN
DOORS OPEN
SPRING CAPSULE SOL (ENERGIZED)
RELEASE SWITCH (ENERGIZED)
DOOR CONTROL SWITCH (ENERGIZED)
CHUTE DEPLOYED
HOOK LOCK SOLENOID (DEENERGIZED)
HOOK LOCK (OPENED)
SHEAR PIN
CHUTE (JETTISONED)
DRAG CHUTE CONTROL HANDLE
DOORS OPEN
SPRING CAPSULE SOL (DEENERGIZED)
RELEASE SWITCH (DEENERGIZED)
DOOR CONTROL SWITCH (DEENERGIZED)
CHUTE JETTISONED
CONTROL PANEL

Figure 1—35

The indicator light is pressed in to determine if the drag chute control circuit is energized. If the light illuminates it indicates that the drag chute control circuit is energized.

CAUTION

To prevent inadvertent jettisoning of the drag chute, the control handle should not be rotated until jettisoning is required.

INSTRUMENTS — FLIGHT.

The turn needle in the turn and slip indicator, the radio compass indicator, the flap position indicator, the voltmeter, the loadmeter, and the pitot heater are operated from the primary bus. The directional indicator, attitude indicator, fuel level indicator, fuel flow indicator, oil pressure indicator and on tandem aircraft the Duplex hydraulic pressure gage are powered from the AC power circuit, and will operate from either the main or alternate inverters. The accelerometer, vertical velocity indicator, altimeter, air-speed indicator, the ball in the turn and slip indicator and the machmeter, installed in the instrument panel, do not require any electrical power. The static vent is located on the lower section of the forward fuselage, while the pitot pressure head is located on the tunnel division of the nose section on A THRU H aircraft, and on a boom on the left wing tip on J and later aircraft. The pitot heater is described in Section IV.

ATTITUDE INDICATOR.

The attitude indicator, Type J-8 (9, figure 1–4; 9, figure 1–5; 11, figure 1–6) shows the attitude of the aircraft in relation to the earth's horizontal plane during any aircraft maneuver, throughout 360 degrees. It is an electrically driven instrument receiving power from the AC bus and will operate from either the main or alternate inverter. The portion of the sphere, which is visible to the pilot during level flight and in dives or climbs up to 27 degrees, is unmarked. Approximately 2½ minutes after system starts, the OFF flag on the upper left section of the indicator face retracts. If the OFF flag requires longer than this to retract, it may indicate the possibility of a malfunction. (Any minor oscillation is cause for rejection of the indicator, and it should be noted in Form 781.) The completely automatic operation eliminates manual caging. Failure of either DC or three phase AC power causes the OFF flag to reappear. Relative motion of the aircraft is indicated on the face of the instrument by movement of the horizontal bar with respect to the miniature aircraft in the center of the dial. Angular displacement of the horizontal bar, with respect to the miniature aircraft, indicates the degree of roll. The actual amount of roll is indicated by the position of the bank index relative to the 10-, 20-, 30-, 60- and 90-degree roll markings on the face of the instrument. When the aircraft exceeds 27 degrees of dive, the horizontal bar is held in its extreme (27 degree) position. At this point, the word DIVE, on the upper portion of the sphere, becomes visible. As the angle of dive increases graduations become visible on the sphere. These graduations are placed at the 70-, 75- and 80-degree intervals; the 85-degree dive indication is reached when the trim indicator coincides with the edge of the bull's eye. When the aircraft exceeds 27 degrees of climb, the horizontal bar is held in its extreme (27 degrees) downward position and any increase in climb is indicated on the sphere. The lower portion of the sphere is marked similarly to the upper with the word CLIMB substituted for DIVE. The horizontal bar may indicate a pitch and/or a bank error in excess of 5 degrees after a loop or during a turn. The J-8 indicator will immediately begin to correct these errors once true gravitational forces are sensed. This characteristic error is commonly called "sluggishness" or "lag" by pilots. In successive loops, the above described error may become increasingly greater and may cause the horizon bar to reach the limit of its movement. This is normal in successive loops and is not indicative of a defective instrument. The J-8 attitude indicator may be caged manually by means of a gyro centering device operated by pulling the cage knob. To cage the gyro, the PULL TO CAGE knob is drawn smoothly away from the face of the instrument. A momentary stop will be felt when the bank caging mechanism is engaged; as the cage knob is pulled further out, the pitch caging mechanism is engaged. As soon as the caging knob reaches the limit of its travel, it should be released quickly.

WARNING

- A slight amount of pitch error in the indication of the Type J-8 attitude indicator will result from accelerations or decelerations. It will appear as a slight climb indication after a forward acceleration and as a slight dive indication after deceleration when the aircraft is flying straight and level. This error will be most noticeable at the time the aircraft breaks ground during the take-off run. At this time, a climb indication error of about 1½ bar widths will normally be noticed; however, the exact amount of error

ENGINE OVERHEAT and FIRE WARNING SYSTEM

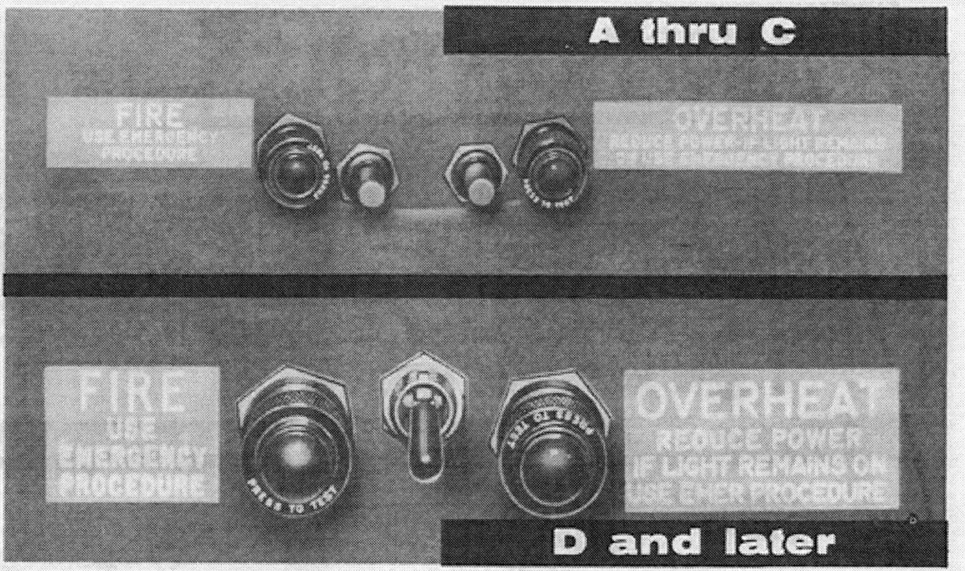

Figure 1—36

will depend upon the acceleration and elapsed time of each individual take-off. The erection system will automatically remove the error after the acceleration ceases.

- A slight reduction in AC or DC power or failure of certain electrical or mechanical components in the system could cause loss of roll indication, and not cause the OFF flag to appear even though the instrument is not functioning properly. Therefore, periodically in flight, the attitude indications should be checked against other flight instruments, such as the standby magnetic compass, or the turn and slip, and vertical velocity indicators.

CAUTION

A violent or hard pull on the caging knob when caging the attitude indicator may damage the instrument. Remember that the indicator cages to the attitude of the aircraft and not to the true vertical. Therefore, the instrument should never be caged to correct in-flight errors, unless the aircraft is in straight and level flight by visual reference to a true horizon.

ALTIMETER.

The altimeter (27, figures 1—4, 1—5 and 1—6) utilizes static pressure and provides the pilot with a constant indication of barometric altitude. Three pointers on the face of the instrument move over a scale graduated in 20 foot increments, with a major division every hundred feet from zero to one thousand feet. The larger pointer indicates hundreds of feet and makes one revolution for each 1,000 feet of altitude. An additional small pointer indicating tens of thousands of feet is painted on a black disc with an extension line terminating in a triangular section. The pointers and barometric scale can be set manually by turning the knob in the lower left corner of the instrument. To determine altimeter error, the pilot sets base altimeter setting on the barometric scale, then notes indicated altitude, which should be compatible with known field elevation.

CAUTION

It is possible to set the altimeter in error by 10,000 feet. This happens when the barometric set knob is continuously rotated after the barometric scale is out of view. The knob can be rotated until eventually the numbers will reappear in the window from the opposite side. If the correct altimeter setting is then established, the altimeter will read approximately 10,000 feet in error.

EMERGENCY EQUIPMENT.

ENGINE OVERHEAT WARNING SYSTEM.

The aircraft has an engine overheat warning system consisting of a set of eight thermal switches, installed in the aft fuselage, and an amber warning light (8, figure 1—4; 7, figure 1—5 and figure 1—36) which is marked OVERHEAT. The system operates automatically and is powered by the primary bus. A button-type of test switch is installed adjacent to the warning light and is marked PRESS TO TEST. If the warning light illuminates when the test switch is depressed, the circuits are complete and the system is functioning properly.

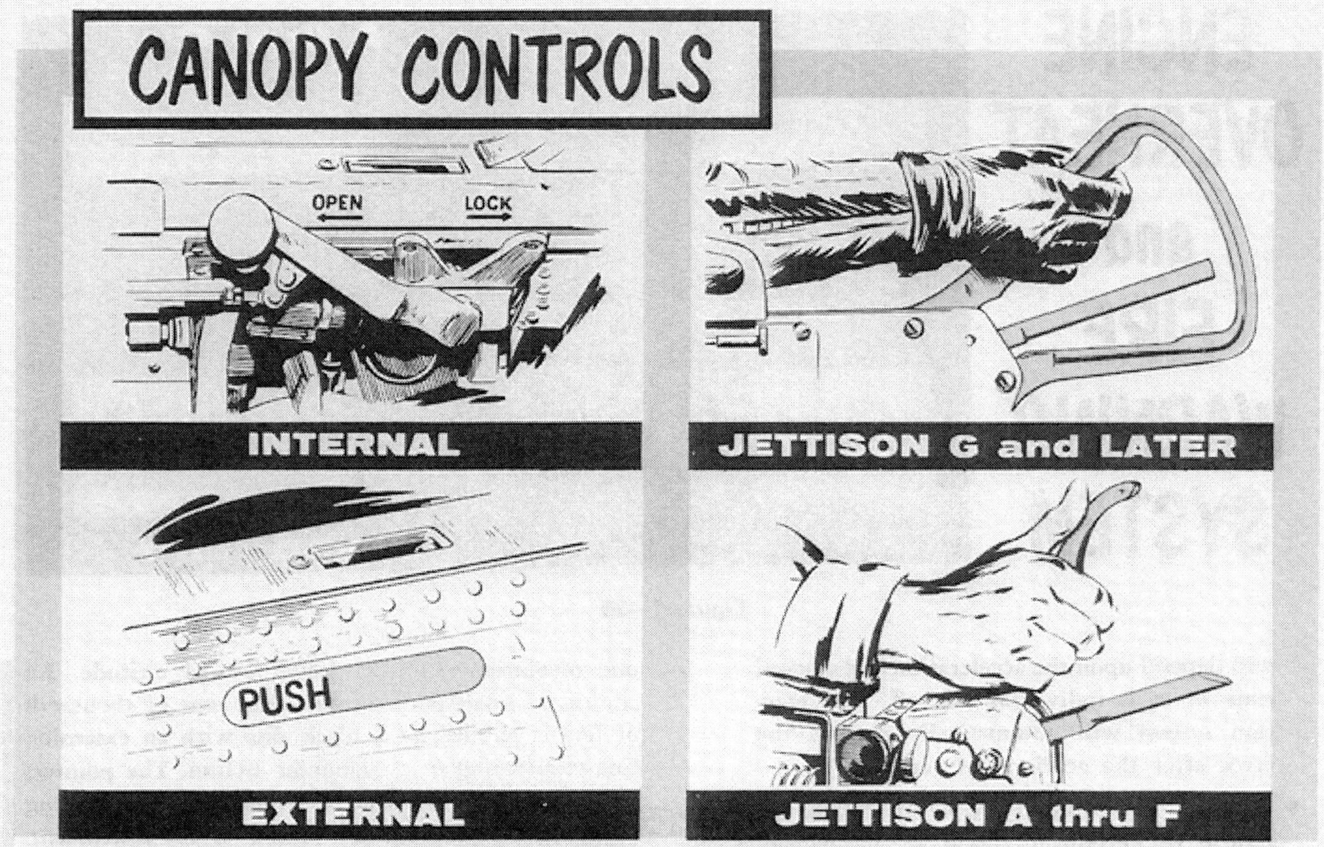

Figure 1—37

ENGINE FIRE WARNING SYSTEM.

An engine fire warning system, consisting of a set of thermal switches in the forward section of the fuselage, and a red warning light (8, figure 1—4 and 7, figure 1—5 and figure 1—36) marked FIRE, is installed in the aircraft. The system operates automatically and is powered by the primary bus. A button-type of test switch is installed adjacent to the warning light and is marked PRESS TO TEST. If the warning light illuminates when the test switch is depressed, the circuits are complete and the system is functioning properly.

WINDSHIELD.

The windshield consists of three transparent panels, set in rubber and mounted in an aluminum frame. The center panel is laminated bullet-resistant glass with an electrical element incorporated between the laminations for de-icing, and de-fogging. Each side panel is made up of two plexiglas panels with an air space between them for de-fogging. Heated and dried air is circulated between the two layers and is controlled by a manually operated switch. A defroster system, with a manual shut-off, directs hot air from the engine compressor to a perforated tube that directs the air over the inside of each side panel. For complete description of the defroster system, see Section IV.

CANOPY.

The canopy is a one piece bubble type made up of two layers of plexiglas with an air space between them. Heated and dried air, controlled by a manually operated switch, is circulated between the two layers to de-fog the canopy. The canopy is manually operated and is hinged to the fuselage in such a way that when it is opened, it moves up and aft in an arc so as to clear the pilot at all times. It is counter-balanced so that opening and closing is accomplished with a minimum effort. The canopy is jettisoned by means of a manually fired initiator that develops a high pneumatic pressure which is carried through a flex hose to the canopy downlock release cylinder which unlocks the canopy and actuates a second initiator. The pneumatic pressure from the second initiator is carried through a flex line to a piston type of thruster which raises the canopy approximately four inches up into the airstream which will force it to the fully open position. In the fully open position, the right canopy mounting arm actuates a microswitch which completes a circuit to the canopy jettison explosive squibs at the four hinge fittings. In the event canopy jettison is attempted and the canopy fails to separate, on aircraft modified by drilling holes in the canopy arms (T.O. 1F-84-763), the canopy will jettison at mimimum airspeeds, since the holes will cause the canopy arms to break under airstream pressure. On unmodified aircraft an airspeed of, approximately, 300 knots IAS at the time the canopy is jettisoned will assure the canopy breaking free from the aircraft in the event the squibs fail to fire. The explosive squibs are normally powered by the battery bus when the battery switch is in the ON position. In the event of battery failure, or when the battery switch is in the OFF position, the explosive squibs are powered by the canopy jettison battery, which is independent of the aircraft electrical system. A canopy squib test circuit is provided to test the squib firing circuits. In the closed position, the canopy is sealed to the cockpit structure by rubber tubes that are automatically inflated by air pressure from the engine compressor. A warning light, powered from the primary bus, is provided to indicate when the canopy is not fully closed and locked. The canopy may be left open during taxiing on smooth ground.

WARNING

The present inspection requirement applicable to the 4.5 volt canopy jettison battery does not determine the ability of the battery to detonate the canopy squibs, therefore the aircraft battery switch should be left in the ON position until the canopy has been jettisoned.

CAUTION

Do not rest arms on longerons while canopy is open as injury will result if canopy unexpectedly closes during taxiing.

CANOPY CONTROL LEVER.

The canopy control lever (10, figure 1–8; 9, figure 1–9) operates a manual lock and has only two positions: OPEN and CLOSE. The CLOSE position mechanically locks the canopy to the fuselage structure. The OPEN position releases the canopy locks and the canopy can then be lifted to the fully open position. When fully open, a lock automatically engages on the side hinges to hold the canopy. These locks are released by pulling the canopy control lever aft, past the OPEN position. This operation assures that the canopy control lever will be in the OPEN position before closing the canopy. When the canopy is jettisoned, the mechanical locks between the canopy and fuselage are automatically opened.

EXTERNAL CANOPY CONTROL.

The external canopy control (figure 1–38) located on the left side of the fuselage below the canopy is used to open or close the canopy from outside the airplane. The forward end is marked PUSH and if pushed in, the control rotates from its flush position so that a handle is available and at the same time

This page intentionally left blank.

WARNING

A type MA-1 seat cushion is provided for use in the seat with a survival kit container when a life raft is not utilized. The forward edge of the packed kit should not be thicker than seven inches. Recent investigations have indicated that some pilots are leaving the standard cushion in the ejection seat when carrying the C-2A life raft. Sitting on both of these items does not provide any added comfort but actually creates a definite injury hazard as well as possibly positioning the pilot where it may be difficult to reach all the controls. Chance of vertebrae injury is increased considerably by the pilot sitting on a thick, compressible mass, such as a combination of the life raft and seat cushion, because when utilizing the ejection seat in this circumstance, it will not exert a direct force on its occupant until the seat has moved two or three inches upward. After this amount of travel, the seat has gathered such momentum that excessive impact is produced when the seat initially lifts the pilot. The chance of injury is also increased considerably under these circumstances if a crash landing cannot be avoided. If a crash landing is made, the combined compressiveness of the two items will permit the pilot to sink down far enough to loosen the shoulder straps allowing him to slump forward. In this position the impact may severely injure his back. The use of additional material under the pilot may raise him to where his arms will not be held by the retainers on either arm rest and they will be subjected to injury due to possible flailing in the windblast. Therefore, the seat cushion should not be left in the seat and should be replaced by the type MA-1 cushion, if the C-2A life raft is being carried.

SHOULDER HARNESS LOCK CONTROL.

A two-position (lock-unlocked) shoulder harness locking control lever (figures 1—39 and 1—40) is installed on the letf side of the pilot's seat. The control handle may be actuated manually or automatically by handgrip action when jettisoning the canopy. On A THRU F aircraft, the reel lock control handle is provided with a latch mechanism for positively retaining the control handle in either the locked (full forward) or unlocked position (full aft) on the quadrant. By pressing down on top of the control handle, the latch is released and the control handle may be moved freely from one position to the other. The inertia reel lock on G and later aircraft may be moved manually or automatically to lock the inertia reel when jettisoning the canopy. On all aircraft, when the control handle is in the unlocked position (full aft) the shoulder harness reel cable will extend to allow the pilot to lean forward in the cockpit and automatically retract as tension is released when the pilot moves back. However, the inertia reel will automatically lock, preventing further extension of the cable, when an impact force of 2 to 3 G's is encountered. When the reel is locked in this manner, it will remain locked until the control handle is moved to the locked position and then returned to the unlocked position. If the control is in the locked position (full forward) while the pilot is leaning forward, as he straightens up, the harness will retract with him, moving into successive locked positions as he moves back against the seat. The locked position is used when a crash landing or ditching is anticipated. This control position provides an added safety precaution over and above that of the automatic safety lock.

A THRU F

VERTICAL ADJUSTMENT CONTROL.

The vertical adjustment control (figure 1—39) with an attached locking pin indicator lever is located on the right side of the pilot's seat. Moving the vertical adjustment control up releases the seat so that it may be raised or lowered to the desired height, and at the same time, the footrests are released from the seat so that they remain in contact with the cabin floor at all times. With the vertical adjustment control in this position there is no "finger" space between the adjustment control and the locking pin indicator. After the seat is at the desired height, the vertical adjustment control is released and the seat and footrest locking pins automatically engage. If the locking pins are fully engaged there will be "finger" space (approximately ⅝ inch) between the vertical adjustment control and the locking pin indicator lever. In the event that the locking pins are not fully engaged (indicated by no "finger" space), jiggle the seat to permit the pins to fully seat.

This page intentionally left blank.

WARNING

After each vertical (height) adjustment of the seat insure that the adjustment handle and latch lever are separated by approximately ⅝ inch (½ to ¾ inch). Failure of the handle and the latch to separate as specified indicates one or more of the four seat locking pins has not locked. Failure of the top pins to lock in place can prevent successful seat ejection.

G and LATER

VERTICAL ADJUSTMENT CONTROL.

The pilot's seat adjustment control (figure 1–40) is a three position momentary contact toggle switch located on the right side of the pilot's seat. The switch actuates the electrically driven seat adjustment and moves the seat in the same direction as switch travel, up, when the switch is in the up position; downward when the switch is held in the down position. Seat height can be adjusted over a range of 3.75 inches of travel. The drive mechanism is self-locking and will operate when the battery switch is ON and the primary bus is energized.

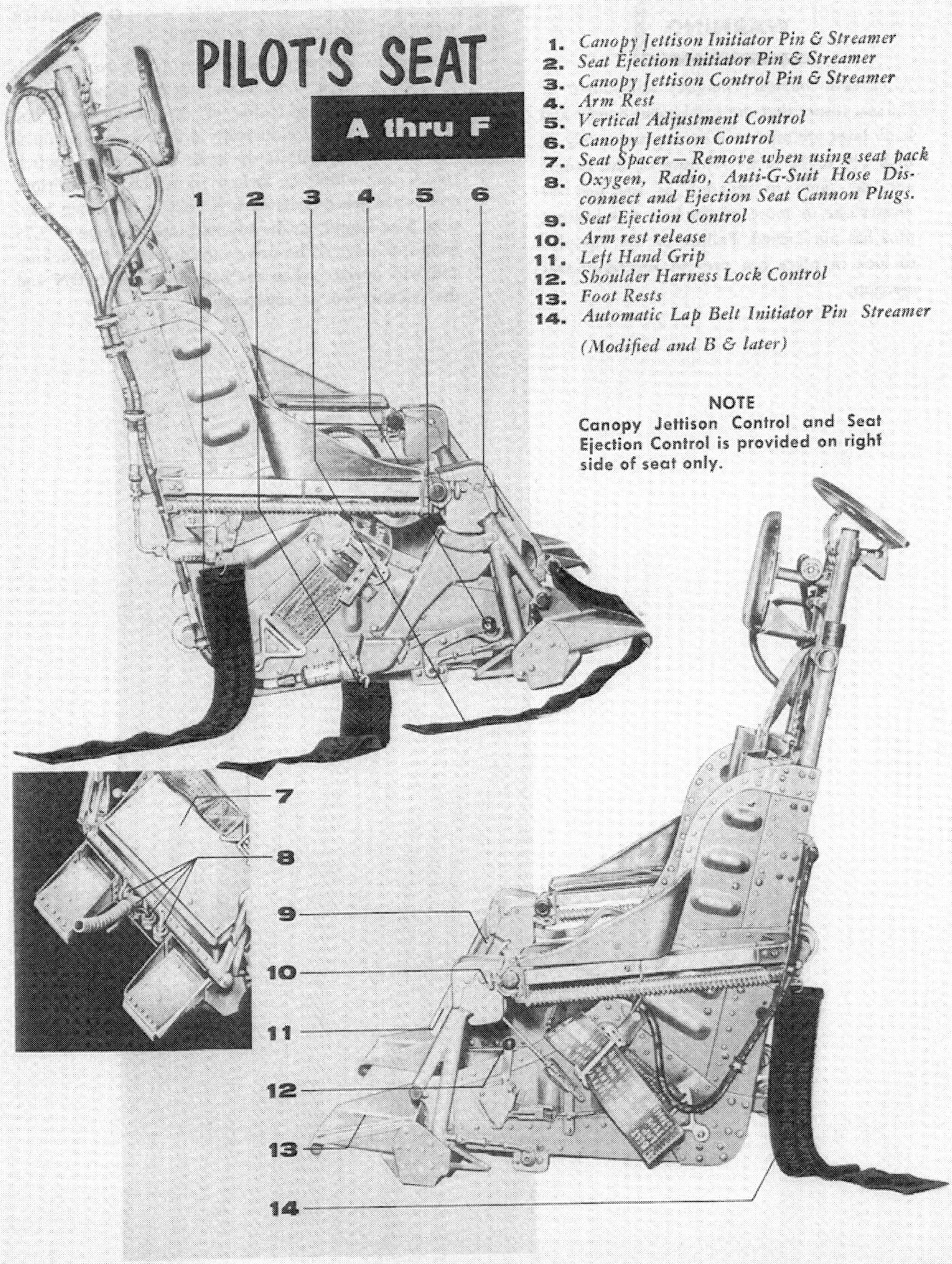

1. *Canopy Jettison Initiator Pin & Streamer*
2. *Seat Ejection Initiator Pin & Streamer*
3. *Canopy Jettison Control Pin & Streamer*
4. *Arm Rest*
5. *Vertical Adjustment Control*
6. *Canopy Jettison Control*
7. *Seat Spacer — Remove when using seat pack*
8. *Oxygen, Radio, Anti-G-Suit Hose Disconnect and Ejection Seat Cannon Plugs.*
9. *Seat Ejection Control*
10. *Arm rest release*
11. *Left Hand Grip*
12. *Shoulder Harness Lock Control*
13. *Foot Rests*
14. *Automatic Lap Belt Initiator Pin Streamer*

(Modified and B & later)

NOTE

Canopy Jettison Control and Seat Ejection Control is provided on right side of seat only.

Figure 1—39

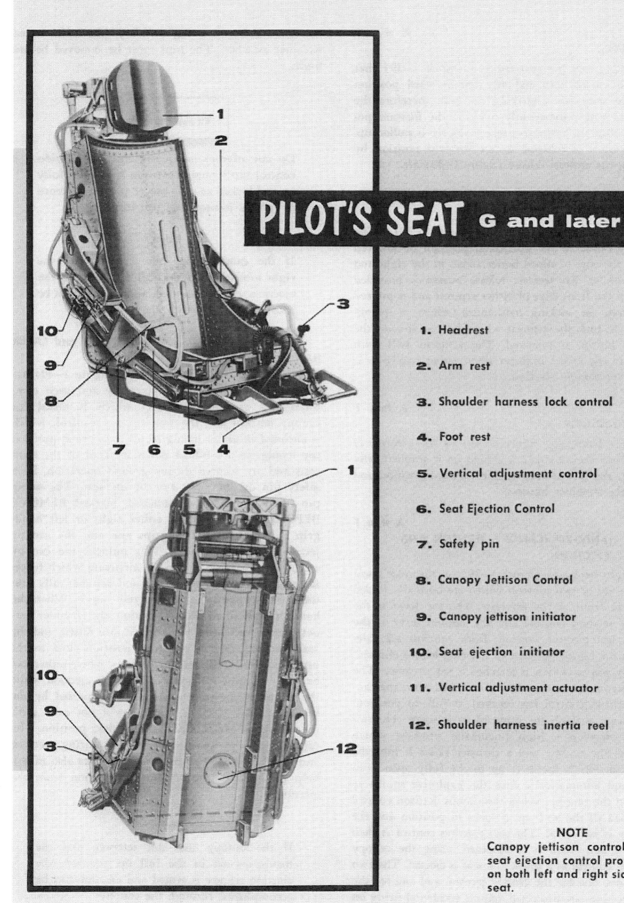

PILOT'S SEAT G and later

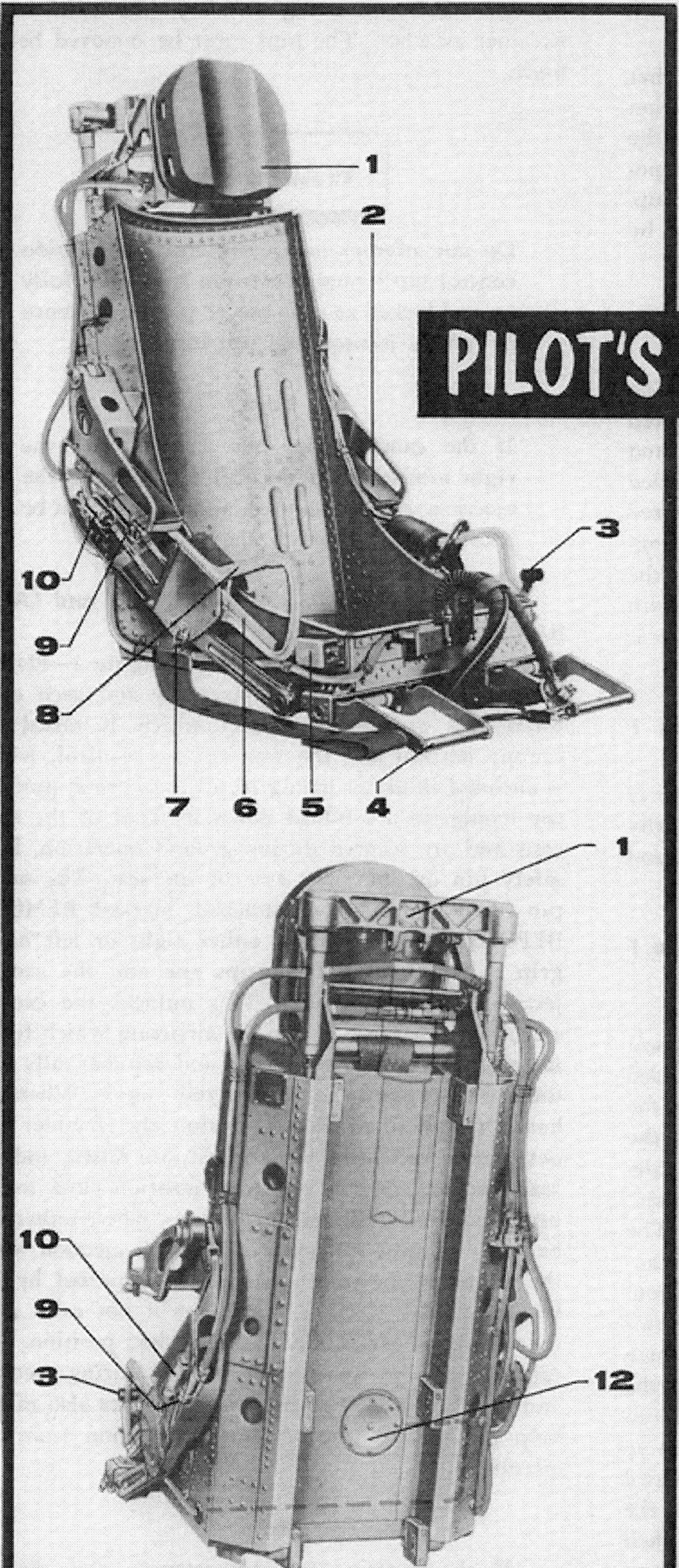

1. Headrest
2. Arm rest
3. Shoulder harness lock control
4. Foot rest
5. Vertical adjustment control
6. Seat Ejection Control
7. Safety pin
8. Canopy Jettison Control
9. Canopy jettison initiator
10. Seat ejection initiator
11. Vertical adjustment actuator
12. Shoulder harness inertia reel

NOTE

Canopy jettison control and seat ejection control provided on both left and right sides of seat.

Figure 1—40

A thru F

ARMRESTS.

The pilot's seat has two armrests (figure 1–39) that may be moved fore and aft. The forward position must be used when ejecting the seat, therefore the armrests will automatically move to the forward position when the respective seat handgrip is pulled up. The armrest is released to the forward position by moving the armrest release (figure 1–39) aft.

G and LATER

ARMRESTS.

The pilot's seat has two armrests (figure 1–40) which may be raised to the horizontal position or lowered out of the way to afford better access to the right and left consoles. An armrest release button is provided beneath the front edge of either armrest and is pressed to release the locking mechanism, which is spring-loaded to lock the armrest when the pressure on the release button is removed. The armrests will both move to the raised position when either handgrip is raised for canopy ejection.

A thru F

LEFT HANDGRIP.

The left handgrip (figure 1–39) is for emergency use. When the handgrip is pulled up it automatically releases the left armrest to the forward position and locks the shoulder harness.

A thru F

RIGHT HANDGRIP (CANOPY JETTISON AND SEAT EJECTION)

The right handgrip (figure 1–39) on the pilot's seat is made up of two controls which are normally folded down in front of the armrests. The top lever is the canopy jettison control and the bottom lever is the pilot's seat ejection control. Both controls are safetied in the folded position by a spring-loaded clip and a safety pin to which is attached a red streamer. The seat ejection control cannot be actuated until the canopy jettison control has reached its full up position. Each lever controls the firing of an initiator. The initiators develop a high pneumatic pressure which unlocks the canopy and a thruster raises it into the airstream which forces it up to the fully open position and automatically fires the explosive squibs to jettison the canopy. When the canopy jettison control is pulled all the way up, it locks in position and the canopy is jettisoned. The seat ejection control is then exposed and when it is squeezed, using the canopy jettison control as a brace, the seat is ejected. The two initiators, one for the canopy jettison and one for the seat ejection, are protected against accidental firing on the ground by inserting a safety pin with a red streamer attached. The pins must be removed before flight.

WARNING

Do not attempt to squeeze the seat ejection control until canopy jettison control is fully up and locked so as to assure proper sequence of canopy jettison and seat ejection.

Note

If the canopy does not jettison, and the right armrest is in the full up position, the ejection trigger is armed, and ejection can be accomplished through the canopy.

G and LATER

HANDGRIPS.

The right and left loop handgrips (figure 1–40) are interconnected to act simultaneously and each combines two controls; the loop which is raised for canopy jettison and the seat ejection control, which is enclosed in either handgrip. In the normal position the handgrips are folded down in front of the armrests and are secured during ground operation, by a safety pin on the right side of the seat. The safety pin has a red streamer attached, marked REMOVE BEFORE FLIGHT. When either right or left handgrip is raised, both handgrips rise and the canopy jettison initiator is fired. This unlocks the canopy and a thruster raises it into the airstream which forces it up to the fully open position and automatically fires the explosive squibs to jettison the canopy. When the handgrips are in the raised position, the shoulder harness inertia reel locks, the armrests are raised, and the seat ejection controls rise to a position close to the upper portion of the handgrip loops, where either can be easily grasped and squeezed for seat ejection while the palms of the pilot's hands are supported by the handgrip. The ejection controls will not raise until the handgrips are in the up and locked position. The raised handgrips serve as leg braces during ejection and together with the armrests, which are also raised, help prevent pilot injury during ejection from the aircraft.

Note

If the canopy does not jettison, and the handgrips are in the full up position, the ejection trigger is armed and ejection can be accomplished through the canopy.

MODIFIED and B and LATER AUTOMATIC SAFETY BELT.

Prior to the use of automatic safety belts in airplanes having ejection seats, instructions were issued that if ejection was necessary below 2000 feet that the safety belt be opened manually prior to ejection. This procedure was necessary to minimize time required for separation of the pilot from the seat after ejection. This procedure was recommended until automatic safety belts could be obtained and installed. It provided best capability to survive even though it was known that manual release of the safety belt may require the pilot to hold himself in the seat prior to ejection (due to negative G on uncontrollable maneuvers) and would place him in a dangerous situation if he changed his mind and decided to crash land the airplane (as he would have no time to relock his safety belt). A type MA-1, MA-3, MA-5 or MA-6 lap type safety belt is installed, which is released automatically after seat ejection by an initiator which is located on the back of the seat. In (A THRU F) aircraft, a ground safety pin is inserted in the initiator to prevent inadvertent action. This pin must be removed prior to flight and re-inserted after each flight. On G and later aircraft, the safety provisions for the belt initiator are provided through the installation of the single ground safety pin in the seat right armrest and additional pins are unnecessary. In operation, the initiator fires when the seat is ejected, supplying gas pressure to a special belt disconnect, thus releasing the belt and shoulder harness. A time delay feature in the initiator insures that the belt will not release until the pilot is entirely clear of the airplane. A key or anchor is attached to the seat belt buckle automatic release mechanism when locking the buckle. This key or anchor is retained in the buckle when the automatic features operate, and releases when the belt is opened manually. Parachutes incorporating automatic rip cord release mechanisms may be used with the automatic seat belts. Manual operation of the system can override the automatic features at any time. For example, it is possible to manually open the lap belt even though the initiator action has started. However, it must be remembered that if the belt is opened manually, the parachute rip cord must be pulled manually. The parachute automatic feature may likewise be overridden by manual operation, even though the automatic parachute rip cord has been actuated. The automatic belt has been thoroughly tested and is completely reliable. UNDER NO CIRCUMSTANCES SHOULD THE BELT BE MANUALLY OPENED PRIOR TO EJECTION REGARDLESS OF ALTITUDE. No matter how fast a pilot's reactions are, he cannot beat the automatic operation and, besides he may not remain conscious during an actual ejection.

Automatic Operation.

Automatic belt opening is accomplished as part of the seat ejection sequence and requires no additional effort on the part of the pilot. When the seat is ejected a firing pin in the initiator is pulled and one second later the initiator is fired, releasing the belt. The time delay feature insures that the belt will not release until the seat and pilot are entirely clear of the airplane. When an automatic parachute is used, an anchor attached to the parachute lanyard is installed over the lap belt swivel link. (See figure 1–40A for correct installation.) During automatic operation, the anchor remains fixed in the release, providing an anchor for the lanyard to the automatic rip cord release which is actuated as the pilot separates from his seat.

With the standardization and installation of automatic safety belts in aircraft, the manual release of the safety belt becomes unnecessary and undesirable for the following reasons:

1. The automatic opening safety belt automatically opens one second after ejection.

2. The escape operation is faster. The automatic opening safety belt eliminates the step of manually opening the safety belt prior to ejection.

3. The manual opening of the automatic safety belt creates a hazard to survival during uncontrollable flight. The opening of the safety belt during uncontrollable flight will mean that the pilot cannot stay in his seat prior to ejection if negative G is incurred.

4. The manual opening of the automatic safety belt creates a hazard to survival if the pilot decides to crash land the aircraft. The pilot will probably not be able to fasten the safety belt and shoulder harness as he will probably be using both hands to control and crash land the airplane.

5. The manual opening of the automatic safety belt will eliminate the automatic opening feature of the automatic opening parachute unless the pilot manually arms his automatic parachute by pulling the arming knobs.

6. Tail clearance is reduced by immediate separation of the pilot from the seat which is likely to occur when the belt is opened before ejection.

7. At high speeds, the peak deceleration due to air loads of the seat and pilot together, approaches the limits of human endurance. Since the deceleration of

AUTOMATIC LAP BELT

Locked Condition

TYPE MA-1

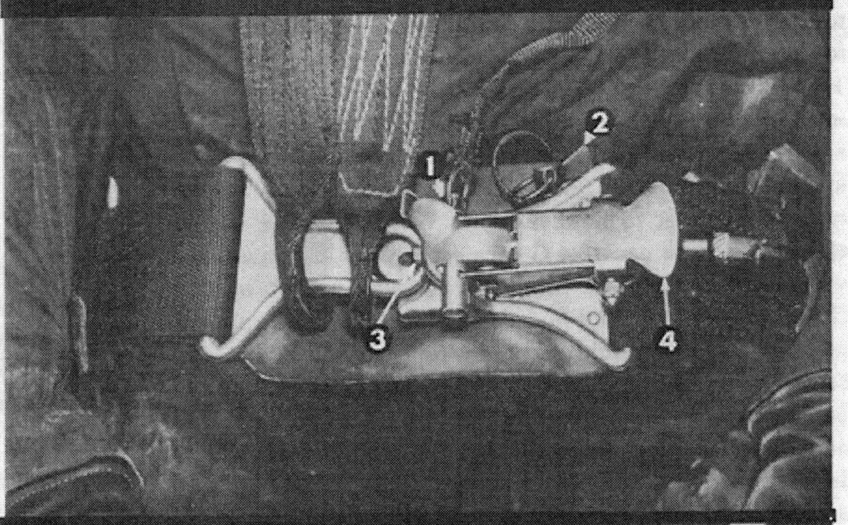

1. PARACHUTE LANYARD KEY
2. SPARE KEY
3. LATCH EXTENDED
4. LOCKHEED TYPE HANDLE EXTENSION

TYPE MA-3

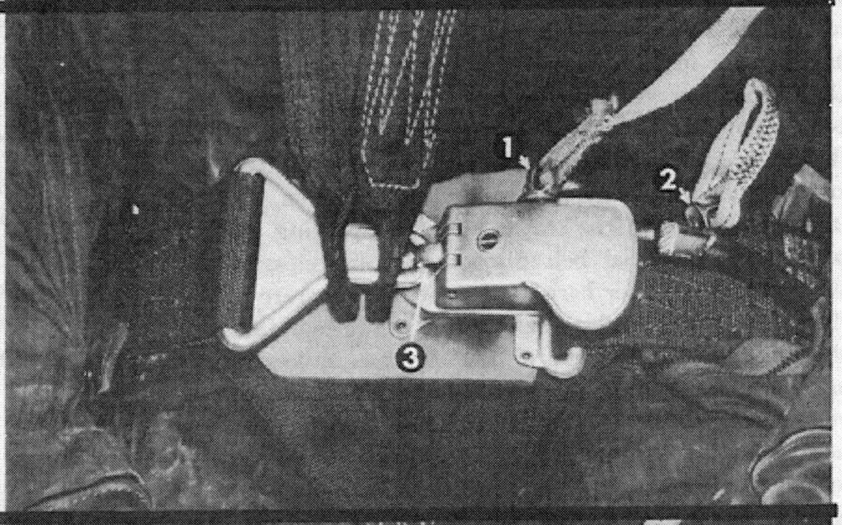

1. PARACHUTE LANYARD KEY
2. SPARE KEY
3. LATCH EXTENDED

TYPE MA-5 or MA-6

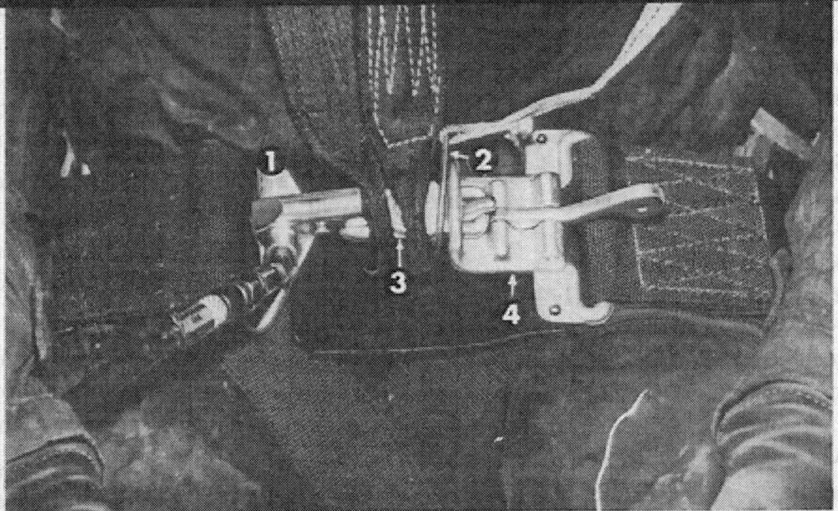

1. AUTOMATIC RELEASE
2. PARACHUTE LANYARD ANCHOR
3. SWIVEL LINK
4. MANUAL RELEASE

Figure 1-40 A (Sheet 1 of 2)

OPERATION

Manually Unlocked

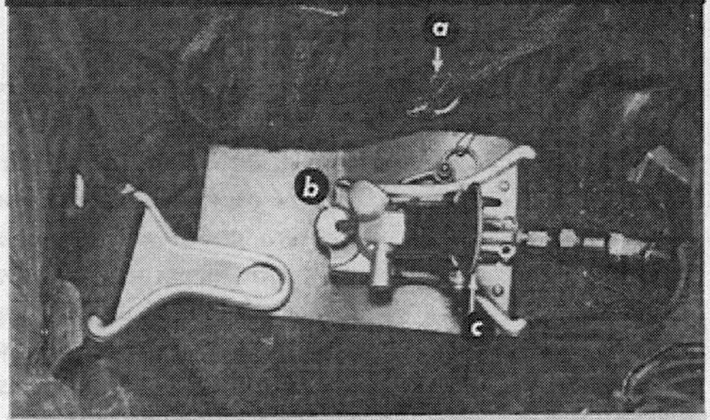

a. Key Ejected
b. Latch Retracted
c. Release Lever Raised

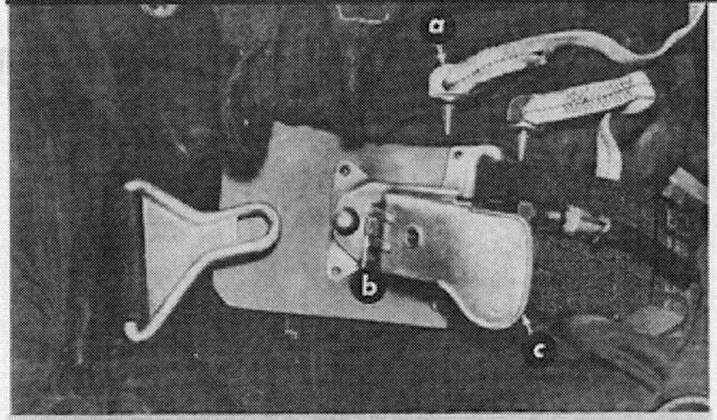

a. Key Ejected
b. Latch Retracted
c. Release Lever Raised

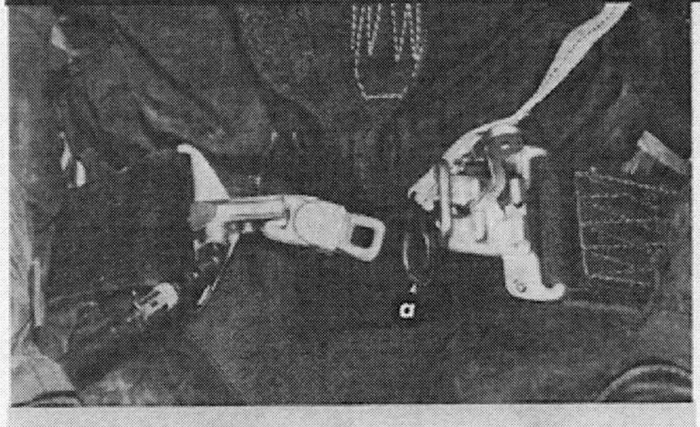

a. Parachute Lanyard Anchor Free

Automatically Opened

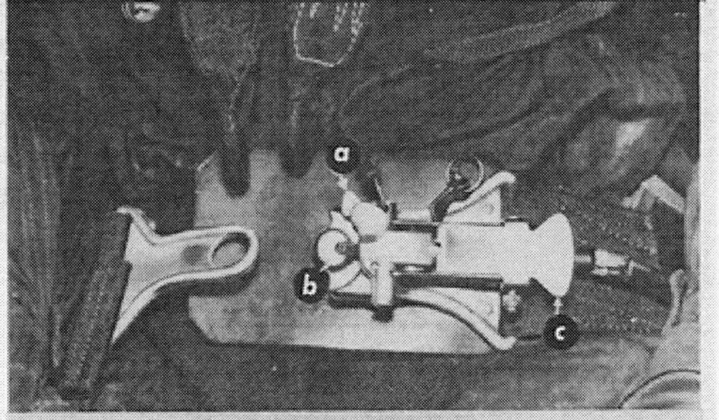

a. Key Retained
b. Latch Retracted
c. Release Lever (Locked)

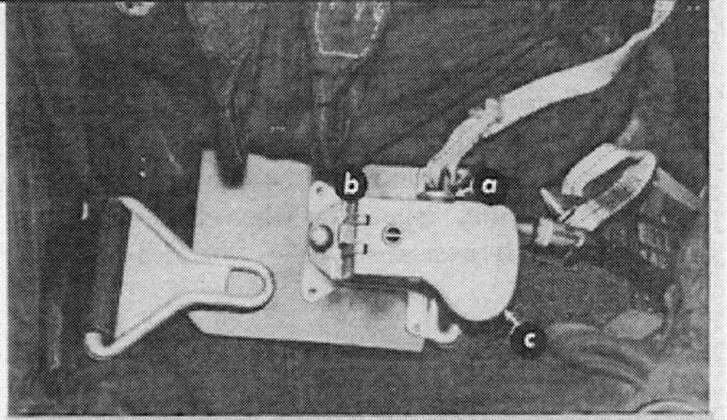

a. Key Retained
b. Latch Pivoted Up (Free To Pivot)
c. Release Lever Down (Locked)

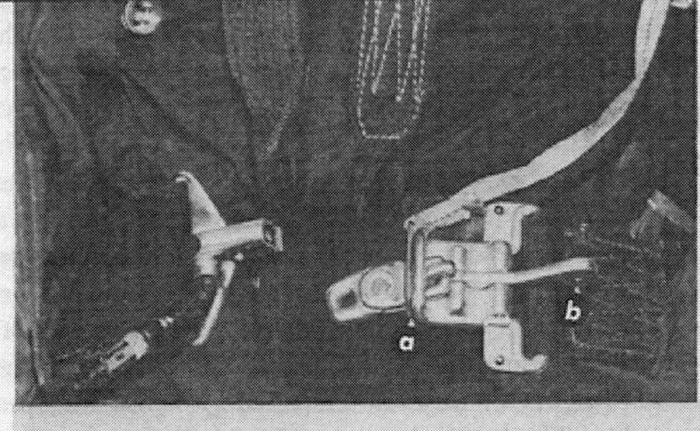

a. Parachute Lanyard Anchor Retained By Shoulder On Swivel Link
b. Manual Release Lever Locked.

Figure 1-40 A (Sheet 2 of 2)

a pilot alone is considerably greater than that of the pilot and seat together, immediate separation at extremely high speeds could result in severe injury to the pilot.

8. Immediate separation of the pilot and seat at extremely high speeds could result in inadvertent opening of the parachute due to the pack being blown open. In this event, fatal injuries will probably be incurred, because of the extremely high opening shock of parachute at this speed, or because of serious damage to the parachute itself upon opening at high speeds.

WARNING

The automatic safety belt initiator ground safety pin with warning streamer is provided for use during maintenance on the ejection system and should not be installed at any other time. Installation of the safety pin does not improve ground safety, since firing of the initiator only actuates the safety belt which would not injure personnel. Failure to remove this pin can result in fatal injury to the pilot, especially when low altitude ejection is necessary.

Manual Operation.

Manual operation of the MA-1 or MA-3 belt is accomplished by pulling up and to the right to release the lap belt tongue and eject the key link. To close the belt, the restraining harness loops are first placed on the lap belt tongue. The tongue is then placed over the latch post on the release mechanism of the belt. The lever is then rotated toward the closed position until it reaches a stop. The key link is then pushed into its receptacle and the lever rotated to the fully closed position. The key link on the MA-1 belt is designed with a wing-like projection which acts as a guide for locating the receptacle and for proper positioning of the key link. When inserting the key link, the wing-like projection should be lined up with a similar protrusion on the release. When fully locked the wing is in position immediately below the protrusion. The key link on the MA-3 belt does not have the wing like projection that is on the key link on the MA-1 belt. On the MA-5 or MA-6 belt, the key link is replaced by a parachute lanyard anchor which is placed over the swivel link before closing the belt.

WARNING

If the lap belt is opened manually, the parachute rip cord must be pulled manually.

Zero Delay Lanyard.

In order to provide an improved low altitude escape capability, a system incorporating a one second safety belt delay (M12 initiator) and a zero second parachute delay (one and zero system) is provided for ejection seat escape systems. This system makes use of a detachable lanyard, installed on the parachute harness (figure 1—40B), that connects the parachute arming ball to the parachute ripcord grip. At very low altitude and airspeed, this zero delay lanyard must be connected, thus providing parachute actuation

D-RING LANYARD

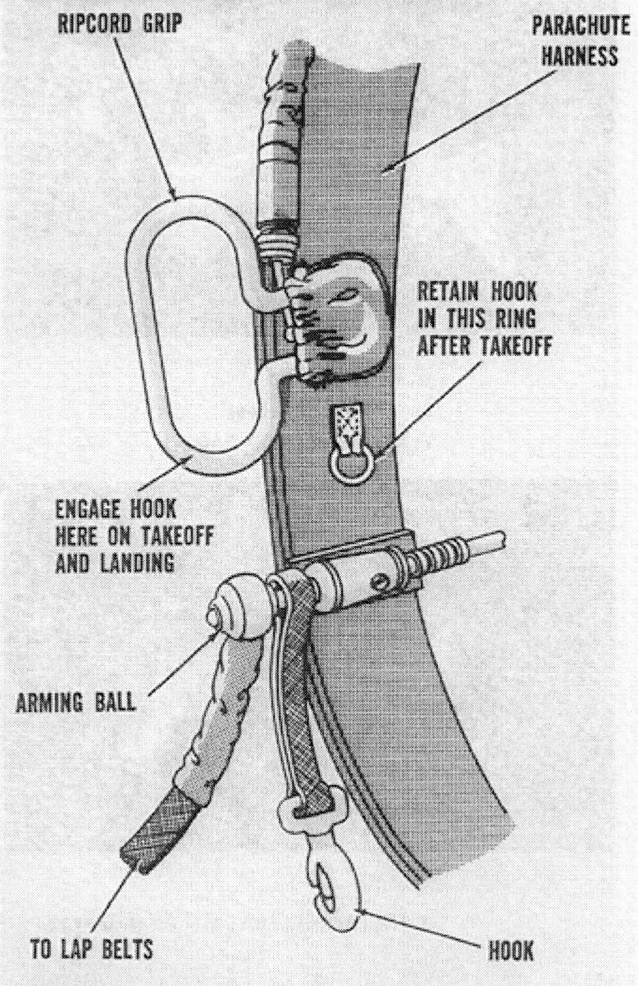

Figure 1—40B

immediately after separation from the ejection seat. At other altitudes and airspeeds, the lanyard *must be disconnected* from the ripcord grip, thus allowing the parachute timer to actuate the parachute below its critical opening speed, and below the parachute timer altitude setting. A ring, attached to the parachute harness, is provided for stowage of the lanyard hook when it is not connected to the parachute ripcord grip. *Hooking and unhooking of the lanyard must be done manually by the pilot.*

WARNING

In the event that the zero delay lanyard is left connected while flying at high altitude, and ejection becomes necessary, the aneroid feature of the parachute will be overridden, and the parachute will be deployed at an altitude where sufficient oxygen is not available to permit safe parachute descent.

- If the zero delay lanyard is left connected while flying at high speed and ejection becomes necessary, the time delay feature of the parachute will be overridden and the parachute will be deployed at high speed, resulting in an extremely high opening shock, in which case injuries will probably be incurred and the parachute seriously damaged.
- If the automatic safety belt is opened manually, the parachute ripcord must be pulled manually.

AUXILIARY EQUIPMENT.

Section IV of this manual contains information on the following auxiliary equipment: heating, pressurizing, ventilating systems; anti-icnig and de-icing systems; communications and associated electronic equipment; lighting equipment; oxygen system; air refueling system; automatic pilot; navigation equipment; armament equipment and miscellaneous equipment.

SERVICING DIAGRAM

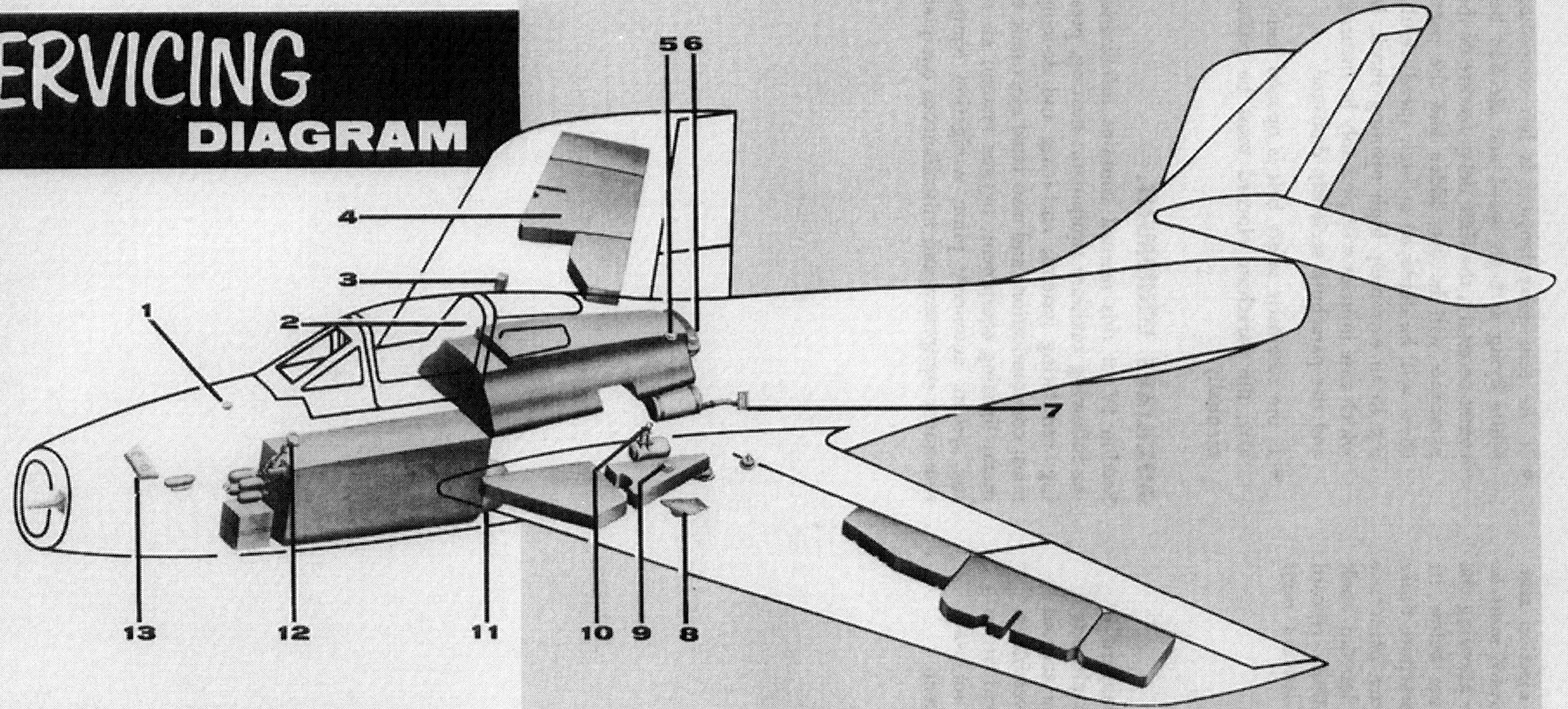

SPECIFICATIONS

Fuel Recommended	MIL-F-5624 GRADE JP-4
Oil	MIL-L-7808 (All Temperature Operation)
Hydraulic Fluid	MIL-O-5606
Alcohol	MIL-A-6091 (System Deactivated)
Oxygen	BB-O-925

1. Oxygen Filler Valve
2. Ground Refueling Receptacle (C and LATER)
3. External Power Receptacle
4. Wing Tank
5. Main Tank
6. Alcohol Tank
7. Oil Tank
8. Air Refueling Receptacle (Air and Ground A and B)
9. Wing Tank
10. Hydraulic Reservoir
11. Starter Air Bottle
12. Forward Tank
13. Pneumatic Ground Service

Figure 1—41

SECTION 2 NORMAL PROCEDURES

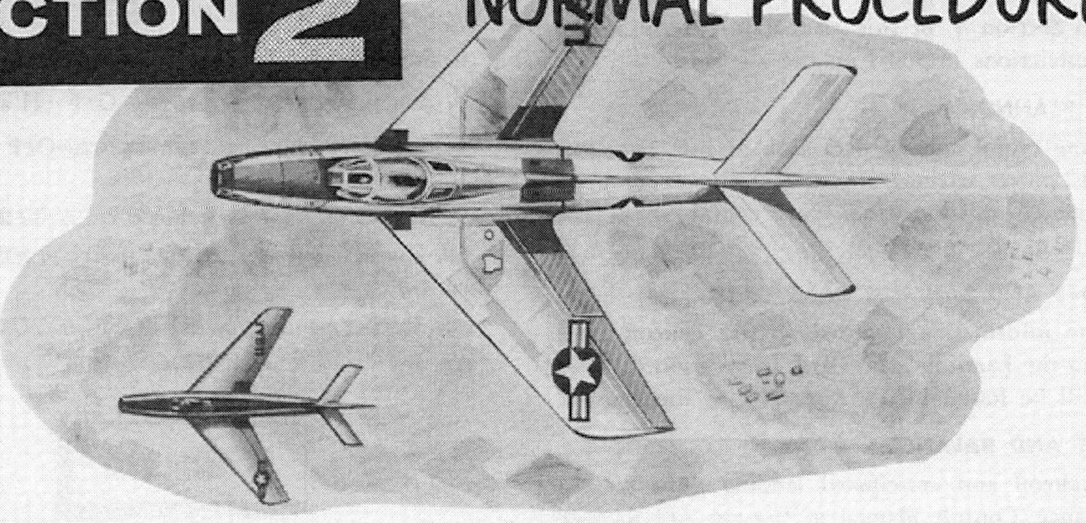

TABLE OF CONTENTS

PREPARATION FOR FLIGHT.

FLIGHT RESTRICTIONS.

Refer to Section V of this manual to determine the flight restrictions imposed on the airplane.

FLIGHT PLANNING.

Determine cruise control data such as fuel required, airspeeds, power settings, etc., as necessary to complete the proposed mission from data contained in the Appendix of the manual.

TAKE-OFF AND LANDING DATA CARD.

Complete information required on the Takeoff Data Card and the Landing Data Card. Illustration of these cards will be found in the Appendix of the manual.

WEIGHT AND BALANCE.

Check takeoff and anticipated landing gross weight and balance. Consult Manual of Weight and Balance Data, T.O. 1-1B-40 for loading procedure. Make sure weight and balance clearance (Form 365F) is satisfactory. Check to see that total weight of fuel, oil, armament, oxygen, and special equipment carried is suitable to the mission to be performed. Refer to Section V for weight limitations for various airplane configurations.

ENTRANCE TO AIRPLANE.

Place a ladder against the left side of the airplane at the cockpit. No external grips or steps are provided. Open the canopy, using the external canopy control to unlock the canopy, then raise it manually using handgrip on canopy skirt.

CAUTION

Do not use the gun sight as a hand hold. Be careful not to kick or drop anything on the control stick grip.

PREFLIGHT CHECK.

BEFORE EXTERIOR INSPECTION.

1. Form 781–Check for status and release of airplane.
2. Check applicable publications–Aboard and current.
3. Canopy jettison safety clip–Check (A THRU G). Clip should be in position over the canopy jettison control handle and the safety pin installed.
4. Canopy jettison and seat ejection control safety pin–Check (H and later).

   Pin should be installed in right handgrip of seat.
5. Safety pin from automatic lap belt initiator–Check removed.

   If pin is installed consult maintenance personnel regarding the status of the ejection system before occupying the ejection seat.
6. Surface control-lock–Release, (if installed).
7. Armament selector switch–OFF (H and later).
8. Auxiliary bomb selector switch–OFF (A THRU G).
9. Rocket selector switch–OFF (A THRU G).
10. Rocket jettison ready switch–NORMAL (A thru G).
11. Rocket arming switch–OFF or FUSE DELAY.
12. Bomb switches–OFF or SAFE.

WARNING

Since takeoff requires a large amount of rearward stick travel, there is a good possibility that the bomb release button could be inadvertently energized by contact with the pilot's clothing or the parachute harness. In order to preclude this possibility, the bomb and rocket selector switches must be placed in the OFF position, and the bomb and rocket arming switches must be in the OFF or SAFE position prior to takeoff.

13. Gun arming switch–OFF.
14. Emergency hydraulic pump switch–OFF.
15. Outboard and inboard pylon jettison switches–OFF and cover guard down.
16. Ato ready switch–OFF.
17. Fuel tank selector–OFF.
18. Landing gear selector handle–DOWN.
19. Emergency landing gear release switch–OFF and cover guard down.
20. Battery switch–ON.
21. Canopy squib A and B circuits–Test (canopy must be fully opened).

WARNING

If the squib test circuits are not functioning properly and the squibs in the rear hinge bolts fire in A and B airplanes (GK airplanes only), the bolts will take a horizontal path. Therefore, be sure that no personnel are in line with these bolts when testing the canopy squib circuits.

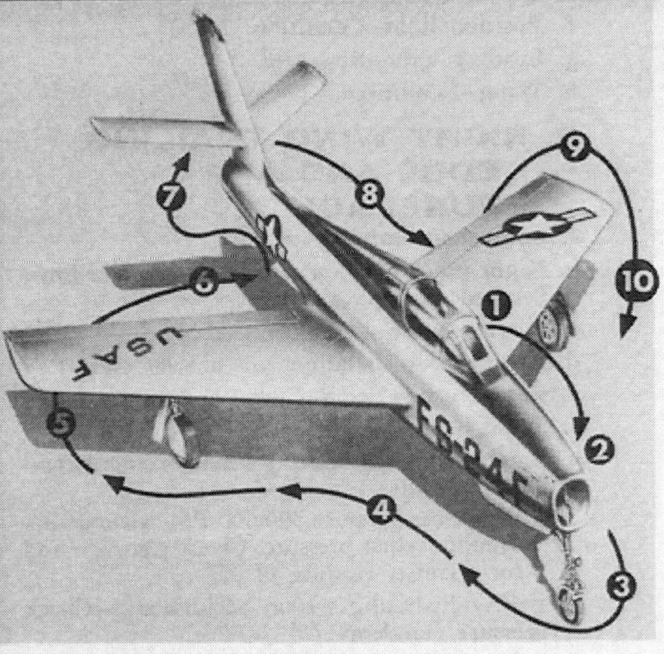

EXTERIOR INSPECTION

1. AFT CANOPY.

a. Canopy jettison hose quick disconnect—Check security.
b. Canopy seal hose and squibs—Check for condition and security.
c. Canopy and canopy frame—Check condition and security.
d. Main fuel tank cap and door—Check index mark and door secured.

WARNING

Be cautious when removing cap when airplane has been serviced through single point refueling receptacle. Fuel under pressure will be released in force.

e. Hydraulic fluid reservoir—Check quantity and cap and door secured.
f. Oil tank—Check quantity and cap installed.
g. Radio access doors—Secured.
h. Left and right wing fuel filler caps—Installed.
i. Forward tank fuel filler cap—Installed (A thru F).

2. FOREWARD FUSELAGE. (Left side and bottom)

a. Wing gun blast tube—Plug installed.
b. Fuselage—General conditions, evidence of fuel and hydraulic leaks.
c. Sucker door and interior—Check condition of door, duct and screen.
d. Static ports—Check condition.
e. Battery lift door—Up and locked.
f. All access doors and gun deck cover—Secured.

3. NOSE SECTION.

a. Intake duct plug and pitot cover—Removed.
b. Air intake ducts—Check.
   Check for foreign objects or cracks.
c. Nose gun blast tubes—Plugs installed.
d. Nose wheel shock strut—Condition and extension.
   Check for visible damage and proper inflation—Strut extension depending on weight of airplane—Use strut extension scale stowed in the nose wheel well.
e. Nose gear safety pin—Remove and stow.
f. Nose wheel tire—Condition.
   Check for proper inflation and evidence of slippage.
g. Nose wheel static ground wire—Grounding.
h. Access doors in wheel well—Closed and secure.
i. Starter air pressure gage—Check pressure (modified).
   Recommended pressure 2500-3000 PSI.
j. Pneumatic pressure for emergency nose gear extension—Check pressure.
   Check pressure 2500-3000 PSI.
k. Nose gear fairing doors—Condition and security.

4. FORWARD FUSELAGE AND RIGHT WING.

a. All access doors and gun deck cover—Secured.
b. Static ports—Check condition.

Figure 2—1 (Sheet 1 of 3)

c. Oxygen filler—Check valve in build up and secure cover.
   Check buildup position (arm horizontal).
d. Sucker door and interior—Check condition of door, duct and screen.
e. Fuselage—General condition evidence of fuel and hydraulic leaks.
f. Wing gun blast tube—Plug installed.
g. Starter air pressure gage—Check pressure (unmodified).
   Recommended pressure 2500-3000 PSI.
h. Gun case ejection chute doors—As required.
i. Right inboard pylon tank or bomb—Check.
   Check for proper installation on pylon. Air pressure in inboard pylon jettison system—1800 (±100) PSI, air pressure in inboard pylon tank jettison system —2000 (±100) PSI.
j. Ground refueling receptacle—Check cap installed door secured (C and later).
k. External power receptacle—Check.
l. Wheel well—Check condition.
   Check for hydraulic and fuel leaks, evidence of wheel rubbing, condition of up stop, doors and fairings.
m. Armament safety override switch—Cover secured.
n. Landing gear strut—Condition and extension.
   Check visible damage and proper inflation. Strut extension approximately 5⅜ inches between scissor hinges.
o. Landing gear tire—Condition.
   Check for blisters, grease or oil, proper inflation and evidence of slippage, short valve stem with internal valve and cap installed.
p. Wheel chock—In place.
q. Landing gear safety clip—Remove and stow.
r. Wing drain and vent holes—Obstructions and leaks.

5. RIGHT WING TIP.

a. Outboard pylon tank or bomb—Check.
   Check for proper installation on outboard pylon and pylon tank filler cover secured. Air pressure in outboard pylon jettison system —1600 (±100) PSI. Air pressure in outboard pylon tank jettison system —2500 (±100) PSI.
b. Front rocket posts—Retracted (A thru F).
c. Front rocket posts—Removed and plugs installed (G and later).
d. Rear rocket doors—Flush and secured (A thru F).
e. Rear rocket posts—Removed and plugs installed (G and later).
f. Position light—Condition.
g. Landing light—Retracted.
h. Wing—Condition.

6. RIGHT WING TRAILING EDGE AND AFT FUSELAGE.

a. Control surfaces—Condition.
b. Right inboard pylon tank filler cap and fins—Secured (if applicable).
c. Fuselage splice cover band—Secured.
d. Ato units—Installation and nozzles clear.
e. Engine access doors—Closed and secure.
f. Speed brake—Condition.
g. Emergency hydraulic system accumulator—Check pressure (non-tandem).
   Check air pressure 500-600 PSI without hydraulic system pressure. Check reservoir rod for quantity reading of 40.
h. Utility hydraulic system accumulator—Check pressure (tandem).
   Check air pressure 500-600 PSI without hydraulic system pressure.
i. Power hydraulic system accumulator—Check pressure (tandem).
   Check air pressure 500-600 PSI without hydraulic system pressure.

7. EMPENNAGE.

a. Fuel tank vent—Condition.
b. Three hydraulic vent ports on right side of the fuselage and one port on right side of vertical fin—Obstructions and hydraulic leaks.
c. Tailpipe dust plug—Removed.
d. Tailpipe—Condition.
e. Empennage—Condition.
f. Tail position lights—Condition.
g. Drag chute check—Complete, doors closed.
   Check drag chute:
   1. Right compartment door—Open.
   2. Both riser cable aft loops—Check loops secured to top of compartment by strap and riser cable end fitting installed in cable hook jaws.
   3. Pilot chute deployment cone—Check centered in rear of deployment bag and no parachute material exposed.
   4. Ground handling pin—Check removed and rip pin on left compartment door properly installed.
   5. Compartment and drag chute—Check free of moisture, fuel, oil and hydraulic fluid.
   6. Right compartment door—Close.
h. Three hydraulic vent ports or left side of fuselage—Obstructions and hydraulic leaks.

Figure 2—1 (Sheet 2 of 3)

FUSELAGE AND WING DRAIN AND VENT HOLES

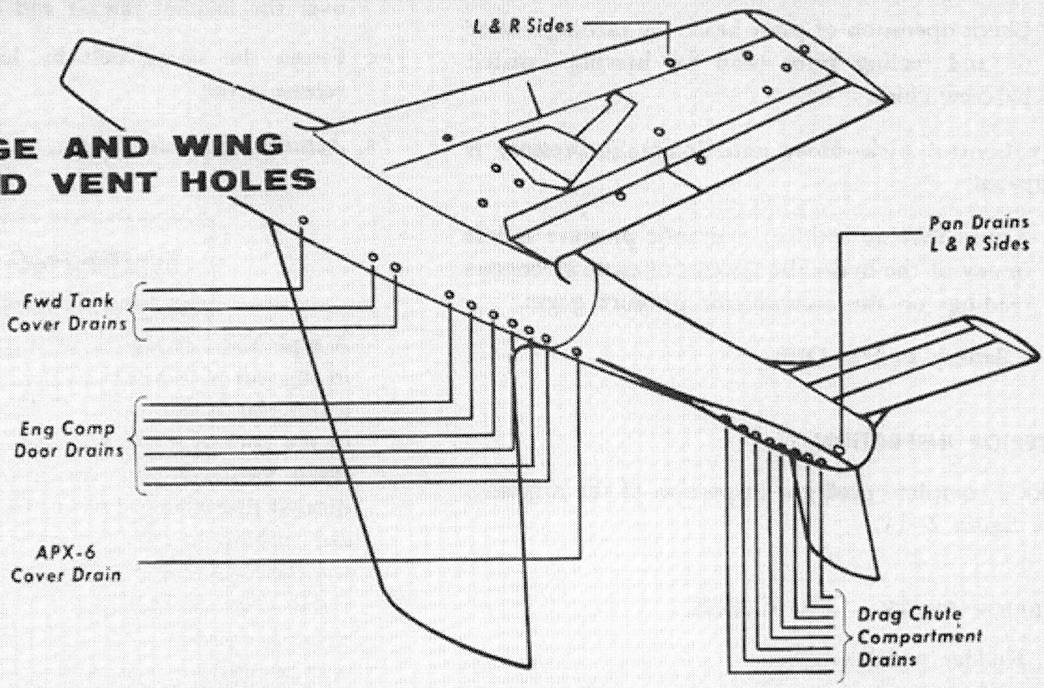

8. AFT FUSELAGE AND LEFT WING TRAILING EDGE.

a. Speed brake—Condition.

b. Engine access doors—Closed and secure.

c. Ato units—Installation and nozzles clear (if applicable).

d. Fuselage splice cover band—Secured.

e. Right inboard pylon tank filler cap and fins—Secured. (if applicable).

f. Control surfaces—Condition.

9. LEFT WING TIP.

a. Wing—Condition.

b. Landing light—Retracted.

c. Position light—Condition.

d. Pitot static boom—Remove cover, condition.

e. Rear rocket posts—Removed and plugs installed (G and later).

f. Rear rocket doors—Flush and secured (A thru F).

g. Front rocket posts—Removed and plugs installed (G and later).

h. Front rocket posts—Retracted (A thru F).

i. Outboard pylon tank or bomb—Check.

Check for proper installation on outboard pylon and pylon tank filler cover secured. Air pressure in outboard pylon jettison system, 1600 (±100) PSI; air pressure in outboard pylon tank jettison system, 2500 (±100) PSI.

10. LEFT WING.

a. Wing drain and vent holes—Obstructions and leaks.

b. Landing gear safety clip—Remove and stow.

c. Wheel chock—In place.

d. Landing gear tire—Condition.

Check for blisters, grease or oil, proper inflation and evidence of slippage, short valve stem with internal valve and cap installed.

e. Landing gear strut—Condition and extension.

Check visible damage and proper inflation. Strut extension approximately 5⅜ inches between scissor hinges.

f. Wheel well—Check condition.

Check for hydraulic and fuel leaks, evidence of wheel rubbing, condition of up stop, doors and fairings.

g. External tanks ground refuel switch—Cover secured C and later).

h. Air refueling valves test switch—Cover secured (C and later).

i. Left inboard pylon tank or bomb—Check.

Check for proper installation on pylon. Air pressure in inboard pylon jettison system —1800 (±100) PSI, air pressure in inboard pylon tank jettison system —2000 (±100) PSI.

j. Gun case ejection chute doors—As required.

k. Power hydraulic system reservoir indicator—Green (tandem).

l. Utility accumulator pressure—Check pressure (non-tandem).

Figure 2—1 (Sheet 3 of 3)

22. Pitot heater check—Climatic.

Check operation of pitot heater by turning heater on and feeling pitot head for heating—assisted by crew chief.

23. Control stick—Move until hydraulic pressure is dissipated.

Assure that no residual hydraulic pressure is left in any of the hydraulic systems to cause erroneous readings on the accumulator pressure gages.

24. Battery switch—OFF.

EXTERIOR INSPECTION.

Make a complete preflight inspection of the airplane. (See figure 2–1).

INTERIOR CHECK — ALL FLIGHTS.

1. Rudder pedals—Adjust.

2. Zero delay lanyard—Hook up.

3. Safety belt, shoulder harness and automatic chute lanyard—Examine and fasten.

Examine for security of adjustment and proper operation of shoulder harness inertia reel lock control. Leave shoulder harness inertia reel lock unlocked.

CAUTION

When adjusting and securing the safety belt on airplanes (A THRU F) make certain belt does not interfere with the canopy jettison linkage.

WARNING

Failure to properly attach the shoulder harness straps to the MA-5 or MA-6 automatic safety belt will prevent separation from the ejection seat after ejection. The following sequence of attachment of the shoulder harness loops, and the automatic parachute lanyard anchor to the safety belt shall be observed. (Refer to Section I.)

a. Place the right and left shoulder harness loops over the manual release end of the swivel link.

b. Place the automatic parachute lanyard anchor over the manual release end of the swivel link.

c. Fasten the safety belt by locking the manual release lever.

4. Armrest and seat—Adjust.

WARNING

A type MA-1 seat cushion is provided for use in the seat with a survival kit container when a life raft is not utilized. The forward edge of the packed kit should not be thicker than seven inches. Recent investigations have indicated that some pilots are leaving the standard cushion in the ejection seat when carrying the C-2A life raft. Sitting on both of these items does not provide added comfort and actually creates a definite injury hazard as well as possibly positioning the pilot where it may be difficult to reach all the controls. Chance of vertebrae injury is increased considerably by the pilot sitting on a thick, compressible mass, such as a combination of the life raft and seat cushion, because when utilizing the ejection seat in this circumstance, it will not exert a direct force on its occupant until the seat has moved two or three inches upward. After this amount of travel, the seat has gathered such momentum that excessive impact is produced when the seat initially lifts the pilot. The chance of injury is also increased considerably under these circumstances if a crash landing cannot be avoided. If a crash landing is made, the combined compressiveness of the two items will permit the pilot to sink down far enough to loosen the shoulder straps thus allowing the pilot to slump forward. In this position, the impact may severely injure his back. The use of additional material under the pilot may raise him to where his arms will not be held by the retainers on either armrest and they may be subjected to injury due to possible flailing in the windblast. Therefore, because of the above, the seat cushion should not be left in the seat and should be replaced by the type MA-1 cushion, if the C-2A life raft is being carried.

5. Seat vertical adjustment control—Locked (5/8 inch gap) (A THRU F).

WARNING

After each vertical (height) adjustment of the seat on airplanes (A THRU F), insure that the adjustment handle and latch lever are separated by approximately 5/8 inch (1/2 to 3/4 inch). Failure of the handle and the latch to separate as specified indicates one or more of the four seat locking pins has not locked. Failure of the top pins to lock in place can prevent successful seat ejection.

6. Ejection seat cannon plug—Secure (A THRU F).
7. Trim switch—Not sticking.

Check for security of mounting on the control stick. Operate the trim switch in all four ON positions and note that it automatically returns to the OFF position when released. If the switch sticks in any of the ON positions, enter this fact with a red cross on Form 781 and do not fly the airplane.

CAUTION

Do not twist the grip as such action may cause the grip to become less secure.

8. LH console circuit breakers—IN (A THRU G).
9. Camera switch—As desired.
10. Gun heater switch—OFF.
11. Hydraulic bypass switch—OFF (non-tandem).
12. Pitch control switch—NORMAL (non-tandem).
13. Alternate trim switch—NORM (non-tandem).
14. Emergency hydraulic pump switch—OFF.
15. Pitch trim switch—NORM (tandem).
16. Pneumatic compressor switch—OFF.
17. Spoiler shutoff switch—ON (Cover down).
18. ARC-34 circuit breaker test light—Test (if installed).
19. ARC-34 emergency battery switch—OFF.

Check that safety wire on switch is not broken.

20. ARC-34 radio—OFF (H and later).
21. Emergency fuel switch—OFF.
22. Engine screen switch—EXTEND or AUTO.
23. Pylon tanks air pressure switches—OFF.
24. Battery bus circuit breaker test light—TEST (A THRU G).
25. Landing flap lever—NEUT.
26. Speed brake switch—As desired.
27. Throttle—CLOSED.
28. External stores jettison press-to-test light—Test (H and later).
29. Hydraulic hand pump—Test (non-tandem).

Test operate to insure pressure, indicated on utility pressure gage.

30. Air refueling receiver switch—Closed (Down).
31. Amplifier override switch—NORMAL (Down).
32. Air refueling receiver light switch—OFF (Down).
33. Fuel shutoff valve switches—UP.
34. Console light switch—OFF (A THRU G).
35. Landing light switch—OFF.
36. Directional indicator switch—NORMAL.
37. Altimeter—Field elevation.

CAUTION

Be sure that the 10,000 foot pointer on the altimeter is set correctly.

38. Drag chute control handle—Full forward.
39. Drag chute circuit breaker test light—Test.
40. Clock, accelerometer—Set.
41. Sight mechanical caging lever—CAGE.
42. Autopilot engaging switch—OFF.
43. Cabin vent selector—PRESSURE or NORMAL.
44. Windshield defroster—OFF.
45. Canopy dry air switch—OFF (Modified, H and later).
46. Radio compass—OFF.
47. ARC-33 radio—OFF (A THRU G).
48. IFF—OFF.
49. Instrument panel light—OFF.
50. Position lights—OFF.
51. Taxi lights—OFF.
52. Console light switch—OFF (H and later).
53. Standby battery—Installed.
54. RH console circuit breakers—IN.
55. Oxygen, radio G-suit—Connect.

CAUTION

Make sure that G-suit hose is properly stored to prevent possible interference with control movement.

56. Oxygen system check—Complete.

Note

The liquid oxygen gage should read between four and four-and-one-half liters when the oxygen system is fully charged. Do not be alarmed that the gage does not read five liters, since it is impossible to charge the liquid oxygen converter to five liters. Use the oxygen duration chart to determine your oxygen duration for the indicated supply.

57. Oxygen diluter lever—As required.

For protection against carbon monoxide gas contamination when taxiing or run-up directly behind another aircraft or during run-up with tail into the wind, the following precautionary procedures will be used:

a. Before starting engine, don oxygen mask and place diluter lever at 100% OXYGEN position.

b. Use 100 per cent oxygen during ground operation and take-off when monoxide gas contamination is suspected.

c. Use NORMAL OXYGEN position when contamination is no longer suspected.

WARNING

The oxygen diluter lever must be returned to the NORMAL OXYGEN position as soon as possible, because the use of the 100 per cent oxygen throughout a long mission will so deplete the oxygen supply as to be hazardous to the pilot.

58. Generator switch—ON.

59. External power connected—Instrument power switch NOR or Battery switch ON and Instrument power switch ALT.

Note

Use external power source whenever practical to conserve the airplane's battery.

60. Alternate inverter—Check if external power is used.

61. Fuel quantity—Check.

62. Attitude indicator—Cage momentarily to insure proper erection.

Note

The attitude indicator should be energized for at least 30 seconds prior to caging.

63. Flight instruments—Check operation.

64. Vertical velocity indicator—Check zero position.

65. All warning lights—Test.

66. Fire warning and overheat test switch—Test.

67. Head airplane into the wind if practicable.

It is best to start the engine with the aircraft headed into the wind as a tailwind will increase exhaust gas temperature and will aggravate any engine fire occurring during starting.

68. Run-up area—Check.

Whenever possible, start and run-up of the engine should be made on a hard surface to minimize the possibility of dirt and foreign objects entering the compressor.

69. Make certain that all personnel are clear of the jet exhaust and intake areas.

INTERIOR CHECK — NIGHT FLIGHTS.

1. Console lights—Check.
2. Instrument panel lights—Check.
3. Position lights—Check.
4. Landing lights—Check.
5. Taxi light—Check.
6. Air refueling receptacle light—Check.
7. Flashlight—Check.

TANDEM

EMERGENCY HYDRAULIC SYSTEM CHECK.

1. External power connected or battery switch ON and instrument power switch ALT.

Note

Use external power source whenever practicable to conserve the airplane's battery.

2. Utility and power hydraulic pressure gages—Zero reading.

3. Emergency hydraulic pump switch—HYD EMER PUMP.

Check power hydraulic gage, should indicate 1550 (±100) PSI.

Note

During control checks, rates of motion of the control stick fore and aft must be equal and from side to side must be equal.

4. Control stick—Full forward.

5. Stabilator operation—Check.

Starting with the emergency hydraulic pressure at 1500 (±100) PSI, control stick should operate smoothly and steadily from full forward to full aft, then to full forward and then the stick action may become restricted during the next aft travel due to the emergency pump limitation. Return control stick to neutral.

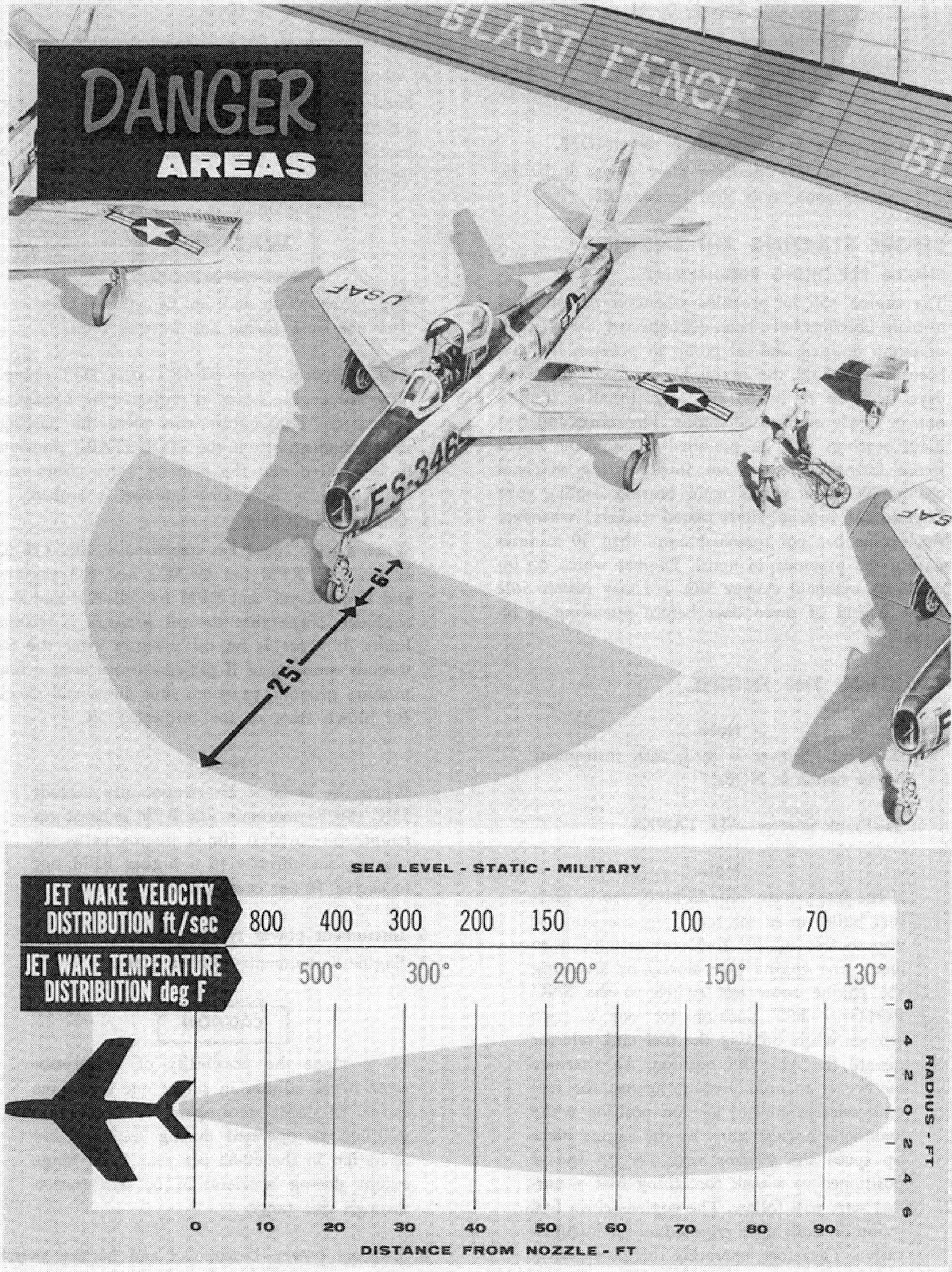
DANGER
AREAS
BLAST FENCE
6'
25'
SEA LEVEL - STATIC - MILITARY
JET WAKE VELOCITY DISTRIBUTION ft/sec
800 400 300 200 150 100 70
JET WAKE TEMPERATURE DISTRIBUTION deg F
500° 300° 200° 150° 130°
RADIUS - FT
6 4 2 0 2 4 6
0 10 20 30 40 50 60 70 80 90
DISTANCE FROM NOZZLE - FT

Figure 2—2

6. Aileron operation—Check.

   Check through complete range at normal rates. Operation should be smooth and steady with no indication of system starvation or reduction in rate of operation.

7. Emergency hydraulic pump switch—OFF.

   Return to OFF position after power hydraulic pressure gage reads 1550 (±100) PSI.

BEFORE STARTING THE ENGINE.

ENGINE PRE-OILING REQUIREMENTS.

The engine will be pre-oiled whenever the oil lines to main bearings have been disconnected, the oil tank of pump drained, the oil pump in pressure line has been disconnected, the engine has been idle for seven days, or prior to initial run after installation of a new or newly overhauled engine. The center and rear main bearings will be pre-oiled through the micro pump fittings (engines not incorporating overhaul change NO. 144, center main bearing cooling tube external and internal silver plated washers) whenever the engine has not operated more than 30 minutes during the previous 24 hours. Engines which do incorporate overhaul change NO. 144 may remain idle for a period of seven days before pre-oiling is required.

STARTING THE ENGINE.

Note

If external power is used, turn instrument power switch to NOR.

1. Fuel tank selector—ALL TANKS.

Note

If the fuel selector should bind, due to pressure build up in the fuel lines, the simplest way to free up the fuel tank selector is to motor the engine over slowly by actuating the engine rotor test switch to the ENG ROTOR TEST position for one or two seconds while holding the fuel tank selector toward the ALL ON position. An alternate method is to hold pressure against the fuel tank selector toward any on position while making a normal start. As the engine picks up speed the selector will free up and if positioned to a tank containing fuel, a normal start will follow. The engine-driven fuel pump depends upon engine fuel for its lubrication. Therefore, operating this pump without fuel may cause pump damage.

2. Throttle—Move to IDLE.

   Move throttle to IDLE to open fuel shutoff valve.

3. Starter switch—START.

   Hold starter switch in START position for approximately three seconds to energize the combustion starter, the engine primer and the ignition timer.

WARNING

The starter switch shall not be actuated more than one time during any starting cycle.

4. Starter switch—STOP START after EGT rising. After the engine starts, as indicated by a definite exhaust gas temperature rise, place the starting switch momentarily in the STOP START position to be assured that the primary valve closes and the circuit to the engine ignition is broken.

5. Oil pressure—Check.

   When engine speed has stabilized at idle (36 to 42 per cent RPM for J65-W-3 and B-3 engines and 42 to 48 per cent RPM for J65-W-7 and B-7 engines), check that the oil pressure is within limits. If there is no oil pressure after the 60 seconds running, or if pressure drops after a few minutes ground operation, shut down and check for blown lines or for congealed oil.

Note

When the ambient air temperature exceeds 32°C (90°F) maintain idle RPM exhaust gas temperature within limits by manually advancing the throttle to a higher RPM not to exceed 50 per cent RPM.

6. Instrument power switch—NOR.
7. Engine instruments—Within limits.

CAUTION

To preclude the possibility of compressor rotor blade failures in stages one and three caused by steady state operation, the engine will not be operated during static ground operation in the 60-82 per cent RPM range except during acceleration or deceleration through this range.

8. External power—Disconnect and battery switch ON (if applicable).

PROCEDURE IN THE EVENT OF FALSE ENGINE START.

An unsuccessful start will result if the exhaust gas temperature reaches maximum allowable for starts or if the exhaust gas temperature does not commence to rise before the engine has accelerated and then decelerated to 17 per cent RPM.

CAUTION

Before attempting a restart, close throttle, allow the engine to stop and be sure all fuel has completely drained from the combustion chamber by observing the drain valve. Restart only if "hot start" conditions have not been exceeded. If the engine fails to start after one attempt, the starter air storage bottle must be recharged. If the engine fails to start after two starting attempts, a 45 minute cooling period is required before making a third attempt. Unlimited single starts at 30 minute intervals may be made thereafter.

WARNING

If a start is attempted and combustion does not occur in the starter, fuel can drain into the pants duct area and possibly into the generator ducts causing a fire hazard. When a false start of the combustion starter occurs, actuate the rotor test switch to scavenge the starter with air. Make a thorough inspection of the compressor inlet for latent fuel, which should be removed before the next start. Wait for completion of engine fuel system drainage before attempting another start. If a start is attempted and the attempt is accompanied by a loud noise for approximately ½ second indicating that combustion has taken place in the starter, and no engine RPM is indicated, the starter has become disengaged from the engine. *Additional start attempts will not remedy this situation. Another start should not be attempted* until the starter has been removed from the engine and the condition responsible for the unlatching has been corrected.

ENGINE GROUND OPERATION.

WARMUP.

No warmup period is required. Takeoff may be made immediately if engine instruments show normal readings. However, some engines will not accelerate properly when they are cold and will therefore require a short warmup period in order that the acceleration limits (See Section V) are met.

CAUTION

During static ground operation, the engine will not be operated in the 60-82 per cent RPM range except during acceleration or deceleration through this range.

Note

In the event condensation in the form of mist or snow is blown into the cockpit from the air conditioning-outlets and is found to be objectionable to the pilot, the cabin vent switch can be positioned to RAM, or a higher temperature selected on the cabin temperature selector.

RPM AND FUEL FLOW FLUCTUATION.

Refer to Section V for RPM and fuel flow fluctuation limits.

TANDEM

GROUND TEST — BEFORE TAXIING.

1. Radio—ON as required.
2. IFF—STANDBY.
3. Emergency hydraulic pump switch—OFF.
4. Engine at IDLE RPM—Check power and utility hydraulic pressures within limits.
5. Brakes—Check operation.
   Check for firm and positive brake pedal feel.
6. Speed brakes—Check operation then IN.
7. Landing flaps—Full down then return to takeoff.
   Lower to full down (100 per cent) then return to takeoff position (50 per cent).
8. Spoiler switch OFF—Control stick full forward. Check both pressure gages, cycle stick, check power and utility hydraulic pressure.
   With spoiler switch OFF, position control stick full forward. With 1500 (±100) PSI indicated on both pressure gages, cycle stick to full limits of aileron and stabilator deflection (rate of one cycle is approximately four to five seconds). Power and utility hydraulic pressure should not drop below 400 PSI. Return control stick to neutral and spoiler switch cover guard down.
9. Rudder—Check for free and correct movement.
10. Trim switch—Actuate and check trim operation.
    Actuate and check trim operation of the stabilator and ailerons. The aileron trim indicator should illuminate at neutral trim. Position the pitch trim

switch in the OFF position while holding the trim switch. Trim actuators should be inoperative. Check NOSE UP and NOSE DN positions, then return pitch trim switch to NORM.

Note

Because of the differential characteristics of some limit microswitches incorporated in the stabilator trim system, the stabilator trim actuators have a dead spot near the full up or down position. Movement of the trim switch to the opposite position momentarily and then returning it to the original position may cause the stabilator trim to become inoperative unless the actuator has moved sufficiently towards the neutral position to reset the limit switch. Therefore, whenever the stabilator trim is in the full up or down position, allow sufficient time by holding the stabilator trim switch in the opposite position for several seconds, before reselecting the original position.

11. Rudder trim switch—Actuate and check rudder trim.

Rudder trim indicator should illuminate at neutral trim.

12. Compressor switch—ON and check pneumatic pressure gage. Return switch to OFF.

Pressure checked by crew chief.

13. Loadmeter—Within range.
14. Voltmeter—Within range.
15. Radios—Check operation.
    a. Radio compass—Check all positions then turn to first desired position.
    b. Command radio—Check necessary frequencies and guard frequencies.
16. No fuel overflowing from vent—Check.

Observed by crew chief.

17. Auxiliary fuel tanks check—Accomplish, then return fuel tank selector to ALL TANKS.
    a. Place fuel tank selector in FWD AUX for 15 seconds. Check fuel quantity forward pump pressure warning light and stabilization of fuel flow.
    b. Place fuel tank selector in WING AUX for 15 seconds. Check fuel quantity, wing pump pressure warning light and stabilization of fuel flow.
    c. Return fuel tank selector to ALL TANKS.
18. Canopy dry air switch—Climatic.
19. Windshield defrost switch—ON.
20. Defroster control—Climatic.
21. Oxygen mask—Properly adjusted and oxygen system operating.

Arrange oxygen hose so as not to interfere with full stick travel.

22. Autopilot check—Complete. (Refer to Section IV.)
23. Autopilot engaging switch—OFF.

Lift autopilot engaging switch out of detent.

24. Pitch trim switch—NORM.
25. Canopy jettison and seat ejection safety pins—Remove, show to crew chief and stow.

There are four pins on airplanes (A THRU F) and one pin on G and later airplanes.

26. Exterior check for hydraulic and fuel leaks—Complete (crew chief).
27. Chocks—Removed.

NON-TANDEM

GROUND TEST — BEFORE TAXIING.

1. Radio—ON as required.
2. IFF—STANDBY.
3. Emergency hydraulic pump switch—NORMAL.
4. Engine at IDLE RPM—Check utility and boost hydraulic system pressure within limits.
5. Hydraulic utility gage—Check while brake pedals are depressed.

Hydraulic pressure will fluctuate when brakes are depressed.

6. Speed brakes—Check operation then IN.

Note

While actuating speed brakes on airplanes incorporating a boost pump shutoff, (modified A THRU E and later airplanes)—Check for loss of pressure in hydraulic boost system, indicated by a temporary drop on boost gage. This indicates boost pump failure.

7. Landing flaps—Full down then return to takeoff.

Lower to full down (100 per cent) then return to takeoff position (50 per cent).

8. Ailerons, rudder and stabilator—Check operation.

Test operate through complete range and check for free and correct movement.

9. Trim switch—Actuate and check trim operation.

Check trim operation of the stabilator and the aileron. Aileron trim indicator light should illuminate at neutral trim. Position the alternate trim switch in the OFF position while actuating the stick trim switch (D and later). Trim actuators should be inoperative. Return alternate trim switch to NORM position.

Note

Because of the differential characteristics of some limit microswitches incorporated in the stabilator trim system the stabilator trim actuators have a dead spot near the full up or down position. Whenver the stabilator trim actuator is in the full up or full down position, movement of the trim switch to the opposite position momentarily and then returning it to the original position may cause the stabilator trim to become inoperative unless the actuator has moved sufficiently towards the neutral position to reset the limit switch. Therefore, whenever the stabilator trim is in the full up or down position, allow sufficient time by holding the stabilator trim switch in the opposite position for several seconds, before reselecting the original position.

10. Rudder trim switch—Actuate and check rudder trim.

Rudder trim indicator should illuminate at neutral trim.

11. Engine—IDLE, emergency hydraulic pump switch NORMAL.

At engine IDLE RPM and with the emergency hydraulic pump switch in the NORMAL position, actuate the hydraulic bypass switch to the momentary BYPASS position. On A THRU E airplanes not modified to the boost pump shutoff valve configuration, both hydraulic pressure gages should show pressure has been dumped, while on A THRU E airplanes modified to the boost pump shutoff configuration and later airplanes, only the boost pressure gages will drop. The emergency hydraulic system should function and the EMER HYD CONTROLS ON indicator should illuminate. Check the ailerons and stabilator for proper movement. Operation should be smooth and steady, within limits.

Note

This operation is a simulated emergency condition and operation is limited by the flow capacity (two gallons per minute) of the electric hydraulic pump.

12. Electrical continuity check—Check.

(GROUND OPERATION ONLY). (For emergency flight operation see Section III.) With stabilator and aileron trim in neutral position and engine at IDLE: Check pitch control switch in NORMAL, place the emergency hydraulic pump switch in the OFF position, bypass switch momentary in BYPASS. Cycle stick to dissipate the emergency hydraulic pressure. While applying 15 to 50 pounds load on the control stick in the aft direction actuate the trim switch to move the control stick aft approximately *one* inch. This check is a continuity check assuring that the electrical system will take over during the normal sequence of failures i.e., normal hydraulic system then emergency hydraulic system.

CAUTION

Do not repeat unnecessarily, or exceed the one inch aft (only) movement, as the actuator may be subjected to jamming loads, due to these artificially induced conditions.

13. Emergency hydraulic pump switch EMERG, hydraulic bypass switch off—Check emergency ON indicator and hydraulic pressure gages.

Light should illuminate and the hydraulic pressure gages should not fluctuate when the control stick is moved. Loadmeter should indicate noticeable increased load.

14. Emergency hydraulic pump switch NORMAL, alternate trim switch NORM, pitch control switch STICK OVERRIDE or ELEC, engine at IDLE—Check stabilator for movement with trim switch.

On unmodified (A THRU E) airplanes, both hydraulic pressure gages should show pressure has been dumped, while on modified (A THRU E) and later airplanes, only the boost hydraulic pressure gage will drop. The EMERGENCY HYD CONTROLS ON indicator should illuminate. The stabilator should not move unless the trim switch is actuated. Otherwise, the airplane should not be flown. However, the control stick can be flexed fore and aft approximately ½ inch which is equivalent to the movement of the linkage to the actuator. Actuate the trim switch to move the stick forward and aft and simultaneously apply sufficient load on the stick to position the actuator control valve in the direction of stabilator motion. Check the ailerons for emergency hydraulic control.

15. Pitch control switch ELEC (D and later), emergency hydraulic pump switch NORMAL, alternate trim switch OFF—Actuate alternate trim switch and apply load on control stick.

Actuate the alternate trim switch (stick switch power auxiliary electrical switch) momentarily to NOSE UP and NOSE DN, and simultaneously apply sufficient load (15 to 50 pounds) on the control stick to position the actuator control valve in the direction of stabilator motion.

CAUTION

Release the alternate trim switch before reaching full-up or full-down stabilator travel to avoid the stabilator actuator from bottoming, which may result in damage to the actuator or to the clutch. This circuit is not protected with limit switches.

16. Control stick neutral, alternate trim switch NORM, emergency hydraulic pump switch NORMAL, hydraulic bypass switch off, and pitch control switch NORMAL—Check flight controls.

Check flight controls for free and correct movement to assure transfer of control to the normal hydraulic system.

17. Compressor switch—ON and check pneumatic pressure gage. Return compressor switch to OFF.

Pressure checked by crew chief.

18. Loadmeter—Within range.
19. Voltmeter—Within range.
20. Radio—Check operation.
    a. Radio compass—Check all positions then turn to first desired position.
    b. Command radio—Check necessary frequencies and guard frequencies.
21. No fuel overflowing from vent—Check.

Observed by crew chief.

22. Auxiliary fuel tank check—Accomplish then return fuel tank selector to ALL TANKS.
23. Canopy dry air switch—Climatic.
24. Windshield defrost switch—ON.
25. Defrost control—Climatic.
26. Oxygen mask—Properly adjusted and oxygen system operating.
27. Autopilot check—Complete.
    (Refer to Section IV.)
28. Autopilot engaging switch—OFF.

Lift autopilot engaging switch out of detent.

29. Pitch control switch—NORMAL.
30. Canopy jettison and seat ejection safety pins—Remove, show to crew chief and stow.

There are four pins on airplanes (A THRU F) and one pin on G and later airplanes.

31. Exterior check for hydraulic and fuel leaks—Complete (crew chief).
32. Chocks—Removed.

TAXIING INSTRUCTIONS.

Remove chocks, release brakes and increase power until airplane starts to move. Once the airplane is moving, taxi at the lowest practicable RPM to conserve fuel and avoid damage from tailpipe blast. Brakes are required for steering as the rudder is ineffective at low speeds and nose wheel steering is not provided. The brakes are quite sensitive and care must be used at all times, if jerky operation is to be avoided. When the airplane is fully loaded with external stores, a much higher RPM is necessary to start the airplane rolling. When heavily loaded, the turning radius must be slightly increased to prevent excessive side loads on the struts and tires. Limit taxiing to a minimum as the airplane range is decreased by the high rate of fuel consumed during taxiing. Fuel consumed during taxiing is approximately 15 to 25 pounds per minute. The canopy may be left open at all speeds when taxiing on smooth ground.

CAUTION

- During taxi operation, the engine will not be stabilized in the 60-82 per cent RPM range.
- Do not rest arms on canopy rails when canopy is open, as injury will result if the canopy unexpectedly closes during taxiing.

1. Turn and slip indicator—Check operation.
2. Directional indicator—Check operation.
3. Attitude indicator—Cage and check position.

Note

The attitude indicator should be energized for at least 30 seconds before caging.

4. Windshield defrosting—As required.

BEFORE TAKEOFF.

CAUTION

On airplanes with fuel booster pumps modified in accordance with T.O. 6J10-3-10-502 dated 15 June 1954, the center of gravity will move approximately one per cent forward of the maximum allowable forward CG position when special stores are carried on the inboard pylons. Furthermore, when operating under this condition, a large unsticking force will be required to raise the nose wheel during take-off.

1. Canopy—Close, canopy open indicator light out.
2. Zero delay lanyard—Hooked up.
3. Engine screens—As desired.

WARNING

Under certain conditions engine screens may have to be retracted during take-off. For further information see ENGINE ICING, Section IX.

4. Attitude indicator—Check setting along with indices.
5. Landing flaps—Check in take-off position.
6. Speed brake switch—IN.
7. Aileron and rudder trim—Take-off position.
8. Pylon tank air pressure switches—OFF.
9. Ato ready switch—ATO READY (for ATO take-off).

   ATO ready warning light should illuminate.
10. Flight controls—Check for free and correct movement.

   Assure that the flight controls have returned to operation on the normal hydraulic system.
11. Engine acceleration check—Completed.
   a. An acceleration check shall consist of a throttle burst from a stabilized engine speed of 47 per cent to 100 per cent. Rapid manipulation of the throttle will not be accomplished. An engine and fuel control should be considered satisfactory and acceptable as long as stall free acceleration can be made within the time limits outlined in Section V.

   Note

   Rapid manipulation of the throttle may result in engine chugs and/or stalls.

   b. If the engine acceleration limits are not met, repeat the acceleration check. If the acceleration limits are not met, after the second check, reject the engine and write up on Form 781.
   c. If the EGT during an acceleration exceeds the maximum limit, retard the throttle slightly until temperature drops below the limiting value and then advance it slowly to avoid exceeding the limit. Avoid further rapid accelerations until the cause of the overtemperature has been determined. Record all operation above the acceleration limit.
12. Engine overspeed check—Completed.

   Refer to Section V for engine overspeed limits.

TAKE-OFF.

1. Release brakes.
2. Throttle—Military power.
3. Go-No-Go speed—Check.

   See figure 2–3 for typical take-off procedure.

ASYMMETRICAL TAKE-OFF.

Asymmetric take-offs can be accomplished without difficulty provided aileron boost is available. In the event of complete hydraulic failure it is imperative that the stores be jettisoned *immediately* as the aileron stick forces will be excessive. The following procedure is recommended for an asymmetric take-off:

a. Take-off speed should be approximately 10 knots higher than the recommended take-off speed for corresponding gross weight as noted in the Appendix.

b. Use brakes in conjunction with the rudder to maintain directional control until the rudder becomes effective at approximately 60 knots.

c. As much as ½ aileron travel may be required to maintain a level attitude.

d. Take-off distances will be increased approximately 15 per cent from that shown in the Appendix due to the use of brakes and the higher take-off speed.

ASSISTED TAKE-OFF.

The effect of ATO on airplane trim is slight as the units are installed near the fuselage center line. No special technique is required. Take-off performance will depend on the speed at which the ATO units are fired during the take-off run. Refer to the Appendix for the ATO cut-in speed for a two or a four ATO unit take-off.

FAILURE OF ASSIST TAKE-OFF UNITS

In the event of failure of one or more of the ato units during take-off, rapid change of trim will not be necessary as the ato units are close to the center line of the airplane. Take-off distance will be increased as noted in the Appendix.

AFTER TAKE-OFF — CLIMB.

1. Landing gear selector handle—UP.

   When definitely airborne move landing gear selector to the UP position, feel for definite engagement in the detent. Leave handle in up position after gear is retracted. Check landing gear position indicators.

   Note

   Do not exceed 190 knots until landing gear is up and locked. The landing gear will retract in approximately 10 seconds at normal temperatures.

TYPICAL TAKE-OFF

1. Release Brakes

2. Throttle-Max Power

3. Go-No-Go Speed-Check

Refer to Appendix for minimum take-off distance required by various combinations of gross weight, pressure altitude and air temperature, also for best climbing speed, rate of and time to climb, and fuel consumption.

If the aircraft is operated under possible conditions of carbon monoxide contamination use 100% OXYGEN during take-off.

The following procedure is recommended to obtain performance illustrated in the Appendix:

a. Visually align aircraft with runway.

b. Advance throttle to take-off rpm. Emergency fuel switch OFF. Check instruments, release brakes and begin take-off run. (See figure 7—1, temperature stabilized curve for engine acceleration.)

c. Maintain directional control by minimum use of brakes until rudder becomes effective at approximately 60 knots.

d. Check airspeed at predetermined points on take-off run to check aircraft acceleration.

e. If using Ato, push ato ignition switch as airspeed reaches the desired values as noted in the Appendix.

f. Leave control stick in neutral until take-off speed is reached. This reduces drag to a minimum.

g. When take-off speed is reached, use necessary stick travel to pull aircraft off the ground.

NOTE

After the landing gear is retracted, a slight aft movement of control stick may be necessary. This slight trim change is caused by the stabilator mechanical advantage shifter moving into the 1.8:1 ratio.

h. Leave throttle in take-off position until the aircraft reaches a safe altitude observing the time limit specified.

CAUTION

If the main fuel system fails during take-off, place the emergency fuel switch in the ON position. This transfers the engine to the emergency fuel system. When this transfer occurs at take-off rpm, a sudden change in fuel flow and rpm will occur. The pilot should recognize what has occurred and advance or retard the throttle smoothly but slowly to regain take-off rpm.

i. After carbon monixide contamination is no longer suspected, place the oxygen diluter lever in the NORMAL OXYGEN position.

WARNING

The oxygen diluter lever must be returned to the NORMAL OXYGEN position as soon as possible because the use of 100 per cent oxygen throughout a long mission will so deplete the oxygen supply as to be hazardous to the pilot.

Figure 2—3

2. Landing flap lever—UP.

Landing flap lever UP after landing gear is fully retracted, and all obstacles are cleared.

3. After flaps are retracted—Return landing flap lever to NEUT.

4. After take-off—Drop ato units (if applicable).

Drop ato units when at a safe altitude, by depressing ato jettison switch, if applicable.

5. Ato ready switch—OFF.

6. Climb to a safe altitude and adjust speed for best climb.

CAUTION

The engine exhaust gas temperature MUST be monitored above 35,000 feet altitude. The throttle may have to be retarded to keep the exhaust gas temperature within specified limits.

7. Engine screen switch—As desired.

8. Pneumatic compressor switch—ON (unmodified airplanes).

CAUTION

The compressor should not be operated for more than 30 minutes continuously to prevent overheating and possible compressor explosion. Therefore, if the storage bottles take an excessive amount of time to refill, the compressor should be shut off.

9. Oxygen diluter lever—Check NORMAL.

10. Armament and gun heater switches—As desired.

11. Pylon tank air pressure switches—OUTB'D or INB'D.

Check external tank feeding (Refer to Section IV.).

12. Engine and electrical indicators—Normal readings.

13. Zero delay lanyard—Unhook from rip cord grip and stow.

Unhook from rip cord grip and stow after passing through the minimum, safe, ejection altitude. See figure 3–4B to determine minimum, safe, ejection altitude for system being used.

14. Cabin vent selector switch—PRESSURE (prior to F-84F-35RE and F-84F-51GK).

15. Cabin vent selector switch—NORMAL (F-84F-35RE and F-84F-51GK and later).

16. Autopilot engaging switch—ON.

Trim airplane for wing-level flight, desired pitch trim and directional heading then engage autopilot, if installed.

CAUTION

Do not engage the autopilot when close to the ground or when near other airplanes. Hold the control stick firmly to prevent any abrupt resultant maneuver that may occur should automatic pilot not function properly. Do not engage at speeds in excess of ~~425~~ 320 knots IAS below ~~20,000~~ 23,500 feet altitude.

17. Altimeter—Set 29.92 IN HG.

Reset altimeter climbing through 23,500 feet to 29.92 IN HG or as required.

18. IFF—Checked.

Note

If positive operation of the normal mode of IFF has not been established during departure with an air traffic control facility, a check should be made with such a facility as soon after take-off as flight conditions permit. This check must be made before entering a radar advisory area. If the IFF is inoperative, consult the appropriate navigation publications.

CLIMB.

The climb characteristics of the airplane permit a high initial rate of climb and sustained climbing speed to the service ceiling. Refer to Appendix for recommended climb speeds, rate of climb, engine RPM and fuel consumption during climb.

WARNING

Since the airplane maintains a nose-high attitude during stalls, airflow to the engine can be critically low when the airplane is in a stall. Therefore, the throttle should be retarded, and rapid throttle advancements should not be attempted until the nose has been lowered, and airspeed is definitely increasing. Should the throttle be advanced before the engine receives sufficient airflow, turbine overtemperature may result.

TYPICAL LANDING APPROACH AND GO AROUND

NOTE

Speeds noted in the landing approach diagram are based on a gross weight of 16,000 pounds on a standard day. Recommended speeds for other gross weights will be found in the Appendix.

NOTE

The taxi light may be used as an indication that the nose gear is down and locked (observed by Mobile Control Observor).

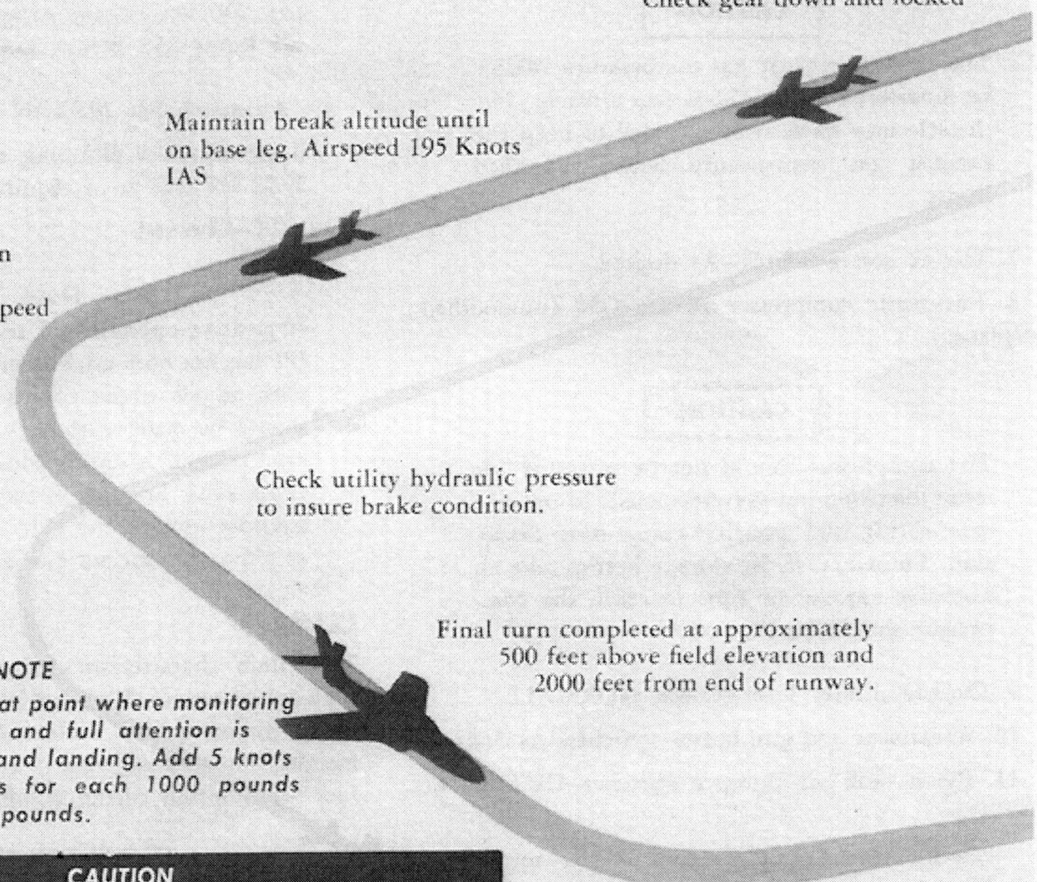
Check gear down and locked
Maintain break altitude until on base leg. Airspeed 195 Knots IAS
Full flaps during turn to final. Return flap lever to NEUT. Airspeed 185 Knots IAS
Check utility hydraulic pressure to insure brake condition.
Final turn completed at approximately 500 feet above field elevation and 2000 feet from end of runway.

NOTE

The term "fence" is that point where monitoring the air speed ceases and full attention is directed to flare-out and landing. Add 5 knots to the above speeds for each 1000 pounds of fuel above 1000 pounds.

CAUTION

With 700 pounds of fuel or less remaining in the main fuel tank, sustained uncoordinated or climbing turns, accelerations, and nose high attitudes can result in booster pump starvation and subsequent engine flame-out. Because of this, the following time limits should not be exceeded while performing any of the above described maneuvers when operating with the fuel tank selector in the ALL TANKS position on aircraft not modified to incorporate the ejector in the main fuel tank.

MAIN TANK FUEL QTY.	TIME LIMIT
700 pounds	1.0 minute
400 pounds	0.5 minute

NOTE

Pitch trim may be used throughout the pattern to lighten stick loads.

Figure 2—4 (Sheet 1 of 2)

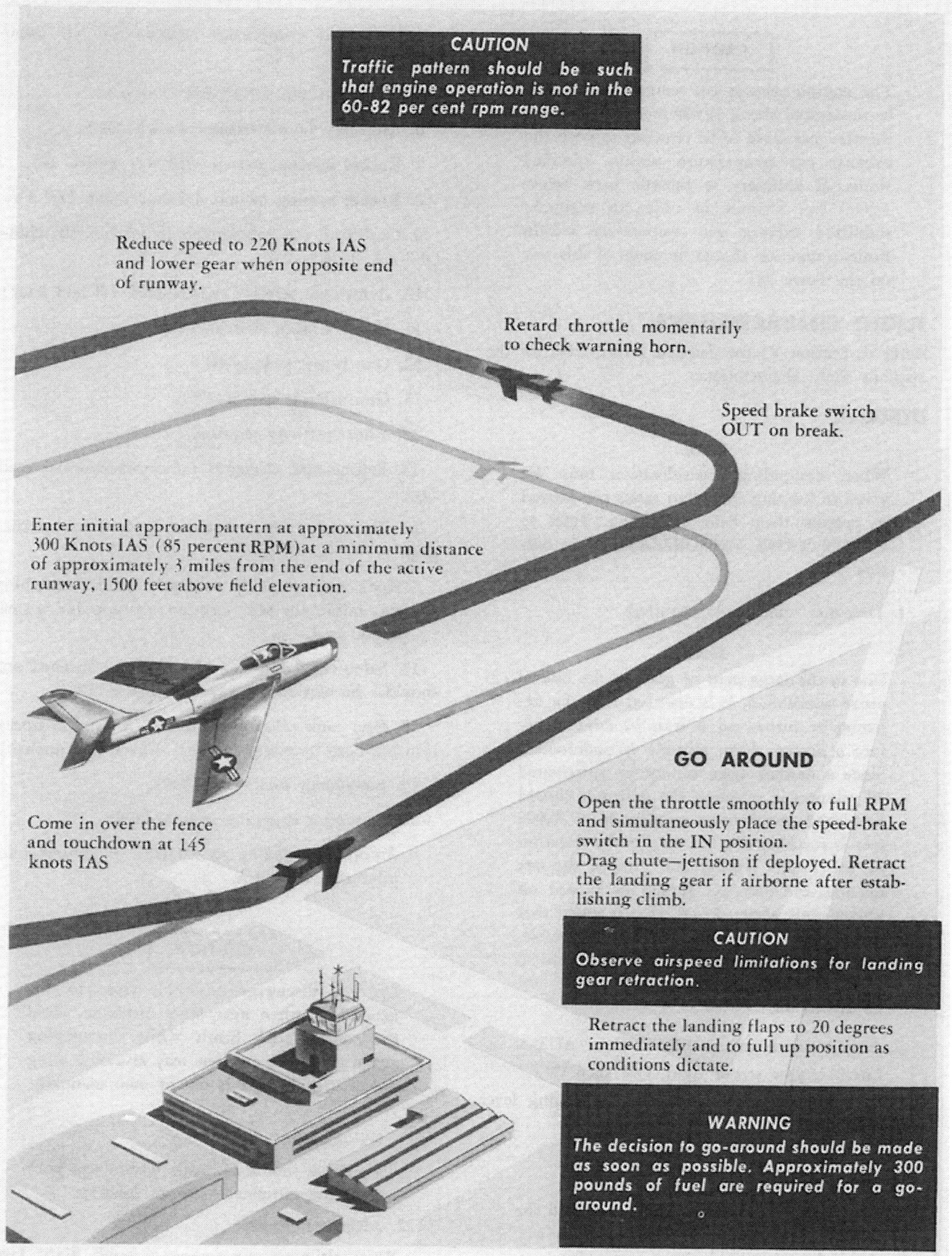
CAUTION
Traffic pattern should be such that engine operation is not in the 60-82 per cent rpm range.
Reduce speed to 220 Knots IAS and lower gear when opposite end of runway.
Retard throttle momentarily to check warning horn.
Speed brake switch OUT on break.
Enter initial approach pattern at approximately 300 Knots IAS (85 percent RPM) at a minimum distance of approximately 3 miles from the end of the active runway, 1500 feet above field elevation.
Come in over the fence and touchdown at 145 knots IAS
GO AROUND
Open the throttle smoothly to full RPM and simultaneously place the speed-brake switch in the IN position.
Drag chute—jettison if deployed. Retract the landing gear if airborne after establishing climb.
CAUTION
Observe airspeed limitations for landing gear retraction.
Retract the landing flaps to 20 degrees immediately and to full up position as conditions dictate.
WARNING
The decision to go-around should be made as soon as possible. Approximately 300 pounds of fuel are required for a go-around.

Figure 2—4 (Sheet 2 of 2)

CAUTION

The engine exhaust gas temperature MUST be monitored above 35,000 feet altitude. The throttle may have to be retarded to keep the exhaust gas temperature within specified limits. If necessary to throttle back below 35,000 feet altitude in order to maintain stabilized exhaust gas temperature within limits a notation should be made of this fact on the Form 781.

FLIGHT CHARACTERISTICS.

Refer to Section VI for detailed information on the airplane flight characteristics.

DESCENT.

Note

When descending, consideration must be given to keeping the pylon tanks pressurized to prevent them from collapsing. Refer to PYLON TANK PRESURIZATION in Section VII.

1. Defroster control—As required.

Note

Due to the large mass of glass in the bulletproof windshield, it is essential that the defroster be turned on at least 30 minutes before a descent from altitude is undertaken. Since a descent often cannot be anticipated 30 minutes in advance, the defroster should be turned on whenever an altitude of 20,000 feet is reached, and left on for the duration of the flight. During the ground support mission, the defroster should be turned on immediately after take-off. If it is found that the defroster air is excessively uncomfortable, the defroster valve should be closed down to some intermediate position, but should not be turned off.

2. Engine screen switch—EXTEND or AUTO. Check engine screen light, ON then OFF.
3. Gun-bomb-rocket, sight mechanical caging lever—CAGE.

CAUTION

If the mechanical caging lever is left in the UNCAGE position during landing or taxiing, the sight mirror or mirror suspension may become damaged due to vibration.

4. Pneumatic compressor switch—OFF (if applicable).
5. Oxygen—100% OXYGEN if desired.
6. Auxiliary bomb selector switch—OFF.
7. Rocket selector switch—OFF (A THRU G).
8. Rocket arming switch—OFF or FUSE DELAY.
9. Outboard and inboard bomb selector switches—OFF (A THRU G).
10. Armament selector switch—OFF (H and later).
11. Bomb arming switch—SAFE.
12. Gun heater switch—OFF.
13. Gun arming switch—OFF.
14. Pitot heat—As required.
15. Engine and electrical indicators—Normal readings.
16. Zero delay lanyard—Hook lanyard to rip cord grip.

    Hook lanyard to rip cord grip prior to reaching the minimum safe ejection altitude for system being used.
17. Safety belt and shoulder harness—Fastened and shoulder harness lock control UNLOCKED.
18. Fuel tank selector—ALL TANKS unless operation has been necessary in AUX. Check fuel quantity.
19. Emergency fuel switch—OFF.
20. Autopilot engaging switch—OFF.

    Autopilot engaging switch OFF or depress autopilot release switch.

CAUTION

Do not disengage autopilot close to the ground or when near other airplanes. Hold the control stick firmly while disengaging autopilot as the airplane may assume a wing down attitude if it is out of trim manually.

21. IFF—Check.

    This check should be made within one hour before the estimated time of landing.
22. Altimeter—Reset.

    Reset altimeter descending through flight level 240 to altimeter setting at point of descent.

BEFORE LANDING.

WARNING

- If the right-hand drag chute door opens during flight there is a possibility of the drag chute deploying inadvertently. Since the shear pin precludes automatic jettisoning of the chute at speeds below 220 knots, consideration should be given to jettisoning the drag chute in event it is not needed to make a safe landing.
- During the landing approach, the hydraulic components; gear, flaps, etc., are initiated independently of each other and far enough apart to avoid excessive demand of the hydraulic system. However, should these components be operated simultaneously, contrary to the approach procedure, hydraulic pressure for stabilator operation may be reduced momentarily.

See figure 2–4 for typical landing approach.

1. Enter initial approach pattern at approximately 300 knots IAS with 85 per cent RPM.

   Enter initial approach pattern at a minimum distance of approximately three miles from the end of the active runway, 1500 feet above field elevation.

WARNING

The traffic pattern should be such that engine operation is not in the 60-82 per cent RPM restricted range.

2. Speed brake switch–OUT on break.
3. Throttle–Retard momentarily to check warning horn.
4. Reduce speed to 220 knots IAS and lower gear when opposite end of runway.
5. Gear down and locked–Check.

   Landing gear will extend in approximately six seconds at normal temperatures.
6. Maintain break altitude and 195 knots IAS until on base leg.
7. Flaps–Full down during turn to final.

   Return flap lever to NEUT, maintain 185 knots IAS.
8. Check brake pedals and utility hydraulic pressure.
9. Final turn completed at approximately 500 feet above field elevation and 2,000 feet from end of runway.
10. Come in over the fence and touchdown at speeds recommended in the Appendix.

LANDING PROCEDURES.

a. When the gear is left extended in flight between landings, a minimum of 15 minutes should elapse between full stop landings to allow enough time for cooling between brake applications.

b. When the gear is retracted in flight between landings, a minimum of 30 minutes should elapse between full stop landing to allow enough time for cooling between brake applications.

c. The full length of the runway should be used during the landing roll in order that the brakes can be used as little and as lightly as possible when bringing the airplane to a stop.

d. Optimum approach and landing speeds should be used or ground roll distances will increase accordingly.

e. Landing with bombs or with fuel in the external tanks requires that a good landing technique be employed to prevent wrinkling or buckling the wings during such landings.

f. It should also be noted that the actual turning moment which depends on the frictional force is the product of the coefficient of friction times the load on the wheel. Therefore excessive braking pressure can result in locking the wheel more easily if the brakes are applied immediately after touchdown than if the same pressure is applied after the full weight of the airplane is on the wheels, and a wheel once locked in this manner right after touchdown will not become unlocked as the load is increased as long as the brake pressure is maintained.

NORMAL LANDING — WITHOUT DRAG CHUTE.

Refer to the Appendix for approach and landing speeds. Ground roll distances in the Appendix are based on maximum braking. The following nose up procedure should be utilized to save tires and brakes. However, this procedure increases landing roll distances by approximately 50 per cent above that required for a maximum braking landing.

1. Touchdown in a nose high attitude.
2. After touchdown, obtain the maximum aerodynamic drag by applying enough aft stick to hold the nose high. Don't drag the tail.

Note

Directional control is good throughout the entire landing roll. However, at speeds below approximately 50 knots, brakes are also required for directional control.

3. Slow down until nose gear touches (approximately 100 knots).

4. Hold full aft stick and apply brakes as required.

CAUTION

Do not apply hard brakes until nose wheel has contacted the ground so as to minimize landing load on the nose wheel strut.

NORMAL LANDING — WITH DRAG CHUTE.

Refer to the Appendix for approach and landing speeds. Ground roll distances in the Appendix are based on maximum braking.

1. Touchdown in a nose-high attitude then let the nose wheel contact the runway.

2. On touchdown, allow nose wheel to settle to the ground immediately and deploy the drag chute.

CAUTION

Due to the faster sink rate and a strong tendency to pitch nose down upon drag chute deployment, all three wheels should be on the ground prior to deployment. Do not push drag chute control handle back in after deploying drag chute as this action will jettison the drag chute.

3. Apply brakes as required until airplane has slowed to taxiing speed.

Note

It is recommended that the drag chute be jettisoned prior to stopping to prevent damage to the chute through contact with the runway or prior to using high power because engine exhaust may burn the shroud lines.

MINIMUM DISTANCE LANDING ROLL ON DRY RUNWAY — WITH OR WITHOUT DRAG CHUTE.

Refer to Appendix for approach and landing speeds and ground roll distances. For a minimum distance landing roll the following procedure is recommended:

a. Turn on to the final approach should be made further out than normal.

b. During final approach, reduce speed to precomputed final approach speed for the airplane gross weight as shown in the Appendix and adjust power so that a 200 to 300 FT/MIN rate of descent can be held.

c. Immediately prior to touchdown, reduce power to IDLE and touchdown at recommended touchdown speed for the airplane gross weight as shown in the Appendix.

d. On touchdown allow the nose wheel to settle to the ground immediately and deploy the drag chute, if installed. Apply braking while gradually applying aft stick. Continue to increase braking so as to hold the nose wheel on the ground as aft stick is applied. Optimum braking is obtained when full aft stick is reached with the nose wheel still on the runway.

Note

It is recommended that the drag chute be jettisoned prior to stopping to prevent damage to the chute through contact with the runway or prior to using high power because engine exhaust may burn the shroud lines.

LANDING ON WET OR ICY RUNWAYS.

The procedure for landing on wet or icy runways is covered in Section IX.

LANDING WITH EXTERNAL LOAD.

Caution must be exercised when landing with external loads such as bombs, fuel in the pylon tanks and rockets because the load applied to the wing structure during such landings may cause wrinkles in the wings unless very smooth landings are made. Landing is made in the normal manner except that the speed will be higher because of the higher stalling speed due to the increased weight.

CROSS WIND LANDING.

The procedure for cross wind landing is the same as for normal landing. However, if the drift appears excessive, the upwind wing may be lowered just before contact. During landing roll, the airplane can be held in a straight path with the rudder until speed reduces to approximately 70 knots. At the lower speeds, direction is controlled by the use of the brakes. Refer to crosswind landing charts in the Appendix of this manual.

EMERGENCY LANDING.

Refer to Section III for procedure in event of an emergency landing.

TOUCH AND GO LANDING.

Touch-and-go landings introduce a significant element of danger because of the many rapid actions which must be executed while rolling on the runway at high speed or while flying in the immediate proximity to the ground. Therefore touch-and-go landings should be made only when authorized and directed by the major command concerned. The following procedure is recommended:

a. Throttle—Increase to 100 per cent RPM while retracting speed brakes.

b. Instruments—Cross check for proper indications. Throttle should be moved smoothly to avoid overtemperature, overspeed, etc.

c. Airspeed—Accelerate to proper airspeed (approximately 140 knots with no external stores, before attempting lift-off).

d. Landing gear—Retract when safely airborne and raise wing flaps at 200 knots.

e. Landing lights—For night go-arounds, retract landing lights as soon as practicable.

GO-AROUND.

See figure 2–4.

WARNING

The decision to go-around should be made as soon as possible. Approximately 300 pounds of fuel are required for a go-around.

CAUTION

On airplanes not modified to incorporate the ejector in the main fuel tank and with 700 pounds of fuel or less remaining in the main fuel tank, sustained uncoordinated or climbing turns, accelerations, and nose high attitudes can result in booster pump starvation and subsequent engine flameout. Because of this, the following time limits should not be exceeded while performing any of the above described maneuvers when operating with the fuel tank selector in the ALL TANKS position on airplanes not modified to incorporate the ejector in the main fuel tank.

MAIN TANK FUEL QUANTITY	TIME LIMIT
700 pounds	1.0 minute
400 pounds	0.5 minute

1. Throttle—Open smoothly to full RPM and simultaneously return speed brake switch to IN.
2. Drag chute—Jettison if deployed.
3. Landing gear—Retract if airborne.

CAUTION

Observe airspeed limitations for landing gear retraction.

4. Landing flaps—Retract as conditions dictate.

AFTER LANDING.

CAUTION

After landing modified airplane, with holes drilled in the canopy arms (T.O. 1F-84-763), the canopy may not be opened until the airplane speed, and/or wind gusts, are less than 40 knots. This restriction is necessary as the canopy arms are subjected to bending, if the canopy is opened at higher airspeeds.

1. Landing flap lever—UP after clearing runway.
2. Speed brake switch—IN after clearing runway.
3. Windshield defroster—OFF.
4. Pitot heater—OFF (if applicable).
5. Radio compass and IFF—OFF.
6. Pylon tank air pressure switch—OFF.

ENGINE SHUTDOWN.

1. Emergency hydraulic pump switch—OFF.
2. Throttle—Move to IDLE.
3. Brakes—Hold until chocks are in place.
4. Throttle—CLOSED.

Note

If the engine RPM has not exceeded 60 per cent during taxiing engine can be shutdown immediately, otherwise run engine at IDLE for one minute prior to shutdown.

5. Landing gear ground safety clips—Install.
6. Fuel tank selector—Turn to OFF after engine stops rotating.

Note

If the fuel tank selector is turned to the OFF position immediately after closing the throttle, the fuel tank selector becomes tightly bound on some airplanes and cannot be moved to any other position without extreme difficulty. Therefore, when stopping the engine, wait until the engine stops rotating before turning the fuel tank selector to the OFF position to avoid fuel pressure buildup between the selector valve and the engine-driven pump. This will also prevent damage to the engine fuel pump due to the possibility of the pump running dry while the engine coasts to a stop.

7. All switches—OFF except generator switch.

CAUTION

The engine must stop rotating before the battery is turned OFF. This prevents hydraulic fluid, pressurized by the windmilling engine, from leaking past the electrically operated landing gear selector valve, which is spring loaded to the neutral position. Fluid leaking past the neutrally positioned valve may release the landing gear downlock.

IF ENGINE FAILS TO STOP WHEN THROTTLE IS CLOSED.

If the engine throttle or fuel control malfunctions so that the engine cannot be stopped, the fuel tank selector should be placed in the OFF position. This action shuts off fuel flow to the engine and the engine will stop without danger of fire.

CAUTION

After using the above procedure for stopping the engine, the fuel tank selector may become tightly bound due to a fuel pressure buildup between the selector valve and the engine-driven pump. This condition should be alleviated and the engine-driven fuel pump checked for damage before attempting to start the engine.

BEFORE LEAVING THE AIRPLANE.

1. Right handgrip canopy jettison control—Install safety pin (A THRU F).
2. Safety pin in right handgrip of seat—Install (G and later).
3. Drag chute control handle—Check in full forward position.

   The full forward position opens the riser cable latch lock so that a repacked chute can be installed in the drag chute compartment.

4. Fill out Form 781.

CAUTION

Make appropriate entries in the Form 781 covering any limits in the Flight Manual that have been exceeded during the flight. Entries must also be made when in the pilot's judgment the airplane has been exposed to unusual or excessive operations such as hard landings, excessive braking action during aborted take-off, long and fast landings and long taxi runs at high speeds, etc.

Note

After landing it is not necessary to disconnect the zero delay lanyard, since it connects the zero delay to the timer knob and is not attached to the lap belt. The parachute may be removed from the airplane with the lanyard in the hooked-up condition.

CONDENSED CHECK LIST.

Your condensed check list is now contained in T.O. 1F-84(25)F-(CL)1-1.

F-84F CONDENSED CHECK LIST

BEFORE EXTERIOR INSPECTION.

1. DD Form 781—Check for status of aircraft.
2. Check applicable publications—Abroad and current.
3. Canopy jettison safety clip—Check (A thru G)
4. Canopy jettison and seat ejection control safety pin—Check (H and later)
5. Safety pin from automatic lap belt initiator—Check removed.
6. Surface control lock—Released.
7. Emergency hydraulic pump switch—OFF.
8. Auxiliary bomb selector switch—OFF (A thru G)
9. Armament selector switch—OFF (H and later)
10. Rocket selector switch—OFF (A thru G)
11. Rocket arming switch—OFF or FUSE DELAY.
12. Rocket jettison ready switch—NORMAL (A thru G)
13. Ato ready switch—OFF.
14. Bomb switches—OFF or SAFE.
15. Gun arming switch—OFF.
16. Fuel tank selector—OFF.
17. Outboard and inboard pylon jettison switches—Cover guard down.
18. Landing gear selector handle—DOWN.
19. Emergency landing gear release switch—Cover guard down.
20. Battery switch—ON.
21. Canopy squib A and B circuits—Test (canopy must be fully open).
22. Control stick—Move until hydraulic pressure is zero
23. Pitot heater check—Climatic.
24. Battery switch—OFF.

EXTERIOR INSPECTION.

1. AFT CANOPY.

a. Oil supply and tank filler cap installed—Check.
b. Hydraulic oil supply, reservoir filler cap and access door—Check.
c. Index mark on main fuel tank filler cap—Check.
d. Alcohol supply tank—Check.
e. Canopy and canopy frame—Check.
f. Canopy—Check.

T.O. 1F-84(25)F-1
15 FEBRUARY 1958
Changed 30 June 1958

1

g. Canopy jettison hose quick disconnect—Check.
h. Air refueling receiver—Check.
i. Wing fuel filler cap—Installed.
j. Forward tank filler cap—Installed (A thru F).

2. FORWARD FUSELAGE (Left Side).

a. Left inboard pylon tank or bomb—Check.
b. Starter air storage bottle—Check.
c. Gun blast tubes—Plugs installed.
d. Gun case ejection chute doors—Check.
e. Static ports—Check.
f. All access doors and gun deck cover—Secured.
g. Battery lift—Check.
h. Sucker doors—Check links.

3. NOSE SECTION.

a. Intake duct plug and pitot cover—Removed.
b. Air intake ducts—Check.
c. Pneumatic pressure for emergency nose gear extension—Check.
d. Nose wheel shock strut—Check.
e. Nose gear safety pin—Removed.
f. Nose wheel tire—Check.
g. Nose gun blast tubes—Plugs installed.
h. Nose wheel static ground wire—Ground contact.
i. Shell casing doors—Check.

4. FORWARD FUSELAGE AND RIGHT WING.

a. All access doors and gun deck cover—Secured.
b. Battery lift—Check.
c. Static ports—Check.
d. Oxygen filler—Secured.
e. Gun blast tubes—Plugs installed.
f. Fuselage and wing drain and vent holes—Check.
g. Gun case ejection chute doors—Check.
h. Ground refueling receptacle—Check (C and later).
i. Right inboard pylon tank or bomb—Check.
j. External power receptacle—Check.
k. Wheel well—Check.
l. Armament safety override switch—Check.

T.O. 1F-84(25)F-1
15 February 1958 **2**

m. Landing gear strut–Check.
n. Landing gear tire–Check.
o. Wheel chock–In place.
p. Landing gear safety clip–Removed.

5. RIGHT WING TIP.

a. Outboard pylon tank or bomb–Check.
b. Front rocket posts–Check (A thru F).
c. Front rocket posts–Check (G and later).
d. Rear rocket doors–Check (A thru F).
e. Rear rocket posts–Check (G and later).
f. Position light–Condition.
g. Landing light–Retracted.
h. Wing–Check.

6. RIGHT WING TRAILING EDGE AND AFT FUSELAGE.

a. Ato units–Check.
b. Engine access doors–Closed.
c. Speed brake–Check.
d. Emergency hydraulic system accumulator–Check (Non-tandem).
e. Utility hydraulic system accumulator–Check (Tandem).
f. Power hydraulic system accumulator–Check (Tandem).
g. Fuselage splice cover band–Check.

7. EMPENNAGE.

a. Fuel tank vent–Check.
b. Three hydraulic vent ports on each side of fuselage and one port on right side of vertical fin–Check.
c. Tailpipe dust plug–Removed.
d. Tailpipe–Check.
e. Empennage–Check.
f. Tail position lights–Check.
g. Drag chute compartment doors–Check.

8. AFT FUSELAGE AND LEFT WING TRAILING EDGE.

a. Speed brake–Check.
b. Engine access doors–Closed.
c. Ato units–Check.
d. Fuselage splice cover band–Check.

9. LEFT WING TIP.

a. Wing—Check.
b. Landing light—Retracted.
c. Position light—Condition.
d. Rear rocket posts—Check (G and later)
e. Rear rocket doors—Check (A thru F)
f. Front rocket posts—Check (G and later)
g. Front rocket posts—Check (A thru F)
h. Outboard pylon tank or bomb—Check.

10. LEFT WING.

a. Wheel well—Check.
b. Fuselage and wing drain and vent holes—Check.
c. Landing gear strut—Check.
d. Landing gear tire—Check.
e. Wheel chock—In place.
f. Landing gear safety clip—Removed.
g. External tanks ground refuel switch—Check (C and later).
h. Air refueling valves test switch—Check (C and later).

INTERIOR CHECK — ALL FLIGHTS.

1. Rudder pedals—Adjust to proper position.
2. Safety belt and shoulder harness—Examine.
3. Armrests and seat—Adjust to desired position.
4. Ejection seat cannon plug—Check for security.
5. Trim switch—Check.
6. Battery bus continuity check—Completed.
7. Drag chute circuit breaker test light—Test.
8. Circuit breakers on left console—Check (A thru G).
9. Hydraulic by-pass switch—Off (Non-tandem).
10. Pneumatic compressor switch—OFF.
11. Spoiler shut-off switch—Cover guard down.
12. AN/ARC-34 emergency battery switch—Check (H and later).
13. AN/ARC-34 circuit breaker test light—Check (H and later).
14. Emergency fuel switch—OFF.
15. Engine screen switch—EXTEND or AUTO.
16. Fuel filter de-icing switch—OFF.
17. Pylon tanks air pressure switches—OFF.

T.O. 1F-84(25)F-1
15 February 1958 4

18. Gun heater switch—OFF.
19. Pitch control switch—NORMAL (Non-tandem).
20. Alternate trim switch—NORM (Non-tandem).
21. Pitch trim switch—NORM (Tandem).
22. Bomb selector switches—OFF (A thru G).
23. Hydraulic hand pump—Test (Non-tandem).
24. Landing flap lever—NEUT.
25. Speed brake switch—Position of speed brakes.
26. Throttle—CLOSED.
27. Air refueling receiver switch—Closed.
28. Air refueling receiver light switch—OFF.
29. Fuel shut-off valve switches—UP position.
30. Landing lights switch—OFF.
31. Directional indicator switch—NORMAL.
32. Altimeter, accelerometer and clock—Set.
33. Gun-bomb-rocket sight mechanical caging lever—CAGE.
34. Drag chute control handle—Full forward position.
35. Instrument power switch—ALT.
36. Generator switch—ON.
37. Oxygen pressure—Check (A and B).
38. Liquid oxygen quantity—Check FULL (C and later).
39. Autopilot engaging switch—OFF.
40. Circuit breakers on right console—Check.
41. Cabin vent selector—PRESSURE (prior to F-84F-35RE and F-84F-51GK).
42. Cabin vent selector—NORMAL (F-84F-35RE and later and F-84F-51GK and later).
43. Defroster control—OFF.
44. Canopy dry air switch—OFF (H and later).
45. All radio controls—OFF.
46. Battery switch—ON.
47. All warning lights—Test.
48. Fuel quantity—Check.
49. Internal and external lights—OFF.
50. Oxygen supply—Connect.
51. Radio connections and G suit—Connect.
52. Oxygen regulator supply lever—ON.

T.O. 1F-84(25)F-1
15 February 1958 5

53. Oxygen system—Check.
54. Flight instruments—Check operation.
55. Attitude indicator—Cage momentarily after inverter starts operating.
56. Vertical velocity indicator—Check zero position.
57. Radio compass—Check.
58. Command radio—Check.
59. Safety pins from canopy jettison and seat ejection initiators—Check removed (A thru F).
60. Emergency escape system safety pin in seat right handgrip—Removed.
61. Oxygen diluter lever—As required.
62. Run-up area—Check.
63. Head aircraft into the wind if practicable.
64. Personnel clear of jet exhaust and intake areas.

INTERIOR CHECK — NIGHT FLIGHTS.

1. Console lights—Check.
2. Instrument panel lights—Check.
3. Position lights—Check.
4. Landing lights—Check.
5. Taxi light—Check.
6. Air refueling receptacle light—Check.
7. Flashlight—Check.

TANDEM

EMERGENCY HYDRAULIC SYSTEM CHECK.

1. External power connected or battery switch ON and instrument power switch ALT.
2. Utility and power hydraulic pressure gages—Check for zero reading.
3. Emergency hydraulic pump switch—HYD EMER PUMP.
4. Control stick—Move to full forward position.
5. Stabilator operation—Check.
6. Aileron operation—Check.
7. Emergency hydraulic pump switch—Return to OFF.

T.O. 1F-84(25)F-1
15 February 1958 6

STARTING THE ENGINE.

1. Fuel tank selector—ALL TANKS.
2. Throttle—Move to IDLE.
3. Starter switch—Immediately actuate to START.
4. Starter switch—STOP START.
5. Instrument power switch—NOR.
6. Oil pressure—Check.
7. Engine instruments—Check within range.
8. External power—Disconnect.

TANDEM

GROUND TEST — BEFORE TAXIING.

1. Emergency hydraulic pump switch—OFF.
2. Hydraulic pressure gage—Check.
3. Brakes—Check.
4. Speed brakes—Test.
5. Landing flaps—Lower to full down then return to take-off.
6. Engine at IDLE RPM—Check power and utility hydraulic pressures.
7. Spoiler switch OFF—Control stick full forward. Check both pressure gages, cycle stick, check power and utility hydraulic pressure.
8. Rudder—Check for free and correct movement.
9. Trim switch—Actuate and check trim operation.
10. Yaw trim switch—Actuate and check rudder trim.
11. Compressor switch—ON and check pneumatic pressure gage. Return compressor switch to OFF.
12. Charge on loadmeter—Check.
13. Voltage reading on voltmeter—Check.
14. Communication equipment—Check.
15. No fuel overflowing from vent—Check.
16. Fuel tank AUX positions check—Completed.
17. Canopy dry air switch—Climatic.
18. Windshield defrost switch—ON.
19. Defroster control—Climatic.
20. Oxygen mask—Properly adjusted and oxygen system operating.
21. Autopilot—Check.

22. Autopilot engaging switch—OFF.
23. AN/APX-6 (IFF)—STANDBY.
24. Emergency hydraulic pump switch—OFF.
25. Pitch trim switch—NORM.

NON-TANDEM

GROUND TEST — BEFORE TAXIING.

1. Emergency hydraulic pump switch—NORMAL.
2. At IDLE engine RPM—Check utility and boost hydraulic system pressure.
3. Hydraulic utility gage—Check while brake pedals are depressed.
4. Speed brakes—Test and return switch to IN.
5. Landing flaps—Full down then return to take-off position.
6. Ailerons, rudder and stabilator—Test operate.
7. Trim switch—Actuate and check trim operation.
8. Rudder trim switch—Actuate and check rudder trim.
9. Engine—IDLE, emergency hydraulic pump switch NORMAL.
10. ELECTRICAL CONTINUITY CHECK—Check.
11. Emergency hydraulic pump switch—EMERG, hydraulic by-pass switch off. Check emergency ON indicator and hydraulic pressure gages.
12. Emergency hydraulic pump switch—NORMAL. Alternate trim switch—NORM: pitch control switch—STICK OVERRIDE or ELEC, engine at IDLE. Check stabilator for movement with trim switch.
13. Pitch control switch ELEC (D and later), emergency hydraulic pump switch NORMAL, alternate trim switch OFF. Actuate alternate trim switch and apply load on control stick.
14. Control stick neutral, alternate trim switch—NORM, emergency hydraulic pump switch—NORMAL, hydraulic by-pass switch—off, and pitch control switch—NORMAL, check flight controls.
15. Compressor switch—ON and check pneumatic pressure gage. Return compressor switch to OFF.
16. Charge on loadmeter—Check.
17. Voltage reading on voltmeter—Check.
18. Communication equipment—Check.
19. No fuel overflowing from vent—Check.
20. Fuel tank AUX positions check—Completed.

T.O. 1F-84(25)F-1
15 February 1958 8

21. Canopy dry air switch—Climatic.
22. Windshield defrost switch—ON.
23. Defroster control—Climatic.
24. Oxygen mask—Properly adjusted and oxygen system operating.
25. Autopilot—Check.
26. Autopilot engaging switch—OFF.
27. AN/APX-6 (IFF)—STANDBY.
28. Emergency hydraulic pump switch—NORMAL.
29. Hydraulic by-pass switch—Off.
30. Pitch control switch—NORMAL.
31. Alternate trim switch—NORM.

TAXIING INSTRUCTIONS.

1. Turn and slip indicator—Check operation.
2. Directional indicator—Check operation.
3. Attitude indicator—Check position.
4. Windshield defrosting—Check.

BEFORE TAKE-OFF.

1. Engine screens—As desired.
2. Attitude indicator—Check.
3. Landing flaps—Check in take-off position.
4. Speed brake switch—IN.
5. Aileron and rudder trim indicator lights—Illuminated.
6. Pylon tank air pressure switches—OFF.
7. Ato ready switch—ATO READY (for ATO take-off).
8. Flight controls—Check for free and correct movement.
9. Engine acceleration check—Completed.
10. Engine overspeed check—Completed.

TAKE-OFF.

1. Release brakes.
2. Throttle—Military power.
3. Go-No-Go speed—Check.

T.O. 1F-84(25)F-1
15 February 1958 9

AFTER TAKE-OFF — CLIMB.

1. Landing gear selector handle—UP.
2. Landing flap lever—UP.
3. After flaps are retracted—Return landing flap lever to NEUT.
4. Pylon tank air pressure switches—OUTB'D or INB'D.
5. After take-off—Drop ato units.
6. Ato ready switch—OFF.
7. Climb to a safe altitude and adjust speed for best climb.
8. D-ring lanyard—Unhook from D-ring and stow.
9. Cabin vent selector switch—PRESSURE (prior to F-84F-35RE and F-84F-51GK).
10. Cabin vent selector switch—NORMAL (F-84F-35RE and F-84F-51GK and later).
11. Engine screen switch—As desired.
12. Pneumatic compressor switch—ON.
13. Autopilot engaging switch—ON.
14. Gun heater switch—Climatic.
15. Oxygen diluter lever—Check. NORMAL.

DESCENT.

1. Auxiliary bomb selector switch—OFF.
2. Rocket selector switch—OFF (A thru G).
3. Rocket arming switch—OFF or FUSE DELAY.
4. Outboard and inboard bomb selector switches—OFF (A thru G).
5. Armament selector switch—OFF (H and later).
6. Bomb arming switch—SAFE.
7. Gun heater switch—OFF.
8. Engine screen switch—EXTEND or AUTO.
9. Fuel tank selector—ALL TANKS unless operation has been necessary in AUX.
10. Gun arming switch—OFF.
11. Gun-bomb-rocket sight mechanical caging lever—CAGE.
12. Emergency fuel switch—OFF.
13. Pneumatic compressor switch—OFF.
14. Autopilot engaging switch—OFF.
15. Oxygen—100% OXYGEN if desired.
16. Safety belt and shoulder harness—Check.
17. D-ring lanyard—Hook lanyard to D-ring.

T.O. 1F-84(25)F-1
Changed 30 June 1958 **10**

BEFORE LANDING.

1. Enter initial approach pattern at approximately 300 knots IAS.
2. Speed brake switch—OUT on break.
3. Throttle—Retard momentarily to check warning horn.
4. Reduce speed to 220 knots and lower gear when opposite end of runway.
5. Gear down and locked—Check.
6. Maintain break altitude until on base leg.
7. Flaps—Full down during turn to final.
8. Utility hydraulic pressure—Check to insure brake condition.
9. Final turn completed at approximately 500 feet above field elevation and 2000 feet from end of runway.
10. Come in over the fence and touchdown at speeds recommended in the Appendix.

GO-AROUND.

1. Throttle—Open smoothly to full RPM.
2. Drag chute—Jettison if deployed.
3. Speed brake switch IN.
4. Landing gear—Retract if airborne.
5. Landing flaps—Retract as conditions dictate.

AFTER LANDING.

1. Landing flap lever—UP after clearing runway.
2. Speed brake switch—IN after clearing runway.
3. Windshield defroster—Off.
4. Pitot heater—OFF (if applicable).

ENGINE SHUT DOWN.

1. Emergency hydraulic pump switch—OFF.
2. Throttle—Move to IDLE.
3. Brakes—Hold until chocks are in place.
4. Throttle—CLOSED.
5. Landing gear ground safety clips—Install.
6. Fuel tank selector—Turn to OFF after engine stops rotating.
7. All switches—OFF except generator switch.

T.O. 1F-84(25)F-1
15 February 1958 11

BEFORE LEAVING THE AIRCRAFT.

1. Right handgrip canopy jettison control—Install safety pin (A thru F).
2. Safety pin in right handgrip of seat—Install (G and later).
3. Drag chute control handle—Check in the full forward position.
4. Fill out DD Form 781.

TAKE-OFF DATA CARD

CONDITIONS

Gross Weight	________LB
Runway Length	________FT
OAT	°F
Pressure Altitude	________FT
Runway Gradient	________%
Wind	________KN
Assist Take-off Units	________
Engine Model	________

TAKE-OFF

Take-off Distance	________FT
Take-off Speed	________KN IAS
Go-No-Go Distance	________FT
Go-No-Go Speed (minimum	________KN IAS

LANDING IMMEDIATELY AFTER TAKE-OFF

Approach Speed	________KN IAS
Landing Ground Roll (without drag chute)	________FT
Landing Ground Roll (with drag chute)	________FT

T.O. 1F-84(25)F-1
15 February 1958

13

LANDING DATA CARD

CONDITIONS

Gross Weight	______ LB
Runway Length	______ FT
OAT	______ °F
Pressure Altitude	______ FT
Runway Gradient	______ %
Wind	______ KN

LANDING

Approach Speed	______ KN IAS
Landing Ground Roll (without drag chute	______ FT
Landing Ground Roll (with drag chute)	______ FT

T.O. 1F-84(25)F-1
15 February 1958

14

SECTION 3 EMERGENCY PROCEDURES

TABLE OF CONTENTS

PROCEDURE

ON ENCOUNTERING ENGINE FLAME-OUT

Complete engine failure, due to damage within the engine rarely occurs. In these cases air starts should not be attempted as fire may result. However, engine flame-outs can occur due to various causes and successful air starts can be made.

The fuel system is designed so that the main tank is kept full until all other fuel is used, provided both wing and forward tank booster pumps are operating. Therefore if a loss of thrust occurs, due to an acceleration or deceleration flame-out, an air start can be successfully made without changing the fuel flow.

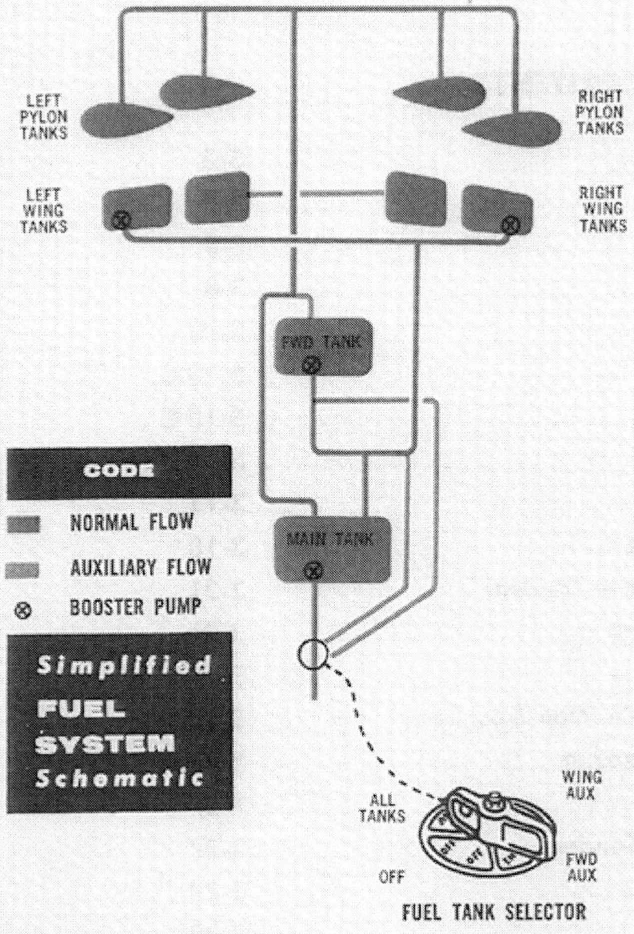

However, if it is suspected that the main tank is empty, an auxiliary fuel flow should be selected before attempting an air start.

If the flame-out is caused by an engine fuel control failure, the emergency fuel flow is selected prior to air starting.

If the engine rpm is below generator cut-in speed the fuel booster pressure warning light will not isolate the fuel system failure, as the fuel tank booster pumps are powered by the secondary bus and the light will remain illuminated as long as the secondary bus is not energized, even though the booster pumps are not defective.

If a flame-out occurs due to rapid engine acceleration or deceleration, sustained inverted flight, or similar reason,

Prepare for an Air Start as follows:

Note

Flight characteristics of the aircraft with a dead engine are normal and rapid trim changes are not necessary.

1. Throttle CLOSED.

   Close throttle to prevent flooding the engine.

   CAUTION! *Do not turn the fuel tank selector to the OFF position, as pressure built up in the fuel lines may make it difficult to place the selector in any other position.*

2. Establish glide speed of 225 knots IAS. This should produce an indicated engine speed between 17 and 21 per cent rpm.

3. Turn off all unnecessary electrical equipment to conserve battery for air starting.

   NOTE: *If time permits, the quantity of fuel remaining in any tank can be determined by positioning the instrument power switch to ALT, then positioning the fuel quantity selector switch to the desired position, and the fuel quantity check switch to the FUEL QTY CHECK position.*

4. Select fuel tank as required.

   CAUTION! *Attempt an air start only if it is suspected that power failure was caused by a fuel system failure or a fuel starvation. If power failure was accompanied by an explosion or similar noise that would indicate damage to the engine, an attempted air start may only result in a fire.*

5. If it is suspected that the flame-out occurred due to failure of the engine fuel control, place the emergency fuel switch in the ON position before attempting an air start.

6. Descend to an altitude of 20,000 feet. Although air starts have been made up to 30,000 feet, air starts at altitudes below 20,000 feet are more positive than attempted starts at higher altitudes.

Figure 3—1.

ENGINE FAILURE.

PROCEDURE ON ENCOUNTERING ENGINE FLAMEOUT.

See figure 3–1 for procedure on encountering engine flameout. When a flameout occurs it is imperative to determine if the reverse current relay is faulty in order to conserve electrical power. Reverse current relay failure after engine flameout can be recognized by any one of the following indications:

a. Fuel booster pressure warning lights not illuminated.

b. Voltmeter indicates any value other than zero.

c. Generator out indicator light not illuminated.

d. Main inverter still operating.

In the event relay failure is suspected accomplish the following steps.

a. Battery switch–OFF.

b. Booster pump circuit breakers–Out.

c. All equipment powered by secondary bus–OFF.

d. Battery switch–ON.

e. Instrument power switch–ALT.

Note

Air starts should not normally be attempted above 20,000 feet. These attempts will shorten the life of the battery making it harder to accomplish air starts at 20,000 feet and below.

COMPLETE ENGINE FAILURE DURING TAKE-OFF.

Before Flying Speed Is Reached.

1. Throttle–CLOSED.
2. Drag chute–Deploy if installed.
3. Fuel tank selector–OFF.
4. External stores–Jettison if runway overrun barrier engagement is not anticipated.
5. Brake to a stop on runway, if possible.
6. If impossible to stop on runway:
   a. Prepare to engage barrier, if available.
   b. If barrier is not available or landing gear is retracted–Jettison canopy (battery switch ON) then turn battery switch OFF.

After Becoming Airborne.

1. Landing gear selector–DOWN.
2. External stores–Jettison, if necessary.
3. Throttle–CLOSED.
4. Fuel tank selector–OFF.
5. Emergency hydraulic pump switch–HYD EMER PUMP (Tandem).
6. Canopy–Jettison (Battery switch ON).

WARNING

Leave battery switch ON until after canopy has been jettisoned as the present inspection requirement applicable to the 4.5 volt canopy jettison battery does not determine the ability of the battery to detonate the canopy squibs.

7. Shoulder harness–LOCKED.
8. Land straight ahead–Change course only enough to miss obstacles.
9. Drag chute–Deploy after touchdown if applicable.
10. Battery switch–OFF on ground contact.
11. Generator switch–OFF on ground contact.

WARNING

Battery switch should be left in the ON position until ground contact is made to insure operation of controls. If sufficient altitude has been obtained before engine failure, follow emergency bailout procedure.

PARTIAL ENGINE FAILURE DURING TAKE-OFF.

1. If not airborne–Abort take-off.
2. Emergency fuel switch–ON.

   Emergency on indicator light should illuminate.

CAUTION

When the emergency fuel system is selected, a sudden change in fuel flow and RPM will occur. The transition from normal to emergency fuel system can be made directly provided the RPM has not dropped below 85 per cent and the altitude is below 6,000 feet. If RPM has dropped below 85 per cent or altitude is above 6,000 feet retard throttle to IDLE, switch emergency fuel switch to ON then advance throttle smoothly, but slowly, to regain take-off RPM.

Note

If the fuel system booster pressure warning light is OFF when power failure occurred, this is indicative of a failed engine fuel control.

3. External stores–Jettison.
4. If power cannot be regained–Fuel tank selector WING AUX or FWD AUX.

Note

If fuel system booster pressure warning light is ON when power failure occurred, this is indicative of a failed main tank booster pump or a restricted fuel supply from the main tank.

5. Go around and land.

COMPLETE ENGINE FAILURE DURING FLIGHT.

Note

Factors affecting the decision to eject or attempt a forced landing are covered in the paragraph "Ejection Versus Forced Landings," in this section.

If complete engine failure occurs during flight, the following procedure is recommended.

1. Throttle—CLOSED.
2. Landing flaps—UP for maximum glide.
3. Speed brake switch—IN for maximum glide.
4. Trim airplane for 225 knots IAS.
5. Attempt an air start.
6. If an air start is impossible—Make a forced landing or abandon the airplane as conditions dictate.

ENGINE FAILURE DURING FLIGHT AT LOW ALTITUDE.

In the event of engine failure during flight at extremely low altitude, and with sufficient airspeed available, the airplane should be pulled up (zoom-up) to exchange airspeed for an increase in altitude. This will allow more time for accomplishing subsequent emergency procedures (air start, establishing forced landing pattern, ejection, etc.).

Note

The point at which climb should be terminated will depend on whether the pilot intends to eject or whether he intends to continue attempting air starts, establish forced landing pattern, etc. In any event, it is recommended that air start be attempted immediately upon detection of engine flameout and repeated as many times as possible during the zoom-up. If the decision is to eject, the airplane should be allowed to climb as far as possible. Ejection should be accomplished while the nose of the airplane is above the horizon but prior to reaching a stall or sink. If the decision is to continue attempting air starts, the climb should be terminated prior to the airspeed dropping below best glide speed in order that engine windmilling RPM will not drop below the minimum required for air start.

In the zoom-up maneuver, more altitude can be gained if external loads are jettisoned. Maximum altitude gain can be achieved by jettisoning external loads prior to zoom-up. The further up the climbing flight path that external loads are jettisoned, the less additional altitude will be gained. Therefore, to attain the most altitude in the zoom-up, the external load should be jettisoned as soon as possible. However, when jettisoning external loads, consideration must be given to several factors such as: sufficient airspeed to allow time for pilot reaction and jettisoning external load; terrain where external load will fall (populated areas, friendy or enemy territory, etc.); type of stores to be jettisoned (special store, conventional bombs, full or empty drop tanks, etc.); controllability of airplane if one or more stores fail to release resulting a dangerous asymmetrical condition at low altitude. Also, of prime importance, are the external load release limits as outlined in Section V. These limits must be observed to prevent damage to the airplane. It is impossible to predict the extent of damage which may occur if the external loads are released outside the established limits because of the number of factors involved. Depending on the emergency, it may be advisable to jettison the external load outside the release limits and risk some damage to the airplane in order to increase the probability of being able to accomplish subsequent emergency procedures. In any event, the decision to jettison or retain external loads must be made by the pilot on the basis of his evaluation of the above factors and conditions existing at the time of the emergency.

PARTIAL POWER FAILURE DURING FLIGHT.

1. Throttle—Retard to IDLE.
2. Emergency fuel switch—ON.
   Emergency on indicator light should illuminate.

Note

The transition from normal to emergency fuel system can be made directly without retarding throttle providing the RPM has not dropped below 85 per cent and the pressure altitude is below 6,000 feet.

3. Throttle—Advance slowly and smoothly to obtain desired RPM.

CAUTION

There is no acceleration or overspeed control in the emergency fuel system, and rapid accelerations should not be made. Small movements of the throttle will result in large power changes.

4. Complete flight on emergency fuel system to preclude flameout.

5. Land as soon as possible.

ENGINE AIR STARTING.

1. Air start switch—Depress momentarily.

   This opens the primer valve and energizes the ignition timer for a period of 15 (±3) seconds.

CAUTION

It is not recommended that the starter switch be used during air starts to increase engine RPM, as damage to the starter may result.

2. Throttle—Immediately move to IDLE position or above, but not exceeding ⅓ open.

   This opens the fuel shutoff valve. A delay in accomplishing this action will waste part of the 15 (±3) second ignition cycle. Maintain until ignition takes place or for 15 seconds maximum. If ignition takes place, monitor the exhaust gas temperature by movement of the throttle and by attitude (IAS) of the airplane. If ignition does not take place close the throttle.

CAUTION

If air start is made on the emergency fuel system, open throttle slowly toward IDLE until EGT indicates engine has started. Continue moving throttle slowly to IDLE, and monitor at all times to avoid excessive speeds and exhaust temperatures. Altitude permitting, maximum exhaust gas temperatures may be controlled or limited by diving the airplane to obtain increased ram air flow.

3. RPM and exhaust gas temperature—When stabilized at idle speed advance throttle as desired.

Note

After the engine has started, momentarily position the starter switch to the STOP START position to assure the primer valve is closed and the ignition timer is deenergized.

4. If start is unsuccessful:
   a. Throttle—CLOSED.
   b. Recheck fuel available.
   c. Reaccomplish engine air start procedure at 15,000 feet or below if initial air start was above 15,000 feet.

5. If start is again unsuccessful:
   a. Throttle—CLOSED.
   b. Fuel tank selector—Any position known to contain fuel.
   c. Attempt another air start.

Note

The ignition timer will not recycle, even though the AIR START switch is pushed again, until its 15 (±3) second period has run out, unless the circuit to the timer is first interrupted. The starter switch must be momentarily tripped to its STOP START position to interrupt this circuit, if a second air start attempt is required during the 15 (±3) second period.

LOSS OF ENGINE OIL PRESSURE.

If an oil system malfunction (as evidenced by high or low oil pressure or excessively low oil quantity) has caused prolonged oil starvation of engine bearings, the result will be a progressive bearing failure and subsequent engine seizure. This progression of bearing failure starts slowly and will normally continue at a slow rate up to a certain point at which the progression of failure accelerates rapidly to complete bearing failure. The time interval from the moment of oil starvation to complete failure depends on such factors as: condition of the bearings prior to oil starvation, operating temperatures of bearings, and bearing loads. A good possibility exists that the engine may operate for 10 minutes after experiencing a complete loss of lubricating oil. Bearing failure due to oil starvation is generally characterized by a rapidly increasing vibration; when the vibration becomes moderate to heavy, complete failure is only seconds away and in most instances the pilot will increase his chances of a successful ejection or power-off landing by shutting down the engine. Since the end result of oil starvation is engine seizure, the following procedures should be observed in an attempt to forestall engine seizure as long as possible.

AT FIRST INDICATION OF OIL SYSTEM MALFUNCTION:

1. Thrust—As required at pilot's discretion.

High thrust settings should be avoided if at all possible, in order to keep bearing loads at a minimum. Upon detection of an oil system malfunction (as evidenced by the oil pressure gage or level warning indication) a minimum thrust setting should be established depending on aircraft configuration, gross weight, and altitude. This setting should be sufficient to maintain level flight and allow for safe approach maneuvers (subsequent variations should be avoided

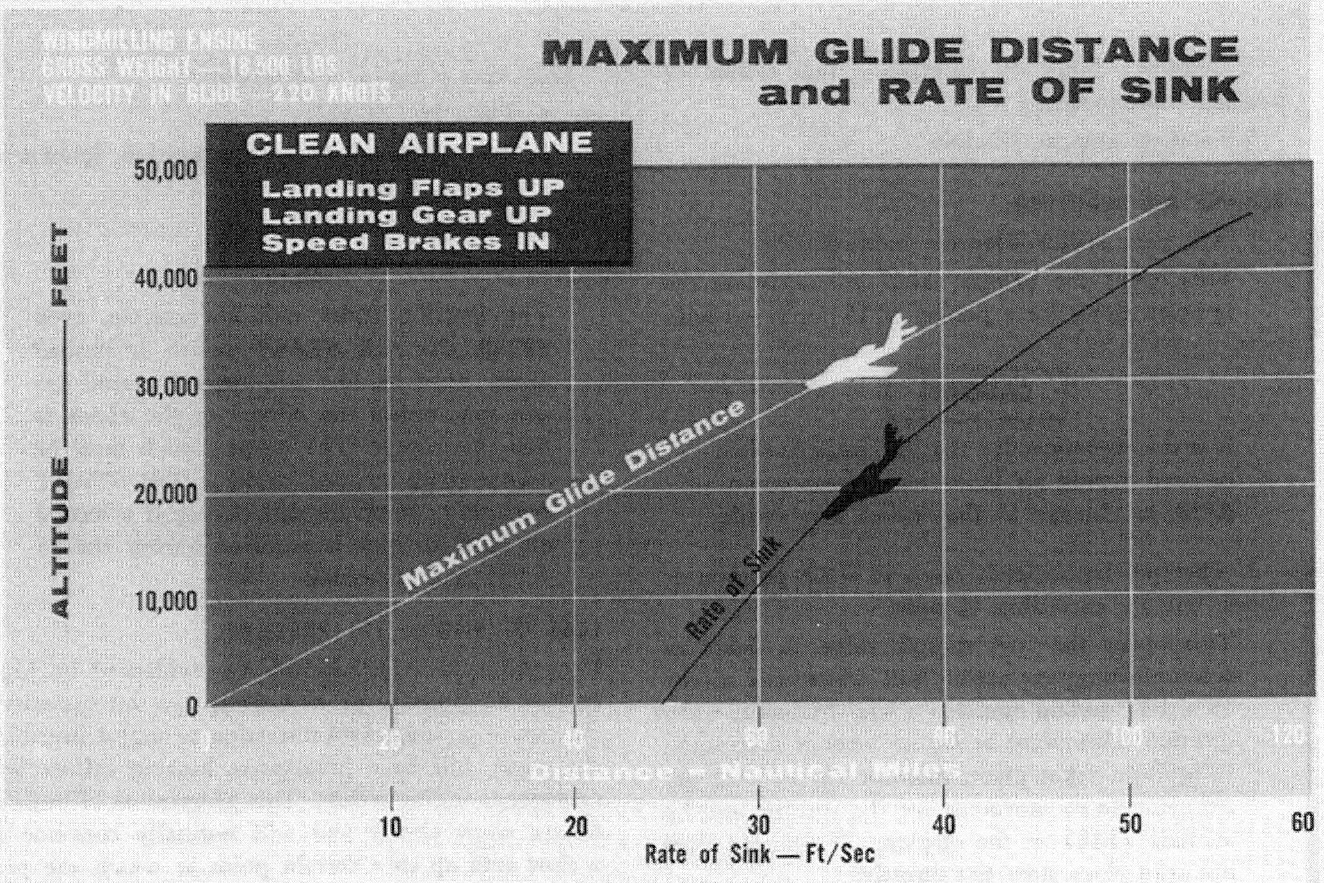

Figure 3—2

if possible). However, if the malfunction has gone unnoticed and has progressed to the point where bearing failure has started, as evidenced by vibration, the throttle should not be retarded. If the throttle is retarded, the resistance to rotation offered by one or more failing bearings may cause further deceleration and complete engine seizure in a very short time.

WARNING

If moderate to heavy vibration occurs, the chance of a successful ejection or power-off landing will be improved by shutting down the engine.

2. External stores not required—Jettison.
3. G forces—Minimize. Avoid all abrupt maneuvers causing high G forces.
4. Land as soon as possible, using the forced landing pattern to insure landing in the event of complete engine failure.

MAXIMUM GLIDE WITH DEAD ENGINE.

For maximum glide distance, trim airplane to maintain recommended glide speed with gear and flaps UP, speed brakes IN. (See figure 3—2.)

EJECTION VERSUS FORCED LANDING.

Normally, ejection is the best course of action with a windmilling or frozen engine, or failure of the flight control hydraulic systems. Because of the many variables encountered, the final decision to attempt a flameout landing, or to eject, must remain with the pilot. It is impossible to establish a predetermined set of rules and instructions which would provide a readymade decision applicable to all emergencies of this nature. The basic conditions listed below, combined with the pilot's analysis of the condition of the airplane, type of emergency, and his proficiency are of prime importance in determining whether to attempt a flame-out landing, or to eject. These variables make a quick and accurate decision difficult. If the decision is made to eject, prior to ejection, if possible, the pilot should attempt to turn the airplane toward an area where injury to persons or damage to property on ground or water is least likely to occur. Before a decision is made to attempt a flame-out landing, the following basic conditions should exist.

a. Flame-out landings should only be attempted by pilots who have satisfactorily completed simulated flame-out approaches in this airplane.

LANDING WITH DEAD ENGINE

The absorption of the initial shock by the extended landing gear has resulted in less damage to the aircraft or injury to personnel, as opposed to wheels-up landing. Emergency landings shall be made on unprepared or prepared surfaces whenever possible with the landing gear extended. Refer to figure 3–2 for maximum gliding distance from various altitudes.

1 Throttle–CLOSED.
2 Fuel tank selector–OFF.
3 External stores–Jettison over uninhabited areas.
4 Safety belt and shoulder harness tightened and inertia reel lock control locked.

CAUTION

The pilot is prevented from bending forward when the inertia reel lock control is in the locked position, therefore, all switches not readily accessible should be "cut" before moving the control to the locked position.

5 Speed brake– IN to increase glide.
6 Landing gear–UP to increase glide.
7 Landing flaps–UP to increase glide.
8 Pneumatic compressors–OFF.
9 Emergency hydraulic pump switch–EMERG at low key point unless needed to maintain control.

WARNING

To avoid any possible hydraulic surge accompanied with erratic control system functioning do not turn on emergency pump when a demand is on the system.

10 Landing gear–DOWN.
11 Flaps–DOWN if available.
12 Speed brake switch–OUT, if available.

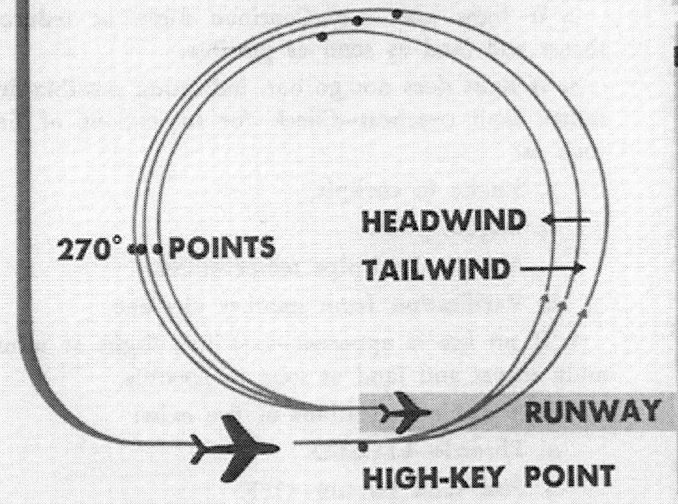

ACTUAL FLAME-OUT DATA

1 Best glide IAS—225 knots (gear up)
2 Rate of descent 2,500 feet/min (gear up)
3 Average time for gear to lock down—seven to eight seconds.
4 Average time for full extension of landing flaps—14 seconds.
5 Recommended altitude for lowering gear—Entirely dependent on altitude available when over field.
6 Approximate distance for rectangular pattern —two x three nautical miles.
7 Hi-Key point 6,000 feet.
8 Average windmill RPM below Hi-Key 14.5 per cent.
9 Average time for Hi-Key to touchdown—1½ minutes.
10 Low Key point 3,000 feet.
11 Base leg (270 degree point) 1,500 feet.
12 Best glide IAS 220 knots (gear down).
13 Rate of descent 4,000 ft/min.
14 Recommended Rate of turn — three deg/sec (gear down).
15 Recommended final IAS 200 knots (maintain this IAS until latter part of final approach).
16 Recommended IAS at touchdown 140 knots.
17 Deploy drag chute after touchdown if installed.

PATTERN ALTITUDE VARIATIONS

1 HiKey point 5,000 to 9,000 feet
2 Low Key Point 2,500 to 4,500 feet
3 270 Degree point 1,000 to 2,000 feet

FLAME-OUT SIMULATION DATA

ALTITUDE	HIGH POWER (Speed Brake Out) *(FLAPS 68 per cent)*	LOW POWER (Speed Brake In) *(FLAPS 54 per cent)*
ABOVE 15,000	88 per cent RPM	
15,000	83 per cent RPM	54 per cent RPM
10,000	83 per cent RPM	54 per cent RPM
5,000	82 per cent RPM	54 per cent RPM
SEA LEVEL	82 per cent RPM	54 per cent RPM

HIGH POWER SIMULATION, SET 83 per cent RPM at high key point, and 82 per cent RPM at low key point.

LOW POWER SIMULATION (WHEN LANDING IS INTENDED) SET 54 per cent RPM at high key point.

Figure 3–3

b. Flame-out landings should only be attempted on a prepared or designated suitable surface of at least 8,000 feet.

c. Approaches to the runway should be clear and should not present a problem during a flame-out approach.

Note

No attempt should be made to land a flamed-out airplane at any field whose approaches are over heavily populated areas, if a suitable area is available to abandon the airplane. If possible, prior to ejection, the pilot should attempt to turn the airplane toward an area where injury or damage to persons or property on the ground or water is least likely to occur.

d. Weather and terrain conditions must be favorable. Cloud cover, ceiling, visibility, turbulence, surface wind, etc., must not impede in any manner the establishment of a proper flame-out landing pattern.

Note

Night flame-out landings, or flame-out landings under poor lighting conditions such as at dusk or dawn, should not be contemplated regardless of weather or field lighting.

e. Flame-out landings should only be attempted when either a satisfactory "High Key" or "Low Key" position can be achieved.

f. If at any time during the flame-out approach, conditions do not appear ideal for successful completion of the landing, ejection should be accomplished. EJECT no later than the "Low Key" altitude.

LANDING WITH DEAD ENGINE.

See figure 3–3.

FLAME-OUT SIMULATION.

The recommended operational procedure for simulated flame-out pattern has been devised to approximate the rate of descent of actual flame-out airplane of 2,500 feet per minute with gear up. Due to the restriction on engine operation in the 60-82 per cent RPM range a compromise between configuration and power setting is required. The flame-out simulation data provided in figure 3–2 will enable the pilot to make either a high power, or low power let-down with a minimum of RPM changes. Recommended airspeeds are 225 knots IAS with gear up and 220 knots IAS with gear down. Changing from gear up to gear down simulation merely requires lowering the landing gear. The low RPM procedure will provide better simulation where actual touchdown is intended because of less thrust during the final stages of the approach.

FIRE.

ENGINE OVERHEAT LIGHT ON.

Illumination of the overheat warning light indicates an overheat condition, or possible fire in the aft section. The exact procedure to follow is dependent on each set of circumstances and depends on altitude, airspeed, length of runway and overrun remaining, location of populated areas, etc. As a guide in making a decision, the following procedures are recommended:

On Take-off Run.

If sufficient overrun is available to abort take-off, proceed as follows:

1. Throttle—IDLE.
2. Brakes—Apply.
3. Drag chute handle—PULL (if installed).
4. Prepare to engage the overrun barrier if available and stop as soon as possible.

Immediately After Becoming Airborne.

If there is not sufficent runway or clear overrun available to abort take-off, proceed as follows:

1. Continue normal take-off to safe ejection altitude.
2. Note the tailpipe temperature and reduce it to within limits (if high).
3. Check for other indications of fire, such as excessive RPM or fuel flow, or fire warning light illuminated.
4. If no fire is apparent, land as soon as possible.
5. If indications of fire exist:
   a. Throttle—CLOSED.
   b. Fuel tank selector—OFF.
   c. Make decision to land or eject.

During Flight.

1. Reduce thrust immediately.
2. If light goes out—Continue flight at reduced thrust and land as soon as possible.
3. If light does not go out, indicating possible fire rather than overheat—Check for indications of fire such as:
   a. Smoke in cockpit.
   b. Engine noise.
   c. Abnormal tailpipe temperatures.
   d. Verification from another airplane.
4. If no fire is apparent—Continue flight at minimum thrust and land as soon as possible.
5. If possible indications of fire exist:
   a. Throttle—CLOSED.
   b. Fuel tank selector—OFF.
   c. Make decision to land or eject.

ENGINE FIRE WARNING LIGHT ON.

Illumination of the fire warning light indicates a fire in the forward engine section. However, some false warnings have occurred. Therefore, it is preferable under unfavorable flight conditions to take precautionary measures before completely shutting down the engine, whenever a fire warning light appears:

On Take-off Run.

1. Throttle—CLOSED.
2. Drag chute—Deploy, if installed.
3. Fuel tank selector—OFF.
4. Prepare to engage runway overrun barrier if available and stop as soon as possible.

Immediately After Becoming Airborne.

1. Maintain maximum thrust and climb immediately to a safe ejection altitude.
2. Level off, then throttle back as far as possible, but still maintain sufficient thrust to sustain flight.
3. If no evidence of fire exists—Proceed cautiously and land as soon as possible.
4. If evidence of fire exists:
   a. Throttle—CLOSED.
   b. Fuel tank selector—OFF.
   c. Make decision to either eject or make an emergency landing.

During Flight.

If altitude and terrain permit.

1. Throttle—CLOSED.
2. Fuel tank selector—OFF.
3. Turn OFF all unnecessary electrical equipment.
4. Check for any evidence of fire:
   a. Presence of smoke in cockpit.
   b. Engine noise.
   c. Abnormal tailpipe temperature.
   d. Verification from another airplane.
5. Do not attempt a restart if indication of fire is present.
6. If a fire persists or even if it goes out—Prepare to abandon the airplane or make an emergency landing.
7. If it is definitely determined that no fire exists, a restart may be accomplished.

If Altitude or Terrain Preclude Shutting Down the Engine Completely.

1. Reduce thrust as much as possible and still maintain flight.
2. Check for evidence of fire.
3. If no evidence of fire exists—Proceed cautiously and land as soon as possible.
4. If evidence of fire exists:
   a. Throttle—CLOSED.
   b. Fuel tank selector—OFF.
   c. Make decision to eject or make an emergency landing.

FIRE WHILE STARTING THE ENGINE.

1. Throttle—CLOSED.
2. Fuel tank selector—OFF.
3. Battery switch—OFF or disconnect external power supply.
4. Leave cockpit immediately.

ENGINE FIRE DURING TAKE-OFF — IF NOT AIRBORNE.

1. Throttle—CLOSED.
2. Drag chute—Deploy, if installed.
3. Fuel tank selector—OFF.
4. Brake to a stop on runway if possible.
5. External stores jettison switch—Depress unless runway overrun barrier engagement is anticipated.
6. If impossible to stop on runway:
   a. Canopy—Jettison (Battery switch ON).
   b. Battery switch—OFF.
7. Leave cockpit as soon as possible.

ENGINE FIRE DURING TAKE-OFF — IF AIRBORNE.

1. If practicable, after passing point of safe abort, climb to a minimum safe ejection altitude.
2. Throttle back enough to maintain altitude.
3. Check for evidence of fire—Smoke, noises, verification from another airplane.
4. If no evidence of fire exists—Land as soon as possible.
5. If evidence of fire persists—Prepare to eject or make an emergency landing.

SMOKE ELIMINATION.

If smoke or fumes enter the cockpit, the cause may be a fire, a fuel leak in the engine compartment, a failed refrigeration turbine or a failed bearing seal.

1. Oxygen regulator diluter lever—100% OXYGEN.
2. Cabin temperature control—Position to highest point.

   If smoke is eliminated the cause is in the refrigeration turbine.
3. If smoke is not eliminated:
   a. Cabin vent selector—RAM.
   b. Defroster control—OFF.
   c. Abort mission and land as soon as possible.

Note

The intake distribution valve will close and the ram air and dump valves will open. The intake of air through the ram air valve will scavenge the air in the cockpit through the open dump valve, and rapidly dissipate the fumes and smoke.

TAKE-OFF AND LANDING EMERGENCIES.

An extensive study of injuries and damages resulting from emergency landings on unprepared surfaces has revealed that a greater injury hazard is presented whenever emergency landings are made with the wheels retracted, regardless of terrain. Increased airspeed or nose-high angle at impact during landings with the wheels retracted is common practice and contributes greatly to the severity of pilot injury and airplane damage. This nose-high attitude causes the airplane to literally "slap" the ground on impact, subjecting the pilot to possible spinal injury. Less injury and less airplane damage resulted when the landings were made with the gear extended because of the absorption of initial shock by the extended landing gear. There were no cases recorded where the airplane flipped or tumbled end-over-end when an emergency landing was attempted with the gear extended. Emergency landings on unprepared surfaces shall be made with the landing gear extended. This also applies to landing short on the runway, or when it is impossible to stop on the remaining runway during either take-off or landing.

ABORT TAKE-OFF.

If sufficient overrun is available to abort take-off, proceed as follows:

1. Throttle—IDLE.
2. Brakes—Apply.
3. Drag chute handle—PULL (if installed).
4. Prepare to engage the overrun barrier if available and stop as soon as possible.

RUNWAY OVERRUN BARRIER ENGAGEMENT.

Runway overrun barrier engagements can successfully be accomplished at any ground speeds from a minimum of approximately 35 knots up to a maximum of 130 knots, but more positive engagement can be assured if the barrier is contacted at a speed as close to minimum as possible. Off center engagements are successful and may be safely accomplished. If target flags are attached to the barrier webbing adapter, it is advisable to avoid hitting the webbing right at the target flag. It is possible for the flag to foul the webbing and prevent engagement. The following technique should be employed to successfully engage a runway overrun barrier, if an engagement becomes necessary.

1. External stores—Jettison.

   Jettison all stores over an open area in those cases of a known emergency prior to landing. Jettisoning the pylon tanks in the case of an emergency on take-off is haardous, and since successful barrier engagement can be accomplished with the tanks still installed, no attempt should be made to jettison them while on the ground.

WARNING

Faulty rigging of barrier may result in an unsuccessful runway overrun barrier engagement.

2. Drag chute—Deploy after touchdown (if installed).
3. Brakes—Avoid excessive use.

   Excessive use of the brakes might result in tire blowout. A blowout could result in a loss of directional control and failure to stay on the runway for barrier engagement.

4. Immediately prior to engaging the runway overrun barrier:
   a. Throttle—CLOSED.
   b. Fuel tank selector—OFF.
   c. Generator switch—OFF.
   d. Battery switch—OFF.

LANDING WITH WHEELS RETRACTED.

In the event it is impossible to get landing gear down, the following procedure is recommended:

1. If time and conditions permit—Jettison all external stores, except empty external tanks, over an uninhabited area.

   Retain pylons when stores are dropped. Less damage will be sustained to the airplane by this method. If time does not permit, jettison external stores with the external stores jettison switch.

WARNING

Be sure that the bomb arming switch is SAFE and the rocket arming switch is OFF or DELAY.

2. Canopy—Jettison (Battery switch ON).

   (Battery switch is ON so that airplane battery power is available to detonate the canopy squibs).

WARNING

If practical, jettison canopy at 300 knots IAS or above to assure removal of canopy from airplane.

3. Safety belt—Tightened.
4. Shoulder harness and inertia reel lock control—Tightened and locked.
5. Landing flaps—Extend as desired.
6. Make normal approach.
7. Before contact with the ground, close throttle and turn fuel tank selector OFF. On contact turn battery switch OFF.
8. Drag chute—Deploy after touchdown (if installed).

Note

Drag chute may be deployed after touchdown the same as for a normal landing. The use of the drag chute may cause less damage to the airplane.

9. Leave cockpit as soon as possible.

WARNING

Use extreme care when leaving the airplane to prevent possible actuation of the seat ejection control.

LANDING WITH MAIN GEAR DOWN — NOSE GEAR UP OR UNLOCKED.

If the nose gear will not extend or lock down as indicated by the landing gear indicators proceed as follows:

Note

The taxi light can be utilized to determine if the nose gear is locked down.

1. If time and conditions permit—Jettison all external stores, except empty external tanks, over an uninhabited area.

   Retain pylons when stores are dropped. Less damage will be sustained to the airplane by this method. If time does not permit, jettison external stores with the external stores jettison switch.
2. Landing gear—DOWN.
3. Canopy—Jettison (Battery switch ON).

   (Battery switch is ON so that airplane battery power is available to detonate the canopy squib).
4. Plan approach to touchdown as near the end of runway as possible.
5. Make normal approach, landing flaps DOWN, and speed brakes OUT.
6. Safety belt—Tightened.
7. Shoulder harness and inertia reel lock control—Tightened and locked.
8. Before contact with the ground—Close throttle and turn fuel tank selector OFF. On contact turn battery switch OFF.
9. Hold the nose wheel off the ground as long as practicable then ease nose down prior to stall and deploy the drag chute if installed.

CAUTION

It is recommended that the drag chute not be used, unless it is necessary for a safe stop, since deploying the chute may cause the nose to contact the runway with great force, thereby causing extensive damage.

10. Do not use brakes unless it becomes absolutely necessary, as this will tend to slam the nose to the ground.
11. Leave cockpit as soon as possible.

LANDING WITH ONE GEAR UP OR UNLOCKED.

Studies of landing gear configurations whenever all gear will not extend reveals that less damage to the airplane and less personal injury will result if the following procedure is observed.

1. If a prepared, hard surface runway is not available, leave any main gear in the extended position.

Note

An extensive study of injuries and damage resulting from emergency landings on unprepared surfaces has revealed that a greater injury hazard is presented whenever emergency landings are made with the wheels retracted, regardless of the terrain.

2. If landing on a prepared ,hard surface runway, leave any two landing gears down and locked and proceed as follows:
   a. External stores—Jettison, except empty external tanks. Empty external tanks will minimize damage to the wing. External tanks which contain fuel should be jettisoned.
   b. Use same procedure as for landing with nose gear up or unlocked.

NOSE GEAR TIRE FAILURE.

Nose Gear Tire Failure on Take-off.

In case of complete nose gear tire failure on take-off run proceed as follows:

Note

If nose gear tire failure occurs at or near nose wheel lift off speed, the pilot may elect to continue take-off in order to reduce the gross weight of the airplane. Controlling the airplane with a flat nose gear tire is much easier at lighter weights.

1. Throttle—IDLE.
2. Drag chute handle—Pull.
3. Brakes—Maximum.

   Even though heavy breaking will increase the load on the nose gear it is considered more important that the airplane be stopped as quickly as possible than to attempt to lighten the nose wheel loading at the expense of a longer roll.
4. Canopy—Open.

MAIN GEAR TIRE FAILURE.

Main Gear Tire Failure on Take-off.

The following procedure is recommended when main gear tire failure is experienced on take-off run.

Note

If insufficient runway remains for stopping the airplane at the time of main gear tire failure the pilot may continue take-off.

1. Throttle—IDLE.
2. Drag chute handle—Pull.
3. Brakes—Use maximum brake away from swerve. Use brakes on flat tire sparingly and with caution.

LANDING WITH FLAT TIRE.

1. Use up excess fuel.
2. If nose wheel tire is flat; make normal landing holding nose wheel off as long as possible.
3. If one main wheel tire is flat; make normal landing on side of runway nearest inflated tire.
4. If both main wheel tires are flat; make normal landing in center of runway and use brakes sparingly and with caution.
5. Drag chute—Deploy after touchdown (if installed).

LANDING WITH UTILITY HYDRAULIC SYSTEM FAILURE.

1. External stores—Jettison if necessary.

   In the event runway barrier engagement is anticipated, all external stores should be jettisoned prior to touchdown.
2. Landing gear selector handle—EMERG DOWN.

   If the landing gear does not lock down (indicated by the landing gear position indicators) hold the selector handle in the EMERG DOWN position for a few seconds.
3. To reduce flight control surface deflection set up a long flat final approach.

   For approach and touchdown speeds refer to the Appendix.
4. Emergency hydraulic pump switch—HYD EMER PUMP.
5. On ground contact:

   a. Drag chute—Deploy, if installed.

   b. Throttle—IDLE (stop cock throttle if necessary).
6. Use recommended emergency braking technique until speed is reduced to "taxi".

Note

With the utility hydraulic system inoperative, approximatelytwice the toe pressure will be required for the brakes to be as effective as when utility pressure is available.

LANDING WITH UNBALANCED EXTERNAL LOAD.

Landing with an unbalanced inboard pylon configuration can be accomplished without any special technique, provided hydraulic power is available for aileron control. However, landing with an unbalanced outboard pylon configuration should only be accomplished if the pilot is sufficiently familiar with the airplane and all other landing factors are normal. With hydraulic pressure available, there is sufficient lateral control to maintain level flight for the critical asymmetric outboard configuration of one full 230 gallon tank if the landing is made at a minimum speed of 180 knots. If the pilot has any doubt about making the landing it is recommended that the outboard stores be jettisoned.

EMERGENCY ENTRANCE

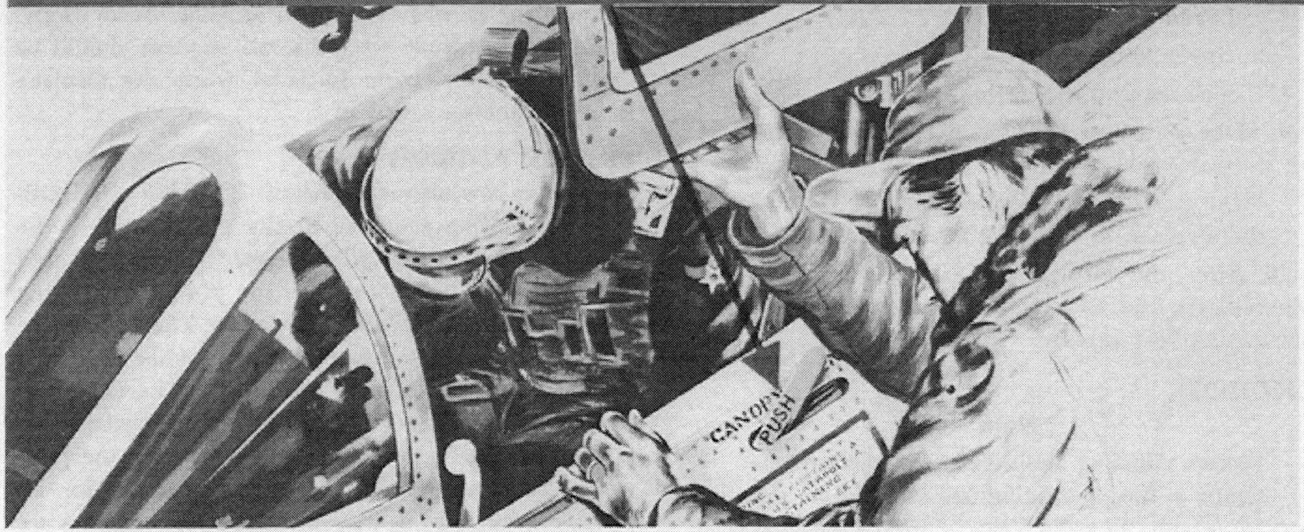

Emergency entrance to the cockpit is made by depressing the forward end of the external canopy control. This action rotates the control so that it can be grasped and pulled forward to unlock the canopy. The canopy is then lifted up with the handgrip on either side of the canopy skirt. If the external canopy control, or the canopy becomes jammed, the pilot should actuate the canopy jettison control to release the canopy from the airplane.

WARNING

If the canopy is released by the pilot, the ground crew should avoid standing in the turtle deck area as the canopy will be lifted high enough for the jettison squibs to release the canopy from the airplane. Extreme caution must be observed, while getting out of the airplane, in order to avoid moving the seat ejection lever, which will be exposed.

Figure 3—4

EMERGENCY ENTRANCE.

See figure 3–4.

DITCHING.

Bailout is preferred to ditching. High speed airplanes tend to skip on contact, therefore, pilot must continue to fly the airplane after touchdown until airplane settles. If a regular wave or swell pattern exists aim the touchdown parallel to the waves and attempt to land on the crest or on the falling side of the wave. More often, the sea surface will be irregular with two or move wave or swell patterns intermingled. In this case, the best compromise is to head into whatever wind may be blowing. Examine the sea to find areas where the intermingling waves cancel out. Aim touchdown for one of these calmer areas. In the event of low altitude or any other reason where bailout is impracticable, ditch the airplane as follows:

1. IFF—Turn to emergency.

Note

The instrument power switch must be in the NOR position to assure operation of the IFF equipment.

2. Oxygen—100% OXYGEN.
3. External stores—Jettison (all).
4. Landing flaps—50 to 75 per cent DOWN.
5. Helmet visor—Down.
6. Canopy—Jettison (Battery switch ON).

WARNING

Be careful not to squeeze the seat ejection control accidentally.

7. If time permits complete the following:
   a. Throttle—CLOSED.
   b. Safety belt—Secured.
   c. Unbuckle parachute and disconnect anti-G and electrical connections.

d. Shoulder harness and inertia reel lock control—Tightened and locked.

e. Landing gear—UP.

f. Speed brakes—IN.

8. Touchdown in a nose high attitude.

9. Make actual touchdown in the same attitude as for normal landing.

Make the "softest" possible landing. Do not stall the airplane at the time of contact.

10. After the airplane has slowed down in the water—Leave the cockpit at once, since fighter airplanes sink very rapidly.

EJECTION.

Note

Factors affecting the decision to eject or attempt a forced landing are covered in the paragraph "Ejection Versus Forced Landings" in this Section.

In all cases of emergency exit in flight, escape should be accomplished by means of seat ejection. This is the safest method of escape in either high speed, or low speed flight, since it precludes the possibility of pilot injury through collision with the tail surfaces. Because of the increasing incidence of vertebrae injuries, occurring to pilots during forced landings of high performance airplanes, more consideration should be given to use of the escape system in preference to forced landings. Ejection should be made whenever possible above 2,000 feet although with the use of the automatic belt and parachute, successful ejection may be accomplished from lower altitudes. The basic ejection procedures are shown in figure 3—5. Study and analyses of escape techniques from airplanes by means of the ejection seat have revealed that:

a. Ejection accomplished at airspeeds ranging from stall speed to 525 knots IAS results in relatively minor forces being exerted on the body, thus reducing injury hazard.

b. The pilot will undergo appreciable forces on the body when ejection is performed at airspeeds of 525-600 knots IAS, and escape is more hazardous in this speed range than at lower airspeeds.

c. Above 600 knots IAS, ejection is extremely hazardous because of the excessive forces to which the body is subjected.

EJECTION ATTITUDES.

On ejection, the seat and pilot will have a component of thrust provided by the airplane. Therefore, if seat ejection is initiated when the airplane is in a climb, the ejected pilot will reach a higher altitude than if he were ejected from an airplane in level flight. The opposite will be true if ejection is accomplished when the airplane is in a nose down attitude. Accordingly, in marginal altitude ejections, the airplane should be "zoomed" and ejection initiated while the airplane is in a climbing attitude.

EJECTION ALTITUDES.

Ejection at low altitudes (below 2,000 feet), is facilitated by pulling the nose of the airplane above the horizon in a "zoom-up maneuver." Ejection should not be delayed when the airplane is in a descending attitude and cannot be leveled out. The chances of successful ejection at low altitudes under this condition are greatly reduced. Because of the numerous combinations of automatic ejection equipment that are possible, a chart, figure 3—4B, provides the pilot with emergency minimum ejection altitudes for all possible combinations of equipment. The altitudes are applicable for *level flight* attitudes. The data are conservative for climbs and optimistic for descending conditions. Data for the "one and zero" system is applicable between 120 knots IAS and the maximum safe ejection speed.

WARNING

Emergency minimum ejection altitudes shown in figure 3—4B were determined through an extensive series of flight tests, and are based on distance above terrain on initiation of seat ejection (initiation of seat ejection is defined as the time the seat is fired). However, human error and equipment malfunctions were not considered in the determination of these altitudes. Therefore, whenever possible, ejection should be initiated at altitudes higher than the minimum shown.

Note

The emergency minimum ejection altitude figures obtained from figure 3—4B during preflight planning, should be used only as a guide. Once an emergency minimum ejection altitude has been determined for a particular configuration of equipment, the decision as to when to eject, or not eject, in an emergency, should not be rigidly determined by the fact that the airplane is above or below the minimum altitude, as determined from these figures. Every emergency will have its particular set of circumstances involving such factors as airplane speed, attitude, and control, as well as altitudes. Based on these figures and the

EMERGENCY MINIMUM EJECTION ALTITUDES

	LEVEL FLIGHT — SPEED 140 TO 300 KNOTS (IAS)					
	2 SECOND PARACHUTE (F-1A Timer)		1 SECOND PARACHUTE (F-1B Timer)		0 SECOND PARACHUTE (Lanyard to "D" Ring)	
	B-4 or B-5 pack C-9 canopy	*B-5 pack C-11 canopy*	*B-4 or B-5 pack C-9 canopy*	*B-5 pack C-11 canopy*	*B-4 or B-5 pack C-9 canopy*	*B-5 pack C-11 canopy*
1 SECOND AUTOMATIC LAP BELT *(M-12 Initiator)*	350	400	200	250	100	150

Figure 3—4B

escape configuration available, a decision should be made before take-off concerning action to be taken in the event of a low altitude emergency.

HIGH ALTITUDE EJECTION.

For a high altitude ejection, the basic ejection procedure shown in figure 3–5 is applicable. The "zoom-up" maneuver is still useful to slow the airplane to a safer ejection speed, or provide more time and glide distance, as long as an immediate ejection is not mandatory. In the event it is mandatory to eject above 30,000 feet and time permits, place the cabin vent selector in the RAM position before initiating ejection procedures. This reduces the hazard of an explosive decompression of the cockpit when the canopy is jettisoned and eliminates possibility of physical incapacitation of the pilot.

WARNING

During high altitude operation the zero delay lanyard MUST NOT be hooked to the parachute rip cord grip. If the zero delay lanyard is hooked to the rip cord grip, the safety feature of the automatic opening parachute will be eliminated. If it becomes necessary to eject at high altitudes, the parachute will open immediately, subjecting the pilot to serious or fatal injury.

EJECTION SPEEDS.

Figure 3–4C includes maximum safe ejection speeds, based on parachute restrictions, for the combination of one second safety belt and zero second parachute or a one second safety belt and one second parachute. The sequence lines (slanting lines) indicate the limits above which the parachute will probably be damaged on opening, or the pilot will probably suffer body injury resulting from parachute opening shock.

WARNING

On aircraft not modified by drilling holes in the canopy arms (T.O. 1F-84-763), it is recommended, if practical, that the aircraft be flown at a minimum speed of 300 knots IAS when jettison of the canopy is initiated. If the canopy is opened and the squibs do not detonate at speeds below 300 knots IAS, acceleration to 300 knots IAS or better may not break the canopy free from the aircraft. If the canopy fails to open by means of the canopy jettison control, the manual unlock system should be used to open the canopy locks. The speed should always be a minimum of 300 knots IAS when the manual unlock system is used to jettison the canopy. The aerodynamic loads at 300 knots IAS should offer a safe assurance of both opening the canopy and shearing the pins at each forward arm. In the further event of seat catapult failure, the aircraft speed after canopy jettison should be reduced to approximately 150 knots IAS, at which time the pilot should manually release his seat belt and leave the aircraft in an inverted position.

This page intentionally left blank.

SAFE EJECTION SPEEDS ALTITUDE VS SEQUENCE (LEVEL FLIGHT)

This graph depicts safe ejection speeds for ideal level flight and average parachute performance conditions only; other ejection altitudes, tumbling, separation delays, variations in parachute opening time, etc., are not included.

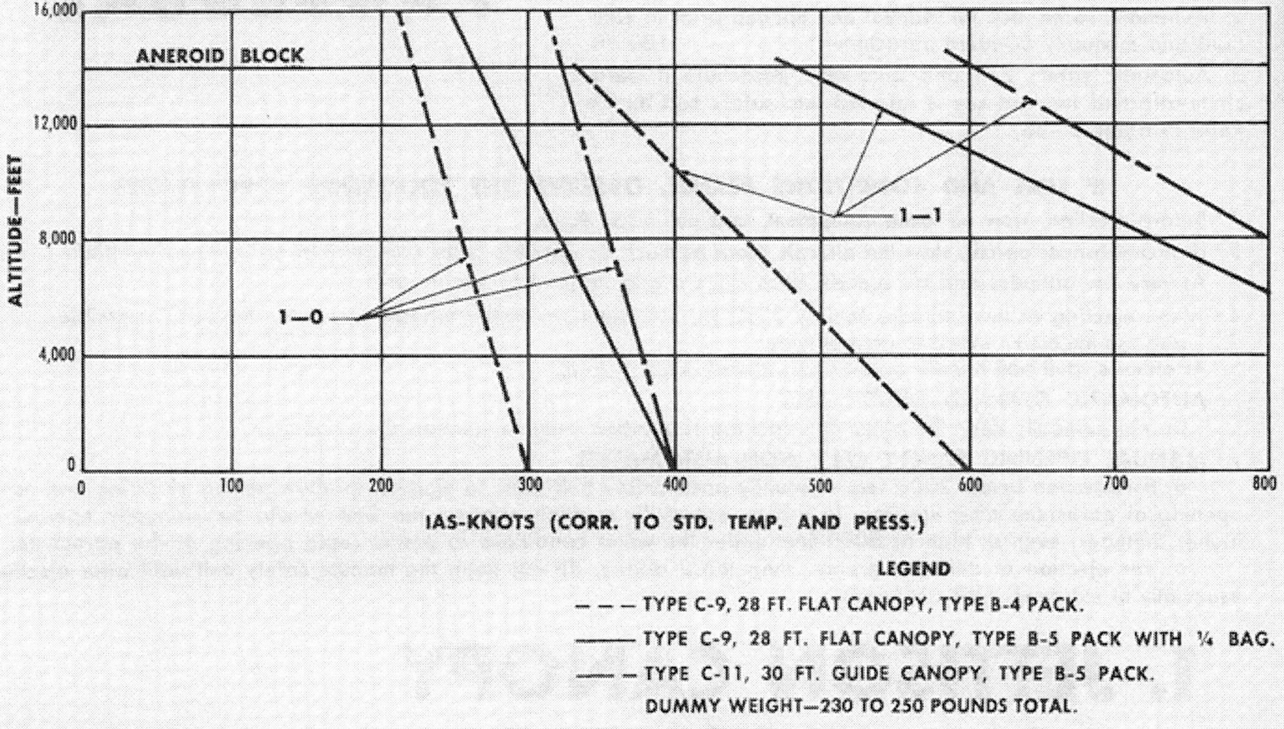

Figure 3—4C

FUEL SYSTEM FAILURE.

Aircraft fuel system failure is indicated by premature illumination of the wing or forward pump pressure warning lights or the fuel system booster pressure warning light. The only indication of an engine fuel system component failure will be a loss of RPM or flame-out. (See figure 3–7).

EXTERNAL TANKS FAIL TO FEED.

1. Receiver door switch—Check closed.
2. AR AMP & CONTR or IFR AMP & CONT circuit breaker—Check in.
3. IFR TEST & FUEL SHUTOFF circuit breaker—Pull (C and later).

IF PYLON TANKS STOP FEEDING.

If the pylon tanks stop feeding prematurely, the cause usually will be due to failure of the tanks to pressurize. The air pressure and vent valves for the tanks are spring-loaded to closed position. Any interruption in electrical current to these valves will cause the fuel transfer from the tanks to stop. On C and later aircraft where the single-point-refueling system is installed, either the AR AMP & CONTR, or the IFR AMP & CONT, circuit breaker must be pushed in for the air-pressure vent valves to be energized. Attempt to obtain fuel transfer as follows:

1. AR AMP & CONTR or IFR AMP & CONT circuit breaker—Check in.
2. Actuate the pylon tank air pressure switches.
3. Cycle the air refueling receptacle door.
4. Execute a series of zooms and dives.
5. Descend to a lower altitude.
6. Cycle the fuel shutoff valves.
7. Pull IFR AMP & CONT circuit breaker (C and later).
8. Pull IFR TEST & FUEL SHUTOFF circuit breaker and reset the IFR AMP & CONT circuit breaker (C and later).

MAIN TANK BOOSTER PUMP FAILURE.

Main tank booster pump failure is indicated by illumination of the fuel system booster pressure warning light (generator out indicator light not illuminated) when operating with the fuel tank selector in the ALL TANKS position. If a main tank booster pump failure is accompanied by power loss (RPM drop or excessive RPM fluctuation), proceed as follows to recover all fuel.

EJECTION PROCEDURE

MINIMUM SAFE EJECTION ALTITUDES

Established minimum safe ejection altitudes for all combinations of automatic or manual seat belts and parachutes must be observed.

1. Manual safety belt and any type parachute 2000 ft.
2. Automatic safety belt (or manual belt opened prior to ejection) and manually actuated parachute 1000 ft.
3. Automatic safety belt and automatic parachute if parachute-attached lanyard key is inserted into safety belt buckle Refer to figure 3—3B.

IF TIME AND CONDITIONS PERMIT, OBSERVE THE FOLLOWING PRECAUTIONS

1. Before ejection, stow all loose equipment, and pull visor down.
2. If circumstances permit, slow the aircraft down as much as possible after canopy jettison, prior to ejection.
3. At very low altitudes and air speeds, hook ripcord grip lanyard to ripcord grip
4. When ejecting at low altitudes (below 2000 ft), pull the nose of the aircraft above the horizon if possible, and use excessive speed to gain altitude.
5. At altitude, pull ball handle on bail-out bottle (Green Apple).
6. **AUTOMATIC OPENING SAFETY BELT**
   Do not manually open the automatic opening safety belt prior to ejection at any altitude.
7. **MANUAL OPENING SAFETY BELT (NON-AUTOMATIC)**
   a. For ejection below 2000 feet, manually open safety belt prior to ejection to allow pulling of D-ring and rapid opening of parachute after ejection. In a dive, especially at high speeds, the belt should be manually opened at higher altitudes, even as high as 5000 feet under the worst conditions to permit rapid opening of the parachute.
   b. For ejection at altitudes greater than those above, do not open the manual safety belt until after ejection, especially at extremely high airspeeds.

1. JETTISON CANOPY
2. ASSUME EJECTION POSITION
3. EJECT

AFTER THE SEAT EJECTS

Aircraft With Automatic Opening Safety Belts:

1. In normal operation the safety belt will separate in one or two seconds after ejection allowing separation from the seat. With the arming lanyard key inserted in the belt the parachute will open automatically at a preset altitude or if below the altitude the parachute will open one or two seconds after separation from the seat.
2. If the safety belt fails to open automatically after one second, manually unfasten the safety belt and kick free of the seat. Then pull the parachute arming lanyard if above 14,000 feet, or the ripcord grip if below 14,000 feet altitude.
3. Manually pull the ripcord grip immediately following separation from the seat for all ejections below 14,000 feet altitude.
4. Kick free of the seat and pull the parachute arming lanyard if wearing an automatic opening parachute WITHOUT arming lanyard key inserted into the safety belt buckle.

If wearing a manually operated parachute, pull the ripcord grip at altitude where normal breathing is possible.

Aircraft Without Automatic Opening Safety Belts:

1. Unfasten safety belt and kick free of seat.
2. If wearing a conventional manually operated parachute, pull the ripcord grip at altitude where normal breathing is possible. If ejection takes place at safe breathing altitude, pull ripcord grip two seconds after separating from seat.
3. If wearing an automatic parachute, pull the arming lanyard manually when kicking free of seat. Parachute will open at a preset altitude. If below the preset altitude the parachute will open one or two seconds after pulling the arming lanyard.

Figure 3—5 (Sheet 1 of 2)

A thru F

Canopy jettison control and seat ejection control provided on right side of seat only.

1 Jettison canopy (battery switch on) by pulling canopy jettison control all the way up until it locks (minimum speed 300 knots IAS, if practicable). Pull up left handgrip (this locks shoulder harness and releases left armrest).

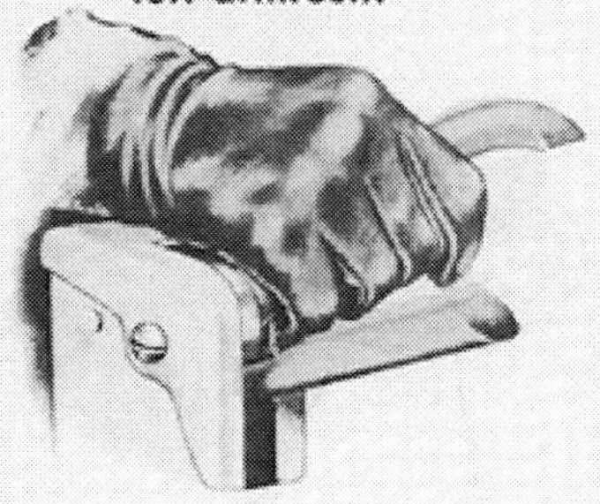

2 Place feet firmly on footrests and brace arms on armrests. Sit erect with head hard against headrest and chin tucked in.

3 Squeeze seat ejection control to eject seat.

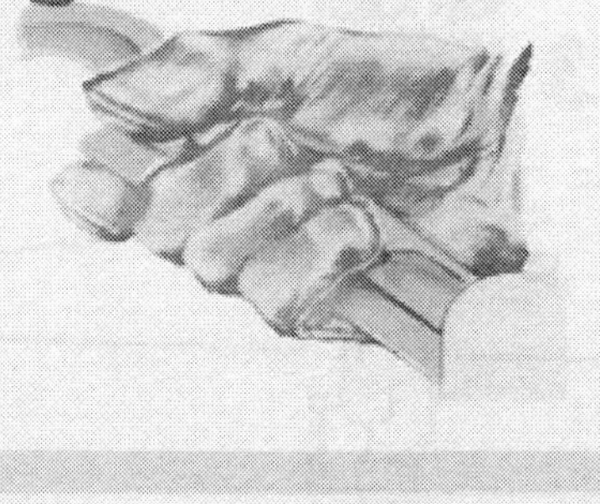

G and LATER

Canopy jettison control and seat ejection control provided on both left and right sides of seat.

1 Jettison Canopy (battery switch on) by pulling both or either canopy jettison control all the way up until it locks (minimum speed 300 knots IAS). Shoulder harness locks automatically.

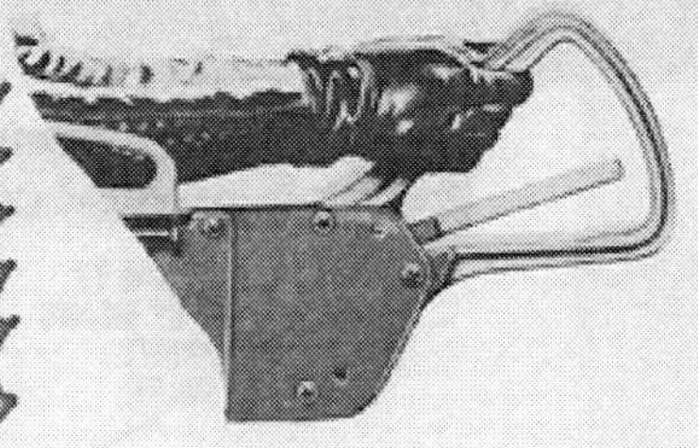

2 Place feet firmly on footrests and brace arms on armrests. Sit erect with head hard against headrest and chin tucked in.

3 Squeeze either seat ejection control to eject seat.

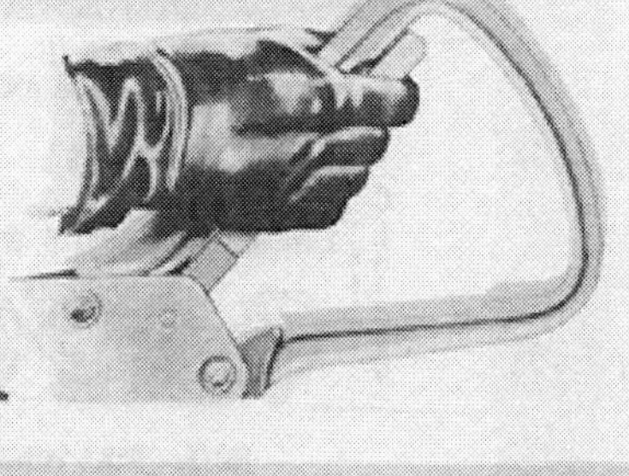

IF CANOPY FAILS TO JETTISON

Attempt to release it manually by opening it and raising it into the airstream (minimum speed 300 knots IAS, if practicable). As a last resort, after all attempts to release canopy have been made, the seat may be ejected through the canopy providing that the canopy is closed and preferrably locked. When using this procedure it is recommended that the seat be adjusted to the bottom position to allow the greatest possible clearance between the top of the head and the top of the head rest. In the event that the canopy opens and does not leave the aircraft and the bow is in the path of ejection, do not eject, use other means to leave aircraft.

WARNING

Do not accidentally squeeze the seat ejection control until ready to eject the seat.

IF SEAT FAILS TO EJECT

Reduce speed to approximately 150 knots IAS, manually release safety belt and leave the aircraft in an inverted position.

Figure 3—5 (Sheet 2 of 2)

FUEL SYSTEM FAILURE

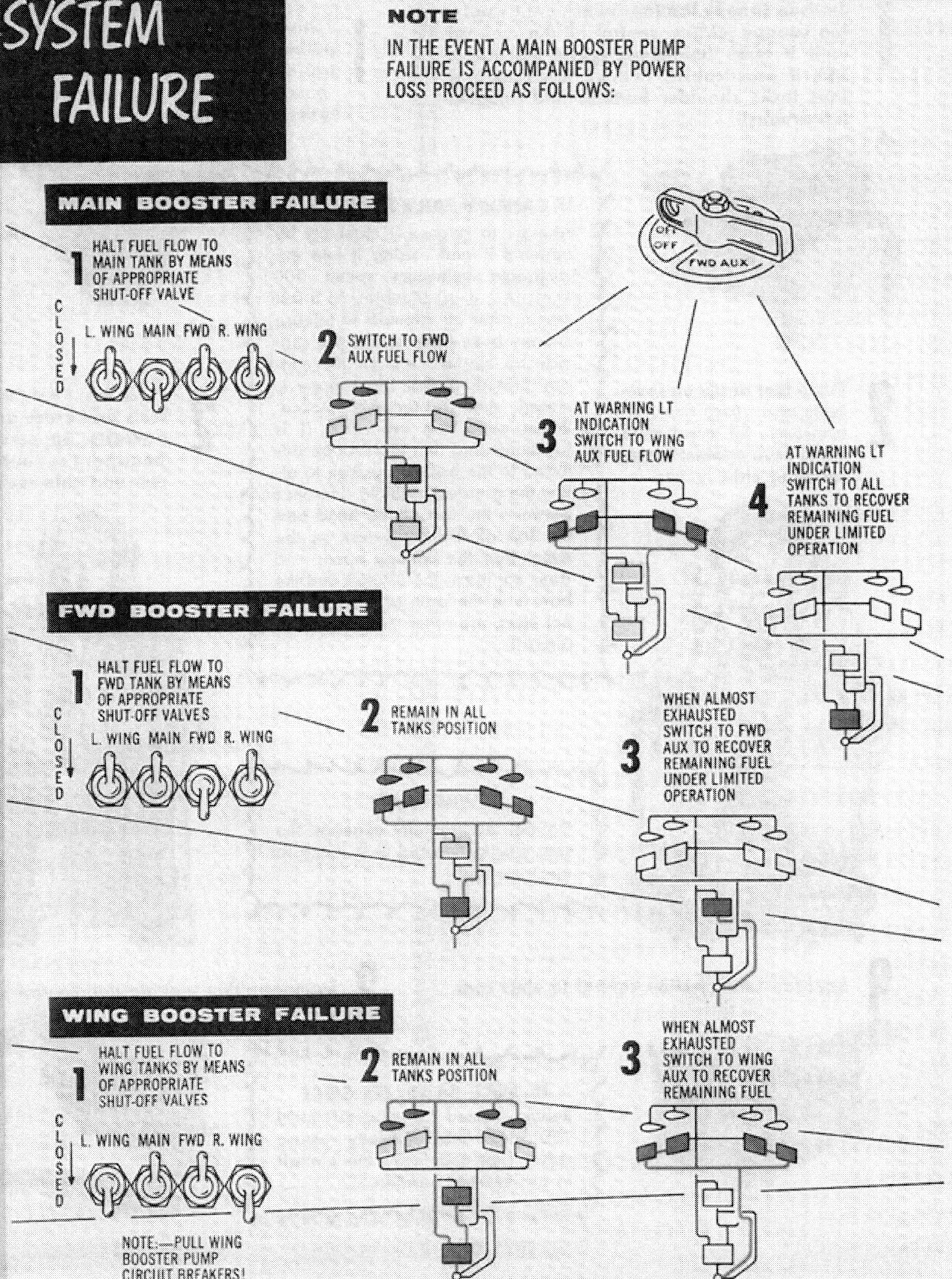
NOTE
IN THE EVENT A MAIN BOOSTER PUMP FAILURE IS ACCOMPANIED BY POWER LOSS PROCEED AS FOLLOWS:
MAIN BOOSTER FAILURE
1 HALT FUEL FLOW TO MAIN TANK BY MEANS OF APPROPRIATE SHUT-OFF VALVE
CLOSED
L. WING MAIN FWD R. WING
2 SWITCH TO FWD AUX FUEL FLOW
OFF
OFF
FWD AUX
3 AT WARNING LT INDICATION SWITCH TO WING AUX FUEL FLOW
4 AT WARNING LT INDICATION SWITCH TO ALL TANKS TO RECOVER REMAINING FUEL UNDER LIMITED OPERATION
FWD BOOSTER FAILURE
1 HALT FUEL FLOW TO FWD TANK BY MEANS OF APPROPRIATE SHUT-OFF VALVES
CLOSED
L. WING MAIN FWD R. WING
2 REMAIN IN ALL TANKS POSITION
3 WHEN ALMOST EXHAUSTED SWITCH TO FWD AUX TO RECOVER REMAINING FUEL UNDER LIMITED OPERATION
WING BOOSTER FAILURE
1 HALT FUEL FLOW TO WING TANKS BY MEANS OF APPROPRIATE SHUT-OFF VALVES
CLOSED
L. WING MAIN FWD R. WING
2 REMAIN IN ALL TANKS POSITION
3 WHEN ALMOST EXHAUSTED SWITCH TO WING AUX TO RECOVER REMAINING FUEL
NOTE:—PULL WING BOOSTER PUMP CIRCUIT BREAKERS!

Figure 3—6 (Sheet 1 of 2).

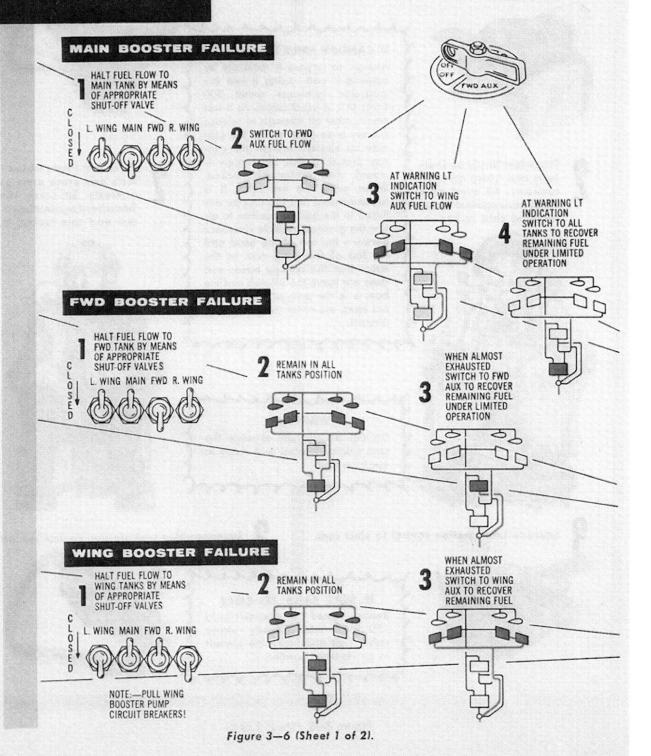

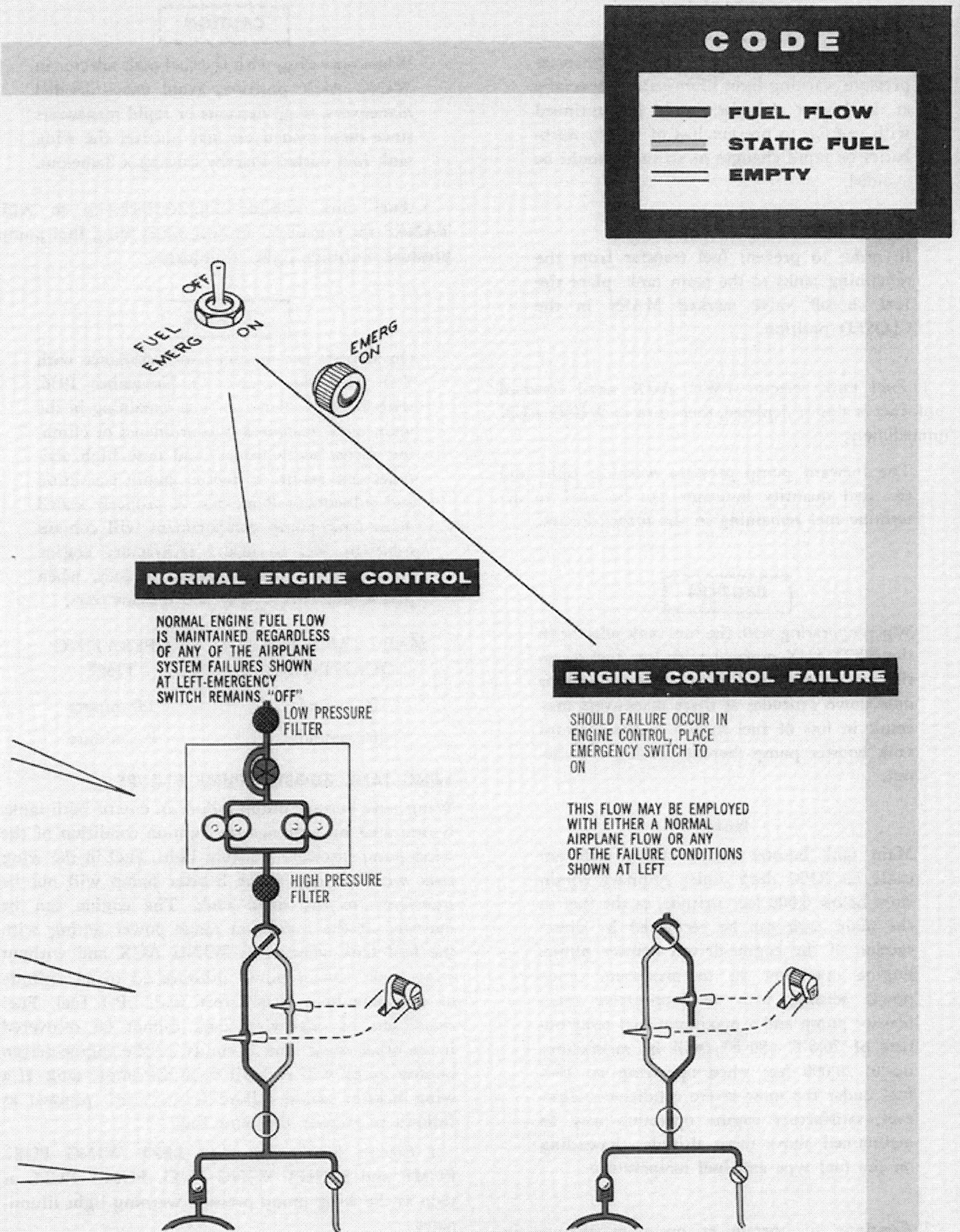
CODE
FUEL FLOW
STATIC FUEL
EMPTY
OFF
FUEL
EMERG
ON
EMERG
ON
NORMAL ENGINE CONTROL
NORMAL ENGINE FUEL FLOW IS MAINTAINED REGARDLESS OF ANY OF THE AIRPLANE SYSTEM FAILURES SHOWN AT LEFT-EMERGENCY SWITCH REMAINS "OFF"
LOW PRESSURE FILTER
HIGH PRESSURE FILTER
ENGINE CONTROL FAILURE
SHOULD FAILURE OCCUR IN ENGINE CONTROL, PLACE EMERGENCY SWITCH TO ON
THIS FLOW MAY BE EMPLOYED WITH EITHER A NORMAL AIRPLANE FLOW OR ANY OF THE FAILURE CONDITIONS SHOWN AT LEFT

Figure 3—6 (Sheet 2 of 2).

CAUTION

Engine operation with the fuel system booster pressure warning light illuminated, especially at the higher altitudes should be continued with caution to prevent loss of thrust. Aerobatics or rapid changes in altitude should be avoided.

Note

In order to prevent fuel transfer from the remaining tanks to the main tank, place the fuel shutoff valve marked MAIN in the CLOSED position.

1. Fuel tank selector—FWD AUX until forward tank fuel is almost depleted, then turn to WING AUX immediately.

The forward pump pressure warning light and the fuel quantity indicator can be used to determine fuel remaining in the forward tank.

CAUTION

When operating with the fuel tank selector in the FWD AUX position with low fuel quantity remaining, avoid uncoordinated turns or nose down attitudes as these maneuvers may result in loss of fuel supply to the forward tank booster pump thereby causing a flame-out.

Note

Main tank booster pump failure will not cause an RPM drop under ordinary conditions below 6,000 feet altitude, as the fuel in the main tank can be recovered by direct suction of the engine-driven booster pump. Engine operation up to maximum range power setting, with an inoperative main booster pump and a maximum fuel temperature of 26.6°C (80°F), will be satisfactory up to 20,000 feet when operating on JP-4 fuel under the most severe conditions. However, satisfactory engine operation may be maintained above these altitudes depending on the fuel type and fuel temperature.

2. Continue to operate at optimum altitude in WING AUX and descend to an altitude below 20,000 feet with JP-4 fuel before wing tank fuel is depleted.

CAUTION

When operating with the fuel tank selector in WING AUX position, avoid uncoordinated maneuvers, steep descents or rapid maneuvers since these maneuvers may uncover the wing tank fuel outlets thereby causing a flame-out.

3. Fuel tank selector—Counterclockwise to ALL TANKS for remainder of fuel when wing fuel pump pressure warning light illuminates.

CAUTION

On aircraft not modified in accordance with T.O. 1F-84-667, dated 15 November 1956, with 700 LBS of fuel or less remaining in the main tank, sustained uncoordinated or climbing turns, accelerations, and nose high attitudes can result in booster pump starvation and subsequent flame-out. A properly sealed main tank pump compartment will contain sufficient fuel to insure satisfactory engine operation, for the following periods, when performing any of the above maneuvers.

MAIN TANK FUEL QUANTITY	OPERATING TIME
700 pounds	1.0 minute
400 pounds	0.5 minute

WING TANK BOOSTER PUMP FAILURE.

Wing tank booster pump failure in one or both tanks is indicated by a premature light-on condition of the wing pump pressure warning light. Fuel in the wing tank with an inoperative booster pump will not be transfered to the main tank. The engine can be operated at the maximum range power setting with the fuel tank selector in WING AUX and without wing tank booster pump pressure up to an altitude of approximately 20,000 feet with JP-4 fuel. Fuel remaining in one wing tank cannot be recovered if the other wing tank is empty, as the engine-driven booster pump will suck air from the empty tank. If a wing booster pump failure is suspected, proceed as follows to recover the most fuel.

1. Circuit breakers marked LEFT WING FUEL PUMP and RIGHT WING FUEL PUMP—PULL as soon as the wing pump pressure warning light illuminates.

This will stop both pumps and prevent the wing tank, with the operating pump, from emptying.

Note

To prevent external fuel from transferring to the wing tanks actuate the L WING and R WING fuel shutoff valves to the CLOSED position.

2. Plan remainder of flight so as to land as soon as possible.

3. Fuel tank selector–ALL TANKS at optimum altitude until the main tank low level warning light illuminates, or if range is critical, until the fuel quantity indicator shows the main tank to be almost empty.

4. Descend to an altitude below 20,000 feet with JP-4 fuel and turn fuel tank selector to WING AUX.

CAUTION

When operating with the fuel tank selector in the WING AUX position, avoid uncoordinated maneuvers, steep descents or rapid maneuvers since these maneuvers may uncover the wing tank fuel outlets thereby causing a flame-out. Should the booster pressure light come on, pull both left and right wing tank booster pump circuit breakers.

FORWARD TANK BOOSTER PUMP FAILURE.

Forward tank booster pump failure is indicated by a premature light-on condition of the forward pump pressure warning light. Fuel in the forward tank can be recovered up to an altitude of approximately 20,000 feet with JP-4 fuel by operating with the fuel tank selector in the FWD AUX position. In event of a forward tank booster pump failure, proceed as follows to recover all fuel.

1. Fuel tank selector–ALL TANKS at optimum altitude until the main tank low level warning light illuminates, or if range is critical, until the fuel quantity indicator shows the main tank to be almost empty.

Note

To prevent external fuel from transferring to the forward tank, actuate the FWD fuel shutoff valve to the CLOSED position.

2. Plan remainder of flight so as to land as soon as possible.

3. Descend to an altitude below 20,000 feet with JP-4 fuel and turn fuel tank selector to FWD AUX.

CAUTION

When turning fuel tank selector from ALL TANKS to FWD AUX do not hesitate when in the WING AUX position as the wing tanks will be empty and a flame-out will result. When operating with the fuel tank selector in the FWD AUX position with low fuel quantity remaining, avoid uncoordinated turns or nose down attitudes as these maneuvers may result in loss of fuel supply to the forward tank booster pump, thereby causing a flame-out.

ENGINE FUEL CONTROL SYSTEM FAILURE.

Failure of the engine fuel control system will be indicated by a loss of thrust with the fuel system booster pressure warning light OFF. Operate as follows:

1. Throttle–Retard to IDLE.
2. Emergency fuel switch–ON.

   If RPM has not dropped below 85 per cent and the pressure altitude is below 6,000 feet the transition from normal to emergency can be made directly without retarding the throttle.

3. Throttle–Advance slowly and smoothly to obtain desired RPM.

CAUTION

There is no acceleration or overspeed control in the emergency fuel system and no rapid accelerations should be made. Small movements of the throttle will result in large power changes.

4. Complete flight on emergency fuel system to preclude engine flame-out.
5. Land as soon as possible.

PYLON TANK JETTISON.

Refer to Section V for recommended speeds for normal release of external tanks.

A THRU G

NORMAL RELEASE.

The pylon tanks are released in the following manner by use of the bomb release switch.

1. Inboard bomb selector switch–SINGLE.

CAUTION

If the inboard bomb selector switch is placed in the ALL position both inboard tanks will drop simultaneously thereby colliding and causing damage to the fuselage.

2. Outboard bomb selector switch—ALL.
3. Bomb release selector switch—MANUAL RELEASE.
4. Depress bomb release switch to drop outboard and left inboard tanks. Depress bomb release switch again to drop right inboard tank.

H and LATER

NORMAL RELEASE.

The pylon tanks are released in the following manner by use of the bomb release switch.

1. Armament selector switch—OUTB'D BOMBS SALVO.
2. Bomb release selector switch—MANUAL.
3. Bomb release switch—Depress momentarily.
4. Armament selector switch—INB'D BOMBS SING.

CAUTION

If the armament selector switch is placed in the INB'D BOMBS SALVO position both inboard tanks will drop simultaneously thereby colliding and causing damage to the fuselage.

5. Depress bomb release switch to drop left inboard tank. Depress bomb release switch again to drop right inboard tank.

EMERGENCY JETTISON.

The inboard or outboard pylon tanks can be jettisoned individually or simultaneously as conditions dictate. To jettison individually, actuate the inboard pylon jettison or the outboard pylon jettison switch to the JETT position. This also jettisons the respective pylons. To jettison the pylon tanks simultaneously, depress the external stores jettison switch. The inboard and outboard pylon tanks will be jettisoned together with the pylons on unmodified -RE aircraft up to F-84F-45RE and all GK aircraft. On F-84F-45RE and subsequent RE aircraft and on aircraft modified to incorporate T.O. 1F-84F-509 only the pylon tanks will be jettisoned and the pylons will be retained. Rockets will be jettisoned on all aircraft if airborne.

ELECTRICAL POWER SUPPLY SYSTEM FAILURE.

COMPLETE ELECTRICAL FAILURE.

If the electrical failure is complete, the emergency hydraulic pump is inoperative, the landing gear will have to be extended by emergency procedure, flaps and speed brakes cannot be selected, the drag chute is not available, and only unboosted brakes are available. (See figure 3–7.)

Fuel System Operation with Complete Electrical Failure.

If the electrical failure is complete, the booster pumps and fuel system indicators will be inoperative and it will not be possible to select the emergency fuel system. At altitudes below 20,000 feet with JP-4 fuel satisfactory engine operation up to maximum range power settings can be maintained with fuel temperatures up to 26.6°C (80°F). Operate fuel system as follows.

1. Fuel tank selector—FWD AUX, until forward tank fuel is used.
2. Fuel tank selector—WING AUX until wing tank fuel is used.
3. Fuel tank selector—ALL TANKS for remainder of fuel in main tank.

WARNING

With complete electrical failure, fuel in the pylon tanks cannot be recovered because the solenoid operated air shut-off valves will be closed, fuel warning lights are inoperative and fuel consumption from the fuel tanks will have to be estimated. Land as soon as possible.

Equipment Available with Complete Electrical Failure.

If the electrical failure is complete, the following systems and instrument readings are available. (See figure 3–7.)

a. Tachometer.

b. Exhaust gas temperature indicator.

c. Canopy jettison.

d. AN/ARC-34 radio.

e. Unboosted brakes.

ELECTRICAL POWER SUPPLY SYSTEM FAILURE

✓ Power available for operation * Power available *only* with battery switch ON

CONTROL OR EQUIPMENT	COMPLETE ELECTRICAL FAILURE	GENERATOR FAILURE	MAIN INVERTER FAILURE	ALTERNATE INVERTER FAILURE
TACHOMETER		No electrical power required		
EXHAUST GAS TEMP. IND.		No electrical power required		
CANOPY JETTISON	✓	✓	✓	✓
AN/ARC-34 RADIO (H & LATER)	✓	✓	✓	✓
DRAG CHUTE		✓		✓
EXTERNAL STORES JETTISON SWITCH		✓	✓	✓
EMERG HYDRAULIC PUMP		✓ *	✓	✓
STABILIZER ACTUATOR (ELECTRIC) (NON-TANDEM)		✓ *	✓	✓
AN/ARC-33 (A THRU G) COMMAND RADIO		✓ *	✓	✓
EMERGENCY FUEL SWITCH & INDICATOR		✓ *	✓	✓
AIR START SWITCH		✓ *	✓	✓
ROCKETS JETTISON		✓ *	✓	✓
FUEL PRESSURE WARNING LIGHT		✓ *	✓	✓
LANDING FLAP SELECTOR		✓ *	✓	✓
LANDING GEAR SELECTOR & BOOST BRAKES		✓ *	✓	✓
LANDING GEAR WARNING LIGHTS and HORN		✓ *	✓	✓
SPEED BRAKE SWITCH		✓ *	✓	✓
INVERTER FAIL WARN LIGHT		✓ *	✓	✓
TURN AND SLIP INDICATOR		✓ *		
ATTITUDE INDICATOR		✓ *	✓	
FUEL FLOW GAGE		✓ *	✓	
FUEL QUANTITY INDICATOR		✓ *	✓	Gage will tend to indicate last reading prior to failure.
DIRECTIONAL INDICATOR		✓ *	✓	
OIL PRESSURE GAGE		✓ *	✓	
HYD. PRESSURE GAGE		✓ *	✓	

Figure 3—7

INVERTER FAILURE.

Inverter failure is noted if the inverter failure indicator light illuminates while the generator out indicator remains out. (See figure 3–7.)

1. Turn the instrument power switch to ALT position.

   If the light does not go out in a few seconds, both inverters have failed.

CAUTION

In the event of failure of the main inverter the directional indicator, the oil pressure gage, hydraulic pressure gage, the fuel flow indicator and the attitude indicator, will tend to remain at their last reading. The above instruments will return to normal operation when the instrument power is placed in the ALT position.

2. Land as soon as possible.

GENERATOR FAILURE.

If generator cuts out of the electrical system, as indicated by the generator out indicator light, accomplish the following:

1. Reduce electrical load.

CAUTION

No attempt should be made to change channels on the command radio as 27 bolts are required for the change. If change in channels is attempted the set may hang up between channels.

2. Instrument power switch–ALT.

   (See figure 3–7 for instruments that are operative.)

3. Generator switch–RESET for a few seconds then return to ON.

   Light should go out and voltmeter read normal voltage.

4. If light remains on–Turn off all electrical equipment possible to conserve battery for necessary electrical operations and land as soon as possible.

monitor engine carefully for high EGT, explosion, loss of RPM, etc

Note

If the generator-out indicator light remains illuminated and the voltmeter and loadmeter readings are normal, a possibility exists that the wire connecting the generator reverse current relay to the coil of the generator ON relay has broken. In this event leave the generator switch in the ON position to permit the generator to remain in the system.

5. If generator power is available–Return instrument power switch to NOR and resume normal operation.

Fuel System Operation During Generator Failure.

In the event of generator failure only, the fuel warning lights will illuminate as they are powered from the primary bus. The fuel tank booster pumps will be inoperative as they are powered from the secondary bus. In this case fuel will not be transferred to the main tank when operating in the ALL TANKS position. The wing and forward tank pressure warning lights will be illuminated when operating in the WING AUX or FWD AUX position. Both the wing or forward and the fuel system pressure warning lights will be illuminated regardless of the position of the main tank fuel shutoff switch as fuel is being sucked from the respective tank by the engine-driven booster pump. Operate the fuel system in the same manner as for complete electrical failure, except that the external tank fuel can be recovered if the respective air pressure switches are in the INB'D PYLON TANKS AIR PRESS or OUT'B PYLON TANKS AIR PRESS position.

CAUTION

Engine operation with the fuel system booster pressure warning light illuminated, especially at the higher altitudes should be continued with caution to prevent loss of thrust. Aerobatics or rapid changes in altitude should be avoided.

Note

Fuel in the pylon tanks can be recovered as the air pressure solenoid shut-off valves are energized by the primary bus.

TANDEM

HYDRAULIC POWER SUPPLY SYSTEM FAILURE.

In the event of utility system pressure failure, all hydraulic systems except the aileron and stabilator will be inoperative. These two control systems will continue to be actuated by the engine-driven power system pump. The rudder will be actuated through the mechanical linkage from the rudder pedals to the rudder. The landing gear can be extended by means of pneumatic pressure. In the event of power hydraulic system failure all hydraulic systems will remain operative. If both utility and power systems are inoperative due to hydraulic pressure failures, the emergency hydraulic system must be selected for aileron and stabilator system operation. A failure in the power hydraulic tubing or reservoir, however, will not be corrected by emergency hydraulic system operation.

TANDEM

FLIGHT CONTROLS SYSTEM FAILURE.

TANDEM

HIGH STICK FORCES.

There is a possibility of an electrical malfunction in the autopilot amplifier box causing excessively high stick forces due to autopilot engagement. In the event such a condition should be encountered during flight, the pilot should turn the instrument power switch to the ALT position to eliminate the autopilot from the system.

TANDEM

UTILITY HYDRAULIC SYSTEM FAILURE.

The power hydraulic system will provide adequate pressure for continued flight. The rudder will be controlled by mechanical linkage and the spoilers will be inoperative. The rudder will buffet slightly at high speed with this type of failure as the rudder system is not irreversible when operating without hydraulic pressure. A landing should be made as soon as possible. When preparing to land it is recommended as an added safety precaution that the emergency hydraulic system be turned ON.

TANDEM

POWER HYDRAULIC SYSTEM FAILURE.

In the event of a line failure in the power hydraulic system, the emergency hydraulic system will be inoperative as these two systems use common lines. If a line failure is suspected in the power hydraulic system, this condition can be further checked by turning the emergency pump on and noting the pressure recorded on the power hydraulic system gage. If the gage indicates that the emergency hydraulic pump has restored operating pressure to the power system, the failure is probably due to power hydraulic pump malfunction. If pressure is not regained a line failure can be suspected. Turn emergency pump off after ascertaining where failure has occurred. The utility hydraulic system will provide adequate pressure for continued flight. However, a landing should be made as soon as possible. When preparing to land it is recommended that the emergency hydraulic system be turned ON if the failure of the power hydraulic system was caused by the power pump. If the failure is such that the emergency system is inoperative, the pneumatic compressor and spoilers should be turned off.

TANDEM

UTILITY AND POWER HYDRAULIC SYSTEM FAILURE.

An engine flame-out will cause both the utility and power hydraulic system pressure to drop considerably. However, the aircraft can be controlled adequately with hydraulic pressure and flow developed by a windmilling engine. Duration of the emergency hydraulic pump when operating on the aircraft's battery will be zero to 15 minutes depending on condition of battery and equipment operating. The emergency hydraulic system should be turned on prior to landing to supplement the windmilling engine-driven pumps.

WARNING

If both utility and power hydraulic pressure gages read zero and engine is operating normally the flight controls must not be moved until the emergency system is turned ON and hydraulic pressure is regained as shown on the power hydraulic pressure gage. If hydraulic pressure is not regained the surfaces will tend to creep in the direction of applied air loads. If hydraulic pressure is not regained abandon aircraft after ascertaining that gage pressure indications are realistic and that hydraulic failure has occurred. This can be accomplished by slightly operating the ailerons to determine if hydraulic pressure is available.

Note

Both utility and power hydraulic pressure gages are powered from the A C bus. In the event of an A C power failure both gages will be inoperative and remain at the last pressure reading.

FLIGHT CONTROLS NON-TANDEM

SYSTEMS OPERATION SCHEMATIC

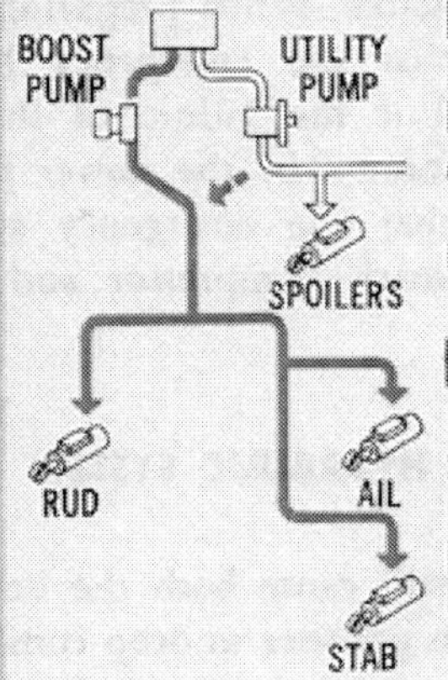

NORMAL OPERATION

BOOST PUMP SUPPLIES PRESSURE FOR HYDRAULIC OPERATION OF RUDDER, AILERONS AND STABILATOR

UTILITY PUMP SUPPLIES UTILITY SYSTEM ONLY

BOOST PUMP FAILURE

IN THE EVENT OF BOOST PUMP FAILURE PRESSURE IS SUPPLIED BY THE UTILITY PUMP FOR HYDRAULIC OPERATION OF THE RUDDER, AILERON AND STABILATOR AS WELL AS THE UTILITY SYSTEM

* NOTE

UNMODIFIED A THRU C ALTERNATE TRIM SWITCH IS NOT PROVIDED. POWER IS SUPPLIED DIRECTLY TO THE TRIM SWITCH SO THAT METHODS OF OPERATION A & B REMAIN AS DESCRIBED. METHOD C, IN THE ABSENCE OF THE SWITCH IS, OF COURSE, NOT AVAILABLE.

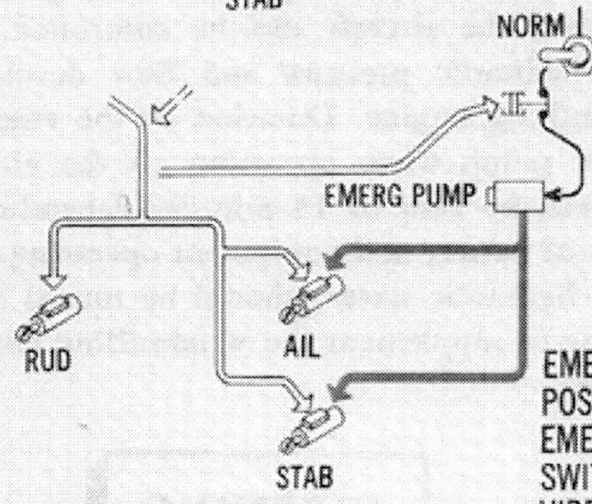

EMERGENCY OPERATION

WITH THE EMERGENCY PUMP SWITCH IN NORMAL POSITION EMERGENCY PRESSURE WILL BE SUPPLIED AUTOMATICALLY TO OPERATE AILERONS AND STABILATOR ON FAILURE OF UTILITY AND BOOST SYSTEMS

RUDDER OPERATION IS THRU MECHANICAL LINKAGE ONLY

EMERGENCY POSITION OF EMERG PUMP SWITCH PROVIDES DIRECT OPERATION OF EMERG SYS

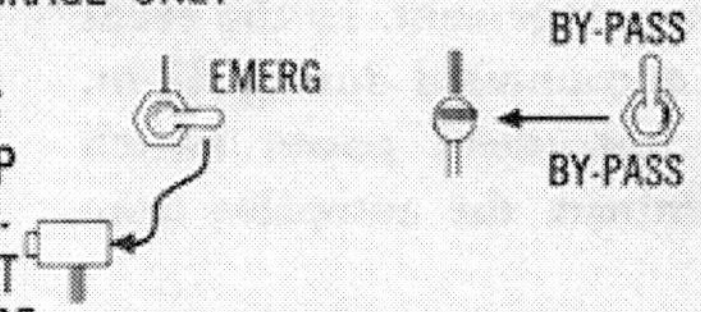

THE BY-PASS SWITCH DUMPS NORMAL PRESSURE TO CHECK EMERG HYD SYSTEM

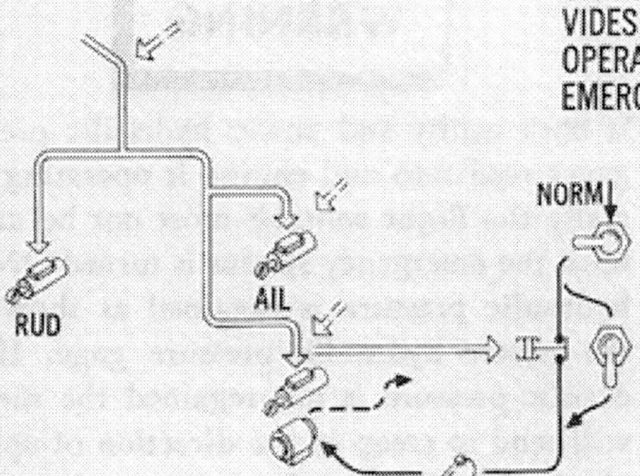

B THE PITCH CONTROL SWITCH PERMITS DIRECT SELECTION OF ELECTRICAL OPERATION

IF EMERG PRESS IS STILL AVAILABLE HYD OPERATION OF AILERONS WILL CONTINUE

ELECTRICAL OPERATION

A WITH ALTERNATE TRIM SWITCH IN "NORM" POSITION FAILURE OF ALL HYDRAULIC SYSTEMS AUTOMATICALLY ENERGIZES THE ELECTRICAL STABILATOR CONTROL SYSTEM WITH CONTROL THROUGH THE TRIM SWITCH

AILERONS AND RUDDER OPERATION IS THROUGH MECHANICAL LINKAGE ONLY.

C IN THE "OFF" POSITION THE ALTERNATE TRIM SWITCH MAY BE EMPLOYED TO CONTROL THE STABILATOR BY MOVING IT TOWARD THE "NOSE D'N" OR "NOSE UP" POSITIONS

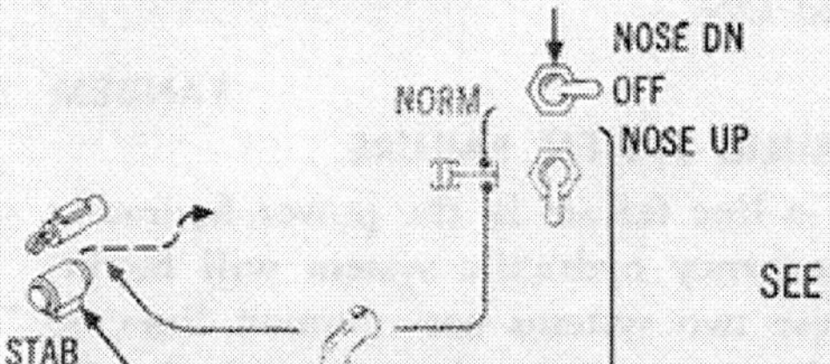

SEE NOTE*

Figure 3—8

TANDEM

TRIM SWITCH FAILURE.

In the event of trim switch failure, which may cause a "runaway" trim or no trim the following procedure is recommended to isolate the trim switch from the circuit:

1. Alternate trim switch (PITCH TRIM)—OFF.
2. Alternate trim switch—NOSE UP or NOSE DN position to trim aircraft about pitch axis. Trim about the roll axis cannot be controlled.

TANDEM

LANDING FLAPS SYSTEM FAILURE.

In the event of utility system failure it will not be possible to lower the landing flaps, since the utility system cannot be operated by the power or emergency hydraulic systems and a hand pump is not installed in the aircraft. In the event of primary bus failure, the flap position indicator will not function and it will not be possible to operate the flap system as the selector valve will remain in the neutral position.

NON-TANDEM

HYDRAULIC POWER SUPPLY SYSTEM FAILURE.

In the event of failure of the boost system hydraulic pump, the utility pump will automatically supply pressure to the control surface actuators. In the event of failure of both the boost and utility hydraulic system pumps, the emergency hydraulic pump will automatically supply hydraulic pressure to the aileron and the stabilator actuators, if the emergency hydraulic pump switch is in the NORMAL position. If the emergency hydraulic system fails, stabilator control is automatically transferred to electrical actuation and is controlled by the control stick trim button, if the pitch control switch is in the NORMAL position, and on D and Later aircraft when the alternate trim switch is in the NORM position. The stabilator can also be controlled with alternate trim switch NOSE UP and NOSE DN positions on D and Later aircraft. The Landing gear is extended for landing by an air pressure system on all aircraft.

Note

The hand pump is basically used during ground maintenance. However, in an emergency, hydraulic systems powered by the utility pump may be actuated by the hand pump. Refer to figure 1—34 for the number of strokes of the hand pump required to operate each system.

NON-TANDEM

FLIGHT CONTROL SYSTEM FAILURE.

NON-TANDEM

STICK LOCK.

There is a possibility of the internal bearings binding in the stabilator actuator causing a momentary hesitation experienced in the control stick. In addition, speed brake operation can cause movements of the transfer valves causing a "stick-lock" condition. In the event of a "stick-lock", reverse the movement of the stick momentarily. If this does not clear the condition, immediately revert to the electrical operation of the stabilator by placing the pitch control switch in the STICK OVERRIDE or ELEC position, and jettison all external stores.

CAUTION

Until such time as the affected transfer valves are replaced by an improved valve, speed brake operation should be avoided whenever a stick-lock condition could become critical; for example, in very close formations or in maneuvers close to the ground.

NON-TANDEM

HIGH STICK FORCES.

There is a possibility of an electrical malfunction in the autopilot amplifier box causing excessively high stick forces due to autopilot engagement. In the event such a condition should be encountered during flight, the pilot should turn the instrument power switch to the ALT position to eliminate the autopilot from the system.

NON-TANDEM

STABILATOR SYSTEM FAILURE.

NON-TANDEM

BOOST PRESSURE FAILURE.

If the boost hydraulic pressure gage drops rapidly on A thru E aircraft not modified to the shut-off valve configuration while the utility gage retains pressure, it is probable that the boost pump drive shaft has sheared, or a leak has developed between the pump and the common line. If the cause is a leak, the utility gage will drop in a short period of time, as the boost pump will pump hydraulic fluid overboard and uncover the utility pump standpipe. However, if the failure is a sheared pump shaft, the utility system will automatically supply pressure to the stabilator, aileron and rudder actuators. When operating all systems from the utility pump, the pressure required for operation of the landing gear, flaps and speed brakes is diverted from the control system and may cause a momentary control stiffness. If the boost gage drops slowly or fluctuates, indicating a leak in the boost system, the utility gage will drop shortly afterwards, with the time element depending on the extent of the leak. On F and later aircraft and aircraft modified to the shut-off valve configuration boost pump failure will not be indicated on the boost gage as the gage will show pressure in the boost system developed by the utility pump. In any event, when the boost gage fluctuates when operating a utility system or drops off, follow the procedure outlined in paragraph under Utility and Boost Pressure Failure.

NON-TANDEM

UTILITY PRESSURE FAILURE.

In the event that the utility system pressure gage drops or fluctuates the probability is that the utility pump drive shaft has sheared or a leak has developed in the utility system. In the event of a leak, hydraulic fluid will be pumped overboard until the level of the utility standpipe is reached. If the leak is in the common system the boost gage will fluctuate or drop shortly thereafter. If the cause is a sheared pump shaft, the boost system will continue to supply pressure to the control actuators while the utility systems will be inoperative. The landing gear will have to be extended by emergency procedure and the landing flaps can be lowered by means of the hydraulic hand pump. In any event, when the utility system gage fluctuates or drops off, follow the procedure outlined in paragraph under Utility and Boost Pressure Failure.

NON-TANDEM

UTILITY AND BOOST PRESSURE FAILURE.

1. In the event both boost and utility pressures fail, the emergency hydraulic system will automatically take over and power the stabilator and ailerons if the emergency hydraulic pump switch is in the NORMAL position. However, if the pressure switch does not function with failure of the normal hydraulic system, the emergency hydraulic system will not take over automatically but it can be actuated by positioning the emergency hydraulic pump switch to the EMERG position. The transition from boost or utility system to emergency hydraulic system will be smooth and automatic due to the emergency system accumulator providing a ready source of pressure, and pressure switches providing circuitry for transfer to the emergency hydraulic system.

2. If the boost and utility hydraulic system pressure gages drop simultaneously, on unmodified A thru E aircraft (not incorporating the shut-off valve), the cause could be an electrical malfunction which causes the dump valve to become energized and dump utility and boost hydraulic pressure. On F thru J aircraft and on modified earlier aircraft only, the boost pressure gage will drop when the shut-off valve is actuated. To isolate this difficulty, the aircraft can be trimmed for straight and level flight, at a safe altitude, and the circuit breaker marked: ROLL & PITCH TRIM & EMERG PITCH on (A thru E) aircraft; ROLL & PITCH TRIM & EMERG on F aircraft; and EMERG HYD CONT on G thru J aircraft; pulled momentarily. Normal hydraulic pressure should return when the circuit breaker is pulled and, in this instance, flight may be continued with the circuit breaker pulled and a normal landing effected as soon as possible. However, since the automatic transfer and the indicating systems are inoperative, the pilot must remain alert to reset the circuit breaker to regain electric power to operate the emergency hydraulic pump or electrical actuator in the event of a subsequent hydraulic pressure failure, as power for the electrical hydraulic pump and the electrical actuator is not available with the circuit breaker out. If pulling the circuit breaker does not cause hydraulic pressure to return to normal, the circuit breaker must be reset to permit control of the aircraft by either the emergency hydraulic system or the electrical actuator on all aircraft, and on A thru F aircraft permit roll and pitch trim circuits to be operative.

WARNING

When the circuit breaker is pulled out, on A thru F aircraft, the roll and pitch trim actuators are inoperative and trim about these axis will have to be overcome manually. A straight in approach should be used for landing to limit control movements to a minimum.

3. If operation on emergency hydraulic system becomes necessary, and a landing can be made within one-half hour, it is recommended the flight be continued on the emergency hydraulic system. To insure that the original failure does not affect the emergency hydraulic system it is recommended that whenever possible, flight be maintained on the emergency hydraulic system for a minimum of 10 minutes prior to landing. If a landing is not contemplated within one-half hour, electrical operation is recommended to avoid excessive use of the emergency hydraulic pump motor. In this event, turn emergency hydraulic pump to OFF *after* positioning the pitch control switch to STICK OVERRIDE or ELEC. Place emergency hydraulic system in operation at least 10 minutes prior

to touchdown. This shall be accomplished by first positioning the emergency hydraulic pump switch to NORMAL and the pitch control switch to NORMAL.

WARNING

Do not attempt electrical operation on aircraft prior to F-84F-40RE or F-84F-46GK with a trim switch failure. In this event, emergency hydraulic system operation is possible; however, aircraft trim must be overcome manually. If the emergency hydraulic system should fail on these aircraft and the trim switch has also failed, the aircraft pitch cannot be controlled. On F-84F-40RE or F-84F-46GK and later aircraft, electrical operation is possible with trim switch failure provided the alternate trim switch (stick switch power auxiliary electric switch) is positioned to OFF before placing the pitch control switch in the ELEC position. In this case longitudinal control must be maintained by selection of the NOSE UP and NOSE DN positions of the alternate trim switch on the left console. On these aircraft an emergency hydraulic system landing is recommended with a trim switch failure with the alternate trim switch in the OFF position. However, in this configuration, aileron and stabilator trim must be manually overcome.

4. A normal landing pattern should not be attempted when making a landing either on the emergency hydraulic system or electric control. A long flat final approach should be set up as described in Para. 5, below so that a safe landing can be accomplished with minimum movement of the stabilator. Although an electric-electric landing is marginal, a satisfactory landing may be made if landing conditions are favorable, however, the final decision as to whether an electric-electric landing is permissible must be decided by the Commander. If an electric landing is to be made, consideration should be given to the following:

a. Experience of pilot in aircraft.

b. Weather—Visibility and wind conditions. Due to poor lateral control, an electric landing is not recommended in strong cross winds or gusty winds.

c. Runway length and condition.

d. Approach to runway.

e. Fuel available.

f. Availability of runway barrier.

5. If electric operation of the stabilator is to be used during flight prior to landing, proceed as follows:

a. Jettison asymmetrical loads.

WARNING

To preclude the possibility of an uncontrollable roll-off in the event of hydraulic failure, asymmetric stores should be jettisoned before switching to STICK OVERRIDE or ELEC and turning the emergency hydraulic pump OFF.

WARNING

If symmetrical external loads are not jettisoned prior to switching to electric control, the loads should not be jettisoned until after the aircraft is returned to emergency hydraulic control. This is to avoid the possibility of an asymmetric loading while flying with manual aileron control in the event that stores on one side of the aircraft fail to release. Electrical operation of the stabilator with unboosted ailerons is inadequate for flight control with asymmetric loads.

b. Place the stabilator in electrical operation with manual aileron control as follows:

1) Pitch control switch—STICK OVERRIDE or ELEC.

2) Emergency hydraulic pump switch—OFF.

CAUTION

Actuate the trim switch to move the stick forward and aft, and simultaneously apply sufficient load on the stick to position the actuator control valve in the direction of stabilator motion.

c. Return aircraft to emergency hydraulic control at least 10 minutes prior to touchdown, as follows:

1) Emergency hydraulic pump switch—NORMAL.

2) Pitch control switch—NORMAL.

Note

When returning to emergency hydraulic control from electric control there may be a few seconds delay in restoration of hydraulic stabilator control. If this occurs, alternate slight forward and aft pressure should be applied to the control stick.

d. Continue to fly aircraft on emergency hydraulic system keeping changes in stabilator setting as small and as gradual as possible. Accomplish a landing as follows:

1) Jettison outboard tanks or any tanks with fuel. Retain empty inboard pylon tanks. In the event runway barrier engagement is anticipated, all external stores should be jettisoned prior to touchdown. However, all external store jettisoning restrictions must be observed.
2) Landing flaps—Down, if possible.
3) Landing gear—Down. Use emergency procedure while still in level flight.
4) Set up a long flat final approach at approximately 500 to 800 feet per minute descent. For approach speeds refer to approach and landing speed figure A-65 or A-151 of the Appendix.
5) Control rate of descent with the throttle. Abrupt changes in attitude are undesirable and should be avoided whenever possible. Changes in stabilator setting should be small with time allowed for the aircraft to respond. When changing power settings make changes smoothly.
6) Do not let airspeed get below recommended final approach speed until just before touchdown.
7) Make landing flare as gradual as possible or, if possible, gradually reduce power and let aircraft settle to contact ground without rounding out.

WARNING

The emergency hydraulic pump has a limited capacity, and abrupt motions of the control stick may cause momentary loss of pressure, with resulting engagement of the electric control. This may appear to the pilot as a stick-lock. Landing on electric control, the same procedure as above should be followed. However, the stabilator response in electrical operation, is too slow to permit round-out and in this case it is necessary to land by gradually reducing power and letting the aircraft settle to the ground. If approach is unsatisfactory, decide on go-around early, using back trim only if absolutely necessary, and only after full throttle is applied.

8) On ground contact, deploy drag chute if installed, reduce throttle to IDLE, and stopcock throttle, if deemed necessary.
9) Use recommended emergency braking technique until speed is reduced to taxi speed.

Note

When landing on either the emergency hydraulic system or in electric control, no boost is available for the brakes, and extremely hard pressure on the pedals is necessary to obtain braking effectiveness.

NON-TANDEM STABILATOR OPERATION WITH ENGINE FAILURE.

In the event of an engine failure the following procedure is recommended:

1. Emergency hydraulic pump switch—OFF.
2. Operate the ailerons and stabilator utilizing pressure from the normal system produced by the windmilling engine. Keep stick movements to a minimum.
3. Make a flame-out pattern landing as described in figure 3–3.

WARNING

If engine failure results in engine seizure the emergency hydraulic system will take over if the emergency hydraulic switch is in the NORMAL position. Duration of the emergency hydraulic pump when operating on the aircraft's battery will be from zero to 15 minutes depending on the condition of the battery and amount of other electrical equipment operating.

NON-TANDEM EMERGENCY HYDRAULIC PUMP FAILURE.

In the event of emergency hydraulic pump failure or if emergency hydraulic operation is not possible and normal hydraulic pressure is not available; the following procedure is recommended.

1. Pitch control switch—STICK OVERRIDE or ELEC.

WARNING

The STICK OVERRIDE or ELEC position must be selected to preclude a "stick-lock" condition. The control stick will remain locked until the stabilator is positioned with the trim switch if emergency hydraulic system failure has caused the electrical system to take over.

2. Fly the aircraft with manual aileron control and electrical stabilator control.

Make an electrical landing as described under UTILITY AND BOOST PRESSURE FAILURE or prepare to abandon the aircraft.

CAUTION

Actuate the trim switch to move the stick forward and aft and simultaneously apply sufficient load on the stick to position the actuator control valve in the direction of stabilator motion.

NON-TANDEM

TRIM SWITCH FAILURE.

NON-TANDEM A THRU C

WITH HYDRAULIC PRESSURE AVAILABLE.

Should trim switch failure occur on A THRU C aircraft, flight can be maintained using normal or emergency system hydraulic power. However, corrections in roll or pitch trim must be overcome manually. It will not be possible to operate the stabilator by means of the electric actuator, as this circuit is completed through the trim switch. Therefore, landings should not be attempted, if the trim switch has failed and the aircraft is on emergency hydraulic operation, because if the emergency system fails it will be impossible to control the aircraft.

NON-TANDEM A THRU C

WITHOUT HYDRAULIC PRESSURE AVAILABLE.

WARNING

If normal and emergency hydraulic pressure fails and the trim switch is inoperative it will be impossible to control the aircraft. Abandon the aircraft.

NON-TANDEM D THRU J

TRIM SWITCH FAILURE.

NON-TANDEM D THRU J

WITH HYDRAULIC PRESSURE AVAILABLE.

In the event of trim switch failure, which may cause a "runaway" trim or no trim, electrical actuation of the stabilator is recommended until preparing to land. This procedure relieves the pilot of maintaining constant stabilator trim loads on the control stick.

1. Alternate trim switch—OFF.
2. Emergency hydraulic pump switch—OFF.
3. Pitch control switch—ELEC.

WARNING

It is mandatory that the STICK OVERRIDE or ELEC position be used whenever the stabilator is operated electrically during flight. Failure to do so will result in a locked stabilator actuator.

4. Alternate trim switch—NOSE UP or NOSE DN to fly the aircraft until preparing to land.

CAUTION

When operating in this manner, aileron trim is inoperative.

5. Emergency hydraulic pump switch—NORMAL prior to landing.
6. Pitch control switch—NORMAL prior to landing.

CAUTION

The ailerons and stabilator will operate from the normal or emergency hydraulic system. However, aileron and stabilator trim will have to be overcome manually during the landing approach. Be prepared for a trim change when the hydraulic system takes over.

NON-TANDEM D THRU J

WITHOUT HYDRAULIC PRESSURE.

1. Alternate trim—OFF.
2. Alternate trim switch—NOSE UP or NOSE DN to fly the aircraft.
3. Fly the aircraft with manual aileron control and electrical stabilator control and make an electrical landing as described under UTILITY AND BOOST PRESSURE FAILURE or prepare to abandon the aircraft.

NON TANDEM

LANDING FLAPS SYSTEM FAILURE.

In the event of hydraulic pressure failure and before emergency extension of the landing gear the landing flaps can be extended by placing the landing flap lever in the DOWN position and by actuating the hydraulic hand pump. If the landing gear has been extended by the emergency system, approximately 80 more strokes of the hydraulic hand pump would be required before the flaps will move as pressure would have to be built up in the down side of the landing gear since the main gear extend by gravity. In the event of primary bus failure, the flap position indicator will be inoperative and it will not be possible to operate the flap system as the selector valve will be in the neutral position.

SPEED BRAKE FAILURE.

In the event of hydraulic failure or primary bus power failure the speed brake cannot be opened. However, if the speed brakes are open when the failure occurs the air loads on the brakes will close them.

LANDING GEAR SYSTEM FAILURE.

LANDING GEAR RETRACTION WHILE ON THE GROUND.

If it becomes necessary to retract the landing gear while the weight of the aircraft is on the main struts proceed as follows:

1. Emergency ground retract switch—Actuate.
2. Landing gear selector handle—UP.

Note

The landing gear will retract only if hydraulic pressure and primary bus power are available.

LANDING GEAR EXTENSION WHILE IN THE AIR.

In the event that the left or right main gear does not extend in flight by means of the main system and utility hydraulic pressure is available the following procedure is recommended:

1. Cycle the landing gear while pulling negative G's or while shaping the aircraft.

Pulling positive G's will normally aggravate this condition since the weight of the wheel rests on the inner landing gear door

Note

The gear cannot be retracted, if desired, once the EMERG DOWN position is selected.

In the event that only the main gear can be locked down with the normal hydraulic system, the possibility exists of back pressure in the return hydraulic lines unlocking the main landing gear whenever the nose gear is lowered by means of the pneumatic system. The main gear will relock itself through normal hydraulic pressure with the landing gear selector handle left in the EMERG DOWN position. Therefore, the landing gear position indicators must be checked to ascertain whether all gears are down and locked after the pneumatic system has been used in the lowering of any gear.

2. If the landing gear will not extend by normal procedures, indicated by the landing gear indicator lights, warning horn or low hydraulic pressure, proceed as follows:

a. Reduce speed to 220 knots IAS or below.

b. Landing gear selector handle—EMERG DOWN.

Note

If the landing gear does not lock down (indicated by the landing gear indicator lights) hold the landing gear selector handle in the EMERG DOWN position for a few seconds.

Note

Once the landing gear is extended by positioning the landing gear selector to the EMERG DOWN position you are committed to a wheels down landing since the selector handle can only be released from the EMERG DOWN position by use of a screwdriver. The possibility of blowing up the hydraulic reservoir due to back pressures when retracting the landing gear also exists. Boost brakes are not available.

c. If necessary, yaw the aircraft to lock the main gear.

d. Landing gear indicator lights—Check to ascertain that all gears are locked down.

Note

Observe the landing gear indicator lights when yawing the aircraft to determine when the spring-loaded downlocks on the main gear are engaged. The nose gear is extended by air pressure and the spring-loaded nose gear downlock is locked when the strut is fully extended as indicated by the nose gear indicator light.

BRAKE FAILURE.

1. Use up the fuel until minimum fuel remains (1000 lbs).
2. Throttle—CLOSED on touchdown.
3. If left brake is out, land on left side of runway.
4. If right brake is out, land on right side of runway.
5. If both brakes are out, land as short as possible at lowest safe speed, landing flaps DOWN, and speed brakes OUT.
6. Drag chute—Deploy after touchdown if installed.
7. Dragging the tail during landing will shorten the roll, if drag chute is not installed.

CONDENSED CHECK LIST.

Your condensed check list is now contained in T.O. 1F-84(25)F-(CL)1-1.

F-84F CONDENSED CHECK LIST

ENGINE FAILURE.

PROCEDURE ON ENCOUNTERING ENGINE FLAME-OUT.

1 Throttle–CLOSED.
2. Establish glide speed of 225 knots IAS.
3. Turn off all unnecessary electrical equipment.
4. Select fuel tank as required.
5. If it is suspected that flame-out occurred due to failure of the engine fuel control, place the emergency fuel switch in the ON position.
6. Descend to an altitude of 20,000 feet or below.

COMPLETE ENGINE FAILURE DURING TAKE-OFF.

Before Flying Speed Is Reached.

1. Throttle–CLOSED.
2. Drag chute–Deploy.
3. Fuel tank selector–OFF.
4. External stores–Jettison if runway overrun barrier engagement is not anticipated.
5. Brake to a stop on runway, if possible.
6. If impossible to stop on runway:
   a. Prepare to engage barrier, if available.
   b. If barrier is not available or landing gear is retracted–Jettison canopy and turn battery switch OFF.

After Becoming Airborne.

1. Landing gear selector–DOWN.
2. External stores–Jettison, if necessary.
3. Throttle–CLOSED.
4. Fuel tank selector–OFF.

T.O. 1F-84(25)F-1
15 FEBRUARY 1958
Changed 30 June 1958 1

5. Emergency hydraulic pump switch—HYD EMER PUMP (Tandem).
6. Canopy—Jettison.
7. Shoulder harness—Locked.
8. Land straight ahead—Change course only to miss obstacles.
9. Drag chute—Deploy, after touchdown.
10. Battery switch—OFF on ground contact.
11. Generator switch—OFF.

PARTIAL ENGINE FAILURE DURING TAFE-OFF.

1. If not airborne—Abort take-off.
2. Emergency fuel switch—ON.
3. External stores—Jettison (if airborne).
4. If power cannot be regained—Fuel tank selector in WING AUX or FWD AUX.
5. Go around and land.

COMPLETE ENGINE FAILURE DURING FLIGHT.

1. Throttle—CLOSED.
2. Landing flaps—UP.
3. Speed brakes—IN.
4. Trim aircraft for 225 knots IAS.
5. Attempt an air-start.
6. If an air-start is impossible—Make a forced landing or abandon the aircraft.

PARTIAL POWER FAILURE DURING FLIGHT.

1. Throttle—IDLE.
2. Emergency fuel switch—ON.
3. Throttle—Advance slowly and smoothly to obtain desired RPM.
4. Complete flight on emergency fuel system.
5. Land as soon as possible.

ENGINE AIR STARTING.

1. Air Start switch—Depress momentarily.
2. Throttle—Immediately move to IDLE or above (not exceeding 1/3 open).
3. RPM and exhaust gas temperature—When stabilized at idle speed advance throttle as desired.
4. If air start is unsuccessful:
    a. Throttle—CLOSED.
    b. Recheck fuel available.
    c. Reaccomplish engine air start procedure at 15,000 feet or below.
5. If start is again unsuccessful:
    a. Throttle—CLOSED.
    b. Fuel tank selector—Any position known to contain fuel.
    c. Attempt another air-start.

LOSS OF ENGINE OIL PRESSURE.

1. Land at first or nearest landing site.
2. If possible reduce to minimum power.
3. Do not vary RPM unnecessarily.
4. External stores—Jettison.
5. Fly as conservatively as possible.
6. Use flame-out pattern for landing.
7. Imminent failure will be indicated by noticeable steadily increasing vibration at which time engine should be shut down.

LANDING WITH DEAD ENGINE.

1. Throttle—CLOSED.
2. Fuel tank selector—OFF.
3. External stores—Jettison.

T.O. 1F-84(25)F-1
15 February 1958 3

4. Safety belt and shoulder harness tightened and inertia reel lock control locked.
5. Speed brake—IN.
6. Landing gear—UP.
7. Landing flaps—UP.
8. Pneumatic compressor—OFF.
9. Emergency hydraulic pump switch—EMERG at low key point unless needed to maintain control.
10. Landing gear—DOWN.
11. Flaps—DOWN if available.
12. Speed brake switch—OUT if available.

ENGINE FIRE DURING FLIGHT.

Engine Overheat Light On.

1. Reduce power immediately.
2. If light goes out—Continue flight at reduced power and land as soon as possible.
3. If light does not go out—Check for indications of fire.
4. If no fire is apparent—Continue flight at minimum power and land as soon as possible.
5. If possible indications of fire exist:
   a. Throttle—CLOSED.
   b. Fuel tank selector—OFF.
   c. Make decision to land or bail-out.

Engine Fire Warning Light On.

On Take-Off Run.

1. Throttle—CLOSED.
2. Drag chute—Deploy.

T.O. 1F-84(25)F-1
15 February 1958 **4**

3. Fuel tank selector—OFF.

4. Prepare to engage runway overrun barrier if available and stop as soon as possible.

Immediately After Becoming Airborne.

1. Maintain maximum power and climb immediately to safe ejection altitude.

2. Level off, then throttle back as far as possible.

3. If no evidence of fire exists—Proceed cautiously and land as soon as possible.

4. If evidence of fire exists:

a. Throttle—CLOSED.

b. Fuel tank selector—OFF.

c. Make decision to either bail-out or make an emergency landing.

During Flight If Altitude and Terrain Permit.

1. Throttle—CLOSED.

2. Fuel tank selector—OFF.

3. Turn off all unnecessary electrical equipment.

4. Check for any evidence of fire.

5. Do not attempt a restart if indication of fire is present.

6. If fire persists or even if it goes out—Prepare to abandon the aircraft or make an emergency landing.

7. If it is definitely determined that no fire exists, a restart may be accomplished.

If Altitude or Terrain Preclude Shutting Down the Engine Completely.

1. Reduce power and still maintain flight.

2. Check for evidence of fire.

3. If no evidence of fire exists—Proceed cautiously and land as soon as possible.

4. If evidence of fire exists:
   a. Throttle—CLOSED.
   b. Fuel tank selector—OFF.
   c. Make decision either to bail-out or make an emergency landing.

FIRE WHILE STARTING THE ENGINE.

1. Throttle—CLOSED.
2. Fuel tank selector—OFF.
3. Battery switch—OFF or disconnect external power supply.
4. Leave cockpit immediately.

ENGINE FIRE DURING TAKE-OFF — IF NOT AIRBORNE.

1. Throttle—CLOSED.
2 Drag chute—Deploy.
3. Fuel tank selector—OFF.
4. Brake to a stop on runway, if possible.
5. External stores jettison switch—Depress unless runway overrun barrier engagement is anticipated.
6. If impossible to stop on runway:
   a. Canopy—Jettison.
   b. Battery switch—OFF.
7. Leave cockpit as soon as possible.

ENGINE FIRE DURING TAKE-OFF — IF AIRBORNE.

1. If practicable after point of safe abort, climb to minimum safe ejection altitude.
2. Throttle back enough to maintain altitude.
3. Check for evidence of fire.
4. If no evidence of fire exists—Land as soon as possible.
5. If evidence of fire persists—Prepare to bail-out or make an emergency landing.

T.O. 1F-84(25)F-1
15 February 1958 6

SMOKE ELIMINATION.

1. Oxygen regulator diluter lever—100% OXYGEN.
2. Cabin temperature control—Position to highest point.
3. If smoke is not eliminated:
   a. Cabin vent selector—RAM.
   b. Defroster control—OFF.
   c. Abort mission and land as soon as possible.

LANDING WITH WHEELS RETRACTED.

1. If time and conditions permit—Jettison all external stores except empty external tanks.
2. Canopy—Jettison.
3. Safety belt—Tightened.
4. Shoulder harness and inertia reel lock control—Tightened and locked.
5. Landing flaps—Extend as desired.
6. Make normal approach.
7. Before contact with ground, close throttle and turn fuel tank selector OFF. On contact turn battery switch OFF.
8. Drag chute—Deploy after touchdown.
9. Leave cockpit as soon as possible.

LANDING WITH MAIN GEAR DOWN — NOSE GEAR UP OR UNLOCKED.

1. If time and conditions permit—Jettison all external stores except empty external tanks.
2. Landing gear—DOWN.
3. Canopy—Jettison.
4. Plan approach to touchdown as near the end of runway as possible.
5. Make normal approach, landing flaps DOWN, speed brakes OUT.
6. Safety belt—Tightened.

7. Shoulder harness and inertia reel lock control—Tightened and Locked.

8. Before contact with ground—Close throttle and turn fuel tank selector OFF. On contact, battery switch—OFF.

9. Hold nose wheel off the ground as long as practicable then ease down prior to stall and deploy drag chute.

10. Do not use brakes unless absolutely necessary.

11. Leave cockpit as soon as possible.

LANDING WITH ONE GEAR UNLOCKED.

1. If prepared hard surface runway is not available leave any main gear in the extended position.

2. If landing on a prepared hard surface runway, leave any two gears down and locked and proceed as follows:

   a. External stores—Jettison, except empty external tanks.

   b. Use same procedure as for landing with nose gear up or unlocked.

LANDING WITH UTILITY HYDRAULIC SYSTEM FAILURE.

1. External stores—Jettison if necessary.
2. Landing gear selector handle—EMERG DOWN.
3. To reduce flight control deflection set up a long flat final approach.
4. Emergency hydraulic pump switch—HYD EMER PUMP.
5. On ground contact:
   a. Drag chute—Deploy.
   b. Throttle—IDLE (stop cock throttle, if necessary).
6. Use recommended emergency braking technique.

RUNWAY OVERRUN BARRIER ENGAGEMENT.

1. External stores—Jettison.
2. Drag chute—Deploy after touchdown.
3. Brakes—Avoid excessive use.

4. Immediately prior to engaging the runway overrun barrier:
   a. Throttle—CLOSED.
   b. Fuel tank selector—OFF.
   c. Generator switch—OFF.
   d. Battery switch—OFF.

LANDING WITH FLAT TIRE.

1. If nose wheel tire is flat; make normal landing holding nose wheel off as long as possible.
2. If one main wheel tire is flat; make normal landing on side of runway nearest inflated tire.
3. If both main wheel tires are flat; make normal landing in center of runway and use brakes sparingly and with caution.
4. Drag chute—Deploy after touchdown.

DITCHING.

1. IFF—Turn to emergency.
2. Oxygen—100% OXYGEN.
3. External stores—Jettison (all).
4. Landing flaps—50 to 75 per cent DOWN.
5. Canopy—Jettison.
6. If time permits complete the following:
   a. Throttle—CLOSED.
   b. Safety belt—Secured.
   c. Unbuckle parachute and disconnect anti-g and electrical connections.
   d. Shoulder harness and inertia reel lock control—Tightened and Locked.
   e. Landing gear—UP.
   f. Speed brakes—IN.
7. Touchdown in a nose high attitude.

8. Make actual touchdown in the same attitude as for normal landing.
9. After aircraft has slowed down in water—Leave cockpit at once.

BAIL-OUT PROCEDURE.

1. Canopy—Jettison.
2. Assume ejection position.
3. Eject.

EXTERNAL TANKS FAIL TO FEED.

1. Receiver door switch—Check CLOSED.
2. AR AMP & CONTR or IFR AMP & CONT circuit breaker—Check IN.
3. IFR TEST & FUEL SHUT-OFF circuit breaker—Pull (C and later).

IF PYLON TANKS STOP FEEDING.

1. AR AMP & CONTR or IFR AMP & CONT circuit breaker—Check in.
2. Actuate the pylon tank air pressure switches.
3. Cycle the air refueling receptacle door.
4. Execute a series of zooms and dives.
5. Descend to a lower altitude.
6. Cycle the fuel shut-off valves.
7. Pull IFR AMP & CONT circuit breaker (C and later).
8. Pull IFR TEST & FUEL SHUT OFF circuit breakers and reset the IFR AMP & CONT circuit breaker (C and later).

MAIN TANK BOOSTER PUMP FAILURE.

1. Fuel tank selector—FWD AUX until forward fuel is almost depleted then turn to WING AUX immediately.
2. Continue to operate at optimum altitude in WING AUX position and descend to an altitude below 20,000 feet, before wing tank fuel is depleted.

T.O. 1F-84(25)F-1
Changed 30 June 1958 10

3. Fuel tank selector—Counterclockwise to ALL TANKS for remainder of fuel.

WING TANK BOOSTER PUMP FAILURE.

1. Pull WING FUEL PUMP CONTR circuit breaker.
2. Plan remainder of flight so as to land as soon as possible.
3. Fuel tank selector in ALL TANKS position at optimum altitude until main tank is almost empty.
4. Descend to an altitude below 20,000 feet and turn fuel tank selector to WING AUX.

FORWARD TANK BOOSTER PUMP FAILURE.

1. Fuel tank selector in ALL TANKS at optimum altitude until main tank is almost empty.
2. Plan remainder of flight so as to land as soon as possible.
3. Descend to an altitude below 20,000 feet, and turn fuel tank selector to FWD AUX.

ENGINE FUEL CONTROL SYSTEM FAILURE.

1. Throttle—IDLE.
2. Emergency fuel switch—ON.
3. Throttle—Advance slowly and smoothly to desired RPM.
4. Complete flight on emergency fuel system.
5. Land as soon as possible.

A thru G

PYLON TANK JETTISON.

NORMAL RELEASE.

1. Inboard bomb selector switch—SINGLE.
2. Outboard bomb selector switch—ALL.
3. Bomb release selector switch—MANUAL RELEASE.
4. Depress bomb release switch momentarily, then depress again.

H and LATER

NORMAL RELEASE.

1. Armament selector switch—OUTB'D BOMBS SALVO.
2. Bomb release selector switch—MANUAL.
3. Bomb release switch—Depress momentarily.
4. Armament selector switch—INB'D BOMBS SING.
5. Depress bomb release switch momentarily, then depress again.

ELECTRICAL POWER SUPPLY SYSTEM FAILURE.

FUEL SYSTEM OPERATION WITH COMPLETE ELECTRICAL FAILURE.

1. Fuel tank selector—FWD AUX—until forward tank fuel is used.
2. Fuel tank selector—WING AUX until wing tank fuel is used.
3. Fuel tank selector—ALL TANKS for remainder of fuel in main tank.

INVERTER FAILURE.

1. Turn instrument power switch to ALT position.
2. Land as soon as possible.

GENERATOR FAILURE.

1. Reduce electrical load.
2. Instrument power switch—ALT.
3. Generator switch—RESET then CN.
4. If light remains on—Turn off all unnecessary electrical equipment possible.
5. If generator power is available—Return instrument power switch to NOR and resume normal operation.

NON-TANDEM

STABILATOR OPERATION WITH ENGINE FAILURE.

1. Emergency hydraulic pump switch—OFF.
2. Operate the ailerons and stabilator utilizing pressure from the normal system produced by windmilling engine.
3. Make a flame-out pattern landing.

NON-TANDEM

EMERGENCY HYDRAULIC PUMP FAILURE.

1. Pitch control switch—STICK OVERRIDE or ELEC.
2. Fly aircraft with manual aileron and electrical stabilator control. Make an electrical landing or abandon the aircraft.

LANDING GEAR RETRACTION WHILE ON THE GROUND.

1. Emergency ground retract switch—Actuate.
2. Landing gear selector handle—UP.

LANDING GEAR EXTENSION WHILE IN THE AIR.

1. Cycle landing gear while pulling negative g's or while shaking the aircraft.
2. If the landing gear will not extend by normal procedures:
   a. Reduce speed to 220 knots IAS or below.
   b. Landing gear selector handle—EMERG DOWN.
   c. If necessary yaw the aircraft to lock the main gear.
   d. Landing gear indicator lights—Check to ascertain that all gears are locked down.

BRAKE FAILURE.

1. Use up fuel until minimum fuel remains (1000 lbs).
2. Throttle—CLOSED on touchdown.
3. If left brake is out, land on right side of runway.
4. If right brake is out, land on left side of runway.
5. If both brakes are out, land as short as possible at lowest safe speed, landing flaps DOWN and speed brakes OUT.
6. Drag chute—Deploy after touchdown.
7. Dragging the tail during landing will shorten the roll, if drag chute is not installed.

This page intentionally left blank.

SECTION 4 Description and Operation of Auxiliary Equipment

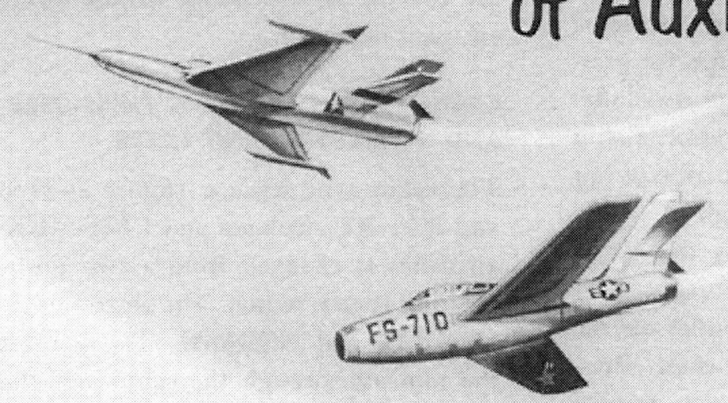

TABLE OF CONTENTS

HEATING, PRESSURIZING AND VENTILATING SYSTEM.

Pressurization, heating and ventilating are combined into an air conditioning system (figure 4–1) which is controlled electrically with power from the primary bus. When the canopy is closed the cabin is sealed by an automatically inflated rubber seal. Air for pressurization, heating, ventilating and canopy seal inflation is obtained from the engine compressor. Cabin temperature is controlled by diverting a portion of hot air from the engine compressor through a turbo-refrigerator, for cooling, before it enters the cabin. Air enters the cabin through two foot registers, a register behind the pilot's seat and an outlet on the right side directly opposite the pilot's chest. Pressurization is maintained for two schedules, normal and combat, by the pressure regulator which releases air from the cabin through a variable opening designed to maintain the proper pressure differential and rate of change of cabin air for the normal schedule. (See figure 4–2.) From sea level to 12500 feet altitude the cabin is unpressurized; from 12500 to 31000 feet altitude normal pressure remains equivalent to atmospheric pressure at 12500 feet; above 31000 feet a constant normal pressure differential of 5.0 PSI is maintained between cabin and outside atmosphere. For the combat schedule; the cabin is unpressurized from sea level to 12500 feet; from 12500 to 21200 feet altitude cabin pressure remains equivalent to atmospheric pressure at 12500 feet; above 21200 feet a constant pressure differential of 2.75 PSI is maintained between cabin and outside atmosphere. The cabin altimeter (figure 1–7) indicates the equivalent cabin altitude. The dump valve operates automatically to relieve excessive cabin pressure and can also be opened to dump cabin pressure if necessary. Outside ventilating air is available to the cabin only if pressurization is shut off.

CABIN VENT SELECTOR — UP TO F-84F-35RE AND F-84F-51GK.

Cabin pressurization is selected by the cabin vent switch (figure 4–3) which is a two position electrical switch marked PRESSURE and RAM. The RAM position opens the ram air valve and dump valves allowing a flow of ram air through the cabin and closes the engine air shut-off valve stopping the flow of engine air to the cabin. Cabin temperature cannot be selected when operating in the RAM position. The PRESSURE position closes the ram inlet and dump valves and the air flow by-pass valve modulates to maintain the temperature called for by the cabin temperature control. The cabin vent switch is powered from the primary bus and is normally left in the PRESSURE position. In the event of primary bus failure cockpit pressurization will be lost due to the canopy seal deflating.

CABIN VENT SELECTOR — F-84F-35RE AND LATER AND F-84F-51GK AND LATER.

The cabin vent selector (figure 4–3) on F-84F-35RE and later RE airplanes and F-84F-51GK and later GK airplanes is changed from a two position to a three position rotary switch. The three positions are: RAM, COMBAT and NORMAL. The RAM position opens the ram air through the cabin and closes the engine air shut-off valve stopping the flow of engine air to the cabin. Cabin temperature cannot be selected when operating in the RAM position. The NORMAL position closes the ram inlet and dump valves and positions the pressure regulator to maintain a maximum pressure differential between the cabin and outside atmosphere of 5.0 PSI. The COMBAT position closes the ram inlet and dump valves and positions the pressure regulator to maintain a maximum pressure differential between the cabin and outside atmosphere of 2.75 PSI. This position is provided for pilot safety to minimize danger resulting from sudden decompression during combat. The cabin vent pressure switch is powered from the primary bus and is normally left in the NORMAL position.

CABIN TEMPERATURE CONTROL.

Cabin temperature is adjusted for pilot comfort by using the cabin temperature control (figure 4–3) which is a three position switch marked COOL, OFF and HEAT, spring-loaded to the OFF position. By holding the cabin temperature control in the COOL or HEAT position, the cabin temperature can be adjusted to the individual needs, if primary bus power is available. Approximately 10 to 15 seconds will be required for the mixing valve to move from the full COOL position to the full HEAT position.

SIDE AIR OUTLET SHUT-OFF.

The side air outlet shut-off (8, figure 1–10; 9, figure 1–11) is located on the right side of the cabin. The quantity of hot or cold air going into the cabin through this outlet may be varied by manually sliding the shut-off over the air outlets.

AIR-CONDITIONING SYSTEM SCHEMATIC

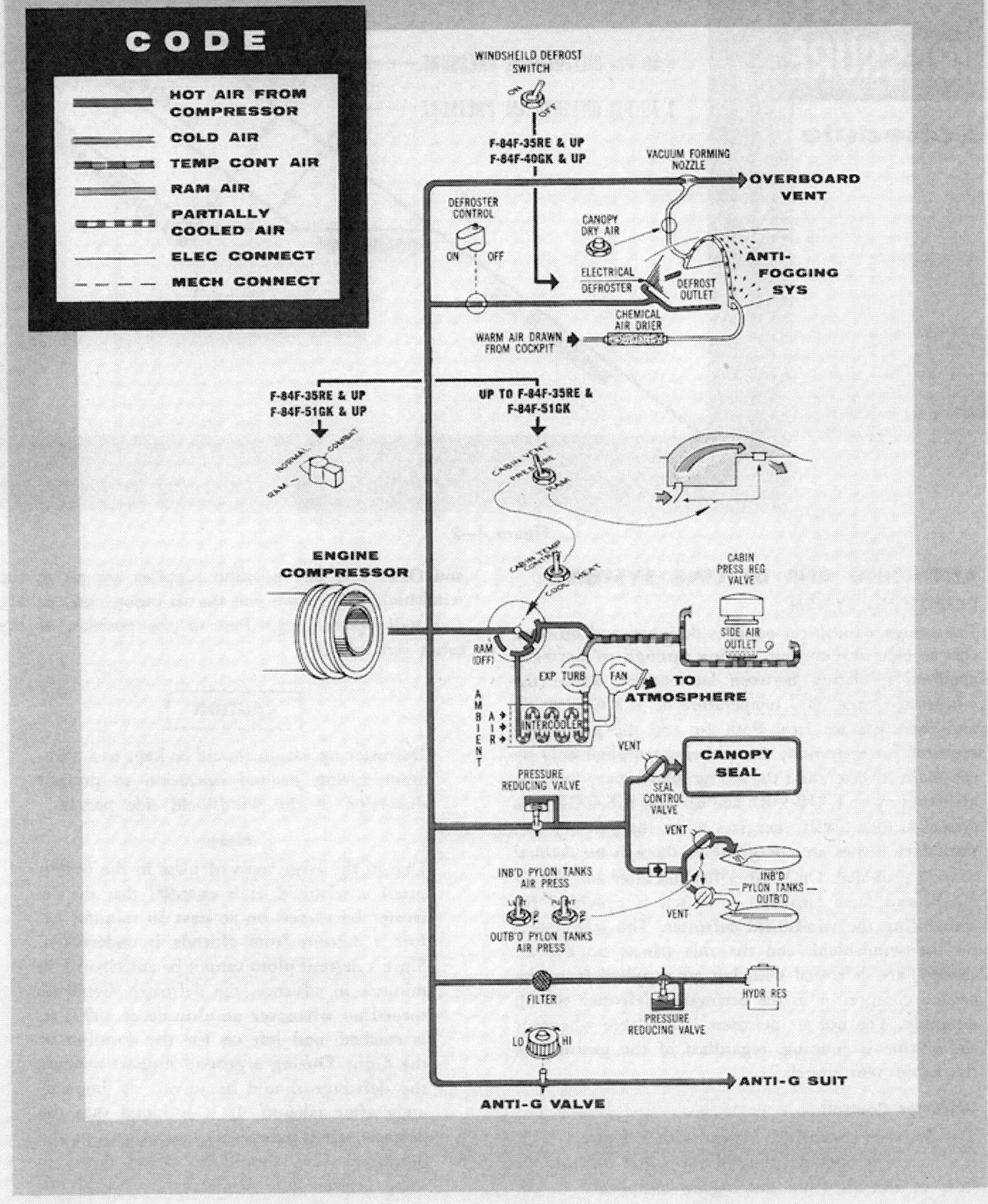
CODE
HOT AIR FROM COMPRESSOR
COLD AIR
TEMP CONT AIR
RAM AIR
PARTIALLY COOLED AIR
ELEC CONNECT
MECH CONNECT
WINDSHEILD DEFROST SWITCH
ON
OFF
F-84F-35RE & UP
F-84F-40GK & UP
VACUUM FORMING NOZZLE
OVERBOARD VENT
DEFROSTER CONTROL
ON
OFF
CANOPY DRY AIR
ANTI-FOGGING SYS
ELECTRICAL DEFROSTER
DEFROST OUTLET
CHEMICAL AIR DRIER
WARM AIR DRAWN FROM COCKPIT
F-84F-35RE & UP
F-84F-51GK & UP
UP TO F-84F-35RE &
F-84F-51GK
NORMAL
COMBAT
RAM
CABIN VENT
PRESSURE
RAM
ENGINE COMPRESSOR
CABIN TEMP CONTROL
COOL
HEAT
CABIN PRESS REG VALVE
SIDE AIR OUTLET
RAM (OFF)
EXP TURB
FAN
TO ATMOSPHERE
AMBIENT AIR
INTERCOOLER
VENT
CANOPY SEAL
PRESSURE REDUCING VALVE
SEAL CONTROL VALVE
VENT
INB'D PYLON TANKS AIR PRESS
INB'D
PYLON TANKS
OUTB'D
VENT
OUTB'D PYLON TANKS AIR PRESS
FILTER
HYDR RES
PRESSURE REDUCING VALVE
LO
HI
ANTI-G SUIT
ANTI-G VALVE

Figure 4—1.

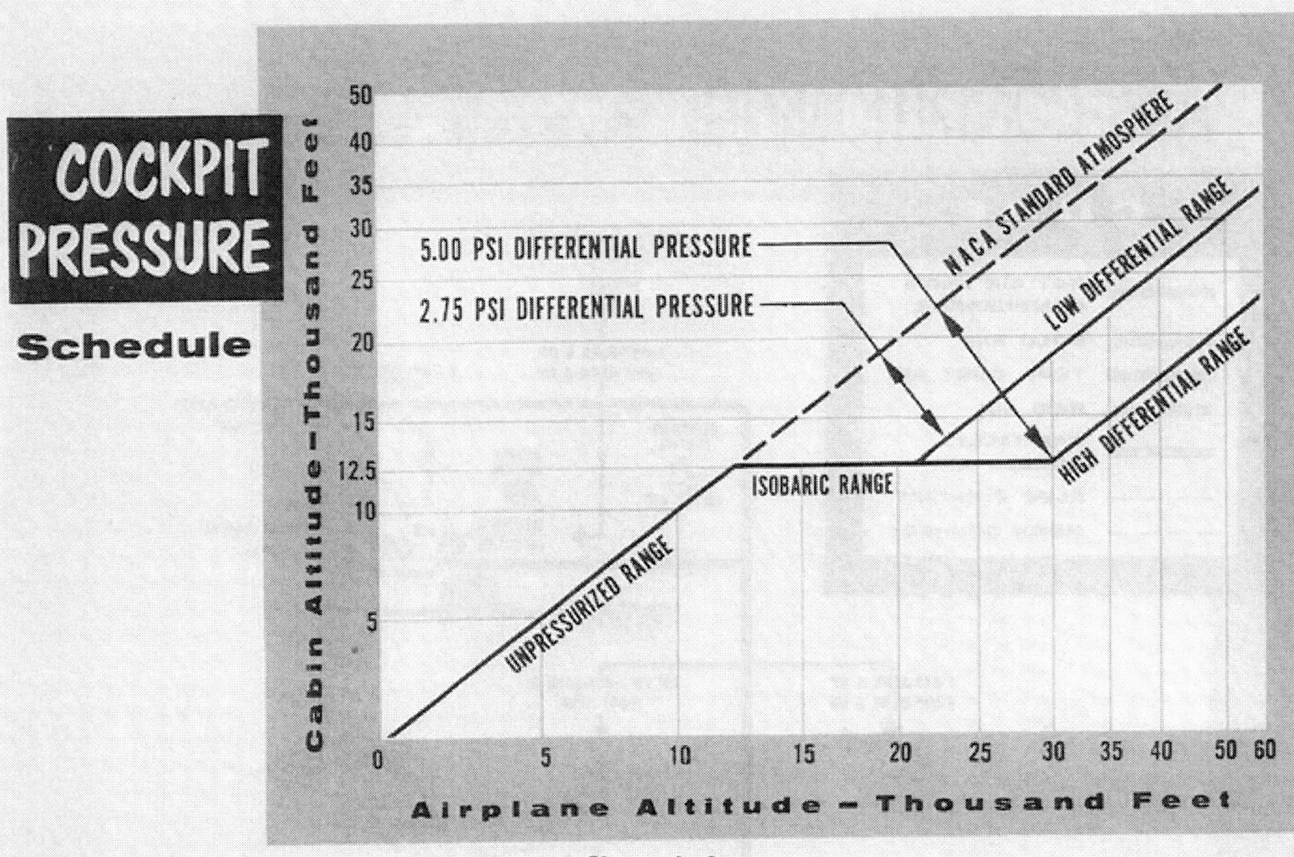

Figure 4—2.

ANTI-ICING AND DE-ICING SYSTEMS.

DEFROSTING SYSTEM.

The center windshield panel is designed and built so that an electrical current, passing through an electrical conductive coating between laminations, keeps the windshield panel at a temperature of 37.8°C (100°F) and therefore ice free. Both a-c and d-c power are required for operation; the a-c being supplied only by the main inverter, and the d-c by the primary bus. On airplanes up to F-84F-35RE and up to F-84F-40GK, the system automatically energizes when the primary and secondary busses are energized, as there is no manual control provided. On F-84F-35RE and later and F-84F-40GK and later airplanes, a switch is provided for controlling the windshield defroster. The side panels on the windshield, and the side panels on the aft canopy, are defrosted with hot air supplied from the engine compressor to the perforated defroster tubing assembly. The hot air defroster is available whenever the engine is running, regardless of the position of the cabin vent switch.

Defroster Control.

The defroster control (5, figure 1—10; 5, figure 1—11), is a manually operated shut-off valve that controls the flow of hot air from the engine compressor to the defroster tubes. The control has two positions: ON and OFF. The ON position supplies hot air to the windshield side panels and the aft canopy side panels, and will operate regardless of the position of the cabin vent switch.

CAUTION

Defroster operation should be kept to a minimum during ground operation to prevent distortion of the windshield side panels.

Note

Due to the large mass of glass in the bulletproof windshield, it is essential that the defroster be turned on at least 30 minutes before a descent from altitude is undertaken. Since a descent often cannot be anticipated 30 minutes in advance, the defroster should be turned on whenever an altitude of 20000 ft. is reached, and left on for the duration of the flight. During a ground support mission, the defroster should be turned on immediately after take-off. If it is found that the defroster air is excessively uncomfortable, the defroster valve should be closed down to some intermediate position, but should not be turned off.

CABIN TEMPERATURE & PRESSURE CONTROLS

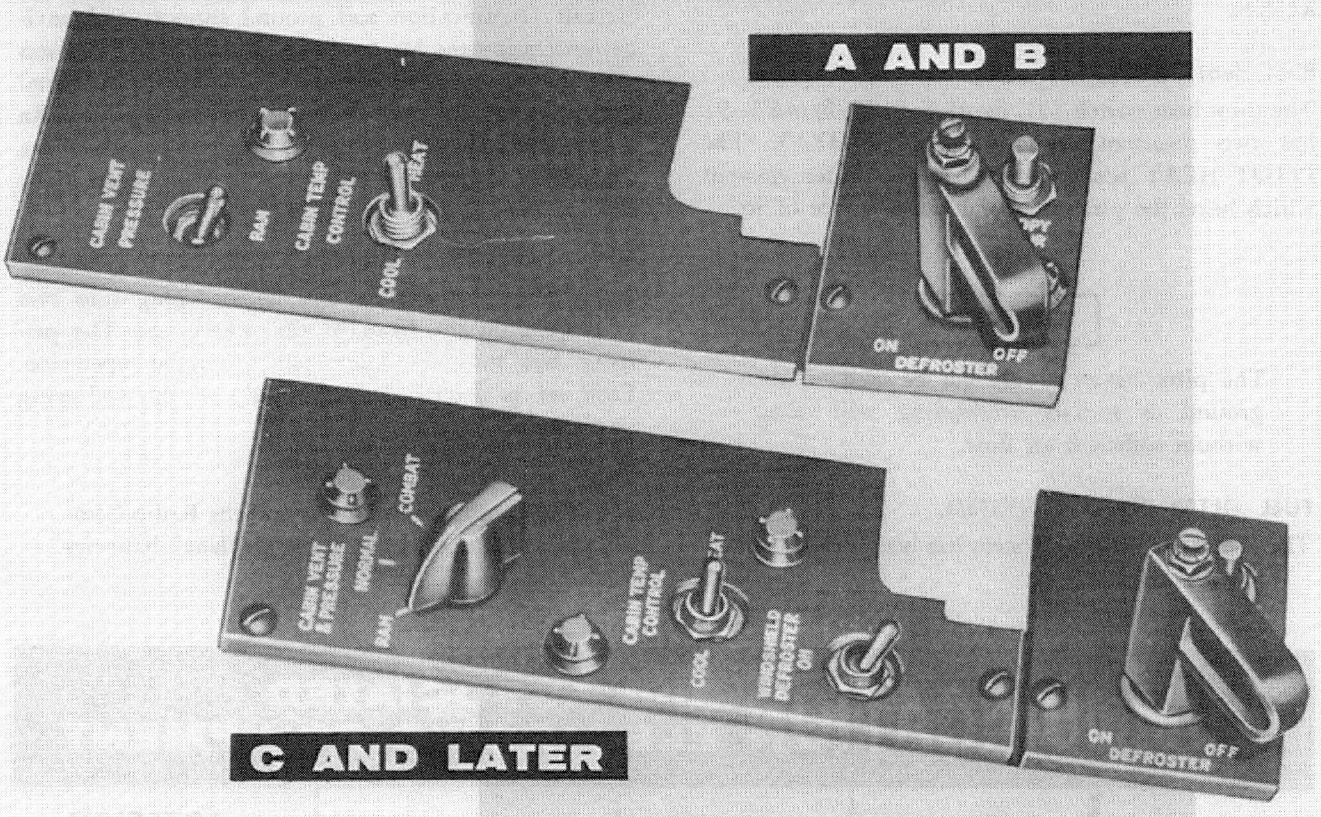

Figure 4—3.

Windshield Defrost Switch.

On F-84F-35RE and later and F-84F-40GK and later airplanes, the windshield defrost switch (figure 4—3) is provided so that the defrosting system can be turned off during ground operation. The switch has two positions; WINDSHIELD DEFROSTER ON and an unmarked off position. The off position opens the a-c power circuit, which in turn automatically opens the d-c power circuit through a relay. The ON position energizes the windshield defroster circuit, if the primary and secondary busses are energized.

ANTI-FOGGING SYSTEM.

The canopy and the windshield side panels are defogged by air, taken from the cabin, passed through a chemical drier, and circulated through the space between the inner and outer surfaces of the panels. The drier must be checked periodically to make sure that the chemical is still active. Circulation is controlled electrically through a normally closed shut-off valve.

Canopy Dry Air Switch (Anti-Fogging).

The canopy dry air switch (6, figure 1—10; 6, figure 1—12) controls the flow of dry air in the anti-fogging system. On A thru G aircraft, it is a momentary push button switch marked CANOPY DRY AIR, and on H and and later aircraft it is at two-position toggle switch marked CANOPY DRY AIR, ON, OFF; actuating the switch for approximately 10 to 15 seconds prior to take-off, opens the normally closed solenoid shut-off valve, which allows dry air to pass through the space between the inner and outer surfaces of the canopy and windshield side panels. Actuating the switch for longer periods is unnecessary, as the space will be filled with dry air and continued use will only replace the dry air with more dry air. When the switch is released the solenoid valve closes and shuts off the flow of dry air to the anti-fogging system. The valve also closes when energy from the primary bus is not available.

PITOT HEATER.

The pitot tube, installed in the duct divider in the nose of the airplane, on A through H aircraft, and on a boom on the left wing tip of J and later aircraft, is electrically heated from the primary bus to keep it ice free.

Pitot Heat Switch.

The pitot heat switch (21, figure 1–8; 18, figure 1–9) has two positions OFF and PITOT HEAT. The PITOT HEAT position turns on the heater element which heats the pitot tube and keeps it free of ice.

CAUTION

The pitot heater should not be used on the ground as serious overheating will occur without sufficient air flow.

FUEL FILTER DE-ICING SYSTEM.

The fuel filter de-icing system has been deactivated.

COMMUNICATION AND ASSOCIATED ELECTRONIC EQUIPMENT.

Communication and associated electronic equipment consists of radio sets to provide airplane-to-airplane and airplane-to-ground communication; radar sets for aircraft identification and ground support and navigation equipment for guidance during flight. All sets are remotely controlled from the cockpit with visual indicators on the console and instrument panel. Antennas are concealed within the airplane and therefore are protected from air loads during high speed flight and are kept free of ice and dirt.

OPERATION OF COMMUNICATION EQUIPMENT.

Insert microphone plug and headset plug into two extensions on the front of the pilot's seat. The primary bus must be energized for radio operation. Each set is described individually in the following paragraphs:

Note

The Command Radio Set and the Radio Compass will operate from the airplane's batteries

COMMUNICATION and ASSOCIATED ELECTRONIC EQUIPMENT

TYPE	DESIGNATION	FUNCTION	RANGE	LOCATION OF CONTROLS
Command Set	AN/ARC-33 or AN/ARC-34	Two-way voice communication	30 miles at 1000 feet altitude to 135 miles at 10,000 feet altitude	AN/ARC-33 R.H. Console (figure 1-10) AN/ARC-34 L.H. Console (figure 1-9)
IFF Set	AN/APX-6 or AN/APX-6A	Automatic identification	100 miles up to 50,000 feet altitude	R.H. Console (figure 1-10)
Radio Compass	AN/ARN-6	Radio navigation	250 miles	R.H. Console (figure 1-10)
Radar Set	AN/APW-11 or AN/APW-11A	Ground support		R.H. Console (figure 1-10)
Tacan	AN/APN-21	To provide an improved tactical air navigational system		Control panel on right console. Course and bearing distance heading indicators on main instrument panel

Figure 4—4

Figure 4–5.

as power is supplied to these sets from the primary bus. When the instrument power switch is in the NOR position, the IFF Radio Set is energized by the main inverter on modified D and later aircraft, and by the alternate inverter on unmodified A thru C aircraft. On all aircraft the IFF Radio Set is inoperative when the instrument power switch is in the ALT position.

A thru G

COMMAND SET — AN/ARC-33.

Radio set AN/ARC-33 is a remotely controlled receiver-transmitter designed to operate, in the 225.0 to 399.9 mc band. A total of 1750 crystal controlled receive and transmit channels are provided within tuning range of the set. Any 20 of these channels may be preset at the radio set control unit so as to be quickly available when desired. Because of the nature of the operating frequencies employed, communication is essentially over line-of-sight distances with a practical maximum of approximately 75 miles. The radio set is designed to operate at altitudes as high as 50000 feet. A guard receiver, constantly tuned to the guard frequency, is incorporated and may be placed in operation along with the main channel receiver, thus making it possible to continually monitor an emergency or command channel while still carrying on communication on another channel. All operating controls are located on the right console (figure 4–5) with the exception of the microphone press-to-talk button which is located on the throttle. A side tone circuit is provided which feeds a side tone signal into the headset, so that the pilot can monitor his own transmitted signals. The radio will operate if the battery switch is in the ON position and the primary bus is energized.

A thru G

Radio On-Off Switch.

A two position toggle switch (figure 4–5) marked ON-OFF located on the radio control panel supplies primary power to operate the radio set when placed in the ON position. A warm up period of approximately 75 seconds is necessary before radio operation becomes satisfactory for reception or transmission.

A thru G

Preset Channel Selector Switch.

The preset channel selector switch (figure 4–5) is used to select any one of the 20 preset channel frequencies desired for radio transmit-receive operation. The indicator windows above the knob will show the channel number selected by the selector knob, however, the channel number will be covered when the function switch is in the G position. A tone will be heard in the headset when changing channels for approximately six seconds until the newly selected channel is ready for operation. Do not attempt transmission until the tone is no longer heard in the headset.

A thru G

Preset Channel Indicator.

The preset channel indicator (figure 4–5) mechanically indicates in the window immediately above the preset channel selector knob, the number of the preset channel selected. When the function switch is placed in the G position the preset channel number shown in the indicator window above the preset channel selector switch will be automatically blanked out.

A thru G

Function Switch.

The function switch (figure 4–5) a four position rotary selector switch marked MAIN, BOTH, G, with a guarded fourth position marked by a white dot, is used to select the type of transmit-receive operation desired. In the MAIN position the main receiver is used and transmission is on the preset channel. In the BOTH position, the radio receives on the main receiver on the preset channel and the guard receiver is tuned to the guard frequency simultaneously while transmission is carried on the preset channel. The G position switches the main receiver and the transmitter over to operate on the guard frequency and cuts off power to the guard receiver, regardless of the position of the preset channel selector switch. The word GUARD will appear in the guard channel indicator window and the channel number in the preset channel indicator window will automatically be blanked out. The guarded white dot position is used to release and extract the preset channel memory cylinder, the preset channel selector switch and the control panel section securing this assembly, from the radio control panel for changing the selection of preset channels, as set up on the preset channel memory cylinder. The release button marked with a white dot must be depressed and held to move the function switch to the guarded white dot position. When the release button is in the depressed position or the function switch is in the white dot position the main receiver and the transmitter will operate on the guard frequency.

CAUTION

When changing the selection of preset channels during flight the memory cylinder should be removed and installed carefully to avoid damage to the cylinder.

A THRU G

Guard Channel Indicator.

The guard channel indicator window (figure 4–5), located above the preset channel indicator window indicates when the main receiver and transmitter are operating in the guard channel. The word GUARD appears in the indicator window automatically when the function switch is in the G position.

A THRU G

Tone Push Button.

The spring-loaded push button marked TONE (figure 4–5) on the radio control panel, switches the radio from receive to tone transmission operation, the preset channel shown in the channel indicator window, when held in the depressed position regardless of whether or not the microphone press-to-talk button is depressed. The 1,020 cycle tone is continuous during voice transmission while the button is held in the depressed position for transmit operation. The radio automatically switches over to receive operation when the button is released.

Note

Tone transmission is usually perceptible over a slightly greater distance and through greater interference than voice transmission. Therefore it is especially adaptable in emergency use for code key operation when interference or jamming makes voice transmission impractical, or as a radio marker signal for radio direction finding.

A THRU G

Volume Control.

The volume control (figure 4–5), a rotary switch marked VOLUME, regulates the volume for the main and guard receivers and the headset volume. However, it cannot increase the volume beyond a predetermined level.

A THRU G

Microphone Press-To-Talk Button.

The microphone press-to-talk button (figure 1–17) located on the engine throttle control, switches the radio set over to transmission operation for voice modulation when held in the depressed position. When the button is released the radio returns to receive operation.

A THRU G

Procedure For Setting Up Channels.

Note

On modified airplanes any one of the 1750 frequencies can be selected by use of the UHF quick manual tuner.

To change preset channels on unmodified airplanes, it is first necessary to know the channel frequency, and the corresponding preset channel number. After these are determined proceed as follows:

1. Push the release pushbutton in, and turn the function switch to the extreme counterclockwise position, where the white dot on the function switch is opposite the white arrow on the panel face.
2. Grasp the preset-channel-selector switch and extract the memory cylinder. Pull the memory cylinder straight out until it is free from the control panel.
3. With one hand, firmly hold the memory cylinder with the longitudinal slot up, and rotate the channel-selector switch to the desired channel position, as indicated by the numbered dial.

Note

When setting up channels on the memory cylinder, hold the cylinder in the horizontal position so that the panel faces the holder, and the longitudinal slot in the cylinder case is up.

4. Place the first tab from the rear of the cylinder opposite the index marked on the cylinder case which corresponds to the first digit of the four-digit channel frequency being set up.

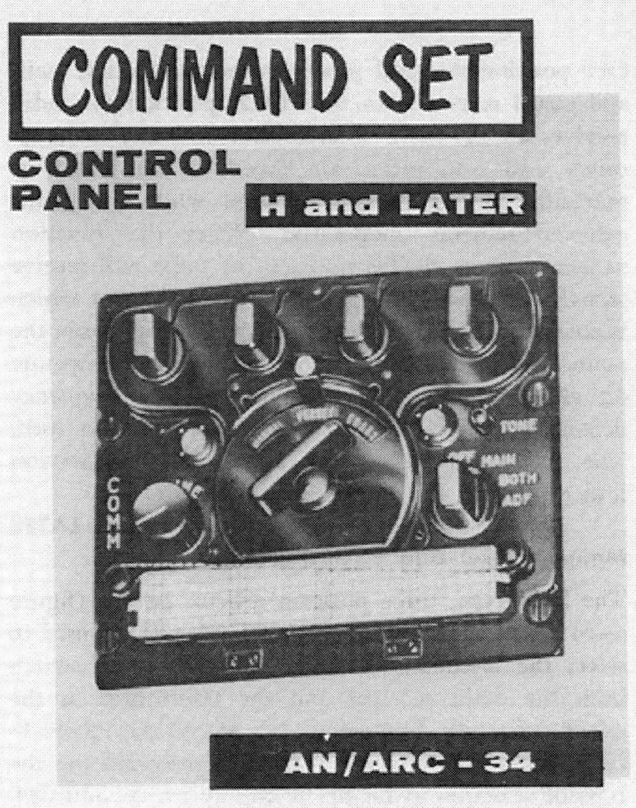

Figure 4—6

5. Place the second tab from the rear of the cylinder opposite the index digit marked on the cylinder case which corresponds to the second digit of the four-digit channel frequency being set up.

6. Place the third tab from the rear of the cylinder opposite the index digit marked on the cylinder case which corresponds to the third digit of the four-digit channel frequency being set up.

7. Place the fourth tab from the rear of the cylinder opposite the index digit marked on the cylinder case which corresponds to the fourth, or decimal, digit of the four-digit channel frequency being set up.

8. After the channel is set up on the memory cylinder, replace it in the control panel as follows:

   a. Make sure the function switch is in the white-dot position.

   b. Insert the memory cylinder in its chamber in the control panel. Slide the cylinder back until its panel is flush with the control panel.

   c. Depress the release pushbutton and place the function switch in either the MAIN or BOTH position.

A THRU G

Starting.

Note

No transmission will be made on emergency (distress) frequency channels except for emergency purposes. For test, demonstration, or drill purposes, the radio equipment will be operated in a shielded room to prevent transmission of messages that could be construed as actual emergency messages.

1. Place the ON-OFF switch in the control panel in the ON position. Allow approximately 75 seconds for radio warm up.

2. Rotate the preset channel selector control knob until the respective number for the desired channel appears in the indicator window above the knob.

Note

When changing channels there will be approximately six seconds delay for channel set up time during which a tone signal will be heard in the headset until the radio is ready for operation on that channel. Do not attempt transmit operation until the tone ceases. The tuning drive motor is protected by a thermal time delay cut-out which opens the motor circuit after approximately one minute of continuous tuning. Therefore, if the tuning motor is unable to come to rest because of some fault in the equipment, or because of constant rotation of the preset channel selector control, the tuning motor circuit is automatically broken after approximately 30 seconds. The motor circuit may be reset by selecting a new channel, then returning to the original channel.

3. Listen for the tone in the headset. Make no attempt to receive or transmit while the tone is heard.

4. Place the function switch in the MAIN, BOTH or G position depending on the operation desired.

5. Adjust the volume control for comfortable level in the headset.

A THRU G

Stopping.

1. Place the ON-OFF switch in the control panel in the OFF position.

Note

Once the radio has been turned off, the 75-second warm-up period and the channel tuning cycle of six seconds must be completed before radio transmission can be resumed.

H and LATER

COMMAND SET AN/ARC-34.

Radio set AN/ARC-34, a remotely controlled receiver-transmitter operating in the 225.0 to 399.9 mc band. A total of 1,750 crystal controlled receive-transmit channels, in increments of one-tenth of a megacycle, are within tuning range of the radio set and may be manually selected by the pilot. Any 20 channels within the radio's total frequency range may be preset in any order for quick selection by the rotary type preset channel selector switch on the radio control panel. Four knobs are provided for manual selection of an operating frequency, so that manual operation does not disturb the preset channel arrangement. Receiver and transmitter tuning is automatically completed after a channel change. Two receivers, a main, and a guard receiver which is constantly tuned to the guard frequency are provided. The guard receiver may be operated along with the main receiver, thus making it possible to monitor the guard channel for command or emergency communication, while still carrying on communication on another channel. Receiver operation is manually selected by a rotary type function switch. Continuous tone transmission is also provided through a tone button which may be used for voice transmission, or as a key for code signals. All radio operating controls are located on the left console (figure 4–6) with the exception of the microphone press-to-talk button, which is located on the throttle. The pilot can monitor his own transmitted signals through a side-tone circuit which feeds from the transmitter into his headset. The transmitter and the main receiver are tuned to the same frequency. The radio will operate if the primary bus is energized. A standby emergency battery may be switched on if the electrical system does not produce sufficient voltage for radio operation. An on-off type switch, which is guarded in the off position by a safety-wired cover guard, is provided to switch the radio to emergency battery power. A circuit breaker press-to-test light is mounted adjacent to the standby battery switch and is used to test the standby battery circuit. The indicator light will illuminate if the circuit is functioning regardless of the position of the standby battery switch.

H and LATER

Function Switch.

The function switch (figure 4–6), a rotary, four position selector switch marked OFF-MAIN-BOTH-ADF is used to turn the AN/ARC-34 radio on or off, or select the type of receiver operation desired. The OFF position shuts off power to the transmitter, main and guard receivers. In the MAIN position the radio receives on the main receiver and both the main receiver and transmitter are tuned to the preset or manually set up channel selected while the guard receiver remains inoperative. When the function switch is in the BOTH position the radio will receive signals simultaneously from the guard receiver which is constantly tuned to the guard channel, and from the main receiver, which with the transmitter will operate on either a preset or manually selected frequency depending upon the type of channel selection used. The ADF position is for an automatic direction finder which is not installed in F-84F aircraft.

H and LATER

Manual-Preset-Guard Selector Switch.

The lever type, three position selector switch (figure 4–6) marked MANUAL-PRESET-GUARD is used to select the method of channel selection or to switch both the main receiver and the transmitter to the guard channel frequency. A MANUAL-PRESET-GUARD switch indicator window is provided and the type of selection as set up by the switch is indicated, while the other positions are visible, but covered by a green shutter. In the MANUAL position, the four tuning knobs at the top of the control panel are used permitting manual selection of any one of the 1750 frequencies in the radio's tuning range for transmit-receive operation. The PRESET and GUARD positions on the MANUAL-PRESET-GUARD switch indicator window will be covered, as will the preset channel indicator window. The indicator windows above each of the four manual tuning knobs are open and provide a direct reading in megacycles and tenths of a megacycle of each frequency manually selected by the knobs. In the PRESET position, the preset channel selector switch is used and transmit-receive operation may be carried on any one of the 20 preset channels as selected by the preset channel selector switch and indicated in the preset channel indicator. The manual tuning knob windows and the MANUAL and GUARD positions on the MANUAL-PRESET-GUARD switch indicator windows will be blanked out. In the GUARD position, the main receiver and transmitter are switched to the guard channel frequency and the guard receiver is inoperative even if the function switch is in the BOTH position. The manual tuning indicator window, the preset channel indicator window and the PRESET and MANUAL positions in the MANUAL-PRESET-GUARD indicator window will be covered.

H and LATER

Tone Push Button.

The spring-loaded push button marked TONE (figure 4–6) switches the radio over from receive to tone transmission on the manually selected or preset frequency that the radio is operating on as long as the depressed position is held. A 1,020 cycle tone is continuous during voice transmission as long as the tone button is depressed. The tone button may also be used as a key for code transmission. The microphone press-to-talk button, regardless of its position has no effect on the operation or function of the tone button.

Note

Tone transmission is usually perceptible over a slightly greater distance and through greater interference than voice transmission. Therefore it is especially adaptable in an emergency for code key operation, when interference or jamming conditions make voice transmission impractical, or to serve as a radio marked signal for direction finding.

H and LATER

Volume Control.

The volume control (figure 4–6) marked VOLUME regulates the headset volume for signals received on both the main and guard receivers. Volume control range is predetermined so that the signal volume may not be reduced below a preset level.

H and LATER

Microphone Press-to-Talk Button.

The microphone press-to-talk button (figure 1–17) located on the engine throttle control, switches the radio set from receive to transmit operation for voice modulation when held in the depressed position. When the button is released the radio returns to receive operation. The microphone press-to-talk button, regardless of its position, has no effect on the function or operation of the tone push button.

H and LATER

Emergency Radio Battery Switch.

On aircraft equipped with the AN/ARC-34 radio, an emergency radio battery and emergency radio battery switch are provided for emergency radio operation. The emergency radio battery switch is located on the left console and is safety-wired and guarded in the off position by a cover-type guard, while the on position is marked ON. The emergency battery is a 26.25 volt wet coil, silver zinc battery located below the aircraft battery and is connected to the radio through a circuit breaker in the battery well.

Remote Channel Indicator.

Airplanes modified in accordance with T.O. 1F-84-786 have a remote channel indicator installed on the main instrument panel. The indicator enables the pilot to see what channel he is selecting on the AN/ARC-33 or AN/ARC-34 radio without looking down at the console.

H and LATER

Emergency Radio Battery Circuit Breaker Check Light.

The emergency battery circuit breaker check light is a press-to-test light mounted adjacent to the AN/ARC-34 radio emergency battery switch and is provided to check the emergency radio battery circuit breaker in the battery well. The check light will function regardless of the position of the emergency radio battery switch.

H and LATER

Starting.

Note

No transmission will be made on emergency (distress) frequency channels except for emergency purposes. For test, demonstration, or drill purposes, the radio equipment will be operated in a shielded room to prevent transmission of messages that could be construed as actual emergency messages.

1. Place the function switch in the BOTH position.

2. Turn MANUAL-PRESET-GUARD selector switch to PRESET position.

3. Rotate preset channel selector knob until desired channel number appears in preset channel indicator window. Allow approximately one minute for equipment warm-up and automatic channel tuning adjustment cycle. At the end of the warm-up period the equipment will be in the standby condition ready to receive signals on the preset command and fixed guard frequencies simultaneously. During transmission periods the pilot should receive his own signals on a side tone received by his headset. A little receiver noise may or may not be heard during non-transmission periods.

This page intentionally left blank.

IFF SET TYPICAL

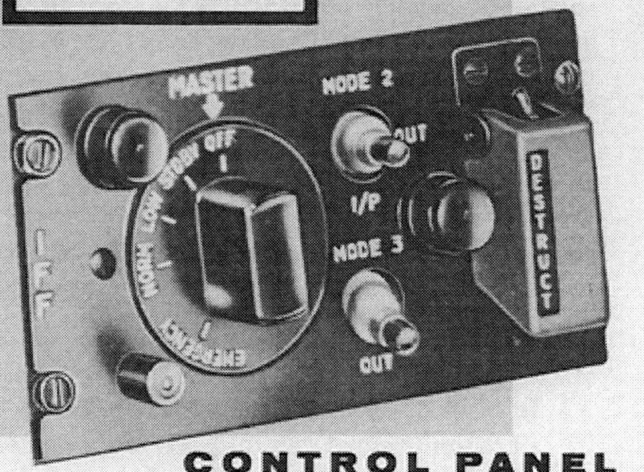

CONTROL PANEL

Figure 4—7

Note

> When selecting a new preset channel there will be a delay of four seconds before the automatic tuning cycle adjusts the set for operation on the newly selected frequency.

4. Adjust the volume control for a comfortable signal level in the headset.

5. Before transmission, check that equipment is functioning and tune-up is complete by listening for voice side-tone, or tone signal when tone button is depressed.

H and LATER

Stopping.

1. Place the function switch in the OFF position.

Note

> Once the radio set has been turned off, the one minute warm-up period must be completed before radio transmission can be resumed.

H and LATER

Operating Precautions.

1. Transmit only when the channel is clear to prevent confusion and loss of messages. The guard frequencies should not be used for transmission unless the message is urgent.

2. Use both receivers for general reception unless signals from one make the desired signals from the other unusable.

3. It is possible to set the preset channel buttons or manual frequency selector knobs for frequencies below 225.0 megacycles. Since this is below the operating frequency range, the automatic tuning mechanism cannot accomplish tune-up. It will operate for approximately 120 seconds and will then be turned off automatically by a protective relay. Under these conditions there will be no side-tone if transmission is attempted and operation of the frequency selection knobs will have no effect. To restore operation, select a channel within the 225.0 to 399.9 megacycle range. Turn the function switch to the OFF position and then to BOTH. After approximately one minute the radio equipment will operate in a normal manner.

UNMODIFIED

IFF SET — AN/APX-6.

The purpose of the IFF Set (figure 4—7) is to enable the aircraft in which it is installed to automatically identify itself as friendly whenever it is properly challenged by suitably equipped friendly surface and airborne radar equipment. A type AN/APX-6 IFF set installed in A THRU G aircraft incorporates three destructors which are inserted in the face of the IFF transponder. On H and later aircraft, an AN/APX-6A type IFF set is installed which is similar to the earlier type but does not incorporate destructor features.

RADIO COMPASS

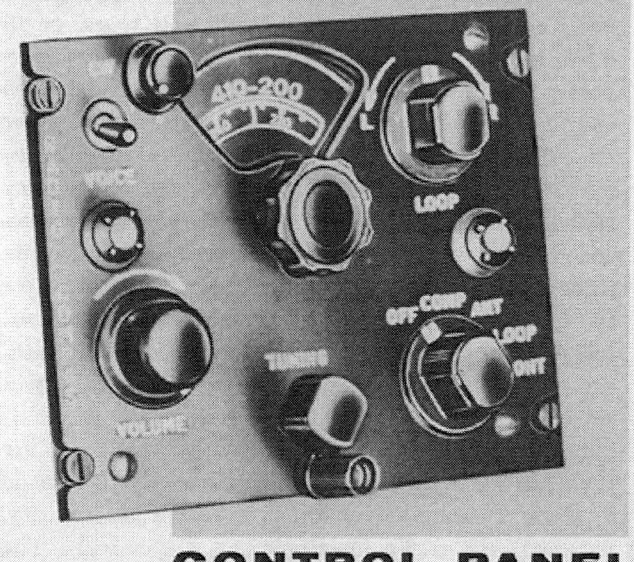

CONTROL PANEL

Figure 4—8

CAUTION

Before take-off make sure that the IFF frequency counters have been set to the proper frequency channels. If the use of destructors is required on AN/APX-6 installations in A THRU G aircraft, check that all three destructors have been inserted in the face of the IFF transponder.

Starting.

1. Rotate the master control to NORM position unless instructed otherwise.

Note

The IFF set is inoperative with the instrument power switch in the ALT position.

2. Set mode 2 and mode 3 control as instructed.

Stopping.

1. Rotate the master control to the OFF position.

Note

On A THRU G aircraft, if the destructor control was operated during the flight, report this fact immediately upon landing so that a new receiver-transmitter may be installed.

MODIFIED

IFF SET — AN/APX-25.

Airplanes modified in accordance with T.O. 1F-84-775 or T.O. 1F-84-783 incorporate an AN/APX-25 radar set. The AN/APX-25 radar identification set (IFF) enables the airplane in which it is installed to identify itself automatically as friendly, whenever it is challenged by the proper signals from other appropriate radar recognition equipment at ground bases, or in other airplanes. The set has two supplementary purposes: (1) It enables specific friendly airplanes to identify themselves apart from numerous other friendly airplanes; (2) provides means for transmitting a special coded signal called the "emergency reply." In operation the AN/APX-25 set receives coded interrogation signals and transmits coded reply signals to the source of the challenging signals where the reply codes are displayed, together with associated radar information (targets, etc.), on the radar indicators. When a radar target is accompanied by a proper reply code from the IFF set, the target is considered friendly. Three modes of operation are provided for response to interrogation signals: mode 1, mode 2, and mode 3 which are used for security, personal, and traffic identifications, respectively. The IFF set provides for two methods of reply coding: Mark X and Mark X SIF. An internal selector switch (set up by ground personnel only) permits the IFF set to be operated in the Mark X, or Mark X SIF, configuration. The Mark X configuration provides for use of the IFF (transponder) control panel only, and selection of reply coding is limited to the one code reply combination preset into the equipment. When using the Mark X SIF configuration, a SIF (selective identification feature) control panel is used in conjunction with the IFF control panel, providing for elaboration of the reply coding through the many code combinations available with the SIF control panel. The radar identification set is powered by the AC and DC secondary busses.

Note

In Mark X operation the SIF (selective identification feature) control panel is eliminated, rendering the code selector dials inoperative. However, the set will still operate in all three IFF modes providing limited preset interrogation and response signals.

Identification Radar Control Panels.

Two radar control panels (figure 4–9) marked SIF and IFF are located on the right console. The IFF or transponder control panel contains two mode switches, an identification of position (I/P) switch, and a fire position master switch whose positions are OFF, STDBY, LOW, NORM, and EMERGENCY. In the STDBY position, the system is inoperative but ready for instant use. In the LOW position, the system operates in partial sensitivity and replies only in the presence of strong interrogation. In the NORM position, the system operates at full sensitivity which provides maximum performance. In the EMERGENCY position, the system replies to all modes of interrogation with a special coded signal to indicate an emergency. The mode 2 switch placarded MODE 2 and OUT, is used by the pilot for personal identification. The mode 3 switch, placarded MODE 3 and OUT is used by the pilot for traffic identification. The identification of position (I/P) switch, placarded I/P, OUT and MIC, is used by the pilot upon request to provide momentary identification of position when held in the I/P position. When placed in the MIC position, the identification of position signals are transmitted when the microphone button is held depressed. The SIF control panel contains two, concentric, rotary, code selector switches which are used to select the specified code signals to be used in mode 1 or mode 3 operation when in Mark X SIF operation. The specified coded signals to be used in mode 2 are preset on the ground and cannot be changed in flight. The rotary code selector switches are marked MODE 1 and MODE 3, and each contain inner and outer knobs for selection of specified code signals. The inner and outer knobs of mode 1, and the outer knob of mode 3, selector switches, are marked 0 through 7 consecutively, as the knobs are turned clockwise. The inner knob of mode 3 code selector switch is marked 0 through 3, consecutively.

Operation of Identification Radar.

1. Rotate master switch to STDBY, to maintain equipment inoperative but ready for instant use.
2. Rotate master switch to NORM to place equipment in operation.

Note

The LOW position of the master switch should not be used except upon proper authorization. Mode 1, the security identification feature, is in operation when the master switch on the IFF control panel is in NORM.

3. Set mode 2 and mode 3 switches OUT unless otherwise directed.
4. Set mode 1 and mode 3 code selector switches on the SIF control panel as directed (when operating in the Mark X SIF configuration).
5. For emergency operation, press dial stop and rotate master switch to EMERGENCY, so that the set will automatically transmit a special coded distress signal in response to interrogation.
6. Rotate master switch to OFF to turn set off.

RADIO COMPASS — AN/ARN-6.

The radio compass AN/ARN-6 (figure 4–8) is an airborne navigational instrument. There are four bands covering a frequency range of 100 to 1,750. The radio compass is capable of providing the following:

1. Automatic visual bearing indication of the direction of arrival of radio signals and simultaneous aural reception of corresponding sound.
2. Aural reception of radio signals, using a non-directional antenna.
3. Aural-null directional indicators of the arrival of radio signals using a loop antenna.

Starting.

1. Turn the function switch to COMP, ANT or LOOP position.

Note

The function switch position marked CONT on the control panel is not used on this installation.

Stopping.

1. Rotate function switch to OFF.

MODIFIED AN/ARN-21 RADIO SET (TACAN).

Airplanes modified in accordance with T.O. 1F-84-775 or T.O. 1F-84-783 incorporate an AN/ARN-21 radio set. The radio set is an airborne portion of the short range air navigation system called TACAN (Tactical Air Navigation). It is designed to operate in conjunction with a surface navigation beacon which enables the airplane to obtain continuous indications of its distance and bearing from any selected surface beacon located within a line-of-sight distance from the airplane up to 200 nautical miles. The bearing information and distance information are displayed on two separate indicators known as the azimuth indicator and the range indicator. A total of 126 channels are provided and spaced one megacycle apart. Transmission frequencies for distance measuring pulses transmitted by the radio set and returned by the ground beacon are in the 1025 to 1150 MC range. The receiver is automatically tuned to a corresponding receiving channel which is divided on either side of the transmitting channels. Channels 1 thru 63 utilize 962 to 1,024 MC and channels 64 thru 126 utilize 1,151 to 1,213 MC. This system enables an equipped airplane to obtain continuous indications of its distance and bearing from any selected TACAN surface beacon located within its range. The radio set utilizes an azimuth indicator mounted on the main instrument panel which indicates the bearing from the airplane to the ground beacon to which it is tuned. While the indicator is searching for the correct bearing, the pointer rotates at a rate which prevents course readings. Distance from the airplane to the ground beacon is indicated in nautical miles on the range indicator which is also mounted on the main instrument panel. While the indicator is searching for the correct range, the rapidily rotating numbers are partially covered by a red flag which warns the pilot against reading incorrect distance indications.

Channel Selector Knob.

The channel selector knob selects the desired navigation beacon channel. The left-hand knob selects the tens and hundred figure of the beacon channel number, and the right-hand knob selects the unit figure of the beacon channel. Combinations of channel settings may be made from 00 to 129, but the equipment operates on channels 01 thru 126 only. The channel frequencies selected are shown on a window marked CHAN.

Power Control Switch.

The power control switch is a three-position switch marked OFF, REC and T/R. The OFF position de-energizes all of the ARN-21 equipment. The REC position places the receiver portion of the equipment into operation so that only bearing information is furnished by the radio set. In the T/R position, the airborne equipment transmits a distance-measuring pulse, and interrogator, a corresponding reply pulse from the ground beacon, to furnish distance information in addition to bearing information.

Volume Control.

A volume control marked VOL is provided for adjusting the volume of the audio identification signal received from the beacon. Clockwise rotation increases the volume.

Operation of AN/ARN-21 Radio Set.

1. Power control switch—REC or T/R.
2. Allow approximately 90 seconds warmup time after power is applied
3. Select the desired beacon channel.
4. Adjust volume to desired level
5. To turn equipment off, position power control switch to OFF.

RADAR SET — AN/APW-11A.

An AN/APW-11A type radar set is installed. These radar sets, have a higher classification than this handbook. Refer to applicable T.O.'s for detailed information.

LIGHTING EQUIPMENT.

INSTRUMENT LIGHTS.

Individual instrument ring lights are installed on each instrument and illuminate the instruments with a red light. Red instrument flood lighting is provided by a spot light mounted on each side of the canopy frame to illuminate placards and can also be used as auxiliary illumination. Thunderstorm lights are provided on some airplanes that produce a bright white light to counteract the brilliance of lightning flashes.

Instrument Panel Light Switches.

The instrument panel lights are controlled by three rheostat switches (figure 4–9). The flight instrument light rheostat is marked FLIGHT, and controls the individual lights on all the flight instruments. The non-flight instrument light rheostat is marked NON-FLIGHT, and controls the individual lights on all non-flight instruments. The red auxiliary light rheostat is marked AUXILIARY, and controls the two spot lights mounted on the canopy frame. All rheostats have two positions OFF and BRIGHT. The FLIGHT and NON-FLIGHT instrument lights are energized by the secondary bus while the AUXILIARY lights are powered from the primary bus.

THUNDERSTORM LIGHTS.

Thunderstorm lights are installed on airplanes AF Serial NOS. 51-1,621, 51-1,622, 51-1,623 and 51-1,624 and are turned on or off with the thunderstorm light switch located above the right console. The switch is a two-position switch marked STORM LTS ON and an unmarked off position. The thunderstorm lights are powered from the primary bus.

CONSOLE AND INSTRUMENT PANEL LIGHTS.

The left console is lighted by three incandescent lights covered with red filters and mounted above the console. The placards on the instrument panel and on the right console are lighted with incandescent lights used in conjunction with plastic panels. These lights are energized by the secondary bus and are controlled by a rheostat switch (15, figure 1–8; 8, figure 1–11) marked CONSOLE LIGHTS with three positions, OFF, DIM and BRIGHT. When placed between the DIM and BRIGHT positions on modified aircraft the landing gear selector handle light brightness is controlled.

INTERNAL & EXTERNAL LIGHTS

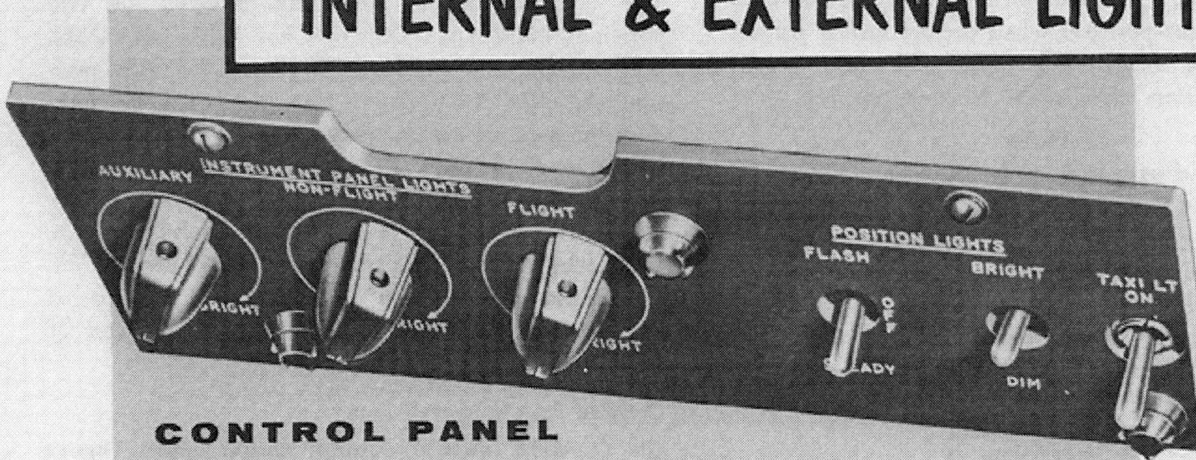

Figure 4–9

COCKPIT LIGHT.

A type C4A cockpit light (9, figure 1–10; 10, figure 1–11) is mounted on the right side of the cabin. The light is provided with an extension cord and it may be removed from its mounting bracket to be used as a portable light. The light is powered by the primary bus and controlled by a rheostat switch located on the light. The rheostat switch controls the intensity of the light for continuous illumination. A push-button type switch on the light may be used for intermittent light use. The light is equipped with a red filter which may be removed and the light used as a white spot light.

LANDING LIGHTS.

A landing light, mounted on an electrically retracted door is installed on each wing tip. The door and light are both actuated with power from the primary bus and controlled by a landing light switch.

Landing Lights Switch.

The landing light switch (34, figure 1–4; 34, figure 1–5; 34, figure 1–6) has three positions EXTEND & ON, OFF and RETRACT. The EXTEND & ON position opens the doors and turns the landing lights on. The OFF position will turn the lights off if the doors are open but the doors will not retract. The RETRACT position retracts the landing lights to the fully closed position.

TAXI LIGHTS.

The taxi light is mounted on the nose wheel strut and is powered by the secondary bus and controlled by an ON-OFF switch (figure 4–9) located above the right console. A switch in the nose wheel well will automatically put the light out when the nose wheel is retracted.

Note

The taxi light will illuminate only when the nose gear is down and locked. Therefore, when viewed from the ground, it may be used as an indication of a safe nose gear condition.

POSITION LIGHTS.

Position lights consist of a red light on the left wing tip, a green light on the right wing tip, a yellow and a white light on the tail, and a white light on the top and the bottom of the fuselage. All lights are powered from the secondary bus and controlled by a position light switch (figure 4–9), having three positions, STEADY, OFF and FLASH. In the STEADY position the wing, tail, and fuselage lights will provide continuous illumination. In the FLASH position the wing and white tail lights will flash alternately with the yellow tail light, and the fuselage lights will be steady. The intensity of the position lights is controlled by a DIM-BRIGHT switch (figure 4–9) adjacent to the position light switch.

OXYGEN SYSTEM.

LIQUID OXYGEN SYSTEM.

The liquid oxygen system consists of a five liter capacity vacuum insulated container, buildup coils, check valves, relief valves and quantity gage. Liquid oxygen is stored in the vacuum container and passes into the buildup coils. Here it evaporates into gaseous oxygen and passes into the oxygen regulator at approximately 70 PSI. The oxygen system from the regulator to the pilot's oxygen mask is identical with previous aircraft. Excessive pressures in the system between the vacuum container and the regulator are relieved through the relief valves and vented overboard. A buildup and vent valve is provided for servicing the oxygen system. It is recommended that this valve be left in the vent position when the aircraft is parked as less loss of oxygen will occur with the valve in the vent position than in the buildup position.

Note

The liquid oxygen quantity gage should read between 4 and 4½ liters when the system is fully charged. Do not be alarmed that the gage does not read 5 liters, since it is impossible to charge the liquid oxygen converter to 5 liters. Use the oxygen duration chart to determine your oxygen duration for the indicated supply.

CAUTION

When installing the canopy cover, care should be exercised to avoid damage to the oxygen filler cover door and the buildup and vent valve when the valve is left in the vent position.

REGULATOR.

An automatic pressure breathing diluter-demand oxygen regulator (figure 4—10) is installed on the right console and includes a low pressure gage and a flow indicator. The regulator automatically supplies the proper mixture of oxygen and air at all altitudes with provisions for positive pressure breathing at high altitudes. Either Type D-1 or Type D-2 automatic pressure demand regulators may be installed. The Type D-2 regulator differs from the late Type D-1 regulator in that the slight positive pressure from 8,000 to 28,000 feet has been removed and a better panel and instrument lighting system has been included. The slight positive pressure has been removed to conserve oxygen. It will be noted in the oxygen duration chart that an increasing number of manhours of oxygen is available above 25,000 feet altitude when operating with the diluter lever in the NORMAL OXYGEN position. This is caused because with increasing altitude, the volume of an equivalent mass of air at sea level increases, the regulator attempts to maintain a constant mass of flow of oxygen to the lungs by increasing the oxygen flow from the oxygen system and decreasing accordingly the amount of air mixed with the oxygen. Beyond the altitude at which 100 per cent oxygen is being used, further expansion of the gas will occur and, unless a pressurized system is used, the lungs cannot expand sufficiently to absorb the normal oxygen consumption. Therefore, even with a pressurized system though not as soon, an altitude will be reached beyond which, less and less mass will be absorbed because of the continually expanding gas.

Regulator Diluter Lever.

The regulator diluter lever (figure 4-10) has two positions, NORMAL OXYGEN and 100% OXYGEN. When the lever is in the NORMAL OXYGEN position, the regulator unit will function to provide automatic mixing of air and oxygen in sea level concentration at all altitudes. In the event that the regulator malfunctions, a pressure relief valve in the regulator unit will protect the pilot from excessive pressure. When the lever is in the 100% OXYGEN position, the automatic air-oxygen mixing feature is by-passed and 100 per cent oxygen is supplied regardless of altitude.

Oxygen Supply Shutoff Lever.

An ON-OFF oxygen supply shutoff lever (figure 4—10) is located on the aft end of the regulator. When the lever is in the ON position, system oxygen is supplied to the regulator unit. When the lever is in the OFF position the oxygen supply to the regulator is shut off. However, the supply valve is lockwired to the ON position. Soft copper wire is used so that the supply may be turned off by the pilot in an emergency.

Oxygen Emergency Toggle Lever.

The oxygen emergency toggle lever (figure 4—10) is provided for emergency operation or to supply maximum pressure for leakage test of the oxygen mask. The emergency toggle lever should remain in the center position at all times, unless an unscheduled pressure increase is required. Moving the toggle lever either to the left or right of its center position to the EMERGENCY position, provides continuous positive pressure to the mask for emergency use. When the toggle lever is depressed in the center position, it provides positive pressure to test the mask for leaks.

CAUTION

When positive pressures are required, it is mandatory that the oxygen mask be well fitted to the face. Unless special precautions are taken to insure no leakage, then continued use of positive pressure under these conditions will result in the rapid depletion of the oxygen supply. In aircraft which employs liquid oxygen, this condition would result in extremely cold oxygen flowing to the mask.

Low Pressure Gage.

The low pressure gage (figure 4—10) incorporated in the oxygen regulator unit records the oxygen pressure being supplied to the regulator. The gage scale reads

OXYGEN REGULATOR

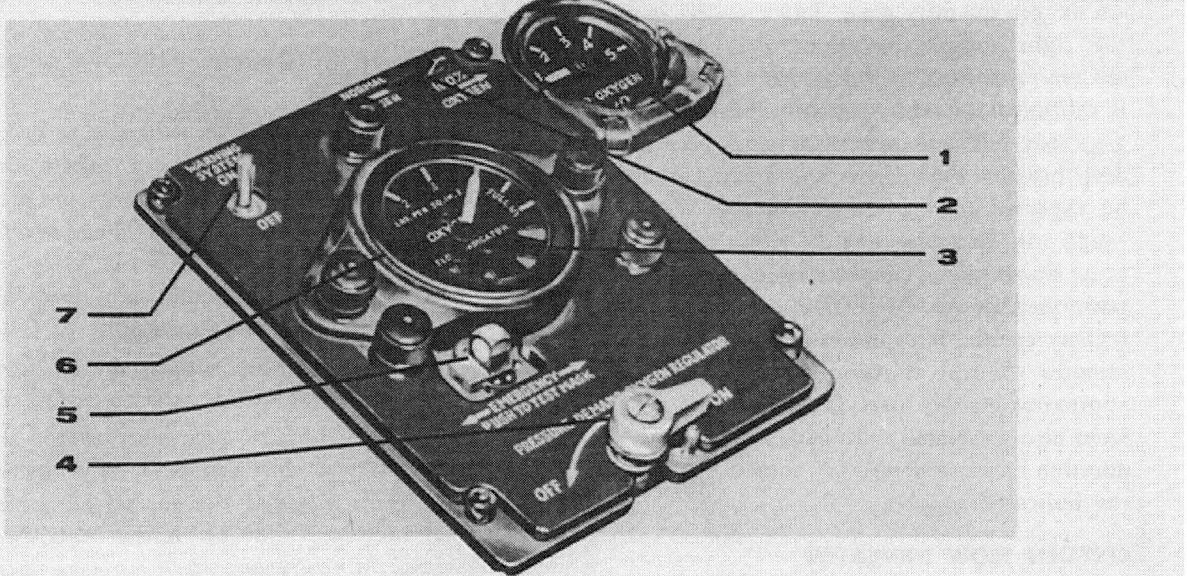

1. *Liquid Oxygen Quantity Gage*
2. *Regulator Diluter Level*
3. *Oxygen Flow Indicator*
4. *Oxygen Supply Shut-off Lever*
5. *Oxygen Emergency Toggle Lever*
6. *Oxygen Pressure Gage*
7. *Oxygen WarningSystem Switch (Deactivated)*

Liquid Oxygen Duration Chart

CREW MEMBER OXYGEN DURATION — HOURS

BLACK FIGURES Indicate Diluter Lever Normal

WHITE FIGURES Indicate Diluter Lever "100%"

1 Crew Member
1 Type A-3 Converter

CABIN ALTITUDE FEET	Diluter Lever	GAGE QUANTITY — LITERS 5	4	3	2	1	BELOW 1
40,000	100% (white)	24.0	19.2	14.4	9.6	4.8	EMERGENCY Descend to Altitude Not Requiring Oxygen
	Normal (black)	24.0	19.2	14.4	9.6	4.8	
35,000	100% (white)	24.0	19.2	14.4	9.6	4.8	
	Normal (black)	24.0	19.2	14.4	9.6	4.8	
30,000	100% (white)	17.3	13.8	10.4	6.9	3.5	
	Normal (black)	17.8	14.2	10.6	7.1	3.5	
25,000	100% (white)	13.3	10.6	8.0	5.3	2.7	
	Normal (black)	16.8	13.4	10.0	6.7	3.4	
20,000	100% (white)	10.1	8.1	6.1	4.0	2.0	
	Normal (black)	19.0	15.2	11.4	7.6	3.8	
15,000	100% (white)	7.8	6.2	4.6	3.1	1.5	
	Normal (black)	23.0	18.4	13.8	9.2	4.6	
10,000	100% (white)	6.6	5.3	4.0	2.6	1.3	
	Normal (black)	23.0	18.4	13.8	9.2	4.6	

Figure 4—10

from 0 to 500 PSI and the operating pressure is approximately 70 PSI. Tolerance and operating conditions can cause this pressure to vary between 65 and 115 PSI in the liquid oxygen system.

OXYGEN QUANTITY GAGE.

An oxygen quantity gage (figure 4–10) is installed on the right console to record the amount of liquid oxygen remaining in the vacuum container. The gage is calibrated to read from 0 to 5 liters. The quantity gage may fluctuate approximately 1 liter while taking deep breaths. Fluctuation and erratic indications may be expected during maneuvers and while flying in rough air. Excessive engine vibration, etc. may also cause fluctuation. After the vent and buildup valve is positioned in the BUILDUP position, the gage will register erratic, false indications for approximately 30 minutes. A full converter will then indicate only approximately 4½ liters due to the vapor loss resulting from heat generated during servicing. Use the oxygen duration chart to determine your oxygen duration for the indicated supply.

OXYGEN FLOW INDICATOR.

A blinker type oxygen flow indicator (figure 4–10) is incorporated in the pressure demand regulator unit. Black and luminescent segments alternately appear through four slots in the indicator face with each breath taken through the oxygen mask.

OXYGEN WARNING SYSTEM.

The oxygen warning system consisting of a warning light on the main instrument panel and an ON-OFF switch on the oxygen regulator panel has been deactivated.

PRESSURE DEMAND OXYGEN MASKS.

Only the type A-13, A-13A or MS22001 pressure demand oxygen mask will be used with the automatic pressure demand oxygen regulator. These masks can be identified by the presence of a gray anodized aluminum exhalation valve (pressure-compensating), which is located in the mask directly below the chin position. Pressure demand masks, when used at altitude, will occasionally produce a distinct vibration in the mask that can be identified by a "wheezing" sound. This condition may be overlooked, in that operational qualities are not disturbed in any manner. If a blocking condition (cannot exhale) occurs during flight, a "sharp" exhalation will usually correct the difficulty. In the event a "sharp" exhalation does not relieve the blocking condition, the mask may be lifted off the face momentarily at the chin section. Extreme caution must be exercised in using this procedure, since the danger of hypoxia increases rapidly above 30,000 feet cockpit altitude. Oxygen masks other than those specified above will not be used with automatic pressuer demand oxygen regulators. Use of unauthorized oxygen masks results in rapid depletion of the aircraft oxygen supply and pressure breathing required at altitude will be lost.

USE OF AUTOMATIC PRESSURE DEMAND REGULATOR.

1. The diluter lever will always be set at the NORMAL OXYGEN position, except in cases where noxious gases are suspected, or pre-breathing of oxygen is deemed necessary. These exceptions are rare, and if the diluter lever is placed in the 100% OXYGEN position, extreme care must be exercised in monitoring the oxygen supply.
2. Turn oxygen supply shutoff lever to the ON position if not already safety wired in the ON position.
3. Press oxygen emergency toggle lever straight in to test mask for leakage at any altitude. Place lever to right or left of the normal off position to provide an increased flow of oxygen only in case of emergency. If emergency use is necessary, however, extreme caution must be used to prevent rapid loss of system pressure through the emergency valve.
4. As breathing through the mask is started, the oxygen flow indicator should start functioning. The proper flow of oxygen will be automatically maintained by the regulator unit.
5. At approximately 10,000 feet (cockpit pressure) the D-1 regulator will supply a "SAFETY PRESSURE" or continuous flow if the pilot's oxygen mask is off or loose fitting. This pressure increases with altitude and will occur in both the NORMAL and 100% OXYGEN positions. The D-2 oxygen regulator has the same characteristics as the D-1 except that the "SAFETY PRESSURE" comes on at approximately 30,000 feet (cockpit pressure).
6. Blow into the mask if the "SAFETY PRESSURE" or flow occurs at cockpit altitudes below those shown above; if the leakage stops the regulator is satisfactory. Numerous instances of the presence of carbon monoxide in the flight compartment of jet aircraft have been suspected. Some of these instances have been brought to light through accident investigations. There are various possibilities by which carbon monoxide may enter the compartment during ground operation; however, as yet, neither the exact concentration nor the exact sources have been determined, except as indicated below. Consequently, the following instructions should be complied with. If the subject aircraft is to be operated under possible conditions of carbon monoxide contamination, such as during

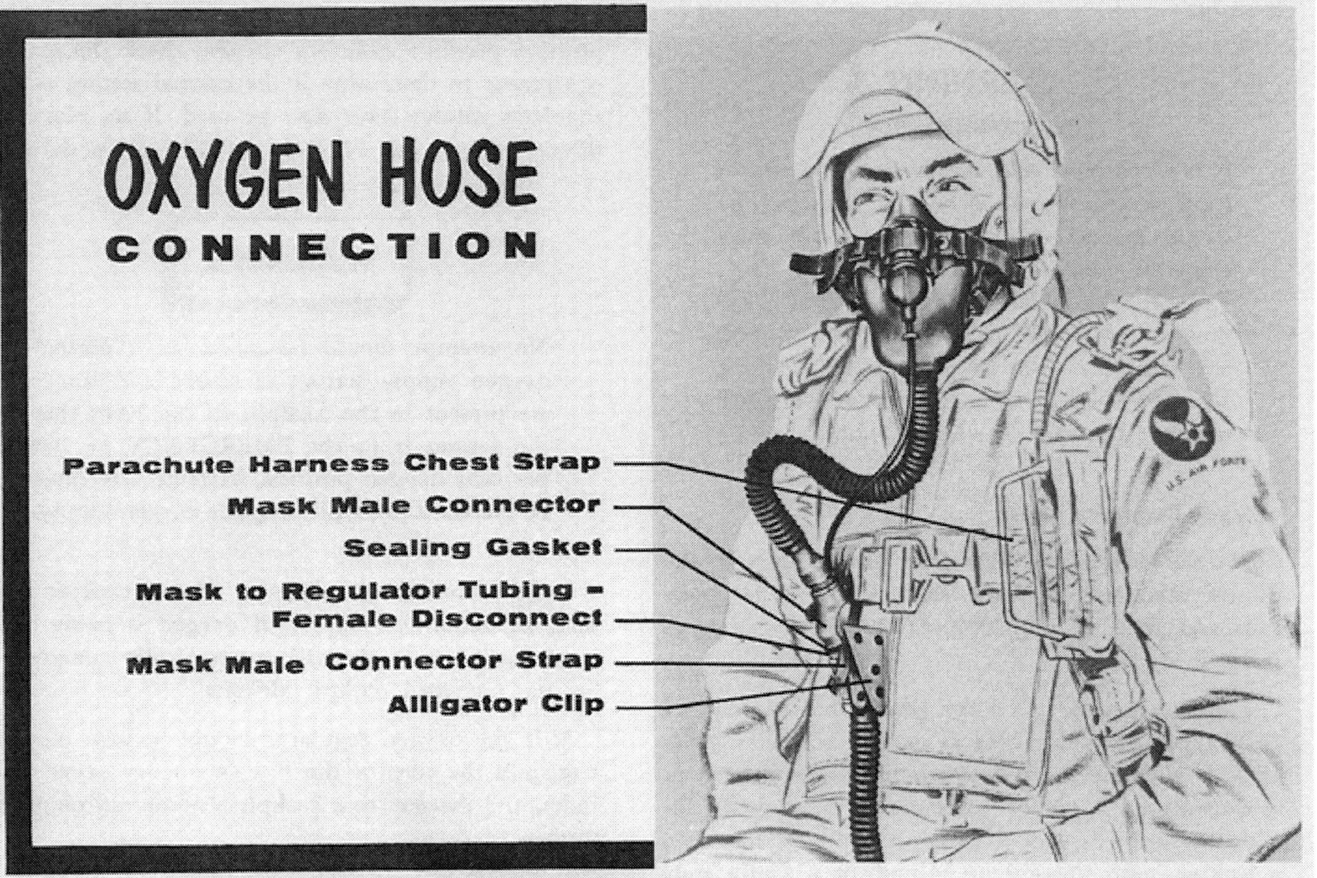

Figure 4—11

"runup" or taxiing directly behind another operating jet aircraft or during "runup" with its tail into the wind the following procedure shall be used:

1. Before starting engine, don oxygen mask, connect hose to oxygen regulator, and place diluter lever at the 100% OXYGEN position.
2. Whenever contamination is suspected, 100 per cent oxygen will be used during ground operation and take-off.
3. After contamination is no longer suspected, place the diluter lever to the oxygen regulator at the NORMAL OXYGEN position.

WARNING

The oxygen diluter lever must be returned to the NORMAL OXYGEN position as soon as possible because the use of 100 per cent oxygen throughout a long mission will so deplete the oxygen supply as to be hazardous to the pilot.

OXYGEN SYSTEM CHECK.

All Flights.

With the liquid oxygen system, the liquid quantity gage should indicate between 4 and 5 liters and the pressure gage on the regulator should read approximately 70 PSI. Tolerance and operating conditions can cause this pressure to vary between 65 and 115 PSI. There should be no evidence of steady venting from the oxygen vent port.

Oxygen Hose Connection.

The oxygen hose will be connected to the parachute chest strap in the following manner to prevent inadvertent opening of the parachute chest strap snap during seat ejection. (See figure 4-11).

1. Attach the connector tie-down strap on the mask male connector to the parachute chest strap by routing the tie-down strap under the chest strap as close to the chest strap snap as possible, up behind the chest strap, then down in front of the chest strap, then around again, and finally snapped to the connector.
2. Connect the mask-to-regulator tubing female disconnect to the mask male connector, listen for the click and check that the sealing gasket is only half exposed.
3. Attach the alligator clip to the end of the mask male connector strap.

WARNING

It is imperative that the connector tie-down strap on the mask male connector not be single looped around the parachute chest strap or attached to the chest strap snap. Failure to double loop the connector tie-down strap around the parachute chest strap will result in the tie-down strap slipping into the chest strap snap and inadvertently opening the snap during bail-out.

Oxygen Regulator Check.

Check the oxygen regulator prior to take-off with the diluter valve first at the NORMAL OXYGEN position and then at the 100% OXYGEN position as follows:

1. Remove mask and blow gently into the end of the oxygen regulator hose as during normal exhalation. If there is a resistance to blowing, the system is satisfactory. Little or no resistance to blowing indicates a faulty demand diaphragm or diluter air valve, a leaking mask-to-regulator tubing, or a faulty ejection seat quick disconnect.

2. With oxygen supply shutoff lever in the ON position, oxygen mask connected to regulator, diluter lever in 100% OXYGEN position, and normal breathing, conduct the following check:

    a. Deflect emergency toggle lever to right or left. A positive pressure should be supplied to mask. Return emergency toggle lever to center position.

    b. Depress emergency toggle lever straight in. A positive pressure should be applied to the mask. Hold breath to determine if there is leakage around mask. Release emergency toggle lever, positive pressure should cease.

3. Return diluter lever to NORMAL OXYGEN.

EMERGENCY OPERATION.

1. Should symptoms of hypoxia be suspected, or if smoke or fumes should enter the cockpit, immediately place the emergency toggle lever in the emergency position. After determining that sufficient oxygen is being received, revert to 100 per cent oxygen, by placing regulator diluter lever in the 100% OXYGEN position and turning the emergency toggle lever off. If it is then ascertained that the 100 per cent oxygen position provides sufficient oxygen, check the oxygen equipment to determine if the normal setting of the regulator diluter lever may be used. If so, place the diluter lever in the NORMAL OXYGEN position.

WARNING

No attempt should be made for a normal oxygen supply setting if smoke and fumes are present in the cockpit. In the event that the system is in the EMERGENCY or 100 per cent oxygen position, extreme care must be exercised in monitoring the oxygen supply.

2. In the event of accidental loss of cockpit pressure, no action is required if oxygen is being used, as the regulator unit will automatically compensate for the increased cockpit altitude.

3. If the oxygen regulator should become inoperative, pull the cord of the H-2 emergency oxygen cylinder, and descend to a cockpit altitude not requiring oxygen.

AIR REFUELING SYSTEM.

The aircraft is equipped with an air refueling system (figure 4–12) which enables the aircraft to be refueled in the air from a tanker using a flying boom. The receiver is located on the left wing upper surface and is concealed by a flush type door, which is hydraulically operated. During the refueling operation, the engine is operated with the fuel tank selector in the ALL TANKS position, so that the engine is fed from the main tank. The forward, wing, main and pylon tanks are filled during the refueling cycle. Once the receiver door in the wing is open and the tanker's boom is inserted in the nozzle, refueling sequence is accomplished electrically through an amplifier, which is powered from the primary bus. When the tanks are full, the fuel flow in the refueling lines is reduced to an amount equal to the engine consumption and the fuel pressure in the lines increases. These changes are noted in the tanker aircraft and a disconnect is effected. An automatic disconnect will be accomplished if the fuel pressure is excessive in the refueling lines, if rough air causes excessive tension on the nozzle, or by any uncontrolled or intentional change in flight attitude of the receiver aircraft wherein a conical angle of 15 degrees from the normal is exceeded. Provision is made so that the forward, main or wing tanks can be isolated, if damaged, from the refueling system. In the

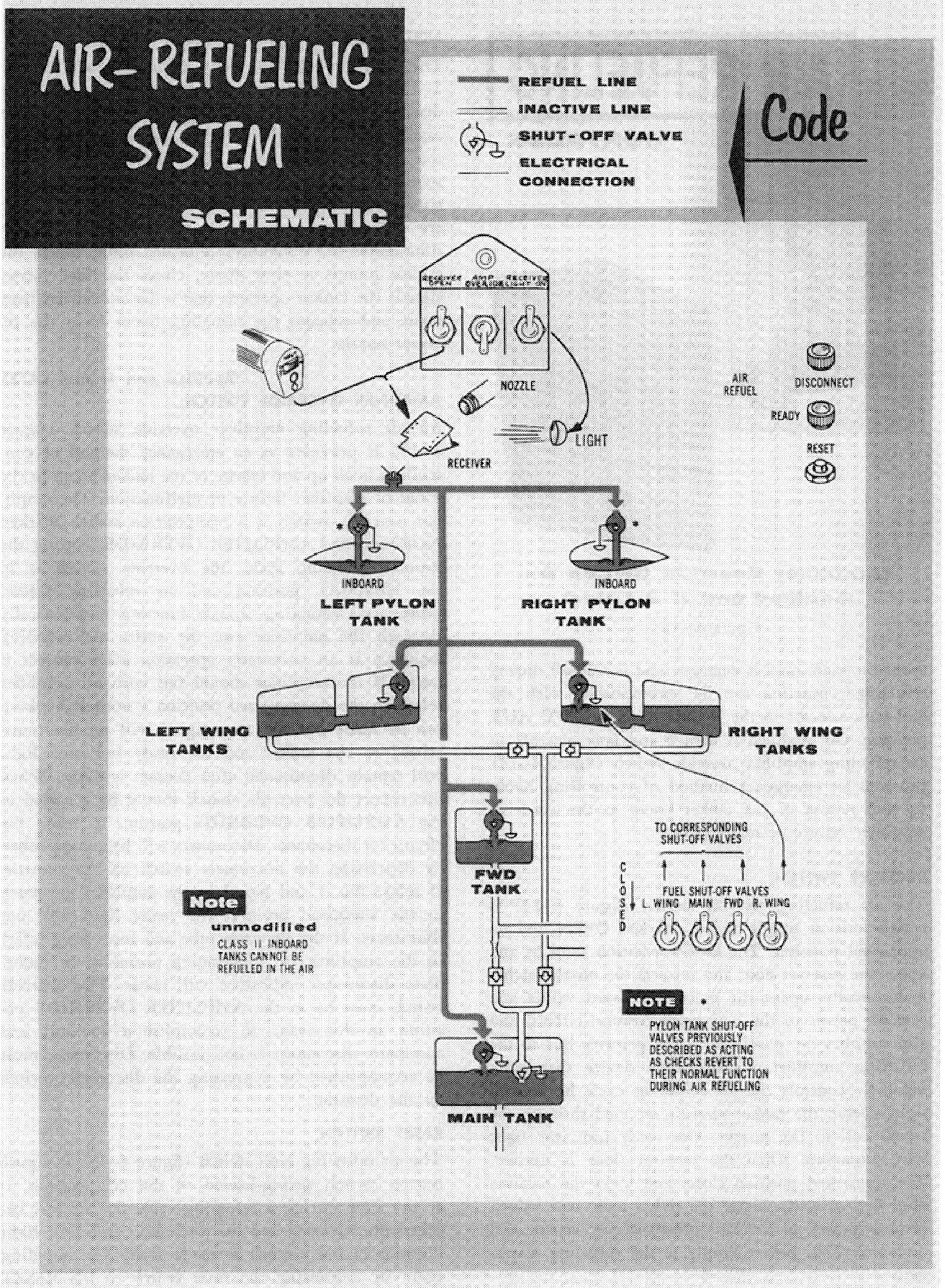
AIR-REFUELING
SYSTEM
SCHEMATIC
REFUEL LINE
INACTIVE LINE
SHUT-OFF VALVE
ELECTRICAL
CONNECTION
Code
RECEIVER OPEN
AMP OVERRIDE
RECEIVER LIGHT ON
NOZZLE
LIGHT
RECEIVER
AIR REFUEL
DISCONNECT
READY
RESET
INBOARD
LEFT PYLON TANK
INBOARD
RIGHT PYLON TANK
LEFT WING TANKS
RIGHT WING TANKS
TO CORRESPONDING SHUT-OFF VALVES
CLOSED
FUEL SHUT-OFF VALVES
L. WING MAIN FWD R. WING
FWD TANK
Note
unmodified
CLASS II INBOARD TANKS CANNOT BE REFUELED IN THE AIR
NOTE
PYLON TANK SHUT-OFF VALVES PREVIOUSLY DESCRIBED AS ACTING AS CHECKS REVERT TO THEIR NORMAL FUNCTION DURING AIR REFUELING
MAIN TANK

Figure 4—12

AIR REFUELING

CONTROLS

(Amplifier Override Switch On Modified and G & Later)

Figure 4—13.

event the main tank is damaged and is shut-off during refueling, operation can be accomplished with the fuel tank selector in the WING AUX or FWD AUX position. On modified A thru F and later aircraft an air refueling amplifier override switch (figure 4—13) provides an emergency method of controlling hook-up and release of the tanker boom in the event of amplifier failure or malfunction.

RECEIVER SWITCH.

The air refueling receiver switch (figure 4—13) is a two-position toggle switch marked OPEN and an unmarked position. The OPEN position unlocks and opens the receiver door and retracts the nozzle latches hydraulically, opens the pylon tank vent valves and cuts off power to the tank-pressurization circuit, and also supplies d-c power from the primary bus to the refueling amplifier, an electronic device that automatically controls the air refueling cycle by control signals from the tanker aircraft received through the signal coil in the nozzle. The ready indicator light will illuminate when the receiver door is opened. The unmarked position closes and locks the receiver door hydraulically, closes the pylon tank vent valves, restores power to the tank-pressurization circuit and disconnects the power supply to the refueling amplifier.

NOZZLE DISCONNECT SWITCH.

The gun-bomb-rocket sight caging switch (figure 1—15) on the throttle control, is used as the nozzle disconnect switch when refueling in the air. The caging switch is in the air refueling circuit only when the receiver switch is in the OPEN position. The switch is depressed to IFR DISCONN if it is desired to end the refueling cycle before the fuel tanks are full. Depressing the nozzle disconnect switch illuminates the disconnect indicator light, causes the tanker pumps to shut down, closes the fuel valves, signals the tanker operator that a disconnect has been made and releases the refueling boom from the receiver nozzle.

Modified and G and LATER

AMPLIFIER OVERRIDE SWITCH.

An air refueling amplifier override switch (figure 4—13) is provided as an emergency method of controlling hook-up and release of the tanker boom in the event of amplifier failure or malfunction. The amplifier override switch is a two-position switch marked NORMAL and AMPLIFIER OVERRIDE. During the normal refueling cycle, the override switch is in the NORMAL position and air refueling system power and actuating signals function automatically through the amplifier and the entire air refueling sequence is an automatic operation after contact is made. If the amplifier should fail with all amplifier relays in the de-energized position a normal hook-up can be made but the start signal will not be transmitted to the tanker and the ready indicator light will remain illuminated after contact is made. When this occurs the override switch should be actuated to the AMPLIFIER OVERRIDE position to ready the circuit for disconnect. Disconnect will be accomplished by depressing the disconnect switch on the throttle. If relays No. 1 and No. 2 in the amplifier are stuck in the energized position the ready light will not illuminate. If the thyraton tube and remaining relays in the amplifier are functioning normally an immediate disconnect indication will occur. The override switch must be in the AMPLIFIER OVERRIDE position, in this event, to accomplish a hook-up, and automatic disconnect is not possible. Disconnect must be accomplished by depressing the disconnect switch on the throttle.

RESET SWITCH.

The air refueling reset switch (figure 4—13) is a push button switch spring-loaded to the off position. If at any time during a refueling cycle the aircraft becomes disconnected and the disconnect indicator light illuminates the aircraft is made ready for refueling again by depressing the reset switch to the RESET

Figure 4—14.

position. The refueling system can also be made ready for refueling by closing then reopening the receiver door.

Note

> Holding the reset button in the depressed position during nozzle contact in a refueling operation will cut off power to the refueling amplifier and cause a disconnect.

RECEIVER LIGHT SWITCHES.

The air refueling receiver light switch (figure 4—13) is a toggle switch with an ON and an unmarked position. The ON position supplies power from the primary bus to illuminate a flood light on the side of the aircraft which lights up the receiver nozzle to aid the boom operator in the tanker during night refueling operations. The light can also be used as a formation light at night.

FUEL SHUT-OFF VALVE SWITCHES.

The same fuel lines are used for transfer of fuel from the external to the internal tanks as are used to supply all tanks during air refueling. The fuel shut-off switches (figure 4—14) provide a means of controlling fuel flow to the individual internal tanks by closing valves at the fuel line entrance to each tank. Each of the four fuel shut-off valves is controlled by a two-position toggle switch. The switches are marked L WING, MAIN, FWD and R WING with an arrow indicating the CLOSED position. During normal operation the switches are left in the up position, which allows fuel to transfer from the external to the internal tanks and also allows all tanks to be refueled from the refueling receptacle. By placing the fuel shut-off valve in the CLOSED position the respective fuel tank will not receive fuel by transfer or air refueling although a slight internal leakage is normal. When the main fuel shut-off switch is in the CLOSED position both valves in the main tank are closed and fuel from neither internal nor external tanks can be transferred. This system is provided to isolate each of the internal tanks, in the event of battle damage to the tank, or failure of the booster pump in the tank. The fuel shut-off valves are actuated by the primary bus.

READY INDICATOR LIGHT.

The ready indicator light (figure 4—13) powered by the primary bus is a green light marked READY and when illuminated indicates that the receiver nozzle is open, power is supplied to the refueling amplifier and the amplifier is ready for the refueling cycle. The ready indicator light will go out when contact is made.

DISCONNECT INDICATOR LIGHT.

The air refueling disconnect indicator (figure 4—13) powered by the primary bus is an amber light marked DISCONNECT. Illumination indicates the tanker's fuel pumps are shut down, the tanker's fuel valves are closed and that the boom nozzle has been disconnected from the receiver. If the disconnect indicator is illuminated the reset switch must be depressed or the receiver door closed and reopened before the refueling system will be ready to make another refueling cycle.

AIR REFUELING SYSTEM OPERATION.

Normal Operation.

(See figure 4—15.)

NON-TANDEM

Operation Without Hydraulic Pressure.

In the event of hydraulic system failure, continuous hand pumping is required during air refueling if automatic refueling is desired. Hand pump actuation at the rate of approximately nine cycles per minute should be adequate for maintaining sufficient pressure in the system to keep the refueling boom latches engaged. In the event hand pumping during the refueling operation is not desired or is impractical, refueling may be accomplished by the following procedure:

1. Open the refueling receptacle by means of hand pump actuation.
2. Make contact with the refueling boom. Hand pump actuation is not required during this operation if the refueling receptacle maintains an open position.
3. Advise tanker by radio communication to begin manual refueling.

AIR REFUELING

1.

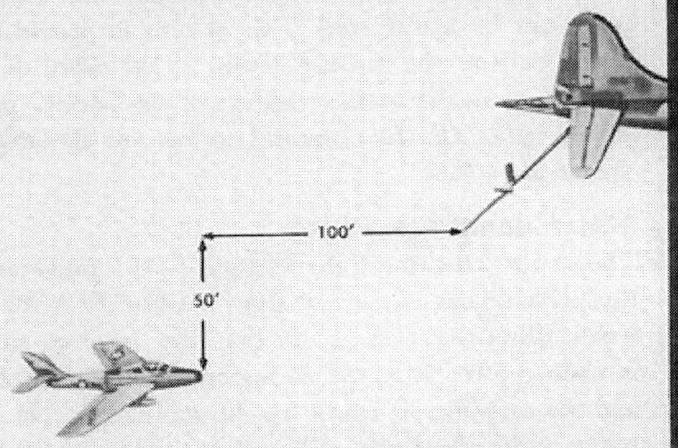

OBSERVATION POSITION

Take a position 100 feet directly behind and 50 feet below the tanker, out of the tanker downwash field, which is known as the observation position. Upon reaching this position the receiver pilot should trim the airplane, stabilize the power setting to maintain position, check the throttle friction lock and open the refueling slipway door. The speed brakes and flaps should be in the retracted position.

Note

Do not open the refueling receptacle at speeds above 350 Knots IAS.

2.

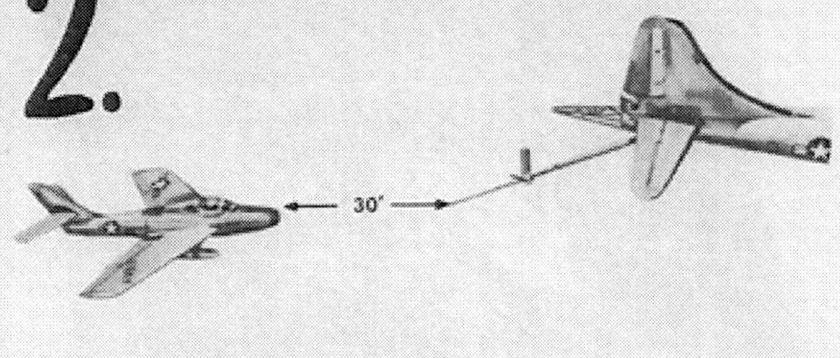

FORMATED POSITION

When going from the observation position to the formated position move forward and upward simultaneously to a position at the proper elevation but approximately 30 feet aft of the end of the boom. Then move forward slowly to contact the boom. It is also possible, though somewhat more difficult, to move from the observation position directly to the contact position. Since the receiver is now in the tanker's downwash, more power and a slightly different pitch trim may be required.

Note

The flaps may be extended 20 percent to aid in contacting the tanker.

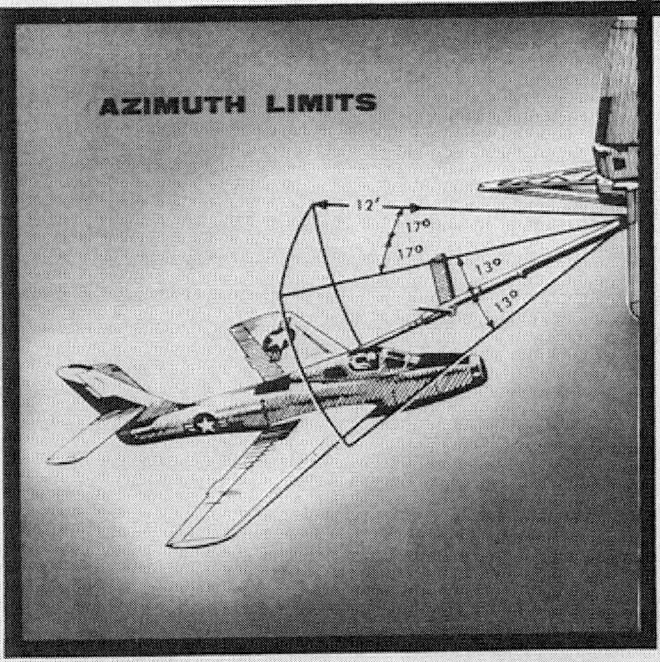

EMERGENCY OPERATION

Figure 4—14A(Sheet 1 of 2)

SYSTEM 3. PILOT TECHNIQUE

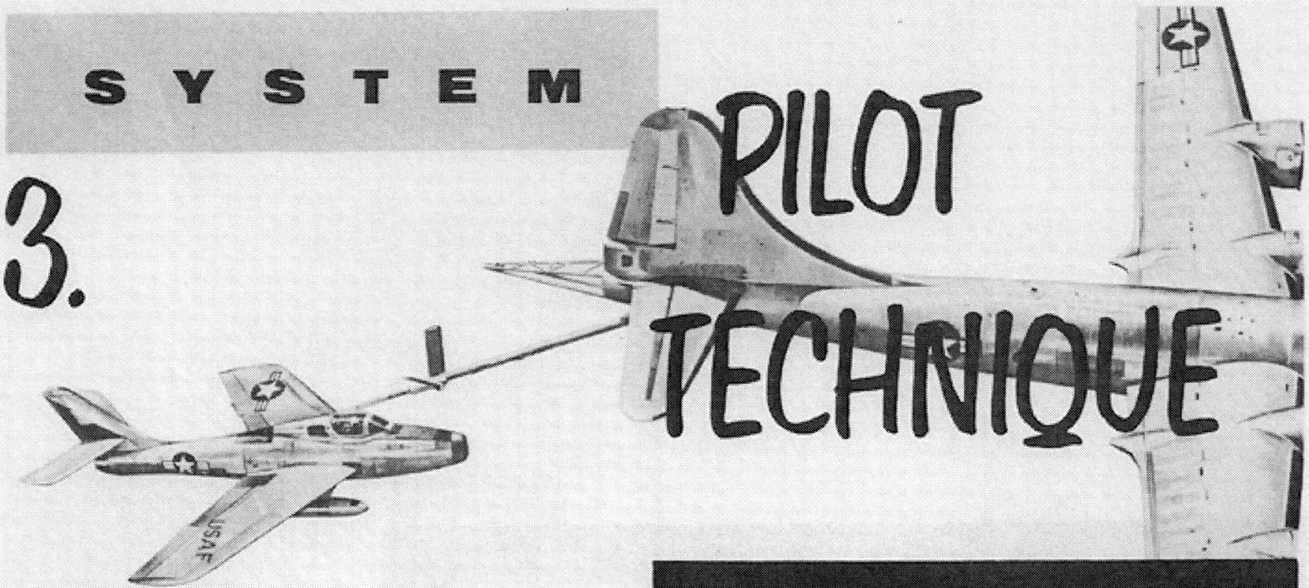

REFUELING

As soon as contact is made, indicated by the nozzle contact indicator light, refueling begins. The receiver pilot should advance the throttle as smoothly as possible to maintain position as the fuel is transferred. The increased power is required to overcome the rate of change of receiver weight, boom effects and the momentum of the fuel being transferred. Prompt throttle action is necessary to avoid dropping backward and downward because of the dynamic effects during transfer. The pilot has to anticipate this power change because it is very difficult to overcome any appreciable aft motion after it becomes established.

Note

After refueling, leave pylon tank air pressure switches in the OFF position until wing tank fuel level drops. The switches are then positioned to the ON posiion. This procedure will purge the fuel transfer lines of air.

CONTACT POSITION

When making contact with the boom the receiver pilot will rely primarily upon his visual observations and verbal instructions from the boom operator. The pilot director lights, located on the bottom of the tanker, are actuated by movements of the boom and therefore provide assistance only when in contact. When the receiver is properly positioned the boom is inserted by the boom operator and contact is made. If contact is broken, the pilot should recycle the amplifier before making another contact.

NORMAL DISCONNECTS

When all the tanks are full the nozzle will automatically disconnect and the boom retract. As soon as contact is broken the receiver pilot should leave the contact position by dropping straight down and aft until out of the tanker's downwash. The refueling door is then closed before increasing speed.

Note

The receiver pilot or the boom operator can effect a disconnect before the tanks are full by depressing the nozzle disconnect switch. Also an automatic disconnect will result if the flight envelope is exceeded in any direction. The limits of this envelope are defined by the boom position relative to the tanker. Disconnect occurs at:

1. Boom extension beyond 45 feet and closer than 33 ft to the tanker.

2. Boom elevation above 25 degrees and below 49 degrees relative to tanker horizontal center line.

3. Boom azimuth of 17 degrees left or right of tanker centerline.

EMERGENCY OPERATION

The word BREAKAWAY is an emergency term and is to be used only as such. When BREAKAWAY is heard over the radio during air refueling operations the following will be accomplished by the indicated crew members:

Tanker pilot: Increase power and climb on course.

Receiver pilot: Actuate the air refueling emergency disconnect switch on the throttle control. Decrease power (chop throttle). Drop aft of tanker. Do not dive under tanker or attempt to turn until well clear of the tanker and boom.

Note

If electrical disconnect fails, mechanical disconnect without damage can be made by the abrupt reduction of power allowing the receiver to slow down and effect a "Brute Force" disconnect.

Figure 4—14A (Sheet 2 of 2)

AIR REFUELING SYSTEM

OPERATION

1. Approach tanker at selected altitude and speed.

2. Fuel tank selector ALL TANKS unless operation is necessary in auxiliary.

3. Pneumatic compressor switch OFF.

4. Pylon tank air pressure switches OFF.

5. Receiver switch OPEN. Ready indicator light on.

6. Receiver light switch ON if refueling operation is at night.

7. Ready indicator light will go out after contact is made.

8. After refueling check disconnect indicator light on.

9. If the disconnect indicator light illuminates before refueling is completed, depress the reset switch.

Note

A disconnect may be accomplished at any time by depressing the disconnect switch.

10. Receiver switch, up position after refueling is completed.

Note

Fuel system operation is normal after the receiver door is closed.

11. Pneumatic compressor switch ON.

12. After refueling, leave pylon tank air pressure switches in the OFF position until wing tank fuel level drops. The switches are then positioned to the ON position. This procedure will purge the fuel transfer lines of air.

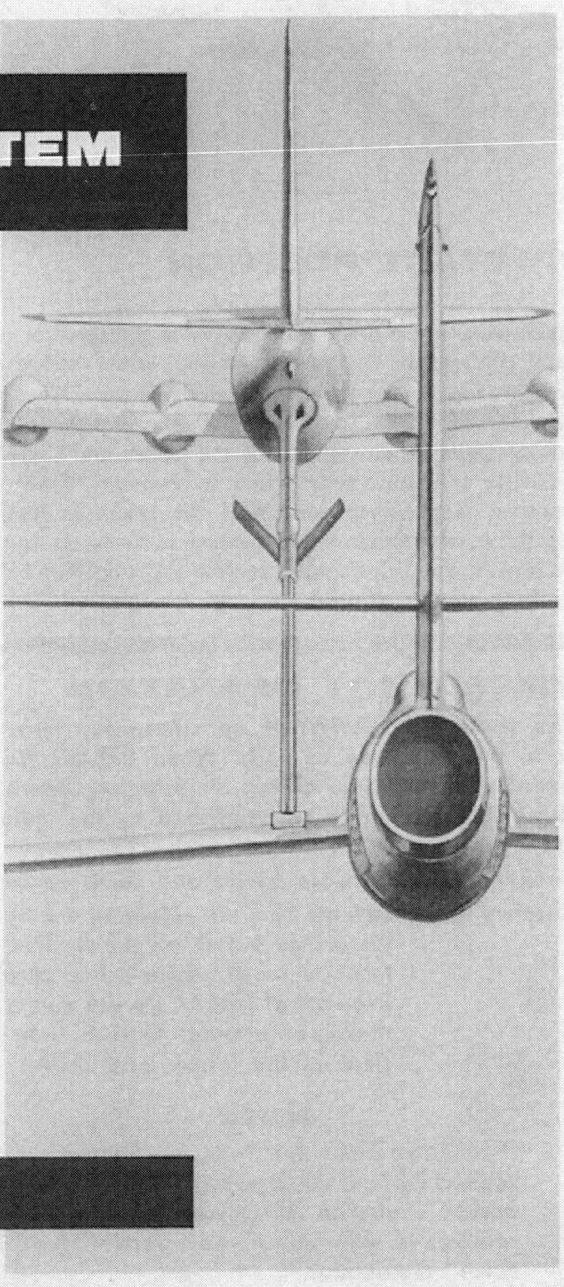

Figure 4—15

4. Maintain contact with the refueling boom by throttle control manipulation. Since the hydraulically operated refueling boom locking latches are ineffective without hydraulic pressure, boom contact must be maintained by constant thrust of the receiver aircraft.

5. Retard throttle to effect disconnect, when advised that refueling is complete.

6. Close refueling receptacle by means of hand pump actuation.

Battle Damage.

If the main, forward or wing tanks have been damaged and it is desired to refuel, these tanks may be isolated from the refueling system by placing the desired fuel shutoff valve in the CLOSED position. Refuel as in normal operation.

AUTOMATIC PILOT.

An MB-2 automatic pilot is installed in some airplanes. The autopilot will automatically hold the airplane on any predetermined course and altitude, change course at will with an exact coordinated turn, maintain the airplane at a selected altitude and pitch trim in laterally straight level flight or at a desired angle of climb or dive up to 45 degrees, and hold a fixed radius in climbing turns and dives. Changes in airspeed are automatically compensated for and control surface displacement is regulated in relation to airspeed and altitude. Automatic control originates in an AC powered gyroscopic unit which includes a vertical and a directional gyro as references. The vertical gyro establishes a flight reference about the lateral and longitudinal axis of the airplane while the directional gyro establishes a reference for the airplane's directional course in relation to the earth's magnetic field. Error signals are transmitted from both the vertical and directional gyros to the autopilot system for any deviation of the aircraft in flight. The autopilot compensates and corrects for these errors through the flight controls. An automatic altitude control senses changes in pressure and when engaged will maintain the airplane at a constant altitude. The automatic pilot can be easily overpowered manually at any time by the human pilot or it can be immediately disconnected by means of the autopilot release switch. If the autopilot is engaged with the aircraft in a wing-level climb or dive, the aircraft will continue on course in the climb or dive. However, if the aircraft is in a turn when the autopilot is engaged, a rapid roll to level attitude will take place as there is no roll synchronization.

INSTRUMENT POWER SWITCH (INVERTER SELECTION).

The instrument power switch (figure 1–7) is a three-position switch marked NOR, OFF and ALT. The NOR position must be selected for autopilot operation. This switch is described in detail in Section I.

FLIGHT CONTROLLER.

All control functions of the autopilot are centered about the flight controller (figure 4–16) which contains the autopilot engaging switch, the roll trim and pitch trim wheels and the turn knob. An automatic interlocking system is provided which prevents the autopilot from being turned on until it is warmed up, or if the "turn" knob is out of the neutral detent position. There is also an electrical circuit provided to automatically disengage the autopilot if the emergency hydraulic pressure pump switch is turned on.

Autopilot Engaging Switch.

The autopilot engaging switch (figure 4–16) is a rotary two-position switch marked OFF and ON. The ON position locks the autopilot to the flight control system. The engaging switch cannot be placed in the ON position unless the instrument power switch is in the NOR position and power is supplied to the autopilot for approximately three minutes, allowing sufficient time for the gyro to reach full RPM and stabilize. It is also impossible to select the ON position if the turn knob is out of the neutral detent position or enough hydraulic pressure is not available to extinguish the emergency hydraulic controls on indicator. The engaging switch will automatically return to the OFF position if AC or DC power supply fails, if the main inverter switch is positioned to the OFF position, if the emergency hydraulic pump switch is turned ON, or if the autopilot release switch on the control stick is depressed.

Pitch Trim Wheel.

The pitch trim wheel (figure 4–16), located on the left and right side of the flight controller and marked DN and UP, controls the pitch attitude of the airplane. If either pitch trim wheel is rotated aft for nose up and forward for nose down trim, the airplane will maintain the selected attitude. The pitch trim is limited to a climb or dive angle of approximately 45 (±5) degrees from the horizontal plane. The wheels may be rotated as rapidly as required, consistent with structural limitations and pilot comfort.

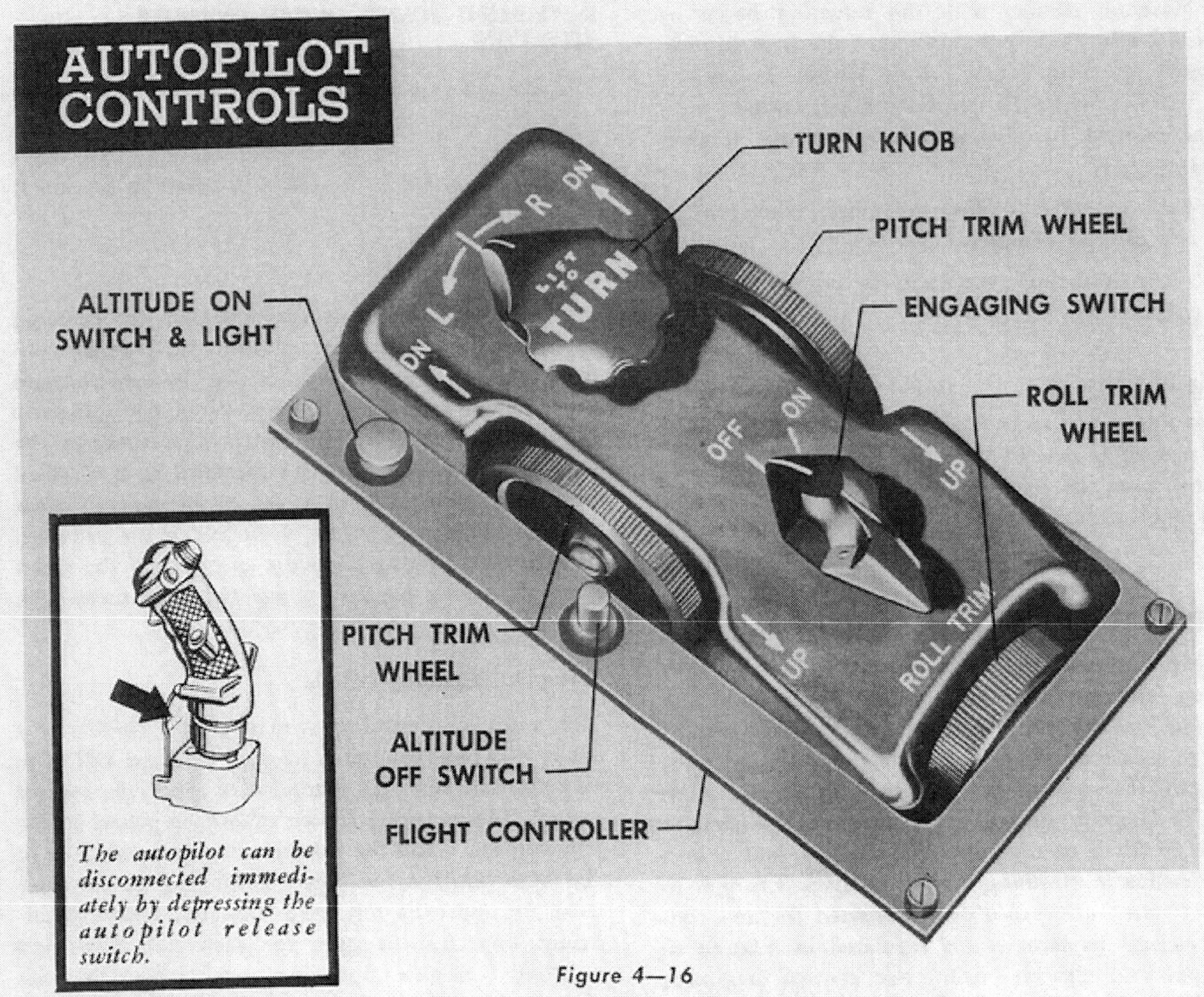

Figure 4—16

Roll Trim Wheel.

The roll trim wheel (figure 4—16) marked ROLL TRIM, controls the lateral trim of the airplane, if the roll trim wheel is rotated clockwise for right wing down or counterclockwise for left wing down, the airplane will maintain a selected trim. The roll trim is limited to approximately ±5 degrees from level flight.

Turn Knob.

The turn knob is a rotary switch marked LIFT TO TURN and has two extreme positions to the left and right of neutral marked L and R. If the turn knob is in the neutral detent position, it must be lifted before it can be turned to the L or R positions. When rotated to the L or R position, the airplane will make a coordinated turn to the left or right. The airplane will lock onto the directional heading it is taking at the time the turn knob is returned to the detent position. The bank angle is governed by the degree of right or left rotation of the turn knob from the neutral position. The bank angle is held to a maximum of 45 (±5) degrees from wing level flight position. The autopilot cannot be engaged if the turn knob is out of the neutral detent position.

Note

When flying an autopilot controlled turn, hold the flight controller turn knob out of detent until the airplane returns to wings-level attitude on the new course. Placing the turn knob in the detent position too soon will cause overshoot and skid condition until return to the "lock-on" course.

AUTOPILOT RELEASE SWITCH.

An autopilot release switch (figure 4—16) is installed on the forward side of the control stick. This is a spring-loaded switch and when depressed, automatically disengages the autopilot from the control system. The autopilot engaging switch, on the flight control, will automatically return to the OFF position when the release switch is depressed.

ALTITUDE CONTROL SWITCHES.

The automatic control consists of two momentary push-button switches (figure 4–16), one marked ALTITUDE ON and the other ALTITUDE OFF. When the altitude on switch is depressed, the automatic altitude control will hold the airplane to within ±30 feet or ±10 per cent of indicated airspeed in feet, whichever is greater in level flight. If engaged in a climb or dive up to 1,500 FPM, the airplane will be stabilized on engaged altitude within ±100 feet. During turns, the altitude should be held to within ±100 feet. The automatic altitude control can be switched off by depressing the altitude off switch, or by disconnecting the autopilot from the control system.

ALTITUDE CONTROL INDICATOR LIGHT.

The altitude control indicator light (figure 4–16) is incorporated with the altitude on switch. The light will illuminate when the altitude on switch is depressed and remains on as long as the automatic altitude control is in operation. Rotating the knurled rim of the light increases or decreases the intensity of the light for pilot comfort.

GROUND TEST BEFORE TAXIING.

WARNING

- Autopilot should be given a thorough check before each flight.
- No airplane with an autopilot write-up should be flown, unless the autopilot is electrically disconnected by pulling the two autopilot circuit breakers in the cockpit, bearing legend AUTOPILOT.

To insure proper operation while airborne, the autopilot should be given a thorough check prior to each flight as follows:

1. With flight controls in neutral and turn knob in detent, turn engaging switch to ON. When the autopilot is engaged on the ground, a slight oscillation of the control may be expected. This is normal and should not be encountered with air loads on the control surfaces when in flight. There should be no appreciable movement of the control stick or rudder pedals. A very slight momentary "kick" is permissible but there should be no movement to a new position.

Note

The engaging switch cannot be moved to the ON position unless the instrument power switch is in the NOR position, and the primary and secondary bus are energized for approximately 3 minutes.

2. Rotate roll trim wheel in each direction and check control stick for corresponding movement. Return ailerons to neutral by rotating roll trim wheel as required.

3. Lift turn knob out of neutral detent. There should be no appreciable movement of the control stick or rudder pedals. Rotate turn knob in each direction and check control stick for corresponding movement. There should be no perceptible rudder movement. Disengage the autopilot.

4. Trim control stick forward approximately two inches, wait five seconds then engage the autopilot. There should be no movement of the stick in pitch. Disengage autopilot. Trim control stick aft approximately two inches, wait five seconds then engage the autopilot. There should be no movement of stick in pitch.

WARNING

If abrupt stick movement occurs when the autopilot is engaged during this pitch synchronization check, the discrepancy must be corrected prior to flight or the autopilot deactivated by pulling the autopilot AC fuse.

Rotate pitch trim wheels in each direction and check control stick for corresponding movement. Return control stick to neutral using pitch trim wheels.

5. With autopilot still engaged, apply overpower momentarily to control stick in all directions and to rudder pedals in either direction. Positive opposition from the autopilot should be felt as overpower force is applied. Controls should return to the original position when overpower force is released.

CAUTION

The overpower check should be performed as rapidly as possible to avoid overheating the clutch mechanism. The clutch should not be forced to slip continuously for periods longer than 20 seconds out of every minute. In the pitch axis the overpower should be held only for a few seconds as the automatic trim will cause the trim to oppose the overpower, and if sustained will give full opposing trim.

6. Depress altitude on switch. Stick movement should not exceed ½ inch. Altitude control indicator light should remain illuminated.

7. Disengage the autopilot by depressing the release switch on the control stick. Autopilot engaging switch (on flight controller) should return to the OFF position. Manually check flight controls for correct operation and freedom of movement.

Note

As a safety feature, the automatic pilot will disengage if the emergency hydraulic pump switch is placed in the ON position. After making this check, manually recheck the flight controls for correct operation and freedom of movement.

WARNING

During preflight check out operations, automatic trim is in operation. At the conclusion of the preflight check out, the controls will be at an undetermined position that may include either extreme. Trim must be checked and reset before takeoff. Aileron and rudder neutral position can be determined by the aileron and rudder neutral trim indicator lights. The control stick should be centered for stabilator neutral.

INFLIGHT OPERATION.

Normal Engagement.

1. Trim the airplane for wing level flight, desired pitch trim and directional heading. Maintain stabilized flight for a few seconds.

Note

Automatic synchronization in pitch and in yaw permits the autopilot to be engaged on any heading and in any pitch attitude up to 45 degrees from level flight, without airplane attitude change. If engagement is made when the airplane is not in a wing level attitude, the airplane immediately will assume a near wing level attitude upon engagement.

2. Prepare to monitor the airplane controls and to overpower the autopilot, if necessary, in case engagement should cause a sudden attitude change as a result of autopilot malfunction during manual flight.
3. Position the engaging switch to ON.

CAUTION

Do not engage the autopilot at speeds above 425 knots IAS below 20,000 feet altitude.

4. After engagement, momentarily check overpower of the autopilot in the yaw, roll and pitch axes. Airplane should return to reference attitude. Control operations should be made at the flight controller while the autopilot is engaged.

CAUTION

The airplane controls should be monitored in autopilot flight whenever in proximity to the ground or other aircraft.

5. Lift the turn knob and adjust the roll trim for wing level flight.

Note

A small wing down error will cause the airplane to turn slightly until balanced by rudder deflection.

6. If necessary, correct the pitch attitude to that desired by means of the pitch trim wheels.
7. For desired change in course, turn knob will produce a coordinated turn up to 45 (±5) degree bank.

Engagement of Automatic Altitude Control.

Automatic altitude control is engaged by depressing the altitude on switch. The indicator light illuminates and stays on while the control is in the autopilot circuit.

Note

Automatic altitude control can be engaged in dive or climb. The airplane will return to and hold the altitude at which the switch is operated. Smoothest "lock-on" of altitude control is made with the airplane in level flight at the desired altitude.

CAUTION

If the altitude control is engaged in a steep dive or climb, the clutch in the altitude unit may slip and the aircraft will level at an altitude below or above the desired altitude at some discomfort to the pilot.

Disengagement of Autopilot.

1. Allow the autopilot to fly the airplane on a straight course for a short time following any autopilot controlled maneuvers. This will insure that the autopilot circuits have stabilized on the present altitude and heading.
2. Monitor the airplane controls, prepare to take immediate corrective action if necessary.
3. Place the engaging switch in the OFF position.

Note

The autopilot can be disengaged by depressing the release switch on the control stick. The engaging switch will automatically return to the OFF position.

Emergency Operation.

In the event any of the following malfunctions occur during flight, corrective action must be taken immediately.

a. Unusually heavy stick forces.

b. Airplane difficult to control.

c. Jerky or erratic controls.

d. Inadvertent engagement.

WARNING

The autopilot may possibly be engaged due to short circuits, even though circuit breakers are pulled.

Corrective Action.

1. Immediately depress the autopilot release switch on the control stick or place the autopilot engaging switch in the OFF position.
2. Instrument power switch—ALT.
3. In the event that G forces prevent the placing of the instrument power switch in the ALT position, place the emergency hydraulic pump switch to EMER PUMP or EMERG until G forces are relieved and instrument power switch is placed in ALT. Return emergency hydraulic pump switch OFF on tandem airplanes or to NORMAL on non-tandem airplanes.
4. If one of the above actions does not correct the malfunction and the airplane is under control, pull the AUTOPILOT circuit breaker located on the right circuit breaker panel.

WARNING

Autopilot may possibly be engaged due to short circuits, even though circuit breakers are pulled.

Note

Do not pull the AUTOPILOT circuit breaker on the left console (A THRU D) or the ALT INV CONT & AUTOPILOT circuit breaker on the right console (E and later) as this will stop the gyro motors on all aircraft and also the alternate inverter on E and later aircraft.

5. Return the inverter switch to the NOR position.
6. Land the airplane as soon as possible.

NAVIGATION EQUIPMENT.

DIRECTIONAL INDICATOR.

A type J-2 directional indicator is installed in the airplane which provides visual indication of the magnetic heading of the airplane. The indication is read on an indicator (6, figure 1–4; 5, figure 1–5; 5, figure 1–6) whose operation is governed by a GYRO whose spin axis is stabilized in a horizontal plane by means of a leveling device and whose orientation in azimuth is slaved to the earth's magnetic meridan by a direction-sensing component, located in the left sta-

This page intentionally left blank.

DIRECTIONAL INDICATOR

CONTROLS

Figure 4—17.

bilizer. The indicator requires both a-c and d-c power. The d-c power is supplied from the primary bus and the a-c power is supplied by the main or alternate inverter. The gyro is free to operate within 85 degrees from level flight in dive and climb, and in right and left bank. At the limits, it strikes mechanical stops, which render the indications on the directional gyro control and the settable dial indicator inaccurate. After return to level flight, errors up to 5 degrees in heading may be introduced; but the gyro will recover its erect and slaved positions automatically, in a period of 5 minutes or less, and thereafter will again resume correct indications until the limits are again exceeded. The flux valve unit of the remote compass transmitter remains pendulous through 30 degrees on both sides of the vertical, in pitch and roll. When these limits are exceeded, or a coordinated turn is being executed, the vertical components of the earth's field are picked up which results in flash signals. Restoration of the airplane to an attitude within these limits renders the flux valve unit pendulous again, and it automatically resumes correct sensing. A thermal switch in the amplifier provides fast slaving and leveling of the directional gyro, during the initial operation of the indicator.

Directional Indicator Slaving Switch.

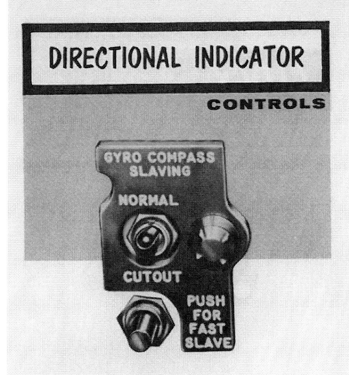

The directional indicator (gyro-compass) slaving switch (figure 4—17) has two positions: NORMAL and CUTOUT. The NORMAL position supplies power to the heating, leveling and slaving systems. The CUTOUT position cuts off the power supply to the control field of the slaving torque motor, and is used when the horizontal lines of magnetic force dip at 84 degrees or more.

Directional Indicator Fast Slaving Switch.

The directional indicator (gyro-compass) fast slaving switch (figure 4—17) marked PUSH FOR FAST SLAVE, is pushed in momentarily to shorten the time required to restore the gyro to its erect and slaved position, after level flight is resumed, following maneuvers in which the gyro has hit the mechanical stops. Approximately three minutes of fast slaving is obtained by depressing the fast slaving switch. The fast slaving switch is also used during initial starting of the indicator to assure both a-c and d-c power are supplied to the system simultaneously.

CAUTION

The fast slaving switch shall not be operated more than once in ten minutes. More frequent use of the switch will damage the torque motors and make the indicator inoperative or inaccurate.

Note

After the fast slaving switch is pressed, a time delay circuit maintains the fast slave action for approximately three minutes. During this three minute interval, any maneuvering of the airplane can induce errors into the equipment. At the completion of the three minutes, the system normally reverts to slow slave and the large errors which have been introduced will remain for a considerable time. Therefore, the fast slaving switch should not be used during flight, except when the airplane can be maintained in straight flight for at least three minutes after the fast slaving switch is depressed.

Starting.

The indicator will operate if the engine is operating and the instrument power switch is in NOR or ALT, if the engine is inoperative and the battery switch is positioned to ON and the instrument power switch is in ALT, or if the primary and secondary busses are energized through the external power receptacle and the instrument power switch is in NOR. Allow three minutes to elapse so that the gyro in the directional gyro control comes up to operating speed, levels and aligns the indication on the settable dial indicator with that sensed by the remote compass transmitter.

Note

It is necessary, for proper operation of the J-2 directional indicator, that the a-c and d-c power supplies to the system be turned on simultaneously.

Operation.

Setting Indicator.

By means of the SET COURSE knob on the indicator, set the dial index for the heading it is desired to fly. It is preferable to set the dial index against the zero bezel index of the indicator, although any index may be chosen.

Using the Indicator — Straight Flight.

After the airplane becomes airborne, the indicator is referred to in the same manner as a magnetic compass.

Using the Compass—In Turns.

Perfect 45, 90 and 180 degree turns can be executed by setting the dial index, with the overlapping pointer against the zero bezel index, then flying the aircraft to align the pointer with the 45 and 90 degree bezel indices on both sides of the zero index, or with the index at 180 degrees. The final heading may be set against the zero bezel index by means of the SET COURSE knob. Another method is to set the dial index for the new heading against any bezel index, then flying the aircraft to align the pointer with that bezel index.

CLOCK.

Unlike previously installed aircraft clocks, this clock (25, figure 1–4; 25, figure 1–5; 25, figure 1–6 and figure 4–18) includes elapsed timing, split-second timing, as well as an 8-day clock winding mechanism. Split-second timing is achieved as follows: Press the right-hand button to start split-second timing; the time traversed by the sweep, split-second needle will be indicated in minutes on the small inner dial, located at the bottom of the instrument face; press the button a second time to stop split-second timing and a third time to return the sweep, split-second needle to zero.

Elapsed time is determined by pressing the left-hand button to start, and reading time traversed on the small inner dial located on the upper portion of the instrument face. Pressing the left-hand button a second time stops elapsed time recording; and for a third time, returns the elapsed time hands to zero.

The button on the lower center (6 o'clock position) of the instrument face is used to manipulate the elapsed time needles for adjustment of elapsed time for refueling, repair, or other conditions, so that an exact record of flying time can be indicated for any desired phase of flight. To assist in adjusting for exact elapsed time, indicators (small holes in the dials) are backed up by a black shield when not in use, or partially backed in black when recording of time has been temporarily arrested. The indicators are backed by a luminous shield when the elapsed time dials are operating. The lower center button is turned to the right when the elapsed time needle is to be arrested; in this event the indicator is partially black. By turning the button to the left elapsed time commences to be recorded and the indicator is luminous. When the upper indicator is partially black, the elapsed time hands have been temporarily stopped; and when completely black, the hands are back to zero. When both upper and lower indicators are completely luminous both elapsed time dials are operating.

Figure 4—18.

PLOTTING BOARD.

A Batori plotting board is provided on modified and E and later aircraft. The purpose of the Batori plotting board is to provide the pilot with a navigational aid for flights of extreme duration. The support sockets for the plotting board are mounted on the top of the upper instrument panel. Stowage provisions consists of two floor mounted brackets on the right side of the cockpit and an elastic strap on the side of the right console just above the main circuit breaker panel.

CAUTION

In order to preclude the possibility of the plotting board holding circuit breakers in the reset position when stowed, assure that it is stowed correctly in the floor mounted brackets.

BATORI COMPUTER.

Provisions are made on the left side of the cockpit for the Batori Computer (2 figure 1–10 and 2 figure 1–11) which is swivel mounted for ready reference. This device is easily accessible to the pilot for one-hand computation of fuel consumption, ground speed, ETA's, etc.

ARMAMENT EQUIPMENT.

This aircraft is equipped to carry guns, bombs, rockets and chemical tanks. The guns are installed internally and the bombs, rockets and chemical tanks are carried as external stores. The gun-bomb-rocket sight is provided, and is used when firing the guns and rockets and to release the bombs at the proper time to be effective on a target. Manual operation is also provided. The pilot in A thru F aircraft is protected from enemy fire by two pieces of armor plate installed behind the seat. On E and later aircraft an armored headrest and plastic shoulder armor is incorporated in the seat, to protect the pilot. A gun camera mounted above the instrument panel records results of fixed gun and rocket firing.

GUN ARMING SWITCH.

The gun arming switch (5, figure 1–8; 19, figure 1–9) has three positions: GUNS, OFF and SIGHT-CAMERA & RADAR. AC power is supplied to the sight gyros and heaters as soon as the secondary bus is energized if the instrument power switch is in the NOR position. The SIGHT-CAMERA & RADAR or GUNS position supplies DC power to the sight tube heaters, relays and control units in addition to the gyros and heaters, and also energizes the gun camera circuit. In addition the GUNS position energizes a relay so that the guns will fire when the stick trigger is actuated and the weight is off the landing gear on A thru G aircraft, or the landing gear handle is moved upward to retract the landing gear on H and later aircraft. The sight is ready for use with the gun arming switch in either the SIGHT-CAMERA & RADAR or GUNS position.

H and LATER

ARMAMENT SELECTOR SWITCH.

The armament selector switch (figure 4–21) is a twelve-position rotary selector switch used to select bombs or rockets to be released and the sequence in which they will be released when the bomb release switch on the control stick is depressed. The armament selector switch may also be used to jettison either bombs or rockets, or all external stores. In its various positions, the switch will function as follows on manual or automatic release when the bomb release switch is depressed.

GUN-BOMB-ROCKET SIGHT.

The A-4 gun-bomb-rocket sight (figures 4–23, 4–24, 4–25) automatically computes the fire control problems for gun fire from fixed guns, for bombing and for air to ground rocket fire. The computed range takes into account the speed of the target. The sight reticle image, consisting of a circle and a central dot, is reflected on an inclined transparent window. The automatic features of the sight enables the pilot to direct his full attention to the selected target, provided he flies the aircraft so that the reticle circle is continuously superimposed on the target. Range data is supplied to the sight, for gunnery operations, by a manual range control, or automatically by an AN/APG-30 radar range unit. The radar system also provides automatic search in range. It automatically locks onto and tracks a target in range and indicates to the pilot when the equipment is tracking a target. On overland targets below 6,000 feet, radar range distance may be reduced by use of the radar range sweep control, to prevent the radar from locking on the ground when target is at low altitudes. Bombs can be released automatically at the proper release point by a mechanism within the sight. Electrical power (28 volt DC) is supplied to the sight and sight heaters from the secondary bus. AC electrical power is supplied from the main inverter and a sight electrical power supply, which are both powered from the secondary bus. If the main inverter is inoperative the sight can not be used in its normal reticle function or as a fixed reticle sight.

Sight Dimmer Control.

The sight dimmer control (figures 4–23, 4–24, 4–25) controls the illumination intensity of the reticle image from DIM to BRIGHT.

ARMAMENT CONFIGURATIONS

ESCORT MISSION

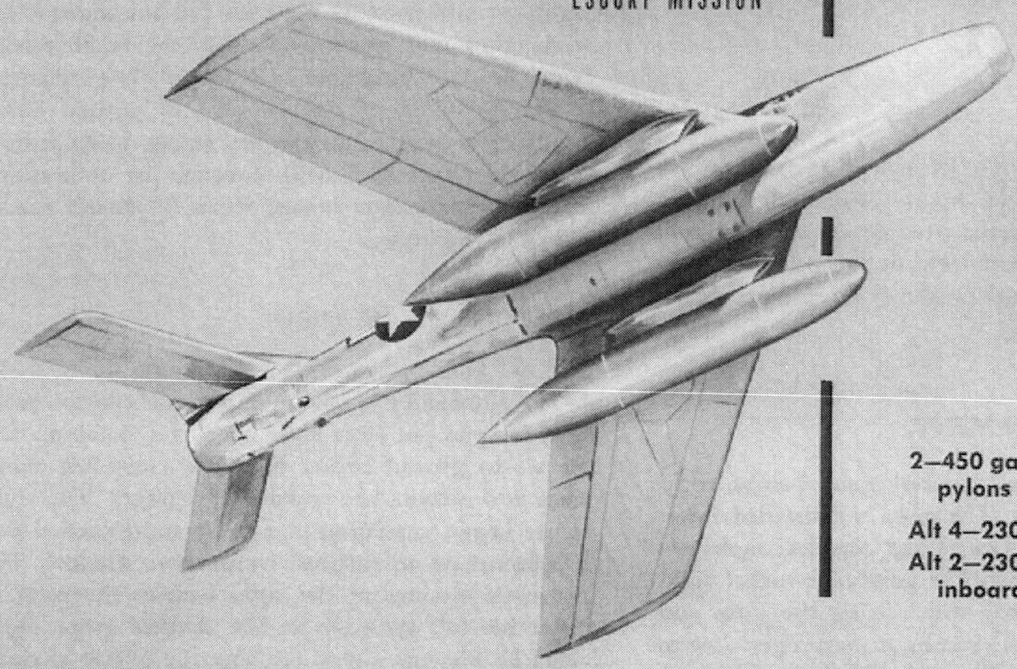

2—450 gal tanks inboard pylons

Alt 4—230 gal tanks

Alt 2—230 gal tanks inboard

GROUND SUPPORT MISSION

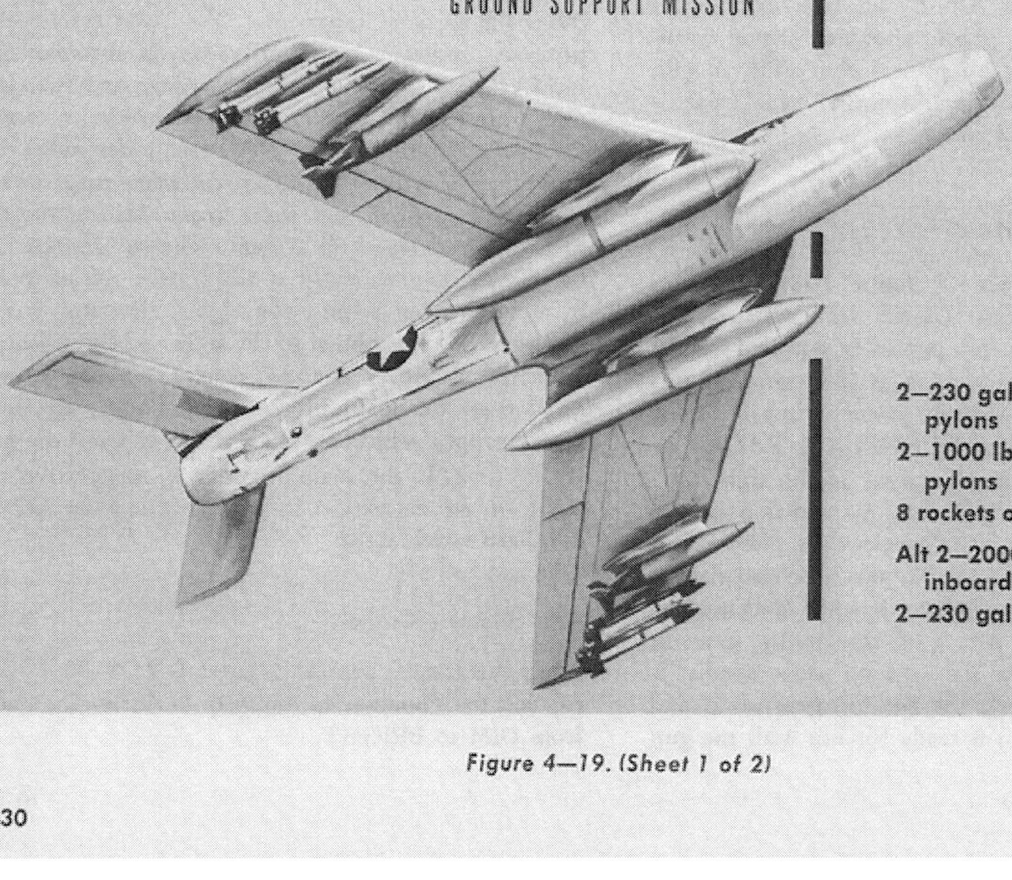

2—230 gal tanks inboard pylons

2—1000 lb bombs outboard pylons

8 rockets outboard

Alt 2—2000 lb bombs inboard

2—230 gal tanks outboard

Figure 4—19. (Sheet 1 of 2)

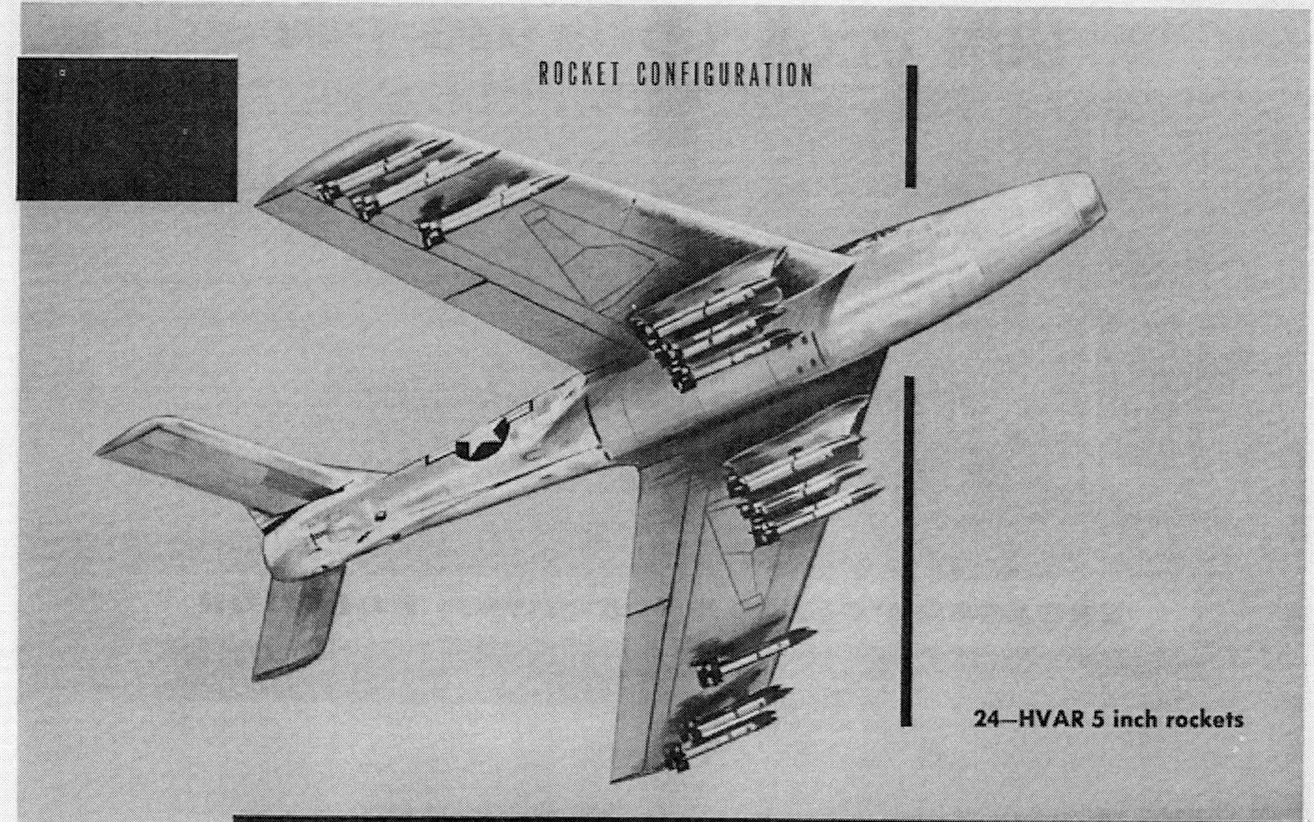

ROCKET FIRING SEQUENCE

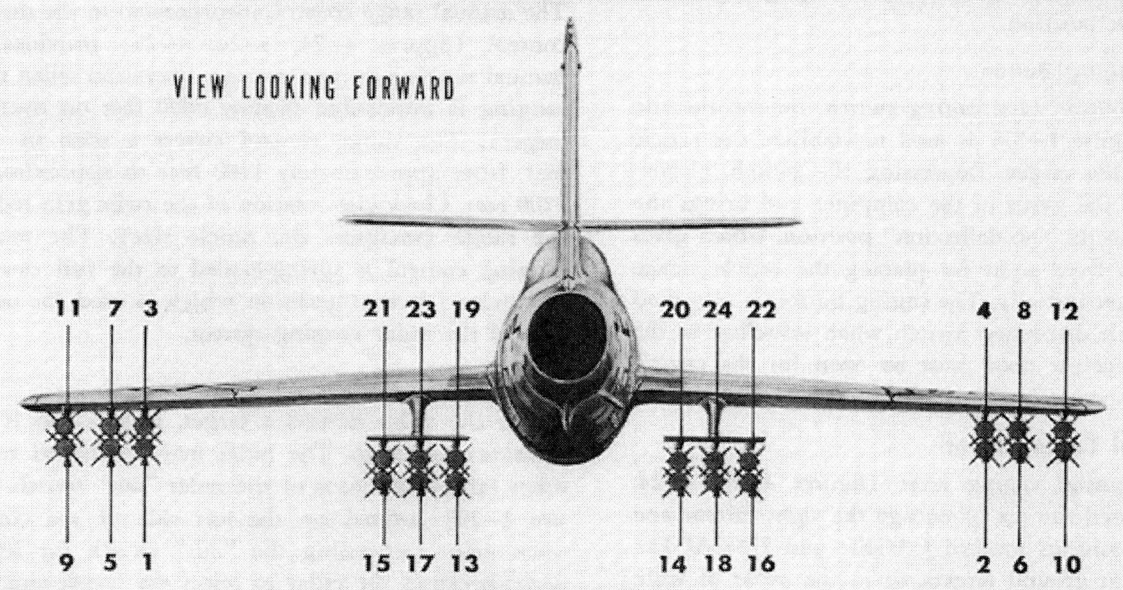

Figure 4—19. (Sheet 2 of 2)

GUN DECK LOCKING PIN

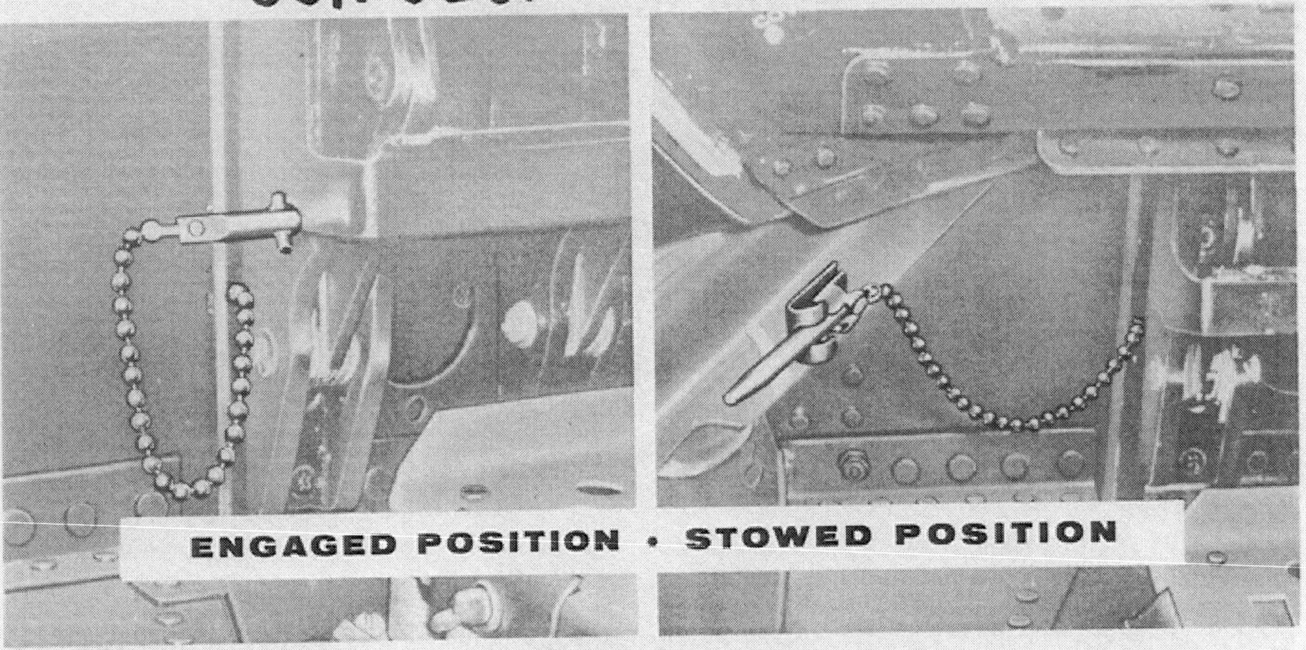

Figure 4—20.

Sight Filament Switch.

The sight filament switch (figures 4—23, 4—24, 4—25) has two positions PRIM and SEC and is used to select one of the dual filaments of the lamp for the sight reticle circle and dot. If either filament goes out, the spare filament may be cut-in by changing the switch to the other position.

Electric Caging Button.

The push-button type caging switch, on the throttle control (figure 1—15) is used to stabilize the reticle image on the target. Depressing the switch, electrically cages the gyros in the computer and brings the sight line to the "no deflection" position, which gives the pilot a fixed sight for placing the reticle image on the target initially. The caging button is also used as the nozzle disconnect switch when refueling in the air. The receiver door must be open for the caging switch to operate in this manner.

Mechanical Caging Lever.

The mechanical caging lever (figures 4—23, 4—24, 4—25) is used to cage or uncage the sight mirror and has two positions marked CAGED and UNCAGED. For firing at ground targets, or in the event of sight failure, the caging lever may be placed at CAGED and the reticle used as a fixed sight. The mechanical caging lever should always be positioned to CAGED during landing, take-off and any ground operation, or for violent air maneuvers when not in use for gunery, in order to protect the sight mirror suspension.

Span Adjustment Lever.

The span adjustment lever (figures 4—23, 4—24, 4—25) is set to the wingspan of the target, in feet, when the gun-bomb-rocket sight is operated with manual range.

Manual Range Control.

The manual range control, incorporated in the throttle control (figures 4—23, 4—24, 4—25) provides for manual ranging during gunnery operation when radar ranging is impossible (below 6000 feet on overland target). The range control covers a span of 1500 feet, from approximately 1200 feet to approximately 2700 feet. Clockwise rotation of the twist grip reduces the range (increases the reticle size). The manual ranging control is spring-loaded to the full counterclockwise (detent) position which is used for operation of the radar ranging system.

Radar "Out" Switch.

When the radar detects a target, it locks on it and measures its range. The radar may be shifted to another target by means of the radar "out" switch (figure 1—29) located on the left side of the control stick grip. Depressing the "out" switch for several seconds causes the radar to reject the target and drift in or out in range.

Radar Range Sweep Control.

The radar range sweep control (figures 4—23, 4—24, 4—24) is a rheostat marked INCREASE with an arrow showing the direction. Turning the control, in the

Figure 4—21.

MIN direction lowers the radar ranging distance to prevent radar from locking on the earth when the aircraft is at low altitudes. Turning the control toward MAX increases the range. During normal operations control should be at MAX.

Bomb-Target-Wind Control.

The bomb-target-wind control (figures 4—23, 4—24, 4—25) has a ROCKET GUN position and a BOMB scale indicating downwind and upwind adjustment. Setting the B-T-W control adjusts the sight to compensate for the components of wind velocity and target motion parallel to the direction of the attacking airplane. The ROCKET GUN position is selected when using the sight for gun or rocket firing.

Rocket Dive Angle Control.

The rocket dive angle control (figures 4—23, 4—24, 4—25) is three separate controls on one dial. One pointer has three positions on the top of the dial; BOMB, GUN and ROCKET. The appropriate position is selected for bombing, gunnery or rocket firing. If the pointer is in either the BOMB or ROCKET position and the radar "out" switch on the control stick is depressed the pointer will automatically return to the GUN position. The second pointer is the rocket setting pointer and has three positions; 5" HVAR, 2.25" SCAR and 2.75" FFAR. Each position has an N and S position. When firing rockets the rocket dive angle control is turned to N for dives up to 40 degrees or to S for dives greater than 40 degrees under the type of rocket being fired. The third pointer has three positions; TR, HI and LO and is used for gunnery only. The TR position is used in a low speed attacking airplane for the purpose of controlling gun fire or rocket fire against a low speed target. The HI position is used when the sight is in a high speed attacking airplane flying against a high speed target. The LO position is used when the sight is in a high speed attacking airplane directing fire against a low speed target.

Target Indicator Light.

The target indicator light (figures 4—23, 4—24, 4—25) is located on the left side of the gun sight. Illumination of the light indicates that the radar set is "locked-on" a target.

Operation.

Starting the Sight.

The sight gyros and heaters start to operate as soon as the secondary bus is energized.

1. Position the gun arming switch to either the GUNS or SIGHT CAMERA & RADAR position which supplies power to the amplifier. Allow approximately five minutes for the amplifier tubes to warm up, and the sight to reach its maximum operating efficiency.
2. Check to see that the reticle image appears on the reflector glass.
3. Check the reticle image for moving the mechanical caging lever from one position to the other. The dot should flicker as the lever is moved, and the circle should change to four circular arcs in the CAGED position.

Stopping.

Position the gun arming switch to the OFF position. This turns off the tube heaters, relays and controls, but leaves the sight operating and ready for use within one minute after repositioning the gun selector switch. The gyros and heaters will operate as long as the secondary bus is energized. Position the mechanical caging lever to CAGED.

GUNNERY EQUIPMENT.

Four .50 caliber machine guns are installed in the gun bay of the fuselage, and one .50 caliber machine gun is installed in the leading edge of each wing. Electric heaters are provided for the guns. The normal load of ammunition for each gun is 300 rounds. All guns are charged manually prior to take-off. A gun camera, installed above the instrument panel operates automatically when the guns are fired, and it may be operated separately. An armament safety switch pre-

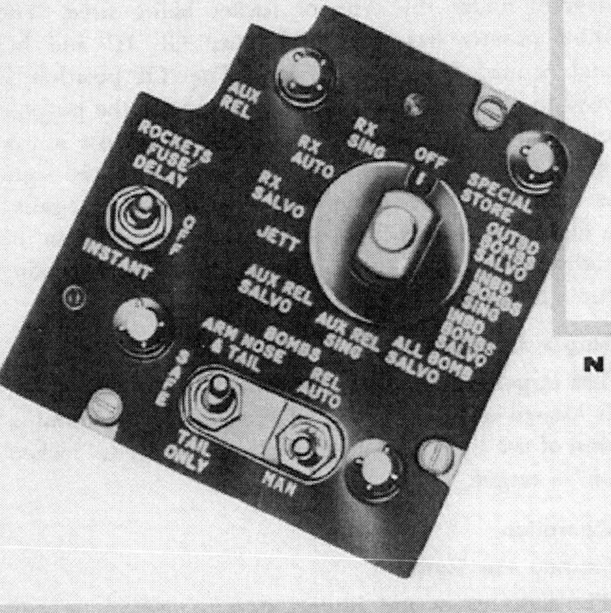

ARMAMENT CONTROL PANEL

H and later

Note When the bomb release selector switch is in the AUTO RELEASE position and the bomb-rocket selector switch is indexed to the OUTB D BOMBS, INB D BOMBS SING, INB D BOMBS SALVO, INB D AND OUTB D BOMBS SALVO, AUX REL SING, AUX BOMB SALVO positions, the source of power for these switch positions is in the secondary bus.

ARMAMENT SELECTOR SWITCH POSITIONS

Position	Power Source	Function
SPECIAL STORE	Primary & Secondary Bus	Equipment has a higher classification than covered by this Handbook.
OUTB D BOMBS	Primary & Secondary Bus	Releases right and left outboard bombs simultaneously when bomb release switch is depressed.
INB'D BOMBS SING	Primary & Secondary Bus	Releases left inboard bomb when bomb release switch is depressed. Circuit transfers to right pylon and right bomb will release when bomb release switch is depressed again.
INB D BOMBS SALVO	Primary & Secondary Bus	Releases right and left inboard bombs simultaneously when bomb release switch is depressed.
INB D & OUTB D BOMBS SALVO	Primary & Secondary Bus	Releases all inboard and outboard bombs simultaneously when bomb release switch is depressed.
AUX REL SING	Primary & Secondary Bus	Releases bombs on left auxiliary rack when bomb release switch is depressed then releases bombs on right auxiliary rack when bomb release switch is depressed again. Auxiliary bomb release indicator light will go out after all auxiliary bombs are released.
AUX BOMB SALVO	Primary & Secondary Bus	Releases auxiliary bombs or fires chemical tanks on both wings simultaneously when bomb release switch is depressed. Auxiliary bomb release indicator will go out when stores are dropped.
JETTISON	Primary & Battery Bus	Will jettison all external stores when bomb release switch is depressed. Jettison indicator light will light. Rockets mounted on retractable posts will not jettison if aircraft is not airborne.
Rx SALVO	Primary & Secondary & Battery Bus	Jettisons all rockets when bomb release switch is depressed.
Rx AUTO	Primary & Secondary Bus	Fires rockets in automatic sequence through intervalometer when bomb release switch is depressed and held.
OFF	No Power	Bomb and rocket circuits not energized, jettison may be accomplished, through external stores jettison switch.

Figure 4—22

vents the guns from being fired or rocket jettisoning while the aricraft is in the static position. On A thru G aircraft the armament safety switch de-energizes the gun and rocket circuit when the landing gear shock struts are compressed by the weight of the aircraft. On H and later aircraft the safety switch disarms the gun and rocket circuits when the landing gear selector is in the DOWN or EMERG DOWN positions. If for any reason the guns must be fired or the rockets jettisoned while the airplane is on the ground, the safety circuit can be by-passed by removing the cover of the armament safety override switch located in the right wheel well. The primary bus must be energized in order to fire the guns.

Stick Trigger.

The gun camera and gun firing circuits are energized by depressing the stick trigger (figure 1–29) located on the control stick grip. There are two definite switch positions. The first position starts the gun camera operating if the gun arming switch is in the SIGHT-CAMERA & RADAR or the GUNS position, and the secondary bus is energized. The second, or fully depressed position fires the guns if the gun arming switch is in the GUNS position; the secondary bus is energized and, on A thru G aircraft, the weight is off the wheels, or on H and later aircraft, the landing gear selector handle is up and out of the DOWN position. If the rocket dive angle control is in either BOMB or ROCKET position in preparation for a bomb or rocket run, and gun firing is required, the rocket dive angle control will automatically return to the GUNS position when the radar "out" switch is depressed.

CAUTION

On A thru G aircraft, if the landing gear shock struts are overinflated so as to be fully extended, it will be possible to fire guns on the ground by depressing the stick trigger, if the gun arming switch is in the GUNS position and the primary bus is energized.

Gun Heater Switch.

The gun heater switch (5, figure 1–8; 23, figure 1–9), located on the left console, is a circuit breaker type switch having two positions: OFF and HEATER. The HEATER position supplies power to the gun heaters from the secondary bus. Gun heat should be used when gunnery is anticipated during cold weather operation and at altitudes above 15000 feet.

Camera Aperture Switch.

The camera aperture switch (5, figure 1–8; 29, figure 1–9) is provided to adjust the camera aperture for existing light conditions so as to improve pictures taken during gunnery. The switch operates from the secondary bus and has three positions: BRIGHT, HAZY and DULL, which refer to light conditions and adjust the camera aperture accordingly.

Gunfire Operation.

With Radar Range.

1. Gun heater switch HEATER.
2. Gun arming switch GUNS. Allow five minutes for amplifier to warm up and stabilize.
3. Set rocket dive angle control to GUN.
4. Set the TR, HI, LO pointer on the rocket dive angle control to suit gunnery target speed.
5. Set B-T-W control to ROCKET-GUN.
6. Check instrument power switch—NOR.
7. Manual range control on throttle in detent.
8. Set mechanical caging lever on sight heat to UNCAGED.
9. Set reticle dimmer control for desired brilliance.
10. When searching for targets, press caging button on throttle control to stabilize the reticle image.
11. When target is located and tracking is started, release caging button. Fly the airplane so that the reticle image is continuously and accurately centered on the target. After the target has been tracked smoothly without slipping or skidding for approximately one second, fire the guns.

With Manual Range.

1. Gun heater switch HEATER.
2. Gun arming switch GUNS. Allow five minutes for amplifier to warm up and stabilize.
3. Set rocket dive angle control to GUN.
4. Set B-T-W control to ROCKET GUN.
5. Check instrument power switch NOR.
6. Set mechanical caging lever on sight head to UNCAGED.
7. Set reticle dimmer control for desired brilliance.
8. Identify target, and set wing span with span adjustment lever corresponding to the target wing span.
9. When searching for targets, press caging button on throttle control to stabilize the reticle image.
10. After the target is close enough for a sphere containing the wing tips to coincide with the framing circle at minimum diameter, release the caging button.
11. Track smoothly, turning manual range control on throttle so that the circle continuously and accurately frames or encloses the target. After the target has been framed and tracked smoothly for approxi-

1 *Target Indicator Light*
2 *Mechanical Caging Lever*
3 *Dive and Roll Indicator*
4 *LABS Control Panel*
5 *Radar Range Sweep Control*
6 *Bomb Target-Wind Control*
7 *Sight Dimmer Control*
8 *Rocket Dive Angle Control*

Figure 4—23

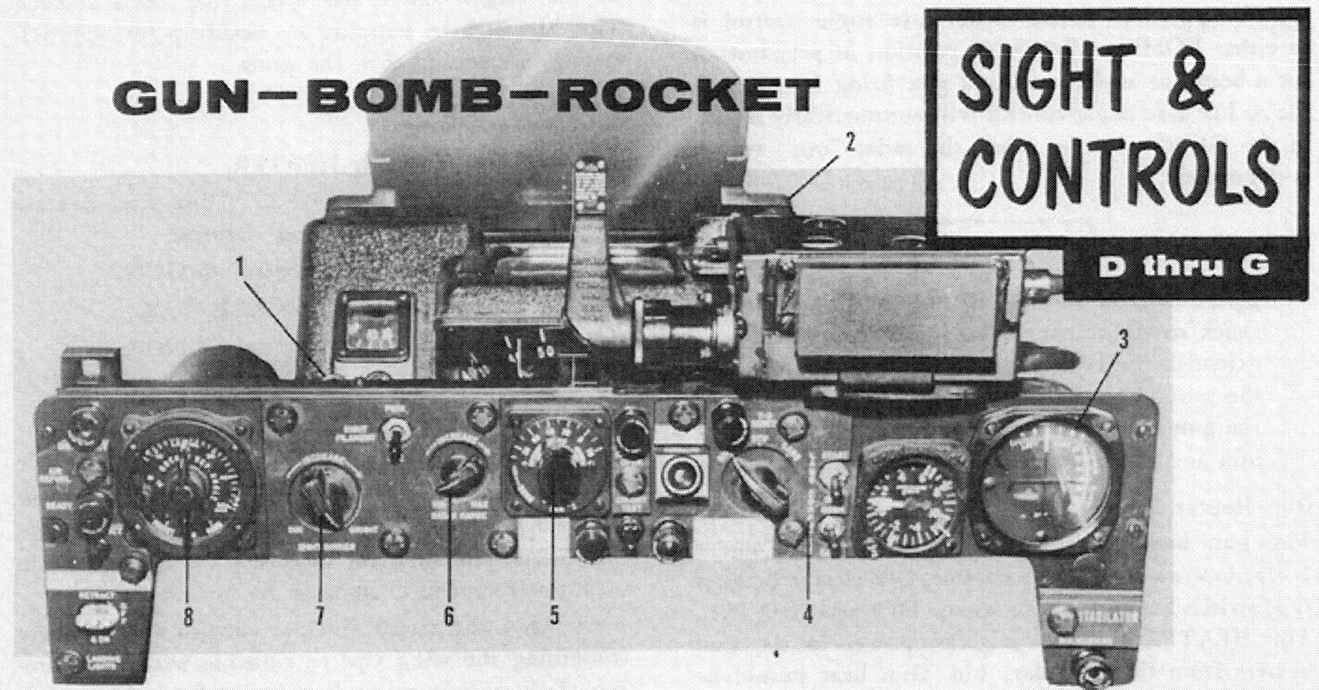

1 *Target Indicator Light*
2 *Mechanical Caging Lever*
3 *Dive and Roll Indicator*
4 *LABS Control Panel*
5 *Bomb Target-Wind Control*
6 *Radar Range Sweep Control*
7 *Sight Dimmer Control*
8 *Rocket Dive Angle Control*

Figure 4—24

1 *Target Indicator Light*
2 *Mechanical Caging Lever*
3 *LABS Control Panel*
4 *Bomb Target-Wing Control*
5 *Radar Range Sweep Control*
6 *Sight Dimmer Control*
7 *Rocket Dive Angle Control*

Figure 4—25.

mately one second, without slipping or skidding, fire the guns.

BOMBING EQUIPMENT.

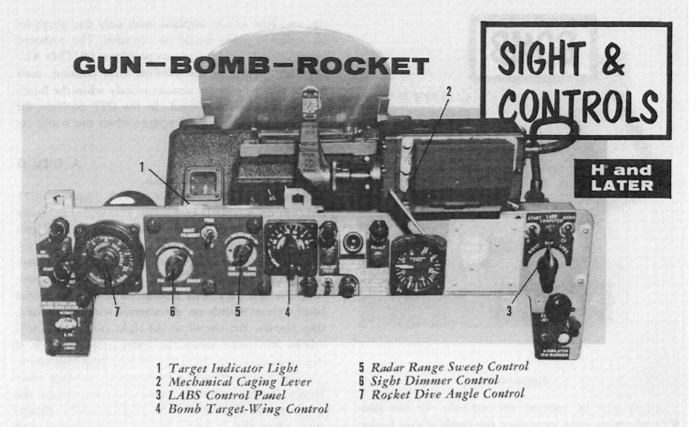

Two jettisonable bomb pylons can be installed under each wing. The pylons are jettisoned by explosive charges in the attaching bolts and are blown away from the aircraft by compressed air cylinders. These cylinders must be charged manually. Each inboard pylon will carry all bombs up to, and including 1,000 pounds, or an external fuel tank. The gun-bomb-rocket sight is used for bomb sighting and automatic bomb release. Controls are provided for normal or emergency release of any external stores carried on the pylons. Normal release may be accomplished automatically or manually, with bombs released singly or simultaneously. The arming condition of the bomb nose and tail fuses is manually selected.

Bomb Arming Switch.

The bomb arming switch (figures 4—22 and 4—24) has three positions; ARM NOSE & TAIL, SAFE, and TAIL ONLY and is energized by the secondary bus on H and later, and unmodified aircraft. On earlier aircraft, and on modified aircraft, power is supplied by the battery bus. When the switch is placed in the ARM NOSE & TAIL position the electric circuit is indexed to arm the bombs for time explosion. When the switch is placed in the TAIL ONLY position, the electric circuit is indexed to arm the bombs for impact explosion. In the SAFE position the bombs are unarmed when dropped.

Note

On aircraft where the circuit is energized by the battery bus, the switch must be in the SAFE position when the aircraft is parked to avoid discharging the battery.

Bomb Release Switch.

The bomb release switch (figure 1—29) is a spring-loaded, button type switch installed on the top of the control stick and energized by the primary bus. When the switch is depressed the bomb rack electric circuit is energized to release the bombs or pylon tanks in accordance with the setting of the bomb release selector switch. The bomb release switch will also fire the rockets, salvo the rockets, or release bombs, depending on the position of the rocket selector switch, or inboard or outboard bomb selector switch respectively, on A thru G aircraft or the position of the armament selector on H and later aircraft.

Bomb Release Selector Switch.

The bomb release selector switch (figures 4—22 and 4—26) is energized from the primary bus and has two positions, which are AUTO RELEASE and MANUAL RELEASE. In the AUTO RELEASE position the

Figure 4—26

bombs will be released automatically by the gun-bomb-rocket sight, providing the bomb release button is depressed by the pilot, and held down until bombs are released. In the MANUAL RELEASE position, the bombs are released manually by depressing the bomb release switch on the control stick, regardless of whether or not the sight is utilized or in operation.

A THRU G

Inboard Bomb Selector Switch.

Release of the bombs installed on the inboard pylons on A THRU G aircraft is controlled by the inboard bomb selector switch (figure 4—26) which is energized from the primary bus. The switch has three positions INBD ALL, OFF and SINGLE. When the switch is in the SINGLE position, the left bomb will release when the bomb release switch is depressed. The right bomb will release when the release switch is depressed again. With the switch placed in the INBD ALL position, both bombs will release simultaneously, when the bomb release switch is depressed. When in the OFF position the bombs on the inboard pylons will not release when the bomb release switch is depressed.

A THRU G

Outboard Bomb Selector Switch.

Release of the bombs installed on the outboard pylons on A THRU G aircraft is controlled by the outboard bomb selector switch (figure 4—26), which is energized from the primary bus. The bombs cannot be released singly, as the pylons are too near the wing tip, and trim of the airplane with only one bomb on an outboard pylon would be excessive. The outboard bomb selector switch has two positions; OUTBD ALL and OFF. When in the OUTBD ALL position, both bombs will be released simultaneously when the bomb release switch is depressed. In the OFF position the outboard bombs will not release when the bomb release switch is depressed.

A THRU G

Auxiliary Bomb Selector Switch.

The auxiliary bomb selector switch (figure 4—27) installed on A THRU G aircraft, is energied from the secondary bus, and is used when fragmentation bomb racks or chemical tanks are attached to the pylons. The auxiliary bomb selector switch has three positions; ALL, OFF and SINGLE. In the SINGLE position, the left rack will release its bombs when the bomb release switch on the control stick is actuated, then transfer the circuit to the right rack, which will release its bombs when the bomb release switch is depressed again. In the ALL position, both fragmentation racks will operate simultaneously to drop their bombs, or the chemical tanks will fire, when the bomb release indicator light (figure 4—27) illuminates when the fragmentation racks are loaded, and remains on until the fragmentation bombs are dropped. The ALL position is used when firing chem-

Figure 4—27

ical tanks. The OFF position must be used when bombs are installed on the pylons.

External Stores Jettison Switch.

The external stores jettison switch (figure 1—7) is a push button type switch recessed in the panel to prevent accidental actuation. The switch is marked EXTERNAL STORES JETTISON, and when depressed will jettison the outboard and inboard pylon tanks together with the pylons on unmodified RE aircraft up to F-84F-45RE and all GK aircraft if the battery bus is energized. On F-84F-45RE and subsequent RE aircraft and on aircraft modified to incorporate T.O. 1F-84F-509 only the pylon stores will be jettisoned and the pylons will be retained. Rockets on rocket posts will be jettisoned on all aircraft if the aircraft is airborne and the primary bus is energized.

This page intentionally left blank.

CAUTION

On unmodified aircraft prior to F-84F-45RE and all GK aircraft the pylons will jettison together with any stores installed if the external stores jettison switch is depressed with the aircraft in the static position. On F-84F-45RE and later RE aircraft and on modified aircraft, only the stores will jettison if the external stores jettison switch is depressed with the aircraft in the static position.

Inboard Pylon Jettison Switch.

The inboard pylon jettison switch (6, figure 1–8; 20, figure 1–9) is guarded with a cover type guard which is marked JETT INBD PYLON and an arrow. When actuated in the direction of the arrow the inboard pylons are jettisoned simultaneously together with any stores carried on them if the battery bus is energized on unmodified A thru D aircraft or if the primary bus is energized on E and later or modified earlier aircraft.

Outboard Pylon Jettison Switch.

The outboard pylon jettison switch (7, figure 1–8; 21, figure 1–9) is guarded with a cover type guard which is marked JETT OUTBD PYLON and an arrow. When actuated in the direction of the arrow the outboard pylons are jettisoned simultaneously together with any stores carried on them if the primary bus is energized on modified and E and later aircraft or if the battery bus is energized on unmodified A thru D aircraft.

Bombing Operation — Manual.

1. Set bomb selector switch as desired.
2. Set bomb arming switch as desired.
3. Set bomb release selector in MANUAL RELEASE.
4. Press bomb release switch on control stick.

CAUTION

Before depressing the bomb release switch, check to determine that all selector switches are in proper position. This is necessary to determine that only the desired circuits will be energized. It is possible, for example, if both the rocket jettison ready switch and the rocket selector switch are both on in A thru G aircraft, to fire some rockets and drop the others when the bomb release switch is depressed.

Bombing Operation Using Gun-Bomb-Rocket Sight.

1. Set rocket dive angle control to BOMB position.

Note

If the radar "out" switch is depressed with the dive angle control in the BOMB position, the control will automatically return to the GUN position.

2. Set B-T-W control to the known or estimated up or downwind velocity plus or minus the estimated target velocity. The reticle will immediately move down to a position just over the nose.
3. Set the mechanical caging lever on the sight head to UNCAGED.
4. Set reticle dimmer control for desired brilliance.
5. Set the bomb release selector switch in the AUTO RELEASE position.
6. Position inboard or outboard bomb selector switch as desired.
7. Fly an approach which will give the desired dive angle during the bombing run. The electrical caging button must be depressed during this maneuver.
8. After a smooth dive has been established with the reticle on the target, release the electrical caging button.
9. Depress the bomb release switch at approximately the bomb drop point.
10. Track very smoothly until the reticle image becomes extinguished. This indicates an automatic bomb release.
11. If manual release is desired, place the bomb release selector switch in the MANUAL RELEASE position, get on the target and track as above. After the reticle image circle becomes extinguished, press the bomb release button. The manual release is not as accurate as the automatic release because of the time lag due to the pilot's reaction time.
12. Computation accomplished during the above prescribed procedure will be good for only one release, i.e., a single bomb, bombs in train with interval control not exceeding ½ second from the first to the last bomb, a pair of bombs released simultaneously or a salvo.

Emergency Operation.

The bombs can be dropped simultaneously unarmed by depressing the jettison external stores switch.

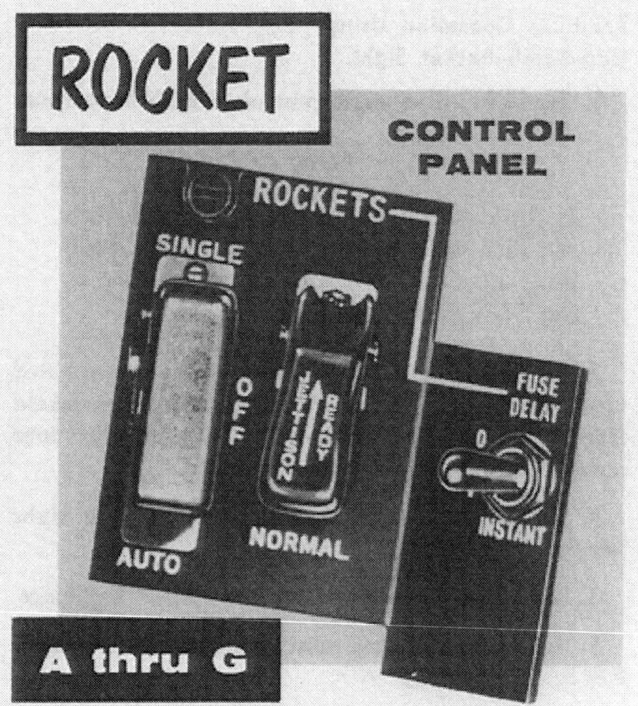

Figure 4—28

CAUTION

On airplanes prior to F-84F-45RE and all GK airplanes the pylons will jettison together with any stores installed if the external stores jettison switch is depressed with the airplane in the static position.

ROCKET EQUIPMENT.

Rocket firing equipment is provided for launching twenty-four 5" HVAR type rockets. Six rockets are carried on three forward and six aft support posts under each wing and six rockets are carried on an adapter on each inboard pylon. Arming solenoids, mounted in the leading edge of each wing and on each pylon, retain the rocket nose fuse arming wires when the rockets are fired in the armed condition. Rocket firing sequence is controlled by a rocket release control. Each forward post is equipped with a jettison mechanism for jettisoning the rockets in an emergency. The fore and aft rocket supports on A thru F aircraft retract automatically after the rockets have been fired or jettisoned. However, on G and later aircraft the rocket support posts are installed only when rockets are to be carried and are not retractable. The forward posts are jettisoned by explosive squib action, while the aft posts remain projecting from the wing after rocket firing or jettison and are later removed on the ground, where flush plugs are installed in place of the posts if rockets are not to be carried on the wing. The rockets may be fired singly or in train. The gun-bomb-rocket sight is used for aiming rockets and the rockets are fired by depressing the bomb release switch. Refer to figure 4—19 for rocket firing sequence.

Rocket Arming Switch.

The rocket arming switch (figures 4—22 and 4—28) is energized by the secondary bus and has three positions: FUSE DELAY, OFF and INSTANT. When the switch is in the INSTANT position the arming wire is retained by the arming solenoid when the rocket is fired. This will arm the rocket for contact detonation by releasing the contact fuse pin. When the switch is placed in the FUSE DELAY or the OFF position, the arming solenoids are de-energized and release the arming wires which remain attached to the rockets. The nose contact fuses will therefore remain in the safe or unarmed position and the rockets will be detonated by the base fuse.

Rocket Release Control.

The rocket release control (figure 4—29) is an electrically operated timing device incorporating an intervalometer, a reset knob and an indicator. The intervalometer automatically selects the next rocket firing circuit in sequence whether firing the rockets singly or in train. The indicator, marked RX TO BE FIRED, informs the pilot by rocket number, the position and location of the next rocket to be fired and enables him to keep check of the number of rockets fired and those remaining on the launchers. The reset knob, turned counter-clockwise, permits the intervalometer to be re-indexed so that any rocket number can be selected for firing depending on the airplane configuration.

A thru G

Rocket Selector Switch.

The rocket selector switch (figure 4—28) is energized by the primary bus and has three positions: OFF, SINGLE and AUTO. When the switch is placed in the SINGLE position, the electrical circuit to the rocket shown by number on the indicator will be energized when the bomb release switch is pressed. When the switch is placed in the AUTO position, the firing circuits to each of the rockets is completed in automatic sequence by the intervalometer when the bomb release switch is pressed and held.

Figure 4—29.

A THRU G

Rocket Jettison Ready Switch.

The rocket jettison ready switch (figure 4—28) is energized by the primary bus and has two positions: NORMAL and JETTISON READY. The NORMAL position is guarded with a red cover guard and is in the rocket firing circuit. The JETTISON READY position indexes the rocket jettison circuit so that the pylon adapter and the rockets on the rocket ports will be jettisoned simultaneously when the bomb release switch on the control stick is pressed. However, if bombs or tanks are installed on the pylons they will not jettison with the rockets on the wings.

Note

Rockets can be jettisoned only if the aircraft is airborne or the struts are overinflated so as to be fully extended.

H and LATER

Armament Selector Switch.

For complete description of this switch refer to BOMBING EQUIPMENT.

External Stores Jettison Switch.

The external stores jettison switch (figure 1—7) is a push button type switch recessed in the panel to prevent accidental actuation. The switch is marked EXTERNAL STORES JETTISON, and when depressed will jettison the outboard and inboard pylon tanks together with the pylon on unmodified aircraft prior to F-84F-45RE and all GK aircraft if the battery bus is energized. On F-84F-45RE and subsequent RE aircraft, and on aircraft modified to incorporate T.O. 1F-84F-509, only the pylon stores will be jettisoned and the pylons will be retained. Rockets on rocket posts will be jettisoned on all aircraft if the aircraft is airborne and the primary bus is energized.

CAUTION

On unmodified aircraft up to F-84F-45RE and all GK aircraft the pylons will jettison together with any stores installed if the external stores jettison switch is depressed with the aircraft in the static position. On F-84F-45RE and later RE aircraft and on modified aircraft only the stores will jettison if the external stores jettison switch is depressed with the aircraft in the static position.

Rocket Firing (Manual).

1. Set selector switch as desired.
2. Set rocket arming switch as desired.
3. Set rocket indicator as desired.
4. Press bomb release switch on control stick.

CAUTION

The lower rocket must always be fired first. When a misfire occurs in the air during "single" round firing the pilot shall check his intervalometer to ascertain which round has misfired, remembering that the number shown on the intervalometer is the rocket number to be fired. Reference to rocket firing sequence will identify the rocket as an upper or lower rocket. If the rocket is a lower rocket, the rocket firing sequence will reveal from which rocket it is suspended. The upper rocket should not be fired. The intervalometer should be positioned upon the next position when the number of the upper rocket of the misfired pair appears on the intervalometer thereby by-passing a double rocket release. (During automatic firing, the pilot has little control over the rockets as they fire at 0.1 second intervals while firing button is held down. If a misfire should occur during automatic firing, experience indicates that if the upper rocket is fired with the lower rocket still attached thereto, and provided the fins are secured per instruction, only slight damage will occur to the aircraft consisting of two superficial scratches on the under surface of the wing and scorched paint. The trajectory of the rocket under double release is immediate nose-over.)

Rocket Fire Operation.

1. Set B-T-W control to ROCKET-GUN.
2. Set rocket dive angle control to type rocket being fired and the expected dive condition; set pointer to S for steep dives of more than 40° or N for dives less than 40°.

Note

If the stick trigger is depressed with the dive angle control in the ROCKET position the sight will automatically return to the GUN position.

3. Set mechanical cage lever on sight head to UNCAGED.
4. Set reticle dimmer control for desired brilliance.
5. Depress the electrical caging button.
6. Fly on the desired approach to the target, until the reticle image lies on the target, then track smoothly.

MISCELLANEOUS EQUIPMENT.

ANTI-G SUIT PROVISIONS.

An air pressure outlet connection on the front of the pilot's seat (figures 1–39 and 1–40) provides for the attachment of the air pressure intake tube of the pilot's anti-g suit. Air pressure for inflation of the anti-g suit bladder is conducted from the engine compressor through a pressure regulating valve (26, figure 1–8 25, figure 1–9) located on the left console which starts functioning when a force of 1.75 g's is applied to the aircraft. A control marked HI and LO allows for adjustment of the rate of inflation of the anti-g suit. In the LO range the valve opens at 1.75 g and then allows 1 psi of air pressure to pass to the suit for every increase of 1 g force thereafter. In the HI range the valve still opens at 1.75 g but delivers 1.5 psi per g force thereafter. The suit will inflate in 0.2 to 2.0 seconds depending on the input pressure.

Fs Fs Fs Fs Fs Fs Fs Fs Fs Fs Fs Fs Fs Fs Fs

T.O. 1F-84(25)F-1EN

SAFETY OF FLIGHT SUPPLEMENT

FLIGHT MANUAL

USAF SERIES
F-84F-25
AIRCRAFT

THIS PUBLICATION SUPPLEMENT T.O. 1F-84(25)F-1. Reference to this supplement will be made on the title page of the basic publication by personnel responsible for maintaining this publication in current status.

COMMANDERS ARE RESPONSIBLE FOR BRINGING THIS SUPPLEMENT TO THE ATTENTION OF ALL AF PERSONNEL CLEARED FOR OPERATION OF SUBJECT AIRCRAFT.

PUBLISHED UNDER AUTHORITY OF THE SECRETARY OF THE AIR FORCE

2 MAY 1960

1. PURPOSE.

To aid the crew member in determining when to disconnect and connect his Zero Delay Lanyard.

2. INSTRUCTIONS.

The following chart contains instructions on when to disconnect and connect the Zero Delay Lanyard.

Fs Fs Fs Fs Fs Fs Fs Fs Fs Fs Fs Fs Fs Fs Fs Fs Fs

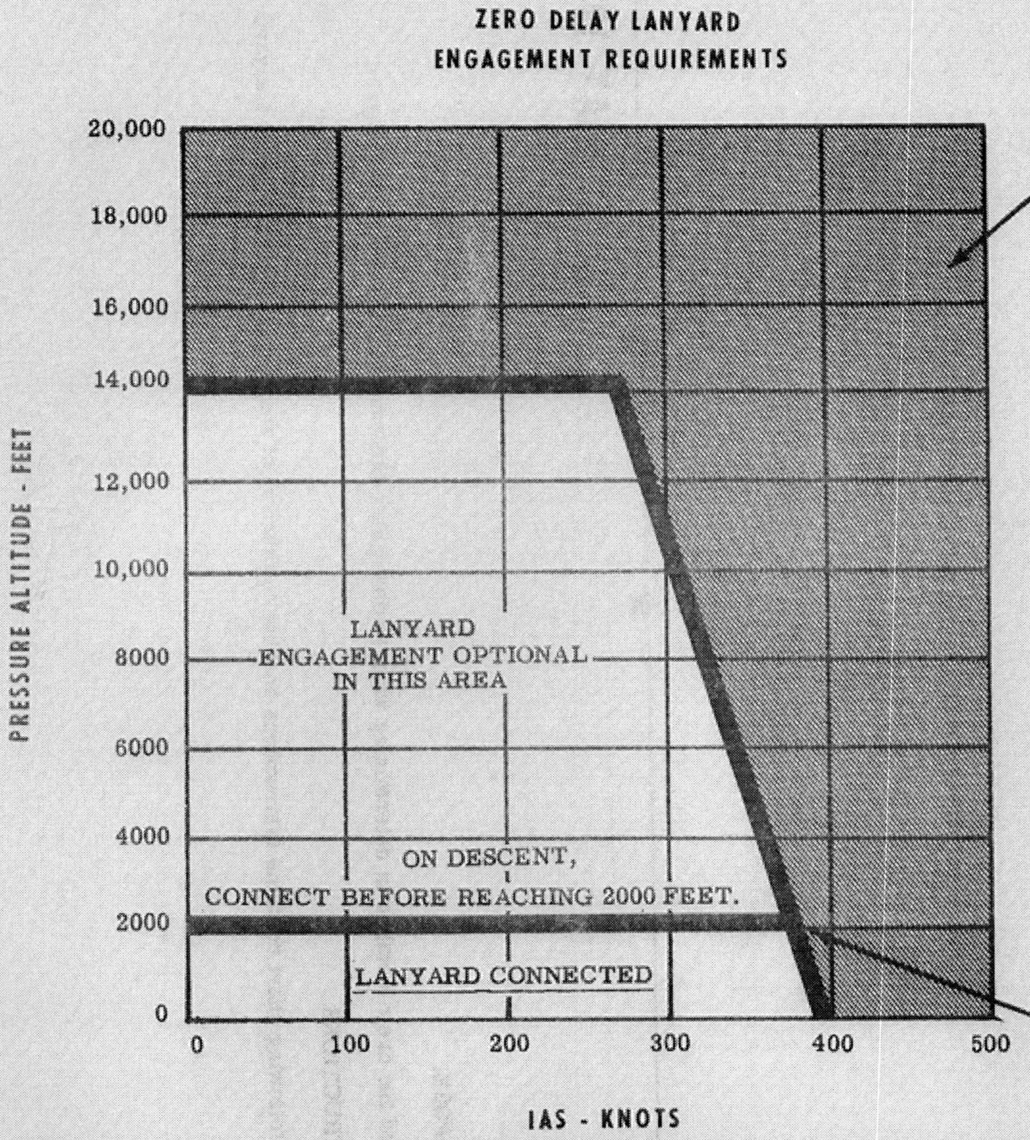

LANYARD DISCONNECTED

NOTE

CHUTE FAILURE LIKELY IN THIS AREA IF EJECTION ATTEMPTED WITH ZERO LANYARD CONNECTED.

WARNING

FOR MAXIMUM SAFETY DURING CLIMB OUT THE LANYARD SHOULD BE DISCONNECTED IMMEDIATELY UPON LEAVING THE "LANYARD CONNECTED" AREA. TO AVOID DISTRACTION DURING CRITICAL PHASES OF FLIGHT, SUCH AS DURING INSTRUMENT OR FORMATION FLIGHT, IT IS PERMISSIBLE TO DELAY DISCONNECTING THE LANYARD TO ANY POINT WITHIN THE OPTIONAL AREA. IF THE AIRCRAFT SHOULD EXCEED THE OPTIONAL AREA, THERE IS NO CAUSE FOR GRAVE ALARM UNLESS THE AIRCRAFT CANNOT BE RETURNED TO THE OPTIONAL AREA, OR THE LANYARD CANNOT BE DISCONNECTED BEFORE EJECTION IS INITIATED. IN ANY EVENT, WHEN CHART LIMITS ARE EXCEEDED, DISCONNECT THE LANYARD AS SOON AS POSSIBLE TO MINIMIZE TIME IN THE UNSAFE AREA.

NOTE

THIS IS NOT A PRESSURE ALTITUDE IT IS ALWAYS 2000 FEET ABOVE THE TERRAIN

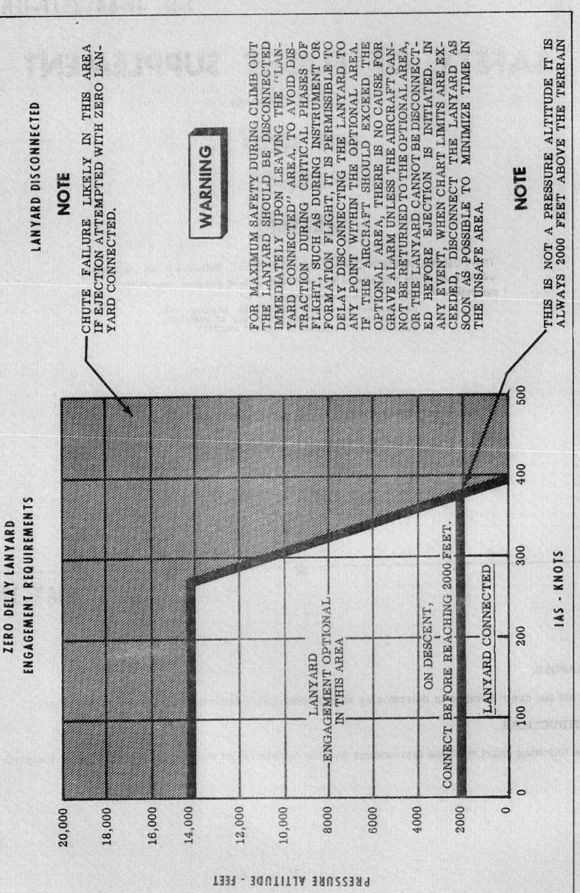

Figure 1. Zero Delay Lanyard Engagement Requirements.

END

SECTION 5 OPERATING LIMITATIONS

TABLE OF CONTENTS

INTRODUCTION.

This section includes the engine and airplane limitations that must be observed for safe and efficient operation. Instrument markings form a part of these limitations. However, they are not repeated in the text and must be referred to on the instrument marking page, figure 5–1. Where necessary, further explanation of the instrument markings are covered in the text of this section, under the appropriate heading. For complete restrictions carefully read the instrument marking page and the explanatory text.

ENGINE LIMITATIONS.

Normal engine limitations are shown in figure 5–1 and in the following paragraphs. Military thrust is obtained by placing the throttle full forward. Maximum continuous thrust (normal rated thrust) is defined as the thrust obtained at approximately three per cent engine RPM below military thrust RPM. Maximum thrust is the same as military thrust since the engine is not equipped with afterburner.

OVER TEMPERATURE LIMITS.

Any start which exceeds 800°C (1,472°F) will require an investigation to determine the cause. A hot section inspection in accordance with the applicable technical order shall be performed after one start, during which the exhaust gas temperature exceeds 900°C (1,652°F). Each hot start will be recorded on the Form 781. The peak temperature should not hold for more than ten seconds. Maximum exhaust gas temperature during engine idling is 660°C (1,220°F) stabilized at 36 to 42

INSTRUMENT MARKINGS

FUEL-GRADE JP4

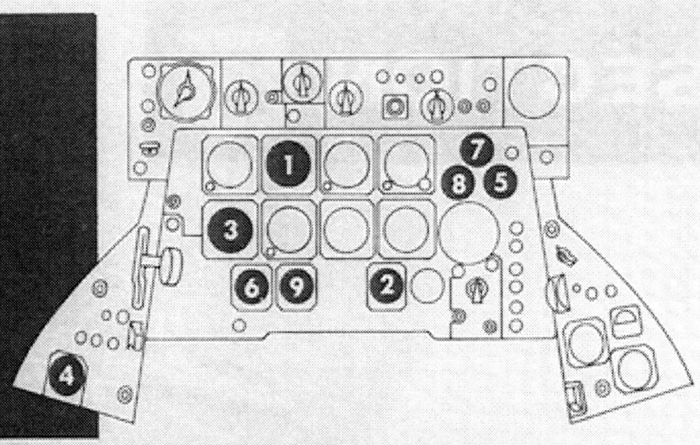

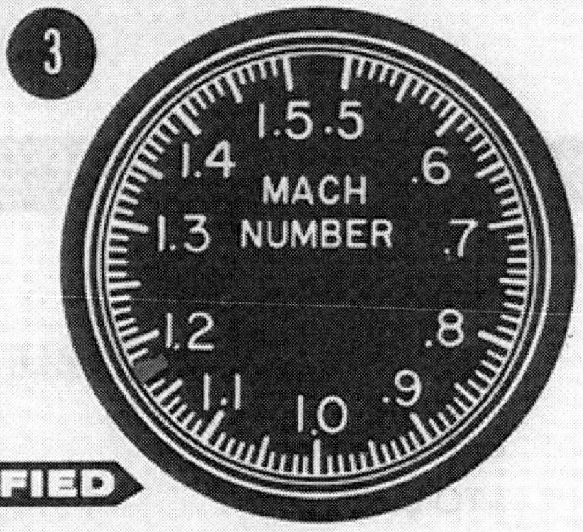

UNMODIFIED

AIR SPEED INDICATOR

- 220 knots *Landing gear down (225 Flaps down)*
- 610 knots *Maximum airspeed*

MACHMETER

- 1.175 *Design Dive Speed*

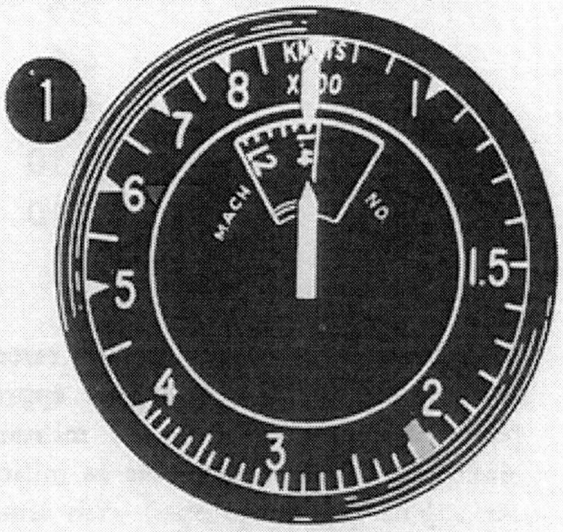

MODIFIED

AIRSPEED AND MACH NUMBER INDICATOR

- 220 Knots Landing gear down (225 Flaps down)
- 610 Knots Maximum airspeed

CAUTION

The red pointer of type ME4 Airspeed Indicator indicates a preset equivalent airspeed in terms of indicated airspeed and therefore increases as altitude increases. This pointer should not be used as a maximum allowable speed reference as the limiting speed is in terms of indicated airspeed.

ACCELEROMETER

- +7.0 G Maximum, clean airplane over Mach No. 0.9 or airplane with combat external stores at any Mach No.
- +8.67 G Maximum, clean airplane under Mach No. 0.9
- −3.0 G Maximum.

Figure 5–1 (Sheet 1 of 3)

4

LANDING FLAP POSITION INDICATOR

- 50% *Take-off*

6

HYDRAULIC PRESSURE UTILITY and POWER SYSTEMS

TANDEM AIRCRAFT

- 1250-1600 PSI Normal full flow
  1450-1600 PSI Static
- 1750-1900 PSI Permissible pressure 5 sec maximum
- 1925 PSI Maximum

9

HYDRAULIC PRESSURE BOOST SYSTEM

NON-TANDEM AIRCRAFT

- 1250-1600 PSI Normal full flow
  1450-1600 PSI Static
- 1750 PSI Maximum

6

HYDRAULIC PRESSURE UTILITY SYSTEM

NON-TANDEM AIRCRAFT

- 1250-1600 PSI Normal full flow
  1450-1600 PSI Static
- 1750-1900 PSI Permissible pressure 5 SEC maximum
- 1925 PSI Maximum

Figure 5–1 (Sheet 2 of 3)

INSTRUMENT MARKINGS

ENGINE INSTRUMENTS

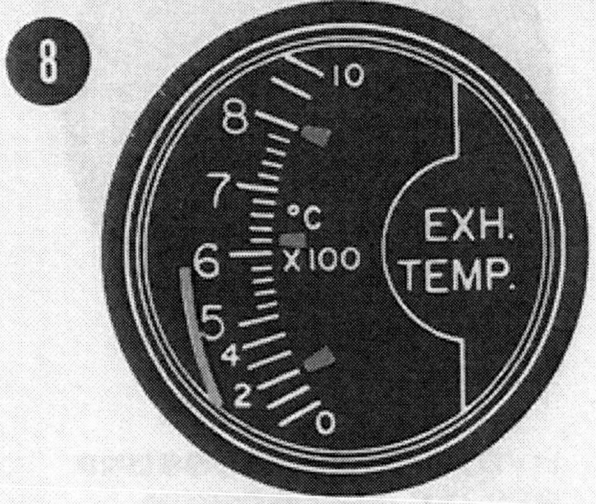

EXHAUST TEMPERATURE INDICATORS

J65-W-3 or B-3 ENGINE

- 200 to 585°C *Continuous operation*
- 200°C *Minimum for flight*
- 620°C *Maximum stabilized for flight (30 min)*
- 800°C *Maximum during starting and acceleration*

J65-W-7 or B-7 ENGINE

- 200 to 595°C *Continuous operation*
- 200°C *Minimum for flight*
- 650°C *Maximum stabilized for flight (30 min)*
- 800°C *Maximum during starting and acceleration*

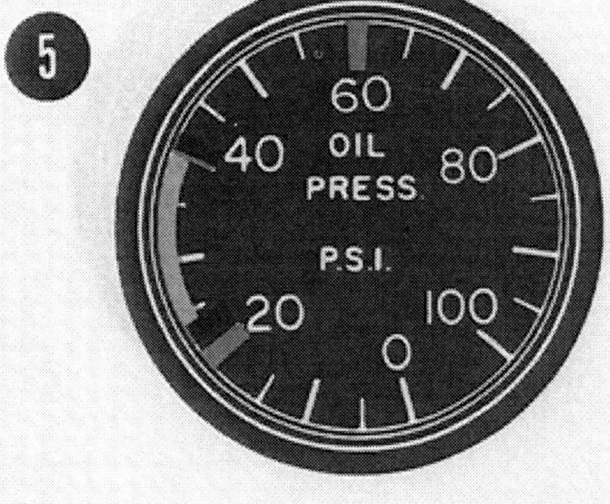

OIL PRESSURE INDICATOR

- 20 PSI *Minimum—below 50% RPM*
- 24 PSI *Minimum—50% RPM and above*
- 40 PSI *Maximum*
- 60 PSI *Maximum for cold start and initial run-up*

TACHOMETER

- 83.0 to 96.5% RPM *Best cruise rpm*
- 101.0% RPM *Maximum (Ground or flight 30 minutes)*
- 60 to 82% RPM *Restricted range see RPM RESTRICTIONS in this section*

Figure 5—1 (Sheet 3 of 3)

per cent RPM for the J65-W-3 and B-3 engines, and 42 to 48 per cent RPM for J65-W-7 and B-7 engines. A stabilized temperature is one that will remain constant for three individual readings taken at 30 second intervals.

The maximum stabilized exhaust gas temperature on take-off and during flight is 620°C (J65-W-B-3 series engines) and 650°C (J65-W-B-7 series engines).

Turbine rotor blade life is progressively shortened each time the engine is operated above the maximum temperature. Continued overtemperature operation may eventually result in turbine rotor blade failure. During takeoff and climb with a cold engine, an effort should be made to monitor the exhaust gas temperature to a point on, or below the maximum steady state allowable, as soon as the airplane is airborne and safety permits. Power may be gradually increased as engine stabilization is approached.

WARNING

- During starting, immediate action should be taken to abort any start in which it appears that the temperature exceeds the normal starting limit, 800°C (1,472°F). Do not attempt to restart the engine after exceeding a starting exhaust gas temperature of 900°C (1,652°F) as turbine wheel replacement is required prior to further flight.
- Since the airplane maintains a nose-high attitude during stalls, airflow to the engine can be critically low when the airplane is in a stall. Therefore, the throttle should be retarded, and rapid throttle advancements should not be attempted, until the nose has been lowered and airspeed is definitely increasing. Should the throttle be advanced before the engine receives sufficient airflow, turbine overtemperature may result.

Note

When the ambient air temperature exceeds 32°C (90°F) maintain idle RPM exhaust temperature within limits by manually advancing the throttle to a higher RPM not to exceed 50 per cent RPM.

OVERSPEED LIMITS.

During an acceleration the RPM may momentarily exceed 101%. If the RPM should exceed 103 per cent, the throttle is to be retarded immediately to stabilize RPM at 100 per cent or below. If the engine exceeds 106 per cent RPM, it is necessary to shut down the engine as soon as possible and perform a hot section inspection and turbine rotor replacement as outlined in the applicable maintenance handbook.

ENGINE ACCELERATION LIMITS WITH SCREENS RETRACTED.

Refer to Section VII for discussion on engine acceleration. Engine acceleration time should be as follows:

ENGINE MODEL	*TYPE FUEL CONTROL*	*ACCEL. TIME SECONDS*	*PER CENT RPM*
J65-B/W3	TJ-L1	8-14	47 to 100
J65-B/W3, B/W7	TJ-L2	14 max	47 to 100

Note

Acceleration time may be increased ½ second for each 1,000 feet increase in altitude. Acceleration limits apply to stall-free acceleration checks, accomplished by advancing the throttle from 47 to 100 per cent RPM in one second. Some engines will not accelerate properly when they are cold and will therefore require a short warmup period in order that the acceleration limits are met.

TRANSIENT TEMPERATURE LIMITS.

For starting and accelerations, exhaust gas temperature should not exceed 800 degrees Centigrade (1,472°F) or remain at its peak value for more than 10 seconds. Following starts and accelerations, the exhaust gas temperature will overshoot the stabilized temperature for a given throttle setting, and several minutes are required to reach the stabilized value. After peaking, the exhaust gas temperature will show a continuous decline until temperature is stabilized. Initial temperature decline rate from the peak will be quite rapid and will become slower and slower as stabilization is approached. Figure 7—1 graphically illustrates the EGT behavior following rapid acceleration of a cold J65-W-B-3 or J65-W-B-7 engine.

Note

The exhaust gas temperature should not stabilize on any point above the maximum steady state value.

This page intentionally left blank.

ENGINE RPM RESTRICTIONS.

Engine operation in the 60 to 82% RPM range can induce compressor blade stresses which may cause rotor blade failures in the number one, and three, stages. When the restricted range is entered, it is not necessary to pass through the complete range before reducing RPM to a point below the restricted range. To preclude the possibility of compressor rotor blade failure caused by steady state operation in the 60 to 82 per cent RPM range the following restrictions will be adhered to:

1. During static ground operation, the engine will not be operated in the 60 to 82 per cent RPM range except during acceleration or deceleration through this range.
2. During taxi operation, the engine will not be stabilized in the 60 to 82 per cent RPM range.
3. Flight operation with the engine stabilized in the 60 to 82 per cent RPM range will be performed only as demanded by operational necessity since operating time in this range is cumulative and has a direct effect on blade life.

ENGINE FLUCTUATION.

Engine fluctuation (at any RPM) is permissible provided that the following is observed:

1. Fuel flow shall not vary a total of more than 500 pounds per hour.
2. Total RPM variation shall not vary more than 1½ per cent.
3. Normal EGT limits are not exceeded.
4. Engine surging does not produce any perceptible movement (or change in movement) of the airplane.

OIL PRESSURE LIMITS.

Time and pressure limitations are based on the minimum oil pressure required to insure an adequate supply of oil to the center and rear main bearings. Minimum allowable oil pressure limits are 20 PSI with engine operating below 50 per cent RPM and 24 PSI with engine operating at 50 per cent RPM and above.

1. Do not operate engine with zero oil pressure longer than one minute after initiating start.
2. Operation below the minimum allowable oil pressure is permissable for a period of one minute or less.
3. Operation below the minimum allowable pressure for a period of over one minute but not exceeding two and one-half minutes requires an engine inspection.
4. Operation below the minimum allowable pressure for a period exceeding two and one-half minutes requires engine removal.

Note

The same time limits shall apply to windmilling engines with zero oil pressure.

ATO UNITS.

The following ATO units are approved for utilization without restriction, except those prescribed in the Inspection, Handling and Storage Technical Order for each type unit.

14AS1000
14DS1000 MK4. Mod 2
14DS1000 M-8
15KS1000
M-15 (16NS1000 or T-60)

AIRSPEED LIMITATIONS.

1. Do not lower the landing gear above 220 knots IAS.
2. Do not lower landing flaps all the way above 225 knots IAS or 50 per cent above 260 knots IAS.
3. Do not extend landing lights above 225 knots.
4. Jettison ato units between 200 and 250 knots IAS to prevent damage to the rear hooks and fuselage.
5. Do not engage the autopilot at speeds in excess of 425 knots IAS at altitudes below 20,000 feet.
6. Dropping fire bombs from the outboard pylons, above 400 knots IAS, will cause the aircraft to lose approximately 60 feet of altitude.

JETTISON EXTERNAL STORES

- *The following procedures and restrictions shall be observed while carrying any of the described external stores. However 250 knots IAS can be considered as a general jettison speed for all external stores.*
- *Jettison speed restrictions shall apply to other external tanks not specifically stated, provided they are of the same type and have similar tank contours and fuel capacities.*

TYPE OF STORE	AIRSPEEDS FOR JETTISON	ATTITUDE FOR JETTISON
1. 230GAL Type II Sutton or Fletcher tanks (on inboard pylons)	200-275 knots IAS any quantity of fuel remaining	Steady
2. 230GAL Type II Sutton or Fletcher tanks (on outboard pylons)	220-310 knots IAS. Tanks must be empty	Steady climb
• 3. 450 GAL Type II Royal Heater tanks (on inboard pylons)	200-275 knots IAS any quantity of fuel remaining	Steady climb
4. 230 GAL Type I tanks with fins (on inboard pylons)	200-500 knots IAS. Release singly, any quantity of fuel remaining	Level, diving, or pulling up, but with no negative "G" force
5. 450 GAL Type II Royal Heater tanks (on inboard pylons)	200-275 knots IAS any quantity of fuel remaining	Steady climb
6. 230 GAL Type IV Royal Heater tanks (on outboard pylons)	230-270 knots IAS any quantity of fuel remaining	Steady climb
7. Inboard pylons (bare)	200-525 knots IAS	Level, diving or pulling up. No negative "G" forces.

• REMARKS While carrying tanks max deceleration 4g safe rate of descent 5000 feet per minute. Max rate of descent 10000 feet per minute

NOTE: All types of tanks can be released when full or nearly full to lighten aircraft in case of an emergency.

WARNING: Jettison of external stores below minimum or above maximum recommended jettison speeds may result in stores striking the aircraft, and causing structural damage.

Figure 5—2

7. Do not open the air refueling receptacle at speeds above 350 knots IAS.

8. Do not deploy the drag chute at speeds in excess of 220 knots IAS as the shear pin will shear; if the aircraft is airborne an excessive high sink rate is experienced with chute deployed.

9. After landing aircraft modified by drilling holes in the canopy arms (T.O. 1F-84-763), do not open a closed canopy until the aircraft speed, and/or wind gusts, are less than 40 knots.

JETTISON EXTERNAL STORES.

See figure 5—2.

PYLON TANKS.

The Type IV, 225 gallon, plastic, external fuel tanks, due to its non-conductivity, is more seriously affected by lightning stroke. Therefore, these tanks should not be flown where lightning hazards exist.

PROHIBITED MANEUVERS.

See figure 5—3.

ACCELERATION LIMITATIONS.

See figure 5—4.

BOMB RACK LIMITATIONS.

Recent accidents resulting from failure of the following bomb racks; S-2, S-2A, S-3, MA-4 and MA-4A, have indicated the necessity for maintaining closer control over these racks. Therefore, until the above mentioned bomb racks are replaced with racks of greater strength, it is necessary that the following instructions be observed to insure that appropriate inspections are accomplished. A notation should be made in the DD Form 781 whenever any of the following conditions are encountered with any of the above mentioned bomb racks installed.

1. Whenever a suspended fuel tank collapses during flight.

2. Whenever any of the following conditions are encountered with more than 1,000 pounds suspended from type S-2, S-2A, MA-4, MA-4A bomb racks, or more than 2,000 pounds suspended from type S-3 bomb racks.

   a. Whenever the aircraft exceeds 90 degrees per second rate of roll.

   b. Whenever the aircraft exceeds 4.5 G.

   c. Whenever any unusual incident occurs in flight or during ground operations which could cause damage to any part of the racks, such as rough taxiing or hard landings.

3. The MA-4 and MA-4A shackles are restricted from use until T.O. 1F-84-669 has been complied with.

PROHIBITED MANEUVERS

1 A thru G AIRCRAFT

Inverted flying or any other maneuver resulting in negative acceleration for longer than one minute may result in engine flameout as the fuel and oil supply is sufficient for one minute of operation in this attitude.

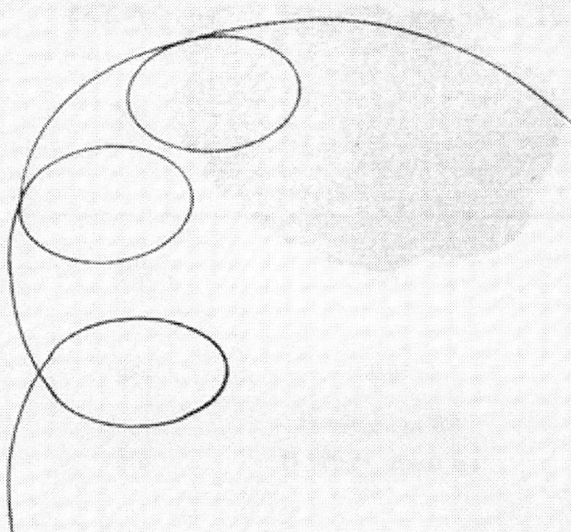

2 H and LATER AIRCRAFT

Inverted flying or any other maneuver resulting in negative acceleration for longer than 15 seconds may result in engine flameout as the fuel supply is sufficient for 15 seconds of operation in this attitude.

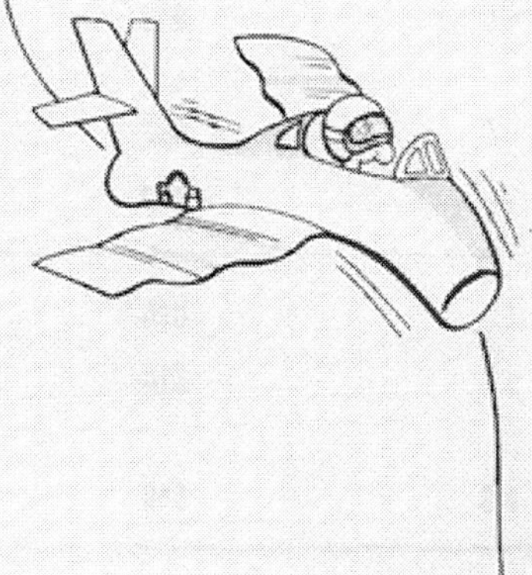

3 *Do not fly in a zero "g" flight configuration for any extended period of time on F-84F-25RE airplanes as hydraulic pressure may be lost due to hydraulic pump cavitation. If loss of pressure occurs, considerable time may elapse before hydraulic pressure is regained.*

4 *Avoid landing the airplane with fuel in the external tanks or with bombs installed. Landing with fuel in the external tanks or with bombs installed requires that a good landing technique be employed to prevent wrinkling or buckling the wings during such landings. Land with a minimum of internal fuel if practicable.*

Figure 5—3

ACCELERATION LIMITATIONS

F-84F SYMMETRIC FLIGHT LIMIT LOAD FACTOR AND SPEED RESTRICTIONS

CONFIGURATION		Maximum Positive G Limit Below 5000 FT	Maximum Positive G Limit Above 5000 FT	Maximum Velocity Knots IAS
CLEAN	Below 0.9 Mach	8.67	7.0	610
	Above 0.9 Mach	7.0	7.0	
TANKS	Two 230 Gallon Type I	7.0	6.0	610
	Two 450 Gallon Type I			
	MB-3 Shackles	6.0	5.0	610
	S-3 Shackles	4.5	4.0 to 15,000 FT 3.0 above 15,000 FT	610
	Two 230 Gallon Type II Inboard Pylons	4.0	4.0	610
	Two 230 Gallon Type II Outboard Pylons	4.0	4.0	350
	Two 450 Gallon Type II	4.0	4.0	610
BOMBS	One 2000 LB	7.0	6.0	610
	Two 2000 LB	7.0	6.0	610
	Two 2000 LB } Two 1000 LB }	7.0	6.0	510
ROCKETS	Single Tier on Wing Posts with or without Double Tier on Inboard Pylon Adapter	5.0	5.0	610
SPECIAL STORE		7.0	6.0	510

NOTES

1. Negative G limit, −3.0 G any configuration, all altitudes, except Type II tanks which are (−2.0 G).
2. Above speed limits apply except when limited to lower values by buffet, or as noted in Section VI of Flight Manual.
3. Abrupt rolling pullouts are prohibited when carrying external stores due to shackle limitations.
4. Abrupt large displacements of the rudder should be avoided. Excessive sideslip angles can lead to dangerous snap maneuvers. External stores and or extended speed brakes tend to further aggravate this condition.
5. When external stores are used in combination, the lower limit shall apply. See Bomb Rack Limitations on page 5-6
6. Type I empty tanks below 5000 feet altitude—clean aircraft limitations apply.

Figure 5—4

OPERATING LIMITS DIAGRAM

GW—18500 LB

* See figure 5—3 for operating limits with external stores.

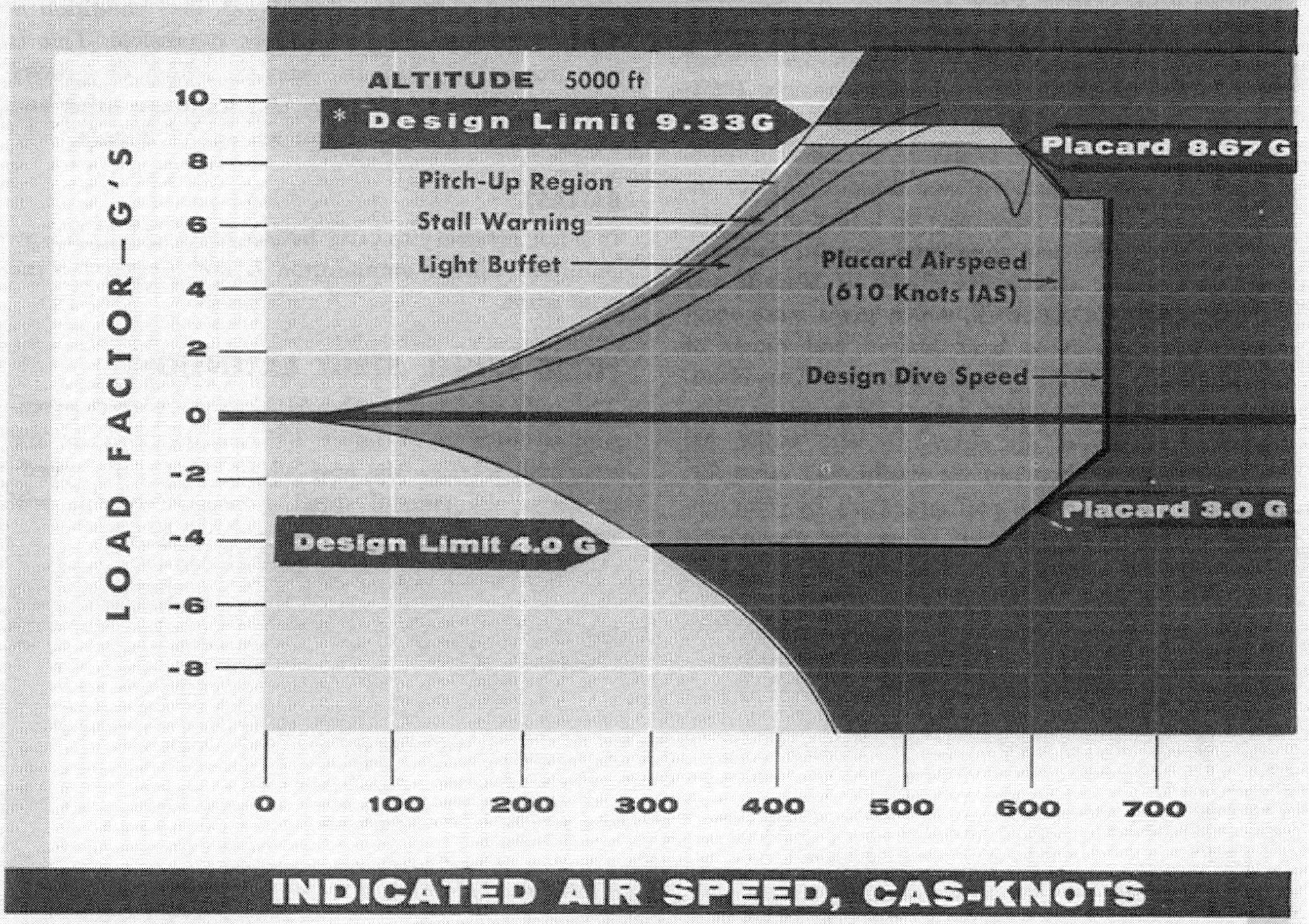

This is an operating limits diagram for the F-84F Aircraft. It shows what speeds and G are possible and allowable for the aircraft. In the light buffet area a very mild buffet is experienced. As the G are increased the buffet intensity increases as denoted by the stall warning area. To keep from exceeding the design limit of 9.33 G it is important that the pilot recognize the heavy buffet region. Heavy buffet is an indication that pitch-up may occur. A more detailed description of the actual operating limits which must be observed by the pilot are given in figure 6—3.

The green area represents the G loads and airspeeds which may be used by the pilot without damage to the aircraft. The aircraft can perform all the maneuvers required within these areas.

Figure 5—5

WEIGHT LIMITATIONS.

MAXIMUM GROSS WEIGHT.

The maximum gross weight of the aircraft has been limited to 27000 pounds, plus the weight of the expendable assist take-off units. The load limiting factor is determined by the main wheel tires which have a maximum loading of 13500 pounds each. The normal gross weight of the aircraft is approximately 18600 pounds. This includes pilot, all internal fuel, oil, all guns and ammunition. Therefore, a total of 8400 pounds of external stores of any description, not including assist take-off units, may be loaded on the aircraft to obtain the maximum gross weight. Each aircraft is equipped with a Handbook of Weight and Balance, Data, T.O. 1-1B-40, which gives more accurate information as to basic weight and center of gravity positions of the individual aircraft. This Handbook should be consulted before each flight. The center of gravity of the aircraft is very stable. All loadings will fall between the established most forward and most rearward positions. Only by some unusual condition in flight of using the expendable load can the center of gravity exceed its limits. If the loading of external stores includes external tanks and fuel on the outboard pylons, this fuel should always be used first, otherwise the center of gravity will exceed its most aft limit, due to taking the fuel from the internal tanks. If a four bomb load condition is used, drop outboard bombs first, if possible. This is not imperative, but for smooth center of gravity travel, is practical. All other conditions are stable and the center of gravity cannot get out of bounds.

BALLAST.

It is not necessary to carry ballast in the gun deck ammunition cans if ammunition is not carried for the nose guns.

NOSE WHEEL STRUT EXTENSION.

The nose wheel strut should have the correct extension as noted in (1, figure 2–1). A bottomed or low strut will increase the nose wheel "unsticking speed" to above the take-off speed. Consequently this will increase the take-off speed and ground roll.

SECTION 6 FLIGHT CHARACTERISTICS

TABLE OF CONTENTS

GENERAL FLIGHT CHARACTERISTICS.

The aircraft is designed for high speed at all altitudes. The swept wing and tail surfaces reduce the effects of compressibility, therefore the high speed is limited only by the available thrust and the total drag of the aircraft. All flight controls are hydraulically operated to provide comfortable control forces and ease of maneuvering. Minor longitudinal trim changes occurring in level flight at high Mach Numbers are easily trimmed out. This, in combination with trimming of the aileron and rudder artificial "feel" systems permits complete trim throughout the speed range.

TANDEM

FLIGHT CONTROLS.

The ailerons and rudder are conventional hydraulically actuated flight controls. The elevators and stabilizer have been combined to form a single surface known as the stabilator. All flight controls are hydraulically actuated and irreversible. The stabilator and ailerons are equipped with tandem hydraulic actuators. The utility hydraulic system provides hydraulic fluid under pressure to the utility side of the stabilator and aileron tandem actuators and the rudder actuator in addition to the landing gear, wing flaps, speed brakes, spoilers and all other hydraulic operated systems. The power hydraulic system provides hydraulic fluid under pressure to the power side of the stabilator and aileron tandem actuators exclusively. Both the utility and power hydraulic systems are completely independent of each other and either system alone is capable of sustaining flight control. Since both hydraulic systems have engine-driven hydraulic pumps, an emergency hydraulic system incorporating an electrically driven hydraulic pump is provided. The emergency pump provides hydraulic fluid under pressure to the power side of the stabilator and aileron tandem actuators. There is a direct mechanical linkage between the control stick and stabilator. However, due to the surface hinge movements, direct manual control without hydraulic pressure is not possible. Although manual operation of the aileron is possible, due to airloads and friction, a high force is required for operation. Rudder operation is adequate through mechanical linkage and is a useful control for picking up a wing. Since the controls are irreversible and airloads on the surfaces cannot be felt, control feel is simulated by artificial feel units in each primary control. The artificial feel unit is a spring capsule designed to give the pilot a sense of control feel by increasing the force required to deflect the controls proportional to the amount of deflection. The feel devices also provide trim control by means of electric actuators which reposition the spring capsules to the selected no-load position. A mechanical advantage shifter is incorporated in the stabilator control system to allow for more control surface deflection in take-off and landing than at high speeds. The shifting is done automatically as the landing gear is retracted or extended. After take-off, when the gear retracts, the stabilator leading edge will move up (aircraft nose down trim change) while the shifter is operating at any fixed aft stick position. Since this movement is relatively slow and the normal aircraft nose up trim requirement is decreasing as the aircraft accelerates, only small corrections need be made by the pilot to compensate for the change in stabilator position. When the gear is extended for landing the procedure outlined in the preceding sentence will be reversed and a slight nose up trim change will be noticed. Failure of the shifter in the one to one ratio during flight makes the aircraft more sensitive in the pitch axis therefore, overcontrolling and porpoising is possible in high speed flight. Failure of the shifter in the 1.8 to 1 ratio in the landing configuration results in approximately nine degrees aircraft nose up stabilator being available for landing. Even with the aircraft at the forward critical loading (CG at 15 per cent MAC) this type of malfunction presents no hazard as long as a touchdown speed of 140 knots or above is utilized. Therefore, the Flight Control Laboratory recommends that in the event of a 1.8 to 1 ratio malfunction in the landing configuration a touchdown speed of 140 knots be used for all loading conditions. In addition to the mechanical advantage indicator light which is illuminated while the ratio is changing or if the system is in the wrong ratio, the pilot can recognize when the mechanical advantage remains in the 1.8 to 1 ratio when the landing gear is extended by the absence of an aircraft trim change.

WARNING

Lateral control with unboosted ailerons is inadequate in high speed, low altitude conditions, under turbulent landing conditions, and with asymmetrical loads installed.

WARNING

Abrupt, large displacements of the rudder should be avoided. Excessive sideslip angles can lead to dangerous snap maneuvers. External stores, and/or extended speed brakes, tend to further aggravate this condition.

MINIMUM FLYING SPEEDS

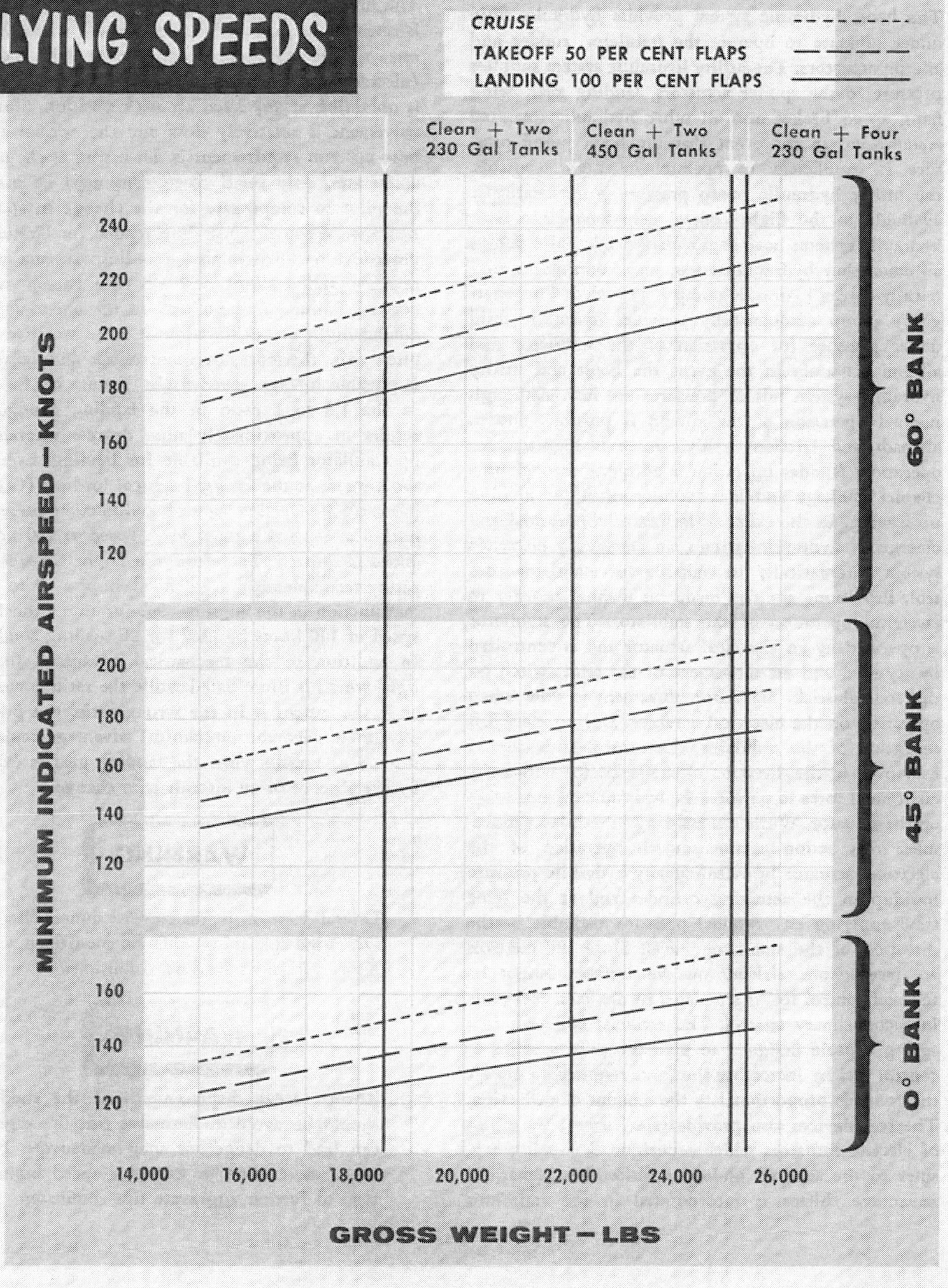

Figure 6–1

NON-TANDEM

FLIGHT CONTROLS.

The boost hydraulic system provides hydraulic fluid under pressure to operate the stabilator, rudder and aileron actuators. The utility hydraulic system supplies pressure to the spoiler actuators, landing gear, wing flaps, speed brakes and all other hydraulic operated systems and in the event that the boost pump pressure is insufficient to operate the flight controls, the utility hydraulic pump pressure is automatically available to the flight control actuators. Since both hydraulic systems have engine-driven hydraulic pumps an emergency hydraulic system incorporating an electrically-driven hydraulic pump is provided. The emergency pump automatically provides hydraulic fluid under pressure for operation of the stabilator and aileron actuators in the event the boost and utility hydraulic system fail or pressures are low. Although manual operation of the aileron is possible, due to airloads and friction, a high force is required for operation. Rudder operation is adequate through mechanical linkage and is a useful control for picking up a wing. In the event of failure of the normal and emergency hydraulic systems, an emergency electrical system automatically is available for stabilator control. Provisions are also made for manual transfer to electrical operation of the stabilator. The stabilator is operated by an electrical actuator and is controlled by forward and aft movement of the trim switch on the control stick. Stabilator movement is slow when operating on the electrical actuator. During electrical operation of the stabilator, the control stick should be moved in the direction of the stabilator with sufficient hand force to position the hydraulic control valve on the actuator. While not assisting in stabilator movement this action assures smooth operation of the electrical actuator by releasing any hydraulic pressure buildup in the actuating cylinder and at the same time applying any residual pressure available in the direction of the stabilator travel. Since the controls are irreversible, airloads on the surfaces cannot be felt and control feel is simulated by artificial feel units in each primary control. The artificial feel unit is a spring capsule designed to give the pilot a sense of control feel by increasing the force required to deflect the controls proportional to the amount of deflection. The feel devices also provide trim control by means of electric actuators which reposition the spring capsules to the selected no-load position. A mechanical advantage shifter is incorporated in the stabilator control system to allow for more control surface deflection in take-off and landing than at high speeds. The shifting is done automatically as the landing gear is retracted or extended. After take-off, when the gear retracts, the stabilator leading edge will move up (aircraft nose down trim change) while the shifter is operating at any fixed aft stick position. Since this movement is relatively slow and the normal aircraft nose up trim requirement is decreasing as the aircraft accelerates, only small corrections need be made by the pilot to compensate for the change in stabilator position. When the gear is extended for landing, the procedure outlined in the preceding sentence will be reversed and a slight nose up trim change will be noticed. Failure of the shifter in the one-to-one ratio during flight makes the aircraft more sensitive in the pitch axis, therefore, overcontrolling and porpoising is possible in high speed flight. Failure of the shifter in the 1.8 to 1 ratio in the landing configuration results in approximately nine degrees aircraft nose up stabilator being available for landing. Even with the aircraft at the forward critical loading (CG at 15 per cent MAC) this type of malfunction presents no hazard as long as a touchdown speed of 140 knots or above is utilized. Therefore, the Flight Control Laboratory recommends that in the event of a 1.8 to 1 ratio malfunction in the landing configuration a touchdown speed of 140 knots be used for all loading conditions. In addition to the mechanical advantage indicator light which is illuminated while the ratio is changing or if the system is in the wrong ratio, the pilot can recognize when the mechanical advantage remains in the 1.8 to 1 ratio when the landing gear is extended by the absence of an aircraft trim change.

WARNING

Lateral control is inadequate under "Boost Off" high speed, low altitude conditions, and under turbulent landing conditions.

WARNING

Abrupt, large displacements of the rudder should be avoided. Excessive sideslip angles can lead to dangerous snap maneuvers. External stores, and/or extended speed brakes, tend to further aggravate this condition.

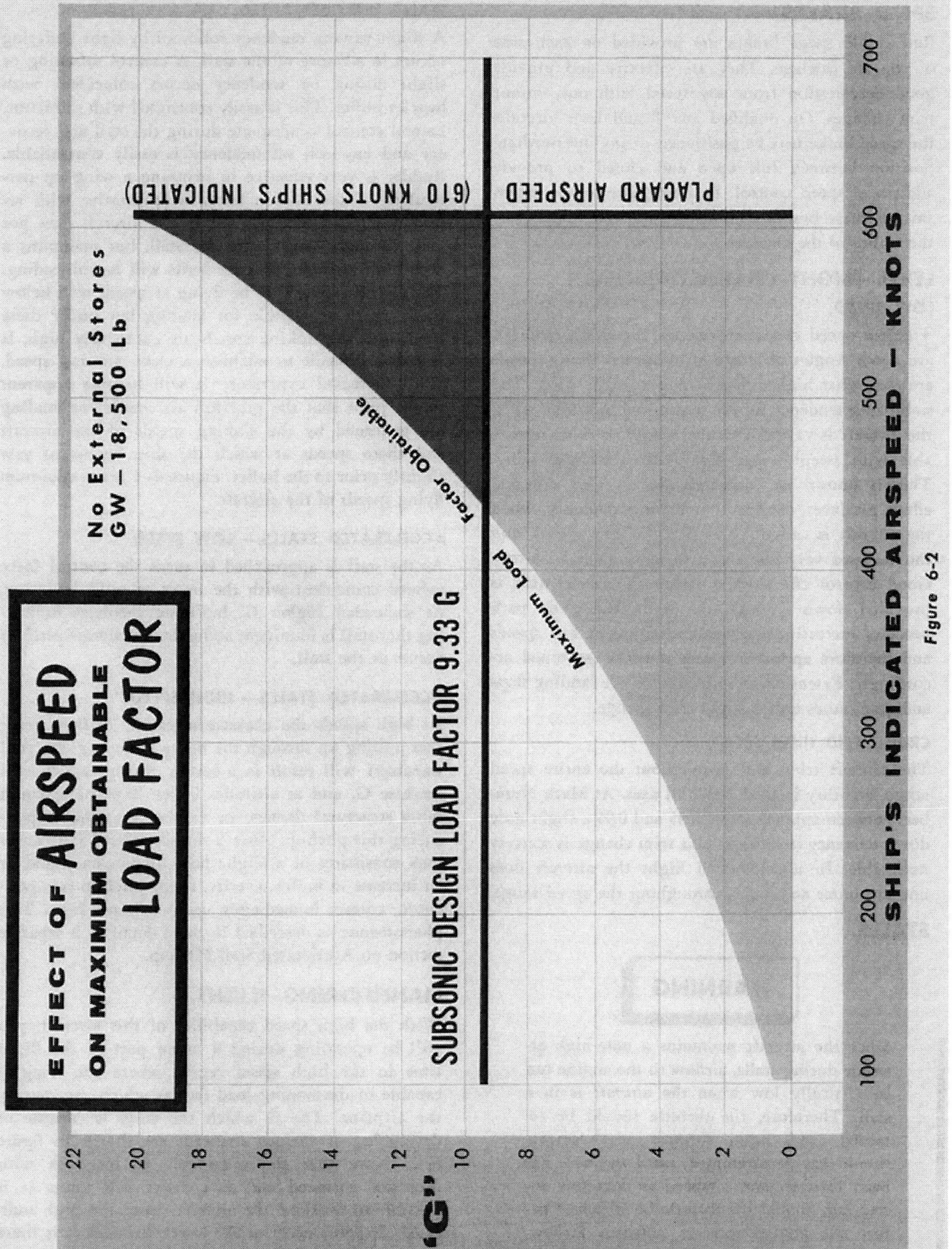
EFFECT OF AIRSPEED ON MAXIMUM OBTAINABLE LOAD FACTOR
No External Stores
GW – 18,500 Lb
PLACARD AIRSPEED (610 KNOTS SHIP'S INDICATED)
SUBSONIC DESIGN LOAD FACTOR 9.33 G
Maximum Load Factor Obtainable
"G"
22
20
18
16
14
12
10
8
6
4
2
0
100
200
300
400
500
600
700
SHIP'S INDICATED AIRSPEED – KNOTS

Figure 6-2

SPEED BRAKES.

Retractable speed brakes are provided on each side of the aft fuselage. They are effective and provide good deceleration from any speed with only minor trim changes. On modified and E and later aircraft, the speed brake may be positioned to any intermediate position between full open and closed to provide additional speed control. Buffeting due to the extension of these brakes is moderate and not limiting to the utility of the airplane.

LEVEL FLIGHT CHARACTERISTICS.

LOW SPEED.

The low speed characteristics and handling qualities are good. Angles of attack at minimum flying speeds are somewhat higher than with a straight wing. The stabilizing tendency for the leading wing to rise, when the aircraft is yawed with the rudder, is more noticeable with swept wings than with a straight wing. This is known as "dihedral effect". This dihedral effect produces considerable rolling tendency when the aircraft is yawed, particularly at low speeds, and can be used very effectively to assist lateral control. Good control effectiveness about all control axes is provided down through the stall. The glide path becomes increasingly difficult to adjust at low speeds and therefore approach speeds must be governed accordingly. Extension or retraction of the landing flaps and gear causes only a slight trim change.

CRUISE AND HIGH SPEED.

The aircraft trims well throughout the entire speed range. Stability is good about all axes. At Mach Numbers between approximately 0.85 and 0.95 a slight nose down tendency is present. This trim change is scarcely noticeable. In unaccelerated flight the aircraft does not encounter any buffet throughout the speed range.

STALLS.

WARNING

Since the aircraft maintains a nose-high attitude during stalls, airflow to the engine can be critically low when the aircraft is in a stall. Therefore, the throttle should be retarded, and rapid throttle advancements should not be attempted, until the nose has been lowered and airspeed is definitely increasing. Should the throttle be advanced before the engine receives sufficient airflow, turbine overtemperature may result.

STALLS UNACCELERATED.

A slight yawing tendency followed by light buffeting occurs in advance of the stall. A control softening or slight nosing up tendency occurs coincident with heavier buffet. This is easily controlled with stabilator. Lateral control is adequate during the stall and recovery and any roll off tendency is easily controllable. Rudder is very effective in bringing a wing up particularly if the aileron boost is inoperative with resultant high stick forces. Since the aircraft does not "fall through" until after the stall, but maintains a nose high attitude, practice stalls will be misleading. The aircraft appears to be flying at speeds well below those which are usable for landing but under these conditions the sinking speeds are excessively high. It is rather difficult to establish a clear cut stall speed. With increased experience it will become apparent to the pilot that the practical air speeds for landing are governed by the sinking speeds of the aircraft and those speeds at which the ship begins to yaw slightly prior to the buffet. Figure 6–1 gives minimum flying speeds of the aircraft.

ACCELERATED STALLS — LOW SPEED.

As the stall is approached in turns the control force softens coincident with the onset of mild buffeting. At somewhat higher G, buffeting increases indicating that stall is imminent and a final moderate pitchup occurs at the stall.

ACCELERATED STALLS — HIGH SPEED.

At high speeds the characteristics are similar except that pulling up through the heavy buffet region (stall warning) will result in a severe pitchup which will increase G, and at altitudes below 25,000 feet could cause structural damage or failure. To avoid experiencing this pitchup, observe the accelerated stall warnings consisting of a slight nose up tendency and/or an increase in buffet severity. If the pitchup is experienced, correct immediately using forward stick. This phenomenon is described in more detail in a separate section on Accelerated Stall Pitchup.

MANEUVERING FLIGHT.

With the high speed capability of this aircraft, you will be operating during a major part of the flight time in the high speed region where the wing is capable of developing load factors which can destroy the airplane. The G which the wing is capable of developing at various airspeeds are shown in figure 6–2. Note that the obtainable G increases with indicated airspeed and that above 430 knots it is possible to overload the aircraft. Since the high indicated airspeeds occur at the lower altitudes it is there that structural overloading is most likely.

MANEUVERING flight characteristics

NO EXTERNAL STORES · GW 18,500

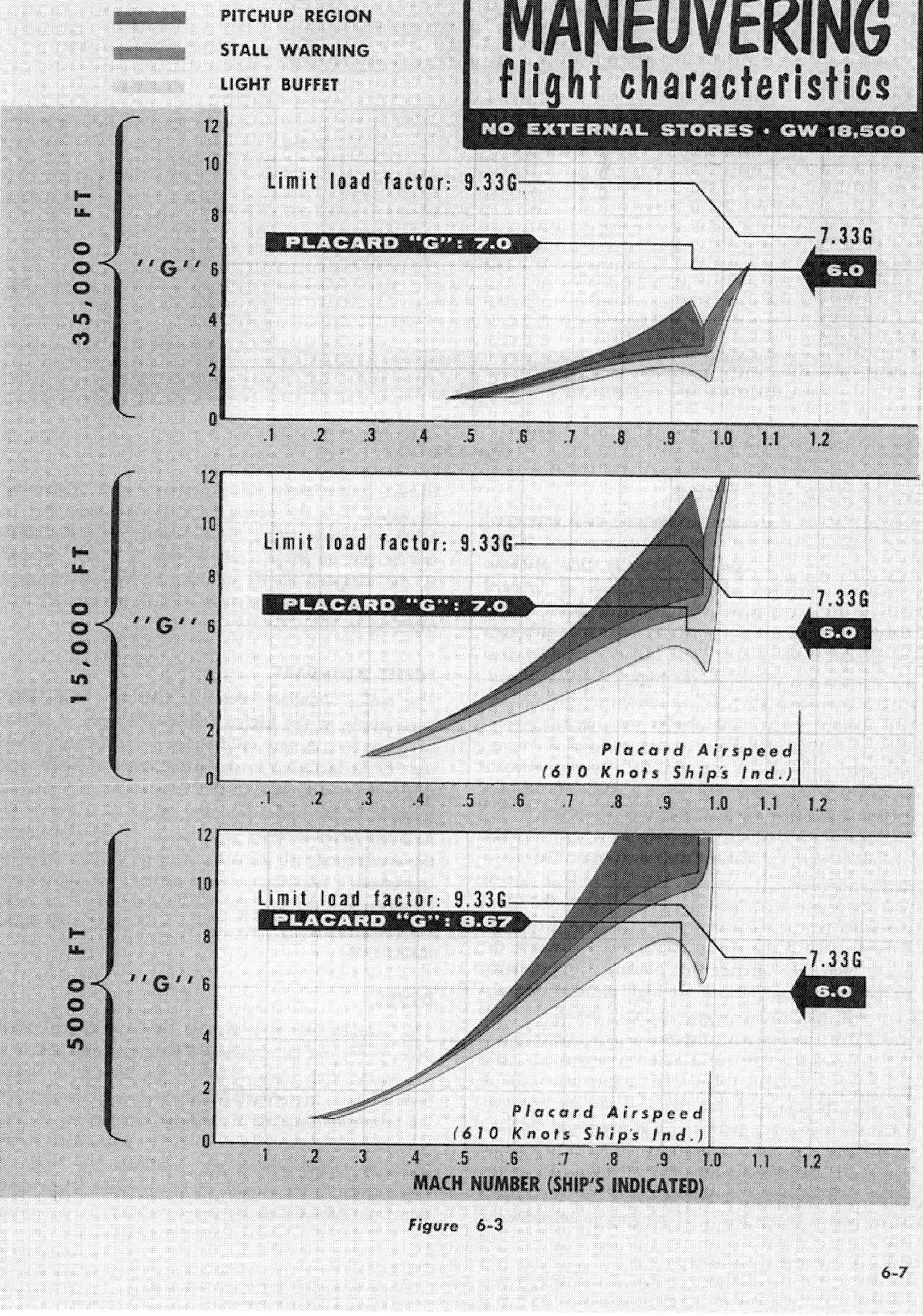

Figure 6-3

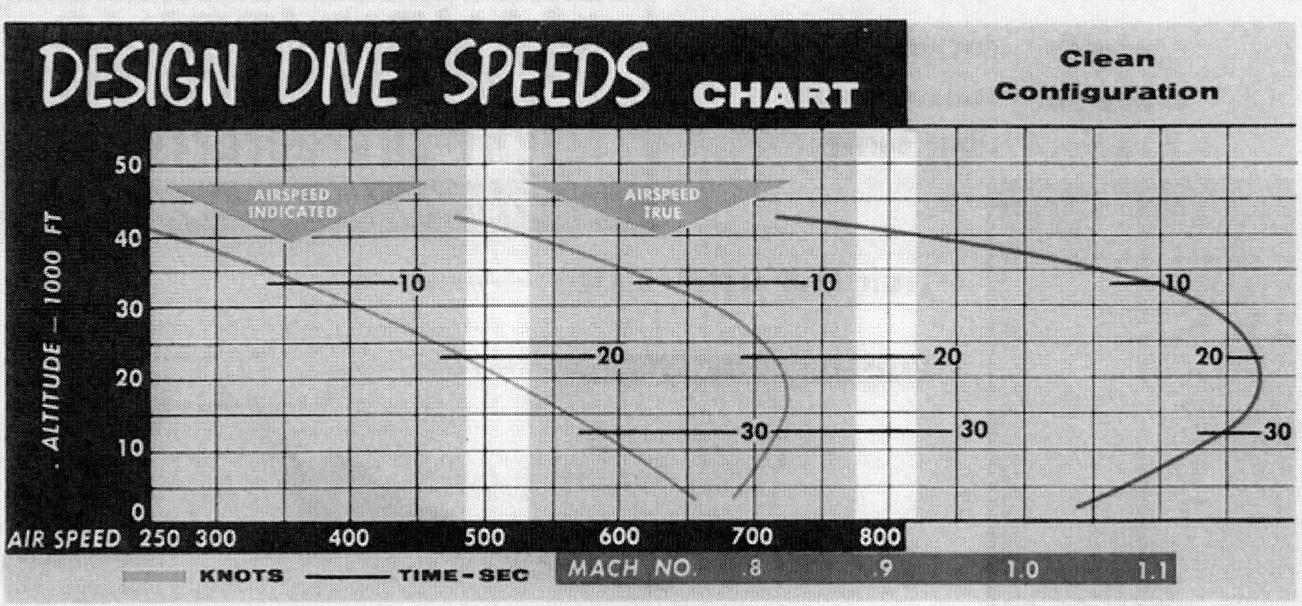

Figure 6—4

ACCELERATED STALL PITCHUP.

The section on high speed accelerated stalls explained that a severe pitchup would be encountered if the stall warning was ignored. Actually this pitchup, which is associated with tip stall and an inboard shift or lift distribution occurs right on down to low speeds. However, at the low Mach Numbers although the aircraft tends to nose up at the stall, the "G" does not increase noticeably. At the higher speeds and consequently at the higher "G" an uncontrollable pitchup will be encountered if the buffet warning is ignored. This pitchup may be severe enough to break the wing. The pitchup is always preceded by a sudden increase in buffet which combined with a moderate control softening provides adequate warning. However, if "G" are applied very rapidly it is possible to pass through the buffet warning without time to correct. The maximum allowable "G" can be applied at high speeds and low altitudes without danger. However, the speed bleeds off very rapidly under high "G" and if the "G" is held on until the aircraft slows down through the heavy buffet the aircraft will pitchup uncontrollably to dangerous load factors. At high altitudes the aircraft will pitchup at correspondingly lower "G". Although the load factors imposed on the aircraft will not be too great, the rapid nose up movement could cause loss of control. Note that at low and medium altitudes the maximum hazard due to pitchup exists because there you may encounter it at just about the limit load factor and the uncontrollable "G" due to pitchup can break the aircraft. *Therefore:* Observe the accelerated stall warning, terminate high "G" maneuvers at or before heavy buffet. If pitchup is encountered, correct immediately using forward stick. Referring to figure 6—3 for example, it can be seen that at 5,000 ft altitude and a Mach Number of 0.90, 8.67G can be put on the aircraft. If this G is not reduced as the airspeed bleeds off, the buffet will become severe at M = 0.80 and at M = 0.78 the aircraft will pitch up to 10½ "G".

BUFFET BOUNDARY.

The buffet boundary occurs at relatively high "G's" particularly at the higher indicated speeds as shown in figure 6—3. A very mild buffet is experienced when the "G" is increased to the initial onset of buffet and this remains mild with further increase in "G" until an increase in the buffet intensity occurs at a higher G. It is the latter increase in intensity which constitutes the accelerated stall warning. This should normally be considered a stopping point to prevent the high Mach Number stall pitchup previously described. The mild buffeting at somewhat lower G's does not limit maneuvers.

DIVES.

The acceleration in a dive is very rapid, and high dive speeds can be obtained. The speeds reached in a 60 degree dive from 40,000 ft are shown in figure 6—4. Dives at high Mach Numbers close to the ground, are prohibited because of the large altitude loss during recovery as shown in figure 6—5. The high Mach Number dive characteristics are excellent. No buffet is experienced as the aircraft picks up speed. The transition from subsonic to supersonic velocity is sometimes accompanied by a very slight, wing dropping, and/or yawing. Both are easily controlled by the pilot. Supersonic characteristics are excellent but reduced control effectiveness should be anticipated. During a high speed dive, engine RPM may increase two to three per cent from the original setting. This is caused by a lag in inlet temperature compensation in the fuel control resulting from the rapid change in the temperature. It may be necessary to control the engine speed in order to avoid exceeding the overspeed limit of 103.0 per cent RPM.

CAUTION

> If the angle of dive is not steep enough to accelerate to supersonic speed and the aircraft stays at approximately Mach 1.0 some aircraft will tend to roll at this transition speed. Recovery is easily accomplished by either increasing speed or slowing down by reducing power and extending speed brakes.

SPINS.

The F-84F aircraft has demonstrated consistently good spinning characteristics with no tendency toward flat or unrecoverable spins. When the aircraft is held in a spin the first three or four turns are oscillatory. These oscillations decrease in amplitude as the spin progresses until after approximately four turns the aircraft's longitudinal axis is at an angle of approximately 60 degrees relative to the ground. The aircraft loses altitude at the rate of 1,000 to 1,200 feet per turn taking approximately two seconds to complete one turn. Demonstration spins indicate that the best recoveries are effected when the corrective action is initiated during that part of the oscillation where the aircraft nose is descending and with speed brakes retracted.

NORMAL SPIN RECOVERY.

To recover from a spin the following procedure is recommended:

1. Apply full back stick and approximately ⅓ ailerons with the spin, then abruptly apply full opposite rudder.

WARNING

> Ailerons held against the spin will prevent recovery. It is imperative that the ailerons be held neutral or slightly with the spin and the stick full back during recovery.

2. When the spin rotation stops neutralize rudder and ailerons and when the pullout is well underway ease the stick forward slowly. Premature forward movement of the stick during recovery attempt will cause the nose to go under excessively which may result in an inverted spin.
3. If the spin is entered when carrying external stores, use the recommended spin recovery procedure. If the spin does not stop in three turns, jettison external stores, and repeat recommended recovery procedure.
4. If a spin is entered with asymmetric external stores it is recommended that the asymmetric store be jettisoned then recommended spin recovery procedure used. If the asymmetric store is outboard of the spin, the spin would be aggravated.

WARNING

> If you get into a spin and are still spinning without any indication of recovery at 8,000 feet (terrain clearance) immediate bail-out is recommended.

EMERGENCY SPIN RECOVERY.

The results of aircraft spin tests have shown satisfactory spin recovery characteristics when proper techniques are followed. These techniques are considered adequate for recovery from most spins. In an emergency, however, if the normal spin recovery techniques have not been successful, the drag parachute may also be used for spin recovery. Use of the drag parachute will probably result in some damage to the drag chute compartment doors because of the large riser angles. However, this should not preclude use of the drag chute in emergencies. After the spin rotation has been stopped and the airspeed starts increasing, jettison the chute in accordance with the procedure set forth in Section 1.

INVERTED SPINS.

As mentioned above, improper recovery procedure from a normal erect spin may cause the spin to become inverted. This is recognized by the negative Gs and the tendency for the pilot's hands and feet to pull away from the controls. The aircraft is easily recoverable from an inverted spin using the following recovery procedure.

1. Neutralize controls.
2. Apply full rudder opposite to the direction of rotation.

WARNING

> The direction of rotation in an inverted spin is easily confused. Therefore, in the inverted

DIVE RECOVERY CHART

EXAMPLE:

If a 6.0g pullout from a 30 degree dive at 600 knots CAS is started at 25000 feet the altitude lost during dive recovery will be 2700 feet.

1. Enter chart at calibrated airspeed of aircraft (A) at start of pullout.
2. Move to the right to curve representing altitude (B) at start of pullout.
3. Sight vertically down to curve representing angle of dive (C).
4. Move to the right to curve representing the load factor to be held in the pullout (D).
5. Sight vertically and read altitude lost during pullout on scale (E).

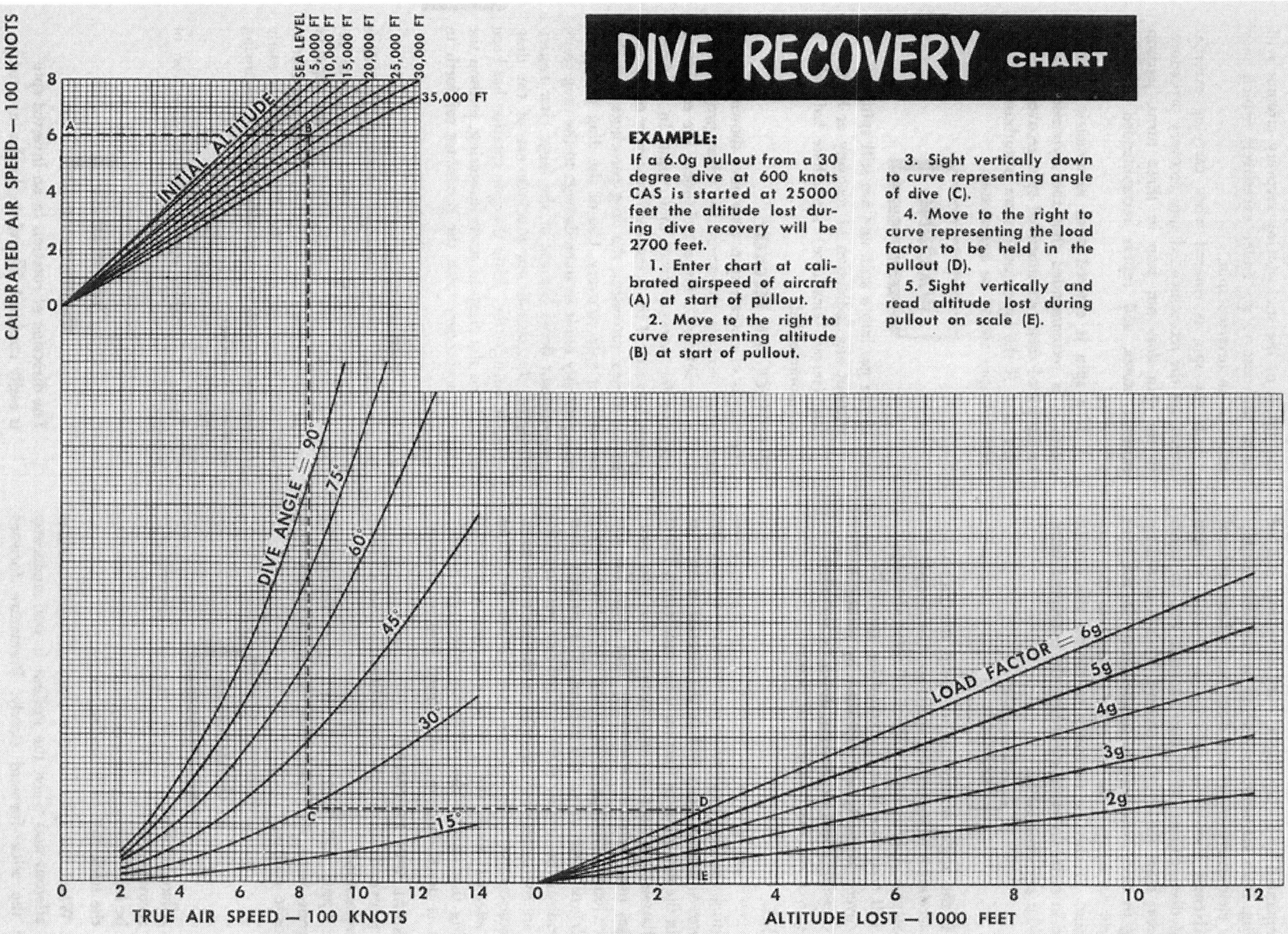

Figure 6—5

spin if the recovery is not accomplished in two turns, the pilot has interpreted the spin direction improperly and the rudder should be reversed.

FLIGHT CHARACTERISTICS WITH EXTERNAL STORES.

Various combinations of external tanks, bombs, rockets and special stores can be flown to very high Mach Numbers with good flight characteristics. The aircraft with external stores is limited from higher speeds only by a slight directional and lateral instability which is easily recognized by the pilot as a "snaking" and roll wobbling effect. External stores tend to increase the severity of buffet during accelerated G maneuvers so that there is ample warning prior to any stall or pitchup.

WARNING

Abrupt, large displacements of the rudder should be avoided with external tanks installed. In this configuration excessive sideslip angles can lead to dangerous snap maneuvers. External stores, and/or extended speed brakes, tend to further aggravate this condition.

WARNING

With unsymmetrical loadings, such as the special store installed, the ship will tend to roll off at high speeds. If severe roll off is encountered, slow down by reducing power. Applying G to the aircraft will increase the roll off tendency.

ASYMMETRIC FLIGHT CHARACTERISTICS.

Flight tests on aircraft not incorporating spoilers have revealed that asymmetrical loading to a maximum of a 450-gallon tank on an inboard pylon is feasible provided that the following precautions are observed. The asymmetrical flight characteristics for those aircraft incorporating spoilers are improved at moderate to high speeds.

INBOARD PYLON.

One Full 450-Gallon Tank or 230-Gallon Tank or Special Store.

1. Since the aircraft will have a tendency to roll into the store on take-off, lateral and rudder trim should be established to counteract this effect before take-off. For the 450-GAL tank configuration approximately 2½ inches of lateral displacement will be required with the rudder trimmed against yaw to the light out position at 165 knots, take-off speed. With the 230-GAL tank or the external store configuration a somewhat less trim setting may be required.

2. During flight with gear and flaps up on non-spoiler aircraft, increase in lateral control is required with increasing speed, the amount of lateral control also dependent on the altitude and the size of the external store. At the higher airspeeds, above 520 knots, 5,000 feet pressure altitude, full 450-GAL tank configuration, available lateral control may run out, however, with spoilers, lateral control will be available at a higher speed. Speeds in excess of those requiring more than ½ stick deflection is not recommended as requirements may suddenly increase with increasing G or gust loads.

WARNING

Whenever loss of hydraulic pressure is encountered, it is recommended that the external store be jettisoned immediately in order to preclude loss of lateral control.

3. With the aircraft in the landing configuration lateral control diminishes with decreasing airspeed until at approximately 150 knots, lateral control becomes negligible. Therefore, approach speed should be maintained above 180 knots with touchdown speed at 165 knots.

OUTBOARD PYLON.

One Full 230-Gallon Tank.

1. Take-off in this configuration is prohibited.

2. The aircraft should not be flown below 180 knots during any phase of flight. Flight tests have revealed that aileron correction between 250 and 460 knots at 5,000 feet pressure altitude should be less than that for the 450-GAL tank inboard configuration.

3. With the aircraft in the landing configuration lateral control diminishes with decreasing airspeed until below approximately 175 knots it becomes marginal. It is recommended, therefore, in the event the full tank cannot be jettisoned, landing be accomplished at the minimum final approach speed of 200 knots and minimum touchdown speed of 180 knots.

4. If on take-off an outboard tank is inadvertently lost, the opposite tank should be jettisoned immediately since lateral control will be insufficient until an airspeed of from 185 to 200 knots is attained.

This page intentionally left blank.

SECTION 7 SYSTEMS OPERATION

TABLE OF CONTENTS

EGT BEHAVIOR Following Cold Engine Acceleration

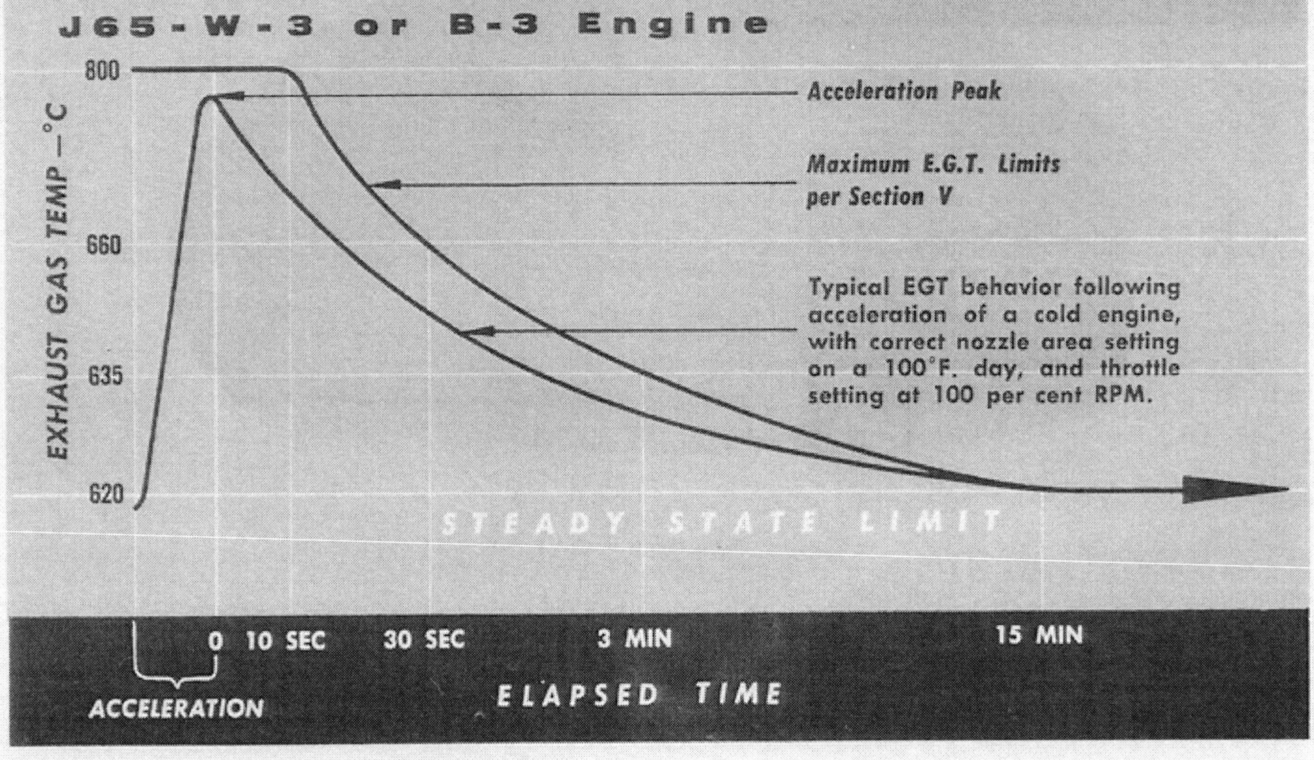

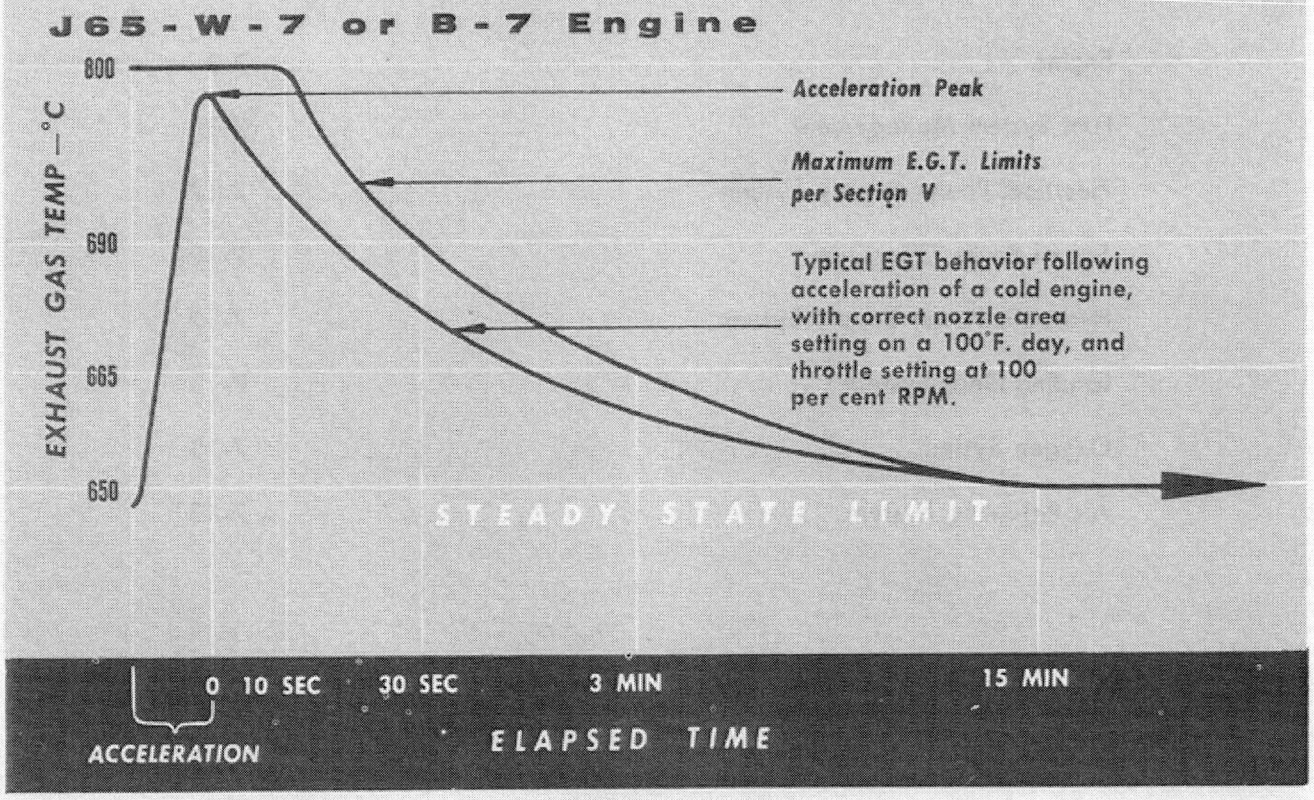

Figure 7—1

ENGINE.

FUEL CONTROL.

The essential purpose of the fuel control is to regulate thrust as a function of throttle control position. Each position of the throttle always corresponds to the same percentage of maximum available thrust. The control also provides protection against overspeed, overtemperature, compressor surge and combustion flame-out. The top speed is established as high as possible without exceeding the maximum allowable turbine stresses, in order to obtain the maximum possible thrust. A small increase in speed above maximum greatly increases the turbine stresses. Furthermore, increasing the speed increases the turbine inlet temperature, and this temperature rise reduces the maximum allowable stresses of the turbine materials. Consequently, any speed increase above maximum will drastically shorten the life of the engine, if not cause immediate failure. In addition to providing steady state speed regulation, the control provides the proper fuel flow during acceleration and deceleration. During an acceleration, the control increases the fuel flow sufficiently to produce the most rapid acceleration of which the engine is capable, without encountering compressor surge or exceeding the maximum temperature limits of the engine. During a deceleration, the control decreases the fuel flow sufficiently to produce a rapid deceleration, yet prevent loss of combustion, or flame-out. A compressor discharge pressure limiter is incorporated in the fuel control which automatically reduces the fuel flow to the engine, whenever the compressor discharge pressure exceeds its limit. This fuel flow reduction will cause a gradual reduction in RPM until the compressor discharge pressure returns to its limit. Reduction in fuel flow will occur at high airspeed, low altitude conditions and should be recognized by the pilot, to avoid unnecessary concern on his part regarding proper functioning of the fuel control. In order to maintain short acceleration times with increasing altitude, and to prevent loss of combustion during deceleration at altitude, it is necessary that the engine idle at progressively higher speeds as the altitude increases. In the event of failure of the main fuel control, the emergency control is provided which will permit safe operation of the engine. The essential requirement of the emergency system is that it be completely reliable. It must be as simple and as uncomplicated as possible. Therefore, the emergency control consists of a simple throttle valve, and during emergency operation fuel flow is a function of throttle control travel only. The compensating features for overspeed, overtemperature, compressor surge, and combustion flame-out have been eliminated to simplify the emergency system as much as possible.

ENGINE CHUGS AND STALLS.

When chugging is encountered during an acceleration there is a momentary RPM hesitation without an exhaust gas temperature rise. When a stall is encountered, the RPM hangs up and then starts dropping as exhaust gas temperature rapidly increases. Engine stall may or may not be accompanied by chugging. If a stall or severe chugging is encountered on the ground, recover by reducing throttle, shut down the engine and investigate. To recover from either a chug or a stall while in flight, reduce throttle. If the stall persists, increase airspeed by reducing altitude. After recovery advance throttle cautiously to desired power. Have this condition investigated as soon as possible.

EXHAUST GAS TEMPERATURE STABILIZATION.

Approximately 8-10 minutes of steady operation at 100 per cent RPM is normally required to obtain stabilized exhaust gas temperature readings after initiating the start of a cold engine. The operating time required for stabilization will be directly affected by the length of time the engine is operated at lower speeds prior to advancing the throttle to 100 per cent RPM, and the ambient temperature at time of run-up. Assume that a start is performed with a cold engine and the throttle is immediately advanced to obtain 100 per cent RPM. As the engine heats up, the turbine casing expands more rapidly than the turbine wheel. The net effect is a temporary increase in the clearance between the rotor blade tips and the shrouds. When the clearance is at maximum, the turbine efficiency is poorest and the exhaust gas temperature is highest. As the engine is run at full RPM, the turbine wheel heats up and expands, the blade tip clearance diminishes, the turbine efficiency improves and the exhaust gas temperature declines until stabilization is achieved. This is graphically llustrated in figure 7—1 under extreme conditions, i.e. rapid acceleration (idle to 100 per cent RPM) of a cold engine immediately following a start on a 100°F day. A cold engine is defined as an engine which has been inactive for a sufficient period of time to allow cooling to ambient temperature.

Note

The criteria illustrated in figure 7—1 are to be used as a guide only and are applicable only to EGT behavior during ground operation of the engine under the conditions mentioned above.

Under normal pre-take-off procedures approximately 5-8 minutes of engine operating time precedes the take-off roll. Exhaust gas temperature stabilization time should be of slight duration depending upon

STABILIZED EXHAUST GAS TEMPERATURE AT 100% RPM FOR GROUND OPERATING INFORMATION:

After 15 Minutes Operation

J-65-W-3 or B-3 ENGINES

AMBIENT TEMP °C	AMBIENT TEMP °F	MAX EGT °C	MIN EGT °C
37.8	100	620	600
32.2	90	618	598
26.7	80	616	596
21.1	70	614	594
15.5	60	612	592
10.0	50	611	591
4.4	40	609	589
−1.1	30	607	587
−6.7	20	605	585
−12.2	10	603	583
−17.8	0	601	581
−23.3	−10	599	579
−28.9	−20	597	577
−33.3	−30	595	575
−40.0	−40	593	573
−45.5	−50	592	572
−51.1	−60	590	570
−53.9	−65	589	569

J-65-W-7 or B-7 ENGINES

AMBIENT TEMP °C	AMBIENT TEMP °F	MAX EGT °C	MIN EGT °C
37.8	100	650	630
32.2	90	649	629
26.7	80	648	628
21.1	70	647	627
15.5	60	645	625
10.0	50	644	624
4.4	40	643	623
−1.1	30	642	622
−6.7	20	641	621
−12.2	10	640	620
−17.8	0	639	619
−23.3	−10	638	618
−28.9	−20	637	617
−33.3	−30	636	616
−40.0	−40	635	615
−45.5	−50	634	614
−51.1	−60	633	613
−53.9	−65	632	612

Figure 7—2

the following variables; length of time the engine is operated and throttle manipulation during the period prior to take-off, ambient temperature, tailpipe nozzle area, etc. The decline rate from peak temperature to stabilization should be more rapid than during ground operation due to the influence of airspeed on engine temperature. Under the most extreme conditions (immediate take-off following cold engine start on a 100°F day), 10-15 minutes may elapse before exhaust gas temperature stabilization is achieved. In such a case, and after the airplane is airborne, the exhaust gas temperature should be monitored, if necessary, to a point on or below the maximum steady state allowable (620°F, J65-W-3, B-3 and 650°C, J65-W-7, B-7). Power may be gradually increased as stabilization is approached. Also to be considered are the factors which influence exhaust gas temperature soaking effects. Lower ambient temperatures encountered as the airplane increases in altitude and characteristics of fuel flow scheduling will account for part of the exhaust gas temperature sag at the medium altitudes. However, at altitudes of 35,000 feet and up, the exhaust gas temperature tends to rise due to the characteristics of the engine and fuel control combination. Both of these later manifestations may occur regardless of how long the engine has been running and should therefore not be confused with exhaust gas temperature stabilization behavior.

EMERGENCY FUEL SYSTEM CHECK.

1. With the engine on the normal system, set the RPM at 90 per cent and place the emergency fuel switch in the ON position. The emergency fuel system warning light should illuminate.
2. If the RPM changes, this indicates that the shift has been made to the emergency fuel system.
3. If there is no RPM change, advance the throttle cautiously to full throttle or 100 per cent whichever comes first. Less than 100 per cent RPM at full throttle on emergency fuel system shows the control has shifted satisfactorily.

Note

If there is no RPM change when shifting to emergency and 100 per cent RPM was obtained either the throttle stop is not set correctly or the control is not shifting from the normal system. This condition should be investigated and corrected.

EMERGENCY FUEL SYSTEM

ESTIMATED FULL THROTTLE RPM

RPM ACCURATE TO ± 2%

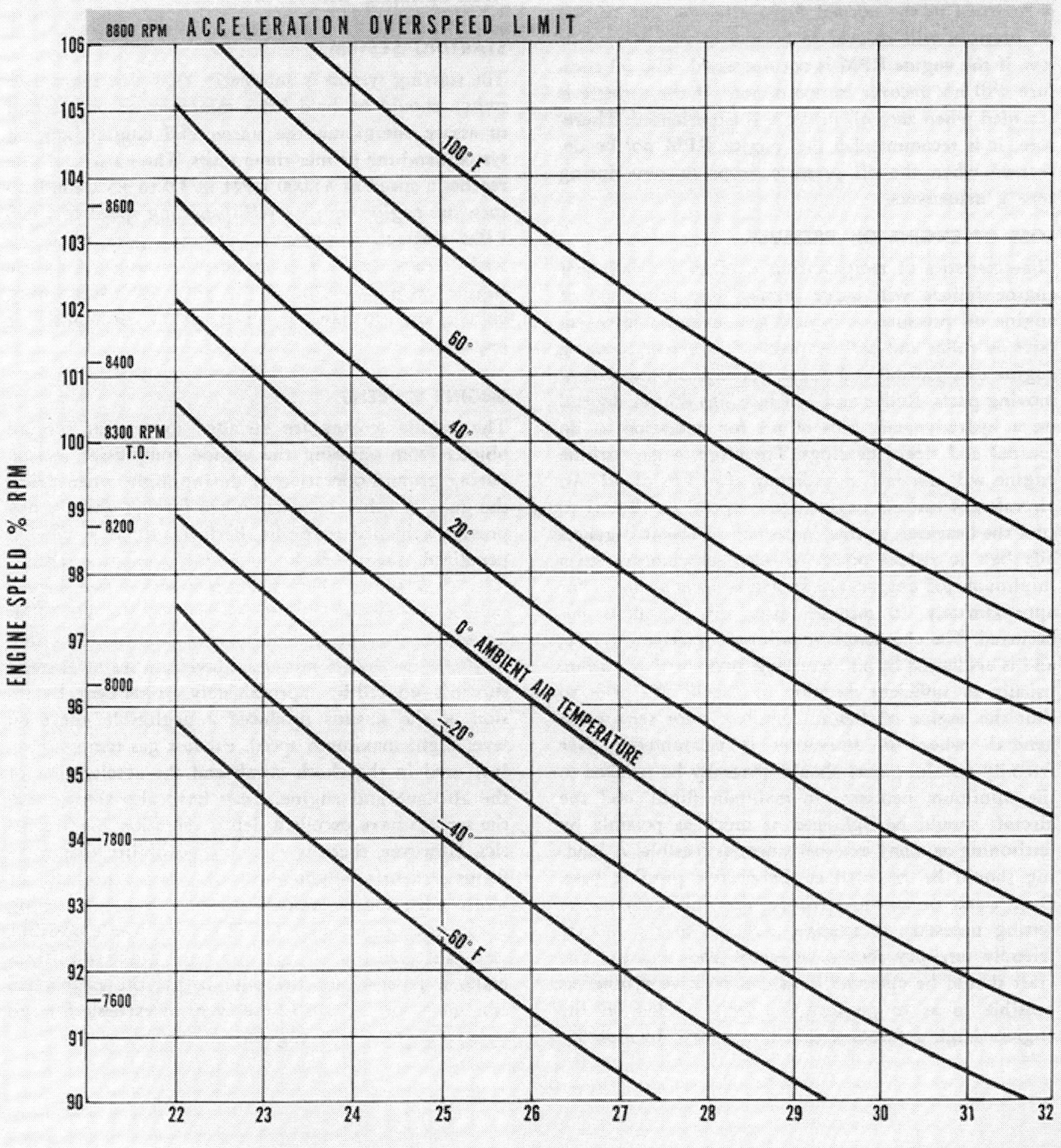

Figure 7—3

4. Retard throttle to IDLE and as RPM passes 60 per cent place the emergency fuel switch in the OFF position. This returns the engine to the normal system.

ENGINE OIL PRESSURE.

Loss of engine oil pressure occurs when the airplane is flown in a zero g attitude and does not return to normal for considerable time, even though the airplane is returned to the normal flight attitude. The engine oil pressure will recover to normal in one minute or less, if the engine RPM is not decreased. The oil pressure will not recover in one minute, if the throttle is retarded when zero oil pressure is experienced. Therefore, it is recommended that engine RPM not be decreased when the oil pressure drops to zero during zero g maneuvers.

LOSS OF ENGINE OIL PRESSURE.

Characteristics of reciprocating engines are such that engine failure will occur immediately after loss of engine oil pressure. A typical gas turbine engine rotates on roller and ball bearings while a reciprocating engine uses journal and sleeve bearings on the critical moving parts. Roller and ball bearings do not depend on a hydrodynamic film of oil for operation as do journal and sleeve bearings. Therefore, a gas turbine engine will not fail immediately after loss of oil. An aircraft gas turbine engine does depend on the oil to cool the bearings, so that in case of oil loss it is generally best to reduce power to keep temperatures to a minimum. J65 engines are known to have operated for approximately 10 minutes after oil starvation has occurred. The characteristic failure is relatively gentle, and is evidenced by an increasing amount of vibration, usually of sufficient duration to permit the pilot to shut the engine off before complete rotor seizure. In general, when oil starvation is encountered over friendly terrain, power should generally be reduced to the minimum necessary to maintain flight, and the aircraft should be lightened as much as possible by jettisoning as many external stores as feasible. A landing should be executed at the nearest possible base. Over water or enemy territory, the minimum power setting necessary to execute a return to the nearest friendly territory should be established, and the aircraft should be operated in as conservative manner as possible so as to produce the least burden on the engine. High g forces should be avoided. In most instances, the ultimate failure of the engine will not occur within 10 minutes, and when it does occur, will be signaled by a steadily increasing vibration. At this time, an engine shut-down should be made to prevent such a destructive failure as would jeopardize a successful ejection or power-off control of the aircraft in a landing attempt. Refer to Section V for limitations when operating without oil pressure.

STARTING SYSTEM.

The starting system is automatic. However, the starter switch should be held for a minimum of one second to assure energizing the starter and engine ignition systems and the primer timer units. The starter turbine reaches a speed of 44,000 RPM in 3.0 to 3.5 seconds to turn the engine up to a self-sustaining speed of 2,000 RPM through a two-stage planetary-gear reduction with a ratio of 22:1. The starter is engaged to the engine with a spline drive with an overrunning clutch, and no jaw or advancement mechanism is required.

ENGINE SCREENS.

The engine screens are installed to prevent foreign objects from entering the engine compressor section during ground operation or during flight when firing the guns or other instances when foreign objects may present a hazard to flight. Performance is somewhat penalized during flight with the screens extended. Test data shows that the extension of the screens reduces the rate of climb between 300 and 475 FPM, depending on altitude, and increases the time to climb to 40,000 feet by 1.5 minutes. Maximum sea level static thrust is reduced by approximately 4.4 per cent. Extension of the screens produced a negligible effect on level flight maximum speed, exhaust gas temperature, fuel used in the check climb and the acceleration of the airplane and engine. Tests have also shown that the screens have excellent debris retention characteristics. However, there is a remote possibility that some forms of debris caught by the screens can peel off the screen edge and be sucked into the compressor during the retraction of the screen. Therefore, where possible, screens should not be retracted until flight conditions make it possible for the pilot to use the established emergency procedures in the event of extreme foreign object damage to the engine.

FUEL SYSTEM MANAGEMENT

SCHEMATIC

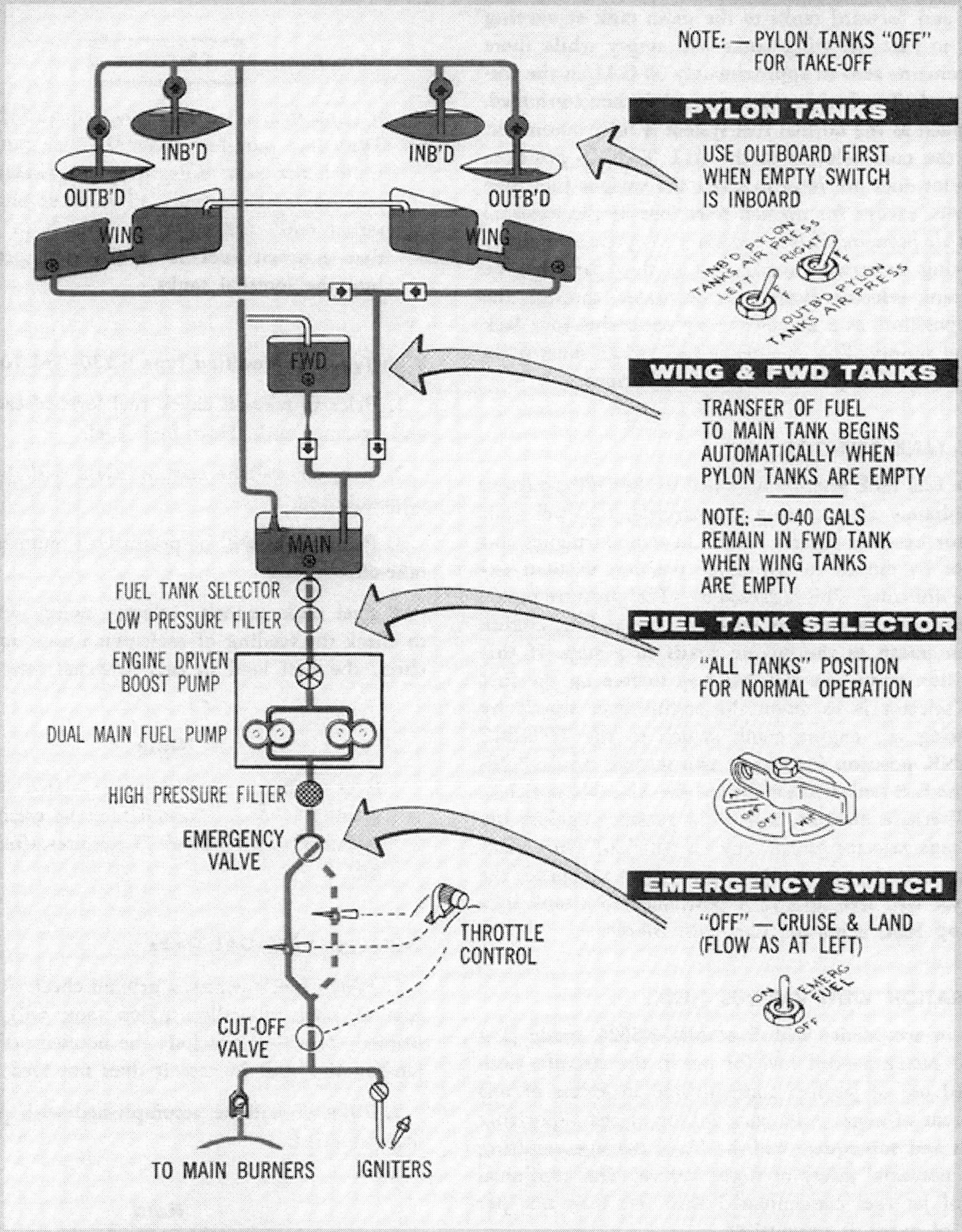

NOTE: THE ENGINE MOUNTED FUEL PRESSURE SWITCH IS BEING DELETED AT ENGINE OVERHAUL ON AIRCRAFT PRIOR TO F-84F-55

Figure 7—4

FUEL SYSTEM MANAGEMENT.

During normal operation fuel is transferred first from the outboard pylons, then the inboard pylons to the main tank so that the pylons can be dropped to reduce drag. Fuel is then transferred simultaneously from the wing and forward tanks to the main tank at varying flows so that the wing tanks will empty while there still remains zero to approximately 20 GAL in the forward tank. The fuel in the main tank is then consumed. Inasmuch as the normal fuel system is fully automatic, with the tank selector in the ALL TANKS position, the pilot does not have to select the various fuel flow patterns, except for manual selections of the external tanks air pressure. If it is necessary to operate on either the wing or forward auxiliary fuel flow patterns, the fuel tank selector should not be turned through the OFF position, as a flame-out may occur due to a lack of fuel supply. The emergency fuel switch must be in the OFF position for all normal operations.

FUEL TANK SELECTOR.

If the fuel tank selector is turned to the OFF position, immediately after closing the throttle, the fuel tank selector becomes tightly bound on some airplanes and cannot be moved to any other position without extreme difficulty. This is caused by a fuel pressure build-up between the fuel tank selector valve and the engine driven pump as the engine coasts to a stop. If this condition exists, the simplest way to free up the fuel tank selector is to motor the engine over slowly by actuating the engine crank switch to the ENGINE CRANK position for one or two seconds while holding the fuel tank selector toward the ALL ON position. An alternate method, is to hold pressure against the fuel tank selector toward any on position, while making a normal start. As the engine picks up speed, the selector will free up and if positioned to a tank containing fuel, a normal start will follow.

OPERATION WITH VARIOUS FUELS.

Fuel in accordance with Spec MIL-F-5624, grade JP-4 is the recommended fuel for use in the aircraft. Continued use of jet fuel contaminated in excess of one per cent of avgas, will cause failure of the vaporizing tubes and subsequent turbine blade damage resulting in a potential safety of flight hazard. The continued use of jet fuel contaminated with less than one per cent of avgas is permissible.

PYLON TANK FEEDING.

All pilots will be certain pylon tanks are feeding immediately after take-off and/or immediately after air refueling. If pilots cannot determine beyond reasonable doubt that pylon tanks are feeding, the mission will be aborted. Check pylon tank feeding as follows:

CAUTION

It is recommended that when operating with Type I or modified Type II pylon tanks installed, the pilot will check individual tank readings at least every 15 minutes until external tanks are emptied. This is to assure that external fuel is transferring properly into the internal tanks.

With Type I or Modified Type II 230-GAL Tanks.

1. Prior to take-off check fuel level of each internal and external tank. Note fuel level.
2. Take-off will be accomplished with pylon tanks unpressurized.
3. Pylon tanks will be pressurized immediately after take-off.
4. Fuel tank quantity selector switch will be used to check the feeding of each pylon tank and to cross-check the fuel load in each internal tank.

Note

It is recommended that the status of fuel feeding of each external tank be established within a maximum of 15 minutes after take-off.

With Type I 450-GAL Tanks.

1. When fuel permits, a ground check of the operation of each 450-gallon pylon tank will be accomplished. This will preclude the necessity of having to jettison the tank in case it does not feed.
2. Take-off will be accomplished with pylon tanks unpressurized.

Note

Certainty that tanks are unpressurized will be accomplished by opening and closing the air refueling receptacle if a ground check of the pylon tanks has been made.

3. Pylon tanks will be pressurized immediately after take-off.

4. Fuel tank quantity selector switch will be used to check the feeding of each pylon tank and to cross check the fuel load in each internal tank.

Note

It is recommended that the status of fuel feeding of each external tank be established within a maximum of 15 minutes after take-off.

With Unmodified Type II or IV 230-GAL Tanks.

These tanks do not afford the pilot an immediate indication of fuel feeding in the cockpit and should be checked for feeding as follows:

Note

This check should be complete within 20 minutes after take-off.

1. Take-off will be accomplished with pylon tanks unpressurized.

2. Place the fuel tanks quantity selector switch in the FWD position.

3. The left outboard pylon tank (if installed) will be pressurized. After a definite indication that the forward tank has refilled, the left outboard pylon tank air pressure switch will be turned to the OFF position.

4. The right outboard tank (if installed) will be pressurized. After a definite indication that the forward tank has refilled, the right outboard pylon tank air pressure switch will be turned to the OFF position.

This page intentionally left blank.

ELECTRICAL LOAD CHART

Major components which require comparatively high amperage while in operation. Components requiring very little amperage are not listed.

Component	Approximate amperage required	Percent of total amperage available (400)
Gun-bomb-rocket Sight heaters	28.0	7
Pitot heater	4.0	1
Flight Instruments	2.4	.6
Forward tank booster pump	28.4	7.1
Wing tank booster pump (2)	27.3	6.8
Main tank booster pump	37.4	9.3
Alternate Inverter	10.0	2.5
Main Inverter	98.5	24.6
Communication Equipment	14.5	3.6
Emergency hydraulic pump	90.0*	
Misc.	12.6	3.1
Total load	**263.1**	**65.6**

*FOR EMERGENCY OPERATION ONLY

Figure 7—5

5. Proceed as in steps 3 and 4 with the left inboard and right inboard tanks, if applicable.

6. If certain all tanks are feeding, pressurize both outboard pylon tanks and when empty pressurize both inboard pylon tanks.

With Type II or IV 450 Gal Tanks.

These tanks do not afford the pilot an immediate indication of fuel feeding in the cockpit and should be checked as follows:

Note

This check should be complete within 20 minutes after take-off.

1. When fuel permits, a ground check of the operation of each 450 GAL pylon tank will be accomplished by placing the fuel tanks quantity selector switch in the FWD position and checking to determine that the tank is feeding.

2. Take-off will be accomplished with pylon tanks unpressurized.

Note

Certainty that tanks are unpressurized will be accomplished by opening and closing the air refueling receptacle if a ground check of the pylon tanks has been made.

3. Place the fuel tanks quantity selector switch in the FWD position.

4. The left inboard pylon tank (if installed) will be pressurized. After a definite indication that the forward tank has refilled the left inboard pylon tank air pressure switch will be turned to the OFF position.

5. The right inboard pylon tank (if installed) will be pressurized. After a definite indication that the forward tank has refilled, the left inboard tank pylon tank will again be pressurized.

IF PYLON TANK STOPS FEEDING.

When a pylon tank stops feeding, there may be a slight rumble accompanied by vibration from below the pilot's seat area. This condition is of short duration and is caused by air surges into the main tank or as the pressure reducing valve, which controls both the cockpit seal and pylon tank air pressure, fluctuates on stabilizing for the new required air flow.

PYLON TANK PRESSURIZATION.

Suction relief valves are not incorporated in Type II external tanks, therefore, the air pressure must be maintained to prevent the tanks from collapsing during high rates of descent. If inboard pylon tanks are retained, the pylon tanks air pressure switches should be left in the INB'D PYLON TANKS AIR PRESS position after the tanks are empty. However, if inboard and outboard tanks are both retained, descents must be made at a slow rate and the pylon tanks air pressure switch alternately placed in the INB'D PYLON TANKS AIR PRESS and OUTB'D PYLON TANKS AIR PRESS position at approximately every 1000 feet of descent in order to equalize the pressure differential and prevent the tanks from collapsing.

PYLON TANK AIR PRESSURE SWITCHES.

When the airplane is serviced on the ground, the feed lines from the wing tanks to the main tank may be filled with air. The trapped air is not bled out until the main tank level drops and allows the main tank shut-off valve, from the wings and forward tanks, to open. If the fuel tank selector is positioned to WING AUX before the wing transfer lines are filled with fuel, the engine will be momentarily starved and a flame-out will result. With the pylon tank air pressure switches OFF enough fuel will be consumed from the main tank during ground operation through take-off to start transfer of fuel from the forward and wing tanks to the main tank. This should bleed the wing and forward transfer lines sufficiently.

ELECTRICAL POWER SUPPLY SYSTEM.

The electrical load chart (figure 7–5) lists the major components which require comparatively high amperage while in operation so that a pilot can ascertain that the loadmeter reading is normal prior to take-off. Components requiring very little amperage are not listed.

SPEED BRAKE OPERATION.

If the landing flap lever is in the DOWN position on non-tandem airplanes, and the speed brakes are extended or retracted, the flaps may momentarily blow up approximately 10 degrees (25 per cent) due to a momentary drop in hydraulic pressure. This conditions can be avoided by placing the flap lever in the NEUT position after the flaps are fully extended.

HYDRAULIC POWER SUPPLY SYSTEM.

TANDEM

EMERGENCY HYDRAULIC SYSTEM.

The emergency hydraulic system can only be tested when the engine is not operating. With external power available or the battery switch ON and the instrument power switch in the ALT position check the utility and power hydraulic pressure gages for zero readings. If gages do not read zero, cycle controls to dissipate pressure. The emergency hydraulic system accumulator should have a charge of 500 to 600 PSI air pressure when the hydraulic pressure is zero. Place the emergency hydraulic pump switch in the HYD EMER PUMP position, the power hydraulic gage should indicate 1550 (±100) PSI. Move control stick full forward. With hydraulic pressure at 1550 (±100) PSI control stick should operate smoothly and steadily from full forward to full aft, then to full forward and then, the stick action may then become restricted during the next aft travel due to the emergency pump limitations. Rate of motion fore and aft must be equal. Return control stick to neutral and observe hydraulic pressure return to 1550 (±100) PSI. Operate ailerons through complete range at normal rates. Operation should be smooth and steady with no indication of system starvation or reduction in rate of operation. Rate of motion from side to side must be equal.

NON-TANDEM

EMERGENCY HYDRAULIC SYSTEM.

Depressing the hydraulic by-pass switch during ground test opens a dump valve in the normal hydraulic system so that hydraulic pressure is relieved on unmodified A thru E aircraft and on modified A thru E and later aircrafts shuts off normal hydraulic pressure to the aileron, rudder and stabilator so the emergency system can be checked for operation. The emergency hydraulic system accumulator consists of a floating piston within a closed cylinder. The compartment on one side of the piston is filled with compressed air while the other side contains hydraulic fluid. When the airplane remains idle for a period of time the hydraulic fluid will lose its pressure and the piston will move to the extreme end of the cylinder. Hydraulic pressure will be regained in the accumulator in approximately 10 minutes during engine operation through a bleed from the normal hydraulic system. However, if the pressure in the boost and utility systems is dumped and the emergency hydraulic system cuts-in, the accumulator will pressurize itself immediately. Therefore, the emergency system is operated prior to take-off so that in the event of failure of the boost and utility systems, a fully pressurized emergency accumulator will be available, if needed, during take-off.

LANDING GEAR SYSTEM.

A vibration, felt after take-off when the landing gear is retracted may be mistaken for engine roughness but could be caused by an unbalanced nose wheel. This vibration would be more noticeable after take-off in the heavier configurations as the take-off speed is higher. Vibration, caused by the nose wheel, will diminish as the wheel coasts to a stop.

USE OF LANDING WHEEL BRAKES.

It is absolutely necessary that airplane brakes be treated with respect. Generally, operating personnel stop the airplane as quickly as possible regardless of the length of the runway, use the brakes consistently for speed up turns, and drag the brakes while taxiing. To minimize brake wear, the following precautions should be observed insofar as is practicable.

1. Use extreme care when applying brakes immediately after touchdown, or at any time when there is considerable lift on the wings, to prevent skidding the tires and causing flat spots. A heavy brake pressure can result in locking the wheels more easily if brakes are applied immediately after touchdown than if the same pressure is applied after the full weight of the airplane is on the wheels. A wheel once locked in this manner immediately after touchdown, will not become unlocked as the load is increased as long as brake pressure is maintained. Proper braking action cannot be expected until the tires are carrying heavy loads. Brakes, themselves, can merely stop the wheel from turning, but stopping the airplane is dependent on the friction of the tires on the runway. For this purpose, it is easiest to think in terms of coefficient of friction which is equal to the frictional force divided by the load on the wheel. It has been found that optimum braking occurs with approximately a 15 to 20 per cent rolling skid; i.e. the wheel continues to rotate but has approximately 15 to 20 per cent slippage on the surface so that the rotational speed is 80 to 85 per cent of the speed which the wheel would have were it in free roll. As the amount of skid increases beyond this amount, the coefficient of friction decreases rapidly so that with a 75 per cent skid the friction is approximately 60 per cent of the optimum and, with a full skid, becomes even lower. There are two reasons for this loss in braking effectiveness with skidding. First, the immediate action is to scuff the rubber, tearing off little pieces which act almost like rollers under the tire. Second, the heat generated starts to melt the rubber and the molten rubber acts as a lubricant. NACA figures have shown that for an incipient skid with an approximate load of 10,000 pounds per wheel, the coefficient of friction on dry concrete is as high as 0.8, whereas the coefficient is of the order of 0.5 or less with a 75 per cent skid. Therefore, if one wheel is locked during application of brakes, there is a very definite tendency for the airplane to turn away from that wheel and further application of brake pressure will offer no corrective action. Since the coefficient of friction goes down when the wheel begins to skid, it is apparent that a wheel, once locked, will never free itself until brake pressure is reduced so that the braking effect on the wheel is less than the turning moment remaining with the reduced frictional force.

2. If maximum braking is required after touchdown, lift should first be decreased as much as possible by raising the flaps and dropping the nose (on tricycle gear airplanes) before applying brakes. This procedure will improve braking action by increasing the frictional force between the tires and the runway.

3. For short landing rolls, a single, smooth application of the brakes with constantly increasing pedal pressure is most desirable. This procedure applies equally well for operation on emergency braking systems.

4. It is recommended that a minimum of 15 minutes elapse between landings where the landing gear remains extended in the slip stream, and a minimum of 30 minutes between landings where the landing gear has been retracted to allow for cooling if brakes are used for steering, crosswind taxiing operation, or a series of landings are performed.

5. The full landing roll should be utilized to take advantage of aerodynamic braking and to use the brakes as little and as lightly as possible.

6. After the brakes have been used excessively for an emergency stop and are in the heated condition, the airplane should not be taxied into a crowded parking area or the parking brakes set. Peak temperatures occur in the wheel and brake assembly from 5 to 15 minutes after a maximum braking operation. To prevent brake fire and possible wheel assembly explosion, the specified procedures for cooling brakes should be followed.

7. The brakes should not be dragged when taxiing, and should be used as little as possible for turning the airplane on the ground.

Figures 7—6 thru 7—8 Deleted

Pages 7-13 thru 7-18 Deleted

SECTION 8

CREW DUTIES

not applicable to this airplane

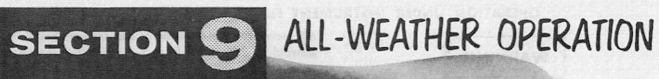

SECTION 9 ALL-WEATHER OPERATION

TABLE OF CONTENTS

INTRODUCTION.

This section contains special or unusual procedures for the F-84F. Instrument flight procedures will be accomplished as outlined in USAF Manuals, Instrument Flying Techniques and Theory of Instrument Flying.

OPERATION UNDER INSTRUMENT FLIGHT CONDITIONS.

WARNING

Investigations of recent accidents have revealed the possibility of engine failure while or after flying through thunderstorms, hailstorms, icing conditions or severe turbulence or even through thick clouds. Engines are in process of being modified to increase shroud clearances in the 8, 9 and 10 stages to a minimum of .070 inch and to .060 inch on the 11 and 12 stages. Aircraft having modified engines installed will *not* be restricted in accordance with the following instructions. Aircraft with unmodified engines installed are placarded to restricting weather flights. Until such time as modified engines are installed the aircraft shall not be flown through any clouds, thunderstorms, hailstorms, icing conditions or severe turbulence (acceleration in excess of 7½g) unless demanded by operational necessity. In the event the aircraft is inadvertently subjected to any of the above conditions an entry should be made in the aircraft DD Form 781. If it becomes absolutely necessary to enter a thunderstorm area or if one is inadvertently entered, follow thunderstorm penetration information. If only icing conditions or precipitation is encountered it is advisable to reduce power as much as possible and still maintain safe flight.

For ease of handling, bank angles should be limited to approximately 30 degrees unless conditions dictate otherwise. At altitudes above 37000 feet with an external load, it becomes necessary to limit the bank angle to 20 degrees to prevent loss of altitude during turns.

When turbulent air is encountered, the aircraft has "snaking" and "dutch roll" characteristics. Without pylon tanks installed, the nose will oscillate from side-to-side several degrees, depending on the degree of turbulence. This condition is accompanied by small bank oscillations. With pylon tanks installed, the same condition exists with a greater amplitude of bank oscillations. The pilot cannot apply sufficient control pressures to stop the condition. For this reason, the attitude indicator should be used for turns instead of the needle and ball. To make a turn, establish the desired bank angle and hold this angle until the desired heading is obtained. On GCA's, hold an average heading on the glide path when the condition is encountered.

All recommended airspeeds in this section should be adhered to during the final stages of any instrument procedure or approach in minimum weather conditions. During some stages of landing approaches, the aircraft is being operated on the flat portion of the power curve, that is, small changes in power give fairly large changes in airspeed. If airspeed is excessive, especially on the glide path in conditions of low ceiling and visibility, with a wet or icy runway, the distance required to stop will be increased.

INSTRUMENT TAKE-OFF AND INITIAL CLIMB.

1. Use normal take-off technique.
2. Use brakes for directional control until rudder becomes effective. Use directional indicator as the primary instrument for heading until the airplane is airborne.

Note

If miniature airplane was lined up with horizon bar on attitude indicator at start of take-off, placing miniature airplane about two horizon bar widths above horizon bar gives a good take-off and initial climb attitude.

3. Take-off at normal speeds (Refer to Appendix).
4. Maintain a constant nose-high, wing-level attitude, after leaving the ground.
5. Landing gear selector UP when definitely airborne and the vertical velocity indicator and altimeter show definite ascent.
6. Landing flap control UP at approximately 190 knots IAS.
7. Climb to a safe altitude and trim for best climbing speed.

INSTRUMENT CLIMB.

1. Maintain take-off attitude until all obstacles are cleared.
2. Accelerate gradually in a climbing attitude, reaching normal VFR climb schedule at 8000-10000 feet.
3. Climbing turns should be limited to a 30 degree bank angle for ease of handling and effective interpretation of the J-8 attitude indicator.

DURING INSTRUMENT CRUISING FLIGHT.

Airspeeds between 250 and 350 knots IAS provide the best handling qualities and allow time for diversions to accomplish other instrument flight functions such as; operate the radio, navigate, etc. Above this range excessive concentration is necessary to maintain pitch attitude.

TYPICAL RADAR RECOVERY PENETRATIONS

DESCENT

Speed Brakes — OUT
IAS — 250 knots
RPM — IDLE

APPROACH ALTITUDE

Landing Gear and Flaps — UP
Speed Brakes — Partial (as required)
IAS — 220 knots
RPM — 85 ± 2%

"GATE" OR TURN-ON

Landing Gear — DOWN
Flaps — 50%
Speed Brakes — IN
IAS — 190 knots
RPM — 85 ± 2%

FINAL APPROACH & GLIDE PATH

Landing Gear — DOWN
Flaps — FULL DOWN
Speed Brakes — OUT
IAS — 165 knots
RPM — 85 ± 2%

Minimum Weather Conditions

Note: The listed time, fuel and distance data are approximate.

ALTITUDE (feet)	TIME (min.)	FUEL (lbs.)	DISTANCE (nautical miles)
40,000	8	97.5	43
30,000	6	78.0	32
20,000	4	59.5	20

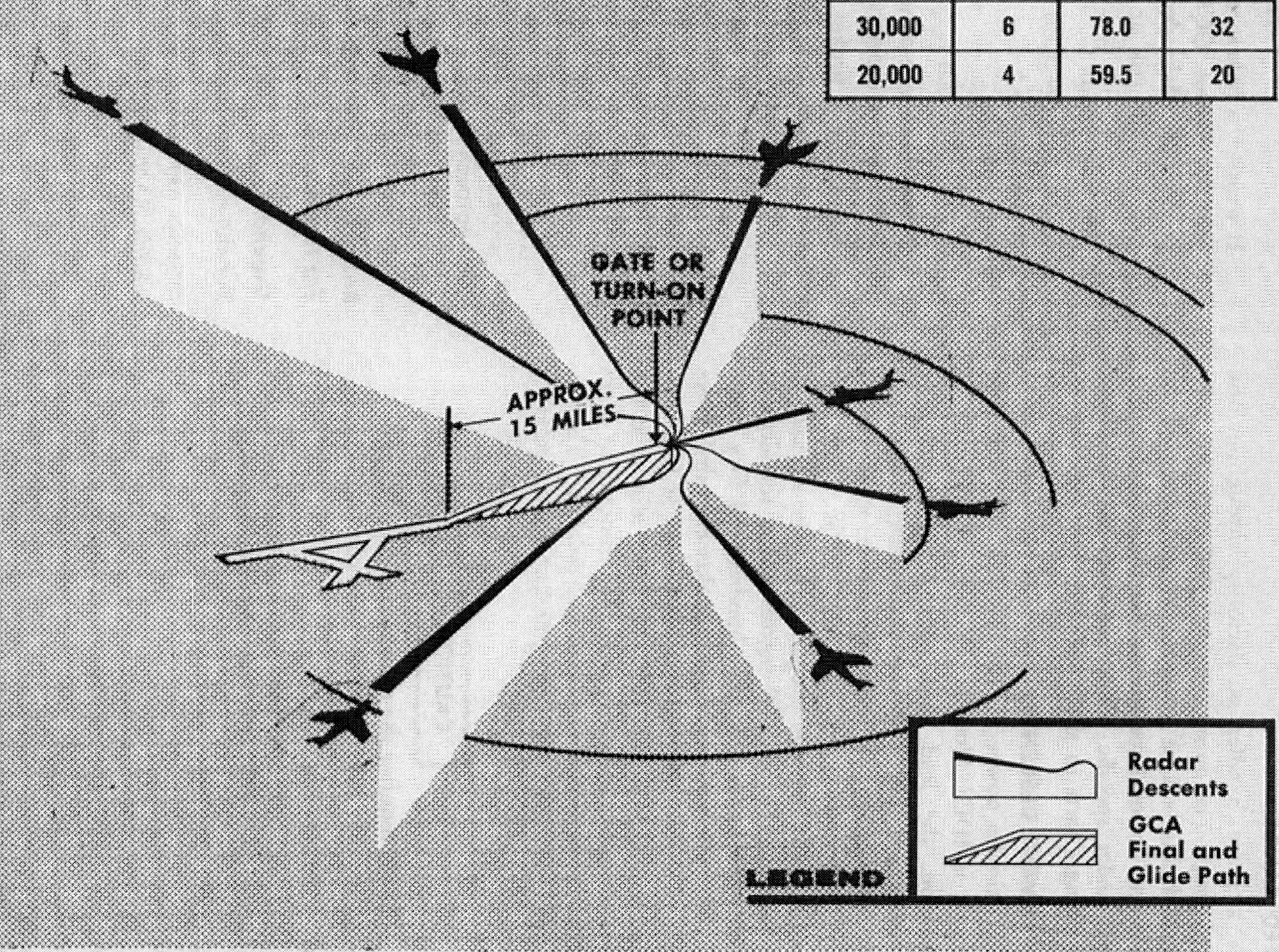

Figure 9—1

SPEED RANGE.

The airplane handles satisfactorily throughout the speed range. The most desirable speed range for individual flights will be governed by the nature of the flight and on local conditions. For the most desirable speed under specific conditions refer to the Flight Operating Charts in the Appendix.

INSTRUMENT DESCENTS.

The optimum power for conservation of fuel during descents is IDLE. However, if descents are made in formation, the lead ship must maintain adequate power above 82% RPM to afford wing ships flexibility of position, otherwise wing ships may periodically be operating in the critical RPM range. Descending turns become progressively more difficult as the bank angle exceeds 30 degrees and or the airspeed exceeds 250 knots IAS. Two hundred and fifty knots IAS, with idle power and speed brakes extended, gives a rate of descent of 5000 feet per minute at high altitudes, and approximately 4000 feet per minute at the lower altitudes. Prior to descending through an overcast use pitot heat and windshield defroster. The defroster should be turned on approximately 15 minutes prior to starting descent.

CAUTION

Closely monitor the altimeter during descents to prevent misreading.

RADAR RECOVERY.

Typical and recommended radar penetration procedures for the F-84F are illustrated in figure 9–1.

IFR LOITERING AND HOLDING PATTERN.

For best loitering indicated airspeeds (fuelwise), which depend upon altitude and weight of the aircraft, refer to the Maximum Endurance Charts in the Appendix. These speeds should be considered as minimum; for better handling it is recommended that 10 knots be added to the listed speeds.

Approximately two to four per cent RPM must be added to the level flight RPM to maintain airspeed during holding pattern turns.

INSTRUMENT APPROACHES.

All instrument approaches will be in accordance with Instrument Flying Techniques and Theory of Instrument Flying Handbooks except for recommended airspeeds.

Radio Range and ADF Penetrations.

Refer to figure 9–2 for jet penetration data. The following is suggested:

1. Maintain initial penetration altitude until reaching the radio fix.

Note

If weather is below minimums for range approach, request GCA prior to beginning penetration. Make decision to proceed to alternate while still at altitude.

2. Throttle—IDLE.
3. Speed brakes—OUT.
4. Maintain 250 knots IAS.
5. Descend outbound one-third (or as locally required) of the initial penetration altitude.
6. Make procedure turn while descending. The 90 degree method is recommended.
7. Retract speed brakes at initial approach altitude, and add power to maintain 220 knots IAS.

CAUTION

Start leveling out at least 1000 feet above minimum penetration (initial approach) altitude.

Note

If the penetration is made on the final approach leg of a radio fix, after reaching final approach altitude inbound, lower gear, 50 per cent of flaps, add power and continue to the station at 175-180 knots IAS.

Radio Range Approach.

For recommended radio range approach see figure 9–3.

Ground Control Approach.

For the recommended GCA procedure, see figure 9–4.

MISSED APPROACH GO-AROUND TECHNIQUES. RADIO RANGE, GCA.

1. 100% power.
2. Speed brakes – IN.
3. Establish climb.
4. Gear—UP.
5. Flaps—UP at 190 knots.
6. Execute established missed approach for particular base.

FLIGHTS IN ICE, SNOW AND RAIN.

The only forward visibility in heavy precipitation is through curved panels of the windshield. Adequate fuel reserve should be allowed for missed GCA approach, due to radar controller's difficulty in maintaining contact with the airplane when precipitation echoes clutter GCA scopes. Icing has marked effect on wings of this aircraft, notably in reduced airspeed and rate of climb. A check of radiosonde information should be made at the point of departure and destina-

TYPICAL RADIO JET PENETRATION

Refer to Pilot's Handbook (Jet, U.S.) for local procedures

If weather is too low for range approach, request GCA, or proceed to alternate while still at altitude

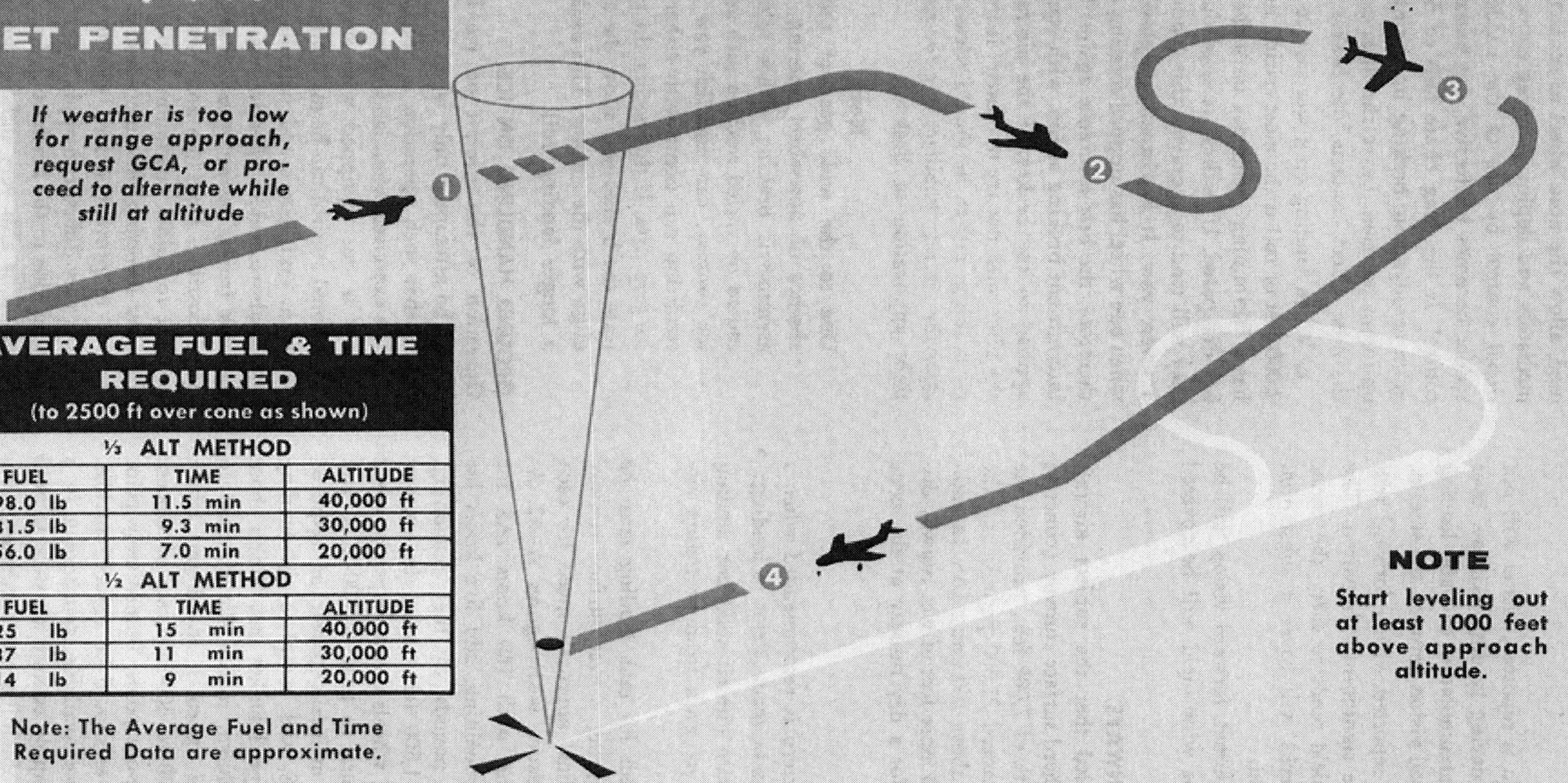

AVERAGE FUEL & TIME REQUIRED
(to 2500 ft over cone as shown)

FUEL	TIME	ALTITUDE
⅓ ALT METHOD		
198.0 lb	11.5 min	40,000 ft
181.5 lb	9.3 min	30,000 ft
156.0 lb	7.0 min	20,000 ft
½ ALT METHOD		
325 lb	15 min	40,000 ft
237 lb	11 min	30,000 ft
214 lb	9 min	20,000 ft

Note: The Average Fuel and Time Required Data are approximate.

NOTE

Start leveling out at least 1000 feet above approach altitude.

1 APPROACH

Approach station at prescribed initial penetration altitude.

2 OUTBOUND

Speed brakes — OUT
IAS — 250 knots
RPM — "IDLE" (or as req.)

Avoid the 60-82% range.

3 PENETRATION TURN

(90° method)

4 *COCKPIT CHECK

Speed Brakes — Partial (as req.)
IAS — 220 knots
RPM — 85±2 per cent
Note:
*If making low approach lower gear before reaching low cone.

Figure 9–2

tion, and the flight should be planned at ice-free altitudes, due to absence of wing and tail de-icing.

Vision in Rain.

1. If mist or light rain is reported, vision will not be significantly affected during landing. Airflow over windshield will prevent accumulation of water leaving streaks or tracks with good vision between the streaks.

2. If moderate rain is reported, vision forward will be possible, but will be significantly impaired. The airflow over the windshield tends to carry the water away but streaks or tracks will cover a substanital portion of the windshield.

3. If heavy rain is reported, forward vision will be virtually impossible. The windshield will be covered with water.

LANDING ON WET RUNWAYS.

Flight tests have revealed that the subject aircraft can be landed on a wet, hard surface runway (concrete or asphalt) with a length of 7,000 feet, provided the gross weight does not exceed 21,600 pounds and the touchdown speed is not above 160 knots IAS. In addition approximately 2,000 more feet of wet runway distance is required than for a dry runway at the same gross weight.

Procedure.

A rectangular traffic pattern is recommended when a wet runway landing is to be made. Prior to landing:

a. The pilot must assure himself sufficient landing pattern spacing so that jet wash from preceding aircraft is at a minimum.

b. The approach speeds for each landing must be computed before attempting a wet surface runway landing. Establish the final approach speed for each gross weight, in accordance with figures A–62. As a rule-of-thumb, starting with 160 knots IAS for 1,000 pounds of fuel remaining, add five knots for each additional 1,000 pounds of fuel remaining. Establish the base leg 1,500 feet above the runway altitude, IAS 225 knots, wheels down, flaps up, speed brakes in. Make the turn to final at 800-1,000 feet about five miles out, maintain power as required to hold 190 knots IAS, and complete the landing checks, full flaps and speed brakes out. When about two miles from touchdown, reduce the airspeed to the predetermined final approach speed and adjust the power so that a 200 to 300-foot rate of descent can be held to the flare-out point. Immediately prior to touchdown, the power should be reduced to idle. If the drag chute is not available, touchdown in a nose high attitude and apply enough aft stick to hold the nose high thereby obtaining the maximum aerodynamic braking. Leave flaps extended and after the nose wheel contacts the runway use brakes intermittently. Do not drag the tail. If the drag chute is to be used, allow the nose wheel to contact the runway immediately and deploy the drag chute. Maintain directional control by use of the rudder and after the rudder becomes ineffective, use brakes for directional control. If the drag chute fails to deploy, use maximum aerodynamic braking by raising the nose to the optimum position. Leave the flaps extended and after the nose wheel contacts, use brakes intermittently.

c. When landing on a wet surface runway, the first 2,000 feet of roll is the most critical in that the aircraft has a "skimming" tendency until the lift of the wing has dissipated. If brakes are used during this period, they will tend to aggravate this condition, resulting in a severe yaw. It is virtually impossible to determine when one wheel has stopped rotating on a wet runway; therefore, the best assurance against blowing a tire is intermittent braking action, with equal pressure being applied to the brakes. If the aircraft starts to yaw, the pilot must not try to "catch" it by use of asymmetric braking; rather, he should release both brakes, and *after* the aircraft stabilizes, *use the rudder,* and, *only* a light application of brakes.

Note

Due to the wide gear of the F-84F and absence of nosewheel steering, the slightest asymmetric braking, either pilot-induced or caused by varied traction such as on a wet or icy runway, can establish yaw angles. This condition can occur with either standard or ice grip tires. If this occurs, the pilot must release the brakes and allow the aircraft to re-align with the runway. This usually results in a longer landing roll.

GROUND HANDLING ON ICE.

Operation of the aircraft on ice is hazardous and should be attempted only when the mission is of the nature that such operation is necessary. Due caution must be exercised when landing or taxiing on ice. The aircraft is not equipped with nose wheel steering. Directional control can be maintained only with wheel brakes at taxi speeds and with brakes and rudder at speeds above rudder effectiveness. Touchdown should be made from a power approach at the minimum safe speed possible. Hold the nose wheel "off" as long as possible to obtain maximum aerodynamic drag. Braking after lowering the nose wheel must be made with caution to prevent sudden yawing and skidding. On ice it is very difficult to apply brakes without skidding the tires, due to the sensitive hydraulic actuated power brakes. It is also very difficult for the pilot to sense that the wheels are skidding. Landings on ice-covered

TYPICAL RADIO RANGE LOW APPROACH

Refer to Pilot's Handbook (East-West) for local procedures

NOTE If fuel is critically low, delay lowering landing gear until over low cone.

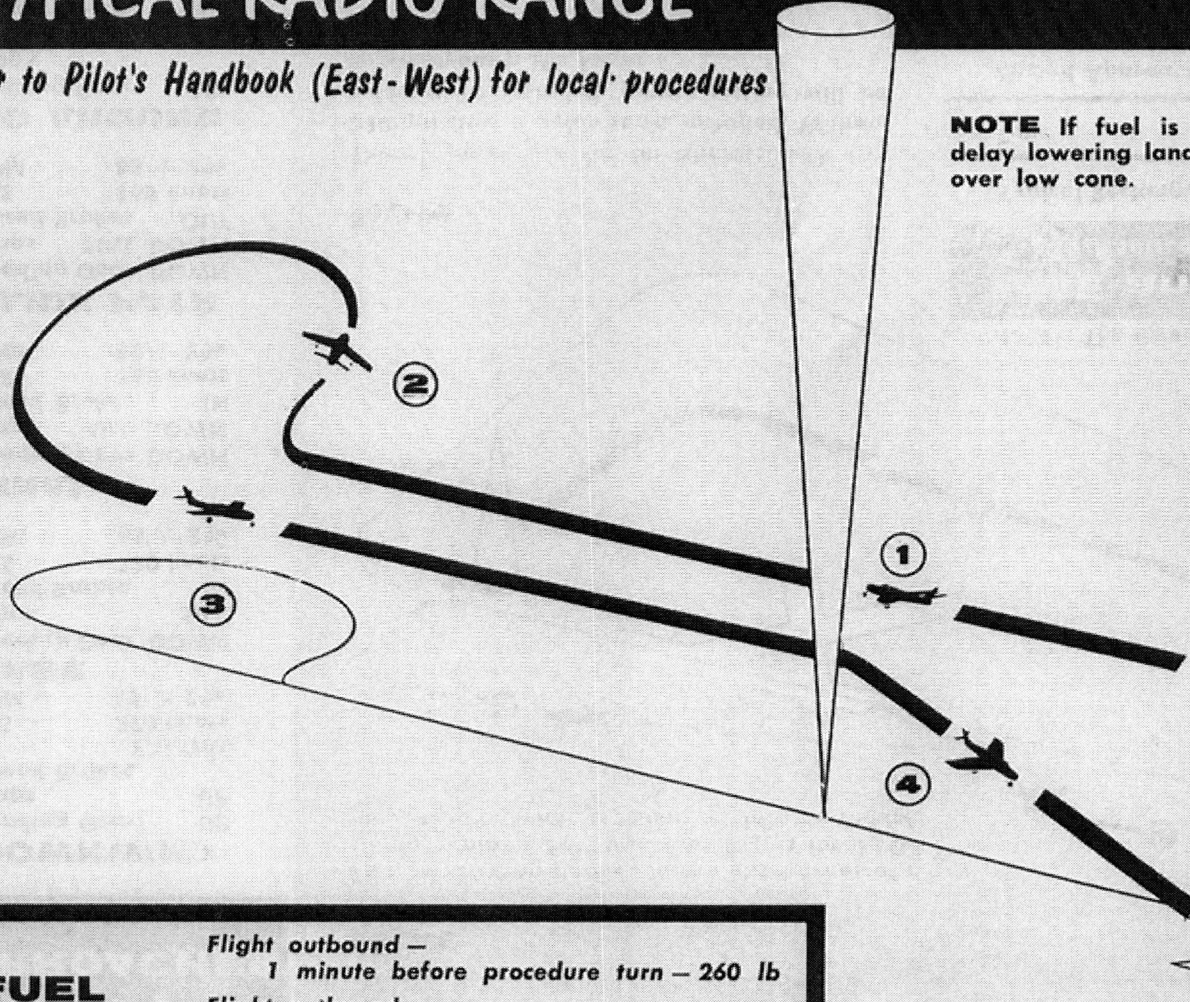

(1) OUTBOUND
Configuration—
Partial Speed Brakes as required
IAS — 220 knots
RPM — 85 ± 2%
Time — 1 minute (or as locally re-required)

(2) PROCEDURE TURN
(90° method)

(3) COCKPIT CHECK
Landing Gear — DOWN
Flaps — 50%
Speed Brakes—IN
IAS—190 Knots
RPM — 85 ± 2%

(4) LOW APPROACH
Flaps— FULL DOWN
Speed Brakes — OUT
IAS — 165 knots
RPM — 85 ± 2%

FUEL REQUIRED

Flight outbound — 1 minute before procedure turn — 260 lb
Flight outbound — 2 minutes before procedure turn — 361 lb
(No conservatism fuel factors included)

Figure 9—3

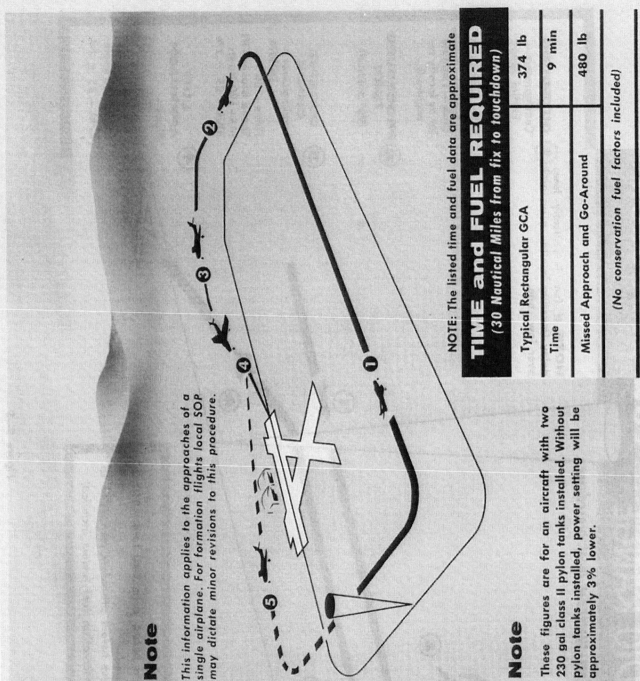

TYPICAL GCA APPROACHES

1 DOWNWIND
Landing Gear UP
Flaps UP
Speed Brakes PARTIAL
IAS — 220 knots
RPM — 85 ± 2%

2 BASE
Landing Gear DOWN
Flaps 50%
Speed Brakes IN
IAS 190 knots
RPM 85 ± 2%

3 FINAL
Landing Gear DOWN
Flaps FULL DOWN
Speed Brakes IN
IAS 165 knots
RPM 85 ± 2%

4 GLIDE PATH
Landing Gear DOWN
Flaps FULL DOWN
Speed Brakes OUT
IAS 165 knots
RPM 85 ± 2%

5 GO AROUND
Landing Gear UP
Flaps UP (at 190 knots)
Speed Brakes IN
RPM 100%

Note

This information applies to the approaches of a single airplane. For formation flights local SOP may dictate minor revisions to this procedure.

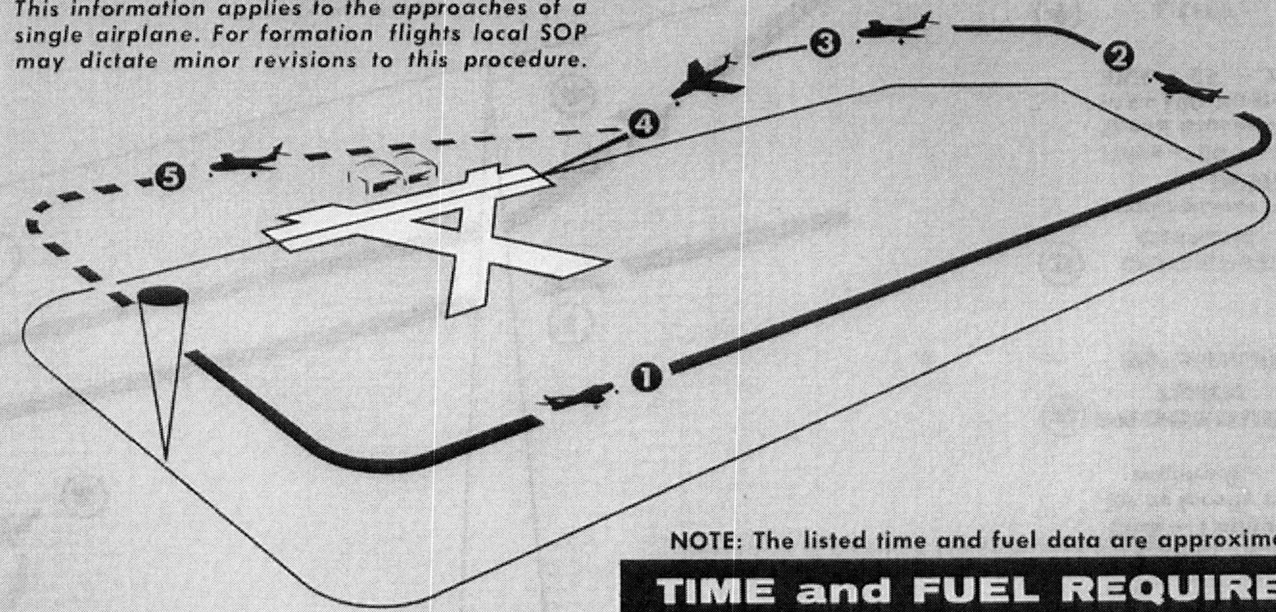

Note

These figures are for an aircraft with two 230 gal class II pylon tanks installed. Without pylon tanks installed, power setting will be approximately 3% lower.

NOTE: The listed time and fuel data are approximate

TIME and FUEL REQUIRED
(30 Nautical Miles from fix to touchdown)

Typical Rectangular GCA	374 lb
Time	9 min
Missed Approach and Go-Around	480 lb

(No conservation fuel factors included)

Figure 9—4

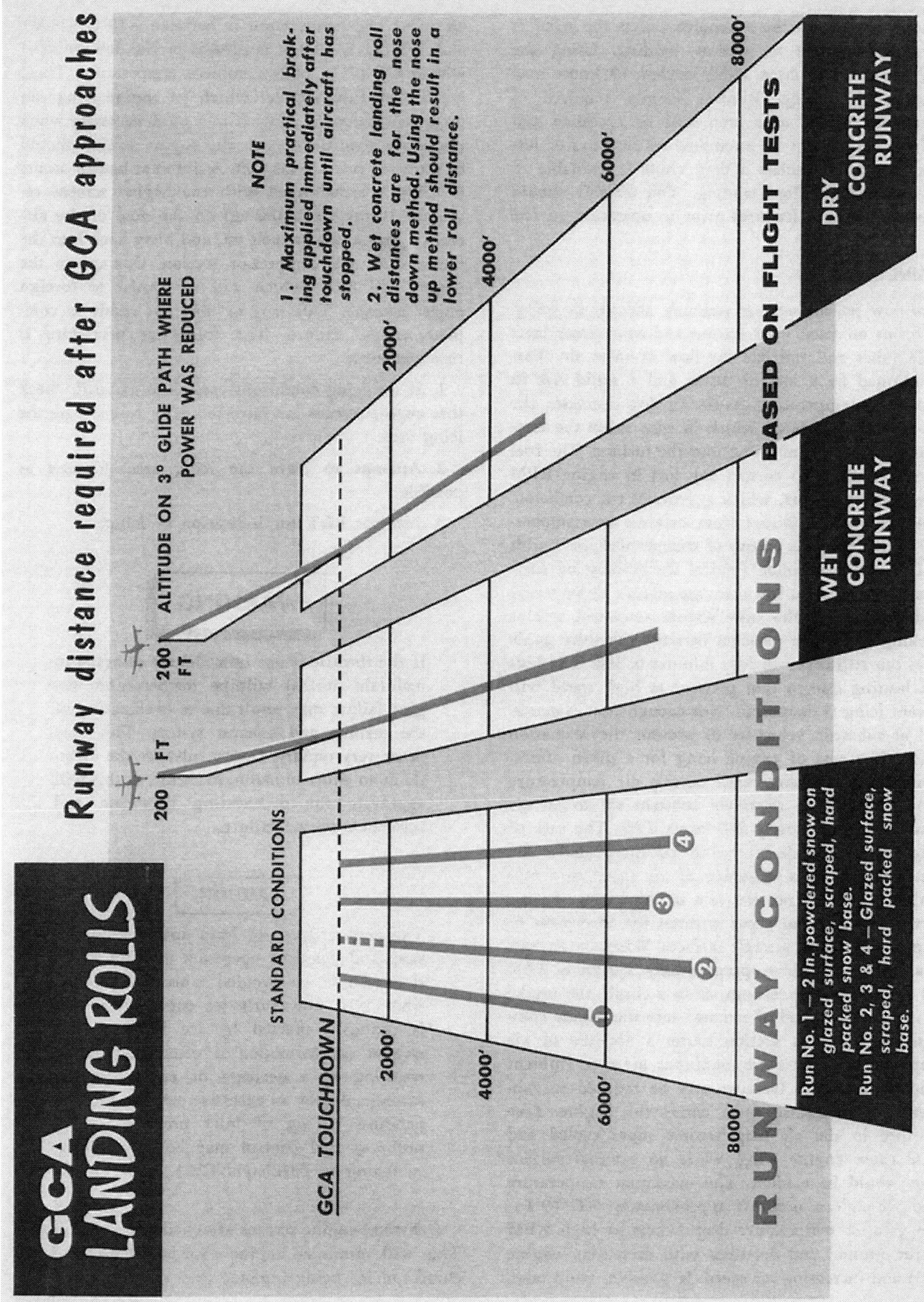
GCA
LANDING ROLLS
Runway distance required after GCA approaches
200 FT
200 FT
ALTITUDE ON 3° GLIDE PATH WHERE POWER WAS REDUCED
NOTE
1. Maximum practical braking applied immediately after touchdown until aircraft has stopped.
2. Wet concrete landing roll distances are for the nose down method. Using the nose up method should result in a lower roll distance.
STANDARD CONDITIONS
GCA TOUCHDOWN
2000'
4000'
6000'
8000'
1
2
3
4
RUNWAY CONDITIONS
Run No. 1 — 2 In. powdered snow on glazed surface, scraped, hard packed snow base.
Run No. 2, 3 & 4 — Glazed surface, scraped, hard packed snow base.
BASED ON FLIGHT TESTS
WET CONCRETE RUNWAY
DRY CONCRETE RUNWAY

Figure 9—5

runways should not be attempted unless the wind is within 10 degrees of runway heading. Using the drag chute when cross winds exceed 10 knots may cause weathervaning. Runway lengths required to bring the aircraft to a stop will be excessive and landings should not be attempted on runways of less than 10,000 feet unless a drag chute is available to aid deceleration after landing. The aircraft should be equipped with ice tires prior to operation on ice.

ENGINE ICING.

Axial-flow jet engines are seriously affected by icing. Ice forms on fixed inlet screens and compressor inlet guide vanes and restricts the flow of inlet air. This is indicated by a loss of thrust and a rapid rise in exhaust gas temperature. As the air flow decreases, the fuel-air ratio increases, which in turn raises the temperature of the gases going into the turbine. The fuel control attempts to correct any loss in engine RPM by adding more fuel, which aggravates the condition. Complete turbine failure from extreme overtemperature may occur in a matter of seconds after ice builds up in the engine inlet. Critical ice buildup on inlet screens can occur in less than one minute under severe conditions. With the inlet screens retracted, serious blocking of the air passages between the inlet guide vanes can still occur in four minutes or less. The idea that heating due to ram pressure at high speed will prevent icing is dangerous. Not enough heat is generated at subsonic velocities to prevent the formation of ice. The rate of engine icing for a given atmospheric icing intensity, with outside air temperature below freezing, is relatively constant up to an air speed of approximately 250 knots TAS. The rate of icing increases with increasing air speed above 250 knots. Therefore, a reduction of air speed to a safe minimum will reduce the rate of inlet icing. Serous inlet duct icing can occur without the formation of ice on the external aircraft surfaces. When jet aircraft fly at velocities below approximately 250 knots TAS, and at high power settings, as in a climb, the intake air is sucked, instead of rammed into the engine compressor inlet. This suction causes a decrease of air temperature. Under these conditions, air at an ambient temperature above freezing may be reduced to subfreezing temperature as it enters the engine. Free moisture in the air may become super cooled and could cause engine icing, while no external surface icing would be evident. The maximum temperature drop which can occur is approximately 5°C (9°F). The greatest temperature drop occurs at high RPM on the ground and decreases with decreasing engine RPM and increasing air speed. If possible, avoid take-off when the temperature is between −10°C (14°F) and 5°C (41°F) if fog is present or the dew point is within 4°C (7°F) of the ambient temperature. These are the conditions under which jet engine icing can occur without wing icing. If take-off is necessary when the above conditions exist, the engine screens should be retracted prior to take-off. A different hazard occurs if icing is encountered with the engine screens retracted. If any ice builds up on the nose of the aircraft it may erode or melt off, and blow back into the intake duct and compressor section. Damage to the engine will occur which will be similar to foreign object damage. This may or may not result in complete engine failure. The following procedure is recommended:

1. If the icing conditions cannot be avoided, check that engine screens are retracted prior to entering the icing area.
2. Attempt to leave the icing area as soon as possible.
3. Monitor EGT for indication of icing.

WARNING

If the throttle is not immediately retarded to maintain normal tailpipe temperatures, engine failure may result due to overheating of the turbine and exhaust system. This may occur very rapidly. Do not advance the throttle in an effort to maintain thrust, as this will aggravate the overheating condition and accelerate engine failure.

CAUTION

The engine does not have anti-icing provisions and therefore operation in icing conditions should be avoided whenever possible since axial flow turbo-jet engine operation is seriously affected by ice buildup. The rate of ice formation is often very rapid, resulting in a decrease of engine airflow accompanied by an excessive exhaust gas temperature. (Icing of inlet pressure sensing probe of fuel control may be corrected by switching to EMERGENCY.)

4. Extend engine screens after icing has terminated. This will minimize engine damage caused by large chunks of ice being ingested into the engine.

TURBULENCE AND THUNDERSTORMS.

CAUTION

> If at all possible, avoid flight through a thunderstorm to prevent damage from hail or icing conditions which are commonly encountered in thunderstorms.

The following factors, singly or in combination, have caused engine flame-outs.

a. Penetration of cumulus buildups with associated high liquid content.

b. Engine icing of either nose cowl or inlet guide vanes.

c. Turbulence associated with penetration can result in angles of attack of plus nine degrees or more causing marginal engine performance.

d. Above 40,000 feet the surge margin of the engine is reduced and there is poor air distribution across the face of the compressor.

CAUTION

> Flying in turbulence, or hail, may increase inlet distortion. At higher altitudes, this distortion can result in engine surge and possible flame-out. However, normal air starts may be accomplished.

Approaching the Storm.

It is imperative that you prepare the airplane prior to entering a zone of turbulent air. If the storm cannot be seen, its proximity can be anticipated by heavy static on the radio compass. Prepare the airplane as follows:

1. A safe comfortable penetration speed for the F-84F airplane in severe turbulence is 275 knots IAS.

CAUTION

> Do not lower gear and flaps as they merely decrease the aerodynamic efficiency of the airplane.

2. Pitot heater—ON.
3. Engine screen—Check RETRACT.
4. Safety belt—Tight (This is important.)
5. Shoulder harness—Locked.
6. Turn radio volume down during severe static conditions.
7. At night, turn cockpit lights full bright to minimize blinding effect of lightning. In addition, turn on thuderstorm lights if aircraft is equipped with them.

Note

> Make every effort to avoid looking up from the instrument panel at lightning flashes. The blinding effect of lightning can be reduced by lowering the seat.

CAUTION

> After leaving the storm, extend engine screens to prevent ice formed on the leading edge of the intake ducts from going through the engine.

NIGHT FLYING.

Before take-off, make certain that all lights function properly. The various rheostat controls afford selection of red lighting contract of the instruments, auxiliary panels and consoles. All cockpit lights should be adjusted to minimum intensity for normal operation to reduce canopy reflections and to permit rapid change over to instrument flight.

When the position light switch is in the FLASH position, the flashing lights cause distracting reflections on the surrounding clouds. The pilot should be aware that a failure of the generator will cause a failure of all cockpit lighting; in addition, all the fuel booster pumps will become inoperative, and the respective warning lights will illuminate.

The taxi light is inadequate, use landing lights for taxiing, but retract before take-off.

If flight through thunderstorm is anticipated, adjust cockpit lights to brightest intensity to prevent momentary blindness from lightning flashes.

The landing lights are aimed for a normal nose-high landing and do not provide adequate runway illumination when a steep power off approach is made. The recommended approach should be made with power and the runway lights should be used as a primary reference for the final approach. The landing lights provide adequate runway illumination once round-out is accomplished.

COLD WEATHER PROCEDURE.

PREFLIGHT.

When the ambient temperature is 0°C (32°F) or lower, use a portable heater to blow hot air into the nose wheel well and starter bottle area. This heating will prevent possible malfunctioning of the starter system. Heat should also be directed toward the engine gear box. Place heater duct in the aircraft air inlet duct for a period of 10-15 minutes. This procedure is necessary to prevent the starter unit from being damaged due to ice seizure of the compressor rotator. Depress the engine rotor test switch for approximately one-half second and listen for audible sound of rotor freedom or observe indication on tachometer. To heat the cockpit, loosen the canopy cover and open the canopy enough so that the heater hose can be inserted into the cockpit. This heating will restore flexibility to the rubber cockpit seal and also will prevent cockpit switch malfunction due to moisture condensation. Remove snow and ice from wings, fuselage, tail and landing gear mechanisms.

WARNING

Depending on the weight of snow and ice accumulated, take-off distances and climb-out performance can be seriously affected. The roughness and distribution of the ice and snow could vary stall speeds and characteristics to an extremely dangerous degree. Loss of an engine shortly after take-off is a serious enough problem without the added, and avoidable, hazard of snow and ice on the wings. In view of the unpredictable and unsafe effects of such a practice, the ice and snow must be removed before flight is attempted.

When conditions are such that mud, snow or slush will freeze, it is recommended that the mud guard be removed from the nose wheel. Use external power for operating and ground check all electrical and radio equipment. Hold battery use to a minimum prior to engine starting. Battery life is reduced to as little as one fifth rated power during extreme cold.

PRESTARTING.

Before the start, one should always manipulate the emergency fuel switch to insure proper operation; otherwise the valves may hang and cause unexpected switching later on. Start engine in normal manner using external power. If there is no oil pressure after 60 seconds running, or if pressure drops after a few minutes ground operation, shut down and check for blown lines or for congealed oil. After start, delay initial movement of controls for a few moments. This is done to permit as much hydraulic fluid as possible to circulate through the pumps. Never turn on electrical equipment except that absolutely needed, until generator shows positive reading. While still in parking area, check out all hydraulic, electrical communication, refueling, lights, defrosting and air conditioning systems and double check that engine screens are extended. Make a final recheck on wings and tail for snow or ice. During taxi-out, check area fore and aft for aircraft. Higher RPM necessary for initial movement and poor braking action on snow require more maneuvering space. Care should be exercised when using full, or near full engine power when aircraft is being runup on chocks as slippage of chocks occurs frequently. Operate wing flaps through several cycles. Turn pitot heater ON if icing conditions are anticipated. Turn heating (pressurizing system), and defroster system ON. Some engines will not accelerate properly when they are cold and will therefore require a short warm-up period in order that the acceleration limits are met. During the runup have aircraft thoroughly checked for leaks. This check should be performed at maximum practicable RPM.

Note

The engine is susceptible to engine hang-up whenever the temperature has dropped five degrees or more at −29°C (−20°F) and usually takes place in the 75 to 80 per cent RPM range. Whenever this occurs, the Bendix fuel control may require an adjustment of the temperature compensating unit.

TAKE-OFF.

WARNING

Prior to take-off make sure all instruments have warmed up sufficiently to insure normal operation. Check for sluggish instruments during taxiing.

Avoid taxiing through loose snow as it may get into brakes and freeze. Pack or remove loose snow from runway prior to take-off. Spacing between elements or aircraft on take-off should not be less than 30 seconds when operating on a snow-covered runway, for aircraft on the take-off roll will leave a hanging wall of powdered snow which requires a few seconds to settle to the runway.

AFTER TAKE-OFF.

After take-off from a snow or slush covered field, operate landing gear and flaps through several complete cycles to preclude their freezing in the UP position. Leave the gear up only after sufficient time has lapsed to remove all slush or moisture. During this time, hold IAS well below maximum permissible gear extension speed. Gear, flaps and all other hydraulic systems will take longer to complete their cycle in cold weather and all limit speeds must be observed.

Note

Landing gear retracting time is from nine seconds at −29°C (−20°F) to 12 seconds at −54°C (−65°F).

LANDING.

Note

In night landings, on snow or ice it is recommended that landing lights not be used for landing due to the intense "bounce" of light the pilot receives from the reflection of the lights on the snow and ice.

Note

Landing gear extension time is from eight seconds at −29°C (−20°F) to 8.5 seconds at −54°C (−65°F).

POST FLIGHT.

If lay-over of several days is expected, remove the battery. Further at temperatures below −29°C (−20°F) remove the battery if lay-over exceeds four hours. Leave canopy slightly open to prevent cracking of transparent areas due to differential contraction. Also air circulation retards frost formation in cockpit. Install wing, empennage and canopy covers and install dust plugs in air intake duct and in tailpipe.

SUMP DRAINAGE.

The fuel tank sump should be drained frequently of condensate. Under prolonged freezing conditions a small amount of ice or snow gets into the fuel tanks each time the aircraft is serviced. When there is sufficient rise in temperature due to placing the aircraft in a hangar or to warmer weather, these crystals melt, resulting in water in the system. Regular and frequent drainage especially under thawing conditions, is the best method of preventing ice in the fuel lines when the aircraft is again subjected to freezing weather. Keeping the tanks as full as possible when the aircraft is parked will also help to reduce moisture condensation.

This page intentionally left blank.

HOT WEATHER AND DESERT PROCEDURES.

BEFORE ENTERING THE AIRCRAFT.

All metal surfaces exposed to the sun are burning hot to touch. Wear gloves to prevent burns. Make all possible ground checks before starting the engine. If operating in sandy country, ascertain that air filters, instrument filters, and oil filters have been cleaned for each flight. Check seals and tires to ascertain that they are not blistered or show other evidence of deterioration.

STARTING THE ENGINE.

Head the aircraft into the wind if practicable. Start in the normal manner. Do not run-up engines to windward of other planes, personnel or ground installations.

Note

When the ambient air temperature exceeds 32°C (90°F) maintain idle RPM exhaust temperature within limits by manually advancing the throttle to a higher RPM but not to exceed 50 per cent RPM.

TAKE-OFF.

If ground is sandy or dusty, avoid taking off in the wake of another aircraft. Place the cabin vent switch in the PRESSURE position unless high humidity causes cockpit to fill up with fog. If so take-off with cabin vent switch in RAM.

Note

Take-off distances will be longer because the air is less dense during warm weather.

AFTER TAKE-OFF.

Do not climb the aircraft at less than flying speed specified in the climb chart.

LANDING.

Because hot air is less dense than cold air, true stalling speed will be greater and additional distance will be required for landing.

PARKING.

If blowing sand is a hazard, close and cover all openings to keep sand out. Cover windshield and canopy to prevent sand scratches. Lay cover on canopy, do not slide. Keep canvas covers on the windshield and canopy whenever the aircraft is parked in the sun. If this is not done, the sun's heat will soften and distort the transparent plastic. Malfunctioning of instruments and communications equipment will also result. If blowing sand is not a hazard, keep canopy and selected access doors open to permit air circulation.

This page intentionally left blank.

APPENDIX 1

PERFORMANCE DATA

TABLE OF CONTENTS

INTRODUCTION.

The flight performance charts in this section provide the pilot with flight test data for flight planning purposes. Two types of charts are included: (1) profile type charts and (2) graphical charts. These charts are described in the following pages. The profile type charts are a supplement to the graphical data and facilitate flight planning by reducing the computations that must be made. These charts are based on the recommended climb and cruise settings shown on the profile for the particular load configuration of the aircraft. This type of presentation gives a direct indication of the fuel and time required to cover a given distance if the recommended settings are maintained. A decrease in weight has been accounted for as fuel is consumed. For cruise at Mach numbers other than those given on the profile charts, the graphical charts should be used for flight planning. For flight planning where accurate results are mandatory, the graphical data should be used. All charts are based on NACA Standard Day conditions. The take-off and landing distance charts contain the temperature correction in graphical form.

ALTIMETER CORRECTION CHART.

Static pressure, which affects the altimeter readings, is not always accurately measured due to the location of the static vent. Additional factors which affect the degree of error in the altimeter are the attitude of the airplane, the airspeed and the installation of external tanks on the inboard pylons. The error in the altimeter reading is greatest at low altitudes and high speeds making it more critical during ground support missions. The altimeter position error correction chart (figure A–1) is provided so that for a given altitude, airplane loading configuration and airspeed the altimeter error can be determined. The error is added algebraically to the altimeter reading to obtain the true pressure altitude.

AIRSPEED INSTALLATION CORRECTION CHARTS.

In order to obtain true airplane speeds several corrections must be applied to the airspeed indicator reading. The first correction is made for the error in the individual instrument. This value is noted on the instrument calibration card and when applied to the instrument reading provides indicated airspeed (IAS). The second correction is for airspeed installation error. This correction is taken from figure A–2, and when applied to the indicated airspeed (IAS) provides calibrated airspeed (CAS). The third correction is taken from figure A–3 and when applied to calibrated airspeed provides equivalent airspeed (EAS). Dividing the equivalent airspeed by the square root of relative density (ratio of ambient to standard sea level density) provides true airspeed (TAS). Vectorially adding wind velocity to true airspeed provides ground speed (GS).

SAMPLE PROBLEM.

For purposes of explaining the use of the airspeed Position Error Correction Chart and the Compressibility Correction Table, consider a clean airplane flying at 25,000 feet at an airspeed indicator reading of 350 knots. Since the airplane is not equipped with an outside air temperature indicator, determine the ambient temperature at 25,000 feet from the Standard Altitude Table (figure A–6) which will be –35°C.

Airspeed Indicator reading	350 knots
Correction for Instrument error (from instrument calibration card)	–2
Indicated Airspeed (IAS)	348 knots
Correction for Installation Error (from airspeed position error correction chart figure A–2)	–8
Calibrated Airspeed (CAS)	340 knots
Correction for Compressibility Error (from compressibility corrrection table figure A–3)	–16
Equivalent Airspeed (EAS)	324 knots
Correction for Air Density (from standard altitude table figure A–6)	x 1.49
True Airspeed (TAS)	483 knots

MACH NUMBER POSITION ERROR CORRECTION CHART.

Due to the inaccuracies of the static pressure position errors will be carried into the Machmeter. The Mach number position error correction chart (figure A–4) indicates the amount of error for various speeds in terms of Mach No.

SPEED CONVERSION CHART.

The speed conversion chart (figure A–5) reflects the Mach No. and the true airspeed at any altitude for a given calibrated airspeed.

TAKE-OFF CHARTS.

Take-off charts are provided for no assist, two ato unit assist and four ato unit assist for the J65-W-3, B-3, W-7 or B-7 engines. Following is a sample of a take-off data card. For the data card form refer to Section II Condensed Check List. Information on this card should be filled out prior to each flight. This data is obtained by referring to the Take-off Chart and the Landing Distance Chart for the subject aircraft configuration.

TAKE-OFF DATA CARD	
Conditions	
Gross Weight	22,300 LB
Runway Length	9,000 FT
OAT	35°C (95°F)
Pressure Altitude	2,000 FT
Runway Gradient	0%
Wind	0 KN
Engine Model	J65-W-3
Assist Take-off Units	0
Take-off	
Take-off Distance	7,250 FT
Take-off Speed	154 KN
Go-No-Go Distance	4,000 FT
Go-No-Go Speed (Minimum)	109 KN
Landing Immediately After Take-off	
Landing Ground Roll (with drag chute)	3,000 FT
Landing Ground Roll (without drag chute)	5,400 FT
Approach Speed	195 KN

TAKE-OFF ACCELERATION CHECK DATA.

F-84F take-off ground roll distance curves, figures A–7, A–8, A–9, A–73, A–74 and A–75, and speed vs distance curves, figures A–10 and A–76 are presented to allow the pilot to determine speed at any distance down the runway (line speed) during the take-off ground roll. Stopping distance curves are presented in figures A–11 and A–77 to determine the minimum distance required to stop from any speed. Using this information, the runway marker (painted stripes, signs, etc.) at which acceleration is checked can be chosen so that the aircraft may be stopped on the remaining runway from normal predicted speed at that point should the take-off be aborted. Figure A–12 presents a graph of airspeed position error correction to be applied in determining cockpit indicated airspeed from any calibrated airspeed presented in figures A–10, A–11, A–76 and A–77.

The purpose of line speed is to provide a means of checking take-off acceleration. A check on the speed at some point reasonably early in the take-off run will allow the pilot to properly monitor his take-off run. It is possible to predict two line speeds for a given line or check point on the runway, (a) normal line speed which assumes normal take-off ground roll distance and (b) minimum line speed which assumes that the entire runway length will be used for the take-off ground roll due to low acceleration. To minimize the number of aborts and assure full utilization of available runway length, the minimum line speed should be used for the acceleration check. If the observed speed at the check point is equal to or more than the minimum line speed, take-off acceleration is acceptable and the take-off should be continued. If it is less than the minimum line speed the take-off should be aborted. Since the check point is chosen so that the airplane can be stopped on the remaining runway from the normal line speed, it will be less difficult to stop from a speed less than minimum line speed.

EXAMPLE:

Determine if a point 4,000 feet down a 9,000 foot runway can be used as a safe acceleration check point. Determine also the normal and minimum line speeds for this check point.

Known.

Take-off gross weight—22,300 pounds.

Configuration—Clean + (two) 230 gallon inboard tanks.

Pressure altitude—2,000 feet.

Runway temperature—plus 35°C (95°F).

Length of available runway—9,000 feet.

Dry hard surface runway conditions.

Zero wind.

Procedure.

Using data obtained from figure A–7 enter figure A–10 at take-off gross weight of 22,300 pounds (A). Proceed right and find normal take-off calibrated airspeed of 154.0 knots CAS (B). Proceed down until guide line intersects ground roll distance of 7,250 feet (D). Follow the curve, down and to the left, to the 4,000 foot distance intersection (G). Read down to find normal line of speed of 119 knots CAS (J).

To determine minimum line speed, again enter figure A–10 at the take-off gross weight of 22,300 pounds (A). Proceed right and find normal take-off airspeed of 154.0 knots CAS (B). Proceed down until guide line intersects the available runway length of 9,000 feet (C). Proceed down and to the left, following the guide lines, to again intersect the 4,000 foot distance (H). This intersection of the guide line at the 4,000 foot point down the runway shows a minimum line speed of 109 knots CAS (K).

Enter figure A–11 at a normal line speed of 119 knots CAS (A), correct for pressure altitude of 2,000 feet (B) and outside air temperature of plus 35°C (95°F) (D) and intersect the stopping distance curve (E). Stopping distance required from normal line speed is 2,825 feet (G).

No conservatism or pilot reaction time is provided in Curve A, figure A–11, as pilot reaction time in deciding to abort the take-off is variable. The average

ALTIMETER POSITION ERROR CORRECTION CHART

FLUSH STATIC SYSTEM

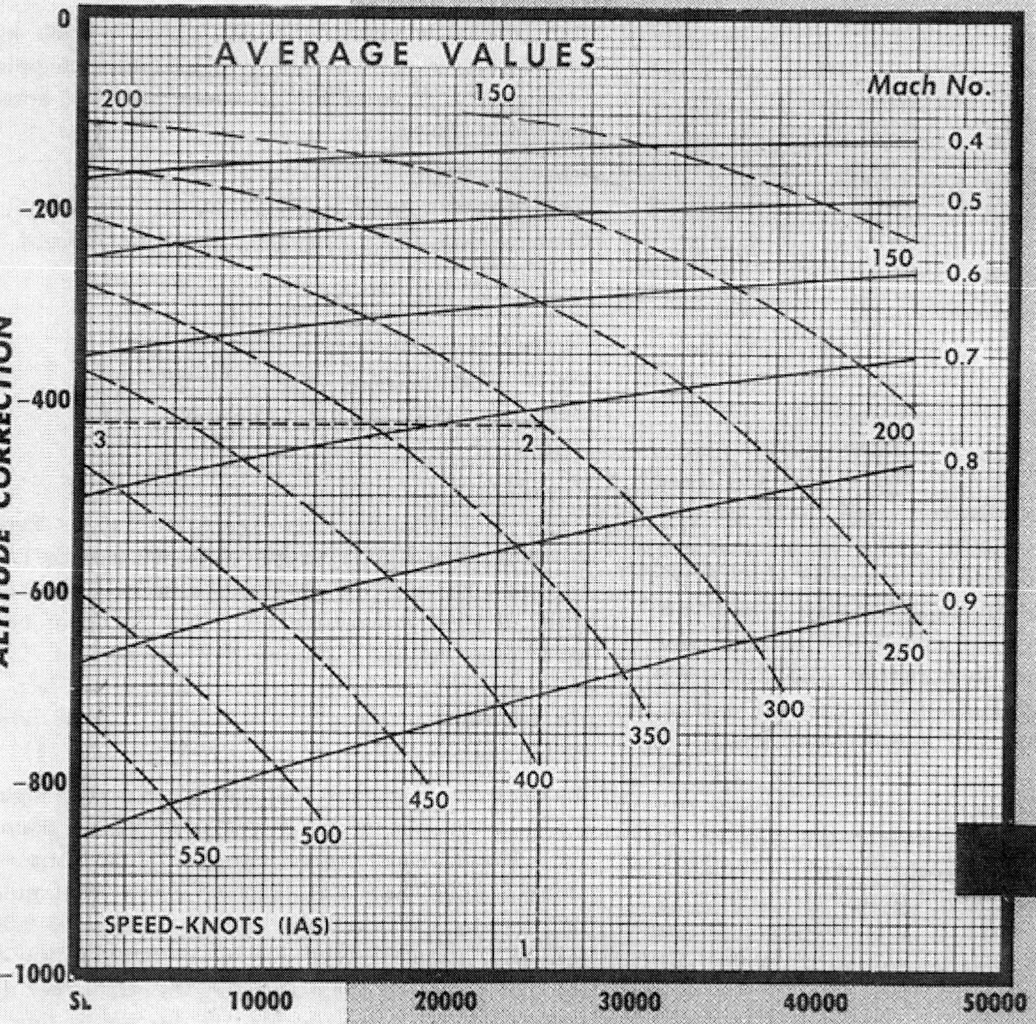

Clean

EXAMPLE

ALTIMETER READS 25,000 ft at 300 knots (IAS) clean airplane.

Enter bottom of chart at 25,000 foot mark (1) and move up to the 300 knot curve (2). Move across and read altimeter correction of −422 feet. Pressure altitude equals 25,000 −422 or 24,578 feet, thus, to hold an assigned pressure altitude of 25,000 feet, the aircraft must be flown at an indicated altitude of 25,422 feet.

Figure A—1 (sheet 1 of 2)

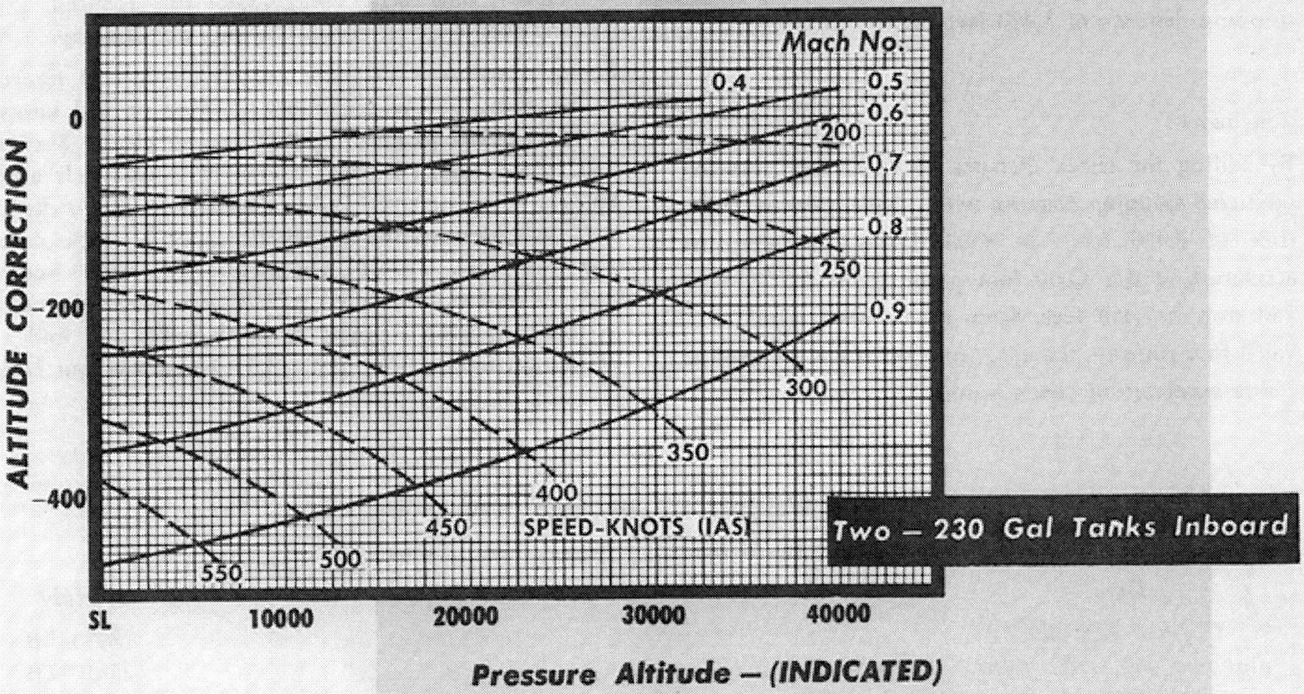

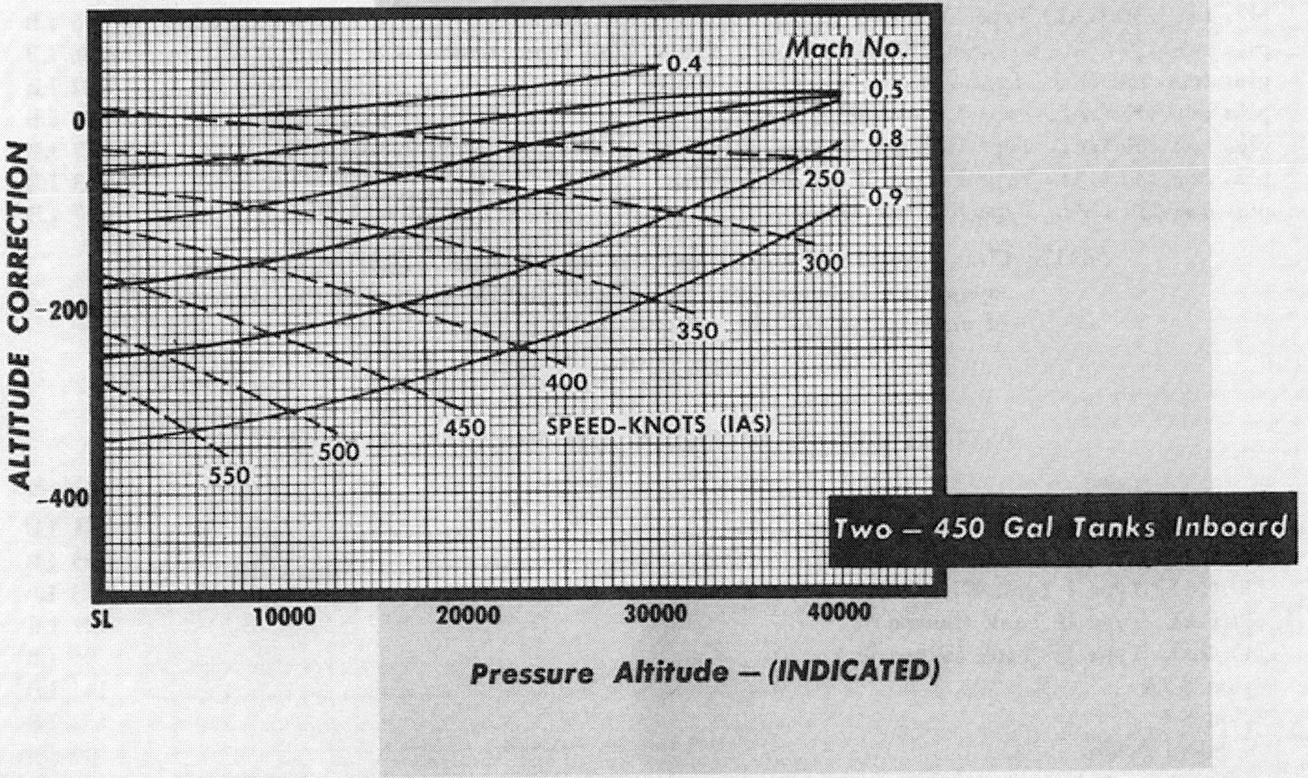

Figure A—1 (sheet 2 of 2)

pilot will use about three seconds to make a decision, cut power, and apply brakes (at 119 knots CAS at 2,000 feet and plus 35°C (95°F) this is 625 feet). Therefore, to facilitate ease of planning, Curve B includes a pilot reaction time of three second—a total stopping distance of 3,450 feet (H).

Conclusion:

By adding the check distance of 4,000 feet plus the predicted stopping distance with normal pilot reaction time of 3,450 feet, the total distance required to accelerate to the 4,000 foot point down the runway and stop is 7,450 feet. Since this is well within the 9,000 foot runway, the 4,000 foot point can be used as a safe acceleration check point.

Note

When determining stopping distance on a wet runway, the speed for which stopping distance is plotted in figure A—11 will be increased by 10 knots CAS before entering the chart. This will compensate for reduced braking friction caused by the wet runway.

Cockpit indicated airspeeds are obtained from figure A—12. Enter at the calibrated airspeed of 119 knots CAS (A) and proceed upward to the clean + two 230 gallon inboard tanks curve (B). Read left and find the correction (C) to be added to obtain a cockpit indicated airspeed for normal line speed. In this case 119 knots CAS + (+4) knots correction = 123 knots IAS. To obtain minimum line indicated airspeed, enter at the calibrated airspeed of 109.0 knots and find a correction to be added of (+3.5) knots. Cockpit IAS = 112.5 knots.

AVERAGE WEIGHTS OF AIRCRAFT AND STORES

Refer to T.O. 1F-84F-5 for Weight and Balance

Item	*Weight*
CLEAN AIRCRAFT	18975 LB
plus two 450 GAL. Type I Tanks With Fins (Full)	25610 LB
plus two 450 GAL. Type II Tanks (Full)	25519 LB
plus two 230 GAL. Type I Tanks With Fins (Full)	22511 LB
plus two 230 GAL. Type II Tanks (Full)	22418 LB
plus two 230 GAL. Type IV Tanks (Full)	22355 LB
plus two 450 GAL. and two 230 GAL. Type I Tanks With Fins (Full)	29146 LB
plus four 230 GAL. Type I Tanks With Fins (Full)	25902 LB
plus one 450 GAL. Type I Tank With Fins (Full)	22293 LB
plus one 450 GAL. Type II Tank (Full)	22247 LB
plus one 230 GAL. Type I Tank With Fins (Full)	20763 LB
plus one 230 GAL. Type II Tank (Full)	20717 LB

NOTE: Clean weight of aircraft includes approximate weight of pilot and equipment, fully serviced internal fuel and oil tanks, full complement of ammunition and six guns.

AVERAGE WEIGHTS OF PYLONS AND EMPTY PYLON TANKS

Item	*Weight*
450 GAL. Type I Tank with Fins	241 LB
450 GAL. Type II Tank	195 LB
230 GAL. Type I Tank With Fins	173 LB
230 GAL. Type II Tank (Sutton)	127 LB
230 GAL. Type IV Tank (Royal Jet)	95 LB
Pylon S-2A	144 LB
Pylon S-3	152 LB
Outboard Pylon	80 LB
Special Store Pylon	167 LB

AIRSPEED POSITION ERROR CORRECTION CHART

FLUSH STATIC SYSTEM

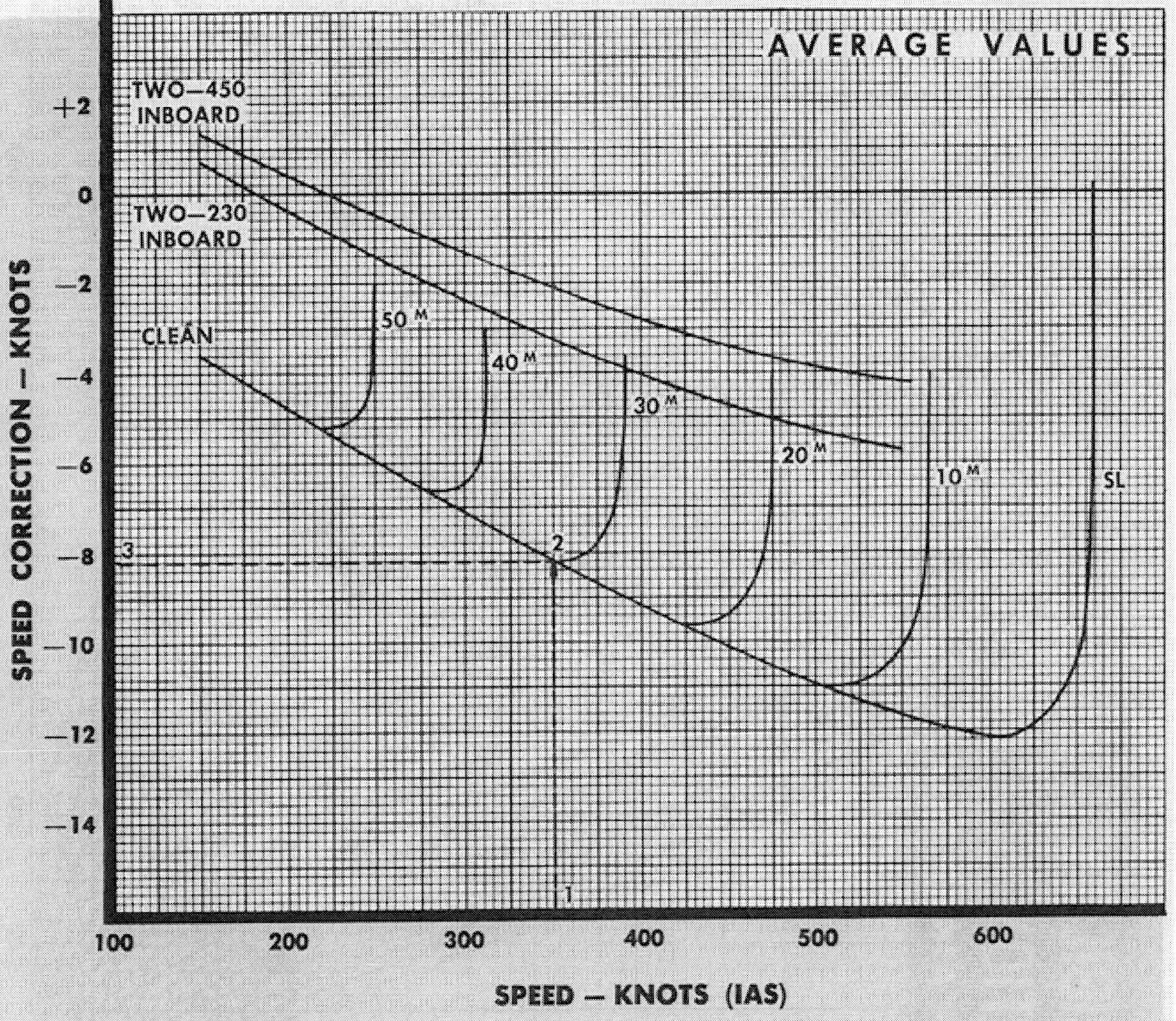

EXAMPLE

Find CAS of clean airplane at 25,000 feet flying at 350 knots IAS.

Enter chart at 350 knots IAS (1) and move up to intersect altitude curve (2). Move across and read correction of −8.2 knots (3). CAS = 350 − 8.2 or 341.8 knots.

Figure A—2

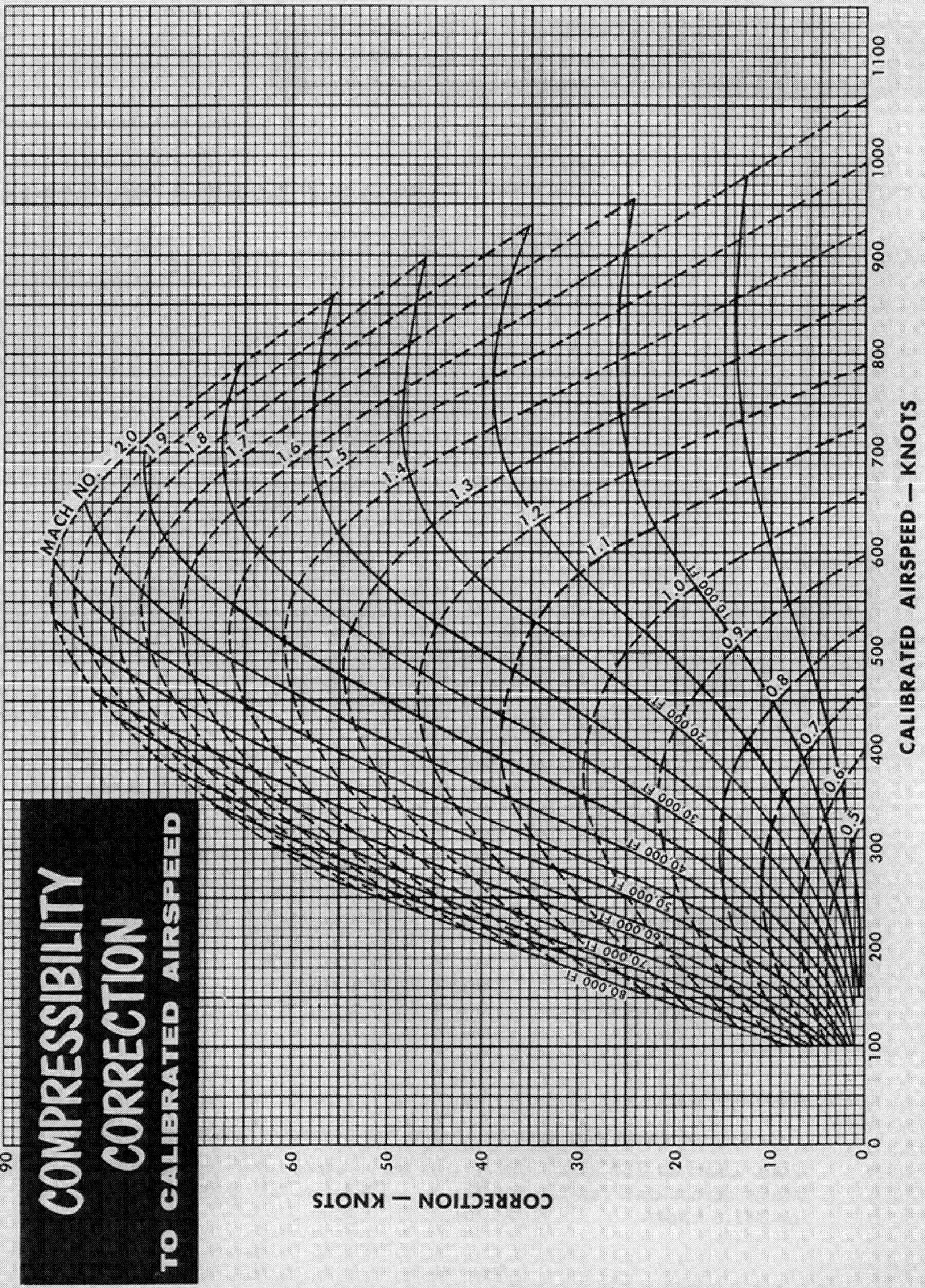
COMPRESSIBILITY
CORRECTION
TO CALIBRATED AIRSPEED
CORRECTION — KNOTS
CALIBRATED AIRSPEED — KNOTS
MACH NO. = 2.0
1.9
1.8
1.7
1.6
1.5
1.4
1.3
1.2
1.1
1.0
0.9
0.8
0.7
0.6
0.5
10,000 FT
20,000 FT
30,000 FT
40,000 FT
50,000 FT
60,000 FT
70,000 FT
80,000 FT
0
10
20
30
40
50
60
90
100
200
300
400
500
600
700
800
900
1000
1100

Figure A—3

MACH NUMBER POSITION ERROR CORRECTION CHART

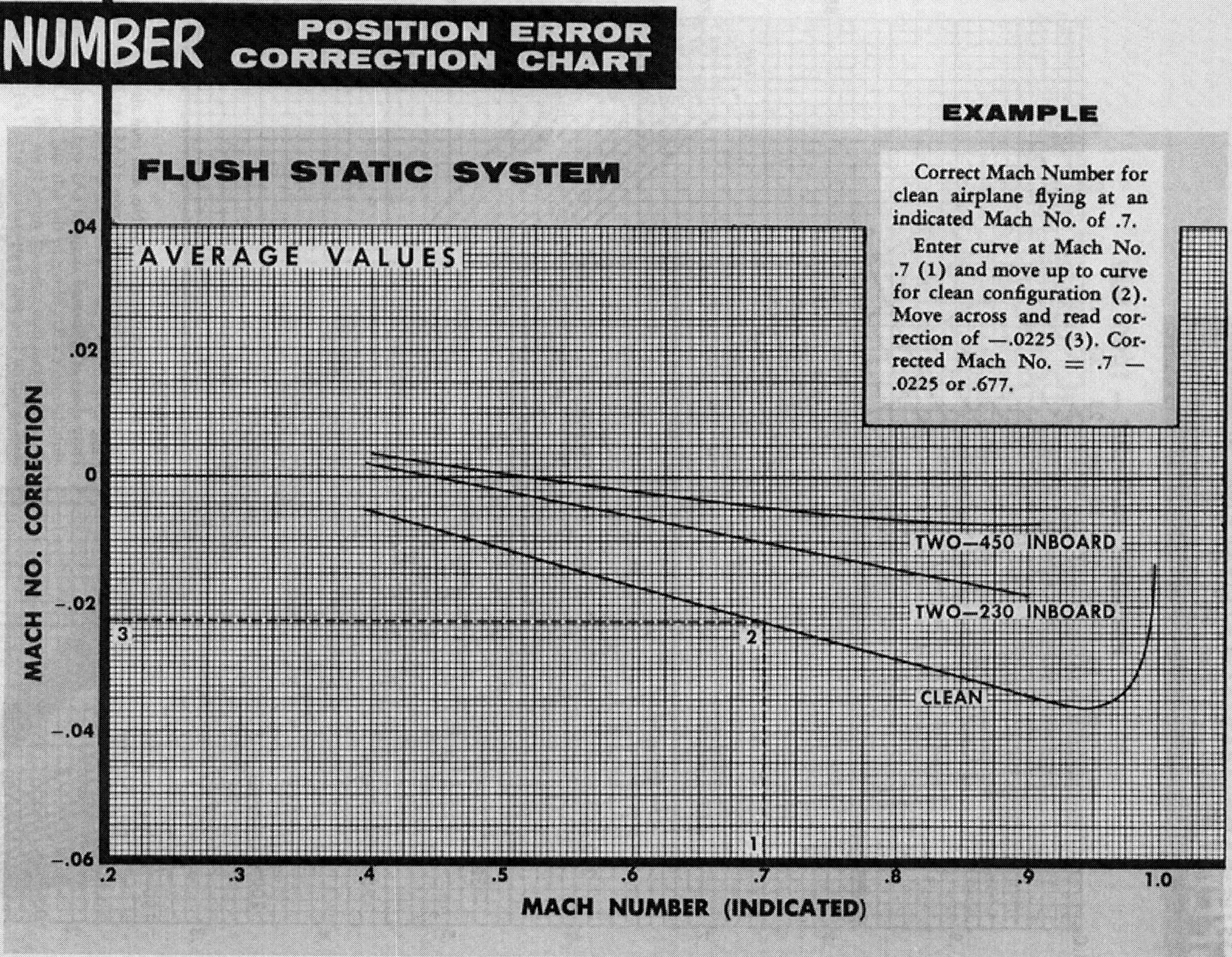

Figure A—4

SPEED CONVERSION CHART

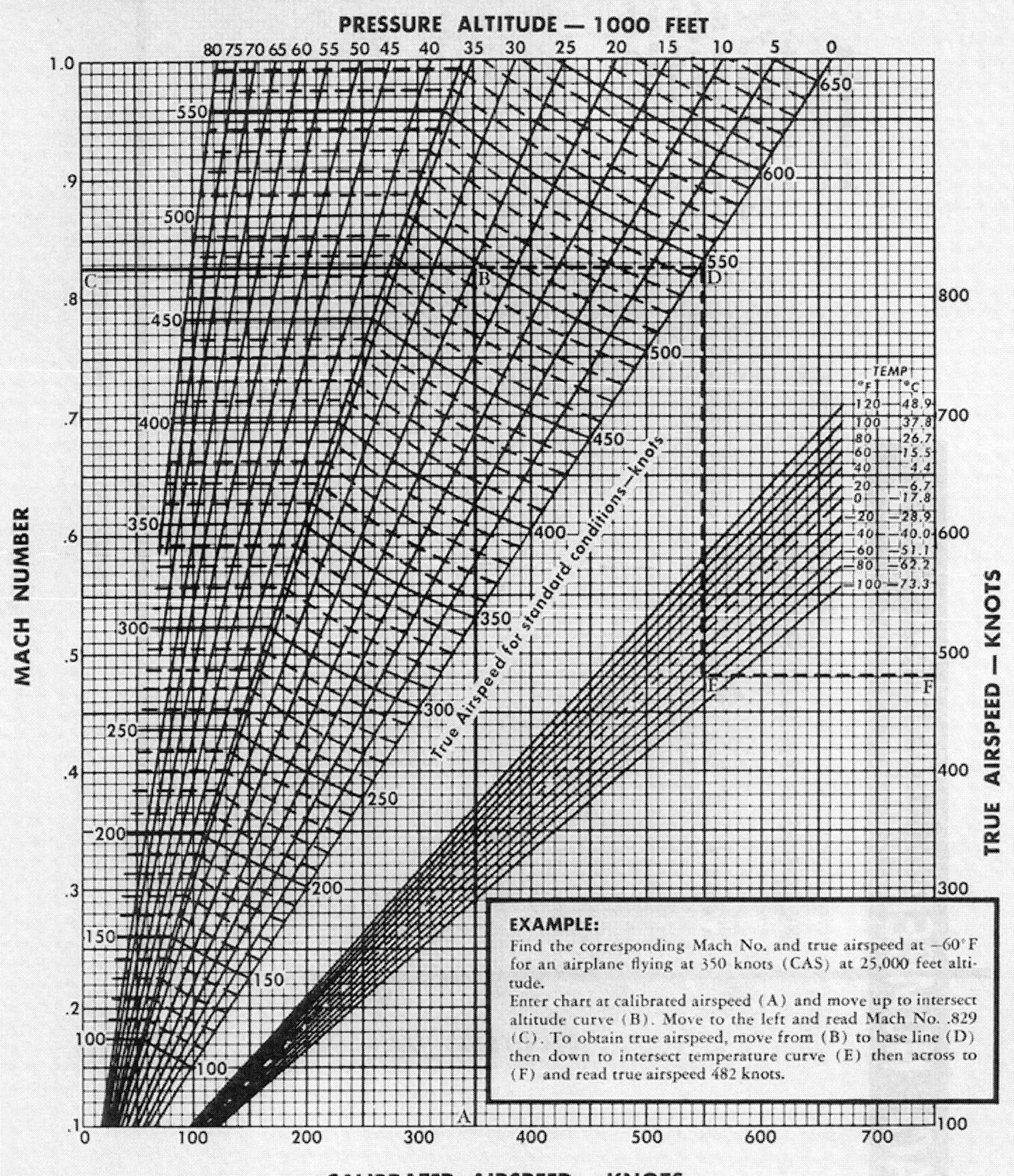

EXAMPLE:

Find the corresponding Mach No. and true airspeed at −60°F for an airplane flying at 350 knots (CAS) at 25,000 feet altitude.

Enter chart at calibrated airspeed (A) and move up to intersect altitude curve (B). Move to the left and read Mach No. .829 (C). To obtain true airspeed, move from (B) to base line (D) then down to intersect temperature curve (E) then across to (F) and read true airspeed 482 knots.

Figure A—5

STANDARD ALTITUDE TABLE

Standard Sea Level Air:
T = 15° C.
P = 29.921 in. of Hg.

W = .07651 lb/cu. ft. ρ_0 = .002378 slugs/cu. ft.
1" of Hg. = 70.732 lb/sq. ft. = 0.4912 lb/sq. in.

This table is based on NACA Technical Report No. 218 a_o - 1116 ft./sec.

Altitude feet	Density Ratio ρ/ρ_0	$\frac{1}{\sqrt{\sigma}}$	Temperature Deg. C	Temperature Deg. F	Speed of Sound Ratio a/a_o	Pressure In. of Hg.	Pressure Ratio P/Po
0	1.0000	1.0000	15.000	59.000	1.0000	29.92	1.0000
1000	.9710	1.0148	13.019	55.434	.997	28.86	.9644
2000	.9428	1.0299	11.038	51.868	.993	27.82	.9298
3000	.9151	1.0454	9.056	48.301	.990	26.81	.8962
4000	.8881	1.0611	7.075	44.735	.986	25.84	.8636
5000	.8616	1.0773	5.094	41.169	.983	24.89	.8320
6000	.8358	1.0938	3.113	37.603	.979	23.98	.8013
7000	.8106	1.1107	1.132	34.037	.976	23.09	.7716
8000	.7859	1.1280	-0.850	30.471	.972	22.22	.7427
9000	.7619	1.1456	-2.831	26.904	.968	21.38	.7147
10000	.7384	1.1637	-4.812	23.338	.965	20.58	.6876
11000	.7154	1.1822	-6.793	19.772	.962	19.79	.6614
12000	.6931	1.2012	-8.774	16.206	.958	19.03	.6359
13000	.6712	1.2206	-10.756	12.640	.954	18.29	.6112
14000	.6499	1.2404	-12.737	9.074	.950	17.57	.5873
15000	.6291	1.2608	-14.718	5.507	.947	16.88	.5642
16000	.6088	1.2816	-16.699	1.941	.943	16.21	.5418
17000	.5891	1.3029	-18.680	-1.625	.940	15.56	.5202
18000	.5698	1.3247	-20.662	-5.191	.936	14.94	.4992
19000	.5509	1.3473	-22.643	-8.757	.932	14.33	.4790
20000	.5327	1.3701	-24.624	-12.323	.929	13.75	.4594
21000	.5148	1.3937	-26.605	-15.890	.925	13.18	.4405
22000	.4974	1.4179	-28.586	-19.456	.922	12.63	.4222
23000	.4805	1.4426	-30.568	-23.022	.917	12.10	.4045
24000	.4640	1.4681	-32.549	-26.588	.914	11.59	.3874
25000	.4480	1.4940	-34.530	-30.154	.910	11.10	.3709
26000	.4323	1.5209	-36.511	-33.720	.906	10.62	.3550
27000	.4171	1.5484	-38.493	-37.287	.903	10.16	.3397
28000	.4023	1.5768	-40.474	-40.853	.899	9.720	.3248
29000	.3879	1.6056	-42.455	-44.419	.895	9.293	.3106
30000	.3740	1.6352	-44.436	-47.985	.891	8.880	.2968
31000	.3603	1.6659	-46.417	-51.551	.887	8.483	.2834
32000	.3472	1.6971	-48.399	-55.117	.883	8.101	.2707
33000	.3343	1.7295	-50.379	-58.684	.879	7.732	.2583
34000	.3218	1.7628	-52.361	-62.250	.875	7.377	.2465
35000	.3098	1.7966	-54.342	-65.816	.871	7.036	.2352
36000	.2962	1.8374	-55.000	-67.000	.870	6.708	.2242
37000	.2824	1.8818	-55.000	-67.000	.870	6.395	.2137
38000	.2692	1.9273	-55.000	-67.000	.870	6.096	.2037
39000	.2566	1.9738	-55.000	-67.000	.870	5.812	.1943
40000	.2447	2.0215	-55.000	-67.000	.870	5.541	.1852
41000	.2332	2.0707	-55.000	-67.000	.870	5.283	.1765
42000	.2224	2.1207	-55.000	-67.000	.870	5.036	.1683
43000	.2120	2.1719	-55.000	-67.000	.870	4.802	.1605
44000	.2021	2.2244	-55.000	-67.000	.870	4.578	.1530
45000	.1926	2.2785	-55.000	-67.000	.870	4.364	.1458
46000	.1837	2.3332	-55.000	-67.000	.870	4.160	.1391
47000	.1751	2.3893	-55.000	-67.000	.870	3.966	.1325
48000	.1669	2.4478	-55.000	-67.000	.870	3.781	.1264
49000	.1591	2.5071	-55.000	-67.000	.870	3.604	.1205
50000	.1517	2.5675	-55.000	-67.000	.870	3.436	.1149

Figure A-6

MB-8 COMPUTER.

The MB-8 computer (figure A—6A) consists of three metal and two plastic discs. The three metal discs are a standard item good for any airplane and are obtainable through regular Air Force channels (refer to foreword). However, they are useless without the plastic "data" discs which contain the airplane performance. The MB-8 computer is designed to solve simple level-flight cruise control problems for jet aircraft. However, exclusive use for preflight planning is not recommended, since under normal conditions the use of the Flight Manual results in far more comprehensive results. The greatest advantage of the computer lies in its simplicity of operation and convenient size: therefore, certain compromises which impose limitations are involved. The computer is designed for an *average* gross weight. This will result in a lower-than-indicated miles per pound of fuel with a subsequent higher rate of fuel flow at the beginning of flight and a higher-than-indicated miles per pound with a subsequent lower rate of fuel flow during the final portion of the flight, giving an average miles per pound as indicated on the computer.

The back or tabulator side of the MB-8 computer shows the cruise data in the MAX RANGE window, listing combinations of fuel remaining at selected pressure altitude. These data can be used as a quick range check for various quantities of fuel remaining at altitude. Range data for both optimum cruise and cruise at constant altitude are given, thereby providing a quick and yet fairly comprehensive picture of the range potential. These data are very similar to the information given in the optimum return profile of the Flight Manual. A second window displays the time, fuel, and distance required for climb or descent, while a third window frames the recommended altitude-speed schedule for these maneuvers. Notice that a black background is used for one configuration, while the other has a white background.

Turn to the front or working side of the computer and begin at the center, working toward the outer edge. *Keep in mind the six factors of range:*

1. Speed (Mach number).
2. Altitude.
3. Fuel.
4. Distance.
5. Time.
6. Wind.

Notice the opposing pie-shaped windows which allow the center plastic discs to show through. The black-bordered window is for the black background data, and vice versa. Notice that the window is divided into *altitudes,* with an index line through the center of the windows. The outer edge of this first (metal) disc is divided into a logarithmic scale labeled "FUEL QUANTITY – POUNDS" (referred to as the fuel disc). With this type of scale, the "1000" mark can mean 1, 10, 100, 1,000, or 10,000 pounds, etc., depending on the magnitude of the other factors in the range problem. Using the tab provided, rotate the fuel disc *counterclockwise* so that the index line for any selected *altitude* passes across the *speed* lines which show through the respective window. The first speed line encountered is the *recommended* speed for maximum range. Further counterclockwise rotation results in passing over increasing speeds until the maximum speed line in the series is reached. In progressing from speed for maximum range to minimum speed, the index passes over a speed line coded as a solid dot with a vertical line passing through it. This is the computer setting for maximum endurance. The speed for maximum endurance is quoted at the extreme right of the *maximum range* speed line. This coded point is used together with the quoted speed to obtain maximum endurance information. Another coded speed line (diamond with a vertical line) is the maximum speed for normal power (maximum continuous thrust). To help understand the position of these speed points on the computer, examine a typical nautical-miles-per-pound-of-fuel (specific range) curve which presents these same speed points graphically.

Note

> The speeds shown on the MB-8 computer are IAS or true Mach number; therefore, any indicated speeds should be corrected for installation error before they are entered on the computer. Because of numerous variables and possible modification of the airspeed indicating system, the indicated airspeed and Mach number are not incorporated in the MB-8 computer.

The second disc (plastic) is a performance data disc around which is placed a logarithmic scale labeled AIR NAUTICAL MILES. Refer to this disc as the distance disc. The placement of the speed lines previously described maintains the proper relationship between the *distance* and *fuel* discs. Note that any specific relationship between the fuel and distance discs for a selected speed-altitude will give the specific range or nautical air miles per pound of fuel; i.e., the air miles at the 1,000 pound mark are actually nautical air miles per 1,000 pounds of fuel. The tab on the distance disc is a special shape with the straight edge of the tab acting as a wiper or cursor

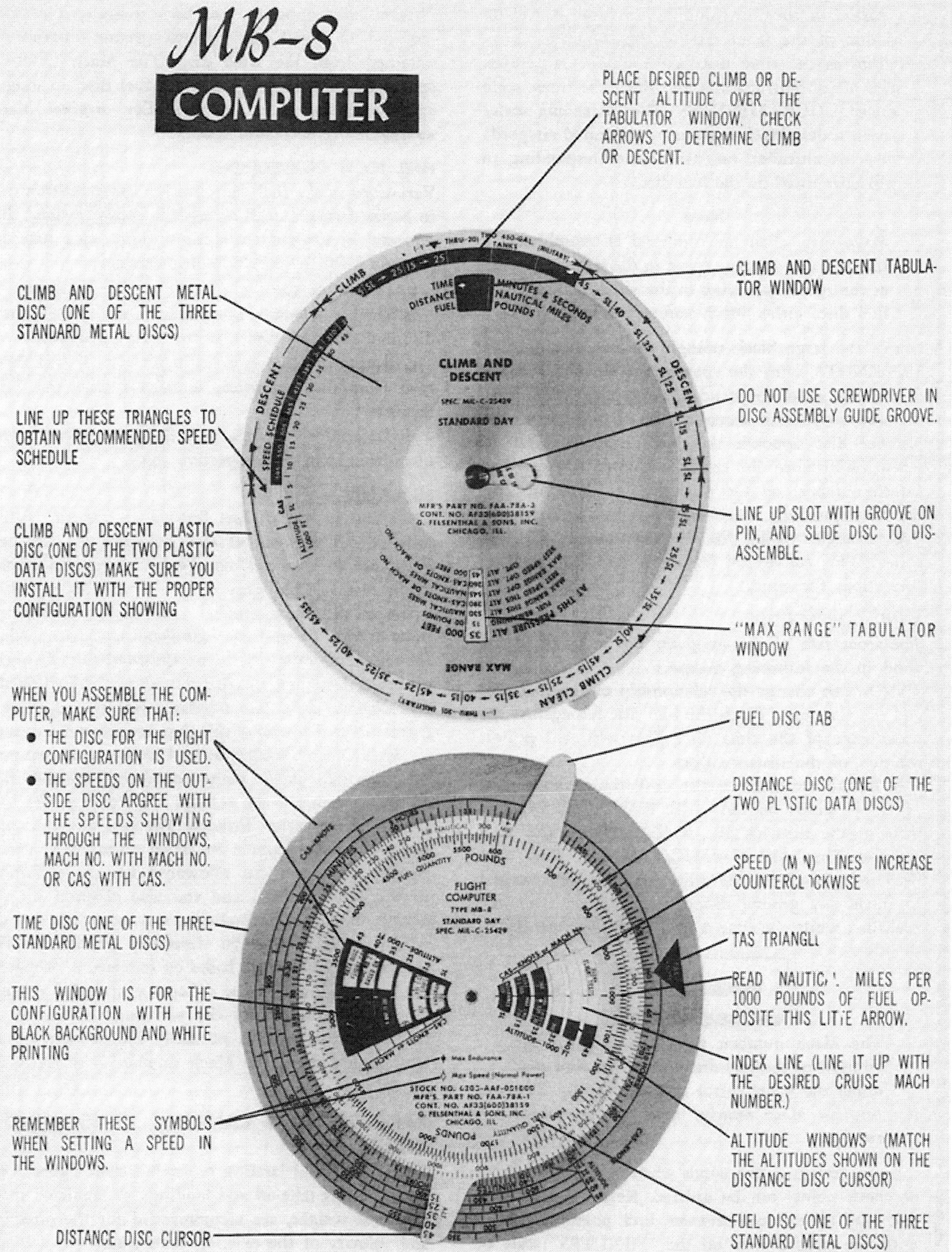
MB-8
COMPUTER
PLACE DESIRED CLIMB OR DESCENT ALTITUDE OVER THE TABULATOR WINDOW. CHECK ARROWS TO DETERMINE CLIMB OR DESCENT.
CLIMB AND DESCENT TABULATOR WINDOW
CLIMB AND DESCENT METAL DISC (ONE OF THE THREE STANDARD METAL DISCS)
DO NOT USE SCREWDRIVER IN DISC ASSEMBLY GUIDE GROOVE.
LINE UP THESE TRIANGLES TO OBTAIN RECOMMENDED SPEED SCHEDULE
LINE UP SLOT WITH GROOVE ON PIN, AND SLIDE DISC TO DISASSEMBLE.
CLIMB AND DESCENT PLASTIC DISC (ONE OF THE TWO PLASTIC DATA DISCS) MAKE SURE YOU INSTALL IT WITH THE PROPER CONFIGURATION SHOWING
"MAX RANGE" TABULATOR WINDOW
WHEN YOU ASSEMBLE THE COMPUTER, MAKE SURE THAT:
• THE DISC FOR THE RIGHT CONFIGURATION IS USED.
• THE SPEEDS ON THE OUTSIDE DISC ARGREE WITH THE SPEEDS SHOWING THROUGH THE WINDOWS, MACH NO. WITH MACH NO. OR CAS WITH CAS.
FUEL DISC TAB
DISTANCE DISC (ONE OF THE TWO PLASTIC DATA DISCS)
SPEED (M N) LINES INCREASE COUNTERCLOCKWISE
TIME DISC (ONE OF THE THREE STANDARD METAL DISCS)
TAS TRIANGLE
THIS WINDOW IS FOR THE CONFIGURATION WITH THE BLACK BACKGROUND AND WHITE PRINTING
READ NAUTICAL MILES PER 1000 POUNDS OF FUEL OPPOSITE THIS LITTLE ARROW.
INDEX LINE (LINE IT UP WITH THE DESIRED CRUISE MACH NUMBER.)
REMEMBER THESE SYMBOLS WHEN SETTING A SPEED IN THE WINDOWS.
ALTITUDE WINDOWS (MATCH THE ALTITUDES SHOWN ON THE DISTANCE DISC CURSOR)
DISTANCE DISC CURSOR
FUEL DISC (ONE OF THE THREE STANDARD METAL DISCS)
CLIMB AND DESCENT
STANDARD DAY
MAX RANGE
FLIGHT COMPUTER
TYPE MB-8

Figure A—6 A

on the third (metal) disc. Mention will be made of this cursor in the discussion of the time disc. The third disc of the front face of the computer is referred to as the time disc. On this disc is printed a series of concentric scales, of which the inner scale is labeled HOURS-MINUTES. The succeeding scales are speed scales (Mach number or calibrated airspeed) for selected altitudes, i.e., altitudes corresponding to the altitudes listed on the fuel disc.

Note

> Make sure when the computer is assembled that this disc displays speed in the same terms as the speed appearing in the window of the fuel disc: either Mach number or IAS.

Notice the large black triangle on this disc, labeled TAS-KNOTS, with the apex at the 60 minute mark of the time scale. For standard atmosphere conditions, the *true* airspeed in knots may be obtained from the distance disc opposite this TAS triangle. That is, when the distance disc cursor is aligned with a speed (Mach number) on the time disc, at some altitude, the corresponding true airspeed is indicated by the TAS triangle (using the distance scale as a speed scale). This conversion feature is useful in making corrections for wind.

WIND CORRECTIONS.

The front face of the computer can be adjusted for wind in the following manner:

a. Do not change the relationship of the fuel disc and the time disc. Pinch the fuel disc tab against the outer edge of the time disc. This will still permit rotation of the distance disc.

b. Rotate the distance disc until the ground speed (TAS + wind) on the "AIR NAUTICAL MILES" scale of the distance disc is aligned with the TAS triangle. The "AIR NAUTICAL MILES" scale then becomes ground nautical miles, and the fuel required to travel any ground distance is obtained from the fuel disc, while the time is read from the time disc.

WARNING

> The Mach numbers now appearing in the window of the fuel disc and on the time disc under the index cursor no longer apply. *The original Mach number must be maintained in flight.*

c. To determine the winds while in flight, a system of check points can be utilized. Rotate the distance disc until the distance between check points is aligned with the elapsed time on the "MINUTES" scale of the time disc. The ground speed is then read on the distance disc opposite the TAS triangle. The fuel required to travel any selected ground distance is obtained from the time disc. The Mach numbers appearing in the window of the fuel disc no longer apply and should be ignored. *The original Mach number must be maintained.*

FUEL FLOW CORRECTIONS.

Variations in the fuel consumption characteristics due to battle damage, small changes in configurations, differences in engines, formation flight, etc., may be accounted for in the following manner:

a. Determine the fuel flow from the flowmeter.

b. Do not change the relationship of the distance disc and the time disc (set from ground speed).

c. Rotate the fuel disc until the rate of fuel flow, read from the flowmeter, is aligned with the TAS triangle.

d. Determine distance and time for selected fuel quantities from the respective discs.

SUMMARY.

Variations in rate of fuel flow of an average magnitude of +5 per cent of that indicated on the computer can be expected on the initial portion of the flight and −5 per cent on the final portion when flying at the Mach number recommended for maximum range. These variations show up plainly when the maximum range shown on the tabulator side of the computer is compared with the range obtained on the fuel distance side. The tabulator side will indicate a greater distance because this data considers the change in airplane gross weight as fuel is consumed, whereas the indicated specific range on the fuel distance side of the computer is an average value and results in a slightly conservative distance. The true airspeeds presented on the computer are based on standard atmospheric conditions. An allowance for the difference in this true airspeed and the true airspeed for the actual atmospheric condition can be made by the wind correction method described previously. The rate of fuel flow is also based on standard atmospheric conditions. However, the difference in fuel flow need not be corrected, since the air range calculated on the flight computer is normally independent of air temperature when the Mach number is correctly indexed.

TAKE-OFF AND LANDING CROSSWIND CHART.

The take-off and landing crosswind chart figure A-6B determines if take-off and landings, at predicted speed for gross weight, are recommended for the direction and velocity of the existing crosswind.

TAKEOFF AND LANDING CROSSWIND CHART

LANDING CONFIGURATION

F-84F-25RE AND LATER AIRCRAFT

ENGINE J65

DATA BASIS: FLIGHT TEST

FUEL GRADE: JP-4

DATA DATE: 1 JULY 1959

FUEL DENSITY: 6.5 LB/GAL

HOW TO USE CHART

GIVEN:

Crosswind 33 knots at 63 degrees to runway. Determine if takeoff speed of 135 knots IAS is recommended for clean airplane.

SOLUTION:

a. Determine coordinates of wind velocity and direction (A).

b. Proceed vertically to intersect grid representing 135 knots IAS (B).

c. This point is in the recommended area.

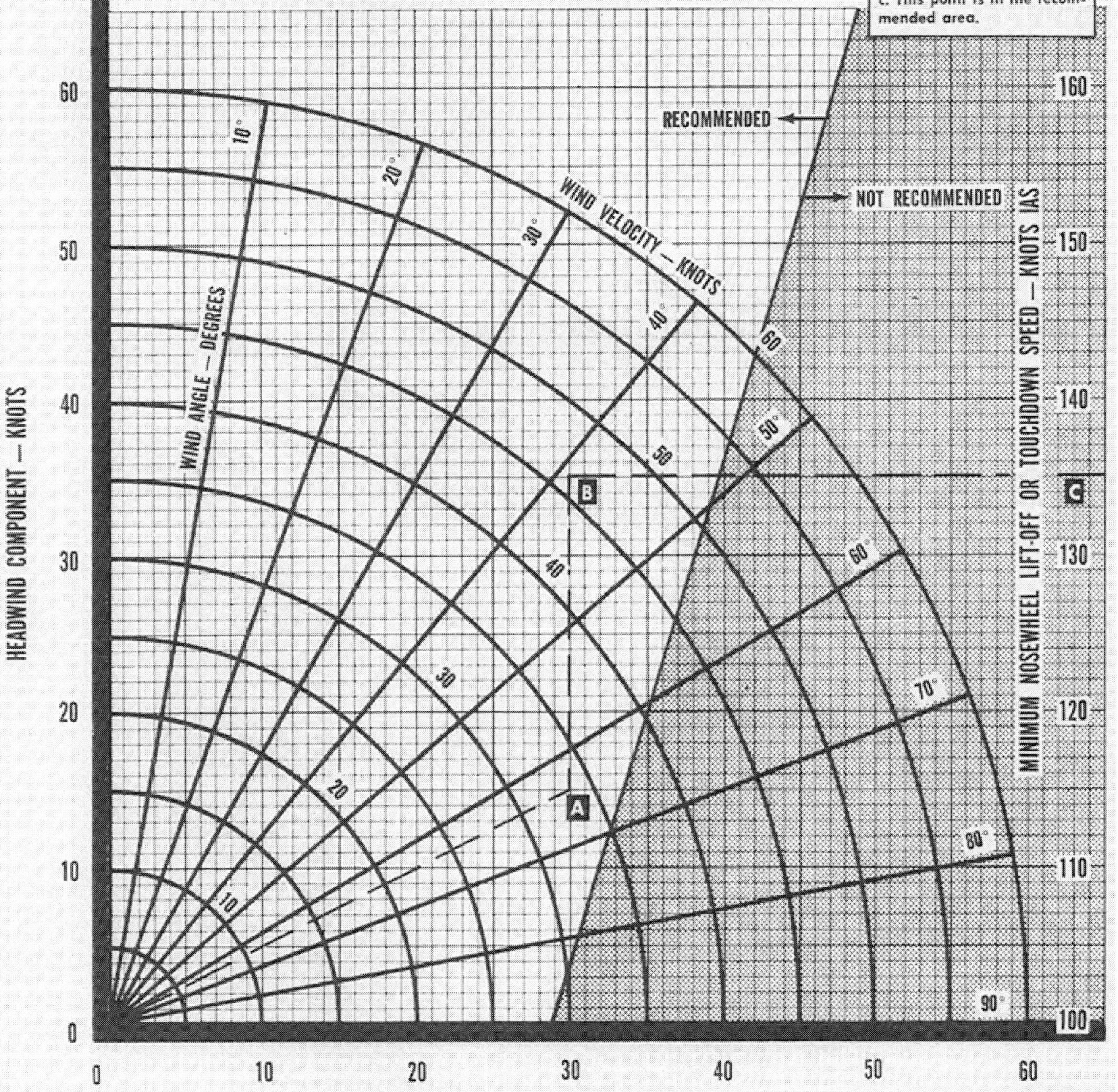

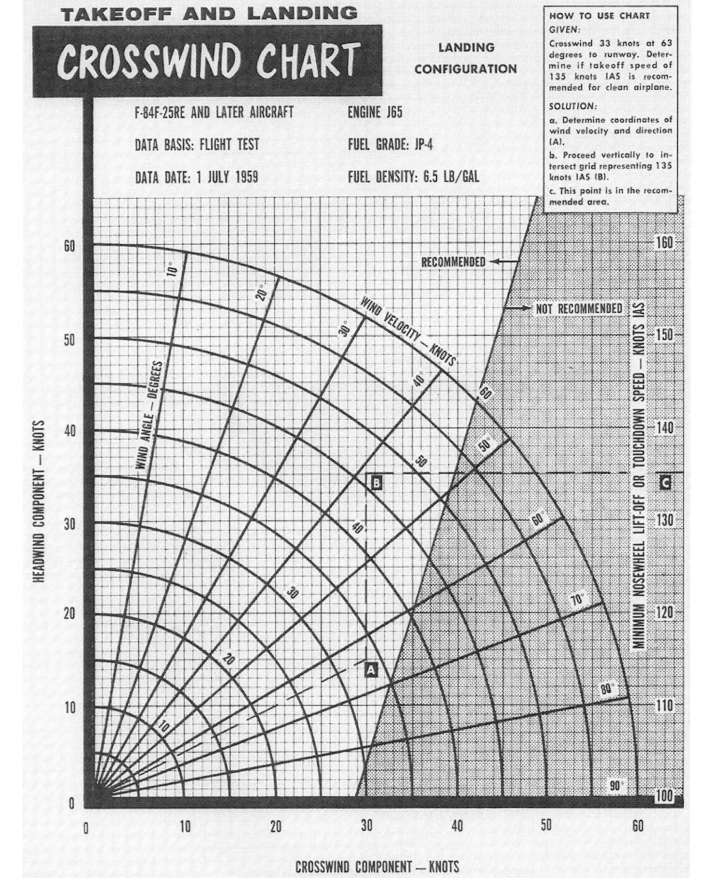

Figure A—6B (Sheet 1 of 2)

This page intentionally left blank.

TAKEOFF AND LANDING CROSSWIND CHART

LANDING CONFIGURATION WITH TWO WING TANKS INSTALLED

F-84F-25RE AND LATER AIRCRAFT

DATA BASIS: FLIGHT TEST

DATA DATE: 1 JULY 1959

ENGINE J65

FUEL GRADE: JP-4

FUEL DENSITY: 6.5 LB/GAL

HOW TO USE CHART

GIVEN:

Crosswind 33 knots at 63 degrees to runway. Determine minimum takeoff speed for given conditions for airplane with two tanks.

SOLUTION:

a. Determine coordinates of wind velocity and direction (A).

b. Proceed vertically to intersect point between recommended and not-recommended areas (B).

c. Minimum speed 145.5 knots IAS

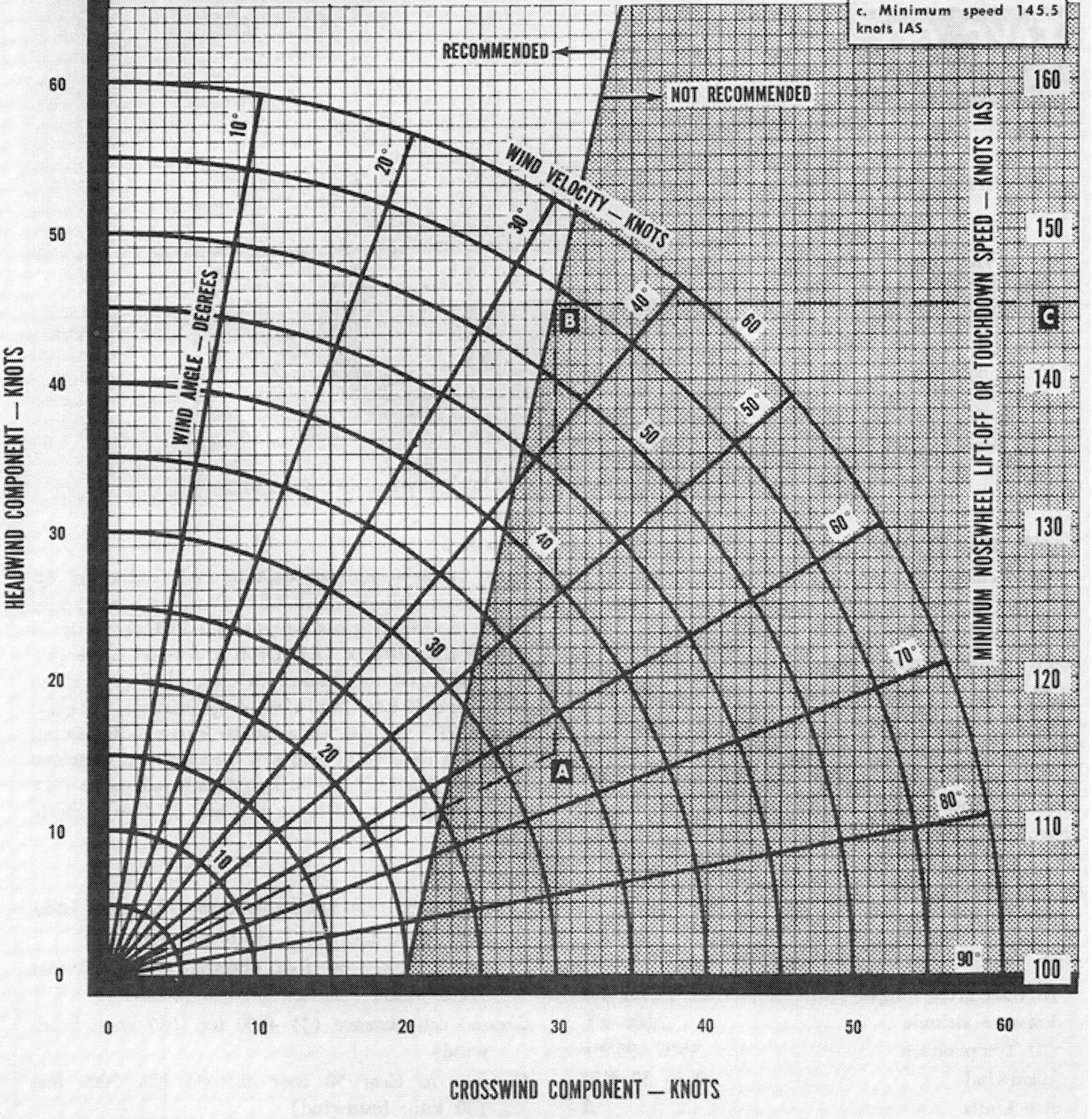

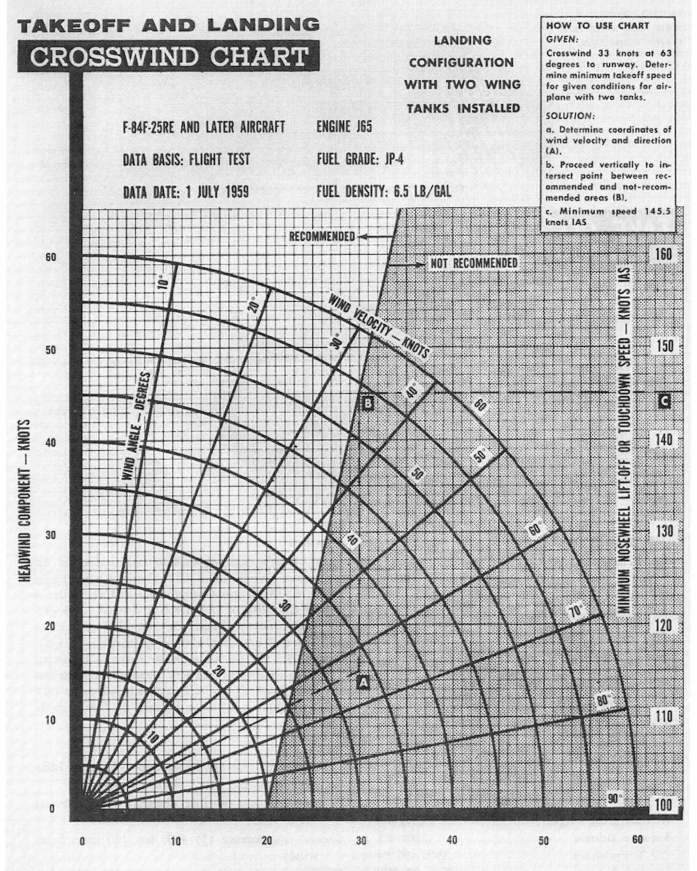

Figure A—6 B (Sheet 2 of 2)

MODEL: F-84F
ENG: J65-W-3, B-3
HARD SURFACE RUNWAY

TAKE-OFF CHART

NO ASSIST

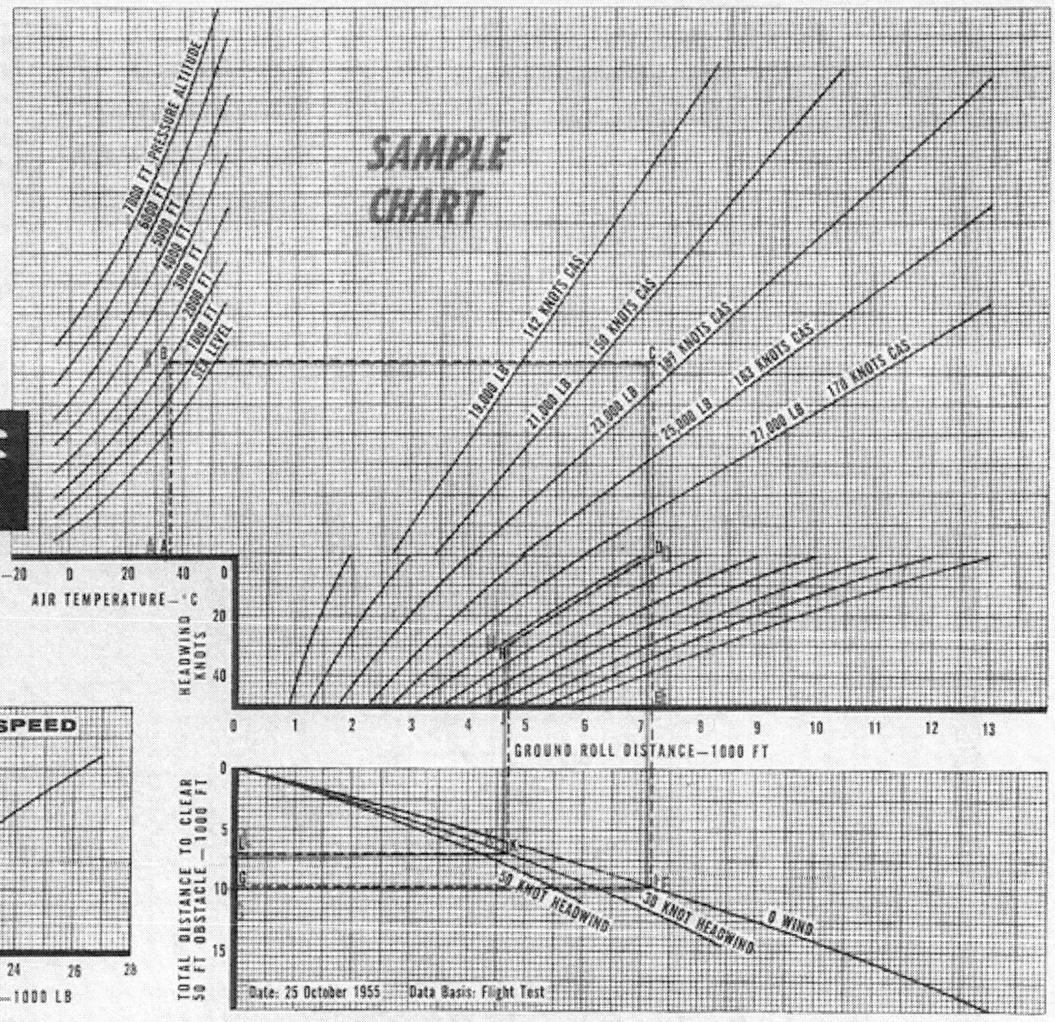

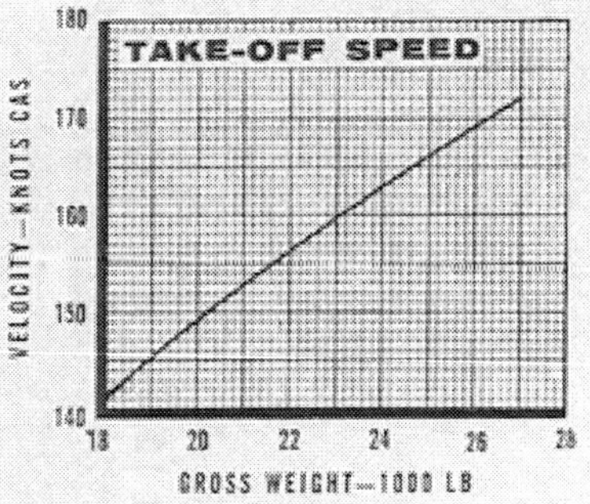

TAKE-OFF CHARTS.

Take-off charts (figures A–7, A–8, A–9, A–73, A–74 and A–75) are provided for no assist, two ato and four ato assist configurations. The charts take into account ambient temperature, pressure altitude and gross weight. Take-off speed, ground roll and distance to clear a 50 foot obstacle with or without a headwind are obtained from the charts. Ato cut-in speeds are plotted on separate charts which are to be used in conjunction with the corresponding take-off chart.

EXAMPLE: NO ASSIST.

Determine take-off speed, ground roll and distance to clear 50 foot obstacle for the following configuration:

Aircraft gross weight	22,300 LB
Pressure altitude	2,000 FT
Air Temperature	35°C (95°F)
Headwind	0 to 30 KN
Ato Units	0

Procedure:

Enter figure at air temperature (A) and move up to pressure altitude curve (B). Continue horizontally across to aircraft gross weight (C) and vertically to ground roll distance (E). Drop to 0 wind curve (F) and horizontal to distance to clear 50 foot obstacle (G). For a 30 knot headwind, drop from aircraft gross weight (C) to (D) then follow curve to headwind (H) then drop to ground roll distance (J). Continue from (J) vertically to 30 knot headwind curve (K) and horizontal to distance to clear 50 foot obstacle (L).

Conclusion:

Take-off speed (C) 154 knots CAS (159 knots IAS)

Ground roll distance (E) 7250 feet (zero wind)

Distance to clear 50 foot obstacle (G) 9650 feet (zero wind)

Ground roll distance (J) 4700 feet (30 knots headwind)

Distance to clear 50 foot obstacle (L) 7000 feet (30 knot headwind)

MILITARY POWER (100% RPM) CLIMB

ALT-FT	CAS-KNOTS	MACH NO.
45,000	220	.81
40,000	245	.81
35,000	275	.81
30,000	300	.79
20,000	340	.74
10,000	380	.68
SEA LEVEL	395	.60

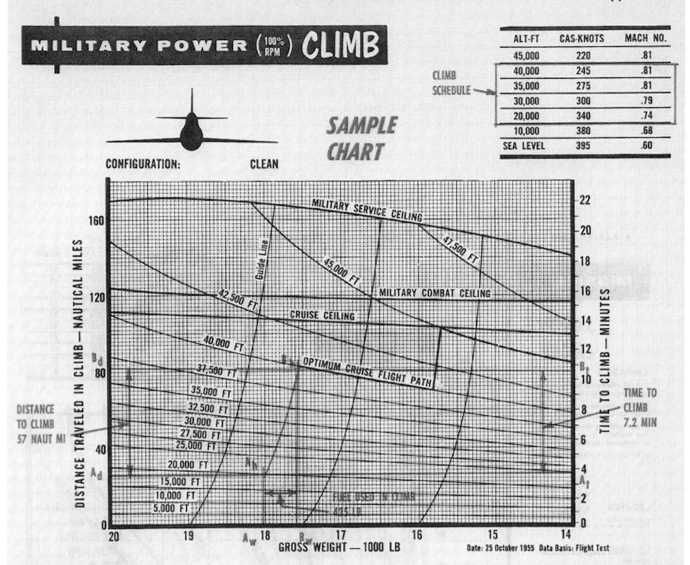

CLIMB SCHEDULE

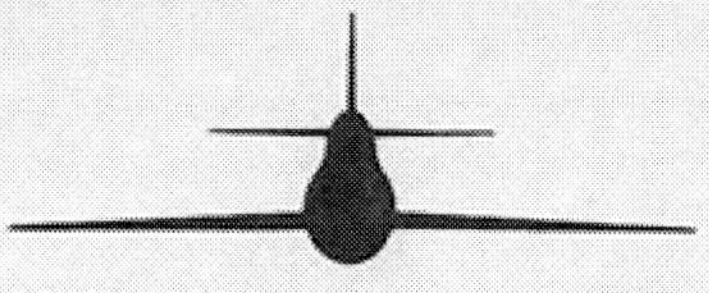

SAMPLE CHART

CONFIGURATION: CLEAN

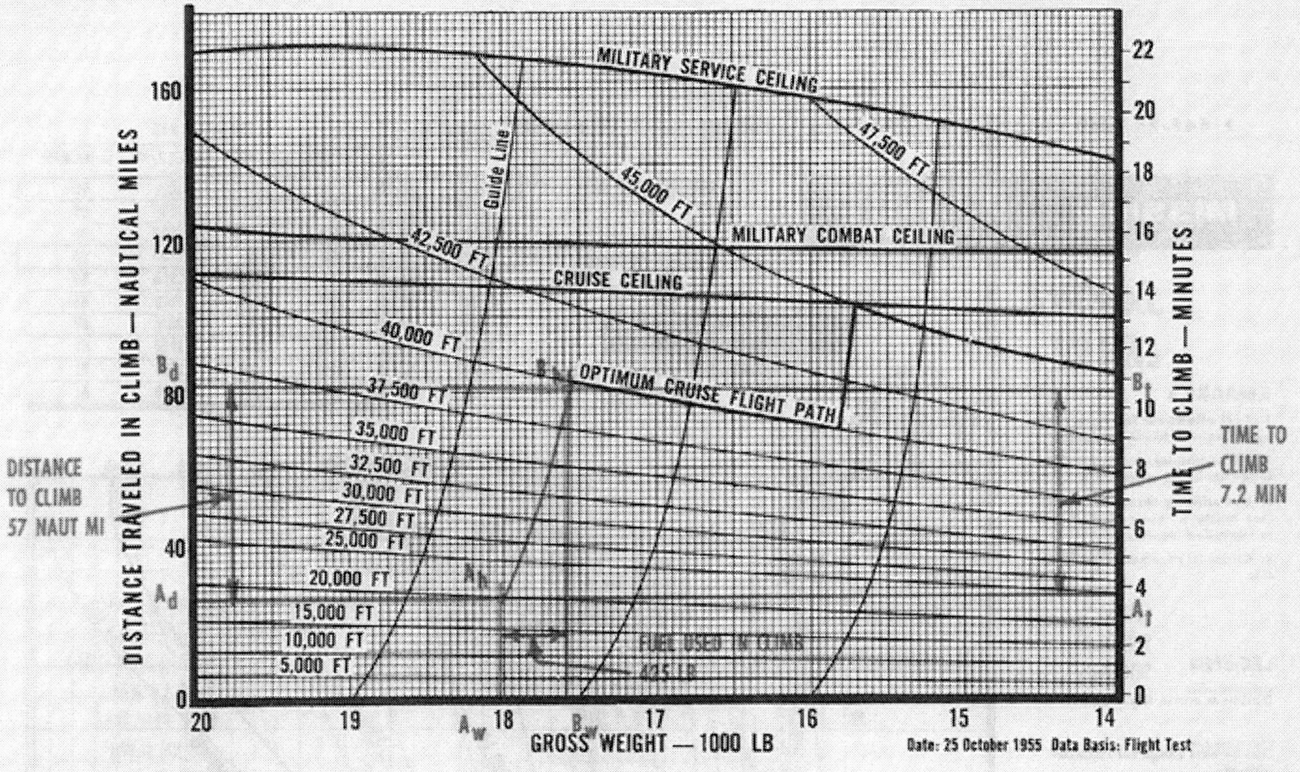

DESCRIPTION.

Climb charts for Military, 98% Military, and Normal Thrust operation based on a recommended climb speed schedule, are shown for each configuration. Time and distance are plotted against gross weight with guide lines to show the reduction in gross weight during climb due to the fuel used. Service ceiling (100 FPM), Combat Ceiling (500 FPM), Cruise Ceiling (300 FPM at Normal Thrust) and optimum cruise flight path are superimposed on the graph.

USE.

To obtain the climb data desired, enter the proper climb chart at the gross weight and altitude at start of climb. Note the time and distance at this point. From this initial altitude point, trace a curve parallel to the guide lines until it intersects the desired altitude at end of climb. Note the time, distance, and gross weight at this intersection. The difference between the initial and final time is the time required to climb. The difference between initial and final values for distance and for gross weight gives, respectively, the distance traveled and fuel used to climb. Since time and distance are zero at sea level, the time required and distance traveled may be read directly for climbs starting at sea level. Fuel used, however, must still be determined by the difference in gross weights. The example shows the fuel used, distance traveled, and time to climb from 15,000 feet to 40,000 feet, using military thrust, clean aircraft, with a gross weight of 18,000 pounds at start of climb.

EXAMPLE:

A_w is initial gross weight (18,000 LB).
A_h is initial altitude (20,000 FT).
A_d is initial distance (26 nautical miles).
A_t is initial time (3.8 MIN).
B_h is final altitude (40,000 FT).
B_w is final gross weight (17,575 LB).
B_d is final distance (83 nautical miles)
B_t is final time (11.0 MIN).
A_w-B_w is fuel used (425 LB).
B_d-A_d is distance traveled (57 nautical miles).
B_t-A_t is time to climb (7.2 MIN).

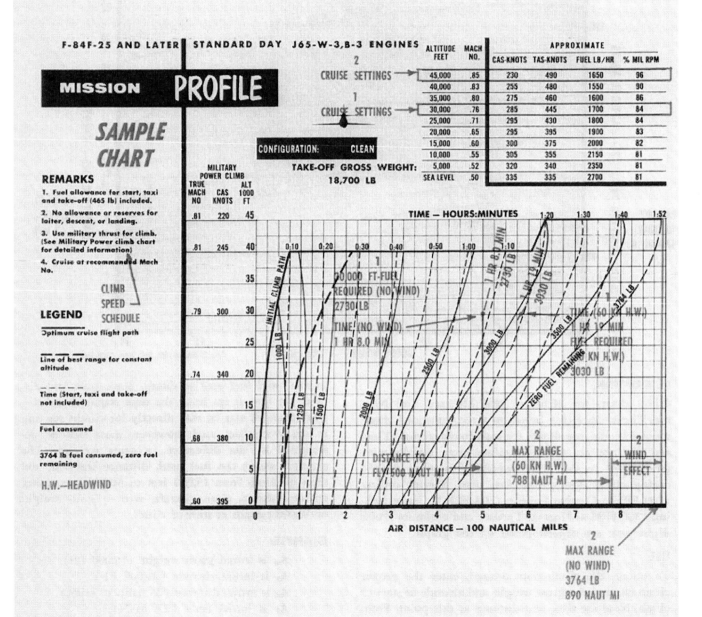

F-84F-25 AND LATER STANDARD DAY J65-W-3,B-3 ENGINES

MISSION PROFILE

SAMPLE CHART

CONFIGURATION: CLEAN

TAKE-OFF GROSS WEIGHT: 18,700 LB

2 CRUISE SETTINGS → 45,000

1 CRUISE SETTINGS → 30,000

ALTITUDE FEET	MACH NO.	APPROXIMATE CAS-KNOTS	TAS-KNOTS	FUEL LB/HR	% MIL RPM
45,000	.85	230	490	1650	96
40,000	.83	255	480	1550	90
35,000	.80	275	460	1600	86
30,000	.76	285	445	1700	84
25,000	.71	295	430	1800	84
20,000	.65	295	395	1900	83
15,000	.60	300	375	2000	82
10,000	.55	305	355	2150	81
5,000	.52	320	340	2350	81
SEA LEVEL	.50	335	335	2700	81

REMARKS

1. Fuel allowance for start, taxi and take-off (465 lb) included.
2. No allowance or reserves for loiter, descent, or landing.
3. Use military thrust for climb. (See Military Power climb chart for detailed information)
4. Cruise at recommended Mach No.

CLIMB SPEED SCHEDULE

LEGEND

Optimum cruise flight path

Line of best range for constant altitude

Time (Start, taxi and take-off not included)

Fuel consumed

3764 lb fuel consumed, zero fuel remaining

H.W.—HEADWIND

MILITARY POWER CLIMB

TRUE MACH NO	CAS KNOTS	ALT 1000 FT
.81	220	45
.81	245	40
		35
.79	300	30
		25
.74	340	20
		15
.68	380	10
		5
.60	395	0

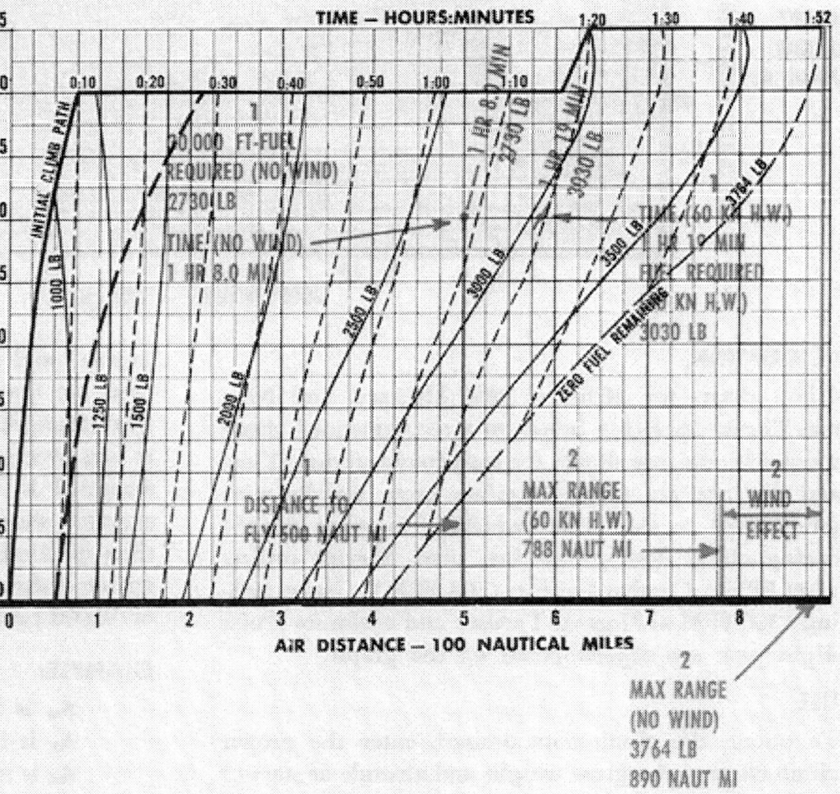

DESCRIPTION.

These charts give the relationship of time, fuel, distance, and altitude to maximum range for no-wind conditions. This relationship is based on a mission sequence of take-off, military thrust climb, and maximum range cruise. The fuel curves include a 465 pound allowance for start, taxi, and take-off, and the fuel used in climb to each altitude, as well as the fuel required for maximum range cruise. The time lines include the time required to climb to cruise altitude but do not include the time to start, taxi, or take-off.

The line labeled "Initial Climb Path" shows the distance traveled during the Military Thrust climb from sea level to cruising altitude, using the climb speed schedule tabulated at the left of the chart.

As an aid to preflight planning, a line of best range for constant-altitude flight appears on the chart. This curve is not a flight path, but a plot of best cruise altitude against distance. For distances greater than those covered by the curve, use step climb procedure for maximum range.

A cruise table gives recommended Mach numbers and approximate operating conditions for cruise at constant altitude. (Cruise-at-constant-altitude data is given for each 5000 feet.)

USE.

The charts may be entered with one or more of the four range factors: time, fuel, distance, and altitude. By entering the chart with the known factors, the others may readily be determined. This is for a no-wind condition. To determine wind effect upon time, fuel, and distance, compute the average true airspeed (TAS)—distance ÷ time, no wind—and apply wind to TAS to obtain ground speed (GS). Then compute the time with wind (distance ÷ GS). Re-enter the profile at the cruising altitude and the computed time with wind to determine the fuel required with wind.

Sample Problem 1.

Using the example shown, find the fuel required, time, necessary speed, and power setting to cruise 500 nautical miles at 30,000 feet with a head wind of 60 knots in the clean configuration.

a. Enter at 500 nautical miles and 30,000 feet to obtain fuel required (no wind) 2730 LB
b. Time (no wind) 1 HR 8.0 MIN (1.13 HR)
c. Average TAS (500 ÷ 1.13) 442 knots
d. Apply wind to obtain G S (442-60) 382 knots
e. Calculate time with 60-knot wind (500 ÷ 382) 1.31 HR (1 HR 19 MIN)
f. Re-enter at cruise altitude at the time with wind. Fuel required with wind. 3030 LB
g. Determine cruise speed from table 0.76 Mach No.
h. Determine cruise power setting from table 84% RPM

Sample Problem 2.

Determine the maximum distance flyable, using clean aircraft with 3764 pounds of fuel and a 60-knot head wind.

a. Enter at 3764 pounds of fuel and obtain maximum air distance at step climb (no wind) 890 (nautical miles)
b. Time (no wind) 1 HR 52 MIN (1.87 HR)
c. Calculate average TAS (890 ÷ 1.87) 476 knots
d. Apply wind to obtain GS (476-60) 416 knots
e. Calculate distance with wind (1.87 × 416) 778 nautical miles
f. Determine step-climb speed from table 0.85 Mach No.
g. Determine cruise power setting from table 96% RPM

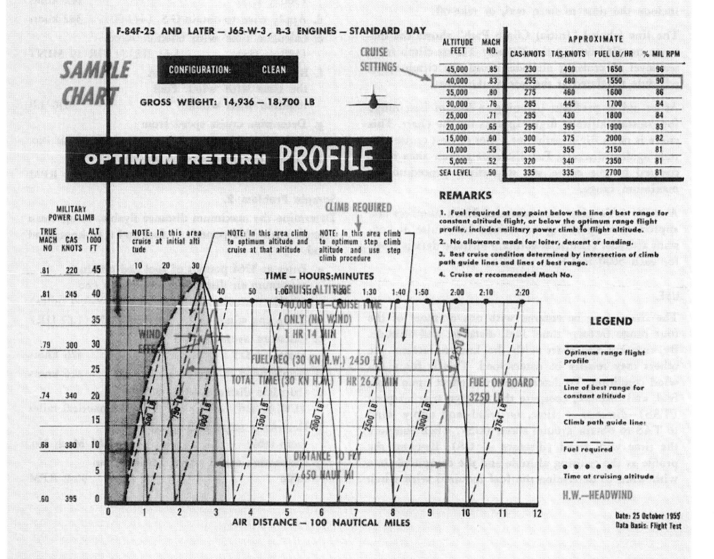

F-84F-25 AND LATER – J65-W-3, B-3 ENGINES – STANDARD DAY

SAMPLE CHART

CONFIGURATION: CLEAN

GROSS WEIGHT: 14,936 – 18,700 LB

OPTIMUM RETURN PROFILE

CRUISE SETTINGS

ALTITUDE FEET	MACH NO.	APPROXIMATE CAS-KNOTS	APPROXIMATE TAS-KNOTS	APPROXIMATE FUEL LB/HR	APPROXIMATE % MIL RPM
45,000	.85	230	490	1650	96
40,000	.83	255	480	1550	90
35,000	.80	275	460	1600	86
30,000	.76	285	445	1700	84
25,000	.71	295	430	1800	84
20,000	.65	295	395	1900	83
15,000	.60	300	375	2000	82
10,000	.55	305	355	2150	81
5,000	.52	320	340	2350	81
SEA LEVEL	.50	335	335	2700	81

REMARKS

1. Fuel required at any point below the line of best range for constant altitude flight, or below the optimum range flight profile, includes military power climb to flight altitude.
2. No allowance made for loiter, descent or landing.
3. Best cruise condition determined by intersection of climb path guide lines and lines of best range.
4. Cruise at recommended Mach No.

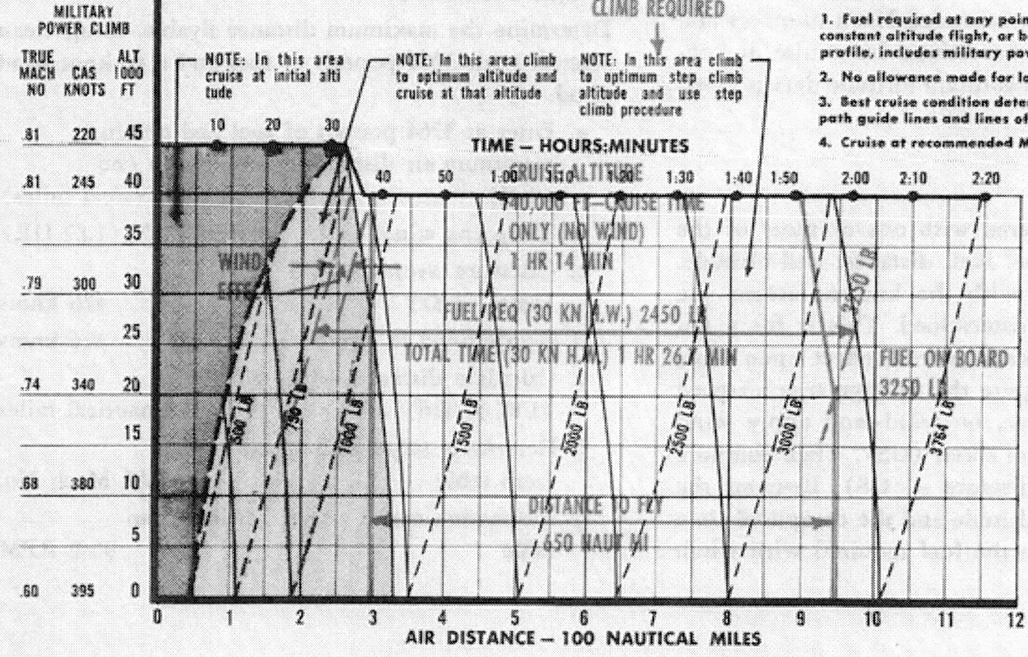

LEGEND

Optimum range flight profile

Line of best range for constant altitude

Climb path guide line

Fuel required

Time at cruising altitude

H.W.—HEADWIND

Date: 25 October 1955
Data Basis: Flight Test

These profiles show the minimum fuel required for maximum distance (no wind) based on an optimum flight path from any starting point within the range of the aircraft configuration. The flight path required is indicated by the different shaded areas and the notes relative to them. The fuel curves are based on a military thrust climb to, and recommended cruise at, the optimum altitude. The military thrust climb speed schedule and recommended cruise settings are tabulated on each chart. No reserve for descent and landing has been included. The time shown at the optimum altitude is cruise time only; it does not include the time required for the climb to optimum altitude or any allowing for descent, loiter, and landing.

USE.

The chart may be entered at the initial altitude with either the fuel on board (to determine the distance available) or with the distance to be flown (to determine the fuel required). The shaded area in which the initial point falls establishes the cruising procedure to be used, as stated in the note relative to the area. The time required to fly the distance is the time at cruise altitude (obtain from profile), plus the time required to climb (obtained from graphical military thrust climb chart).

The effect of wind must be applied to obtain the actual fuel and time to fly the distance. A close approximation can be obtained by considering the head or tail wind for the time it requires to complete the flight (neglecting the difference in wind at the lower altitudes, since comparatively little time is spent during the climb phase).

From example shown, determine the fuel and time required to return to a base 650 nautical miles away. The aircraft is at 20,000 feet with 3250 pounds of fuel on board in the clean configuration (gross weight 18,186 LB). A 30 knot head wind is assumed.

a. Enter profile at 3250 pounds of fuel at 20,000 feet to establish a starting point which is 950 nautical miles from zero fuel (no wind). End point 650 nautical miles away is 300 nautical miles. Fuel required no wind (3250 LB – 900 LB) = 2350 LB.

b. In this area note that a climb is required and a step-climb procedure is followed.

c. By following the climb guide lines, the Initial Cruise altitude is 40,000 feet.

d. Cruise time (no wind) 1 HR 53 MIN to zero fuel less cruise time at 300 nautical miles or 39 MIN = 1 HR 14 MIN.

e. From the military power climb chart for the clean configuration – time to climb = 7.3 MIN.

f. Total time (no wind) (d + e) 1 HR 21.3 MIN (1.355 HR).

g. Average Cruise Speed (distance ÷ total time) 480 knots.

h. Ground Speed (TAS–headwind) 450 knots.

i. Total time (with wind) (distance ÷ GS) 1 HR 26.7 MIN.

j. Cruise time (with wind) (i – e) 1 HR 19.4 MIN.

k. Enter profile at initial cruise time–j, 33.6 MIN, and read fuel remaining at end of cruise with wind: 800 LB.

l. Fuel used with wind in cruise (fuel on board–fuel remaining) 2450 LB.

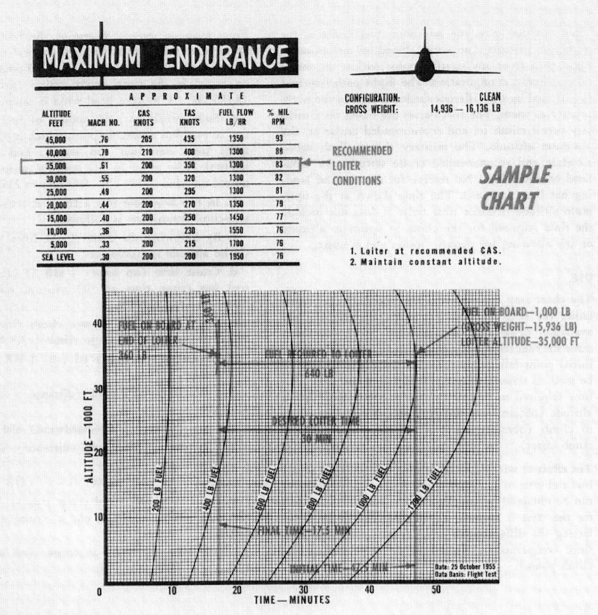

MAXIMUM ENDURANCE

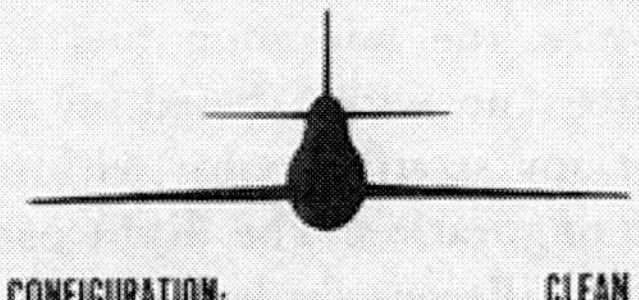

CONFIGURATION: CLEAN
GROSS WEIGHT: 14,936 — 16,136 LB

ALTITUDE FEET	MACH NO. (APPROXIMATE)	CAS KNOTS	TAS KNOTS	FUEL FLOW LB/HR	% MIL RPM
45,000	.76	205	435	1350	93
40,000	.70	210	400	1300	88
35,000	.61	200	350	1300	83
30,000	.55	200	320	1300	82
25,000	.49	200	295	1300	81
20,000	.44	200	270	1350	79
15,000	.40	200	250	1450	77
10,000	.36	200	230	1550	76
5,000	.33	200	215	1700	76
SEA LEVEL	.30	200	200	1950	76

RECOMMENDED LOITER CONDITIONS

SAMPLE CHART

1. Loiter at recommended CAS.
2. Maintain constant altitude.

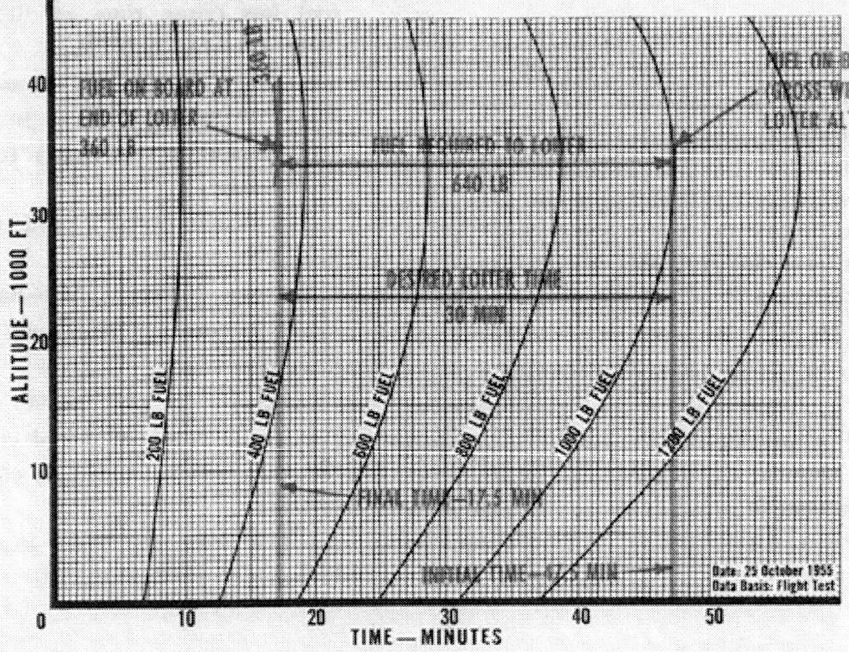

DESCRIPTION.

These profiles show the maximum time available for the fuel on board when loitering at a constant altitude. The recommended calibrated airspeed (CAS) and the approximate operating conditions are tabulated on each chart for the average gross weight.

USE.

To determine the time available for a given amount of fuel: Enter the chart at the amount of fuel on board at the start of loiter and the flight altitude; note the initial time. Re-enter the chart at the amount of fuel on board at the end of the endurance flight (initial fuel on board less fuel to be used) and read the final time. The difference between the initial and final time is the time available to loiter at constant altitude.

To obtain the fuel required to loiter a given time: Enter the chart at the amount of fuel on board at the start of loiter and flight altitude; note the initial time. Re-enter the chart at the time at the end of loiter (initial time less time to loiter) and read final fuel on board. The difference between the initial and final fuel on board is the fuel required to loiter.

From the example shown, determine the fuel required to loiter at 35,000 feet with no external load for 30 minutes. The fuel on board at start of loiter is 1000 pounds (gross weight = 15,936 pounds).

a. Initial time at 1000 pounds and 35,000 feet	47.5 MIN
b. Final time (47.5 – 30.0)	17.5 MIN
c. Fuel on board at end of loiter (17.5 MIN at 35,000 feet)	360 pounds
d. Fuel required to loiter (1000 – 360)	640 pounds
e. Recommended loiter CAS	200 knots

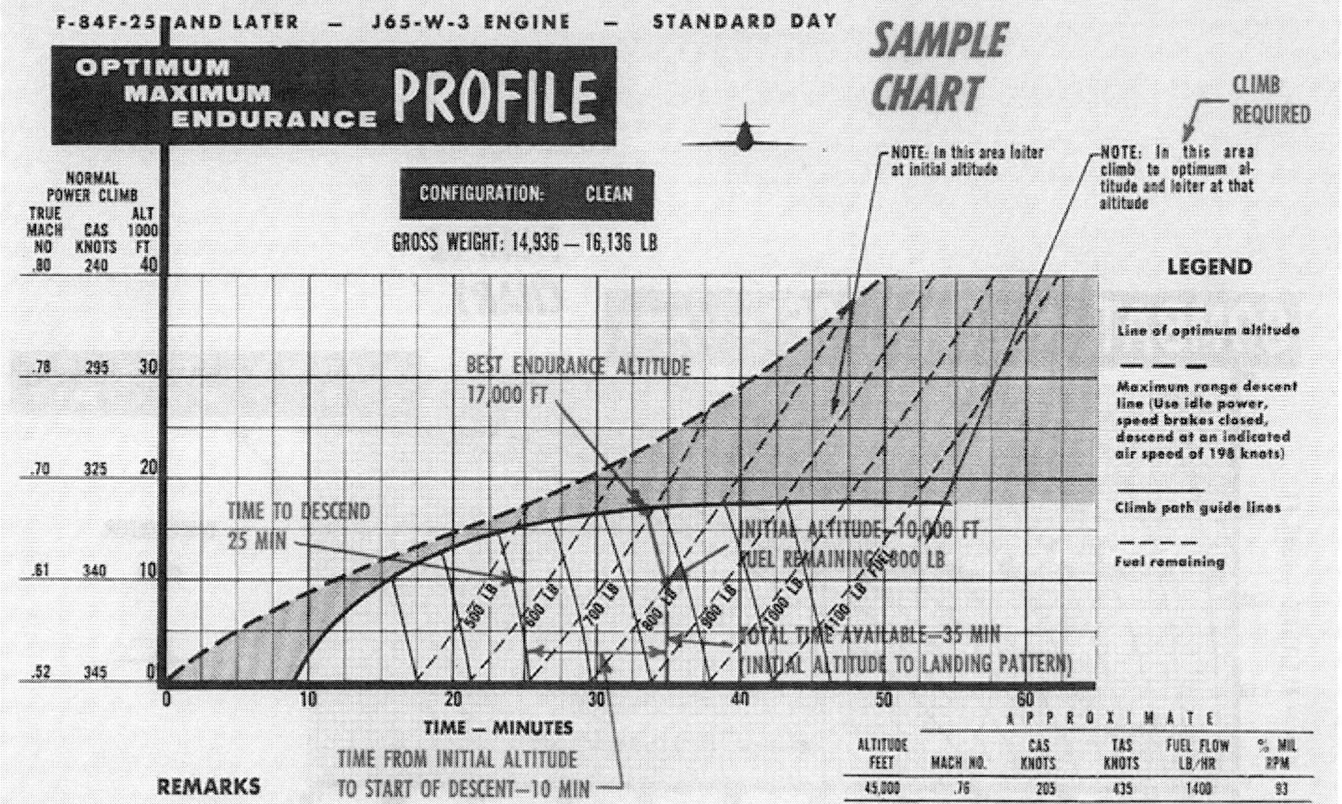

ALTITUDE FEET	MACH NO. (APPROXIMATE)	CAS KNOTS	TAS KNOTS	FUEL FLOW LB/HR	% MIL RPM
45,000	.76	205	435	1400	93
40,000	.70	210	400	1300	89
35,000	.61	200	350	1300	83
30,000	.55	200	320	1300	82
25,000	.49	200	290	1300	81
20,000	.44	200	270	1350	79
15,000	.40	200	250	1450	77
10,000	.36	200	230	1550	76
5,000	.33	200	215	1700	76
SEA LEVEL	.30	200	200	1950	76

RECOMMENDED LOITER CONDITIONS (20,000 and 15,000 FEET)

REMARKS

1. Maximum range descent. (Use idle power, speed brakes closed) Descend at 200 knots CAS.
2. Loiter at recommended CAS
3. Use normal power for climb. (See Normal Power climb chart for detailed information.) — CLIMB SCHEDULE
4. No allowance or reserve made for landing.

Date: 25 October 1955 Data Basis: Flight Test

DESCRIPTION.

These profiles give the maximum time in the air for the fuel remaining, based on an optimum flight path, from any starting altitude. The flight path required is indicated by the different shaded areas and the notes relative to them. Time and fuel lines shown are based on a normal thrust climb to best endurance altitude, loiter at the altitude, and a maximum-range descent to sea level (no reserve for landing). The loiter speed schedule is tabulated below the chart.

USE.

The chart may be entered at the initial altitude with either the fuel remaining (to determine the time available) or the time desired (to determine the fuel requirment). The shaded area in which the initial point falls establishes the flight path to be used, as stated in the note relative to the area.

From the example shown, determine the time available and necessary flight path to remain aloft with 800 pounds of fuel remaining at 10,000 feet in the clean configuration.

a. Enter profile at 10,000 feet and 800 pounds of fuel remaining to establish starting point. Total time available is 35 MIN.

b. In this area, note that a climb is required.

c. By following the climb guide lines, the best endurance altitude is 17,000 feet.

d. Descent time from 17,000 feet to sea level is 25 minutes.

e. Elapsed time from start of climb to start of descent is 10 minutes (35-25).

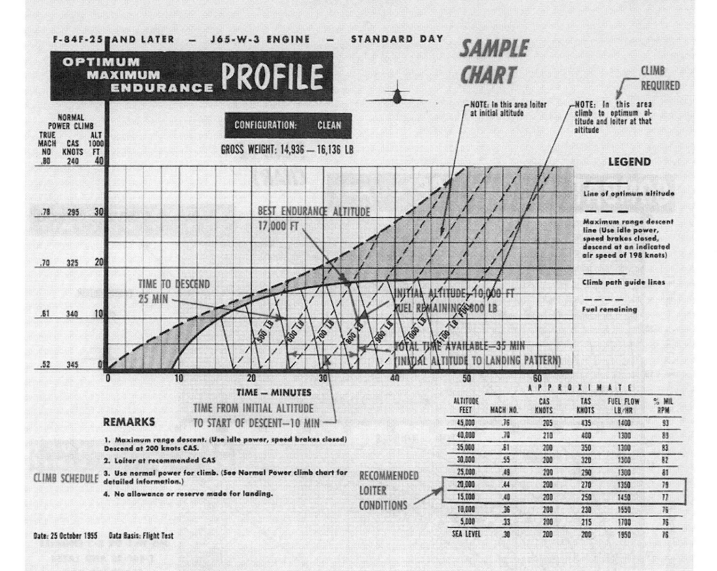

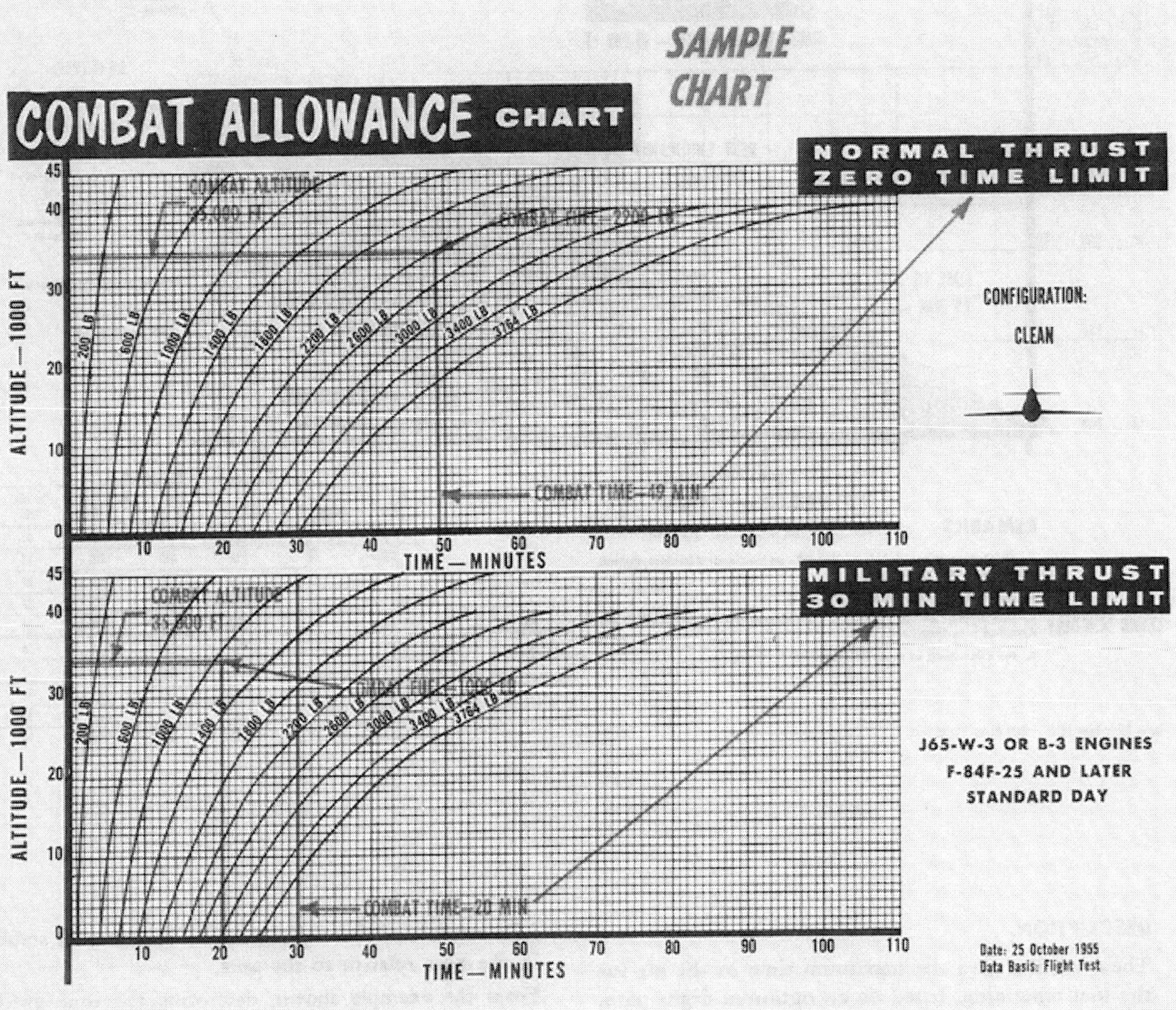

DESCRIPTION.

The Combat Allowance chart shows the relationship between time and fuel with changes in altitude at Military and Normal Thrust settings. Combat time or fuel may be determined from this chart for a given thrust setting. The time limitations for Military Thrust operation are shown. Normal thrust does not have a time limitation.

USE.

Enter the chart at the combat altitude and the fuel quantity to be used for combat to obtain the time available. Enter at the altitude and time available for combat to obtain the fuel required.

Using the example shown, obtain the time available for a combat fuel allowance of 2,200 pounds at Normal Thrust and 1,000 pounds at Military Thrust at 35,000 feet.

Military thrust — 20 min
Normal thrust — 49 min

DESCENT.

Two types of descents are shown for all configurations, recommended descent and maximum range descent, both at idle power with speed brakes closed. Distance, time, fuel used, and rates of descent are shown in figures A–58 thru A–63 and A–144 thru A–149.

LANDING DISTANCES.

Landing ground roll distances and total distances to clear a 50-foot obstacle are shown in figures A–65, A–66, A–151 and A–152 in the clean and externally loaded configurations. The distances are computed for a condition with flaps down and speed brakes closed. The following conditions are also considered: dry hard-surface runway; temperature, −20°C thru 60°C (4°F thru 140°F) pressure altitude, sea level through 6000 feet; gross weight, 15,000 through 23,000 pounds; and head winds, 0 through 50 knots. The recommended indicated airspeeds for the approach, over a 50-foot obstacle, and touchdown are listed on the graphs. Following is a sample of the landing data card. For the data card form refer to Section II Condensed Check List. Information on this card should be filled out prior to each flight. This data is obtained by referring to the Landing Distance Chart.

LANDING DATA CARD

Conditions	
Gross Weight	16,500 LB
Runway Length	9,000 FT
OAT	35°C (95°F)
Pressure Altitude	2,000 FT
Runway Gradient	0%
Wind	0 KN
Landing	
Landing Ground Roll (without drag chute)	4,000 FT
Landing Ground Roll (with drag chute)	2,300 FT
Approach Speed	170 KN

NAUTICAL MILES PER POUND OF FUEL GRAPH.

Cruise data (no wind) throughout the speed range from maximum endurance to Military Thrust are shown on the Nautical Miles per Pound of Fuel graphs. (See figures A–67 thru A–72 and A–153 thru A–158). Several weights for each configuration are given at altitudes of sea level, 15,000, 25,000, 30,000, 35,000, and 40,000 feet. Each graph includes specific range (nautical miles per pound), fuel flow, and RPM. Also included are curves of recommended cruise Mach number, maximum endurance, and Normal and Military Thrust. Specific range is plotted against Mach number.

Cruising range is calculated from the Nautical Miles per Pound of Fuel graphs on a fuel increment basis. The smaller the increment of fuel used in the calculation, the greater the accuracy of the range; therefore, if a high degree of accuracy is desired, several increments should be used.

To obtain the cruising range for an increment of fuel, use the following steps: (If several increments of fuel are used, repeat the steps shown for each increment. The sum of the individual ranges is the total cruising range.)

a. Select the proper graph for the aircraft configuration and altitude.

b. Determine the average weight of aircraft for the increment of fuel being considered.

c. Enter the graph at this average weight and the desired Mach number, or desired RPM, to obtain specific range (nautical miles per 1000 pounds of fuel).

d. The specific range multiplied by the amount of fuel (pounds ÷ 1000) equals cruising range.

e. Determine the approximate fuel flow and RPM at the Mach number and average weight. When there is a wind to be considered, multiply the specific range found in step c. by the range factor (ground speed divided by true airspeed) to obtain the specific range for wind. Proceed with steps d. and e. to complete the problem.

SUMMARY.

Check your flight plan during the actual flight to determine whatever deviations exist. These deviations may be applied to the reserve expected at the destination. The most important factors to consider are:

Fuel used during start, taxi, and take-off. (The profile allows 587 pounds for this phase.)

Wind effect.

Deviation from the recommended climb schedule.

Deviation from the recommended cruise settings.

Variation in engine performance.

Navigational errors, formation flight, and fuel actually aboard at take-off.

SAMPLE PROBLEM.

This sample problem combines the use of the charts and graphs in this section to plan a mission.

A combat mission is to be flown using two 230 gallon drop tanks which are to be dropped prior to combat. Prepare a flight plan based on the following data:

Distance to combat area .... 450 nautical miles
Winds .... 30 knot head wind to combat area
45 knot tailwind from combat area
Time required at Military
Thrust for combat .... 10 MIN
Total fuel on board .... 6754 LB

TAKE-OFF.

Obtain the take-off distance from figure A–7.

Altitude .... Sea level
Temperature .... 15°C (59°F)
Gross weight .... 22,204 LB
Wind .... None
Ground roll distance .... 4000 feet
Total distance to clear 50-foot obstacle .... 5500 feet
Take-off speed .... 156 knots IAS

INBOUND TO TARGET.

The inbound leg to the combat area may be determined directly from the Mission Profile chart for two 230 gallon drop tanks. The profile includes a 587 pound fuel allowance for start, taxi, and take-off, as well as fuel required for climb and cruise.

a. Distance .... 450 nautical miles
b. Cruise altitude .... 40,000 feet
c. Fuel required (no wind) from profile .... 3130 LB
d. Time (no wind) from profile .... 1 HR
e. Average TAS (a ÷ d) .... 450 knots
f. Ground speed (e – 30 knots) .... 420 knots
g. Time with wind (a ÷ f) .... 1 HR 4 MIN
h. Fuel required (with wind) from profile 3250 LB
i. Cruise speed .... 0.81 Mach NO.
j. Cruise power setting .... 95% RPM
k. Military Thrust climb speed schedule .... (See figure A–81)

COMBAT ALLOWANCE.

The tanks are dropped during cruise, prior to entering combat. From the Combat Allowance chart (figure A–55), obtain the fuel required for combat at 40,000 feet.

Combat–Military Thrust (10 minutes) .... 450 LB

Determine the weight and fuel remaining at the end of combat.

Take-off, climb, and cruise fuel .... 3250 LB
Combat fuel .... 450 LB
Total fuel used .... 3700 LB

Fuel remaining (6754 pounds–3700 pounds) .... 3054 LB
Two 230 gallon drop tanks and pylons .... 514 LB
Gross weight (22,204 pounds–3,700 pounds–514 pounds) .... 17,990 LB

RETURN.

Assume the return is started from 250 nautical miles from a base at an altitude of 25,000 feet. Fuel on board 3054 pounds. Use Optimum Return Profile for the clean configuration. (Figure A–37.)

a. Enter profile at 3054 LB and 25,000 FT to establish a starting point which is 920 nautical miles from zero fuel (no wind). End point 250 nautical miles away is 670 nautical miles. Fuel required (no wind) 3054 LB – 2090 LB = 964 LB.
b. Cruise Altitude .... 40,000 FT
c. Distance required to fly .... 250 nautical miles
d. Cruise time .... 27 MIN
e. Time to climb Military Thrust–Fig. A–13 Clean (25,000 FT to 40,000 FT) gross weight 17,990 LB .... 5 MIN
f. Total time (d + e) .... 32 MIN
g. Average TAS (c ÷ f) .... 469 KN
h. Ground Speed (g + tail wind) .... 514 KN
i. Total time (with wind) (c ÷ h) .... 29 MIN
j. Cruise time with wind (i – e) .... 24 MIN
k. Enter profile at initial cruise time–j and read fuel at end of cruise with wind 2150 LB
l. Fuel to cruise with wind (initial fuel on board–k) .... 904 LB
m. Cruise Speed .... 0.83 Mach NO.
n. Cruise power setting .... 94% RPM
o. Fuel remaining over base (3054 LB–904 LB) .... 2150 LB

DESCENT.

Obtain the fuel required to descend to a sea level base from 40,000 feet. (See figure A–58.)

Recommended descent fuel required .... 55 LB
Fuel reserve for landing (2150 pounds – 55 pounds) .... 2095 LB
Aircraft weight for landing .... 17,301 LB

LANDING.

Use Landing Distances chart (figure A–65) for clean configuration without drag chute.

Altitude .... Sea level
Temperature .... 15°C (59°F)
Gross weight .... 17,031 LB
Ground roll distance .... 4250 FT
Total distance over 50-foot obstacle .... 6900 FT
Approach speed .... 180 knots IAS
Over 50-foot obstacle speed .... 160 knots IAS
Touchdown speed .... 145 knots IAS

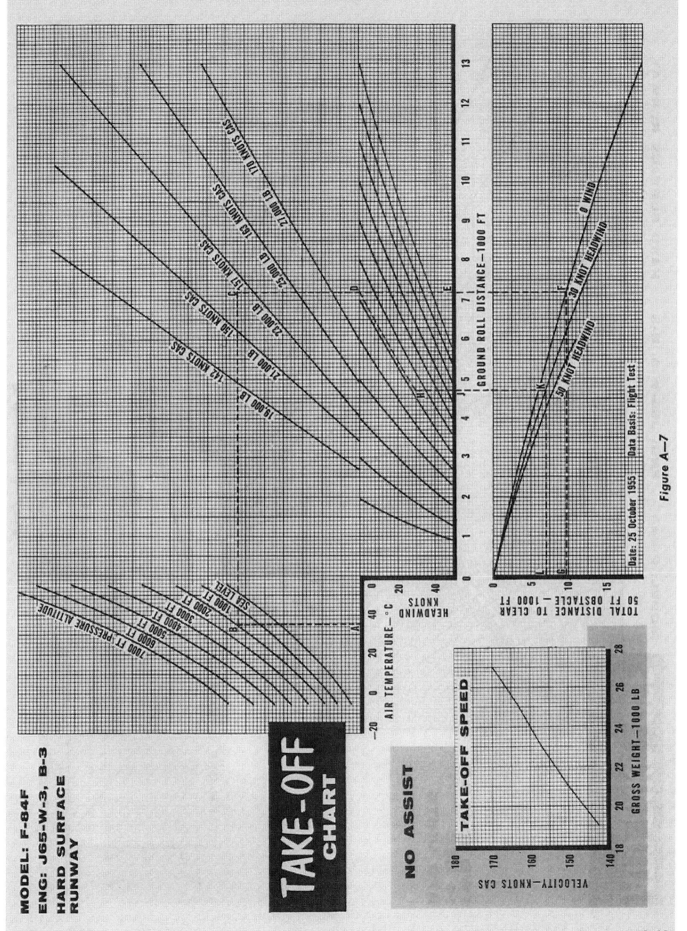

MODEL: F-84F
ENG: J65-W-3, B-3
HARD SURFACE RUNWAY

TAKE-OFF CHART

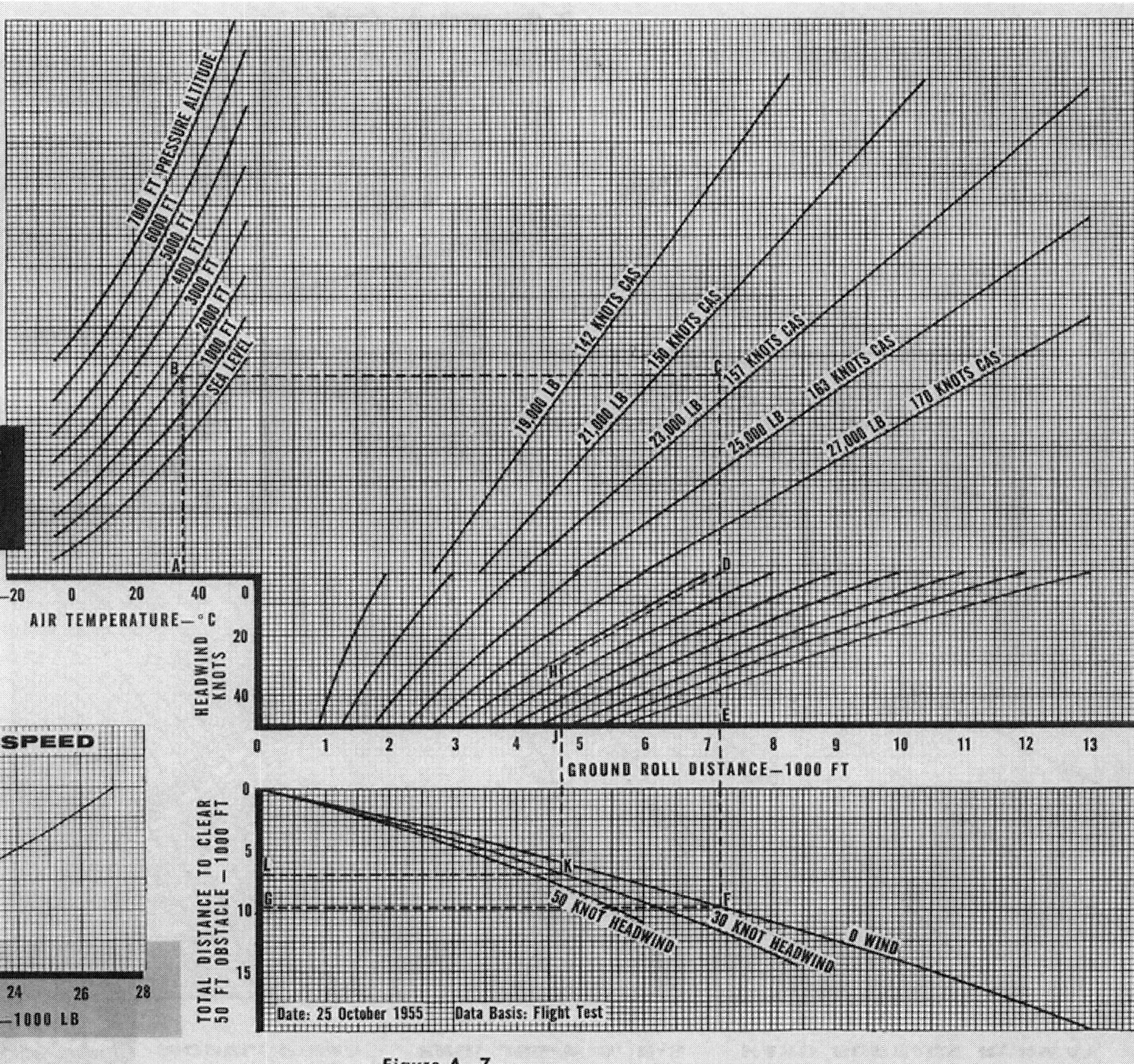

NO ASSIST

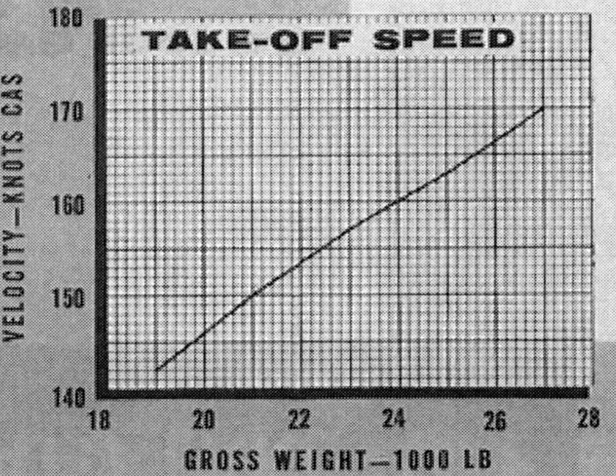

Figure A—7

ATO TAKE-OFF CHART

MODEL: F-84F ENG: J65-W-3, B-3 HARD SURFACE RUNWAY

TWO 1000 LB
14 SEC
ATO UNITS

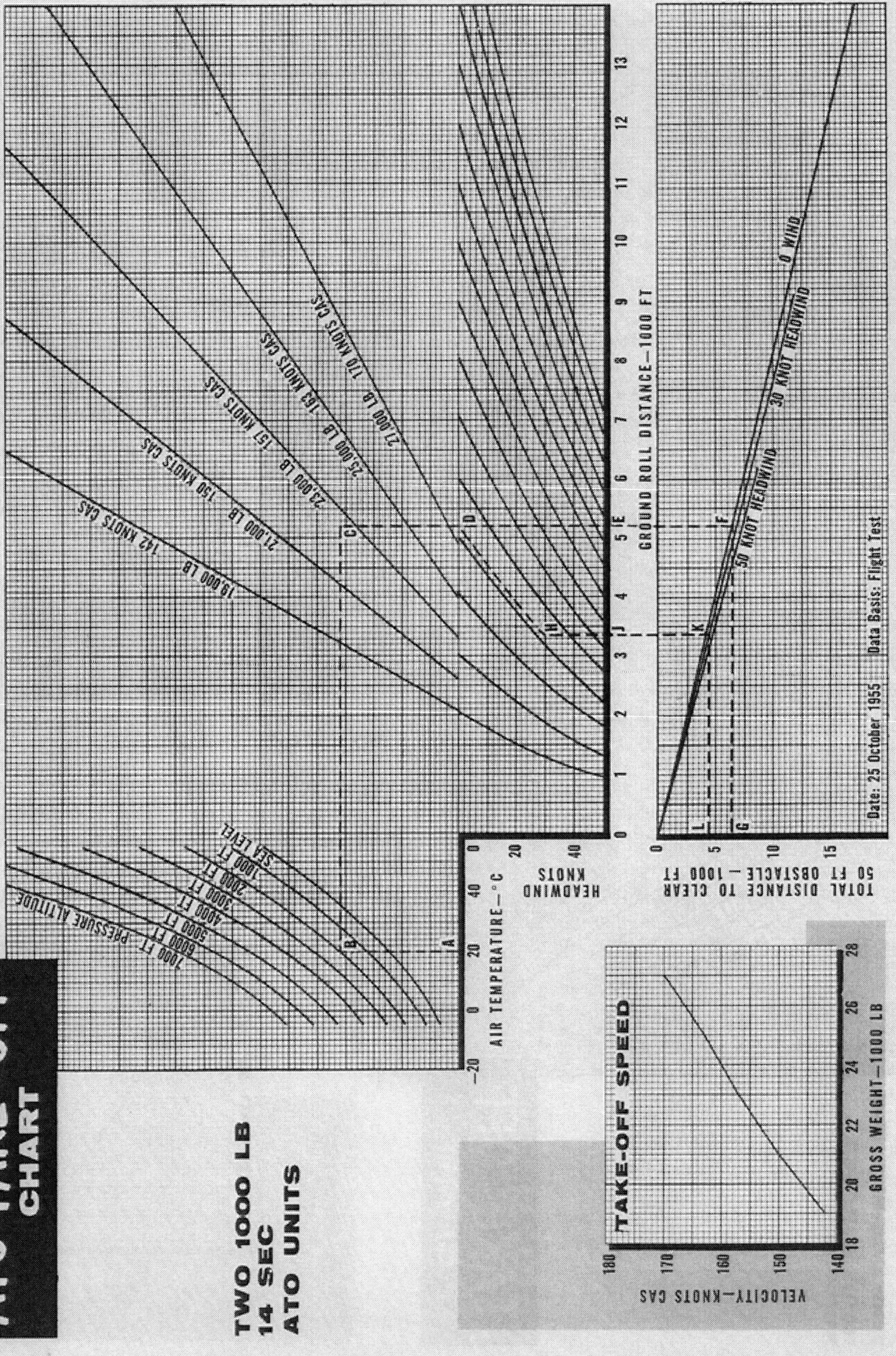

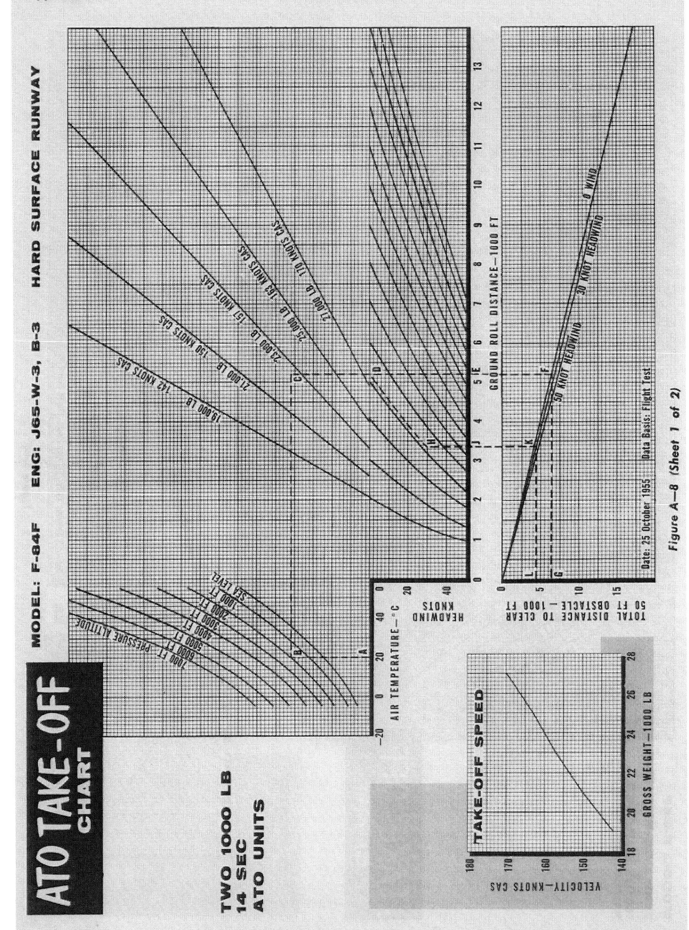

Figure A—8 (Sheet 1 of 2)

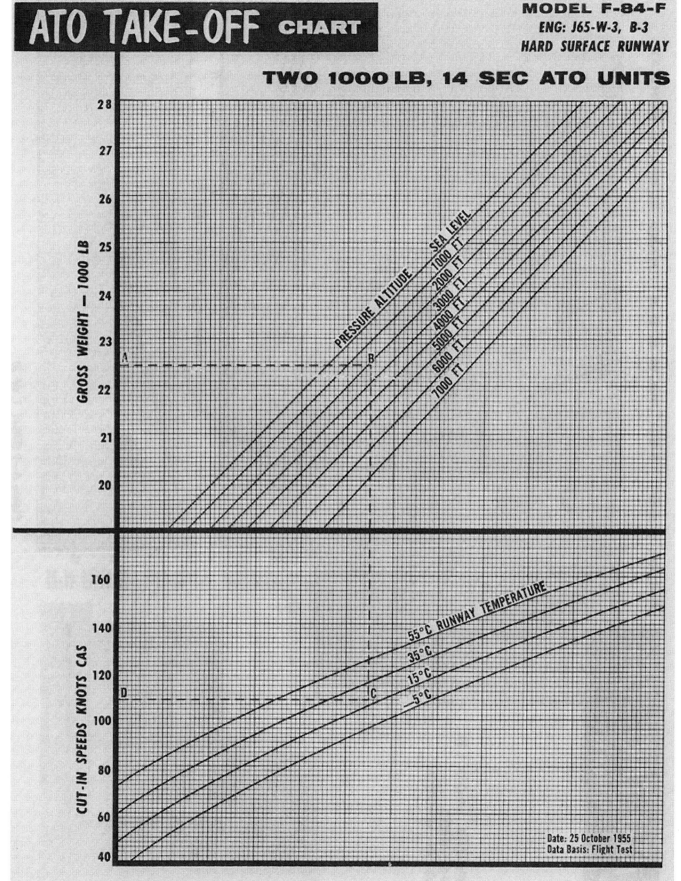

ATO TAKE-OFF CHART

MODEL F-84-F
ENG: J65-W-3, B-3
HARD SURFACE RUNWAY

TWO 1000 LB, 14 SEC ATO UNITS

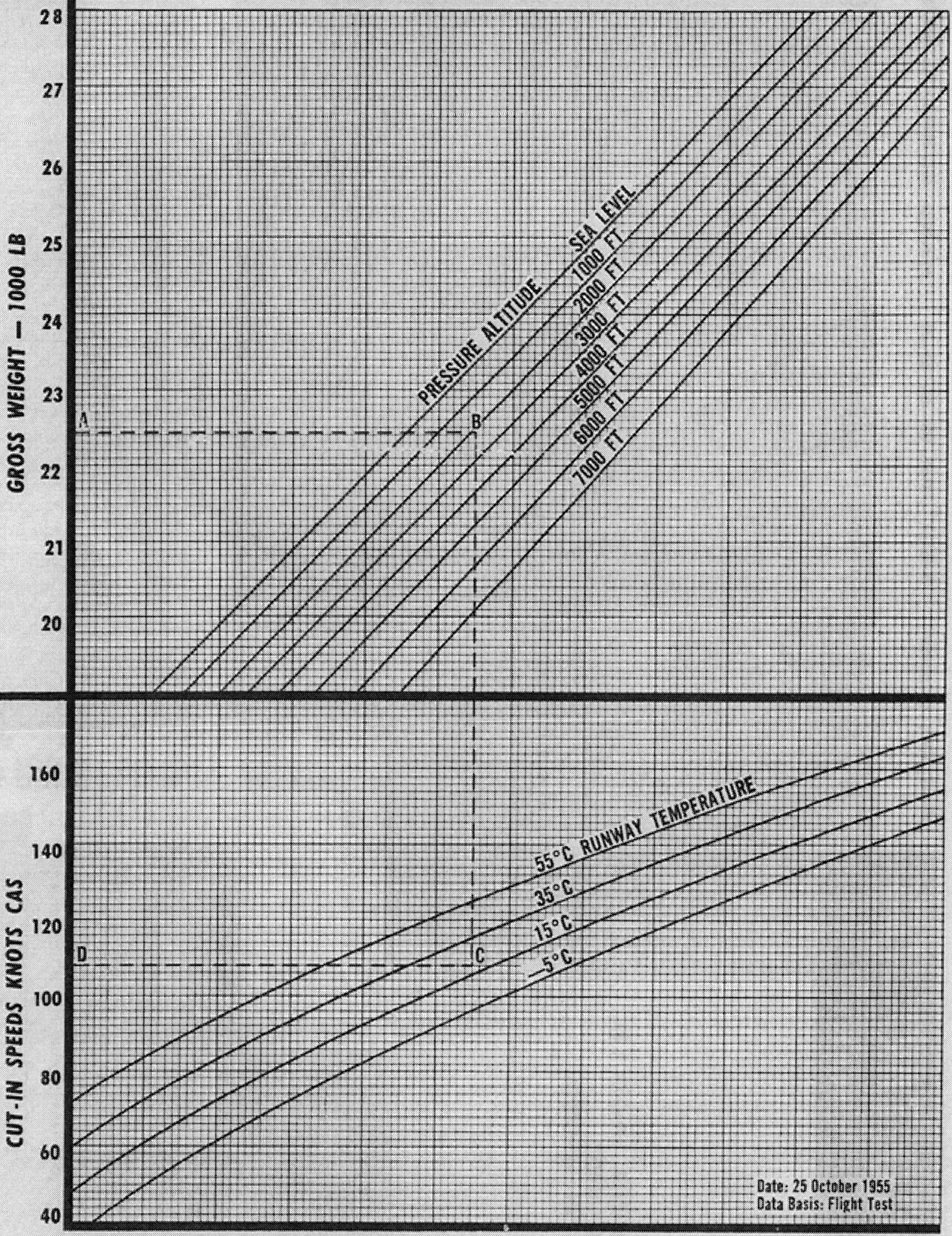

Figure A—8 (Sheet 2 of 2)

ATO TAKE-OFF CHART

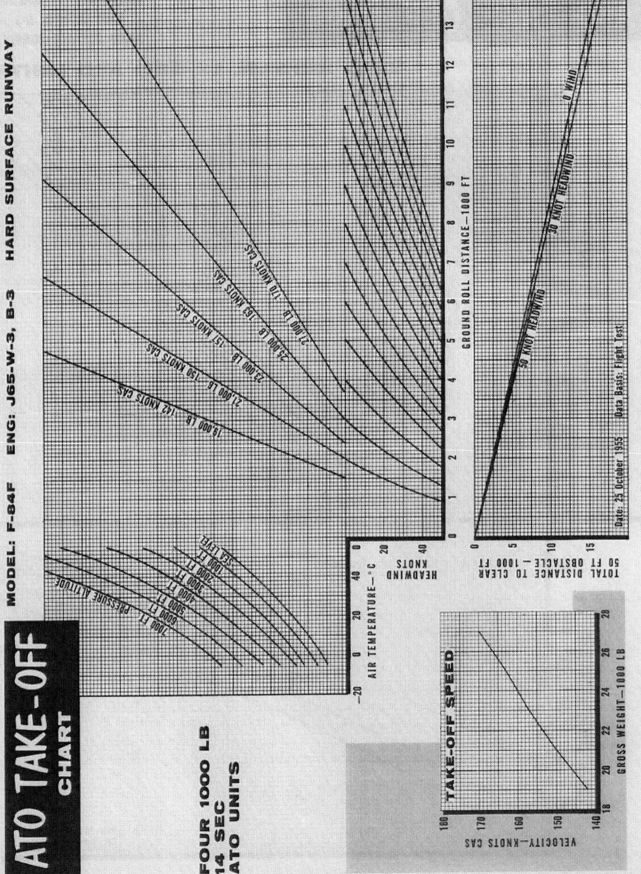

MODEL: F-84F ENG: J65-W-3, B-3 HARD SURFACE RUNWAY

FOUR 1000 LB 14 SEC ATO UNITS

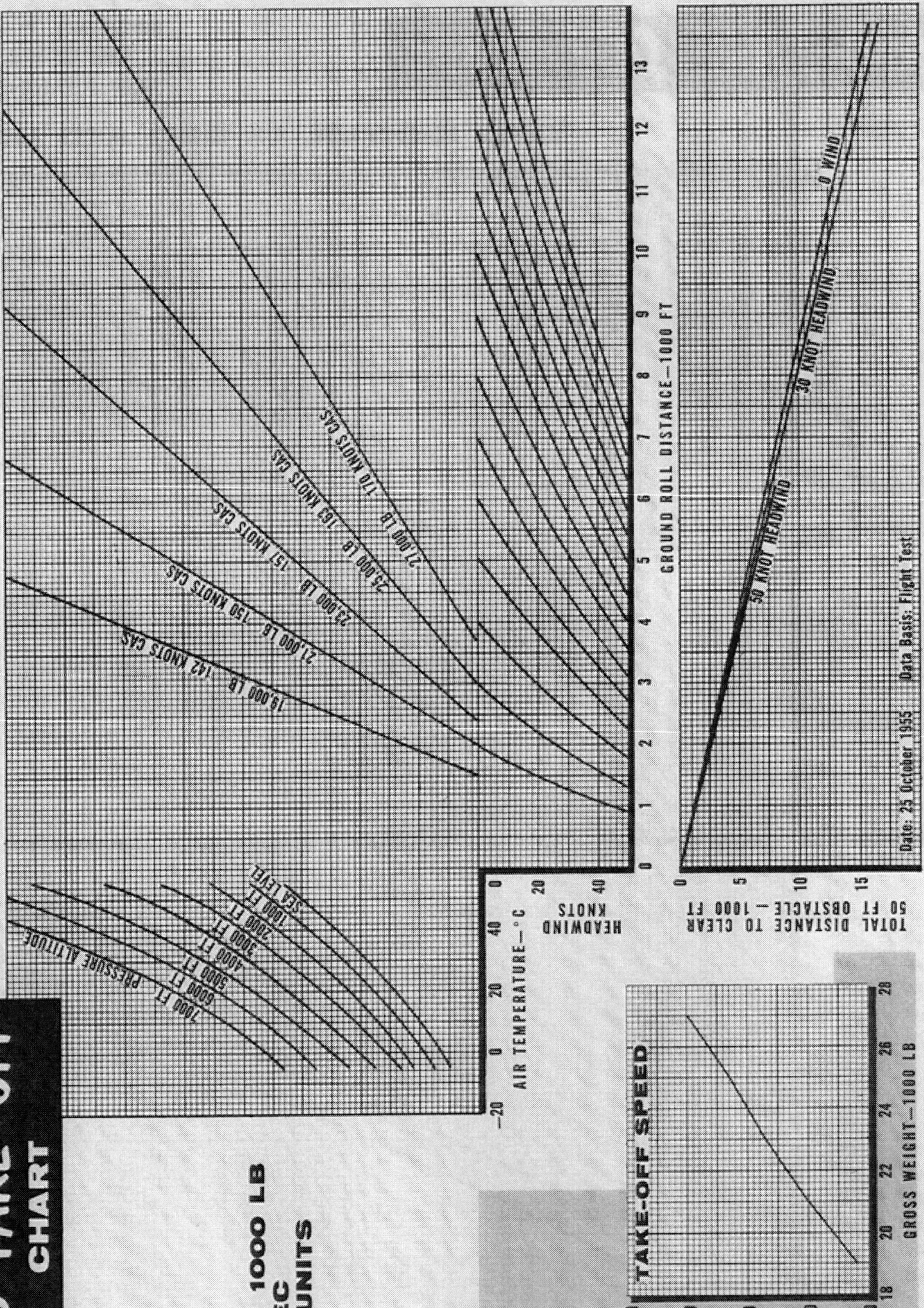

Figure A—9 (Sheet 1 of 2)

ATO TAKE-OFF CHART

MODEL F-84-F
ENG: J65-W-3, B-3
HARD SURFACE RUNWAY

FOUR 1000 LB, 14 SEC ATO UNITS

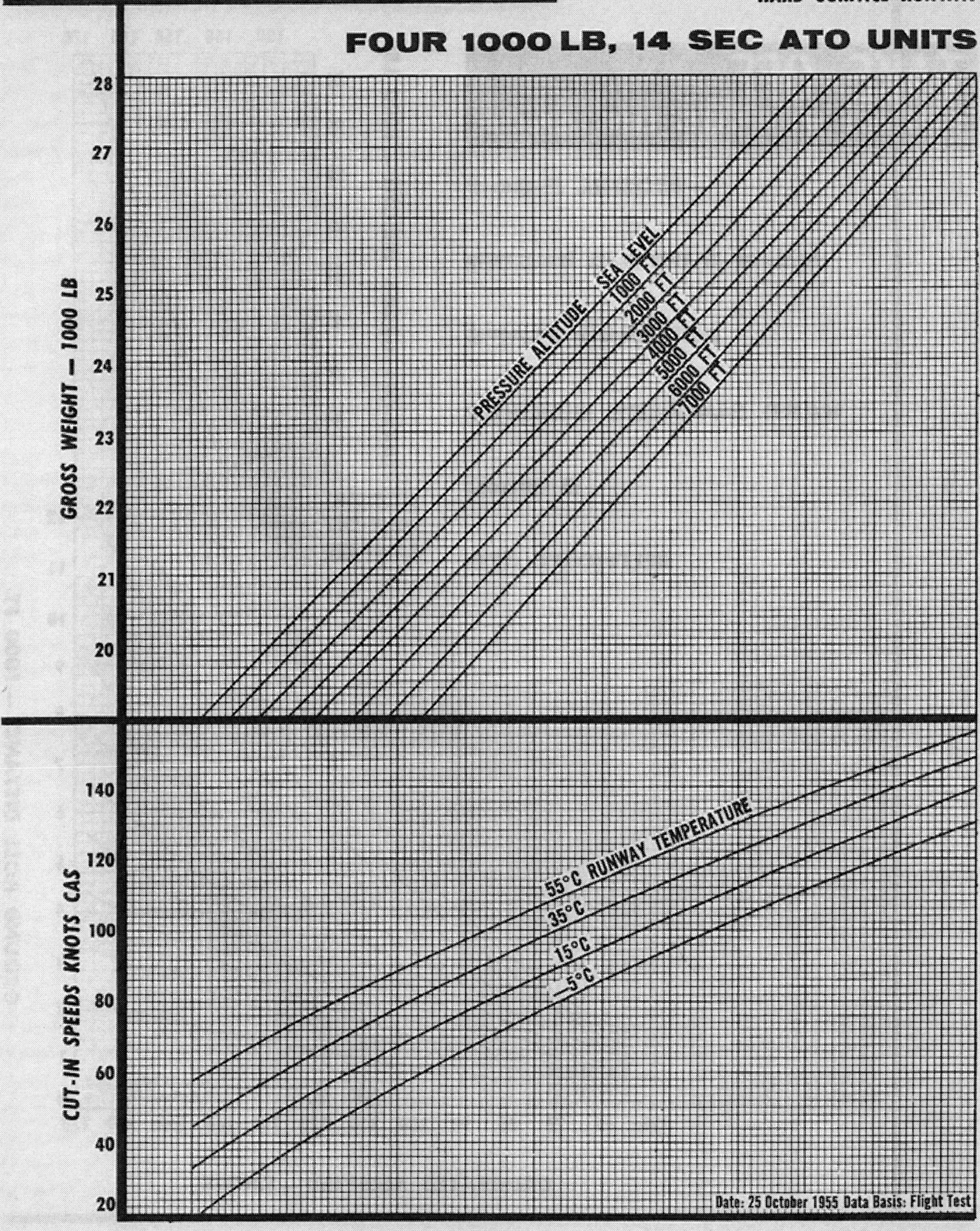

Figure A—9 (Sheet 2 of 2)

TAKE-OFF
SPEED VERSUS DISTANCE

MODEL F-84-F

J65-W-3 OR B-3 ENGINE

20° WING FLAPS

HARD SURFACE RUNWAY

0 WIND

WITHOUT ATO

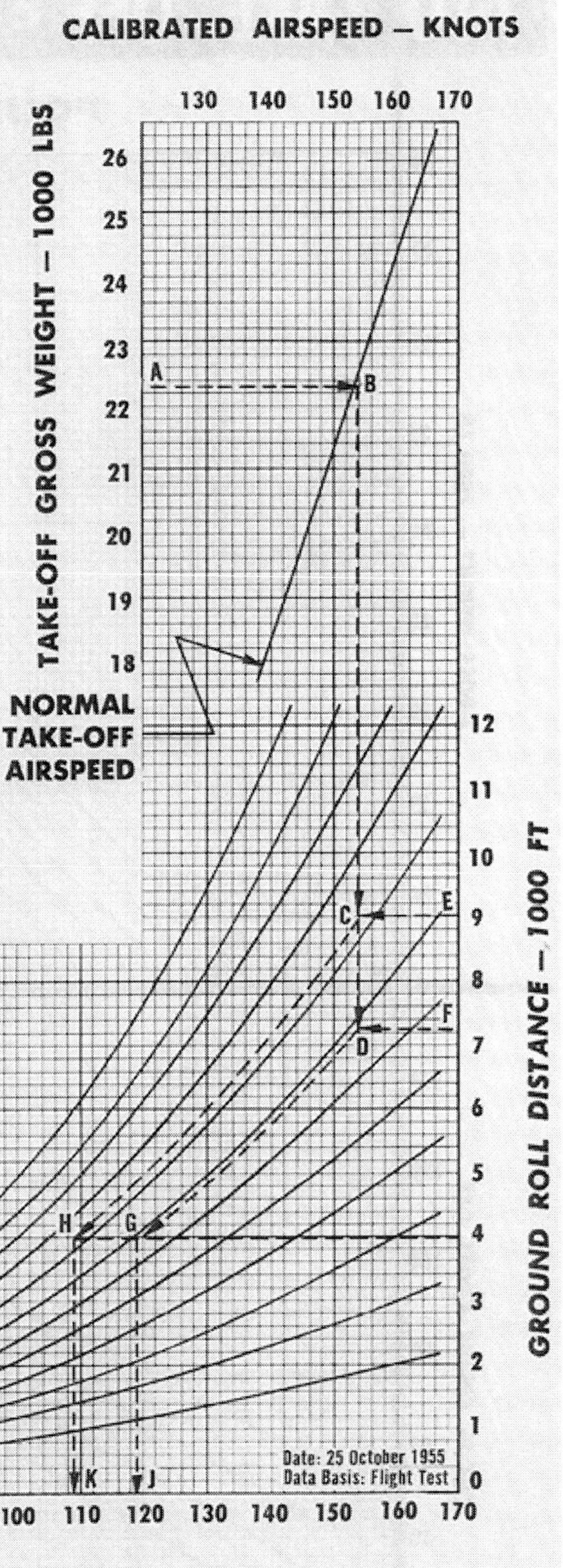

Figure A—10

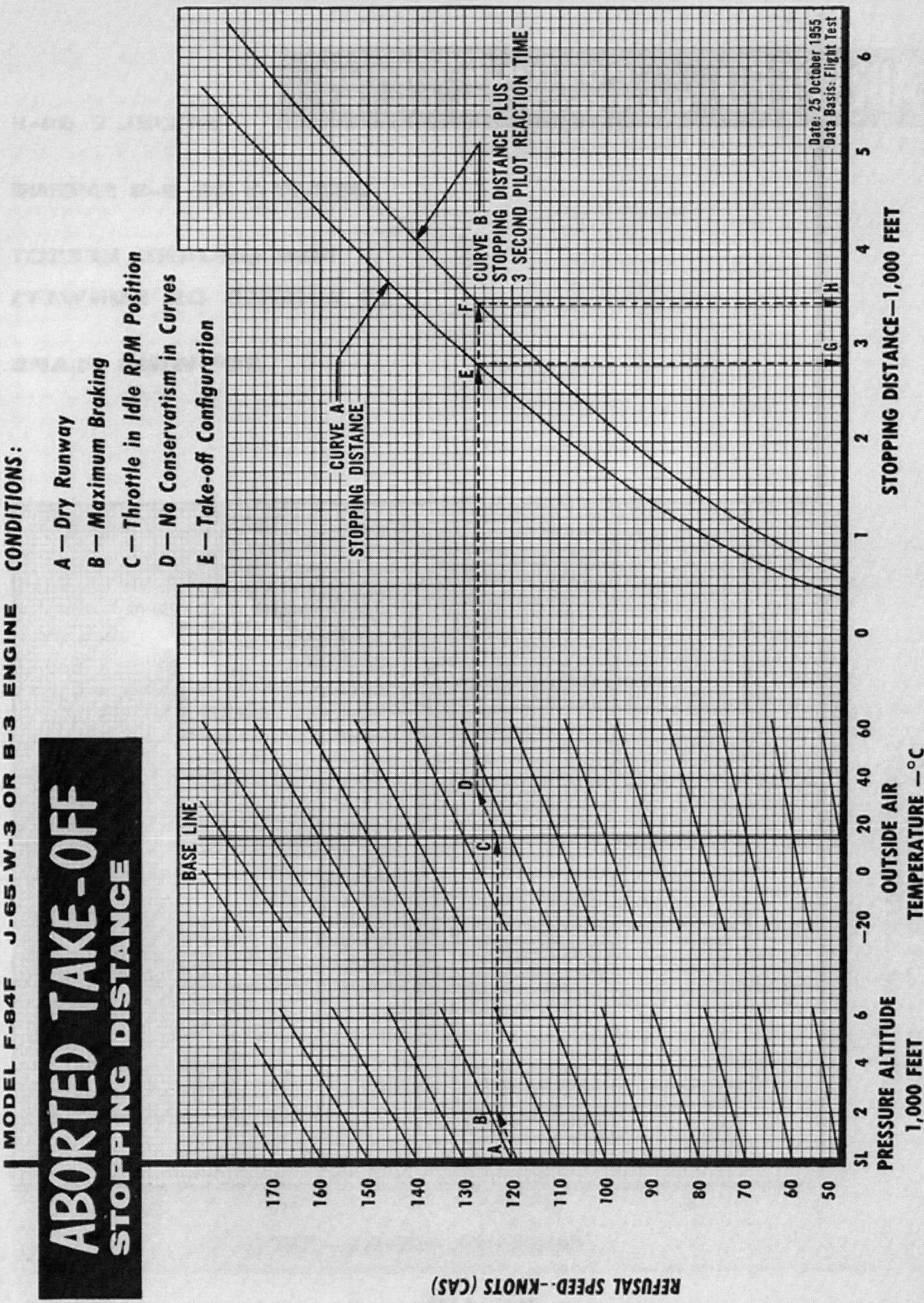
MODEL F-84F J-65-W-3 OR B-3 ENGINE
ABORTED TAKE-OFF
STOPPING DISTANCE
CONDITIONS:
A — Dry Runway
B — Maximum Braking
C — Throttle in Idle RPM Position
D — No Conservatism in Curves
E — Take-off Configuration
BASE LINE
CURVE A
STOPPING DISTANCE
CURVE B
STOPPING DISTANCE PLUS
3 SECOND PILOT REACTION TIME
Date: 25 October 1955
Data Basis: Flight Test
REFUSAL SPEED–KNOTS (CAS)
PRESSURE ALTITUDE
1,000 FEET
OUTSIDE AIR
TEMPERATURE —°C
STOPPING DISTANCE—1,000 FEET
A
B
C
D
E
F
G
H

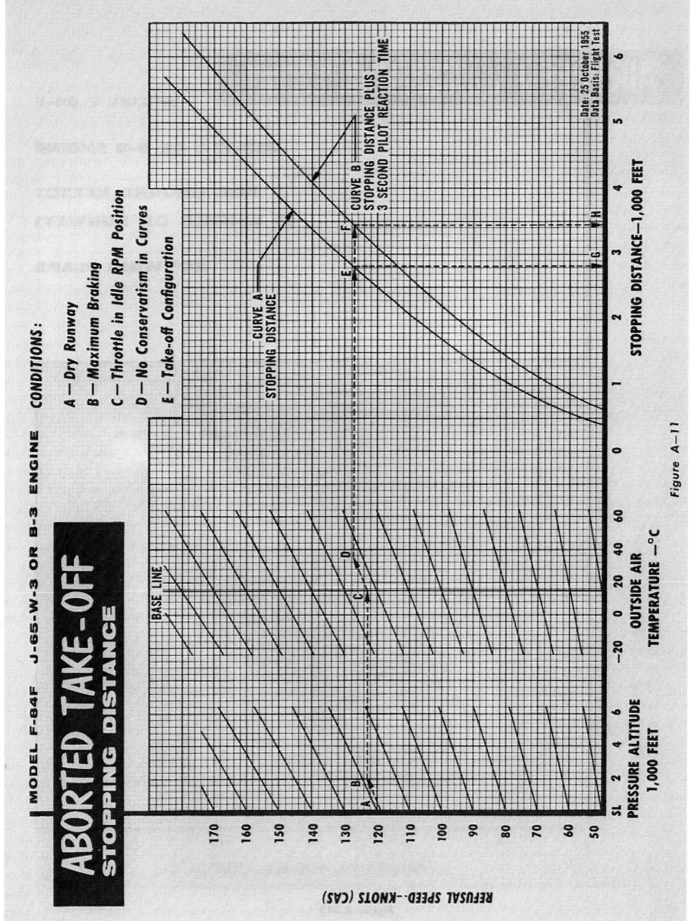

Figure A–11

AIRSPEED POSITION ERROR CORRECTION CHART

MODEL F-84-F

J65-W-3 OR B-3 ENGINE

FOR GROUND EFFECT

(3 WHEELS ON RUNWAY)

20° WING FLAPS

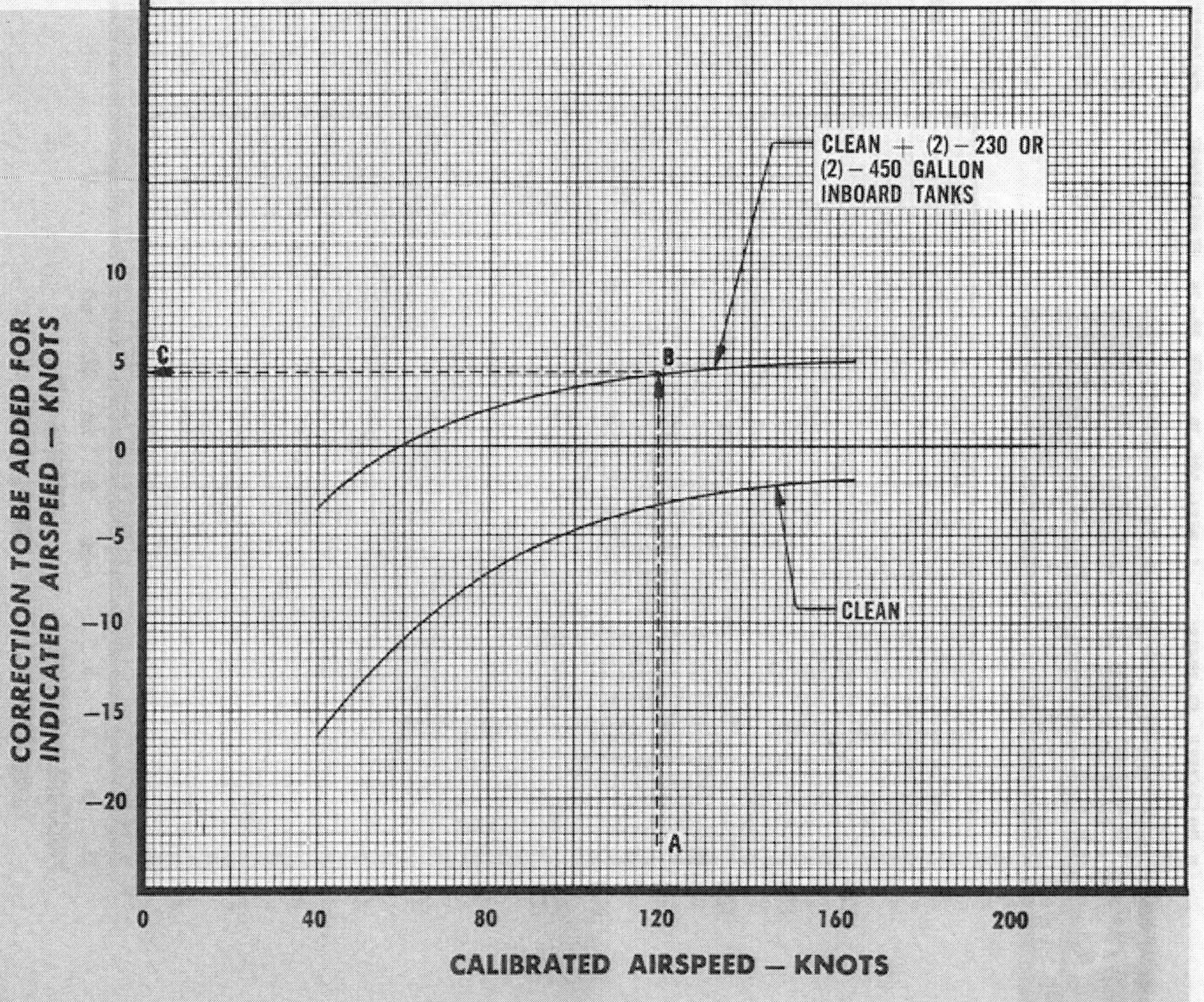

Figure A—12

F-84F-25 AND LATER • J65-W-3 ENGINE • STANDARD DAY •

MILITARY POWER (100% RPM) CLIMB

CONFIGURATION: CLEAN

ALT-FT	CAS-KNOTS	MACH NO.
45,000	220	.81
40,000	245	.81
35,000	275	.81
30,000	300	.79
20,000	340	.74
10,000	380	.68
SEA LEVEL	395	.60

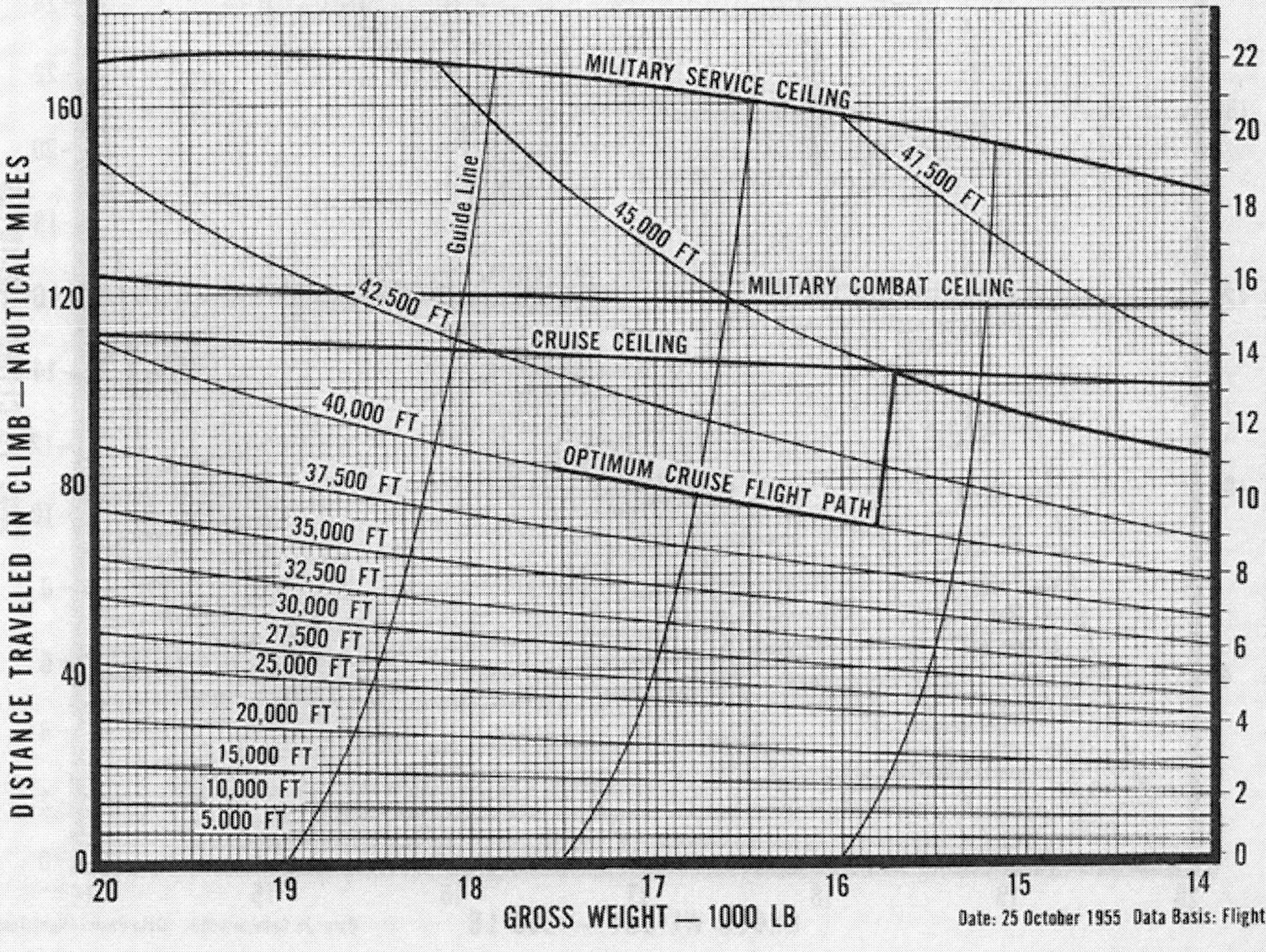

Date: 25 October 1955 Data Basis: Flight Test

Figure A—13

F-84F-25 AND LATER • J65-W-3 OR B-3 ENGINES • STANDARD DAY •

98% MILITARY RPM CLIMB

CONFIGURATION: CLEAN

ALT-FT	CAS-KNOTS	MACH NO.
45,000	220	.81
40,000	250	.81
35,000	275	.81
30,000	300	.79
20,000	335	.73
10,000	360	.64
SEA LEVEL	380	.57

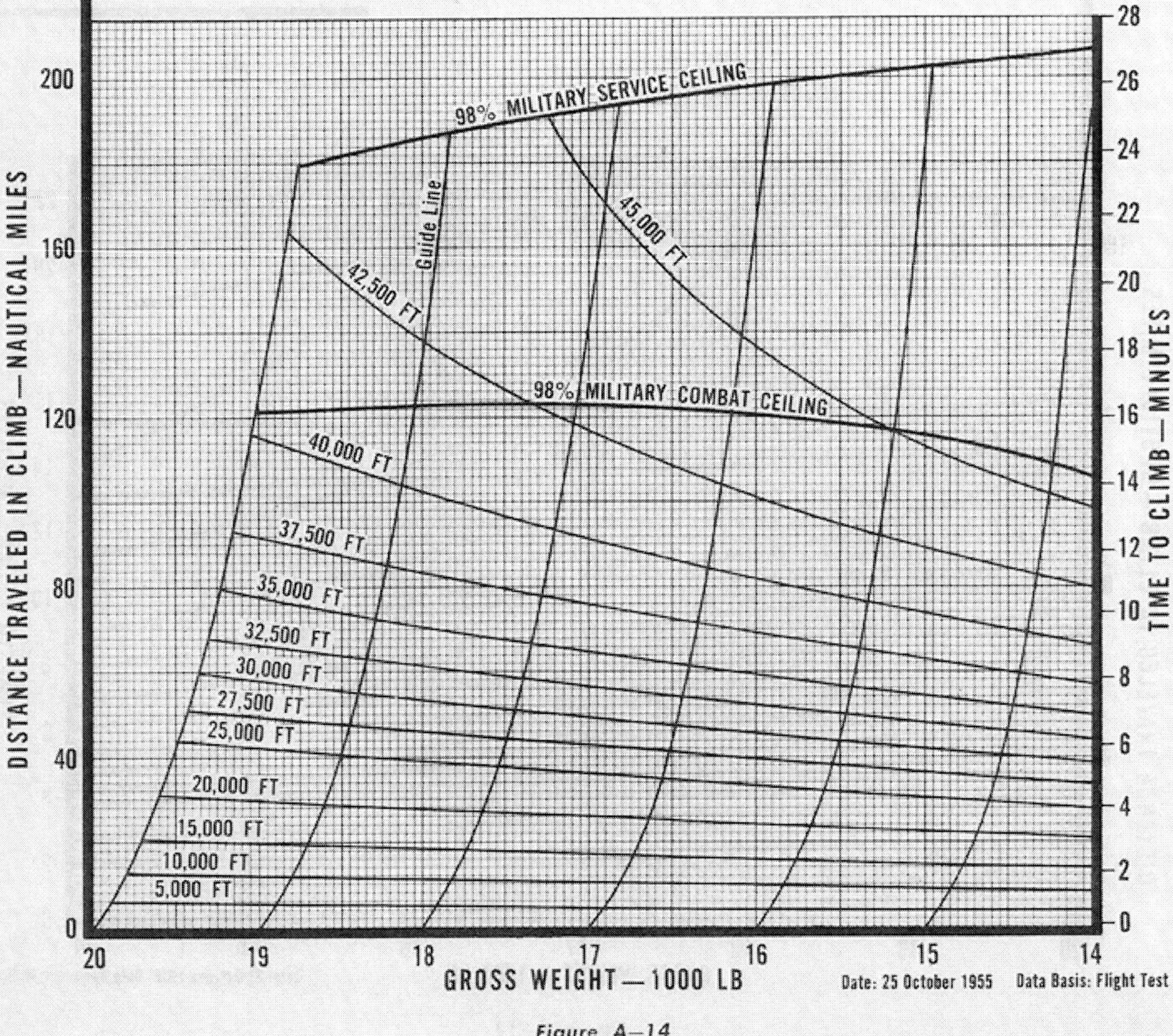

Date: 25 October 1955 Data Basis: Flight Test

Figure A–14

F-84F-25 AND LATER • J65-W-3 ENGINE • STANDARD DAY •

NORMAL POWER (96% RPM) CLIMB

CONFIGURATION: CLEAN

ALT-FT	CAS-KNOTS	MACH NO.
45,000	215	.80
40,000	240	.80
35,000	270	.80
30,000	295	.78
20,000	325	.70
10,000	340	.61
10,000	345	.52

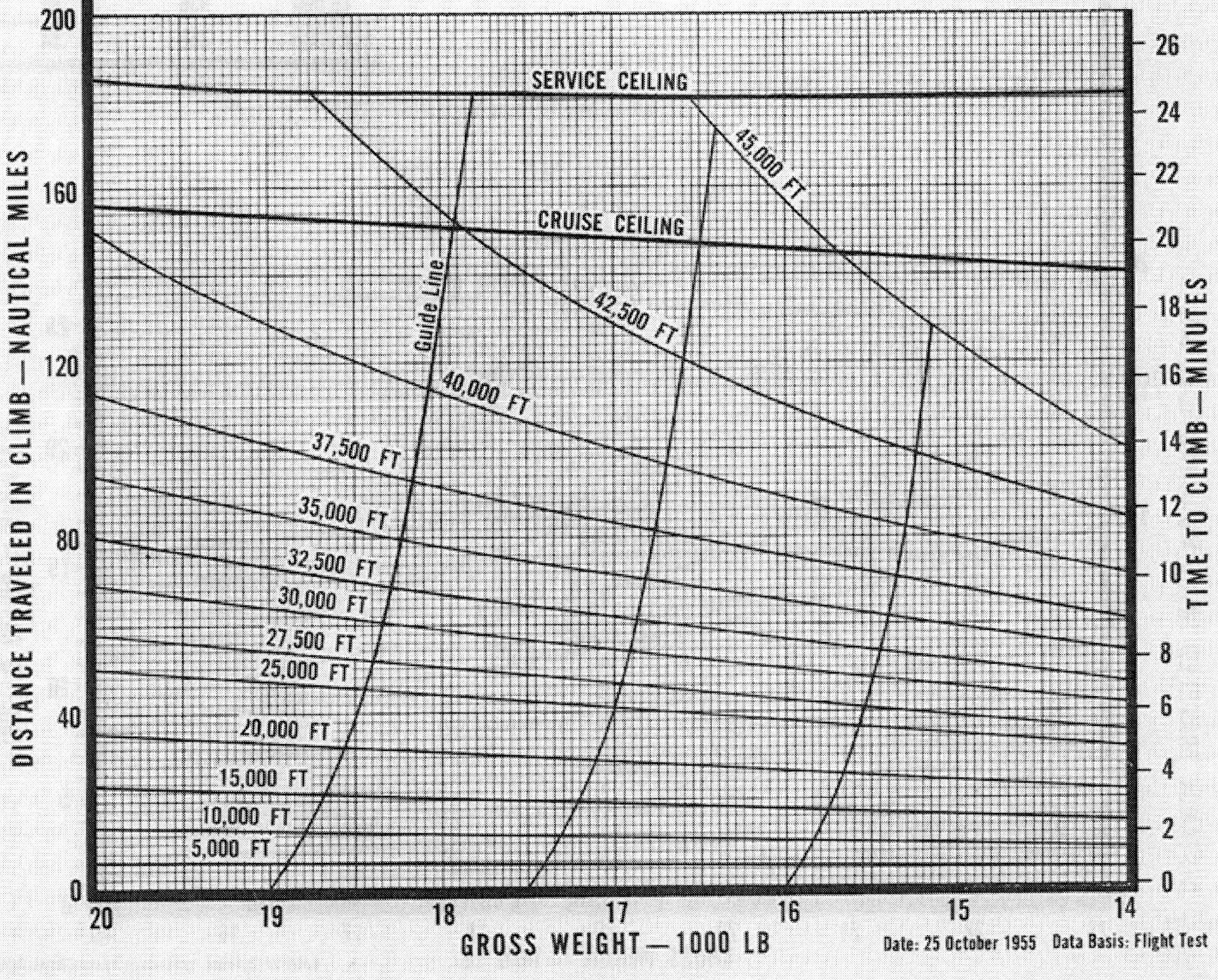

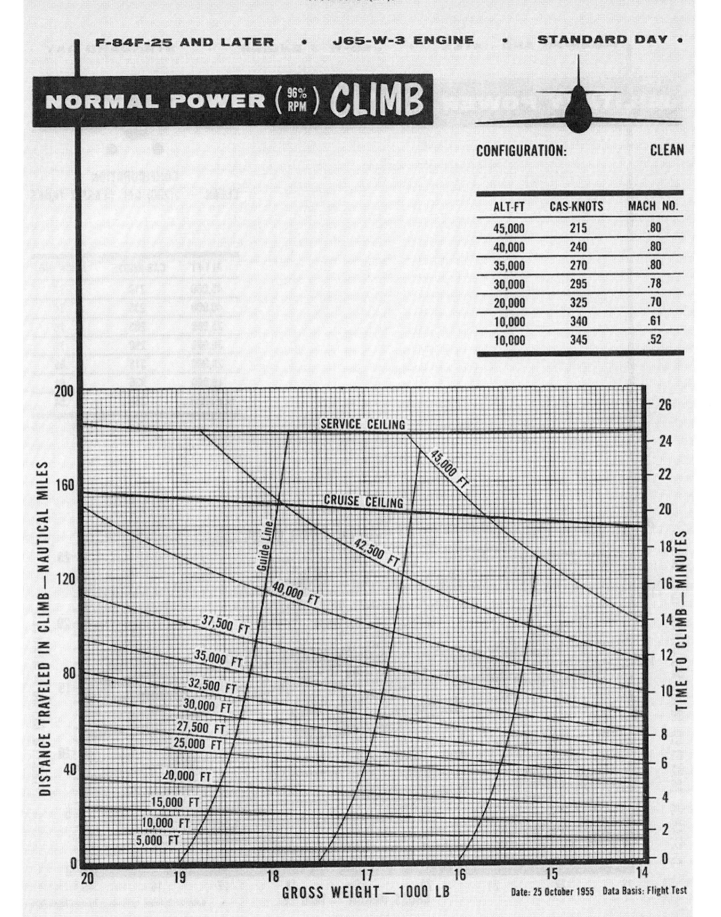

Date: 25 October 1955 Data Basis: Flight Test

Figure A–15

F-84F-25 AND LATER • J65-W-3 ENGINE • STANDARD DAY •

MILITARY POWER (100% RPM) CLIMB

CONFIGURATION
CLEAN + 2(230) GAL CLASS I TANKS

ALT-FT	CAS-KNOTS	MACH NO.
45,000	210	.78
40,000	235	.78
35,000	265	.78
30,000	290	.76
20,000	315	.68
10,000	335	.61
SEA LEVEL	360	.54

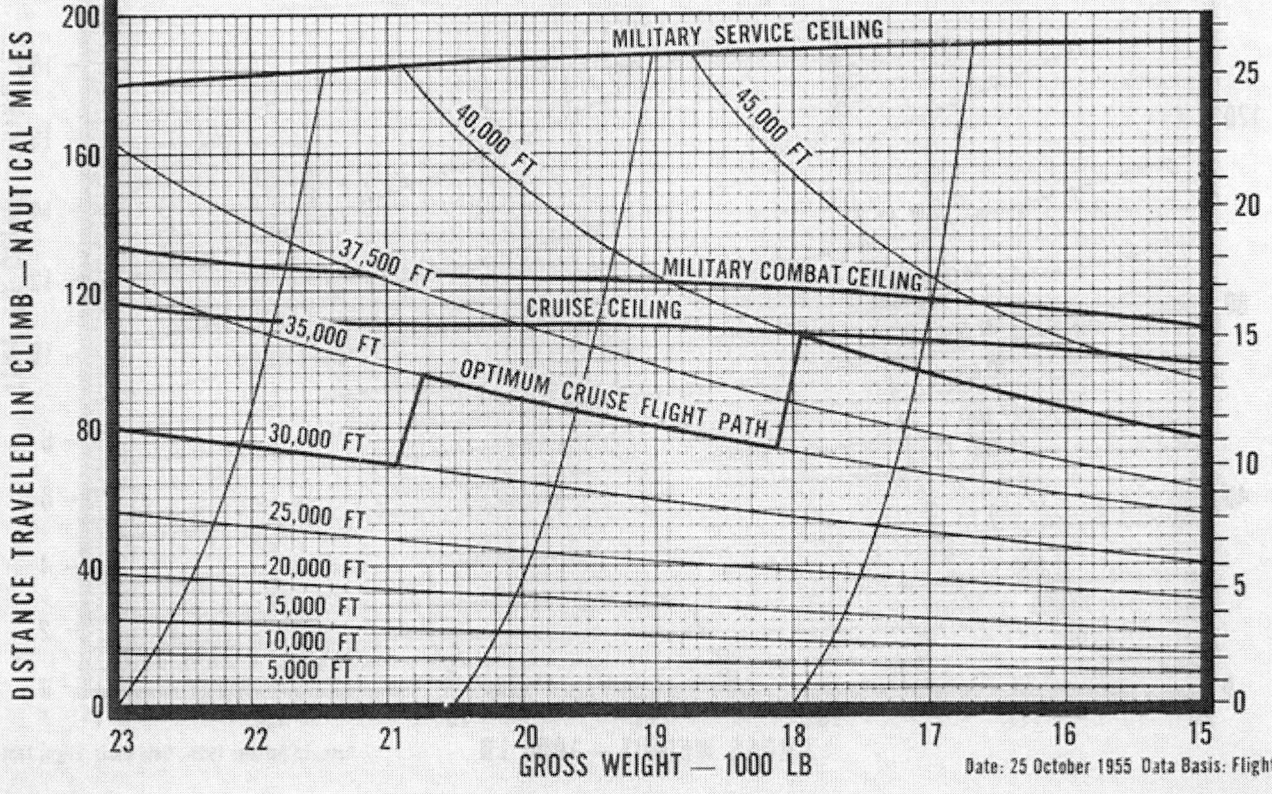

Date: 25 October 1955 Data Basis: Flight Test

Figure A–16

F-84F-25 AND LATER • J65-W-3 OR B-3 ENGINES • STANDARD DAY •

98% MILITARY RPM CLIMB

CONFIGURATION
CLEAN + 2(230) GAL CLASS I TANKS

ALT-FT	CAS-KNOTS	MACH NO.
45,000	210	.78
40,000	235	.78
35,000	265	.78
30,000	275	.73
20,000	300	.65
10,000	320	.58
SEA LEVEL	340	.51

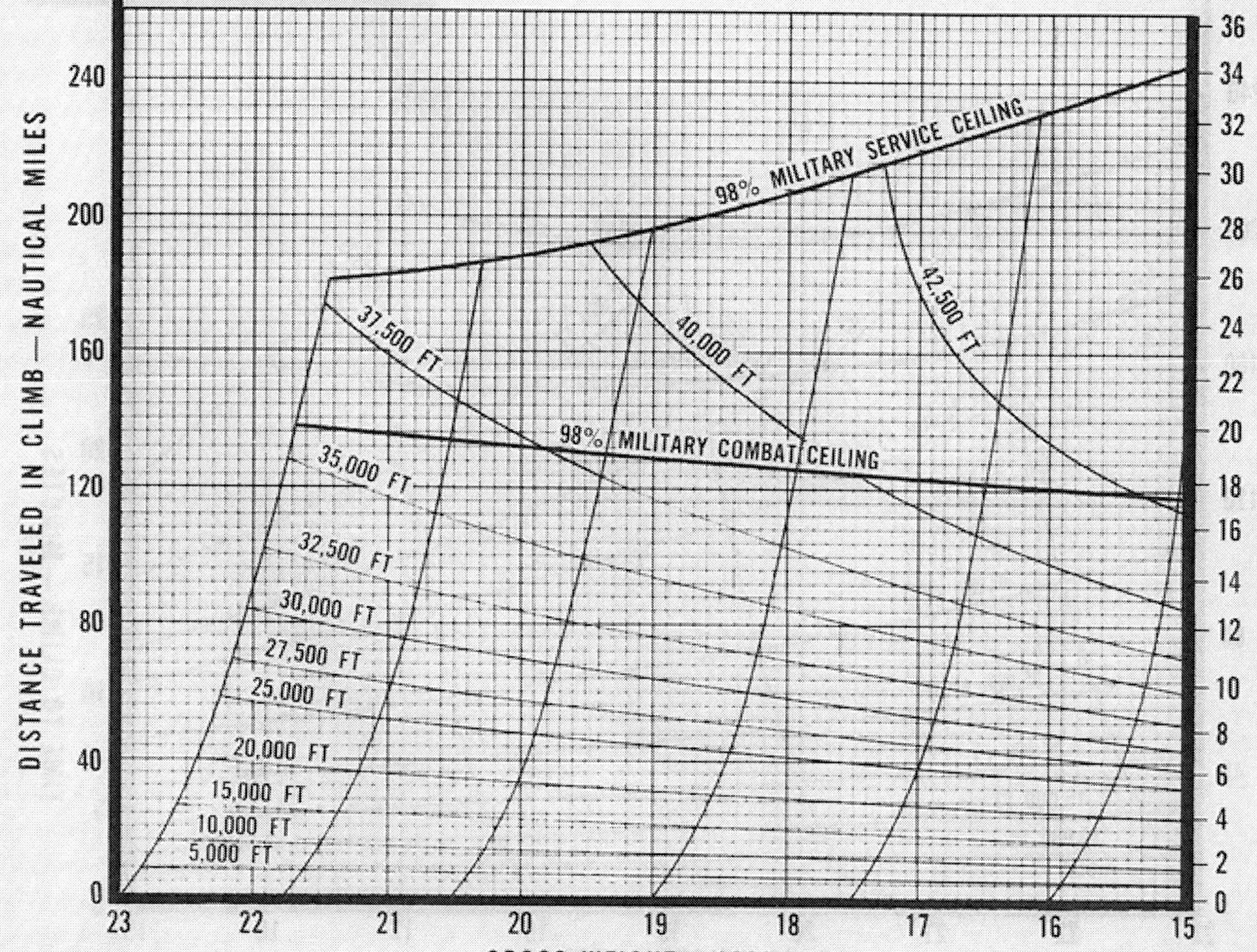

Date: 25 October 1955 Data Basis: Flight Test

Figure A—17

F-84F-25 AND LATER • J65-W-3 ENGINE • STANDARD DAY •

NORMAL POWER (96% RPM) CLIMB

CONFIGURATION
CLEAN + 2(230) GAL CLASS I TANKS

ALT-FT	CAS-KNOTS	MACH NO.
45,000	210	.78
40,000	235	.78
35,000	265	.78
30,000	275	.73
20,000	295	.64
10,000	305	.56
SEA LEVEL	320	.48

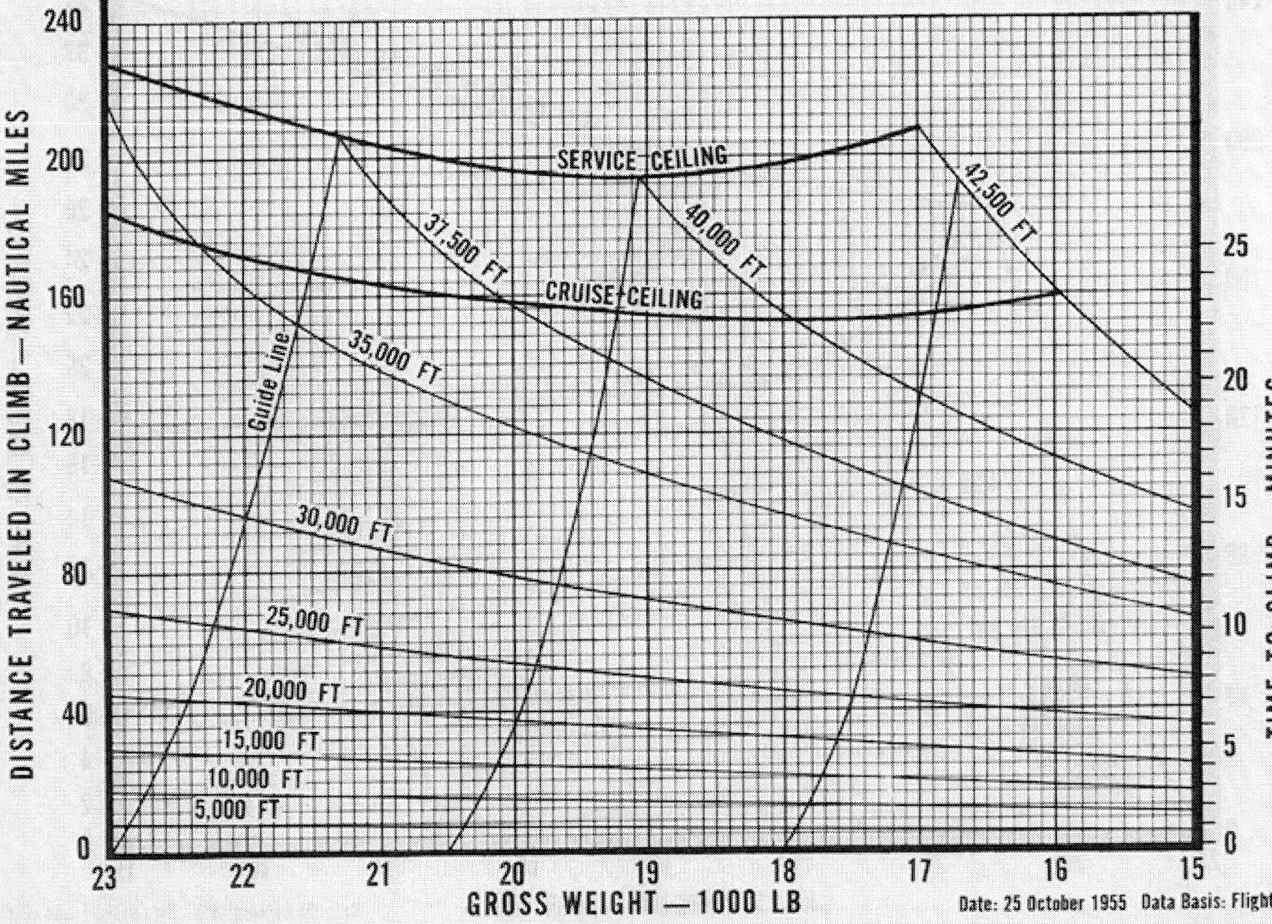

Date: 25 October 1955 Data Basis: Flight Test

Figure A—18

F-84F-25 AND LATER • J65-W-3 ENGINE • STANDARD DAY

MILITARY POWER (100% RPM) CLIMB

ALT-FT	CAS-KNOTS	MACH NO.
45,000	210	.78
40,000	235	.78
35,000	265	.78
30,000	275	.74
20,000	300	.66
10,000	320	.58
SEA LEVEL	335	.51

CONFIGURATION CLEAN + 2(450) GAL CLASS I TANKS

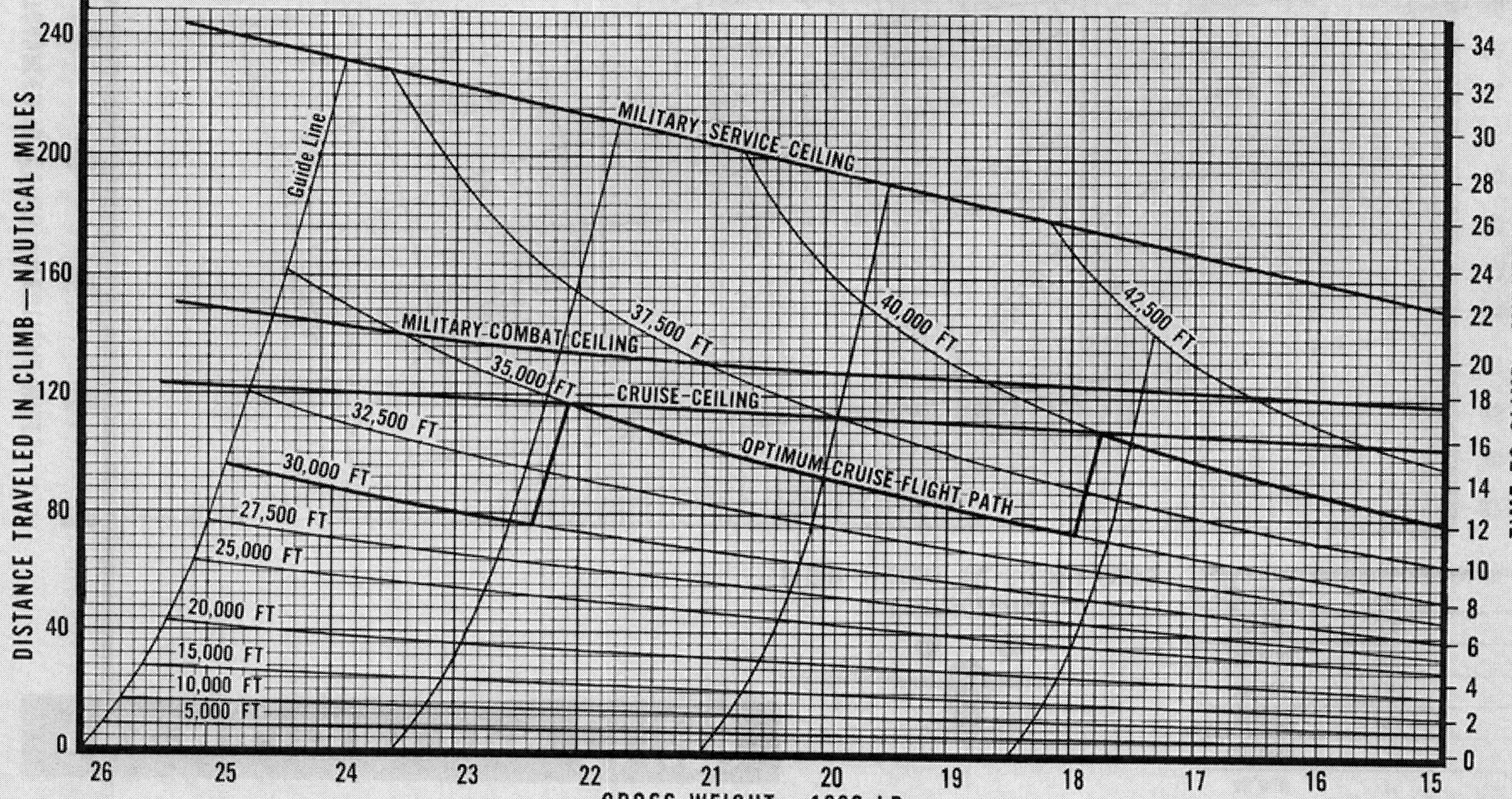

Date: 25 October 1955 Data Basis: Flight Test

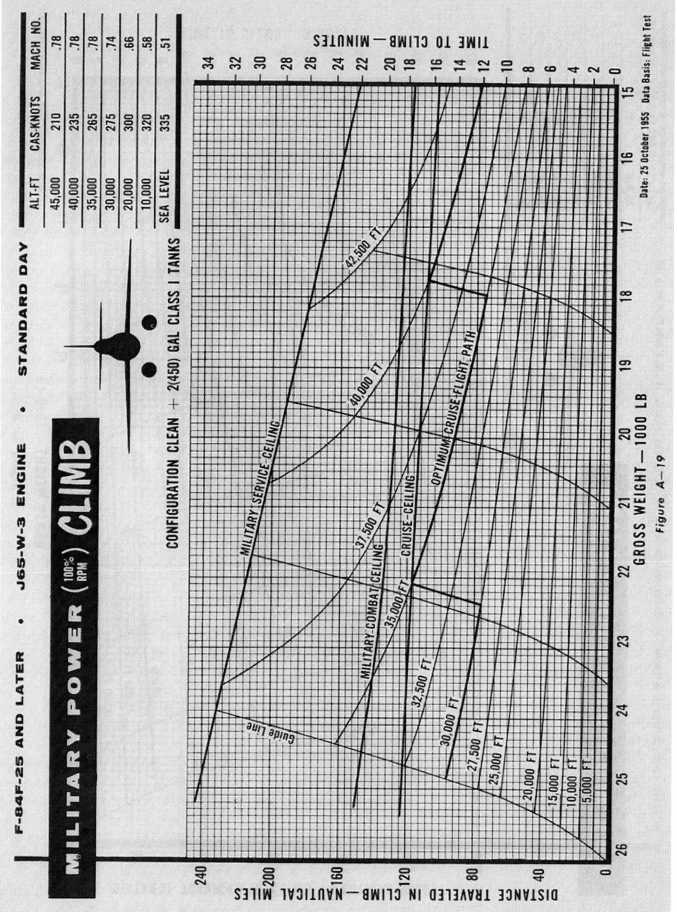

Figure A—19

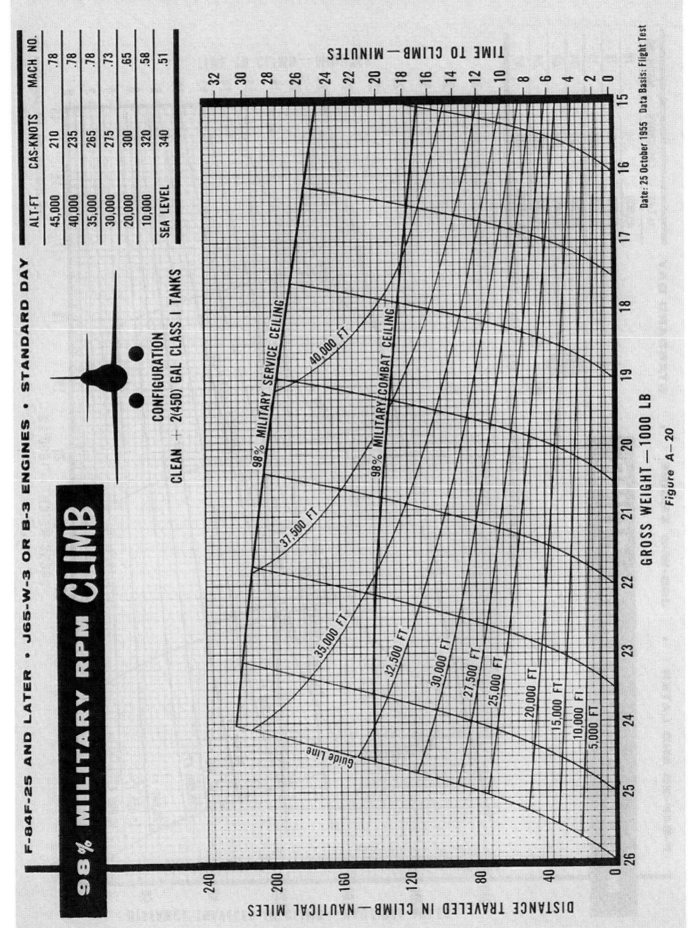

F-84F-25 AND LATER • J65-W-3 OR B-3 ENGINES • STANDARD DAY

98% MILITARY RPM CLIMB

CONFIGURATION
CLEAN + 2(450) GAL CLASS I TANKS

ALT-FT	CAS-KNOTS	MACH NO.
45,000	210	.78
40,000	235	.78
35,000	265	.78
30,000	275	.73
20,000	300	.65
10,000	320	.58
SEA LEVEL	340	.51

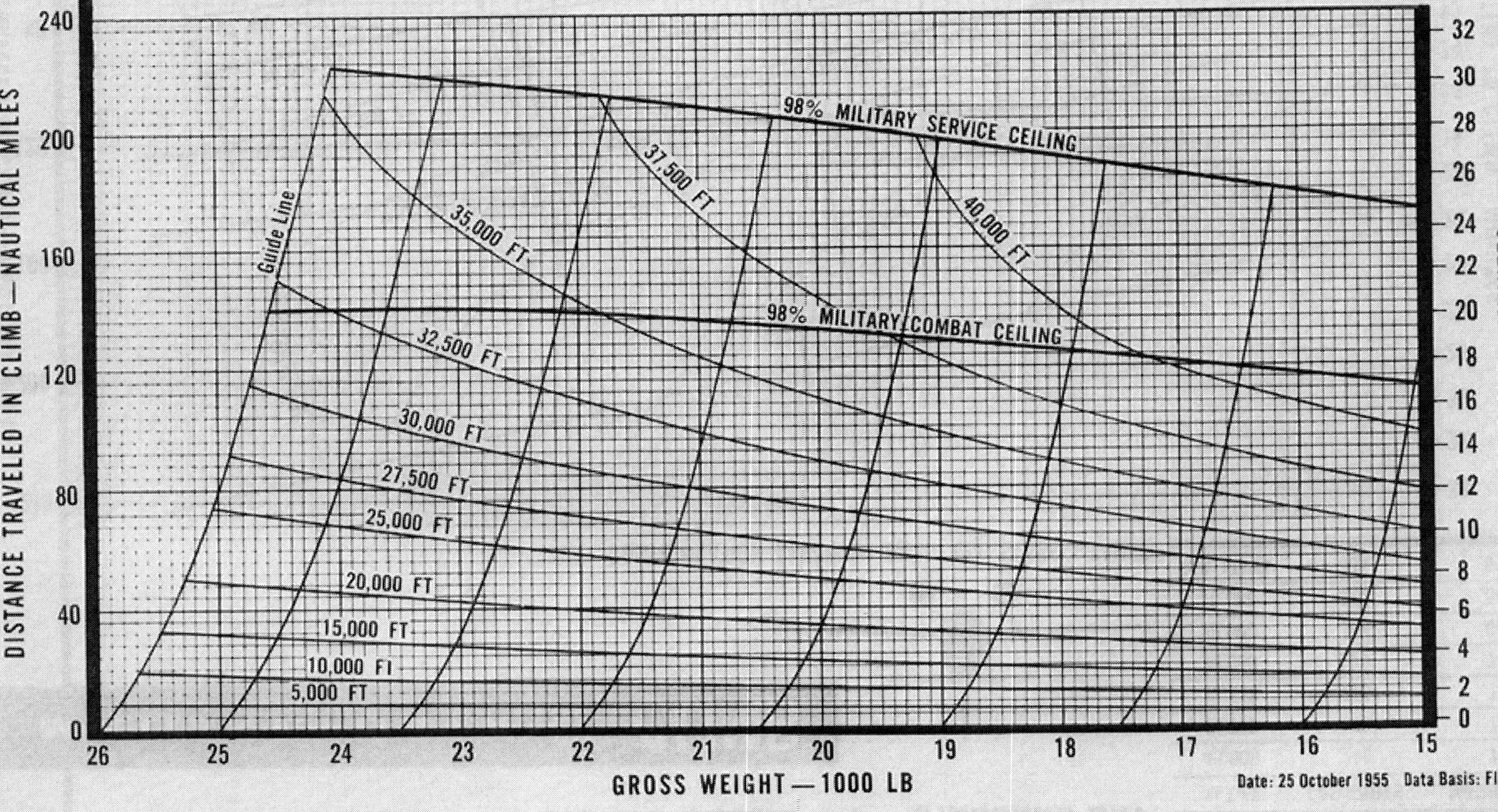

Date: 25 October 1955 Data Basis: Flight Test

Figure A—20

F-84F-25 AND LATER • J65-W-3 ENGINE • STANDARD DAY

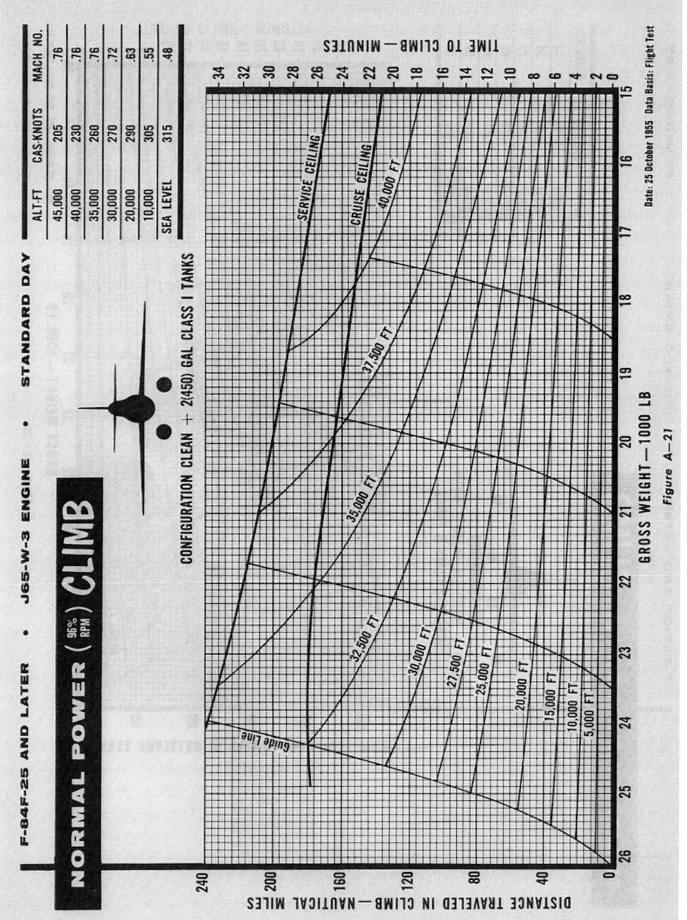

NORMAL POWER (96% RPM) CLIMB

ALT-FT	CAS-KNOTS	MACH NO.
45,000	205	.76
40,000	230	.76
35,000	260	.76
30,000	270	.72
20,000	290	.63
10,000	305	.55
SEA LEVEL	315	.48

CONFIGURATION CLEAN + 2(450) GAL CLASS I TANKS

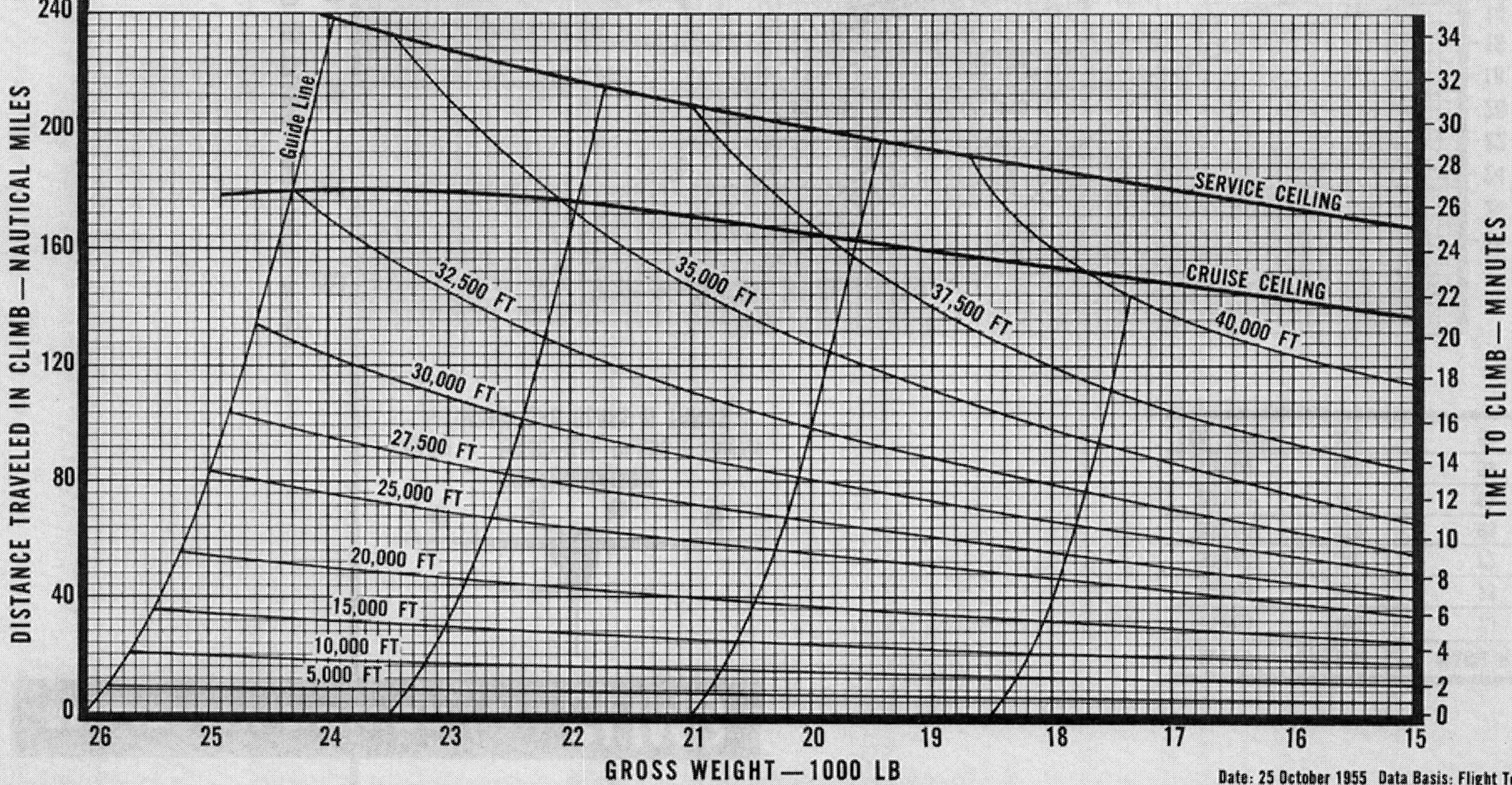

Date: 25 October 1955 Data Basis: Flight Test

Figure A—21

F-84F-25 AND LATER • J65-W-3 ENGINE • STANDARD DAY •

MILITARY POWER (100% RPM) CLIMB

CONFIGURATION
CLEAN + 2(230) GAL CLASS I TANKS
+ 2(230) GAL CLASS II TANKS

ALT-FT	CAS-KNOTS	MACH NO.
45,000	195	.73
40,000	220	.73
35,000	245	.72
30,000	255	.69
20,000	285	.62
10,000	305	.56
SEA LEVEL	330	.50

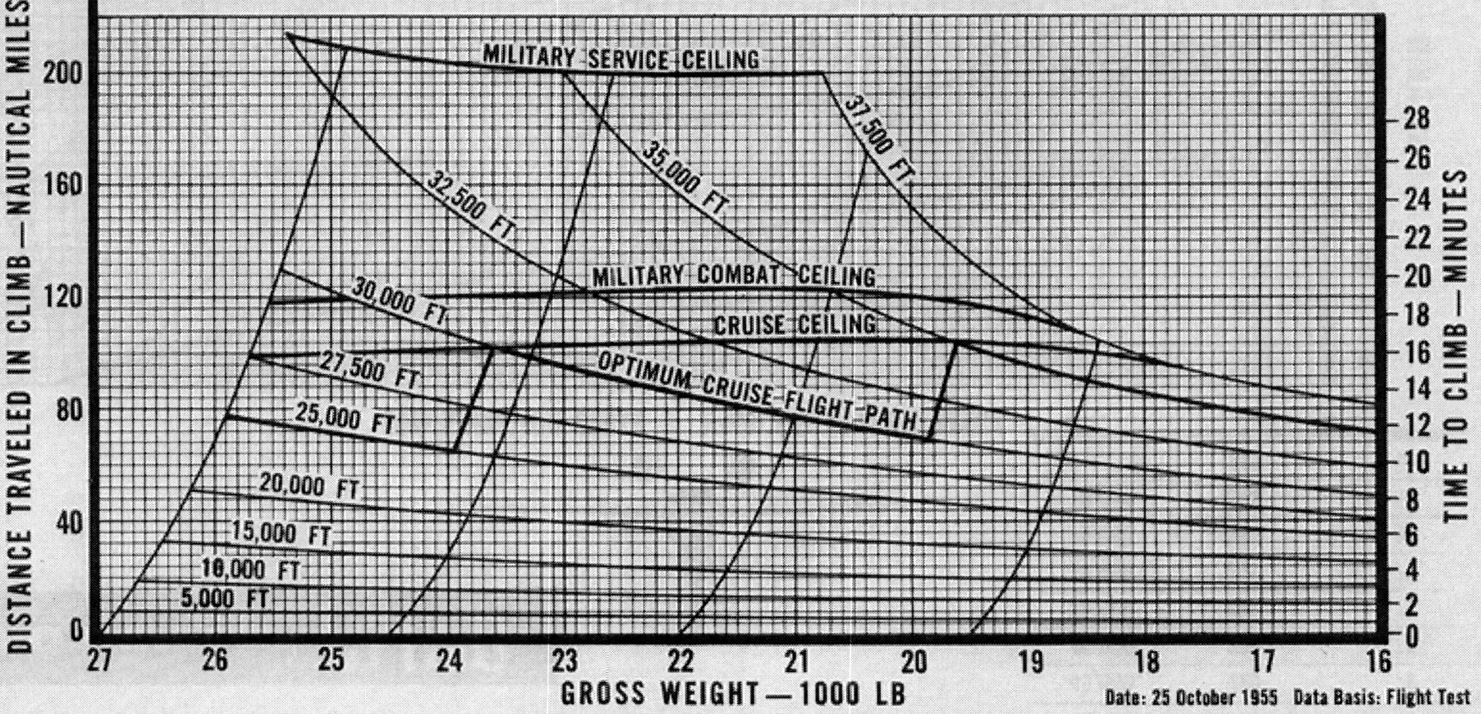

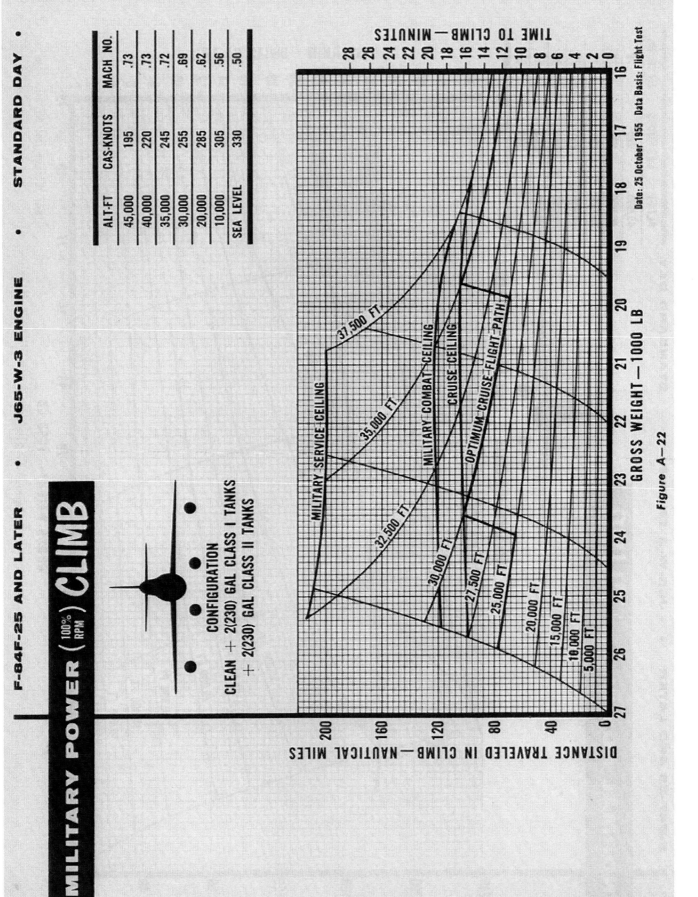

Date: 25 October 1955 Data Basis: Flight Test

Figure A—22

F-84F-25 AND LATER • J65-W-3 OR B-3 ENGINES • STANDARD DAY •

98% MILITARY RPM CLIMB

CONFIGURATION: CLEAN + 2(230) GAL CLASS I TANKS + 2(230) GAL CLASS II TANKS

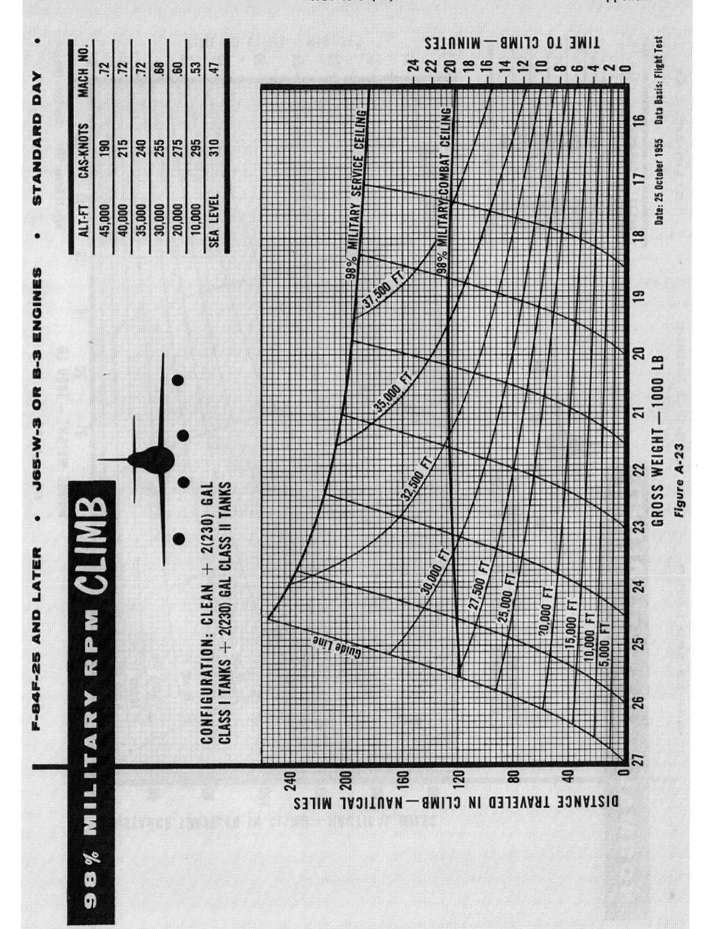

ALT-FT	CAS-KNOTS	MACH NO.
45,000	190	.72
40,000	215	.72
35,000	240	.72
30,000	255	.68
20,000	275	.60
10,000	295	.53
SEA LEVEL	310	.47

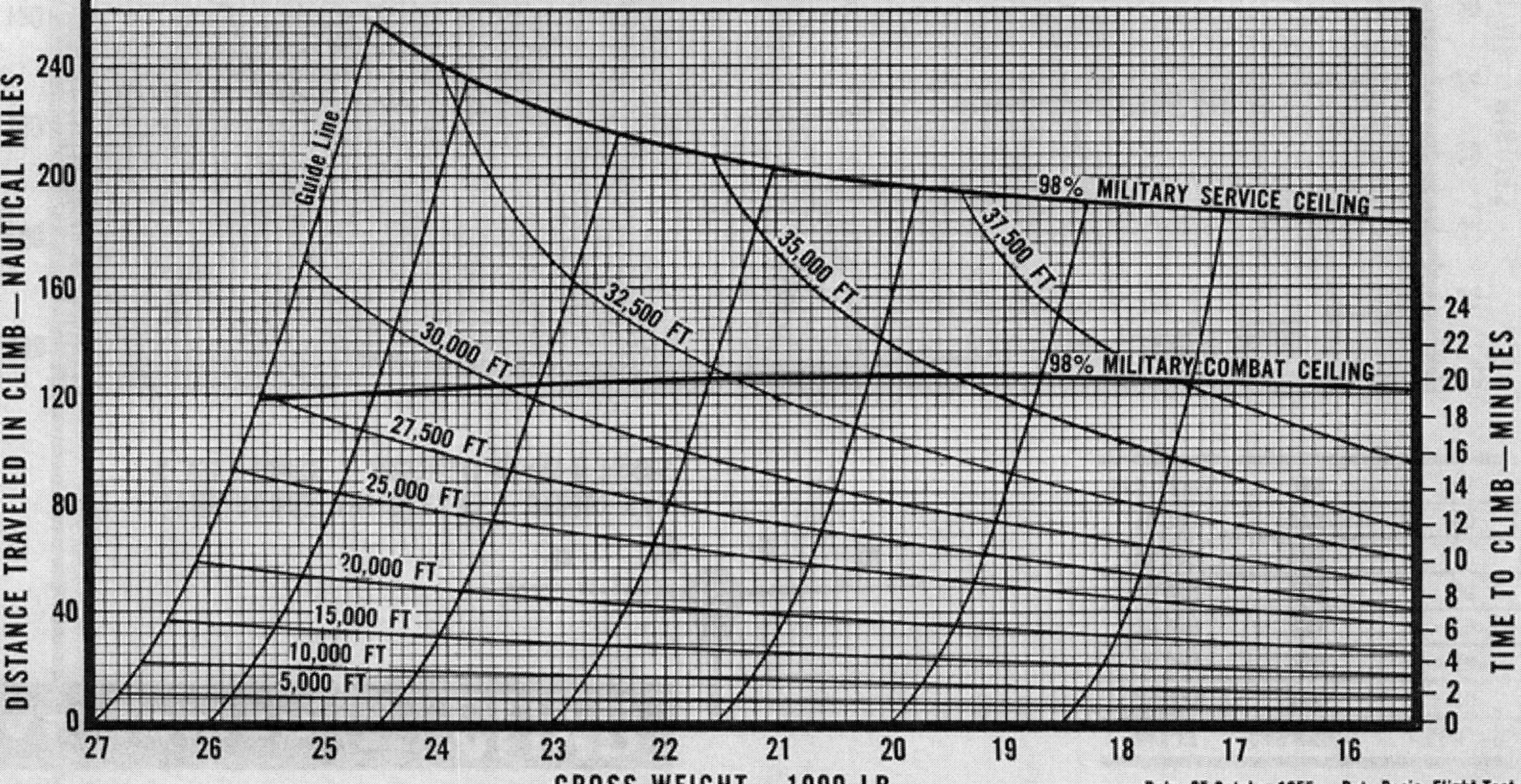

Date: 25 October 1955 Data Basis: Flight Test

Figure A-23

F-84F-25 AND LATER • J65-W-3 ENGINE • STANDARD DAY •

NORMAL POWER (96% RPM) CLIMB

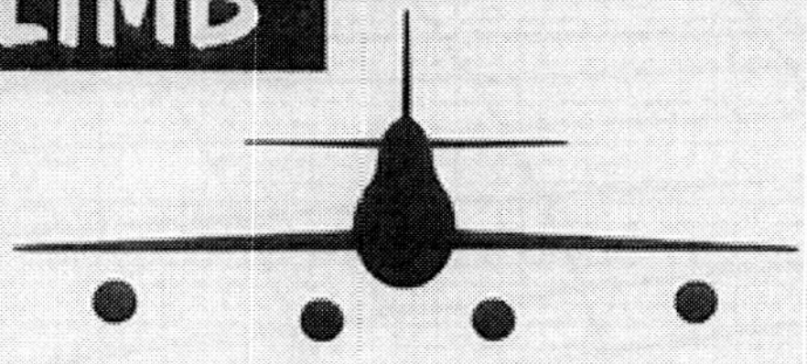

CONFIGURATION: CLEAN + 2(230) GAL CLASS I TANKS + 2(230) GAL CLASS II TANKS

ALT-FT	CAS-KNOTS	MACH NO.
45,000	185	.70
40,000	210	.70
35,000	235	.70
30,000	245	.66
20,000	260	.57
10,000	275	.50
SEA LEVEL	285	.43

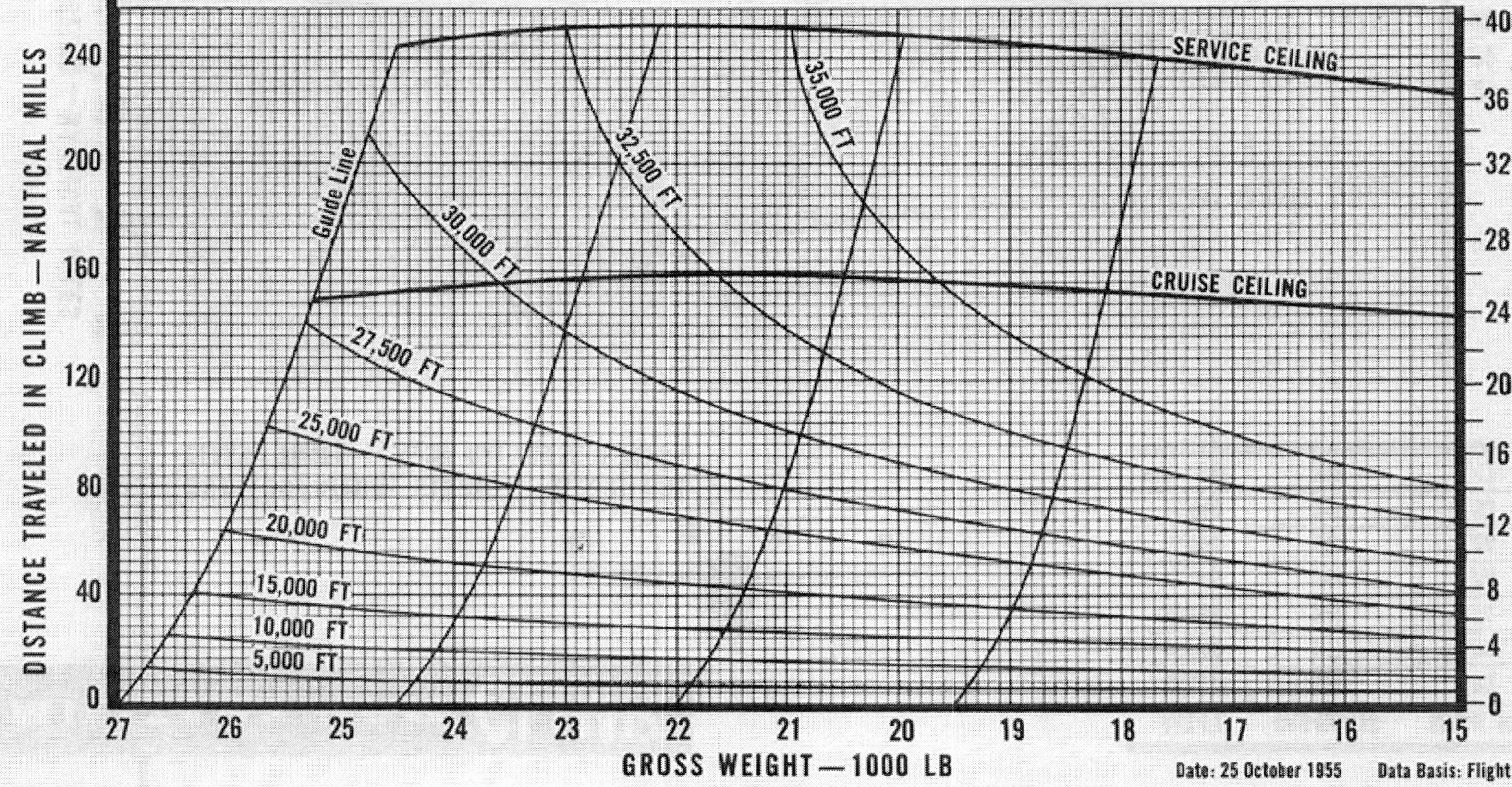

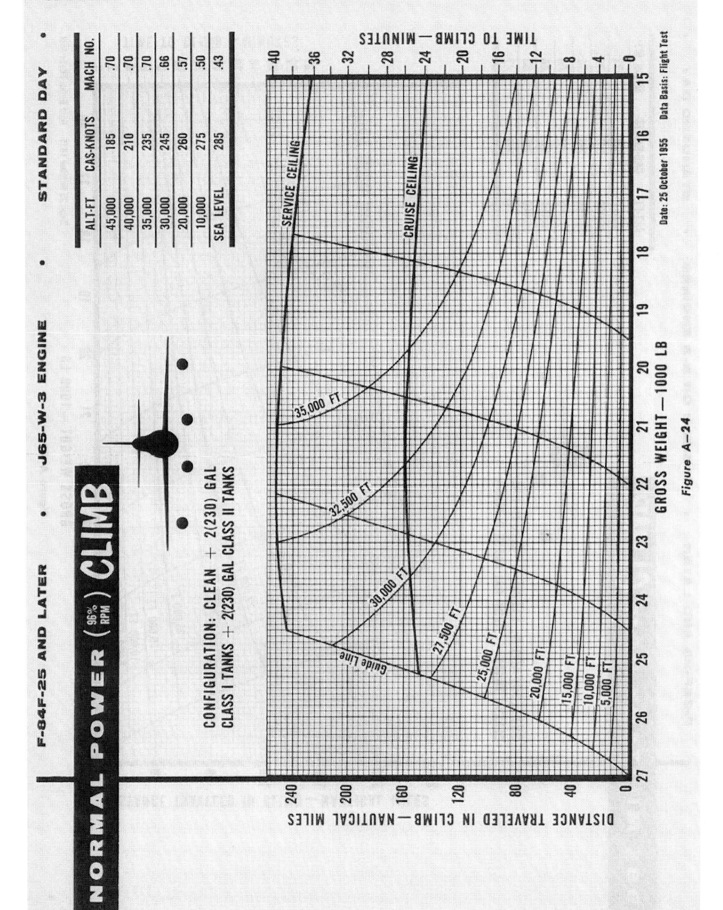

Date: 25 October 1955 Data Basis: Flight Test

Figure A–24

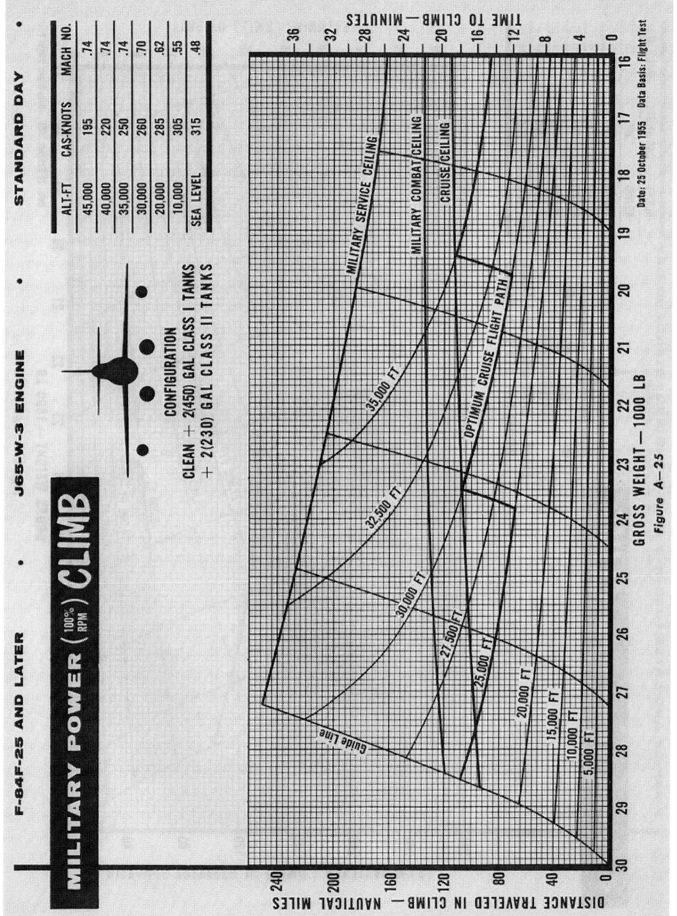

F-84F-25 AND LATER • J65-W-3 ENGINE • STANDARD DAY •

MILITARY POWER (100% RPM) CLIMB

CONFIGURATION
CLEAN + 2(450) GAL CLASS I TANKS
+ 2(230) GAL CLASS II TANKS

ALT-FT	CAS-KNOTS	MACH NO.
45,000	195	.74
40,000	220	.74
35,000	250	.74
30,000	260	.70
20,000	285	.62
10,000	305	.55
SEA LEVEL	315	.48

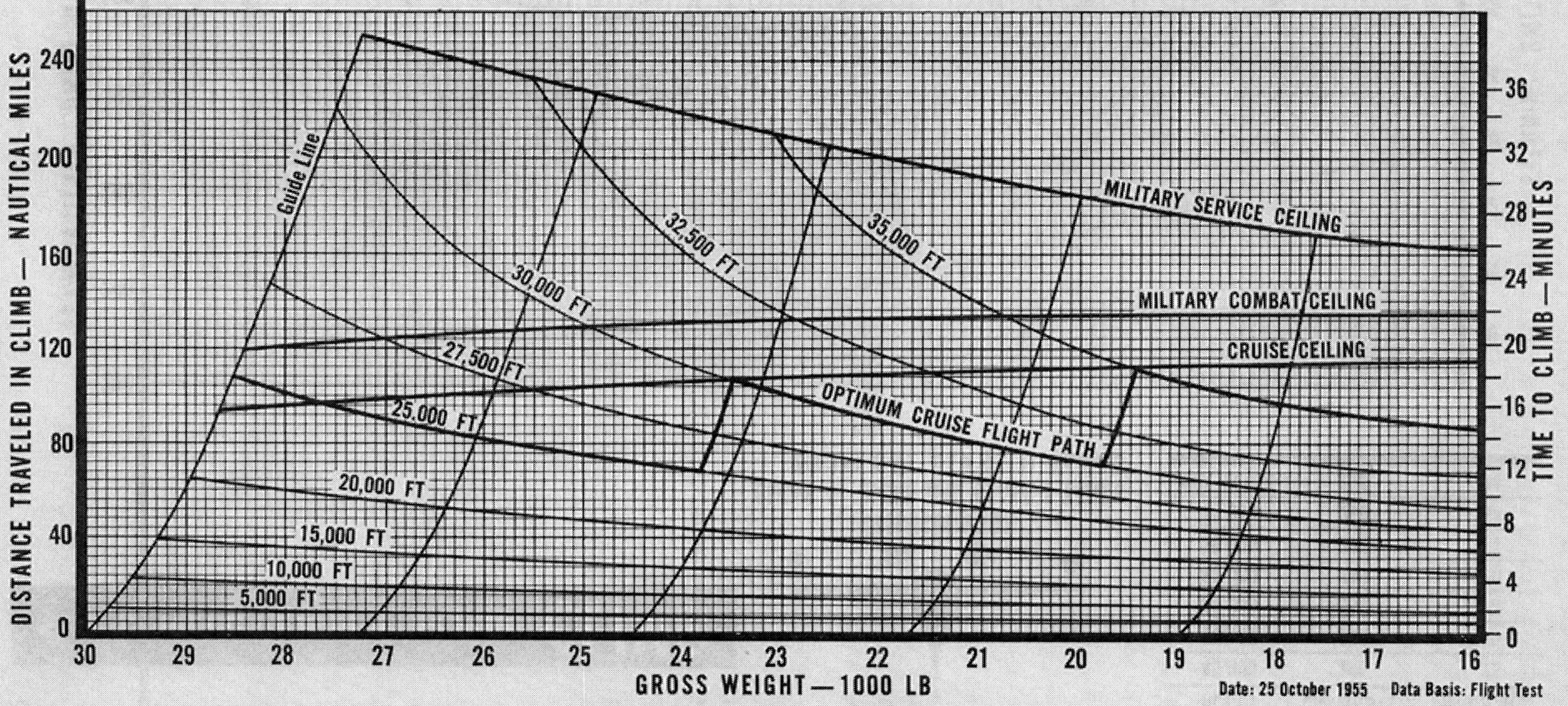

Date: 25 October 1955 Data Basis: Flight Test

Figure A—25

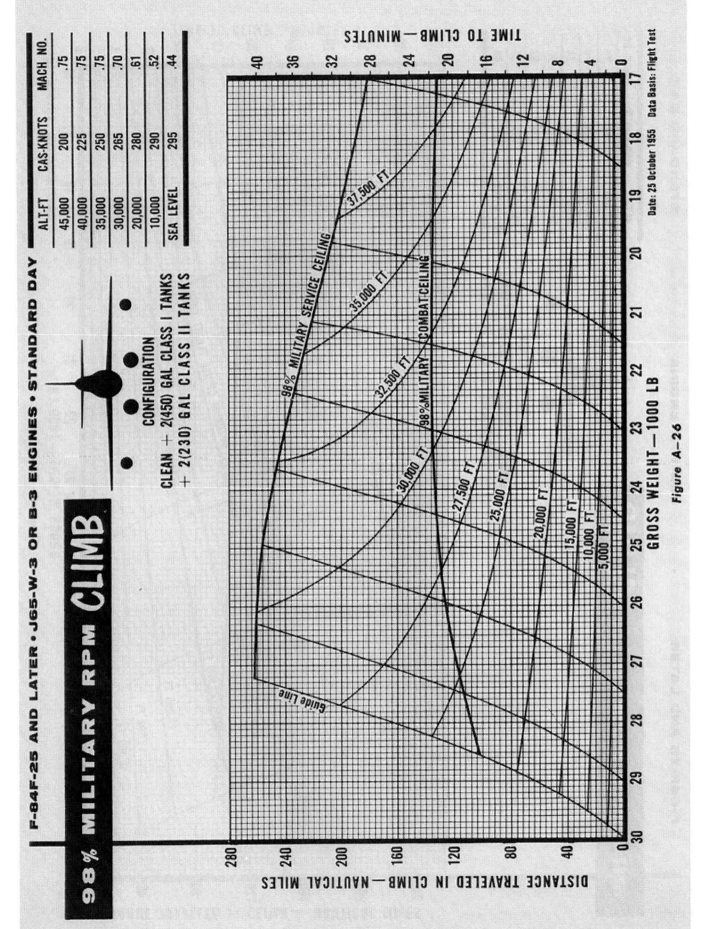

F-84F-25 AND LATER • J65-W-3 OR B-3 ENGINES • STANDARD DAY

98% MILITARY RPM CLIMB

CONFIGURATION
CLEAN + 2(450) GAL CLASS I TANKS
+ 2(230) GAL CLASS II TANKS

ALT-FT	CAS-KNOTS	MACH NO.
45,000	200	.75
40,000	225	.75
35,000	250	.75
30,000	265	.70
20,000	280	.61
10,000	290	.52
SEA LEVEL	295	.44

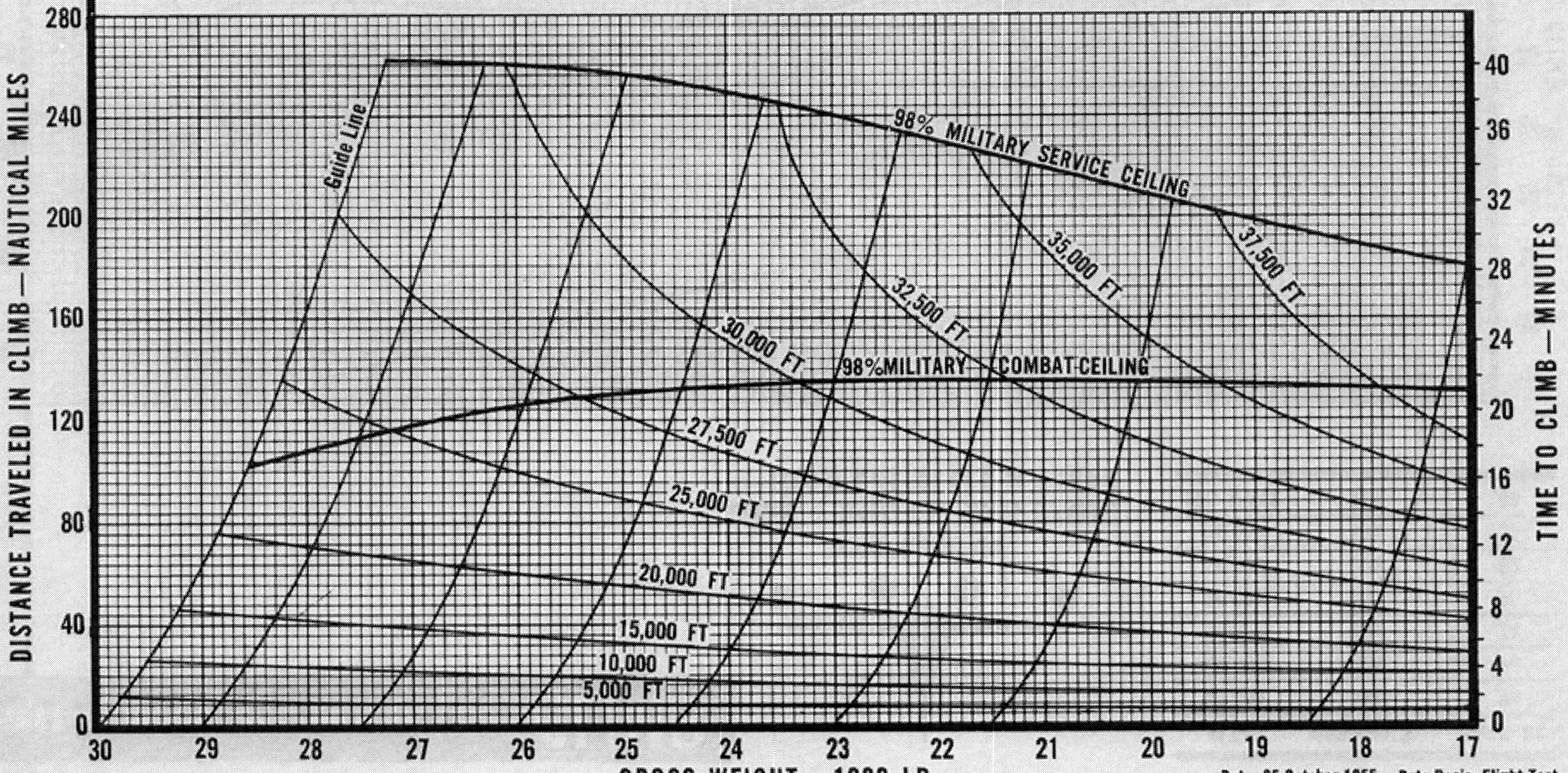

Date: 25 October 1955 Data Basis: Flight Test

Figure A–26

F-84F-25 AND LATER • J65-W-3 ENGINE • STANDARD DAY •

NORMAL POWER (96% RPM) CLIMB

CONFIGURATION
CLEAN + 2(450) GAL CLASS I TANKS
+ 2(230) GAL CLASS II TANKS

ALT-FT	CAS-KNOTS	MACH NO.
45,000	200	.74
40,000	225	.74
35,000	250	.74
30,000	260	.69
20,000	275	.59
10,000	280	.51
SEA LEVEL	285	.43

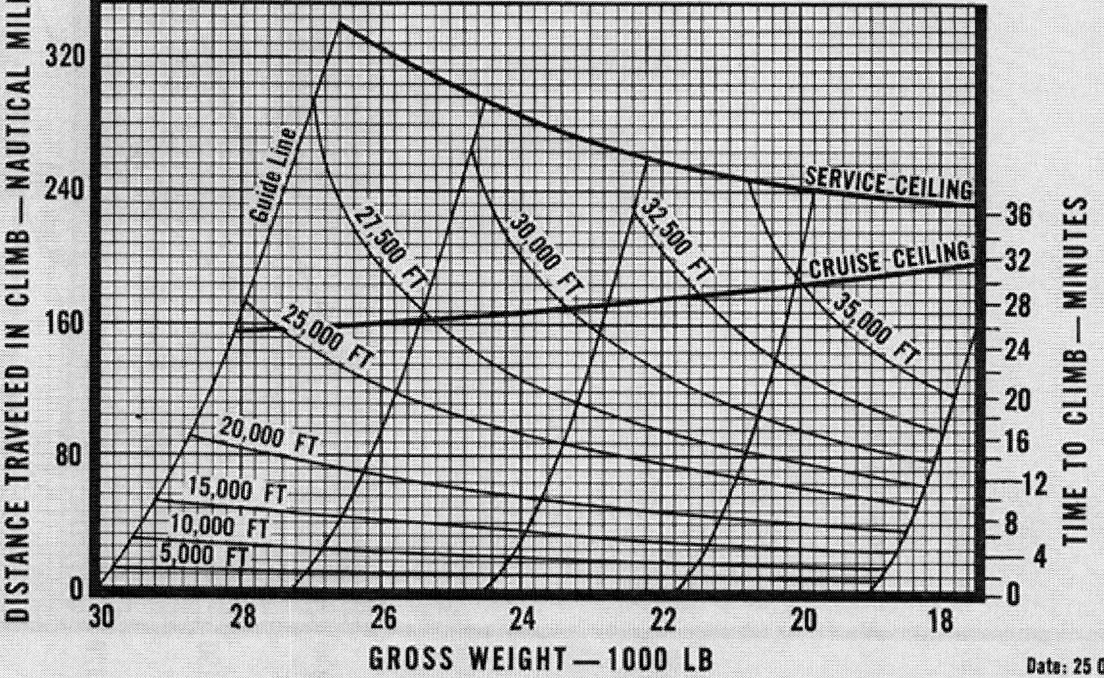

Date: 25 October 1955 Data Basis: Flight Test

Figure A-27

F-84F-25 AND LATER • J65-W-3 ENGINE • STANDARD DAY •

MILITARY POWER (100% RPM) CLIMB

CONFIGURATION
CLEAN + 2(230) GAL CLASS I TANKS
+ 2 1000 LB BOMBS + 4 HVARS

ALT-FT	CAS-KNOTS	MACH NO.
45,000	175	.66
40,000	200	.66
35,000	220	.66
30,000	245	.64
20,000	260	.57
10,000	285	.51
SEA LEVEL	305	.46

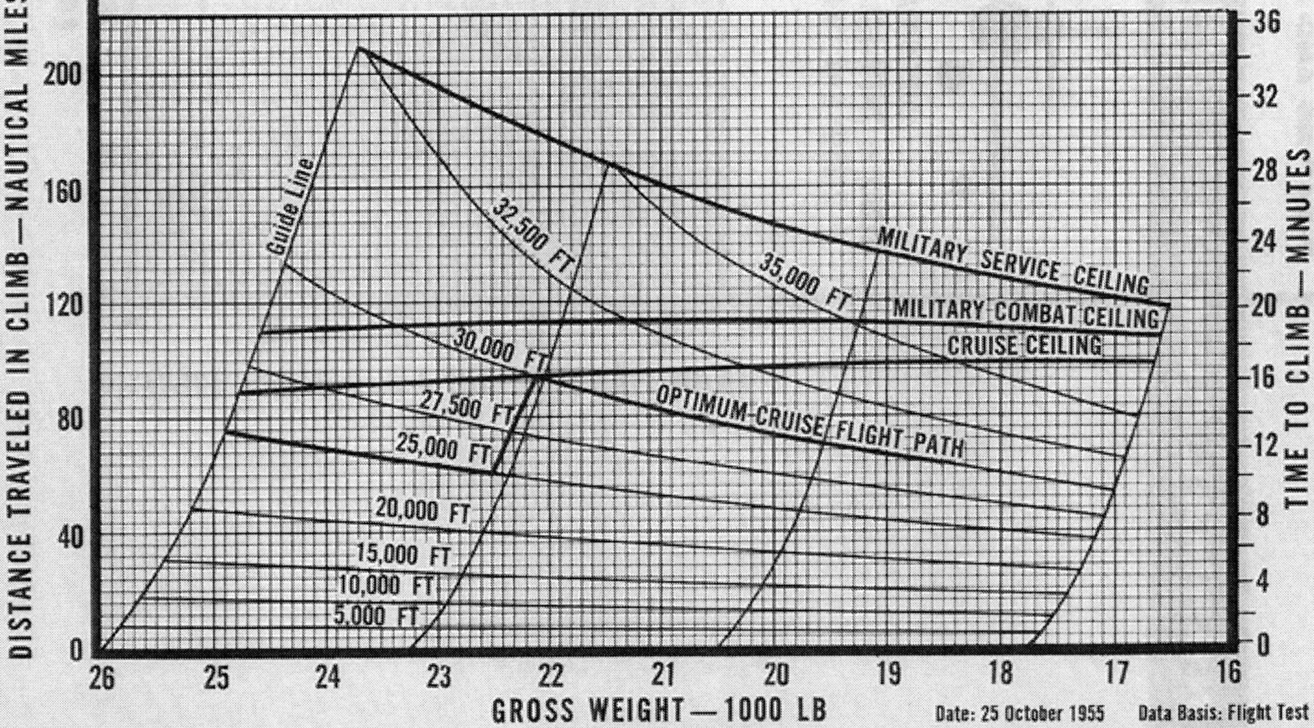

Date: 25 October 1955 Data Basis: Flight Test

Figure A–28

F-84F-25 AND LATER • J65-W-3 OR B-3 ENGINES • STANDARD DAY •

98% MILITARY RPM CLIMB

CONFIGURATION
CLEAN + 2(230) GAL CLASS I TANKS + 2 1000 LB BOMBS + 4 HVARS

ALT-FT	CAS-KNOTS	MACH NO.
45,000	180	.67
40,000	200	.67
35,000	225	.67
30,000	240	.64
20,000	260	.57
10,000	275	.50
SEA LEVEL	290	.44

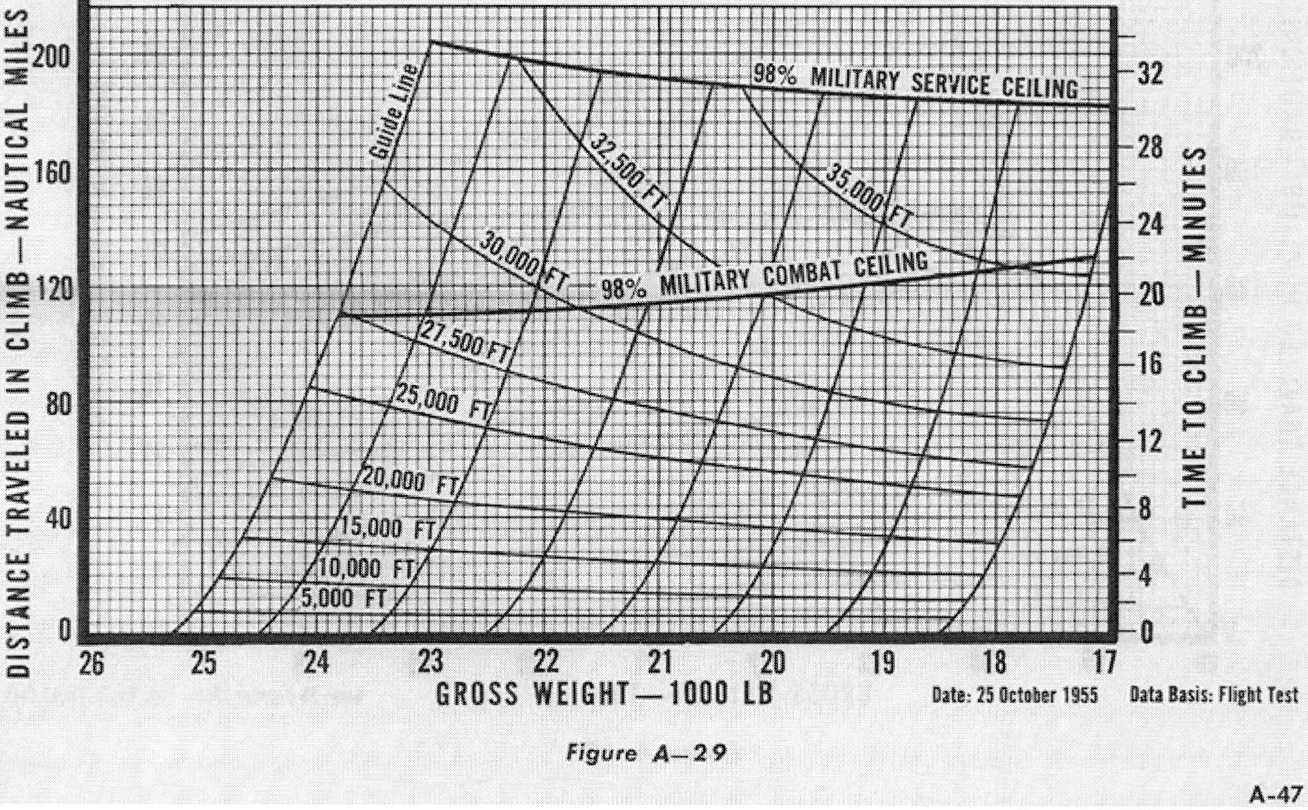

Figure A—29

F-84F-25 AND LATER • J65-W-3 ENGINE • STANDARD DAY •

NORMAL POWER (96% RPM) CLIMB

CONFIGURATION
CLEAN + 2(230) GAL CLASS I TANKS
+ 2 1000 LB BOMBS + 4 HVARS

ALT-FT	CAS-KNOTS	MACH NO.
45,000	175	.66
40,000	195	.66
35,000	220	.67
30,000	230	.62
20,000	250	.55
10,000	265	.48
SEA LEVEL	280	.42

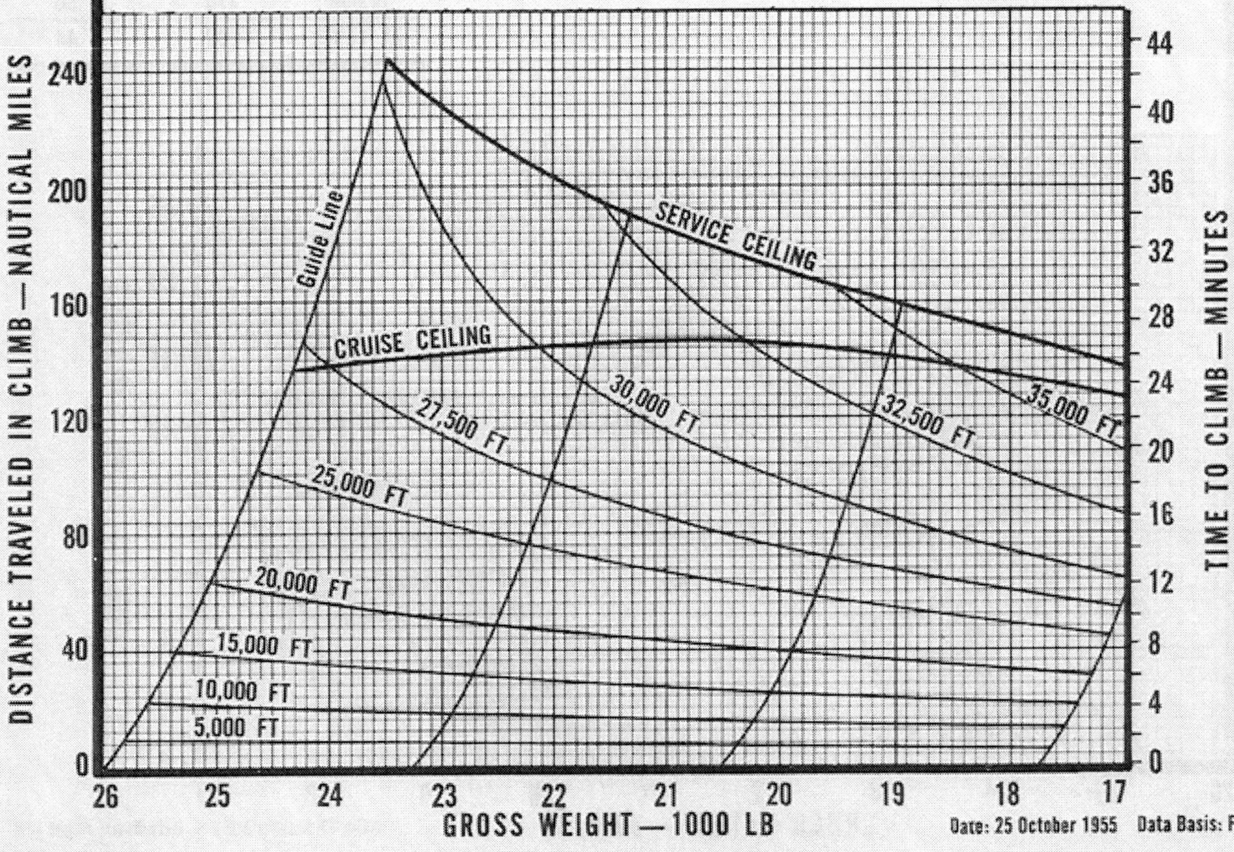

Date: 25 October 1955 Data Basis: Flight Test

Figure A–30

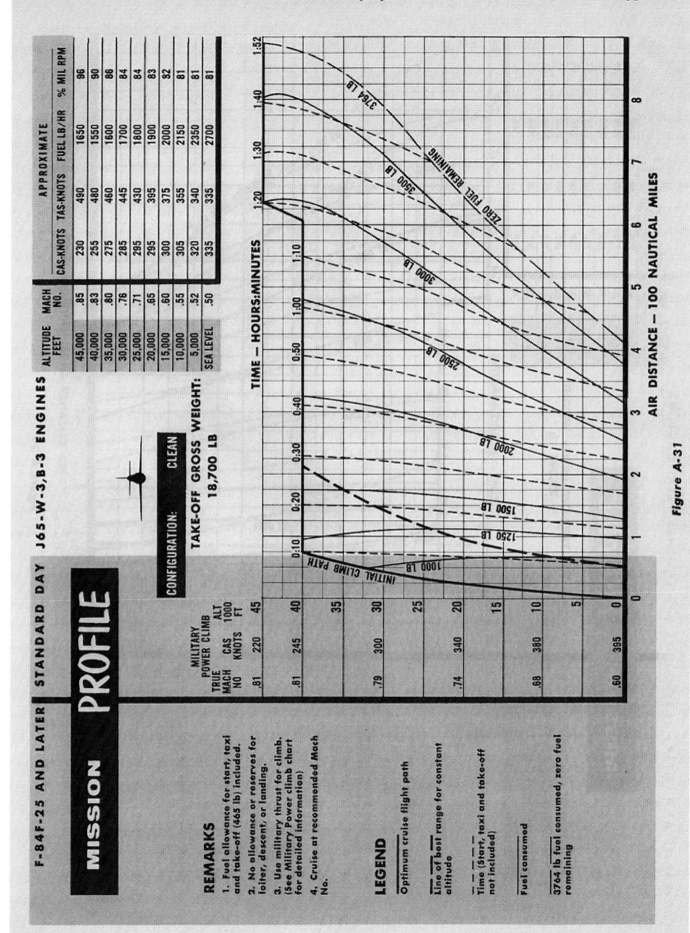

F-84F-25 AND LATER | **STANDARD DAY** | **J65-W-3,B-3 ENGINES**

MISSION PROFILE

CONFIGURATION: CLEAN

TAKE-OFF GROSS WEIGHT: 18,700 LB

ALTITUDE FEET	MACH NO.	APPROXIMATE CAS-KNOTS	APPROXIMATE TAS-KNOTS	APPROXIMATE FUEL LB/HR	APPROXIMATE % MIL RPM
45,000	.85	230	490	1650	96
40,000	.83	255	480	1550	90
35,000	.80	275	460	1600	86
30,000	.76	285	445	1700	84
25,000	.71	295	430	1800	84
20,000	.65	295	395	1900	83
15,000	.60	300	375	2000	82
10,000	.55	305	355	2150	81
5,000	.52	320	340	2350	81
SEA LEVEL	.50	335	335	2700	81

REMARKS

1. Fuel allowance for start, taxi and take-off (465 lb) included.
2. No allowance or reserves for loiter, descent, or landing.
3. Use military thrust for climb. (See Military Power climb chart for detailed information)
4. Cruise at recommended Mach No.

LEGEND

Optimum cruise flight path

Line of best range for constant altitude

Time (Start, taxi and take-off not included)

Fuel consumed

3764 lb fuel consumed, zero fuel remaining

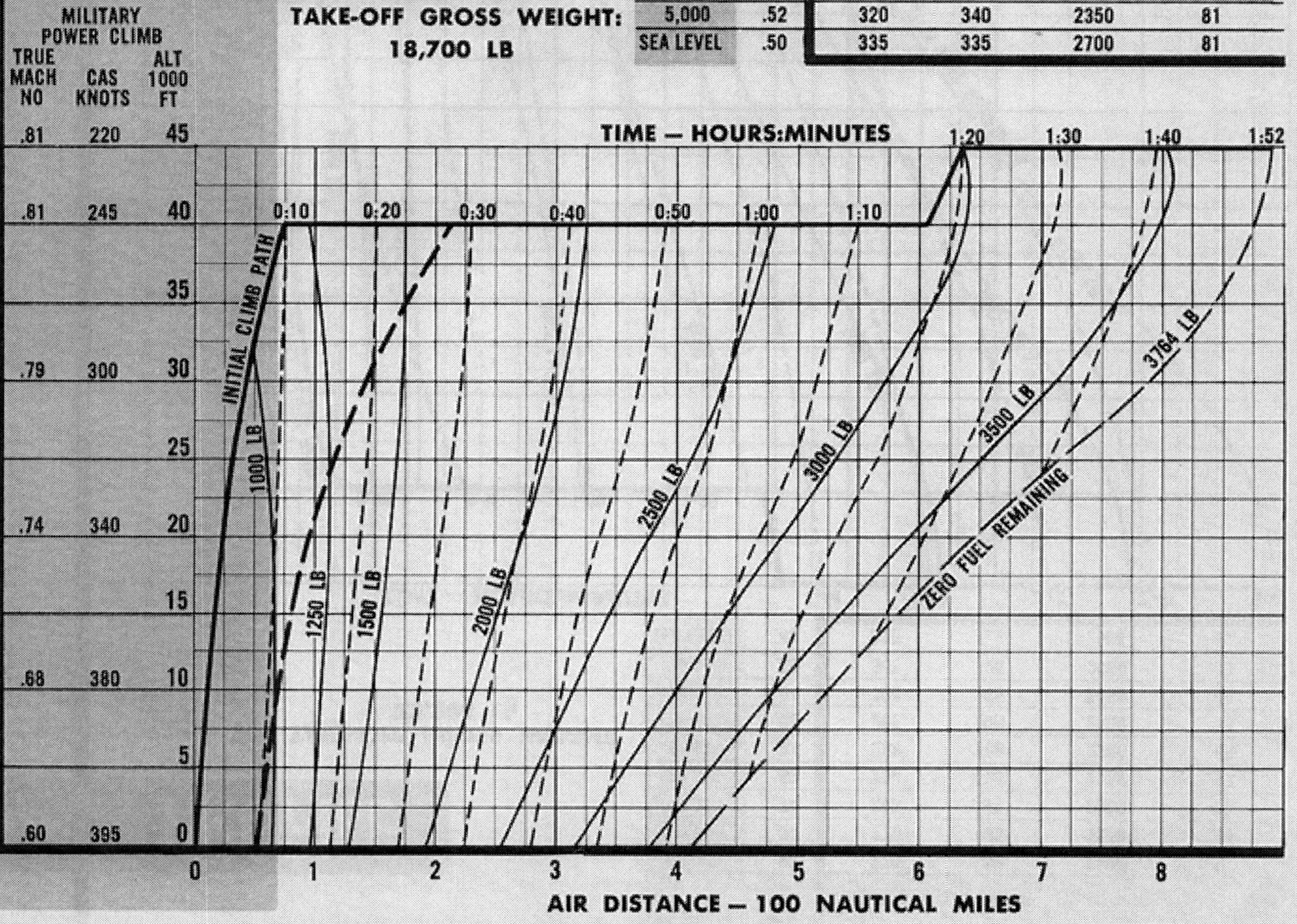

Figure A-31

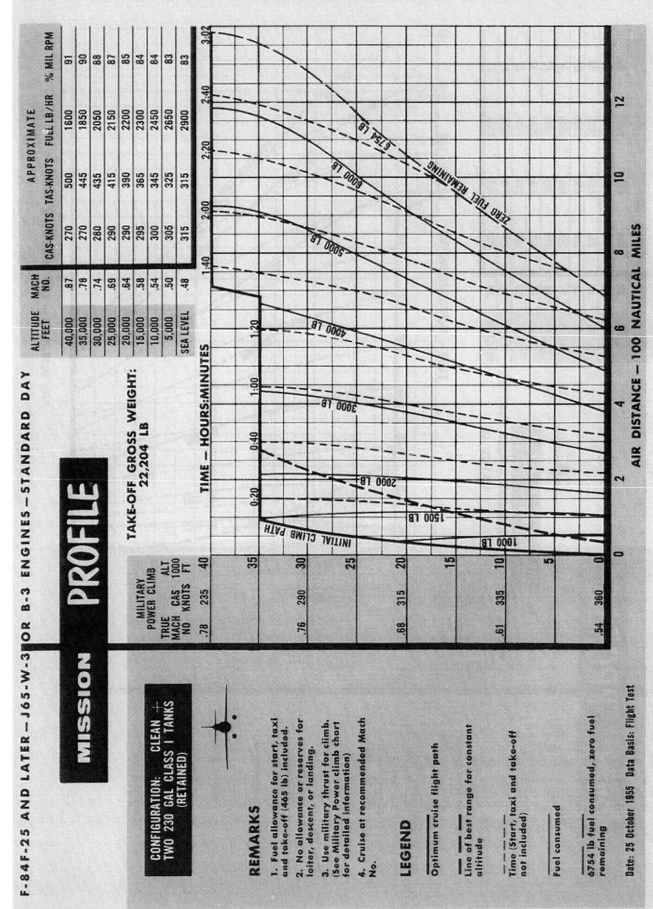

F-84F-25 AND LATER — J65-W-3 OR B-3 ENGINES — STANDARD DAY

MISSION PROFILE

ALTITUDE FEET	MACH NO.	APPROXIMATE CAS-KNOTS	APPROXIMATE TAS-KNOTS	APPROXIMATE FUEL LB/HR	APPROXIMATE % MIL RPM
40,000	.87	270	500	1600	91
35,000	.78	270	445	1850	90
30,000	.74	280	435	2050	88
25,000	.69	290	415	2150	87
20,000	.64	290	390	2200	85
15,000	.58	295	365	2300	84
10,000	.54	300	345	2450	84
5,000	.50	305	325	2650	83
SEA LEVEL	.48	315	315	2900	83

TAKE-OFF GROSS WEIGHT: 22,204 LB

CONFIGURATION: CLEAN + TWO 230 GAL CLASS I TANKS (RETAINED)

MILITARY POWER CLIMB

TRUE MACH NO	CAS KNOTS	ALT 1000 FT
.78	235	40
		35
.76	290	30
		25
.68	315	20
		15
.61	335	10
		5
.54	360	0

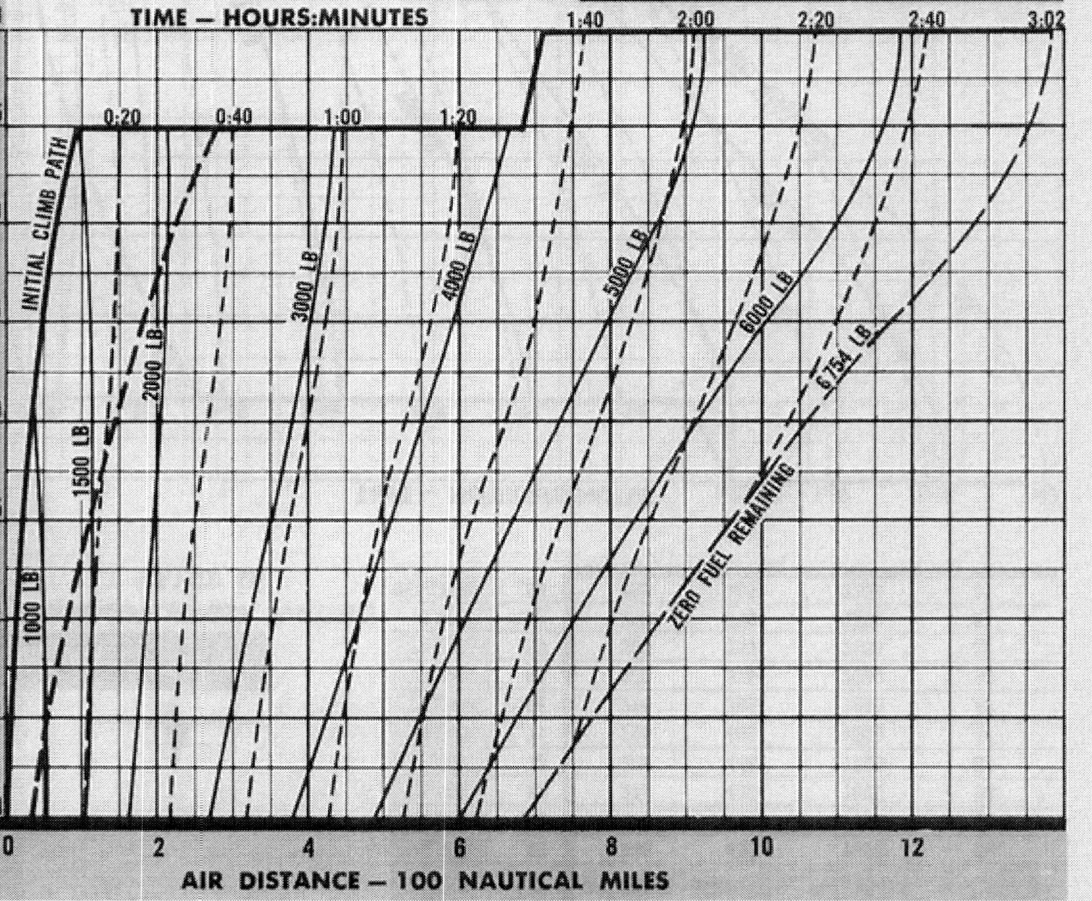

REMARKS

1. Fuel allowance for start, taxi and take-off (465 lb) included.
2. No allowance or reserves for loiter, descent, or landing.
3. Use military thrust for climb. (See Military Power climb chart for detailed information)
4. Cruise at recommended Mach No.

LEGEND

Optimum cruise flight path

Line of best range for constant altitude

Time (Start, taxi and take-off not included)

Fuel consumed

6754 lb fuel consumed, zero fuel remaining

Date: 25 October 1955 Data Basis: Flight Test

Figure A-32

STANDARD DAY F-84F-25 AND LATER — J65-W-3 OR B-3 ENGINES

MISSION PROFILE

CONFIGURATION: CLEAN + TWO 450 GAL CLASS I TANKS (RETAINED)

TAKE-OFF GROSS WEIGHT: 25,226 LB

ALTITUDE FEET	MACH NO.	APPROXIMATE CAS-KNOTS	TAS-KNOTS	FUEL LB/HR	% MIL RPM
35,000	.79	270	455	2150	93
30,000	.75	280	445	2200	90
25,000	.71	300	425	2350	89
20,000	.66	305	405	2450	87
15,000	.61	310	380	2550	86
10,000	.56	310	360	2650	85
5,000	.52	315	340	2850	84
SEA LEVEL	.48	320	320	3050	83

REMARKS

1. Fuel allowance for start, taxi and take-off (465 lb) included.
2. No allowance or reserves for loiter, descent, or landing.
3. Use military thrust for climb. (See Military Power climb chart for detailed information)
4. Cruise at recommended Mach No.

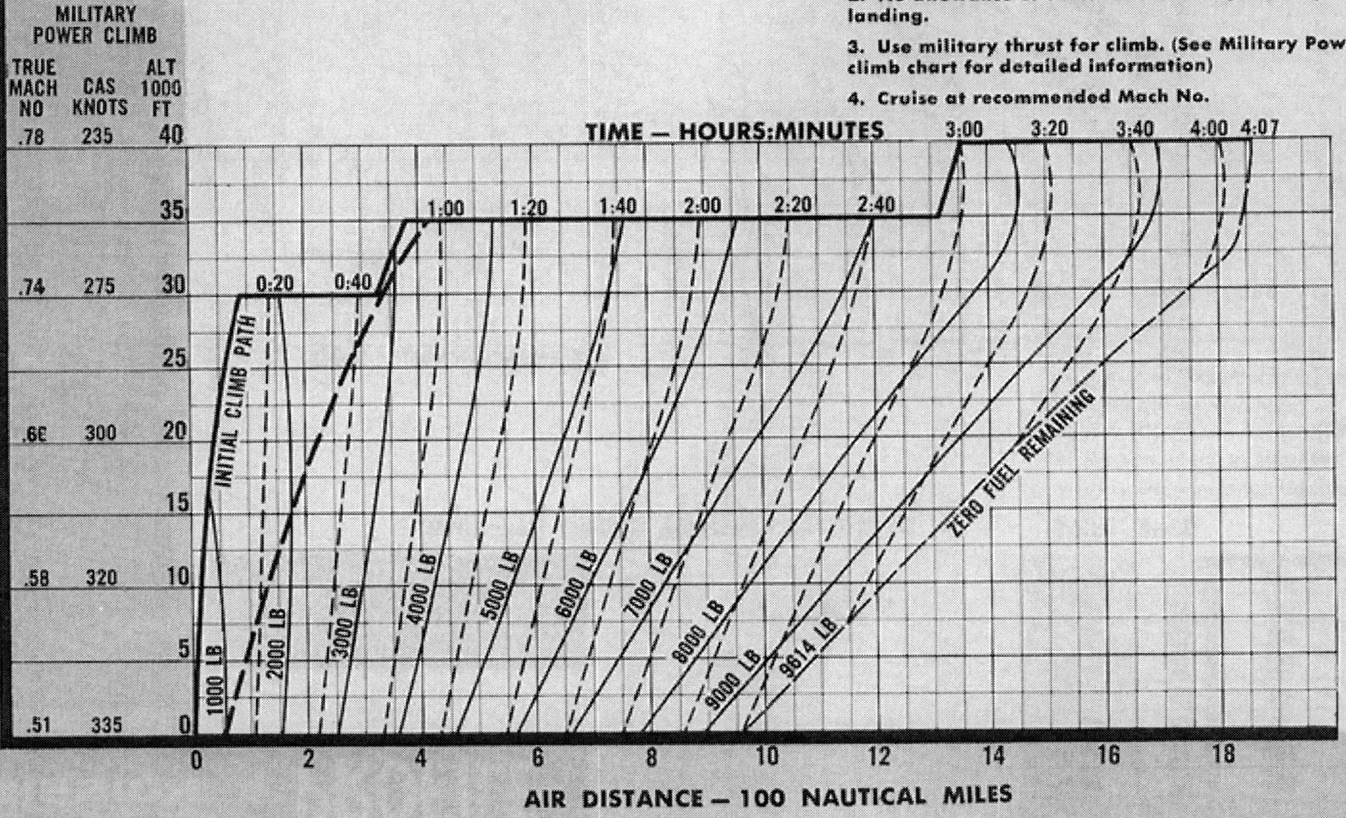

LEGEND

Optimum cruise flight path

Line of best range for constant altitude

Time (Start, taxi and take-off not included)

Fuel consumed

9614 lb fuel consumed, zero fuel remaining

Date: 25 October 1955
Data Basis: Flight Test

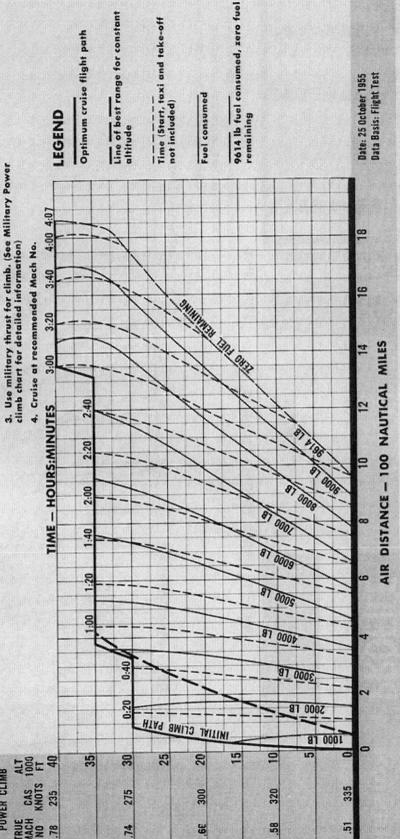

Figure A-33

STANDARD DAY — F-84F-25 AND LATER — J65-W-3 OR B-3 ENGINE

MISSION PROFILE

CONFIGURATION: CLEAN +
TWO 230 GAL CLASS I TANKS +
TWO 230 GAL CLASS II TANKS
(RETAINED)

ALTITUDE FEET	MACH NO.	APPROXIMATE CAS-KNOTS	APPROXIMATE TAS-KNOTS	APPROXIMATE FUEL LB/HR	APPROXIMATE % MIL RPM
35,000	.73	250	420	2050	92
30,000	.71	265	420	2400	92
25,000	.67	280	405	2500	90
20,000	.63	290	385	2600	89
15,000	.58	295	365	2750	87
10,000	.53	295	340	2850	86
5,000	.49	300	320	3000	85
SEA LEVEL	.47	310	310	3250	80

TAKE-OFF GROSS WEIGHT: 25,693 LB

REMARKS

1. Fuel allowance for start, taxi and take-off (465 lb) included.
2. No allowance or reserves for loiter, descent, or landing.
3. Use military thrust for climb. (See Military Power climb chart for detailed information)
4. Cruise at recommended Mach No.

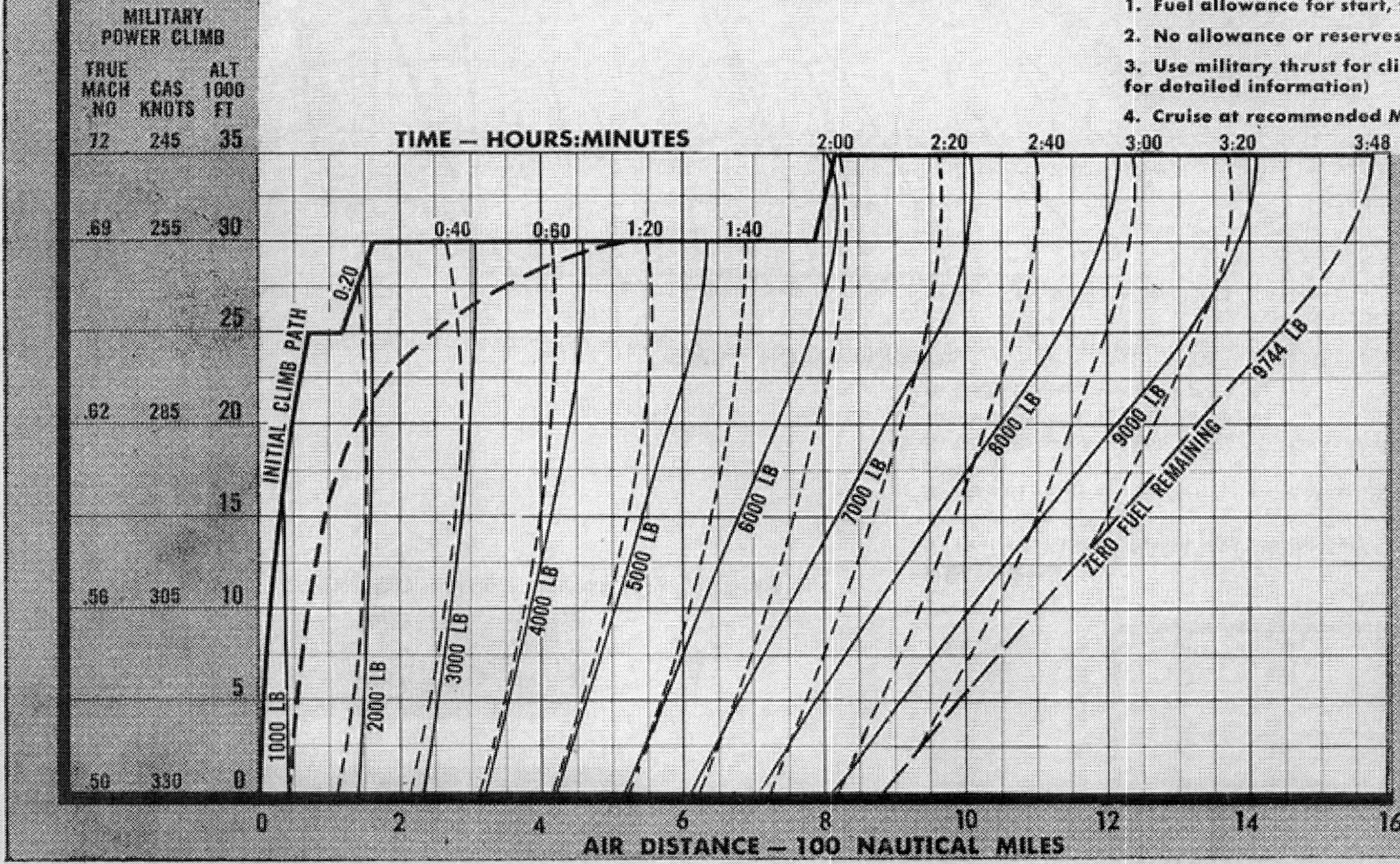

LEGEND

Optimum cruise flight path

Line of best range for constant altitude

Time (Start, taxi and take-off not included)

Fuel consumed

9744 lb fuel consumed, zero fuel remaining

Date: 25 October 1955
Data Basis: Flight Test

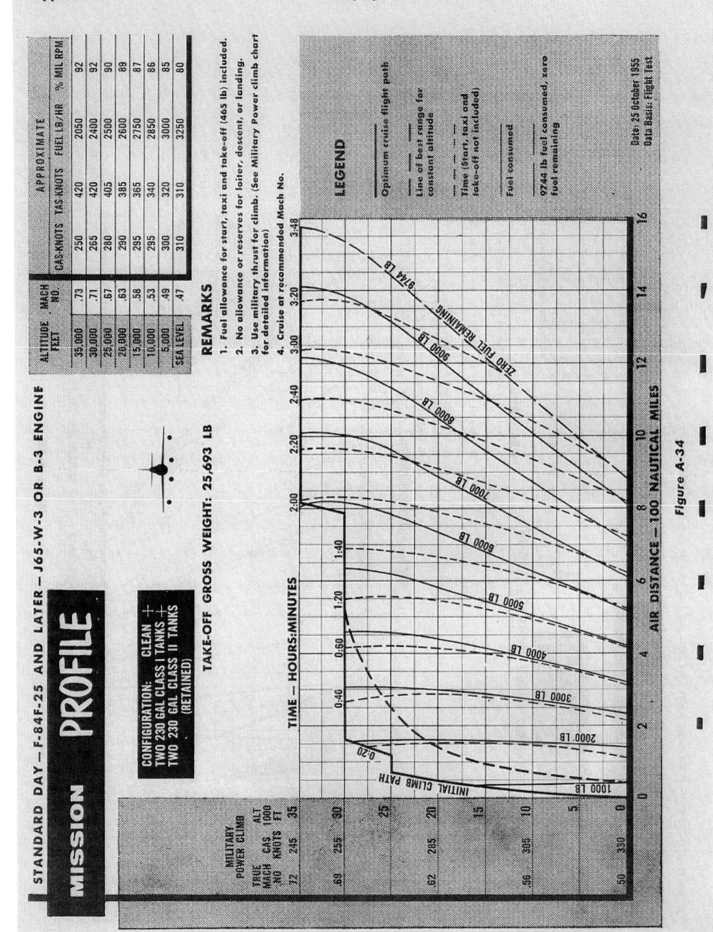

Figure A-34

STANDARD DAY — F-84F-25 AND LATER — J65-W-3 OR B-3 ENGINES

MISSION PROFILE

TAKE-OFF GROSS WEIGHT: 27,000 LB

CONFIGURATION: CLEAN + TWO 450 GAL CLASS I TANKS + TWO 230 GAL CLASS II TANKS (RETAINED)

ALTITUDE FEET	MACH NO.	APPROXIMATE			
		CAS-KNOTS	TAS-KNOTS	FUEL LB/HR	% MIL RPM
35,000	.71	240	435	2100	94
30,000	.75	280	440	2500	94
25,000	.74	305	445	3200	94
20,000	.66	305	400	2900	90
15,000	.60	305	380	3000	89
10,000	.56	305	355	3100	87
5,000	.51	310	330	3250	86
SEA LEVEL	.47	310	310	3350	85

REMARKS

1. Fuel allowance for start, taxi and take-off (465 lb) included.
2. No allowance or reserves for loiter, descent, or landing.
3. Use military thrust for climb. (See Military Power climb chart for detailed information)
4. Cruise at recommended Mach No.

LEGEND

Optimum cruise flight path

Line of best range for constant altitude

Time (Start, taxi and take-off not included)

Fuel consumed

10,889 lb fuel consumed, zero fuel remaining

MILITARY POWER CLIMB

TRUE MACH NO	CAS KNOTS	ALT 1000 FT
.74	250	35
.70	260	30
		25
.62	285	20
		15
.55	305	10
		5
.48	315	0

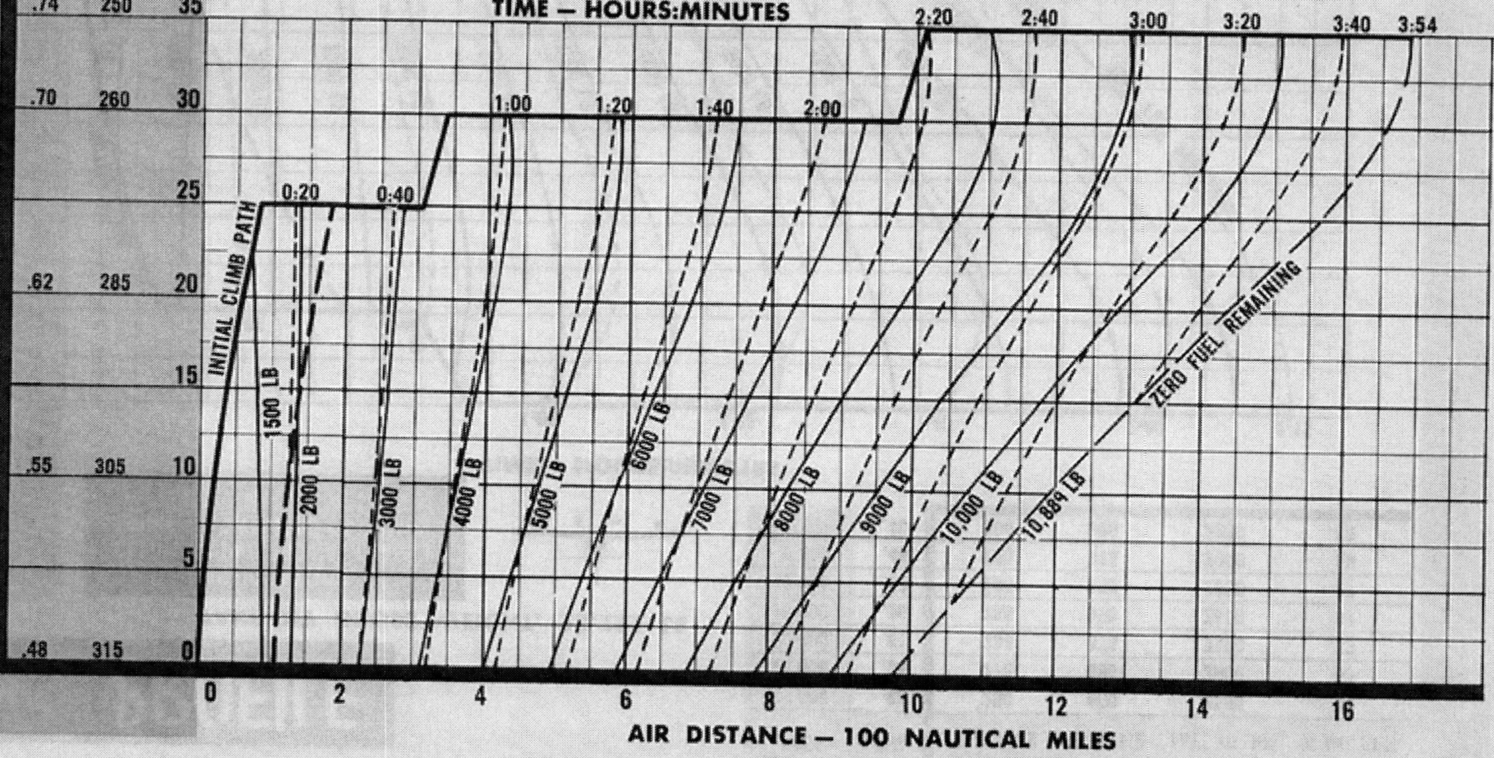

Date: 25 October 1955
Data Basis: Flight Test

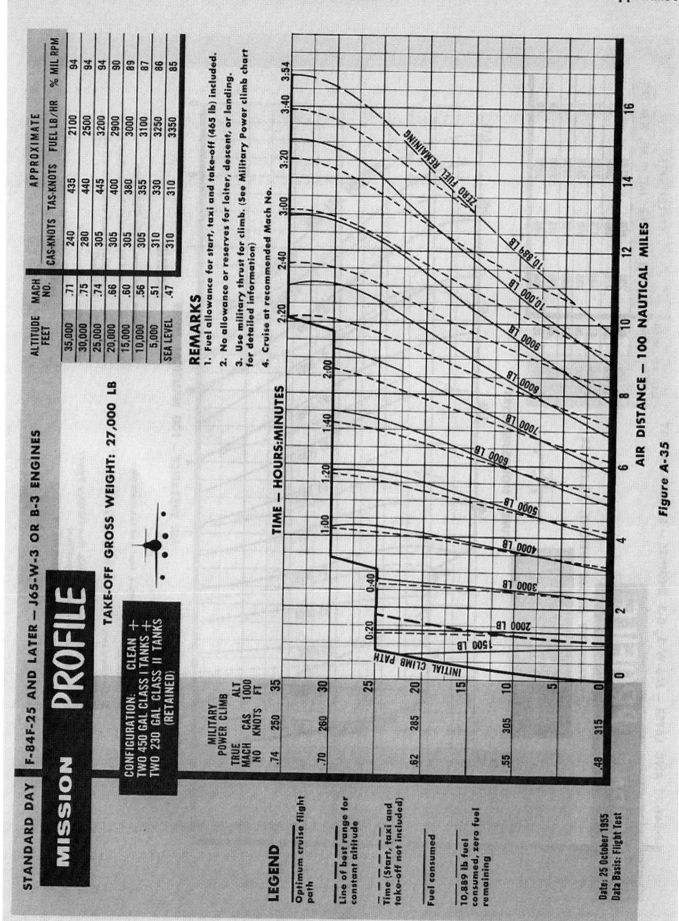

Figure A-35

F-84F-25 AND LATER — J65-W-3, B-3 ENGINES — STANDARD DAY

MISSION PROFILE

TAKE-OFF GROSS WEIGHT: 25,252 LB

CONFIGURATION:
CLEAN + TWO 230 GAL CLASS I TANKS + TWO 1000 LB BOMBS + FOUR HVARS — (RETAINED)

ALTITUDE FEET	MACH NO.	APPROXIMATE CAS-KNOTS	TAS-KNOTS	FUEL LB/HR	% MIL RPM
30,000	.68	250	400	2500	94
25,000	.65	270	390	2600	92
20,000	.61	280	375	2700	89
15,000	.56	285	355	2800	87
10,000	.52	285	330	2850	87
5,000	.48	290	315	3000	86
SEA LEVEL	.46	300	300	3200	85

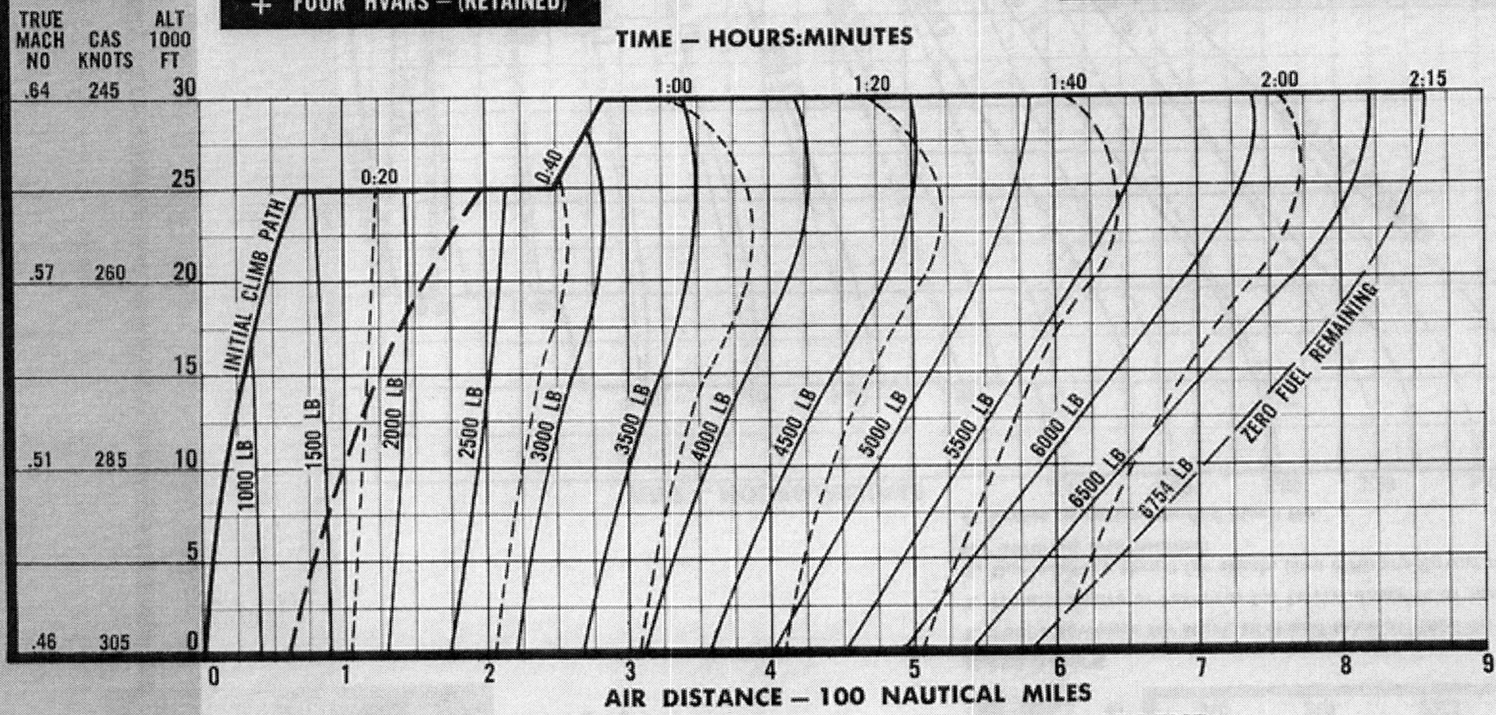

REMARKS

1. Fuel allowance for start, taxi and take-off (465 lb) included.
2. No allowance or reserves for loiter, descent, or landing.
3. Use military thrust for climb. (See Military Power climb chart for detailed information)
4. Cruise at recommended Mach No.

LEGEND

- Optimum cruise flight path
- Line of best range for constant altitude
- Time (Start, taxi and take-off not included)
- Fuel consumed
- 6754 lb fuel consumed, zero fuel remaining

Date: 25 October 1955
Data Basis: Flight Test

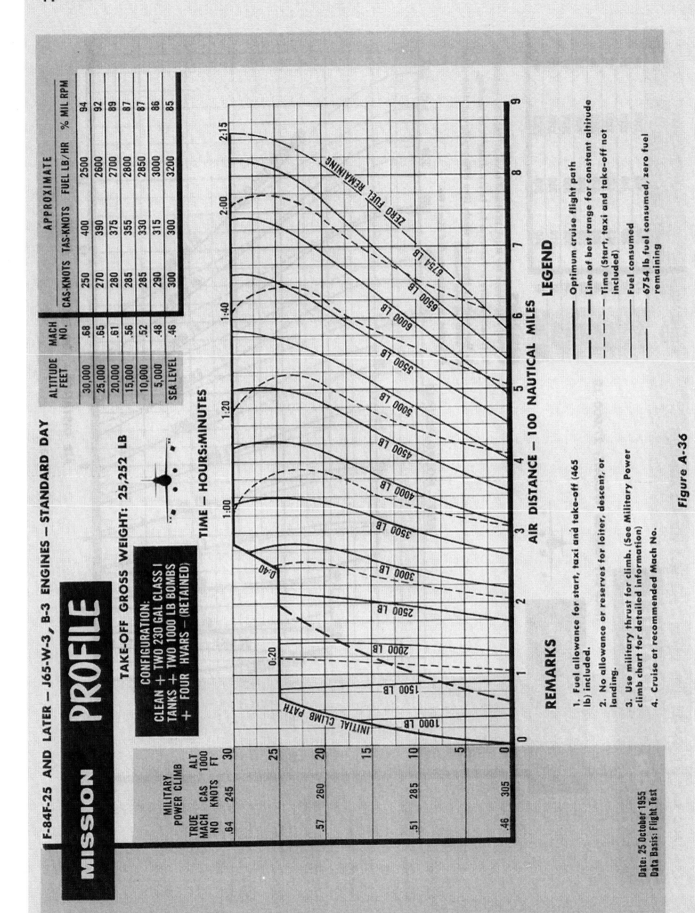

Figure A-36

F-84F-25 AND LATER — J65-W-3, B-3 ENGINES — STANDARD DAY

CONFIGURATION: CLEAN

GROSS WEIGHT: 14,936 — 18,700 LB

OPTIMUM RETURN PROFILE

ALTITUDE FEET	MACH NO.	APPROXIMATE CAS-KNOTS	TAS-KNOTS	FUEL LB/HR	% MIL RPM
45,000	.85	230	490	1650	96
40,000	.83	255	480	1550	90
35,000	.80	275	460	1600	86
30,000	.76	285	445	1700	84
25,000	.71	295	430	1800	84
20,000	.65	295	395	1900	83
15,000	.60	300	375	2000	82
10,000	.55	305	355	2150	81
5,000	.52	320	340	2350	81
SEA LEVEL	.50	335	335	2700	81

REMARKS

1. Fuel required at any point below the line of best range for constant altitude flight, or below the optimum range flight profile, includes military power climb to flight altitude.
2. No allowance made for loiter, descent or landing.
3. Best cruise condition determined by intersection of climb path guide lines and lines of best range.
4. Cruise at recommended Mach No.

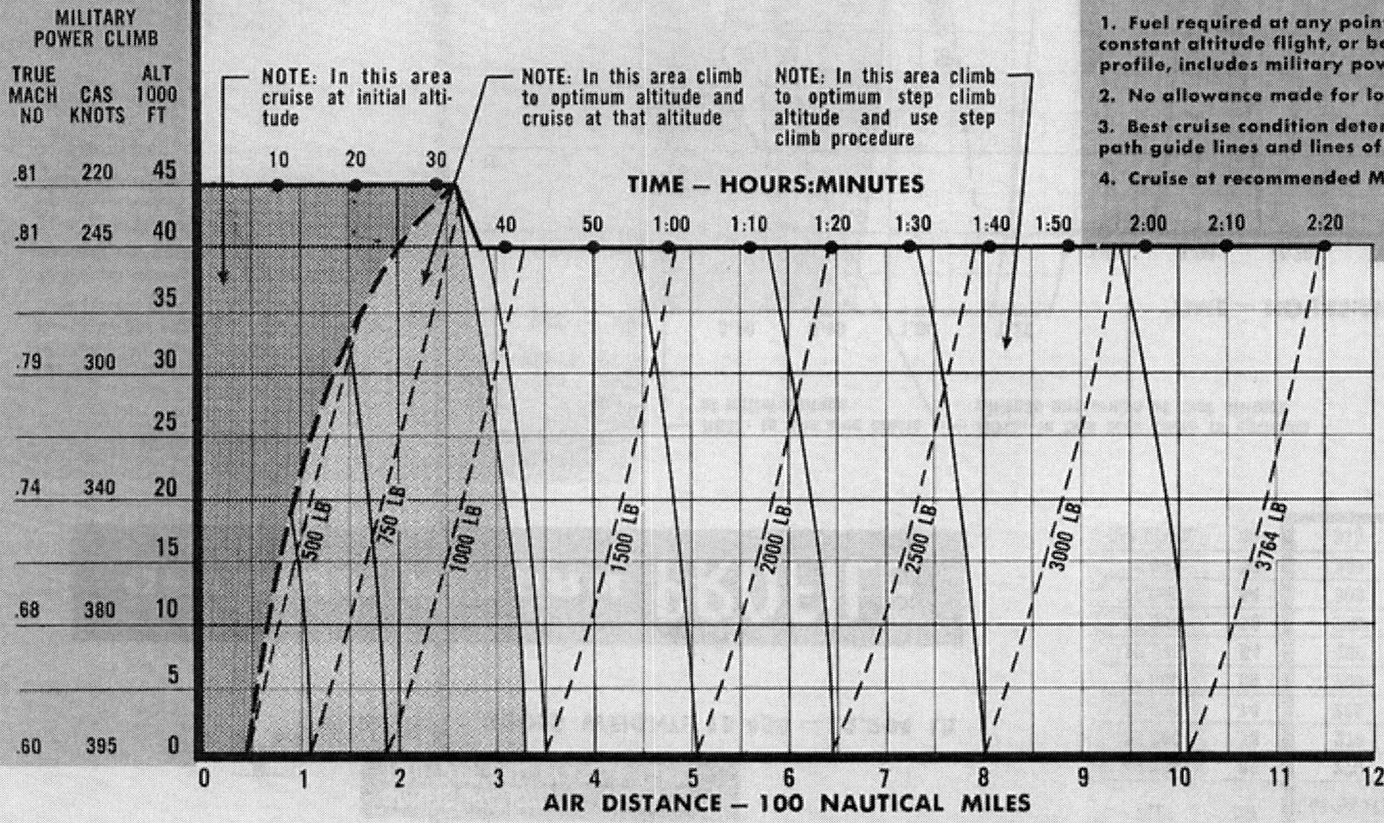

LEGEND

Optimum range flight profile

Line of best range for constant altitude

Climb path guide lines

Fuel required

Time at cruising altitude

Date: 25 October 1955
Data Basis: Flight Test

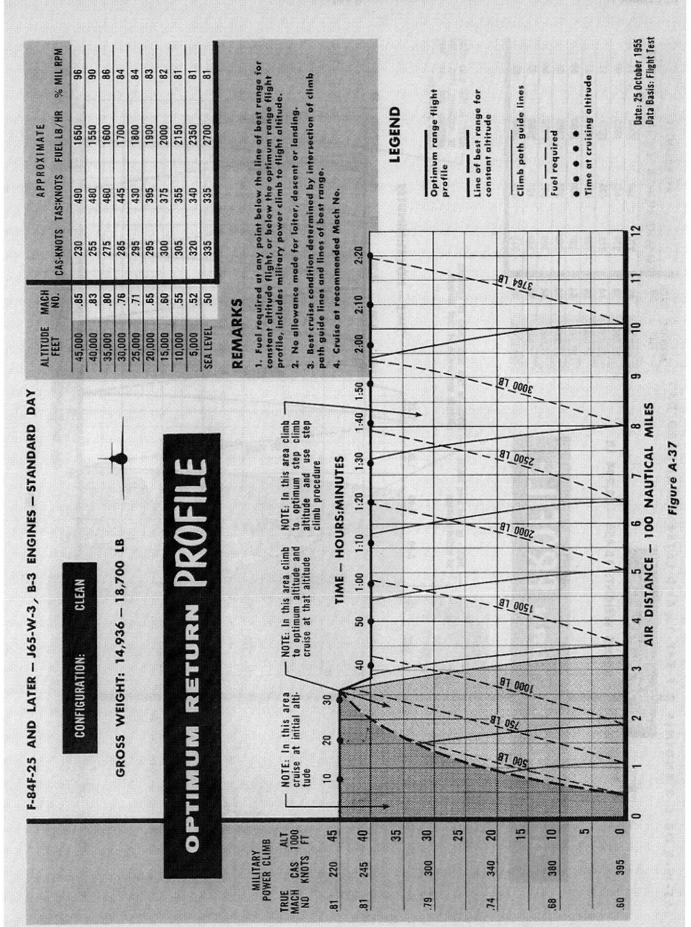

Figure A-37

J65-W-3 OR B-3 ENGINES — F-84F-25 AND LATER — STANDARD DAY

CONFIGURATION: CLEAN + TWO 230 GAL CLASS I TANKS

GROSS WEIGHT: 15,450 — 22,204 LB

OPTIMUM RETURN PROFILE

ALTITUDE FEET	MACH NO.	APPROXIMATE CAS-KNOTS	TAS-KNOTS	FUEL LB/HR	% MIL RPM
40,000	.87	270	500	1600	91
35,000	.78	270	445	1850	90
30,000	.74	280	435	2050	88
25,000	.69	290	415	2150	87
20,000	.64	290	390	2200	85
15,000	.58	295	365	2300	84
10,000	.54	300	345	2450	84
5,000	.50	305	325	2650	83
SEA LEVEL	.48	315	315	2900	83

REMARKS

1. Fuel required at any point below the line of best range for constant altitude flight, or below the optimum range flight profile, includes military power climb to flight altitude.
2. No allowance made for loiter, descent or landing.
3. Best cruise condition determined by intersection of climb path guide lines and lines of best range.
4. Cruise at recommended Mach No.

LEGEND

- Optimum range flight profile
- Line of best range for constant altitude
- Climb path guide lines
- Fuel required
- Time at cruising altitude

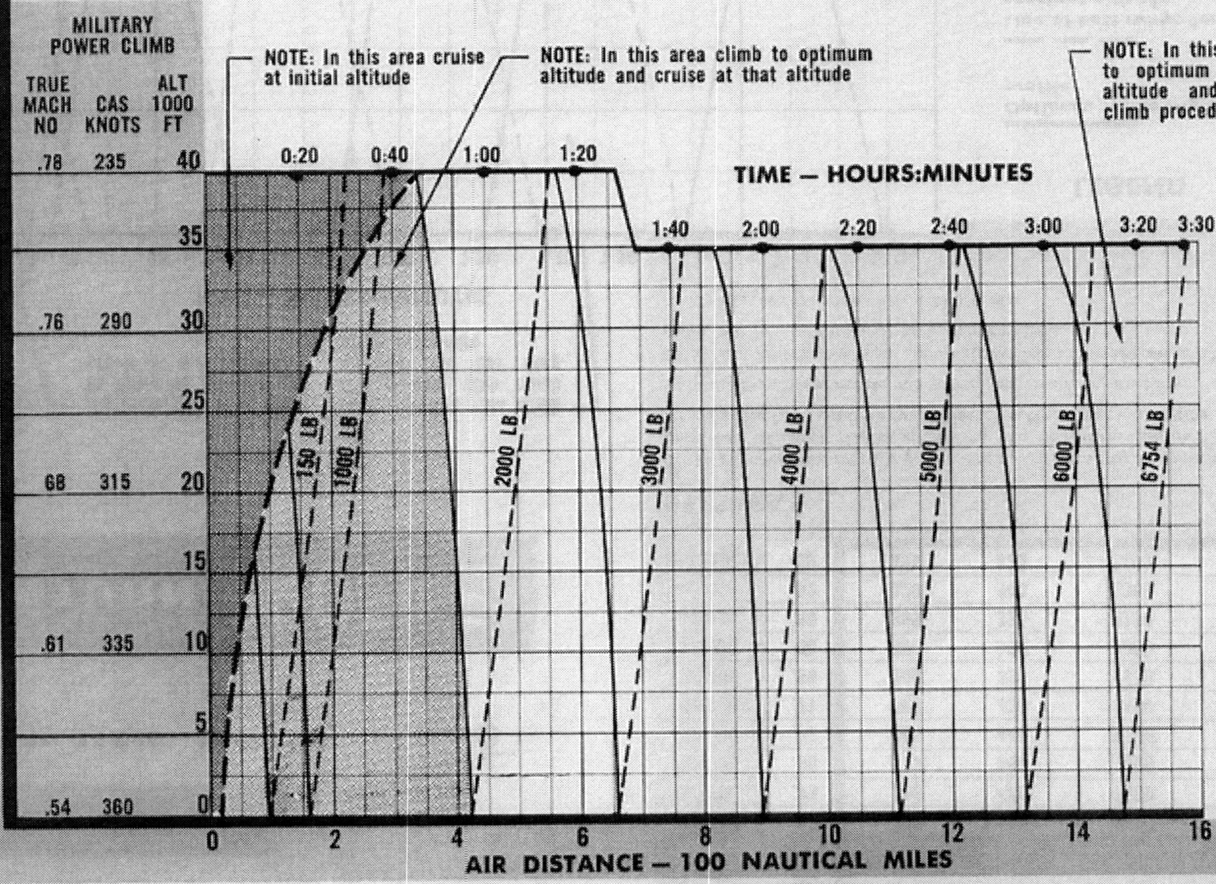

Date: 25 October 1955 Data Basis: Flight Test

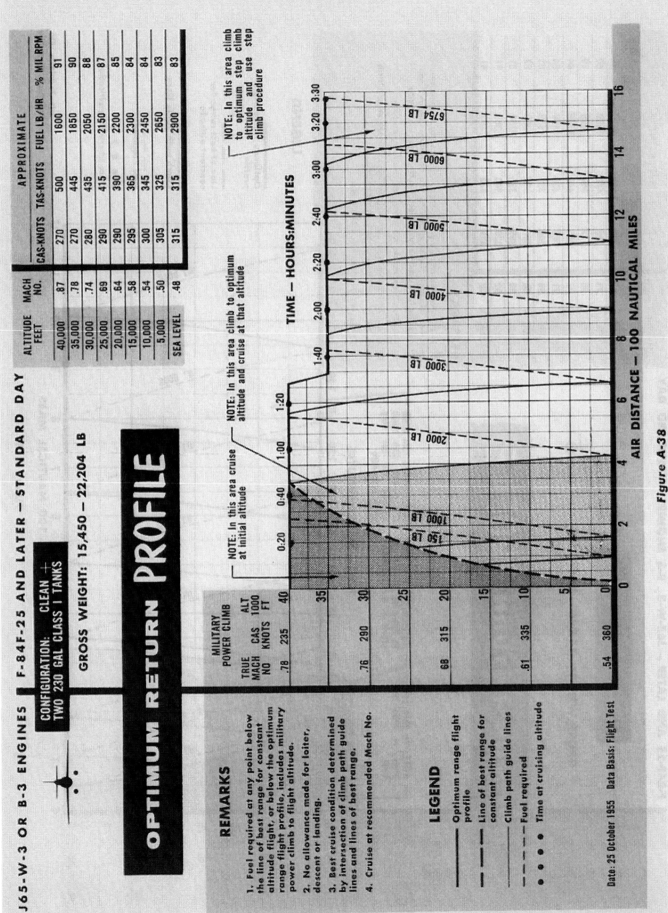

Figure A-38

F-84F-25 AND LATER — J65-W-3 OR B-3 ENGINES — STANDARD DAY

OPTIMUM RETURN PROFILE

CONFIGURATION: CLEAN
TWO 450 GAL CLASS I TANKS

GROSS WEIGHT: 15,612 — 25,226 LB

ALTITUDE FEET	MACH NO.	APPROXIMATE CAS-KNOTS	APPROXIMATE TAS-KNOTS	APPROXIMATE FUEL LB/HR	APPROXIMATE % MIL RPM
35,000	.79	270	455	2150	93
30,000	.75	280	445	2200	90
25,000	.71	300	425	2350	89
20,000	.66	305	405	2450	87
15,000	.61	310	380	2550	86
10,000	.56	310	360	2650	85
5,000	.52	315	340	2850	84
SEA LEVEL	.48	320	320	3050	83

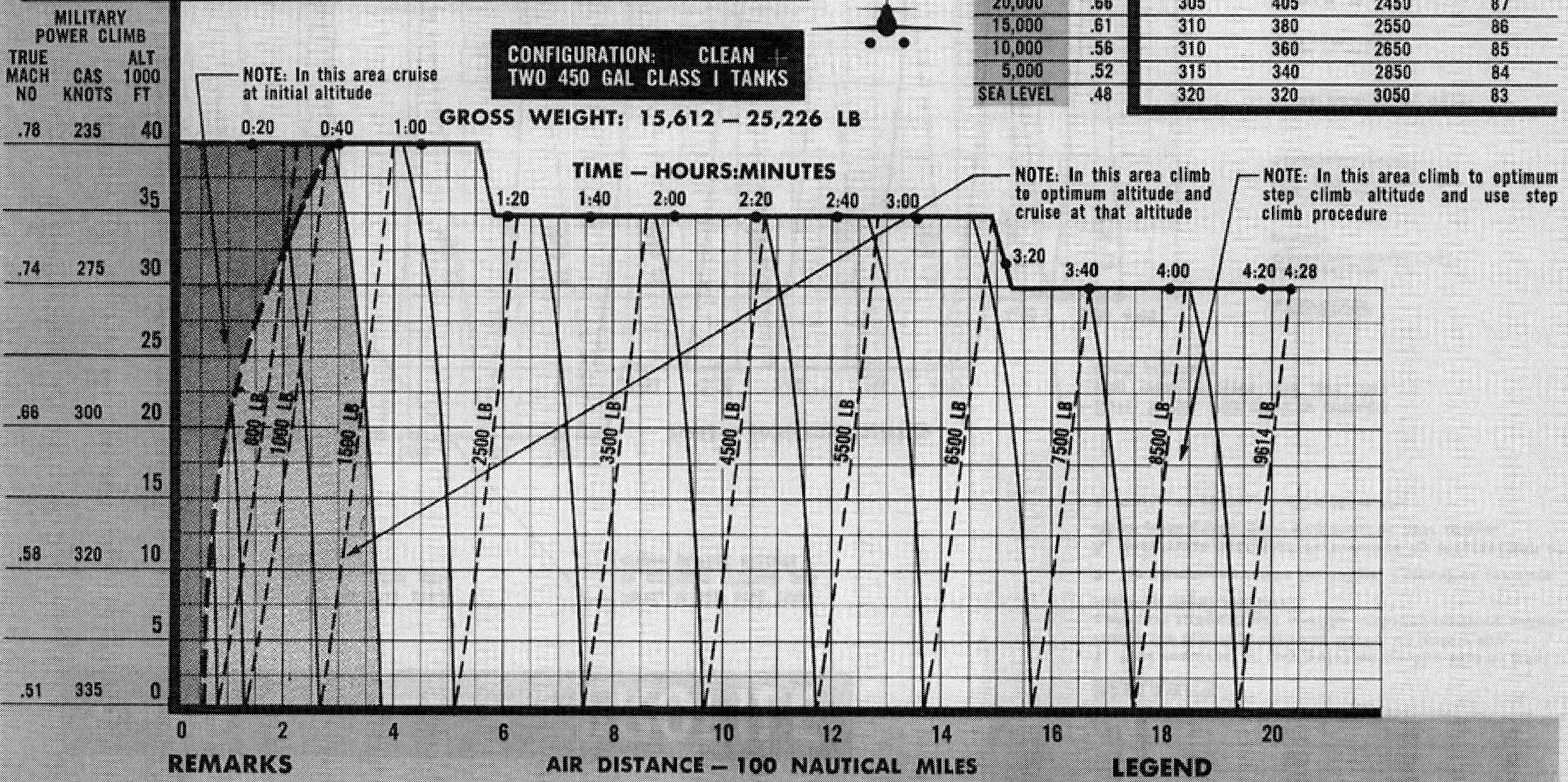

REMARKS

1. Fuel required at any point below the line of best range for constant altitude flight, or below the optimum range flight profile, includes military power climb to flight altitude.
2. No allowance made for loiter, descent or landing.
3. Best cruise condition determined by intersection of climb path guide lines and lines of best range.
4. Cruise at recommended Mach No.

LEGEND

- Optimum range flight profile
- Line of best range for constant altitude
- Climb path guide lines
- Fuel required
- Time at cruising altitude

Date: 25 October 1955
Data Basis: Flight Test

Figure A-39

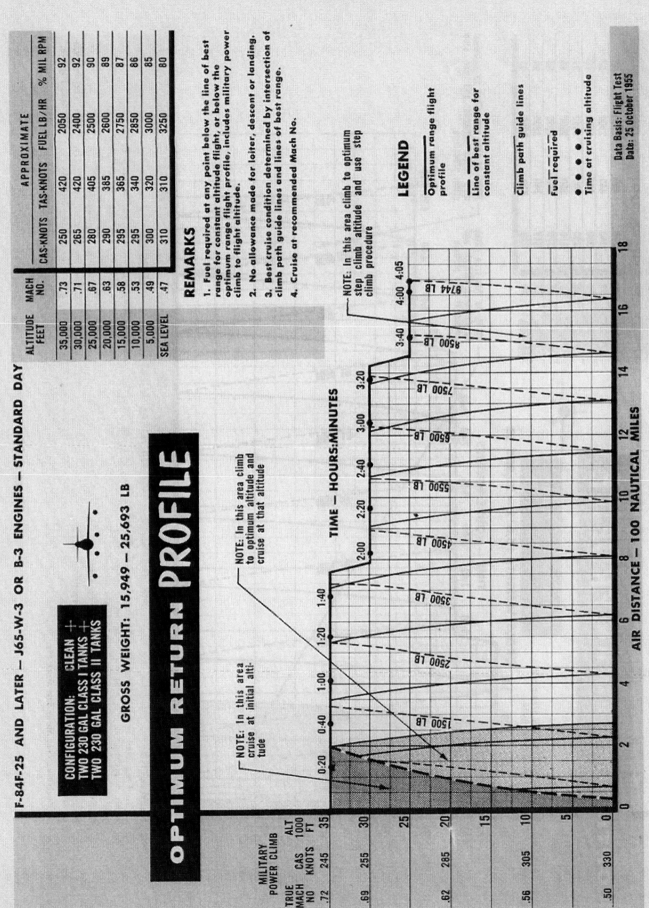

F-84F-25 AND LATER — J65-W-3 OR B-3 ENGINES — STANDARD DAY

CONFIGURATION: CLEAN
TWO 230 GAL CLASS I TANKS
TWO 230 GAL CLASS II TANKS

GROSS WEIGHT: 15,949 — 25,693 LB

OPTIMUM RETURN PROFILE

ALTITUDE FEET	MACH NO.	APPROXIMATE CAS-KNOTS	TAS-KNOTS	FUEL LB/HR	% MIL RPM
35,000	.73	250	420	2050	92
30,000	.71	265	420	2400	92
25,000	.67	280	405	2500	90
20,000	.63	290	385	2600	89
15,000	.58	295	365	2750	87
10,000	.53	295	340	2850	86
5,000	.49	300	320	3000	85
SEA LEVEL	.47	310	310	3250	80

REMARKS

1. Fuel required at any point below the line of best range for constant altitude flight, or below the optimum range flight profile, includes military power climb to flight altitude.
2. No allowance made for loiter, descent or landing.
3. Best cruise condition determined by intersection of climb path guide lines and lines of best range.
4. Cruise at recommended Mach No.

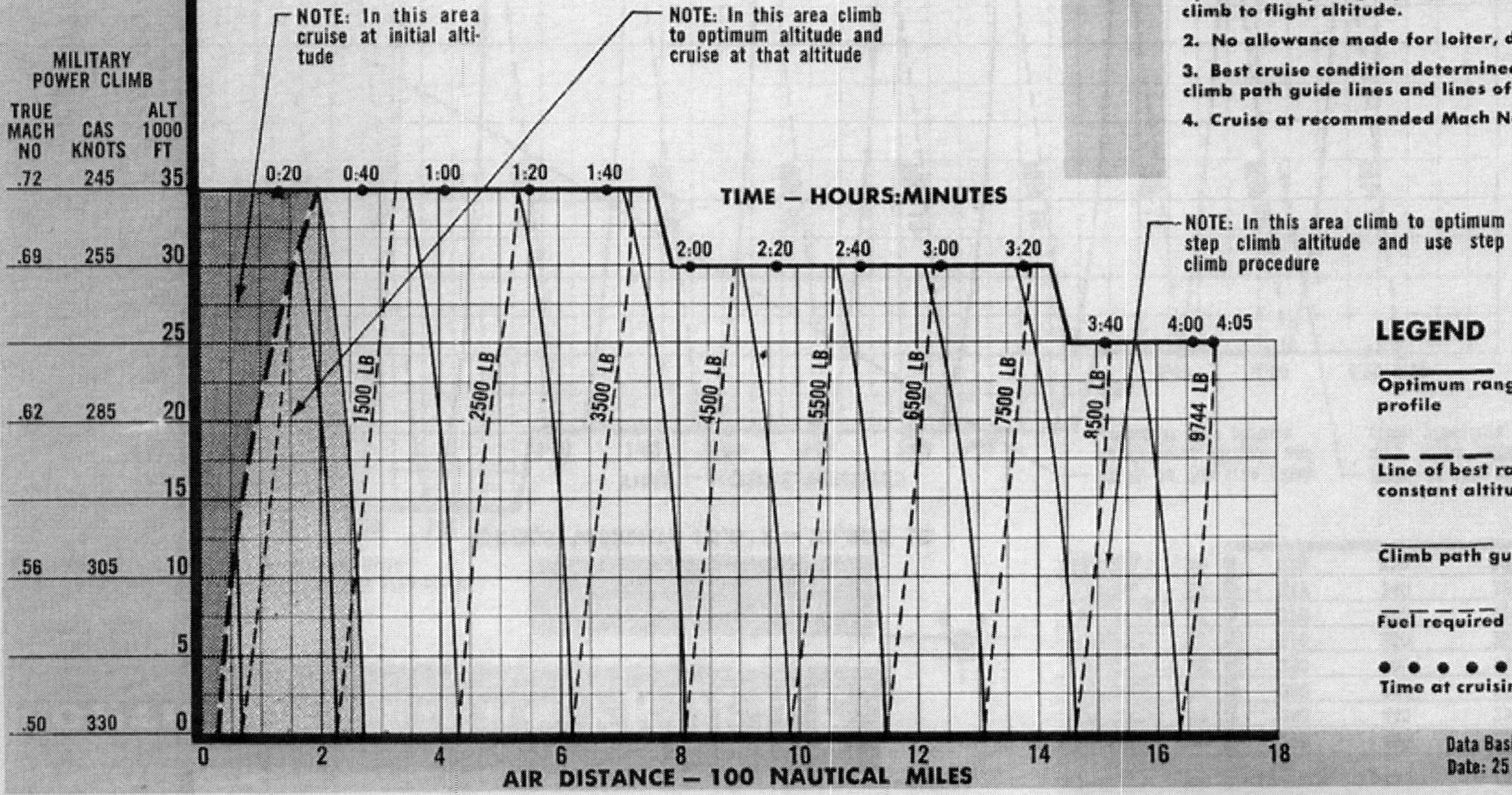

LEGEND

Optimum range flight profile

Line of best range for constant altitude

Climb path guide lines

Fuel required

Time at cruising altitude

Data Basis: Flight Test
Date: 25 October 1955

Figure A-40

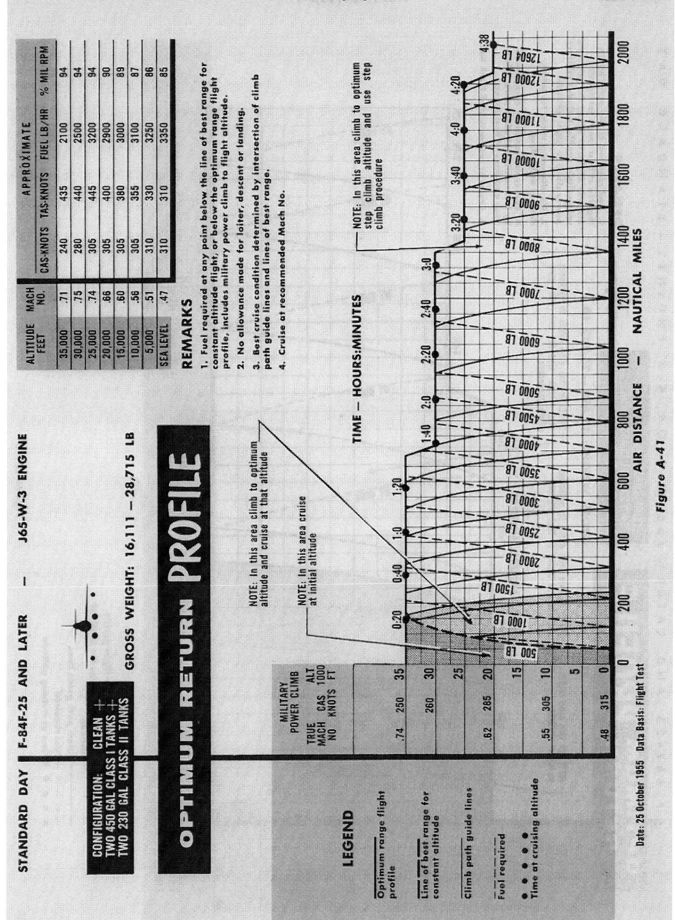

STANDARD DAY | F-84F-25 AND LATER — J65-W-3 ENGINE

CONFIGURATION: CLEAN
TWO 450 GAL CLASS I TANKS
TWO 230 GAL CLASS II TANKS

GROSS WEIGHT: 16,111 — 28,715 LB

OPTIMUM RETURN PROFILE

ALTITUDE FEET	MACH NO.	APPROXIMATE CAS-KNOTS	TAS-KNOTS	FUEL LB/HR	% MIL RPM
35,000	.71	240	435	2100	94
30,000	.75	280	440	2500	94
25,000	.74	305	445	3200	94
20,000	.66	305	400	2900	90
15,000	.60	305	380	3000	89
10,000	.56	305	355	3100	87
5,000	.51	310	330	3250	86
SEA LEVEL	.47	310	310	3350	85

REMARKS

1. Fuel required at any point below the line of best range for constant altitude flight, or below the optimum range flight profile, includes military power climb to flight altitude.
2. No allowance made for loiter, descent or landing.
3. Best cruise condition determined by intersection of climb path guide lines and lines of best range.
4. Cruise at recommended Mach No.

LEGEND

Optimum range flight profile

Line of best range for constant altitude

Climb path guide lines

Fuel required

Time at cruising altitude

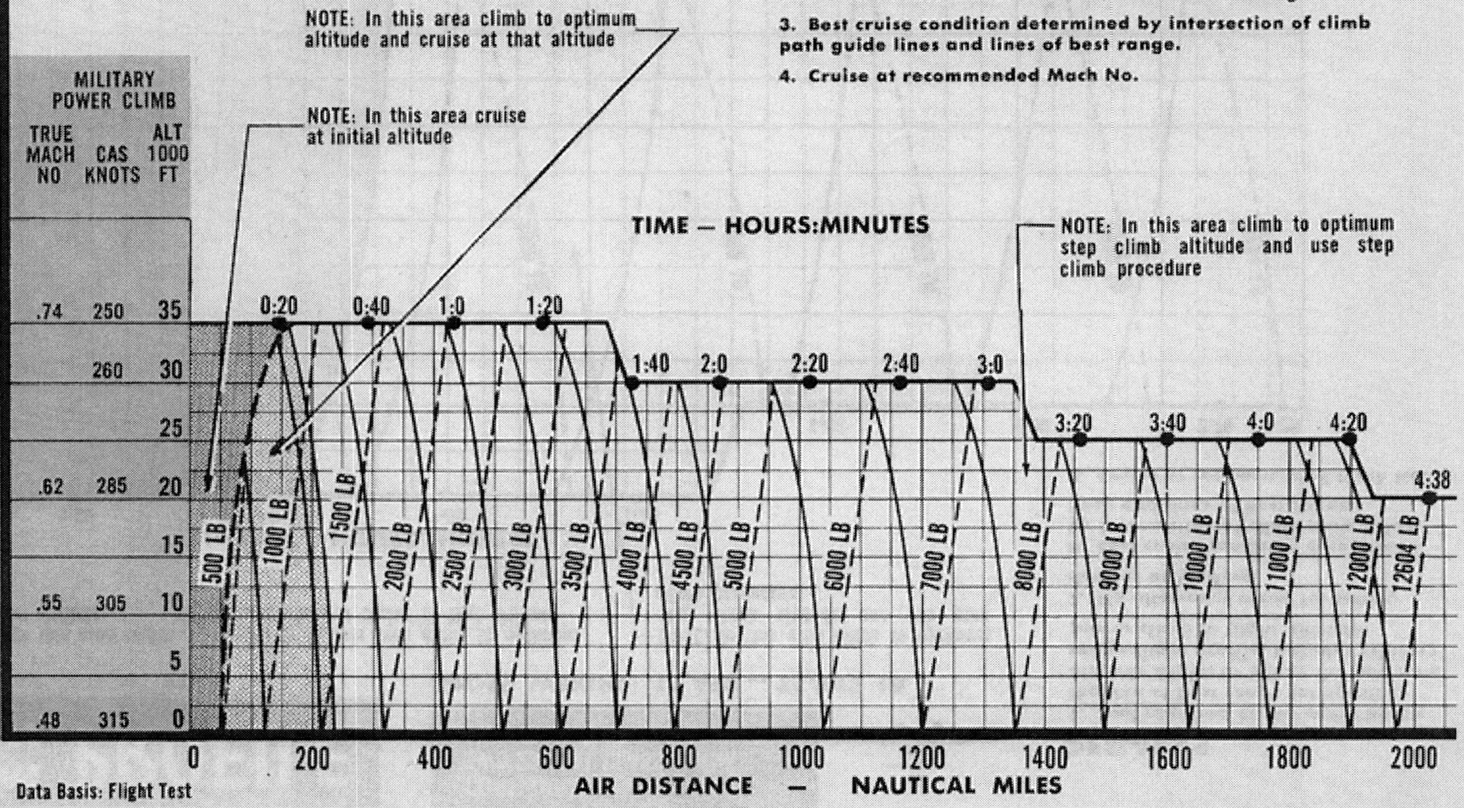

Date: 25 October 1955 Data Basis: Flight Test

Figure A-41

F-84F-25 AND LATER — J65-W-3 OR B-3 ENGINES — STANDARD DAY

OPTIMUM RETURN PROFILE

CONFIGURATION: CLEAN + TWO 230 GAL CLASS I TANKS + TWO 1000 LB BOMBS + 4 HVARS

GROSS WEIGHT: 18,498 — 25,252 LB

REMARKS

1. Fuel required at any point below the line of best range for constant altitude flight, or below the optimum range flight profile, includes military power climb to flight altitude.
2. No allowance made for loiter, descent or landing.
3. Best cruise condition determined by intersection of climb path guide lines and lines of best range.
4. Cruise at recommended Mach No.

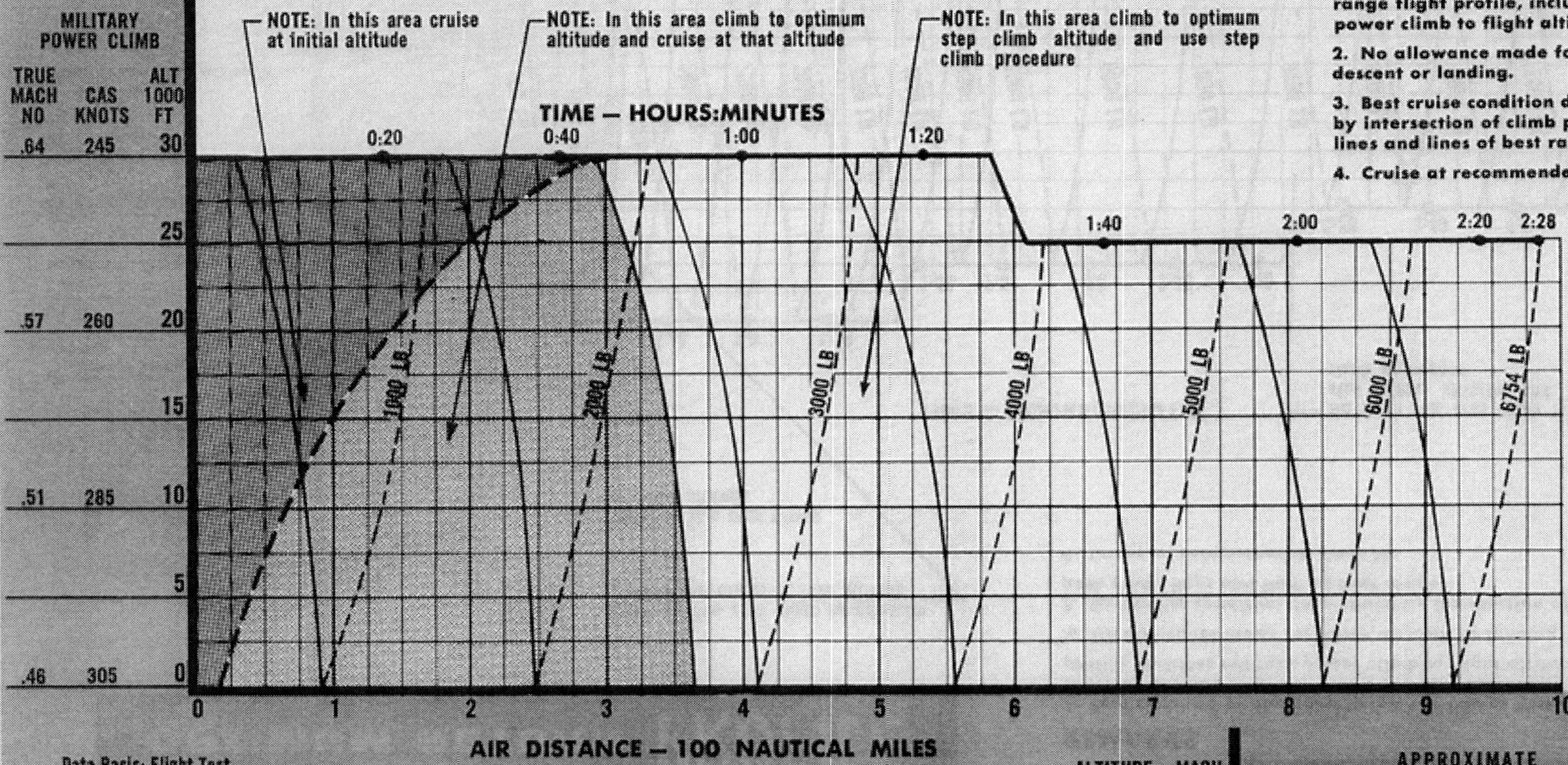

Data Basis: Flight Test
Date: 25 October 1955

ALTITUDE FEET	MACH NO.	APPROXIMATE CAS-KNOTS	TAS-KNOTS	FUEL LB/HR	% MIL RPM
30,000	.68	250	400	2500	94
25,000	.65	270	390	2600	92
20,000	.61	280	375	2700	89
15,000	.56	285	355	2800	87
10,000	.52	285	330	2850	87
5,000	.48	290	315	3000	86
SEA LEVEL	.46	300	300	3200	85

LEGEND

- ——— Optimum range flight profile
- — — — Line of best range for constant altitude
- ——— Climb path guide lines
- - - - - Fuel required
- ● ● ● ● Time at cruising altitude

Figure A-42

F-84F-25 and later • J65-W-3 Engine • Standard Day •

MAXIMUM ENDURANCE

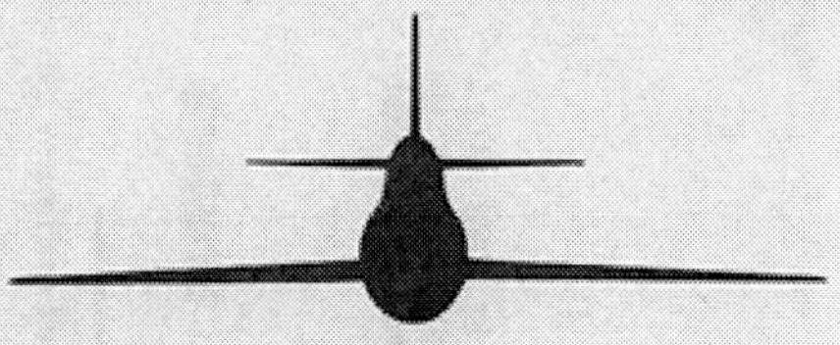

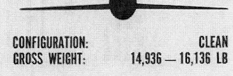

CONFIGURATION: CLEAN
GROSS WEIGHT: 14,936 — 16,136 LB

ALTITUDE FEET	MACH NO.	CAS KNOTS	TAS KNOTS	FUEL FLOW LB HR	% MIL RPM
	APPROXIMATE				
45,000	.76	205	435	1350	93
40,000	.70	210	400	1300	88
35,000	.61	200	350	1300	83
30,000	.55	200	320	1300	82
25,000	.49	200	295	1300	81
20,000	.44	200	270	1350	79
15,000	.40	200	250	1450	77
10,000	.36	200	230	1550	76
5,000	.33	200	215	1700	76
SEA LEVEL	.30	200	200	1950	76

1. Loiter at recommended CAS.
2. Maintain constant altitude.

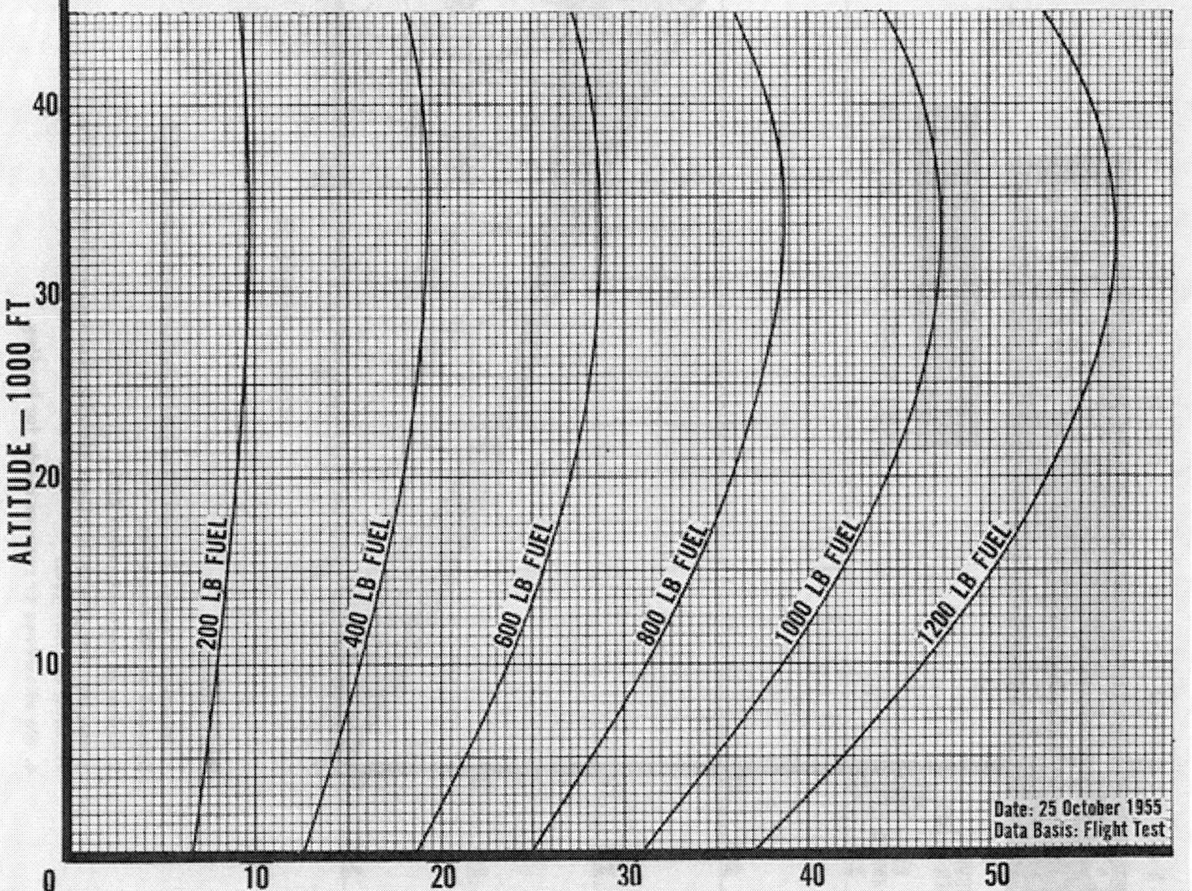

Figure A—43

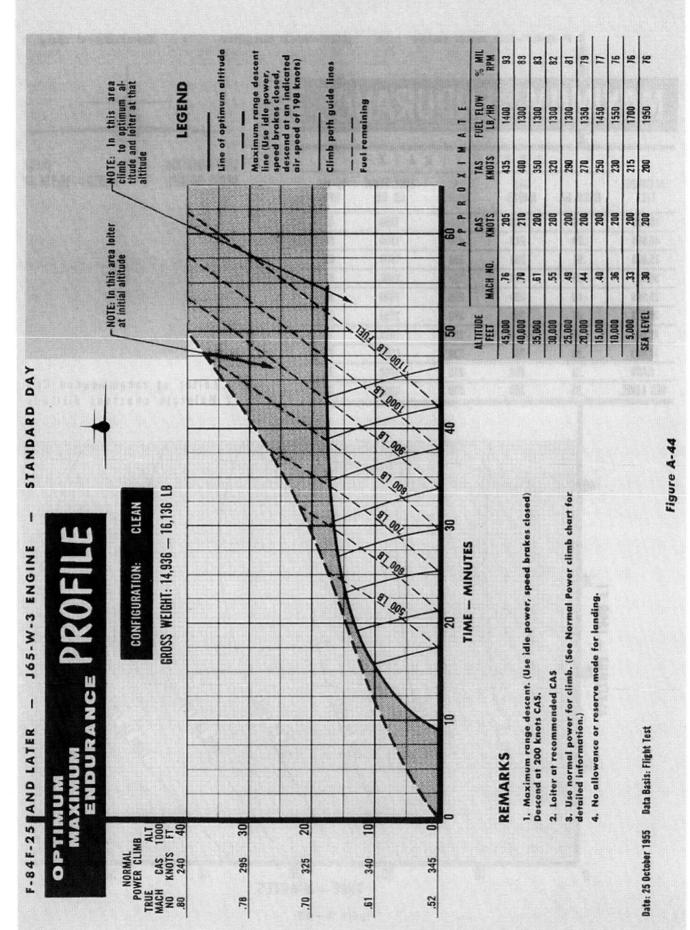

F-84F-25 AND LATER — J65-W-3 ENGINE — STANDARD DAY

OPTIMUM MAXIMUM ENDURANCE PROFILE

CONFIGURATION: CLEAN

GROSS WEIGHT: 14,936 — 16,136 LB

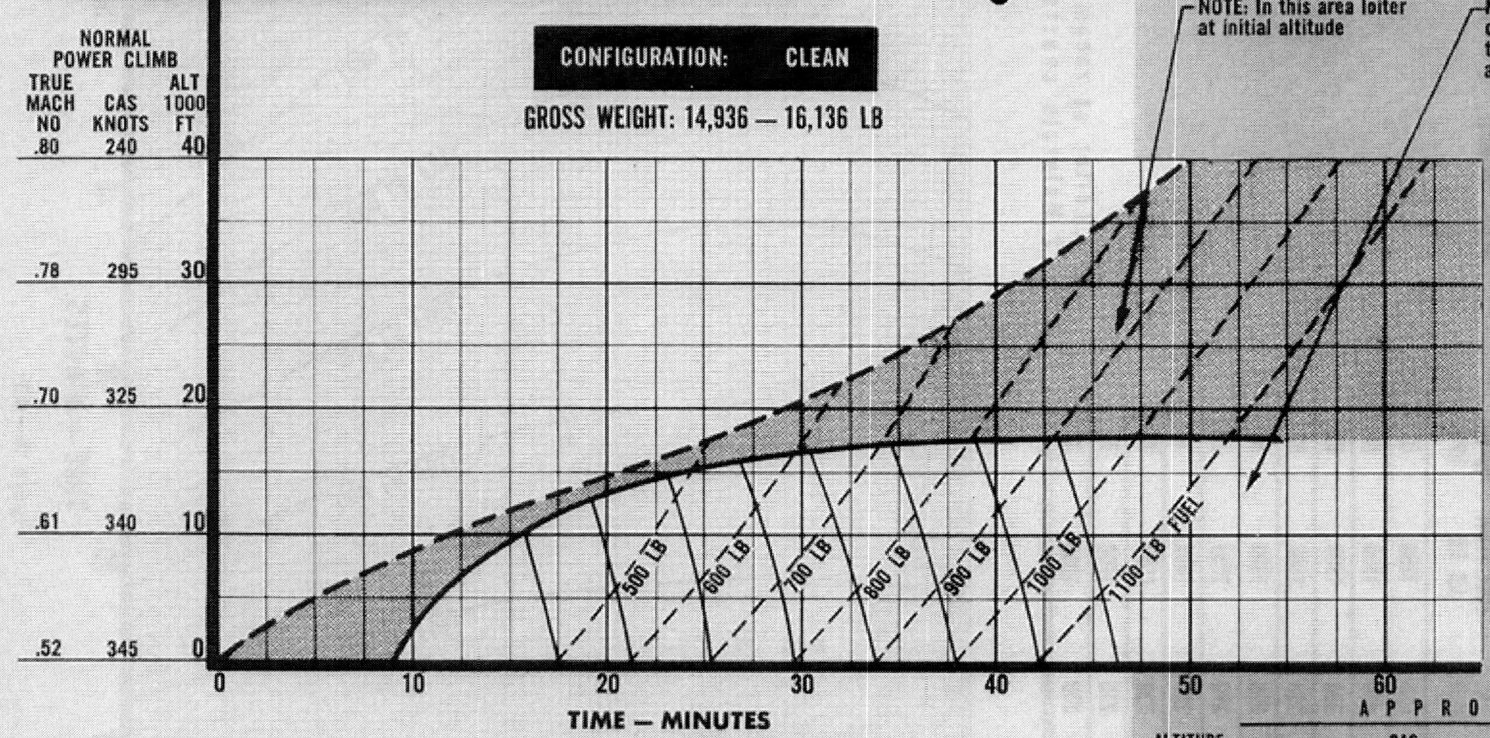

LEGEND

Line of optimum altitude

Maximum range descent line (Use idle power, speed brakes closed, descend at an indicated air speed of 198 knots)

Climb path guide lines

Fuel remaining

ALTITUDE FEET	APPROXIMATE MACH NO.	CAS KNOTS	TAS KNOTS	FUEL FLOW LB/HR	% MIL RPM
45,000	.76	205	435	1400	93
40,000	.70	210	400	1300	88
35,000	.61	200	350	1300	83
30,000	.55	200	320	1300	82
25,000	.49	200	290	1300	81
20,000	.44	200	270	1350	79
15,000	.40	200	250	1450	77
10,000	.36	200	230	1550	76
5,000	.33	200	215	1700	76
SEA LEVEL	.30	200	200	1950	76

REMARKS

1. Maximum range descent. (Use idle power, speed brakes closed) Descend at 200 knots CAS.
2. Loiter at recommended CAS
3. Use normal power for climb. (See Normal Power climb chart for detailed information.)
4. No allowance or reserve made for landing.

Date: 25 October 1955 Data Basis: Flight Test

Figure A-44

F-84F-25 AND LATER
J65-W-3 ENGINE
STANDARD DAY

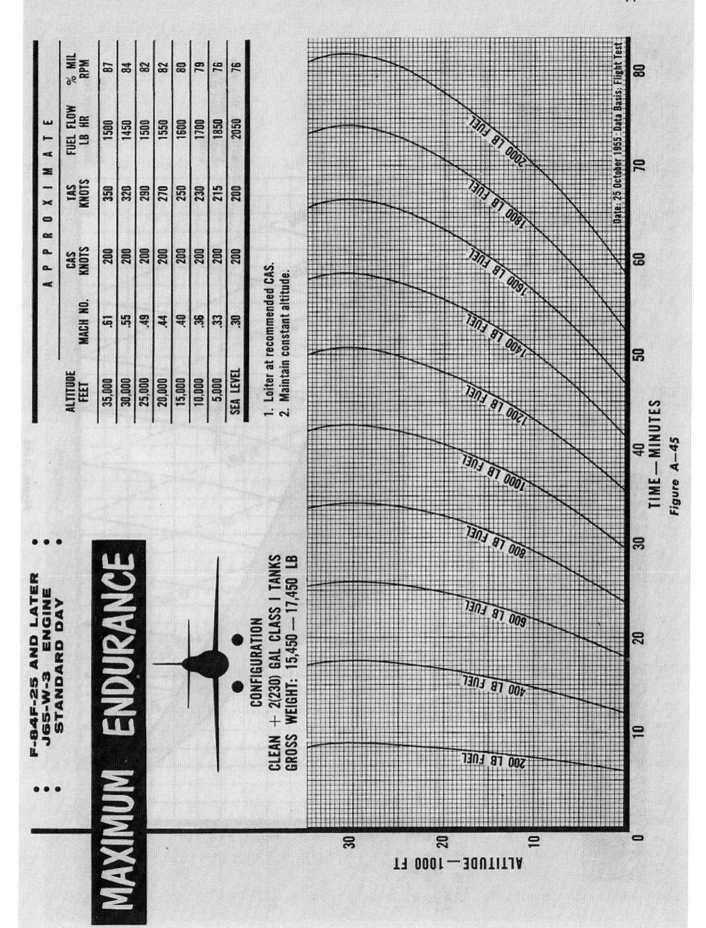

MAXIMUM ENDURANCE

CONFIGURATION
CLEAN + 2(230) GAL CLASS I TANKS
GROSS WEIGHT: 15,450 — 17,450 LB

	APPROXIMATE				
ALTITUDE FEET	MACH NO.	CAS KNOTS	TAS KNOTS	FUEL FLOW LB HR	% MIL RPM
35,000	.61	200	350	1500	87
30,000	.55	200	320	1450	84
25,000	.49	200	290	1500	82
20,000	.44	200	270	1550	82
15,000	.40	200	250	1600	80
10,000	.36	200	230	1700	79
5,000	.33	200	215	1850	76
SEA LEVEL	.30	200	200	2050	76

1. Loiter at recommended CAS.
2. Maintain constant altitude.

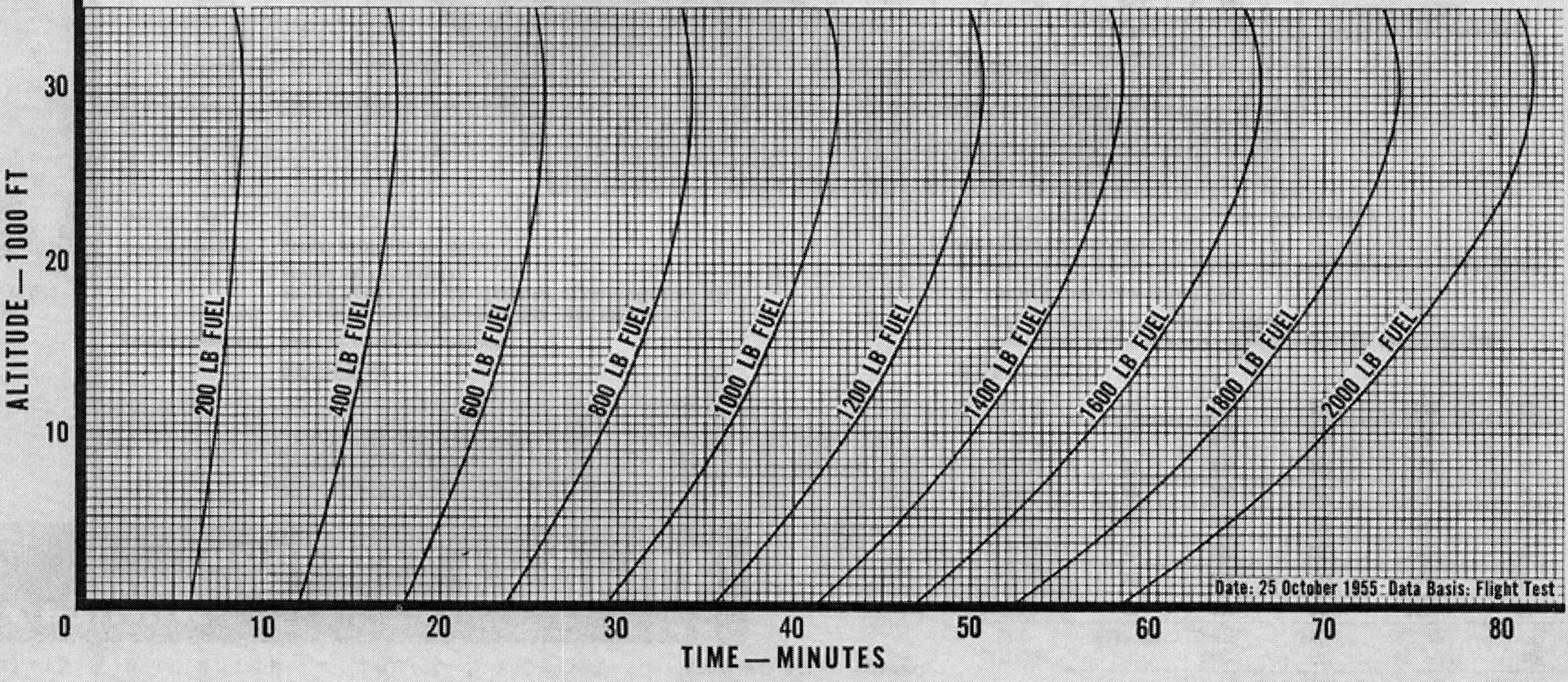

Figure A—45

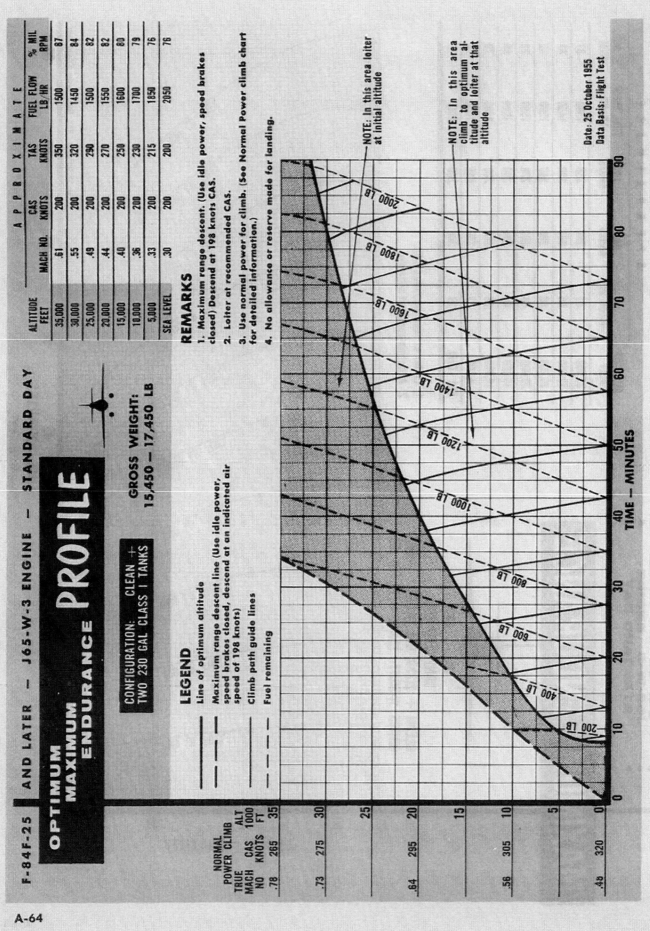

F-84F-25 AND LATER — J65-W-3 ENGINE — STANDARD DAY

OPTIMUM MAXIMUM ENDURANCE PROFILE

CONFIGURATION: CLEAN + TWO 230 GAL CLASS I TANKS

GROSS WEIGHT: 15,450 — 17,450 LB

ALTITUDE FEET	APPROXIMATE MACH NO.	CAS KNOTS	TAS KNOTS	FUEL FLOW LB/HR	% MIL RPM
35,000	.61	200	350	1500	87
30,000	.55	200	320	1450	84
25,000	.49	200	290	1500	82
20,000	.44	200	270	1550	82
15,000	.40	200	250	1600	80
10,000	.36	200	230	1700	79
5,000	.33	200	215	1850	76
SEA LEVEL	.30	200	200	2050	76

LEGEND

- (solid heavy line) Line of optimum altitude
- (heavy dashed line) Maximum range descent line (Use idle power, speed brakes closed, descend at an indicated air speed of 198 knots)
- (thin solid line) Climb path guide lines
- (thin dashed line) Fuel remaining

REMARKS

1. Maximum range descent. (Use idle power, speed brakes closed) Descend at 198 knots CAS.
2. Loiter at recommended CAS.
3. Use normal power for climb. (See Normal Power climb chart for detailed information.)
4. No allowance or reserve made for landing.

NORMAL POWER CLIMB

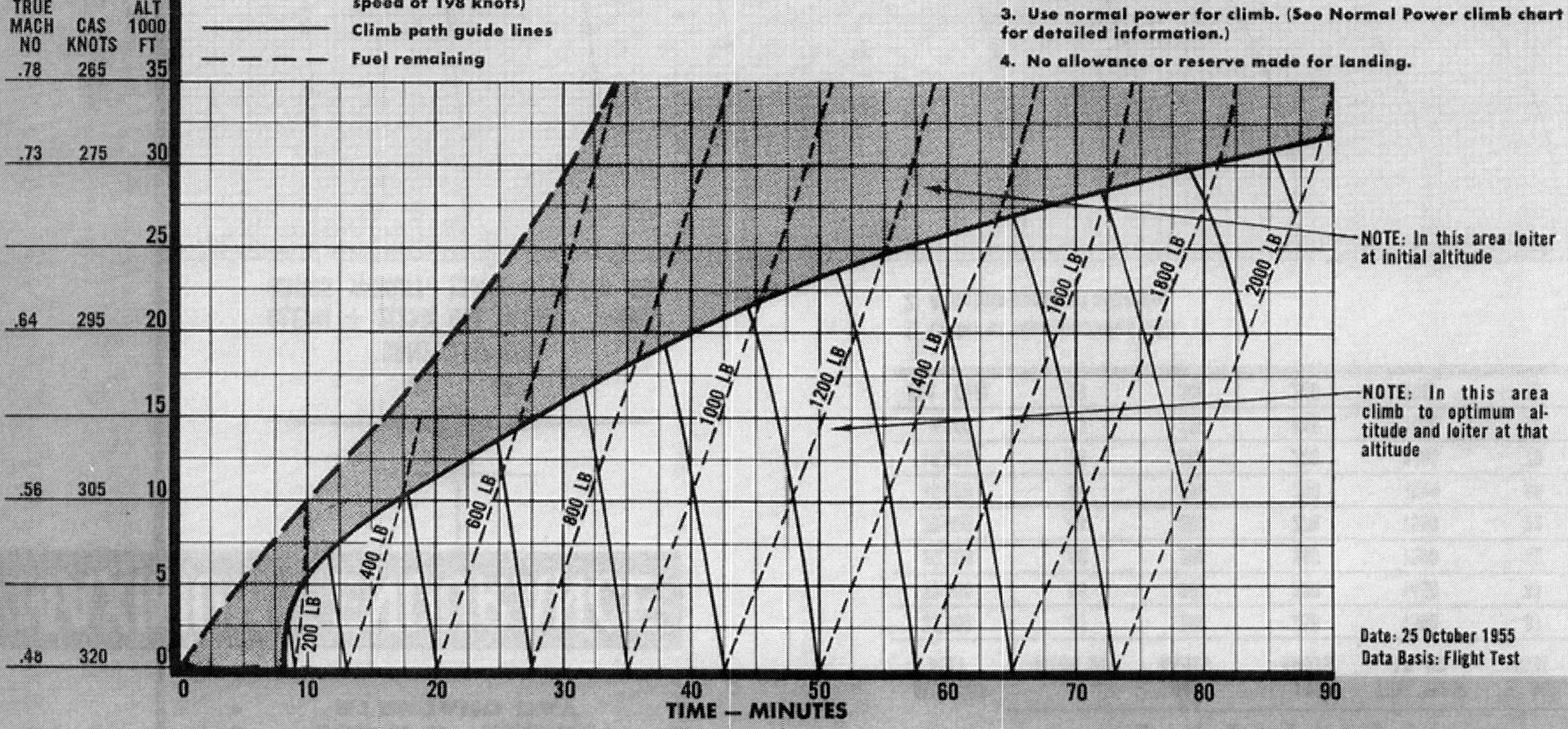

Figure A-46

F-84F-25 and later • J65-W-3 Engine • Standard Day •

MAXIMUM ENDURANCE

CONFIGURATION
CLEAN + 2(450) GAL CLASS I TANKS
GROSS WEIGHT: 15,612—18,612 LB

	APPROXIMATE				
ALTITUDE FEET	MACH NO.	CAS KNOTS	TAS KNOTS	FUEL FLOW LB HR	% MIL RPM
35,000	.61	200	350	1600	88
30,000	.55	200	320	1550	84
25,000	.49	200	290	1550	82
20,000	.44	200	270	1600	81
15,000	.40	200	250	1650	80
10,000	.36	200	230	1750	79
5,000	.33	200	215	1900	77
SEA LEVEL	.30	200	200	2100	76

1. Loiter at recommended CAS.
2. Maintain constant altitude.

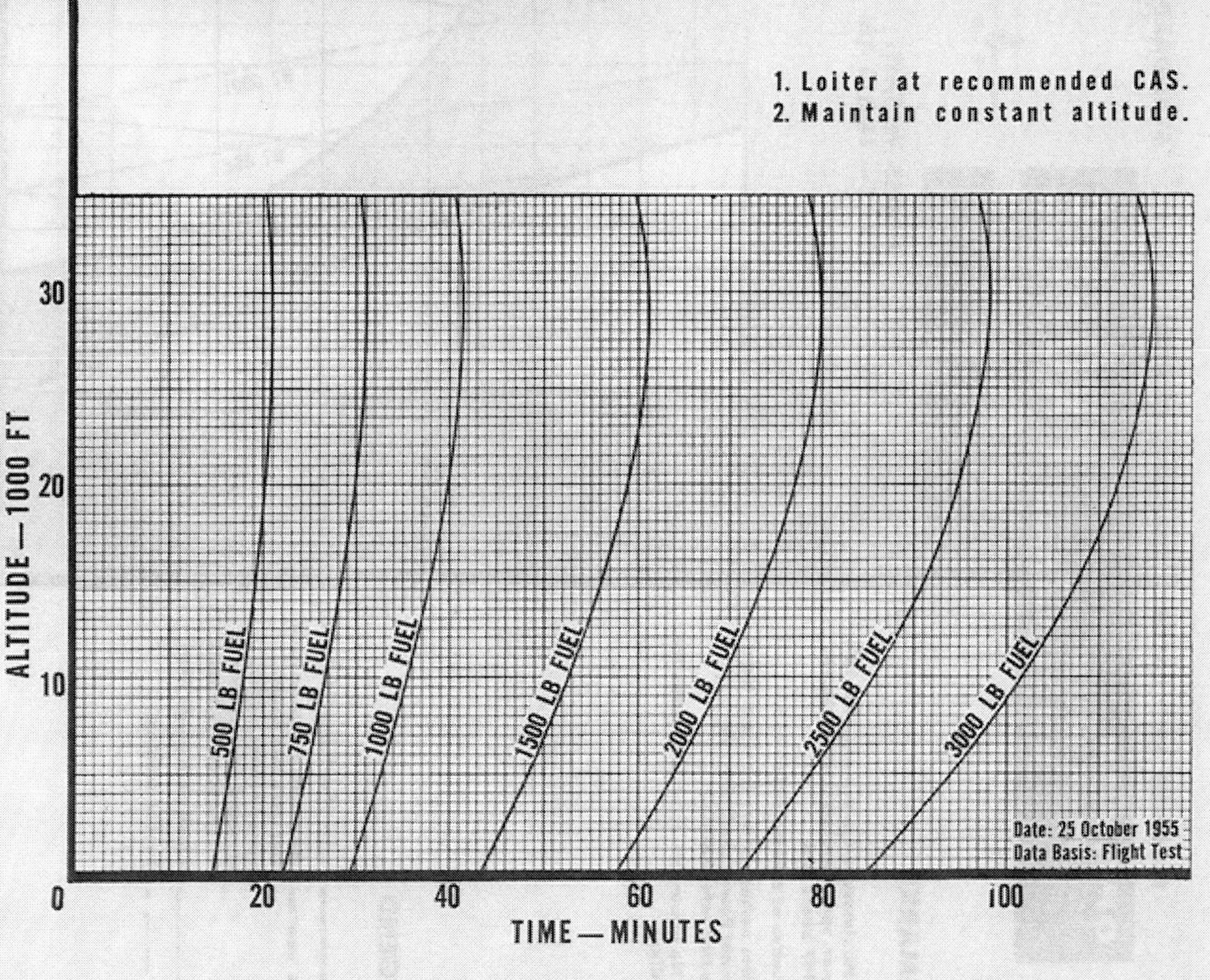

Figure A—47

F-84F-25 AND LATER | J65-W-3 ENGINE — STANDARD DAY

OPTIMUM MAXIMUM ENDURANCE PROFILE

CONFIGURATION: CLEAN + TWO 450 GAL CLASS I TANKS

GROSS WEIGHT: 15,612 — 18,612 LB

ALTITUDE FEET	APPROXIMATE MACH NO.	CAS KNOTS	TAS KNOTS	FUEL FLOW LB/HR	% MIL RPM
35,000	.61	200	350	1600	88
30,000	.55	200	320	1550	84
25,000	.49	200	290	1550	82
20,000	.44	200	270	1600	81
15,000	.40	200	250	1650	80
10,000	.36	200	230	1750	79
5,000	.33	200	215	1900	77
SEA LEVEL	.30	200	200	2100	76

REMARKS

1. Maximum range descent. (Use idle power, speed brakes closed) Descend at 197 knots CAS.
2. Loiter at recommended CAS.
3. Use normal power for climb. (See Normal Power climb chart for detailed information.)
4. No allowance or reserve made for landing.

LEGEND

- (solid heavy line) Line of optimum altitude
- (heavy dashed line) Maximum range descent line (Use idle power, speed brakes closed, descend at an indicated airspeed of 197 knots)
- (thin solid line) Climb path guide lines
- (thin dashed line) Fuel remaining

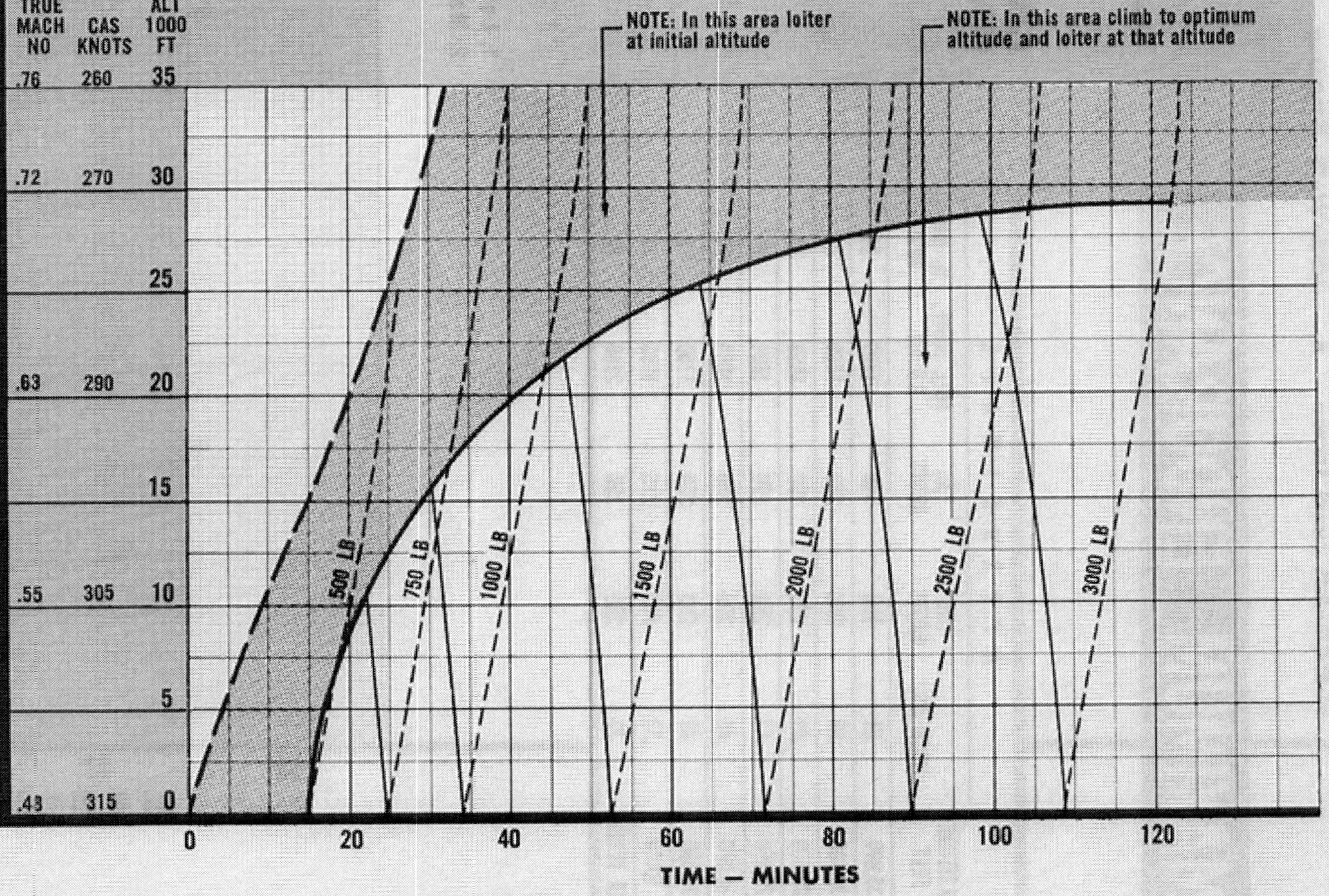

Figure A-48

Date: 25 October 1955
Data Basis: Flight Test

F-84F-25 and later • J65-W-3 Engine • Standard Day •

MAXIMUM ENDURANCE

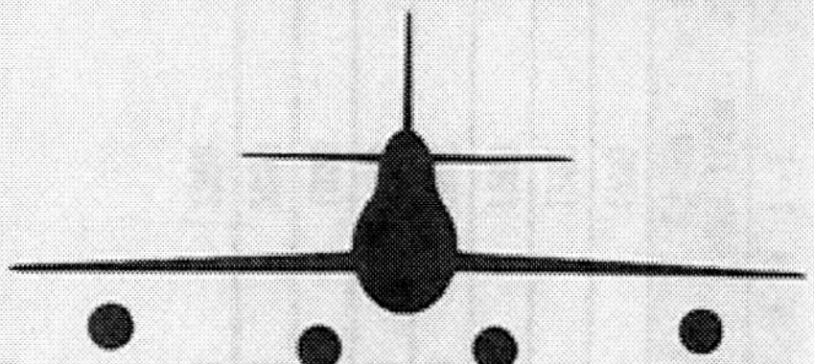

CONFIGURATION
CLEAN + 2(230) GAL CLASS I TANKS
+ 2(230) GAL CLASS II TANKS
GROSS WEIGHT: 15,949 — 19,449 LB

		APPROXIMATE			
ALTITUDE FEET	MACH NO.	CAS KNOTS	TAS KNOTS	FUEL FLOW LB HR	% MIL RPM
35,000	.61	200	350	1750	89
30,000	.55	200	320	1750	87
25,000	.49	200	290	1700	85
20,000	.44	200	270	1800	84
15,000	.40	200	250	1850	81
10,000	.36	200	230	1900	80
5,000	.33	200	215	2050	79
SEA LEVEL	.30	200	200	2200	78

1. Loiter at recommended CAS.
2. Maintain constant altitude.

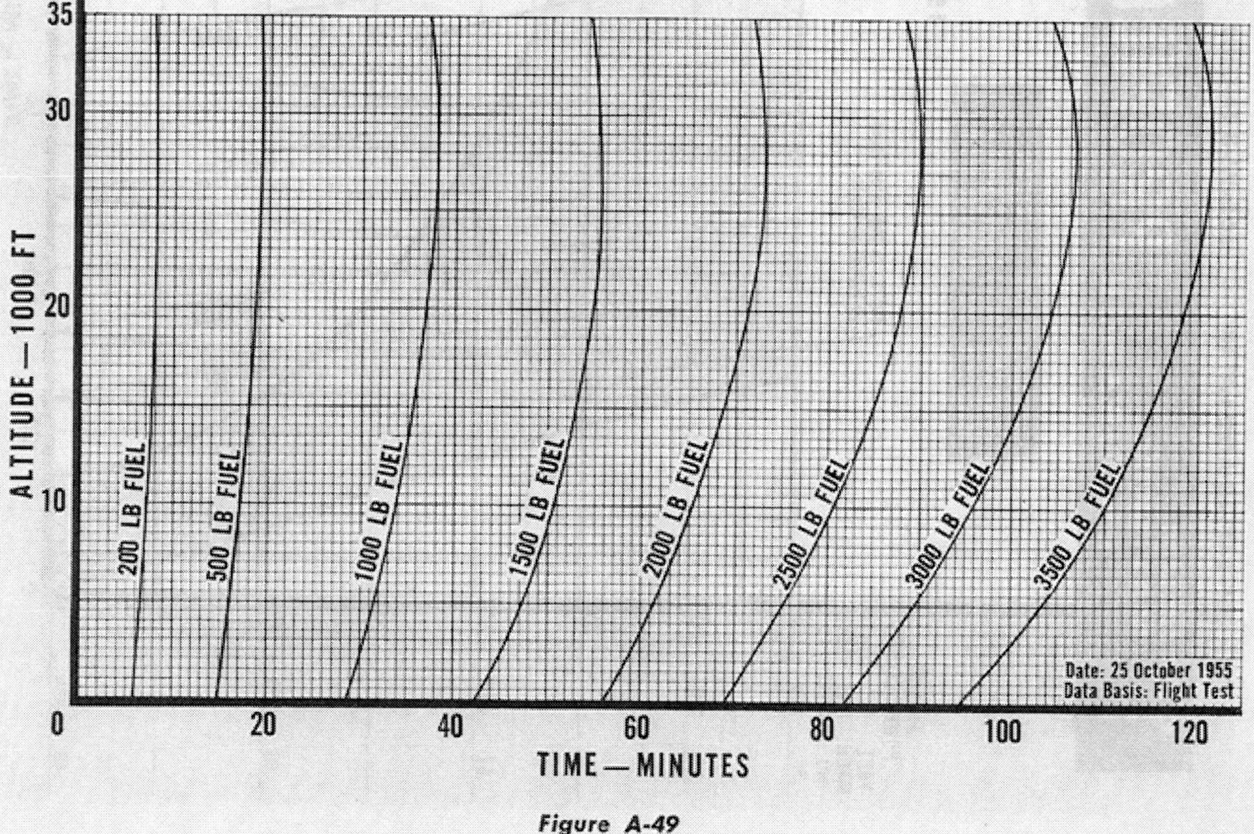

Figure A-49

F-84F-25 AND LATER — J65-W-3 ENGINE — STANDARD DAY

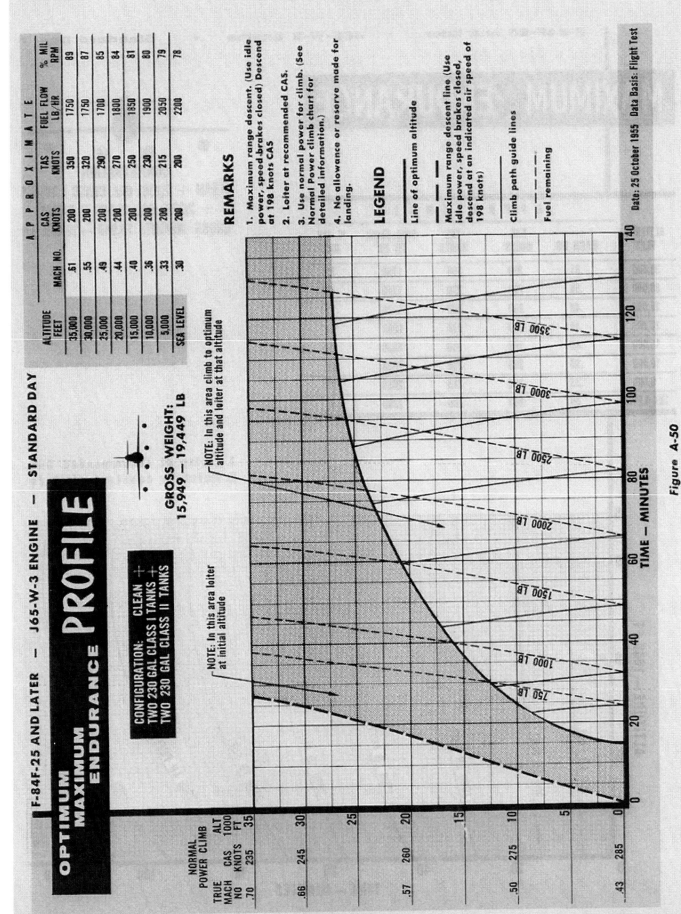

OPTIMUM MAXIMUM ENDURANCE PROFILE

CONFIGURATION: CLEAN +
TWO 230 GAL CLASS I TANKS +
TWO 230 GAL CLASS II TANKS

GROSS WEIGHT: 15,949 — 19,449 LB

ALTITUDE FEET	APPROXIMATE MACH NO.	CAS KNOTS	TAS KNOTS	FUEL FLOW LB/HR	% MIL RPM
35,000	.61	200	350	1750	89
30,000	.55	200	320	1750	87
25,000	.49	200	290	1700	85
20,000	.44	200	270	1800	84
15,000	.40	200	250	1850	81
10,000	.36	200	230	1900	80
5,000	.33	200	215	2050	79
SEA LEVEL	.30	200	200	2200	78

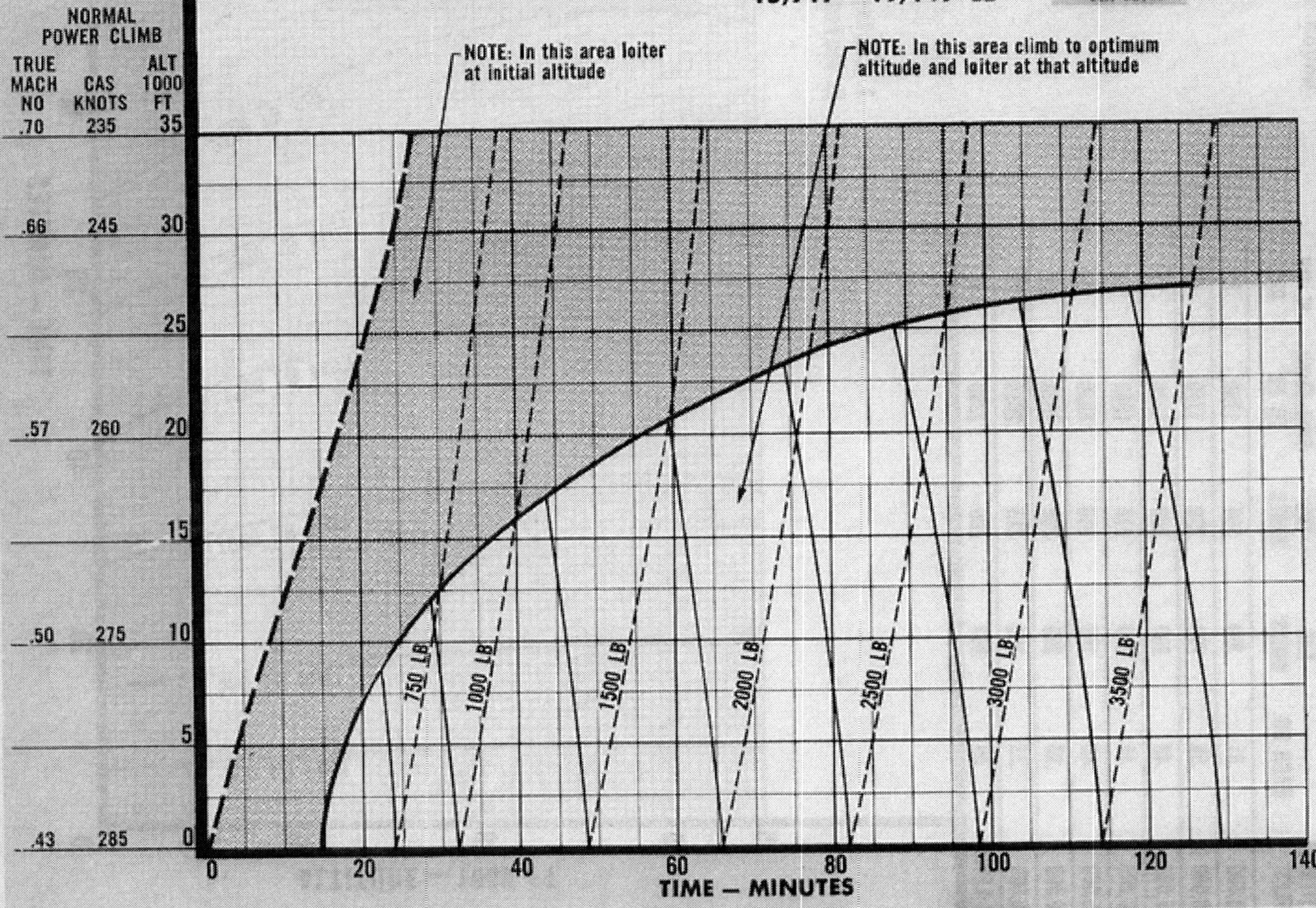

REMARKS

1. Maximum range descent. (Use idle power, speed brakes closed) Descend at 198 knots CAS
2. Loiter at recommended CAS.
3. Use normal power for climb. (See Normal Power climb chart for detailed information.)
4. No allowance or reserve made for landing.

LEGEND

Line of optimum altitude

Maximum range descent line (Use idle power, speed brakes closed, descend at an indicated air speed of 198 knots)

Climb path guide lines

Fuel remaining

Date: 25 October 1955 Data Basis: Flight Test

Figure A-50

- F-84F-25 AND LATER
- J65-W-3 ENGINE
- STANDARD DAY

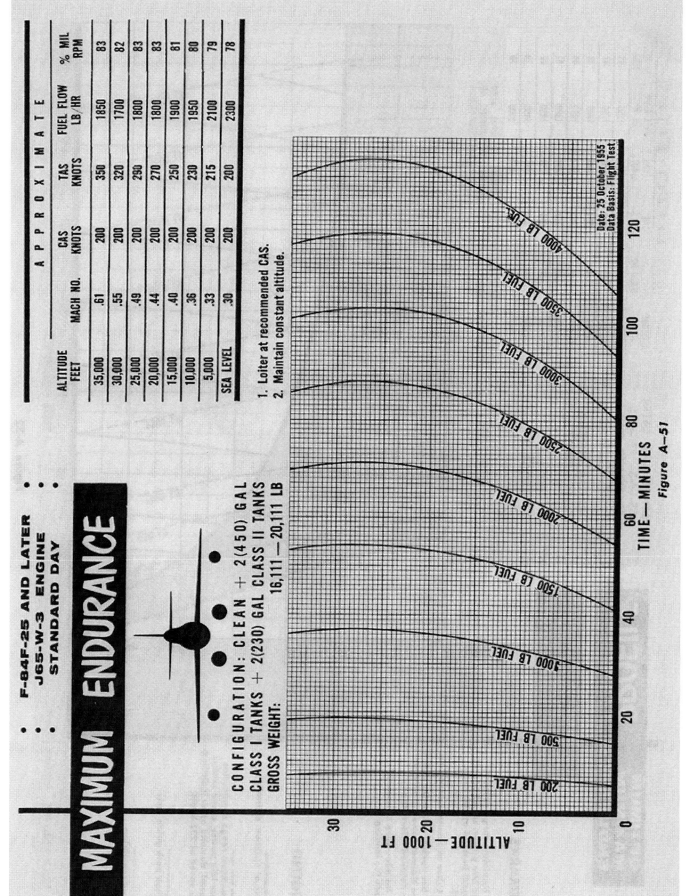

MAXIMUM ENDURANCE

CONFIGURATION: CLEAN + 2(450) GAL CLASS I TANKS + 2(230) GAL CLASS II TANKS

GROSS WEIGHT: 16,111 — 20,111 LB

		A P P R O X I M A T E			
ALTITUDE FEET	MACH NO.	CAS KNOTS	TAS KNOTS	FUEL FLOW LB/HR	% MIL RPM
35,000	.61	200	350	1850	83
30,000	.55	200	320	1700	82
25,000	.49	200	290	1800	83
20,000	.44	200	270	1800	83
15,000	.40	200	250	1900	81
10,000	.36	200	230	1950	80
5,000	.33	200	215	2100	79
SEA LEVEL	.30	200	200	2300	78

1. Loiter at recommended CAS.
2. Maintain constant altitude.

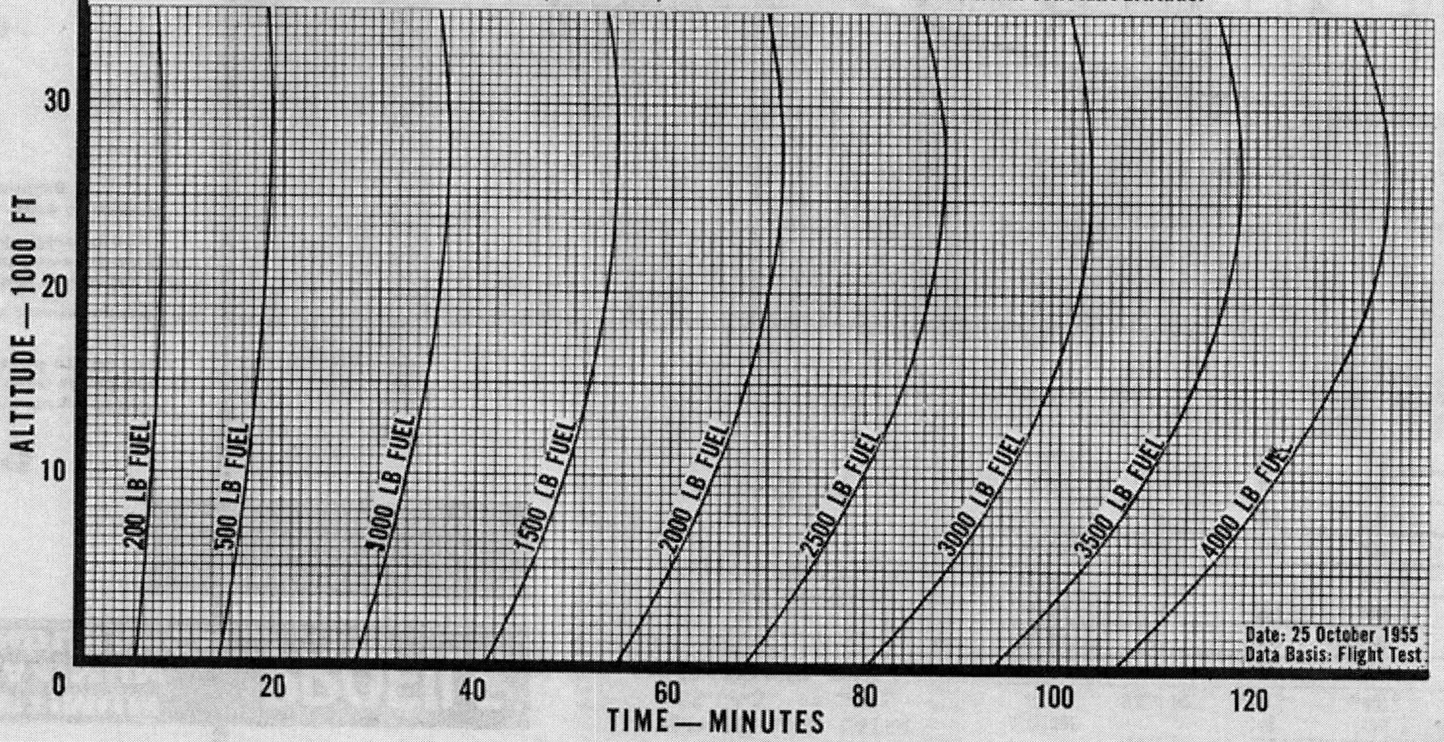

Figure A–51

OPTIMUM MAXIMUM ENDURANCE PROFILE

CONFIGURATION: CLEAN + TWO 450 GAL CLASS I TANKS + TWO 230 GAL CLASS II TANKS

F-84F-25 AND LATER
J65-W-3 ENGINE
STANDARD DAY

GROSS WEIGHT:
16,111 — 20,111 LB

ALTITUDE FEET	APPROXIMATE MACH NO.	CAS KNOTS	TAS KNOTS	FUEL FLOW LB HR	% MIL RPM
35,000	.61	200	350	1850	83
30,000	.55	200	320	1700	82
25,000	.49	200	290	1800	83
20,000	.44	200	270	1800	83
15,000	.40	200	250	1900	81
10,000	.36	200	230	1950	80
5,000	.33	200	215	2100	79
SEA LEVEL	.30	200	200	2300	78

REMARKS

1. Maximum range descent. (Use idle power, speed brakes closed) Descend at 200 knots CAS.
2. Loiter at recommended CAS.
3. Use normal power for climb. (See Normal Power climb chart for detailed information.)
4. No allowance or reserve made for landing.

LEGEND

Line of optimum altitude

Maximum range descent line (Use idle power, speed brakes closed, descend at an indicated air speed of 202 knots)

Climb path guide lines

Fuel remaining

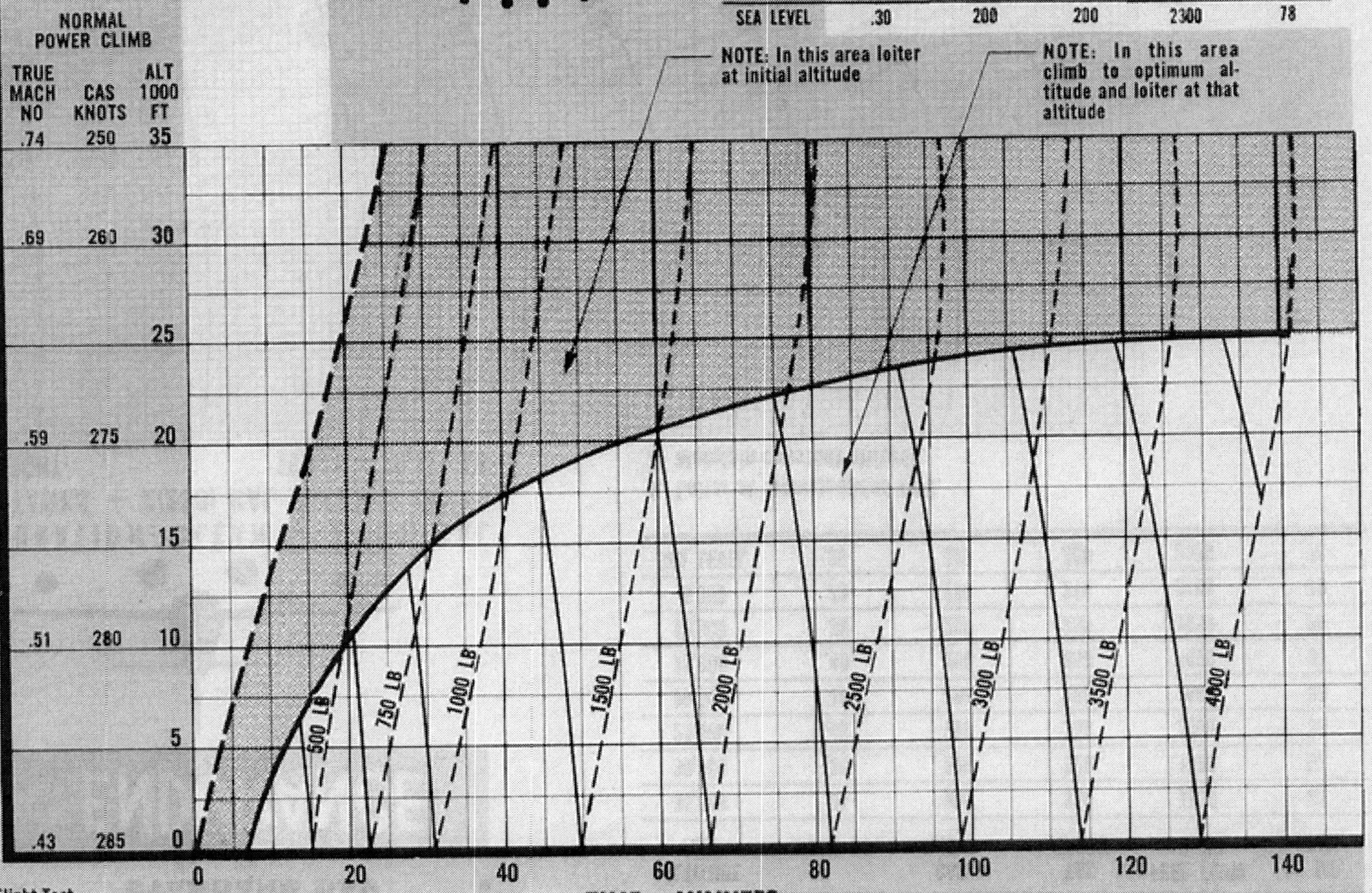

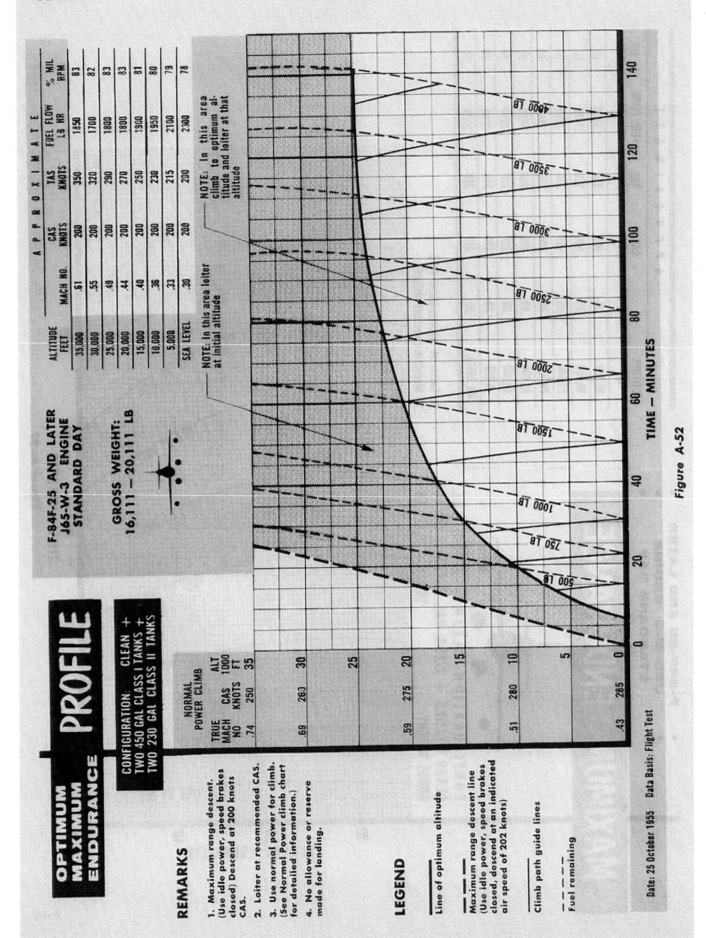

Date: 25 October 1955 Data Basis: Flight Test

Figure A-52

F-84F-25 and later • J65-W-3 Engine • Standard Day •

MAXIMUM ENDURANCE

CONFIGURATION
CLEAN + 2(230) GAL CLASS I TANKS + 2 1000 LB BOMBS + 4 HVARS
GROSS WEIGHT: 18,498 — 20,498 LB

	APPROXIMATE				
ALTITUDE FEET	MACH NO.	CAS KNOTS	TAS KNOTS	FUEL FLOW LB/HR	% MIL RPM
35,000	.61	200	350	2150	96
30,000	.55	200	320	2050	90
25,000	.49	200	290	2050	88
20,000	.44	200	270	2050	85
15,000	.40	200	250	2050	83
10,000	.36	200	230	2150	81
5,000	.33	200	215	2250	81
SEA LEVEL	.30	200	200	2400	80

1. Loiter at recommended CAS.
2. Maintain constant altitude.

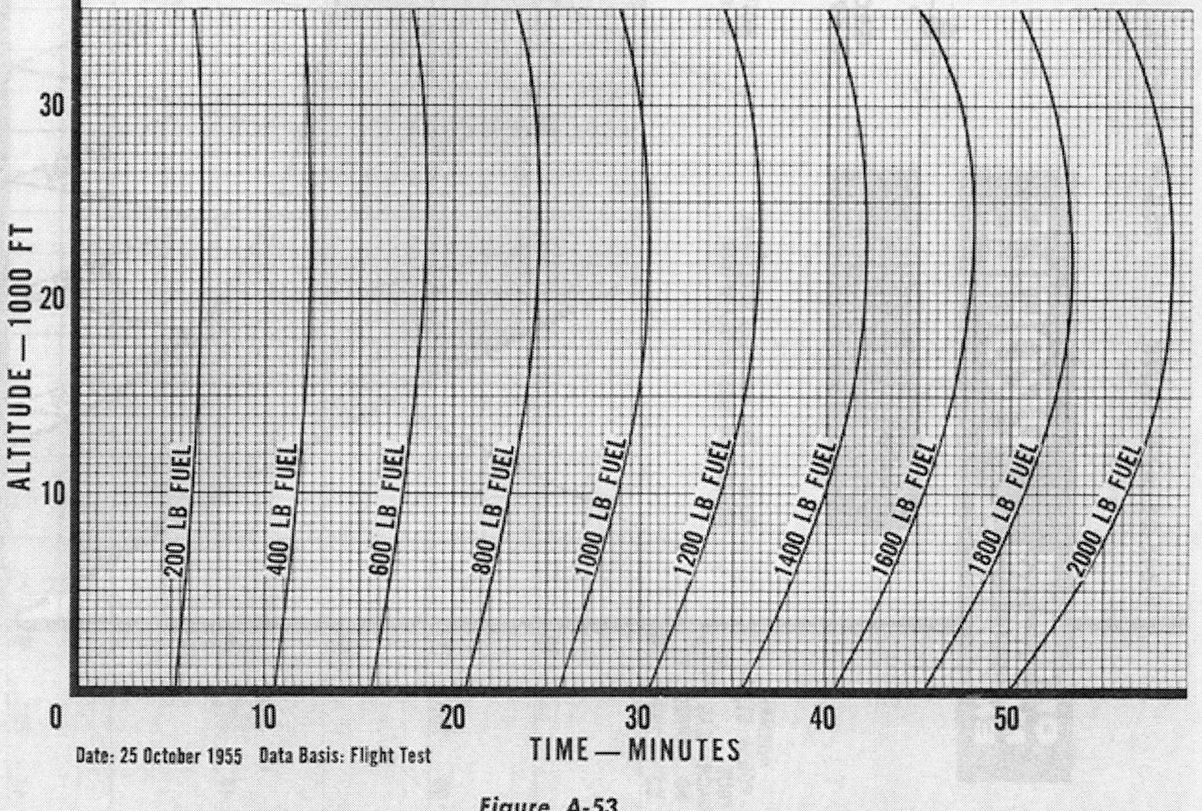

Date: 25 October 1955 Data Basis: Flight Test

Figure A-53

F-84F-25 AND LATER — J65-W-3 ENGINE — STANDARD DAY

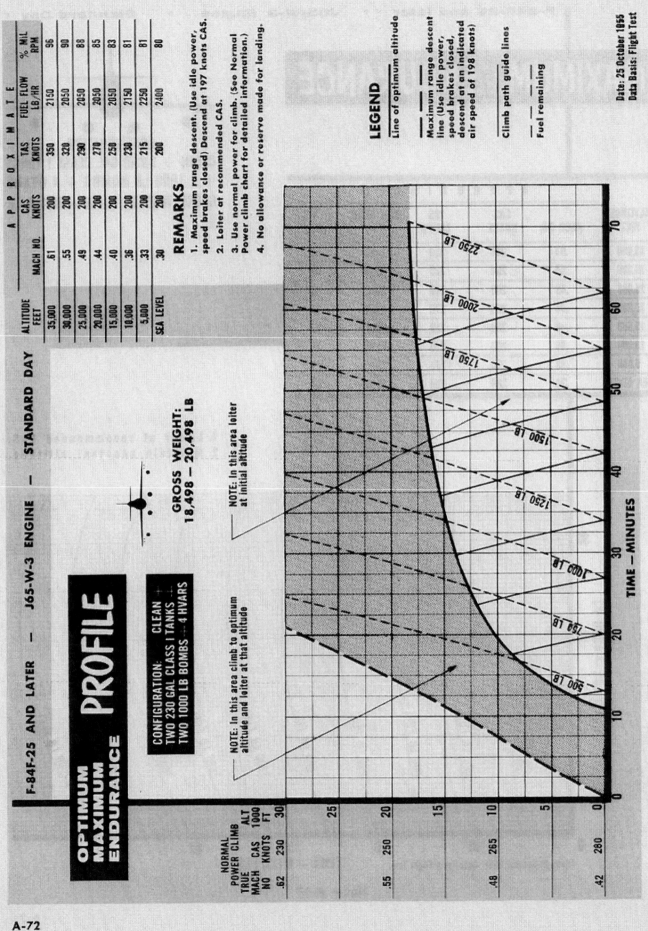

OPTIMUM MAXIMUM ENDURANCE PROFILE

CONFIGURATION: CLEAN
TWO 230 GAL CLASS I TANKS
TWO 1000 LB BOMBS 4 HVARS

GROSS WEIGHT:
18,498 — 20,498 LB

ALTITUDE FEET	APPROXIMATE MACH NO.	CAS KNOTS	TAS KNOTS	FUEL FLOW LB/HR	% MIL RPM
35,000	.61	200	350	2150	96
30,000	.55	200	320	2050	90
25,000	.49	200	290	2050	88
20,000	.44	200	270	2050	85
15,000	.40	200	250	2050	83
10,000	.36	200	230	2150	81
5,000	.33	200	215	2250	81
SEA LEVEL	.30	200	200	2400	80

REMARKS

1. Maximum range descent. (Use idle power, speed brakes closed) Descend at 197 knots CAS.
2. Loiter at recommended CAS.
3. Use normal power for climb. (See Normal Power climb chart for detailed information.)
4. No allowance or reserve made for landing.

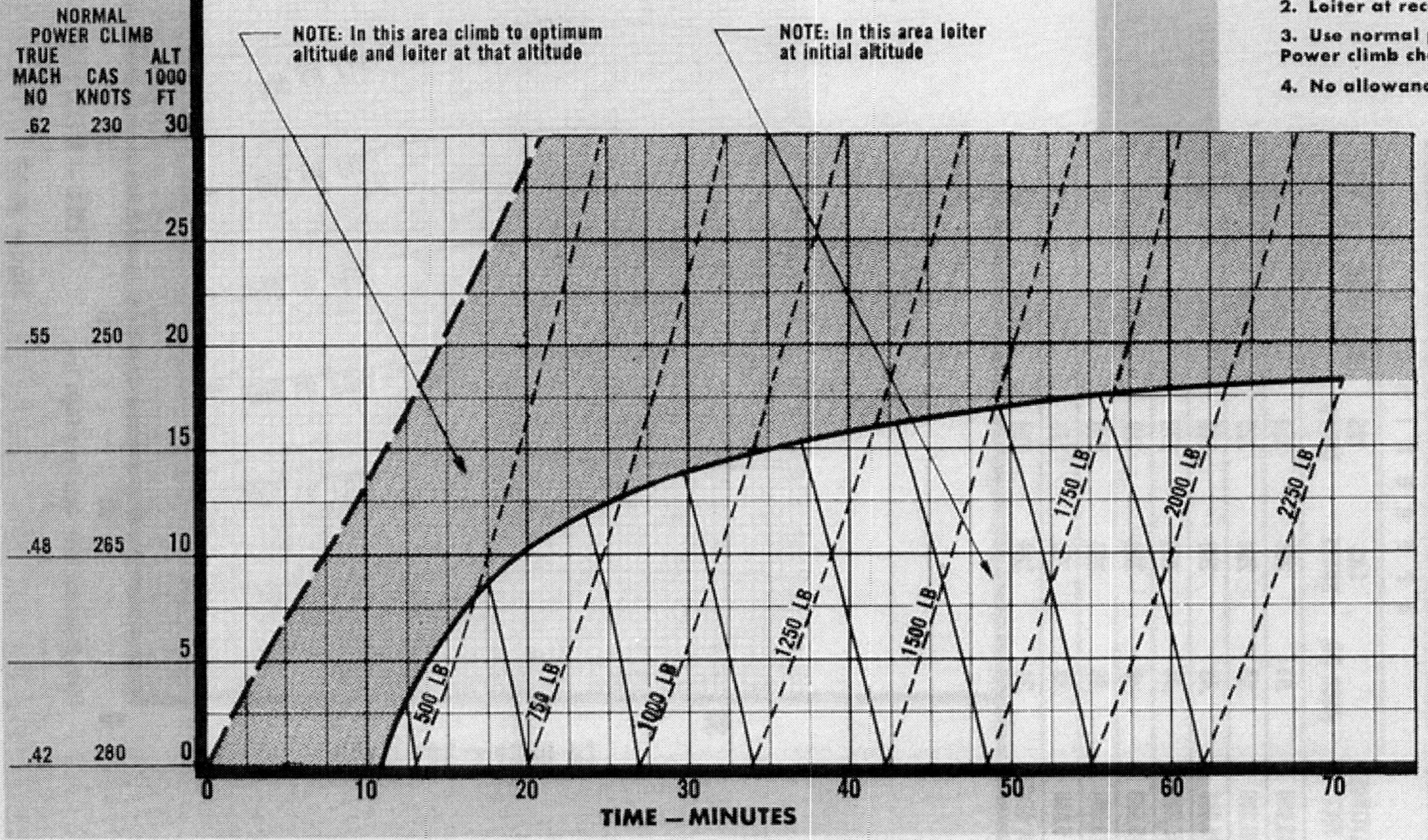

LEGEND

Line of optimum altitude

Maximum range descent line (Use idle power, speed brakes closed, descend at an indicated air speed of 198 knots)

Climb path guide lines

Fuel remaining

Date: 25 October 1955
Data Basis: Flight Test

Figure A-54

COMBAT ALLOWANCE CHART

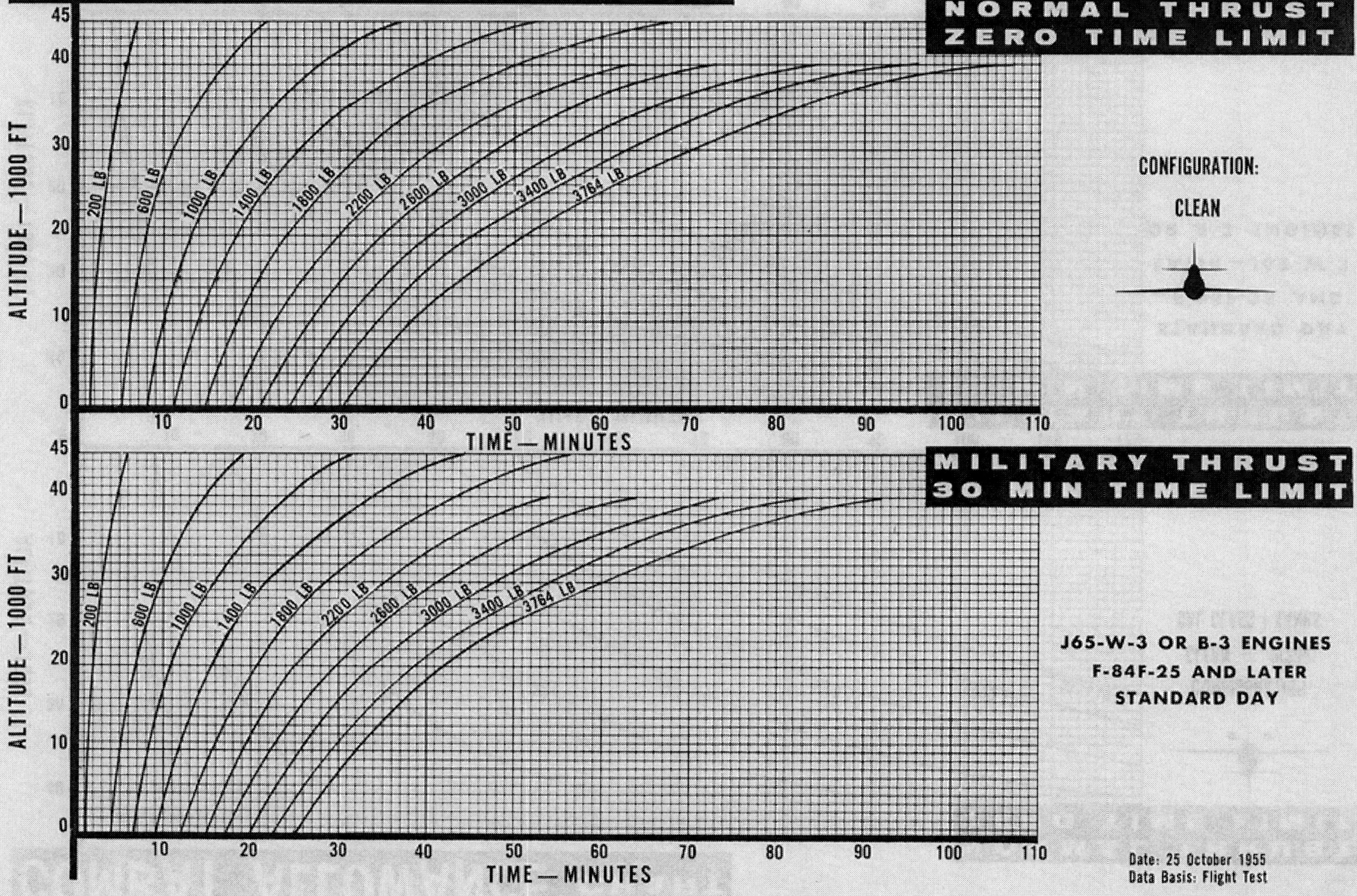

Figure A-55

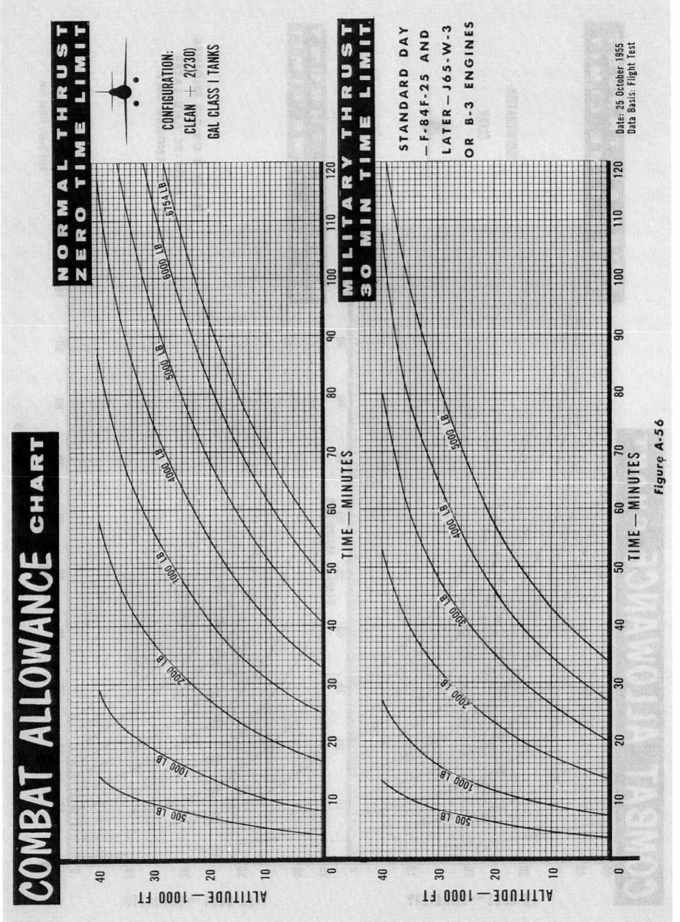

COMBAT ALLOWANCE CHART

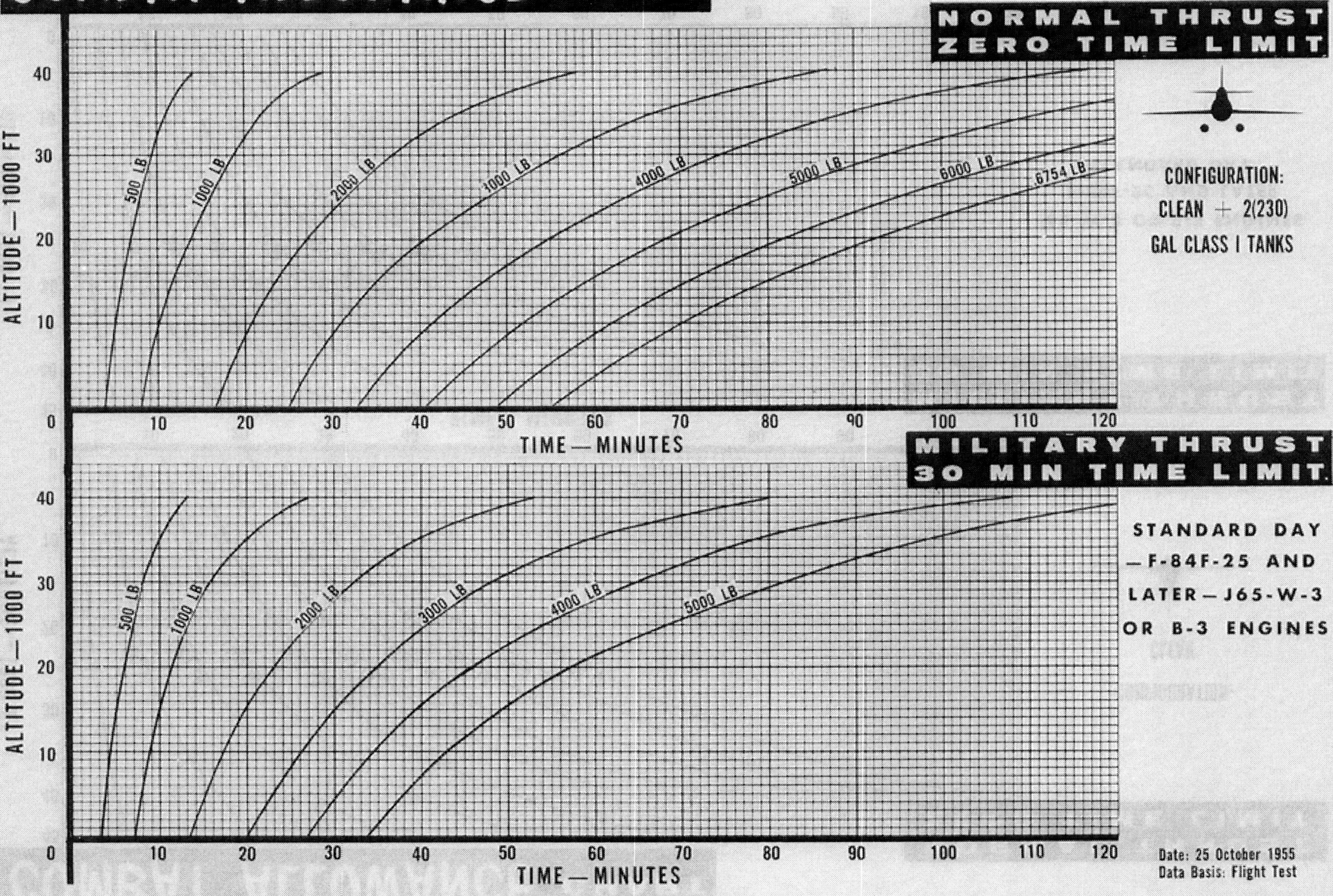

Date: 25 October 1955
Data Basis: Flight Test

Figure A-56

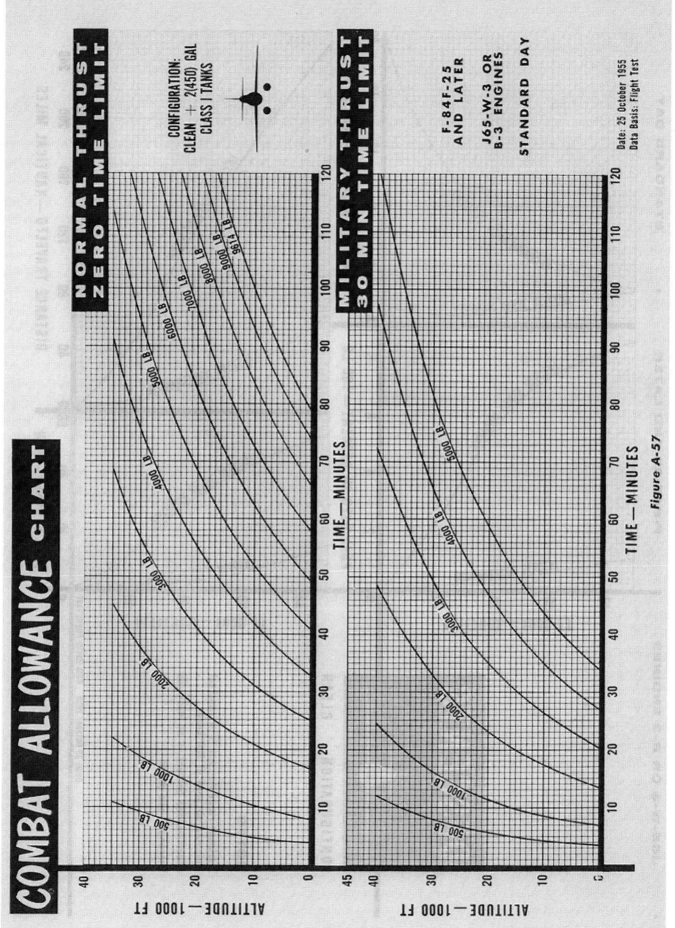

COMBAT ALLOWANCE CHART

NORMAL THRUST ZERO TIME LIMIT

MILITARY THRUST 30 MIN TIME LIMIT

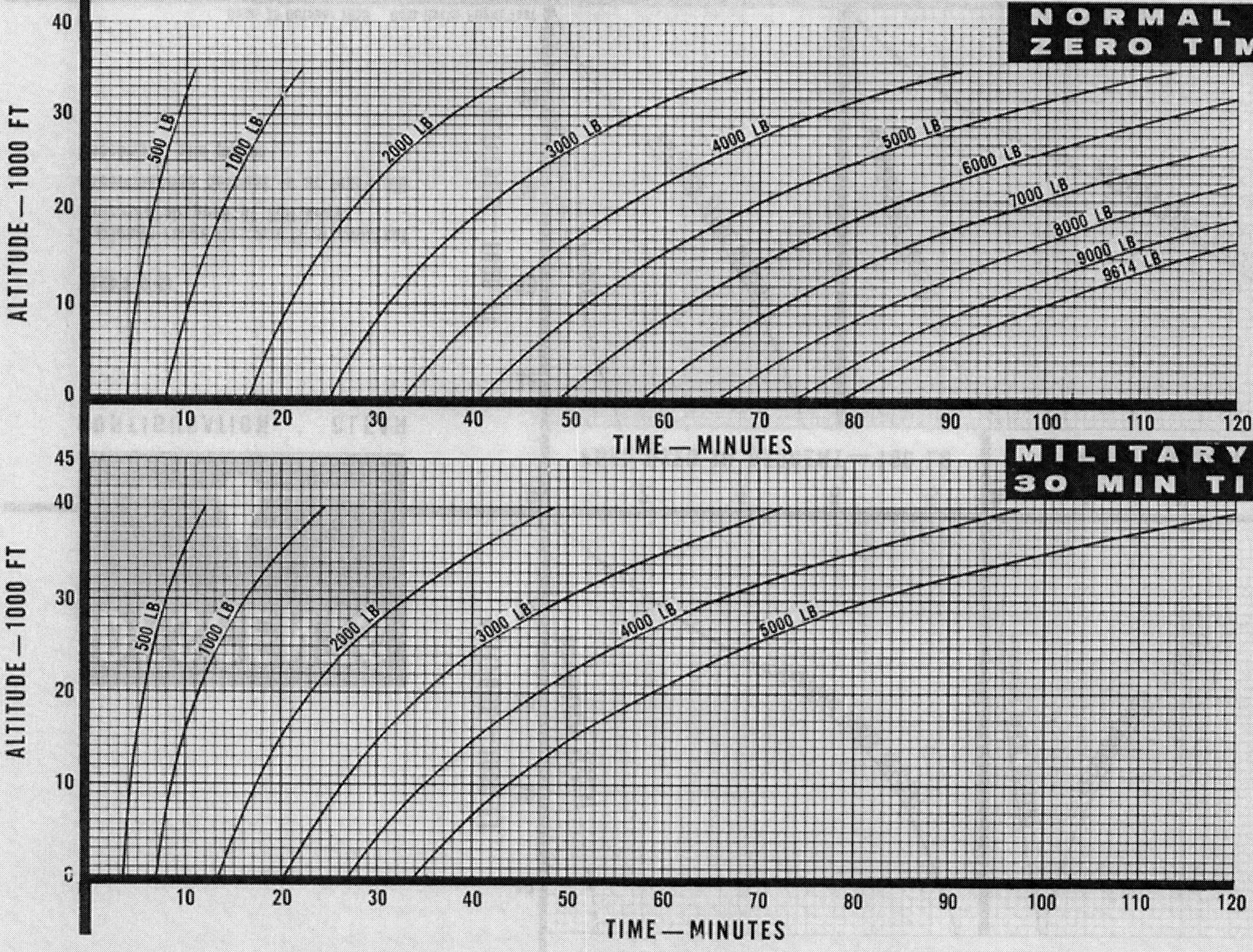

CONFIGURATION:
CLEAN + 2(450) GAL
CLASS I TANKS

F-84F-25
AND LATER

J65-W-3 OR
B-3 ENGINES

STANDARD DAY

Date: 25 October 1955
Data Basis: Flight Test

Figure A-57

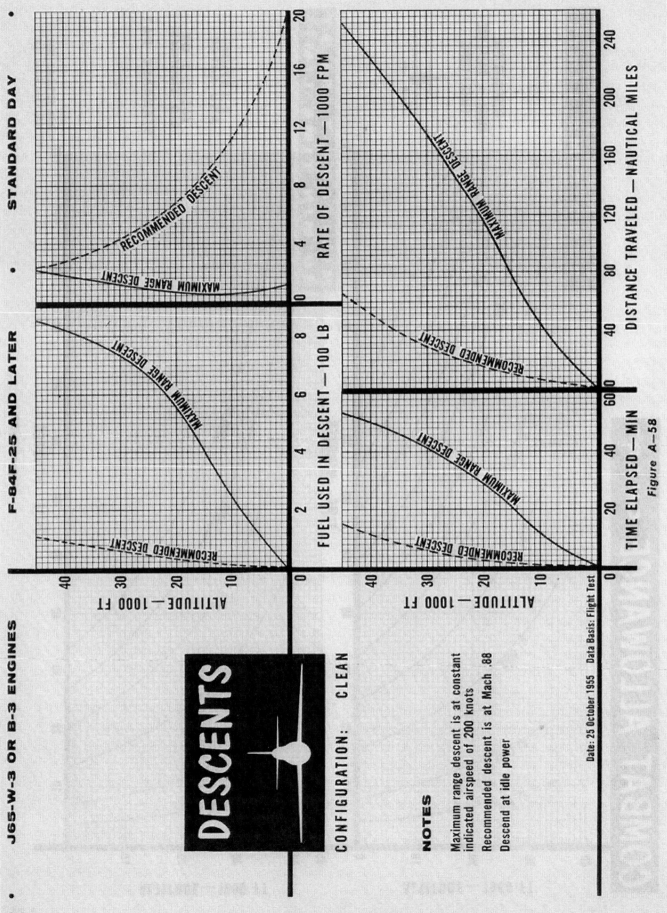

J65-W-3 OR B-3 ENGINES • F-84F-25 AND LATER • STANDARD DAY

DESCENTS

CONFIGURATION: CLEAN

NOTES

Maximum range descent is at constant indicated airspeed of 200 knots

Recommended descent is at Mach .88

Descend at idle power

Date: 25 October 1955 Data Basis: Flight Test

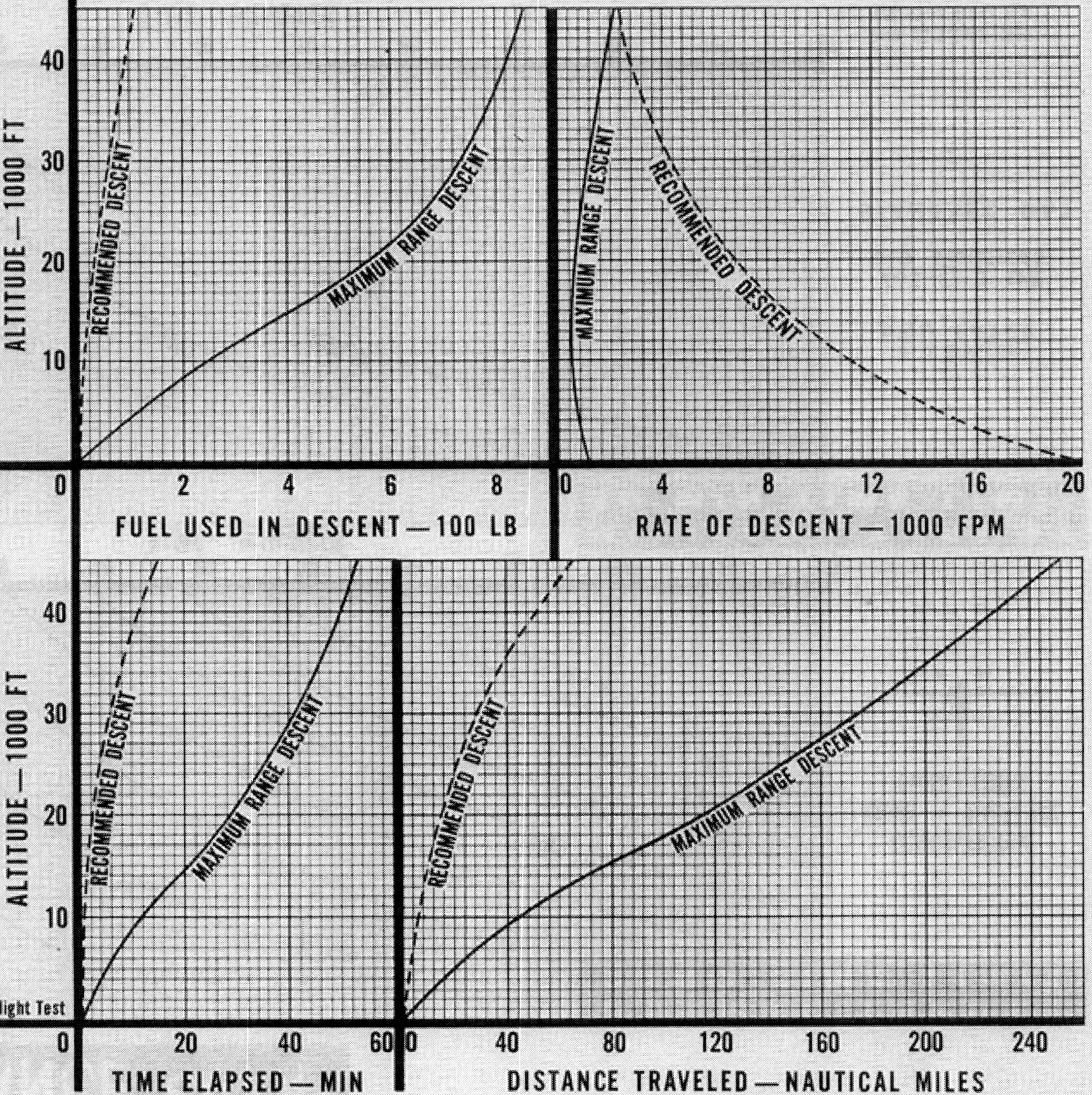

Figure A—58

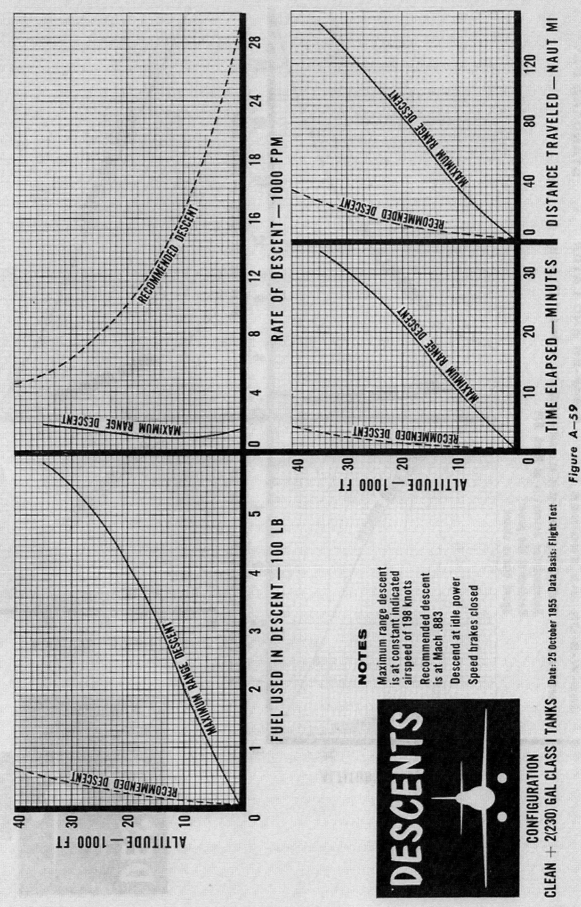

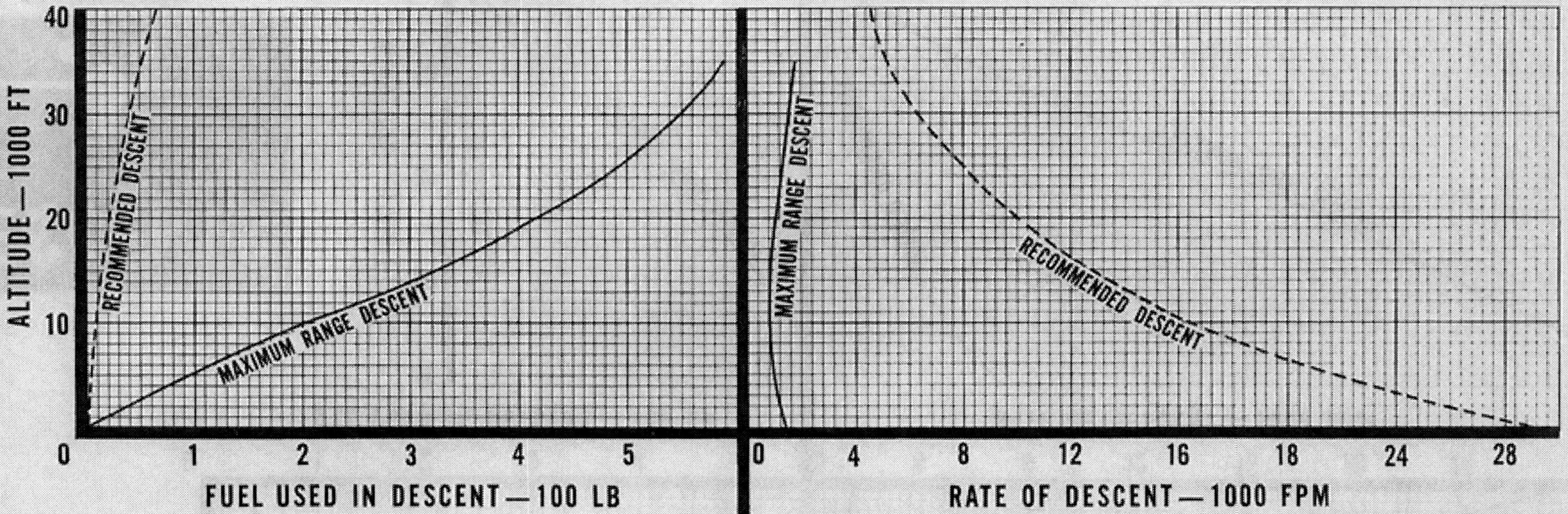

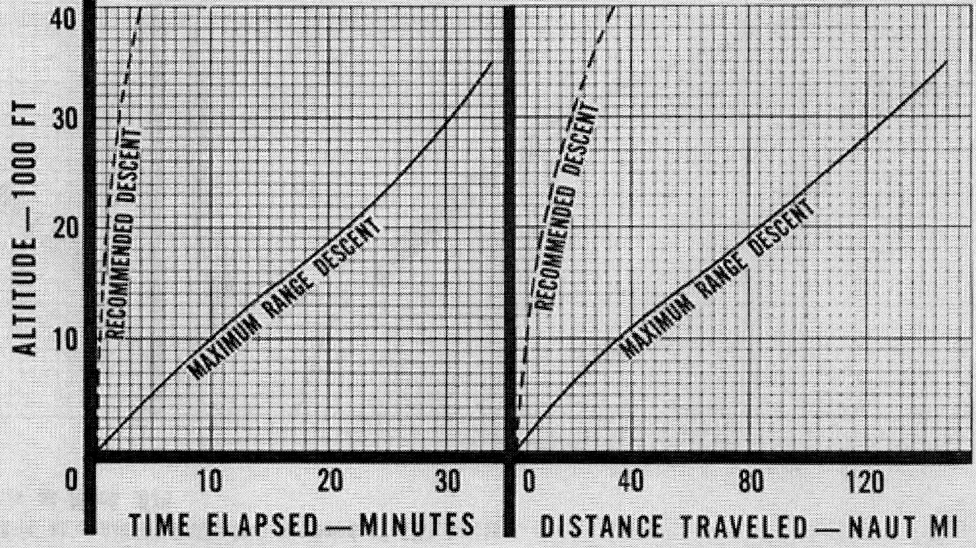

CONFIGURATION

CLEAN + 2(230) GAL CLASS I TANKS

NOTES

Maximum range descent is at constant indicated airspeed of 198 knots

Recommended descent is at Mach .883

Descend at idle power

Speed brakes closed

Date: 25 October 1955 Data Basis: Flight Test

Figure A–59

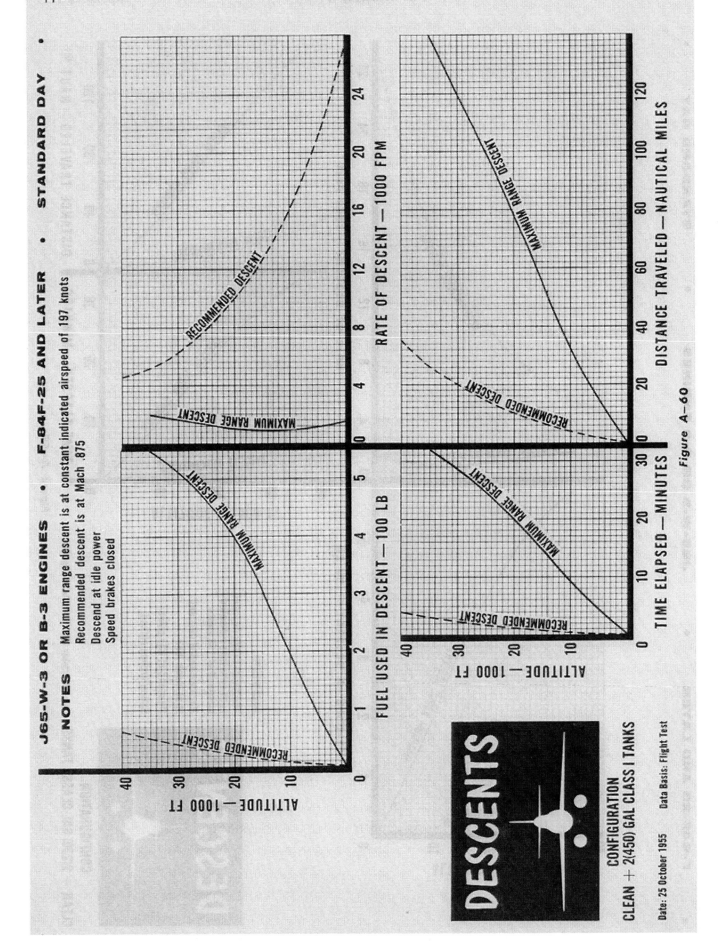

J65-W-3 OR B-3 ENGINES • F-84F-25 AND LATER • STANDARD DAY •

NOTES Maximum range descent is at constant indicated airspeed of 197 knots
Recommended descent is at Mach .875
Descend at idle power
Speed brakes closed

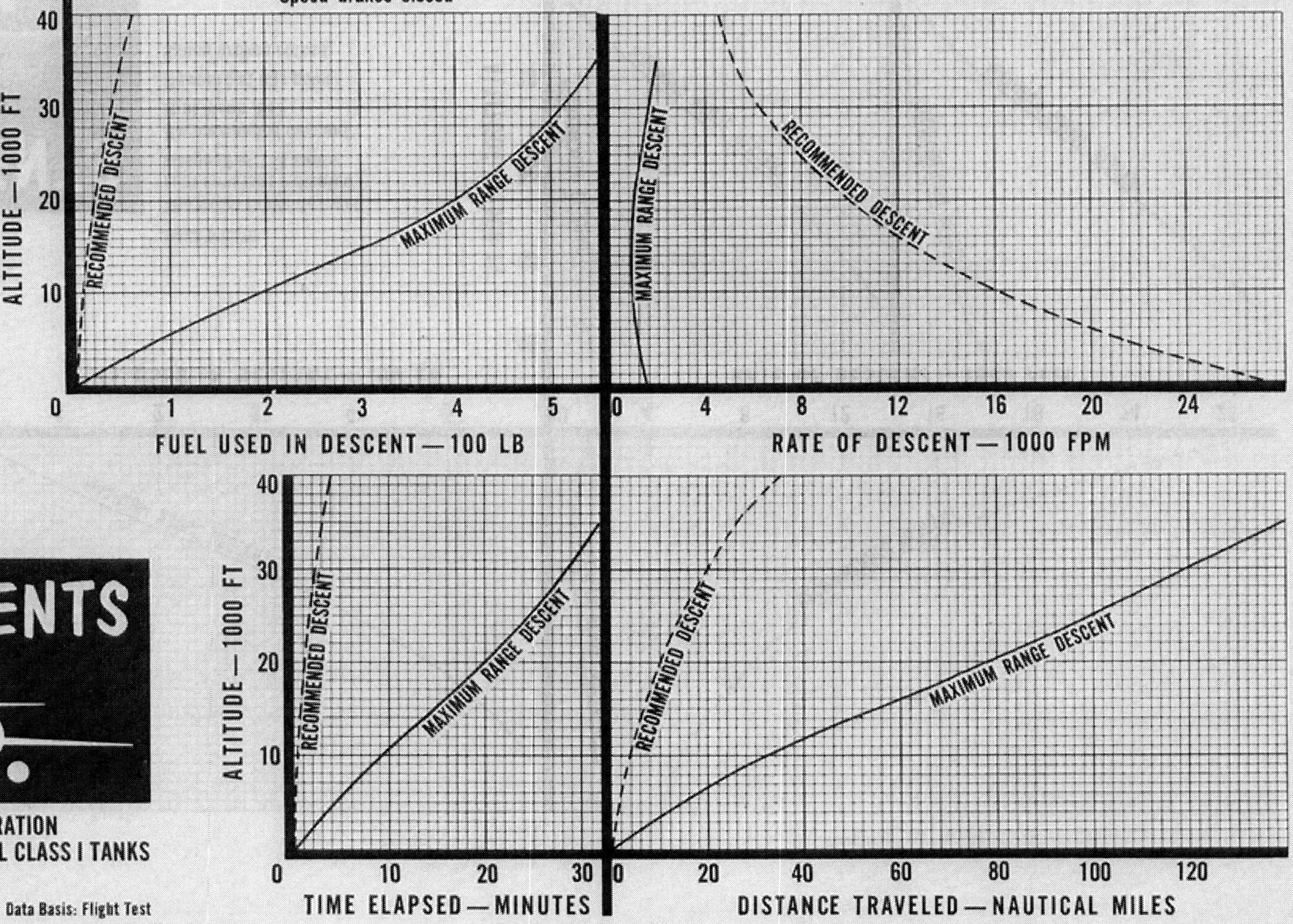

CONFIGURATION
CLEAN + 2(450) GAL CLASS I TANKS

Date: 25 October 1955 Data Basis: Flight Test

Figure A–60

F-84F-25 AND LATER • J65-W-3 OR B-3 ENGINES • • STANDARD DAY •

NOTES

Maximum range descent is at constant indicated airspeed of 198 knots

Recommended descent is at Mach .89

Descend at idle power

Speed brakes closed

Date: 25 October 1955 Data Basis: Flight Test

CONFIGURATION: CLEAN + 2(230) GAL CLASS I TANKS + 2(230) GAL CLASS II TANKS

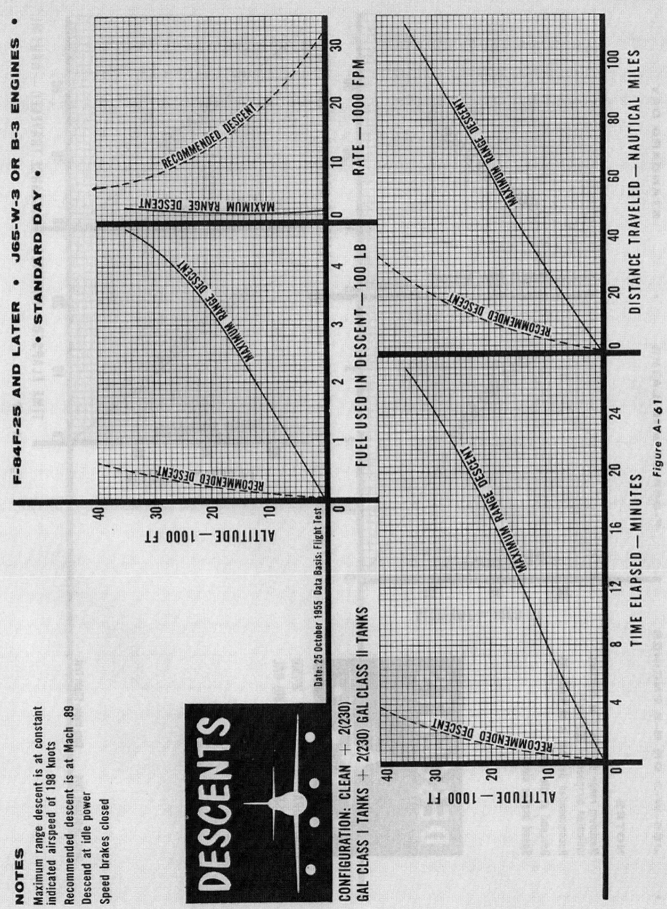

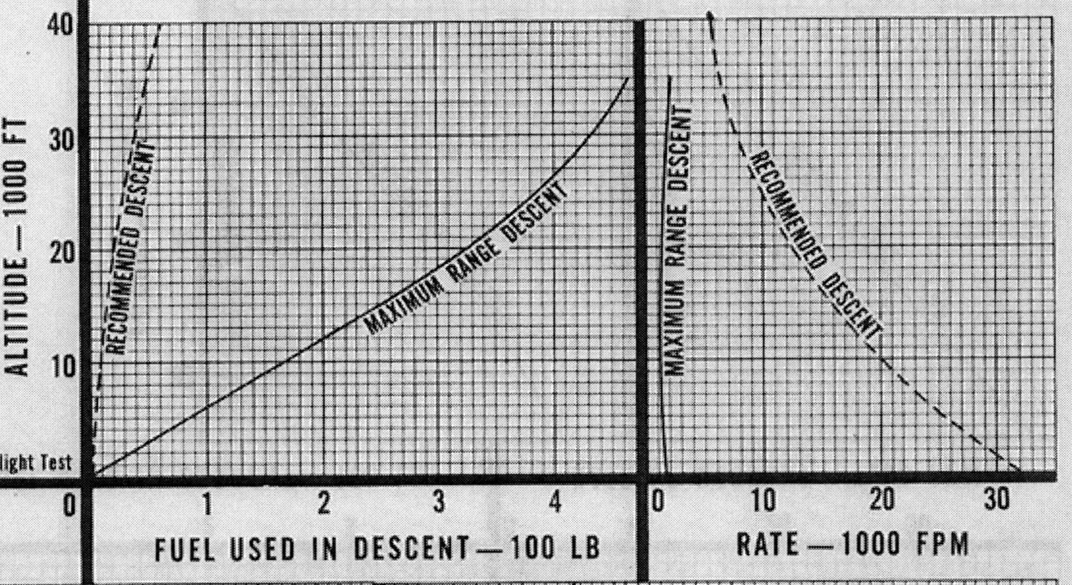

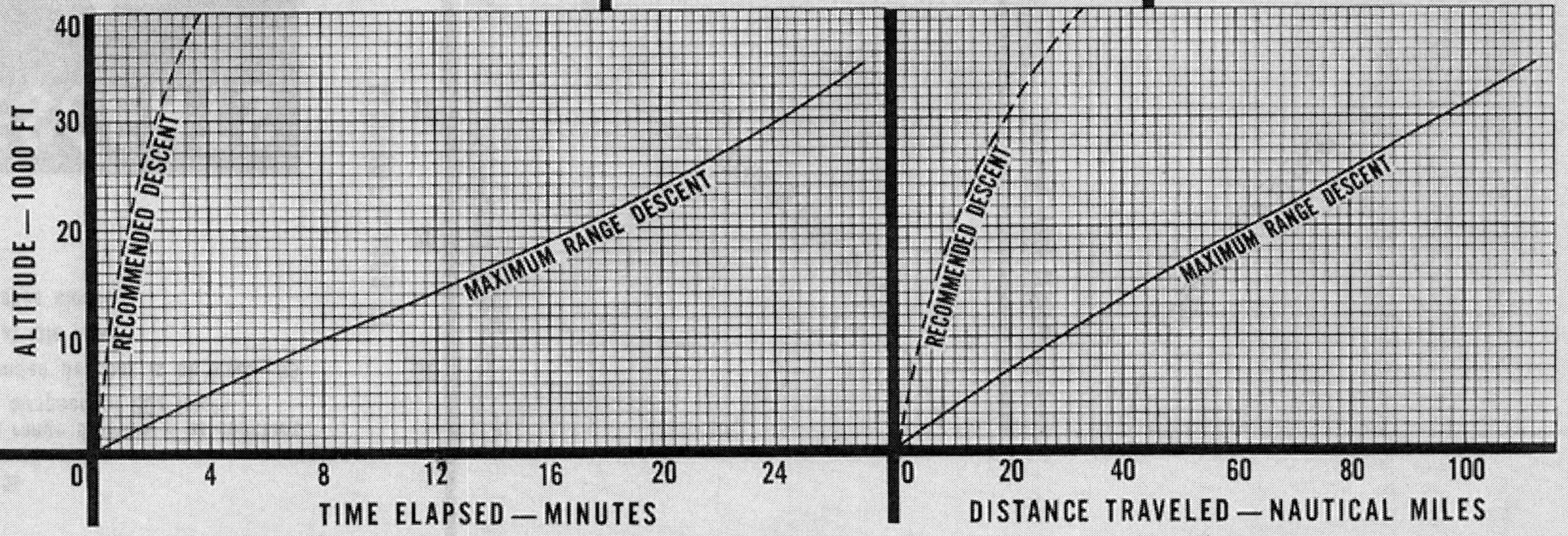

Figure A-61

J65-W-3 OR B-3 ENGINES • F-84F-25 AND LATER • STANDARD DAY

NOTES

Maximum range descent is at constant indicated airspeed of 202 knots

Recommended descent is at Mach .89

Descend at idle power

Speed brakes closed

DESCENTS

CONFIGURATION: CLEAN + 2(450) GAL CLASS I TANKS + 2(230) GAL CLASS II TANKS

Date: 25 October 1955 Data Basis: Flight Test

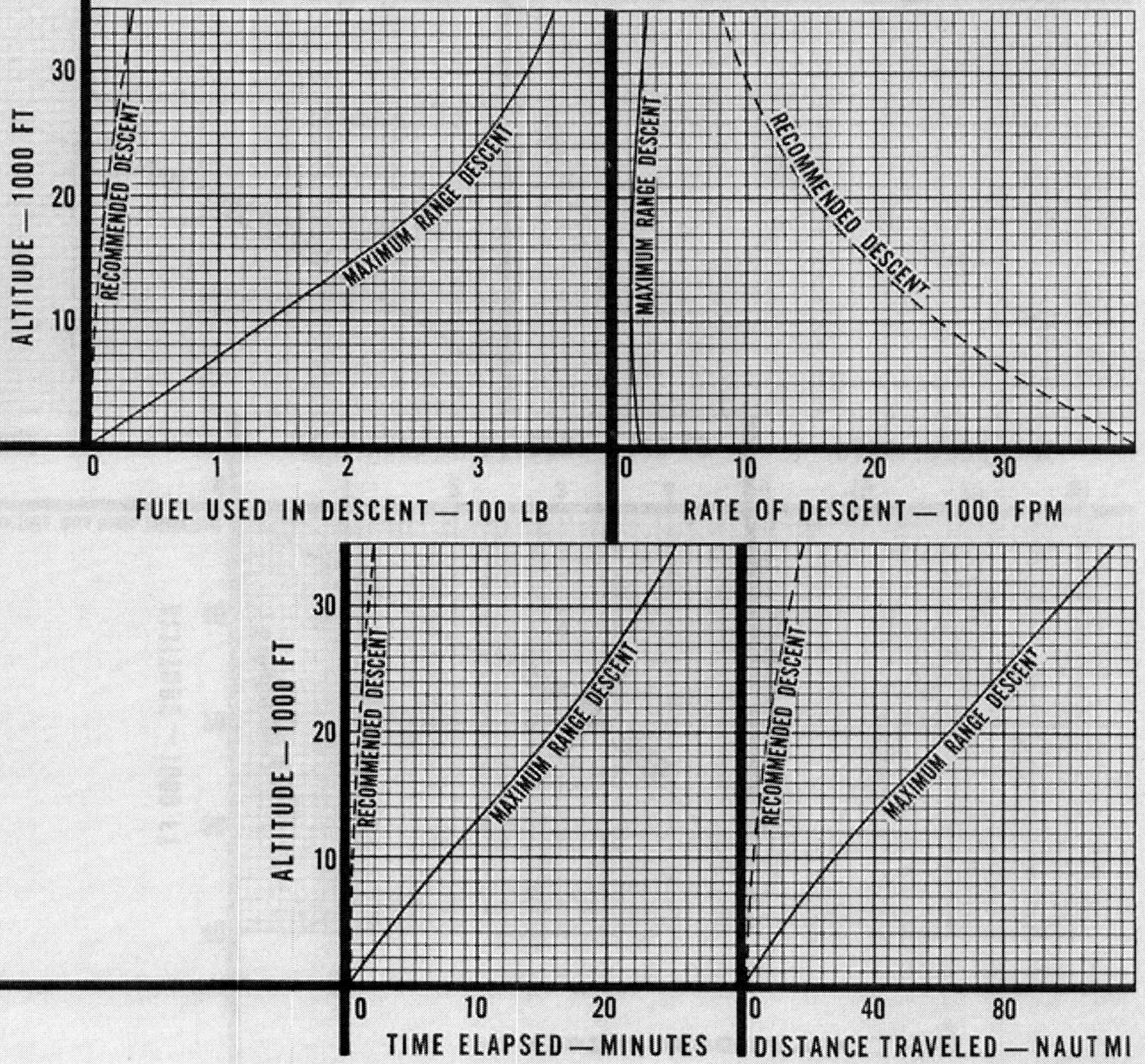

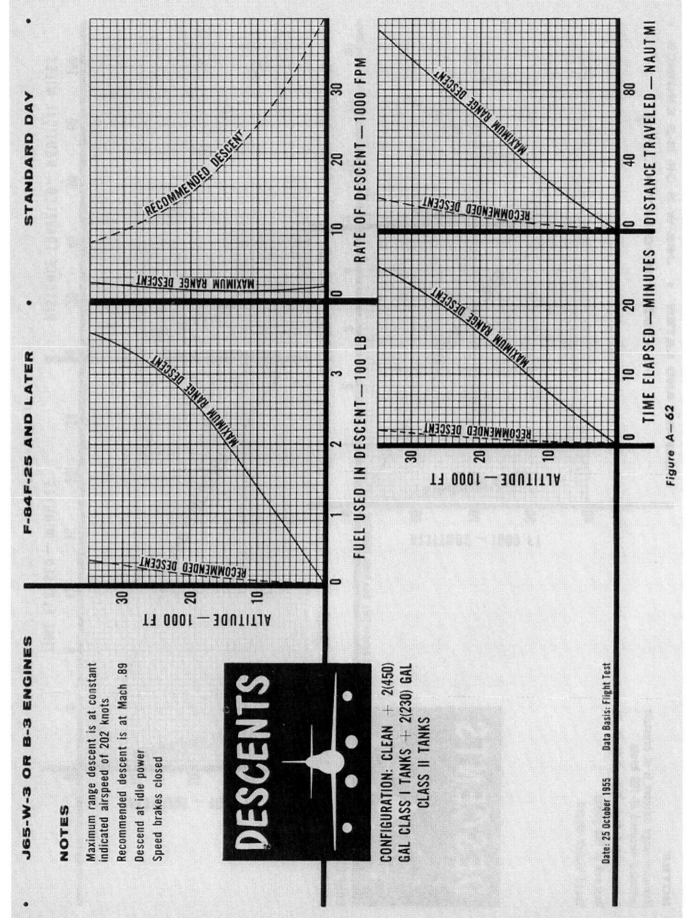

Figure A-62

F-84F-25 AND LATER • J65-W-3 OR B-3 ENGINES • STANDARD DAY •

NOTES

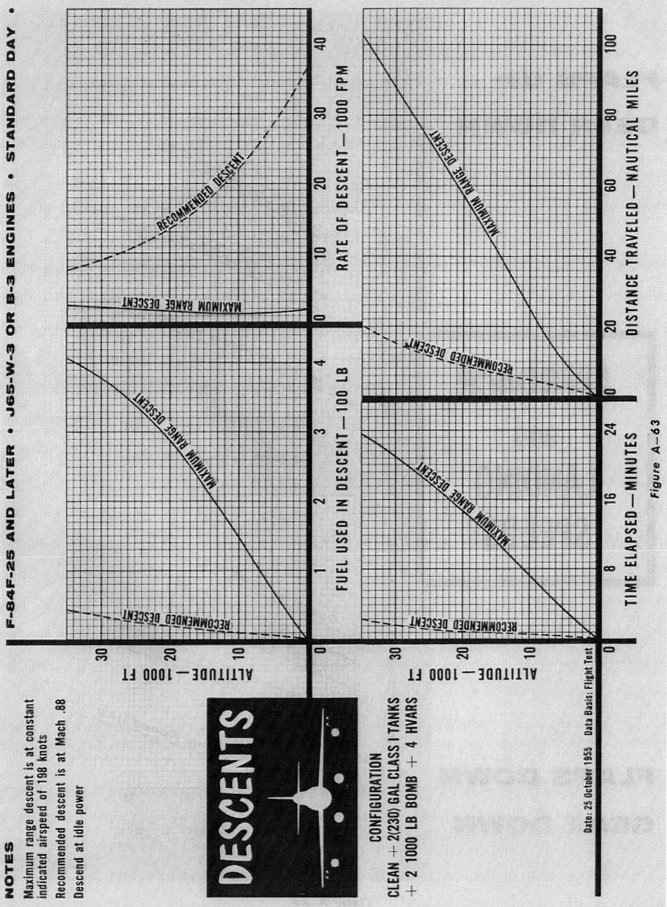

Maximum range descent is at constant indicated airspeed of 198 knots

Recommended descent is at Mach .88

Descend at idle power

CONFIGURATION
CLEAN + 2(230) GAL CLASS I TANKS + 2 1000 LB BOMB + 4 HVARS

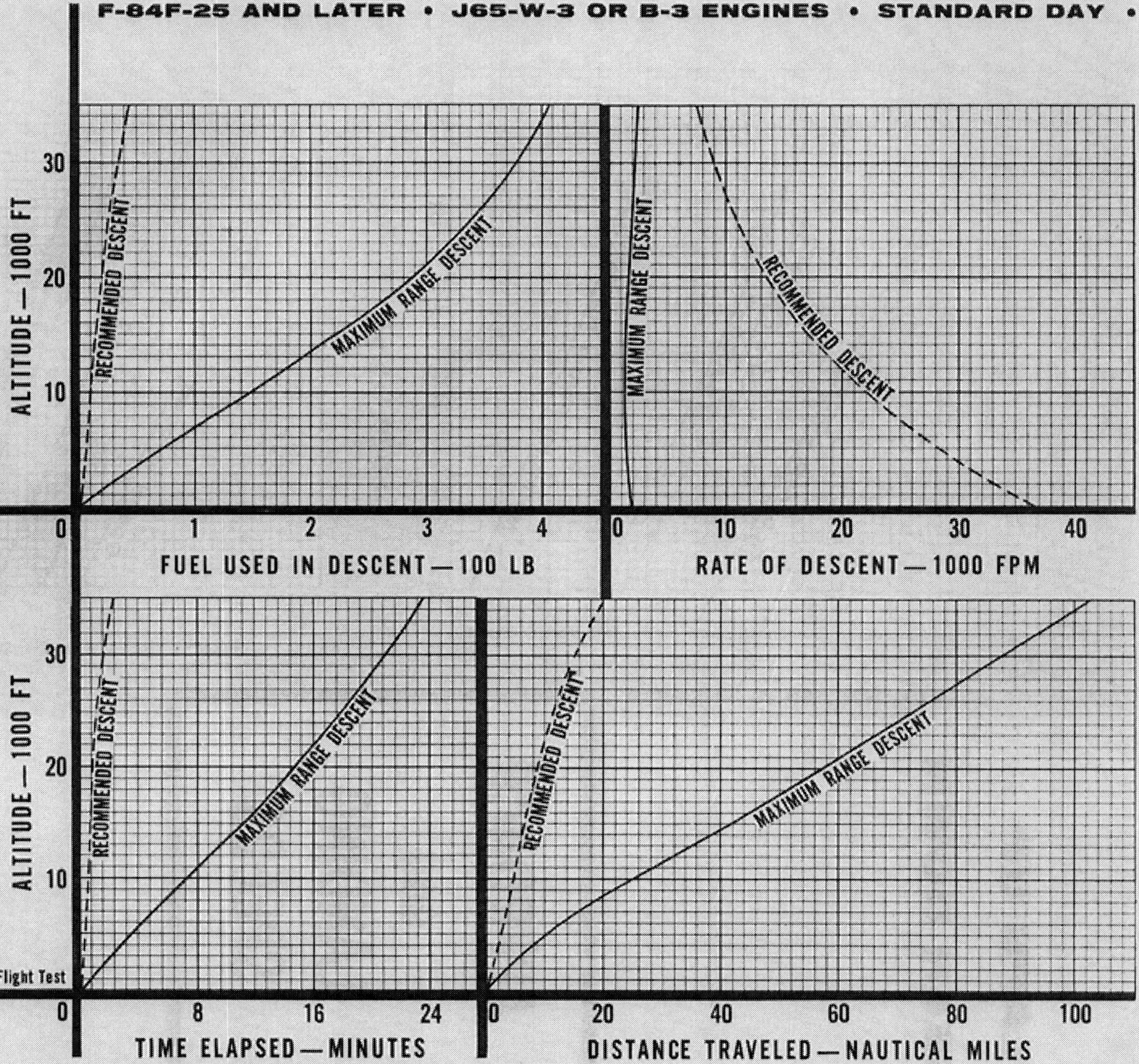

Date: 25 October 1955 Data Basis: Flight Test

Figure A–63

APPROACH and LANDING SPEEDS

FLAPS UP GEAR DOWN

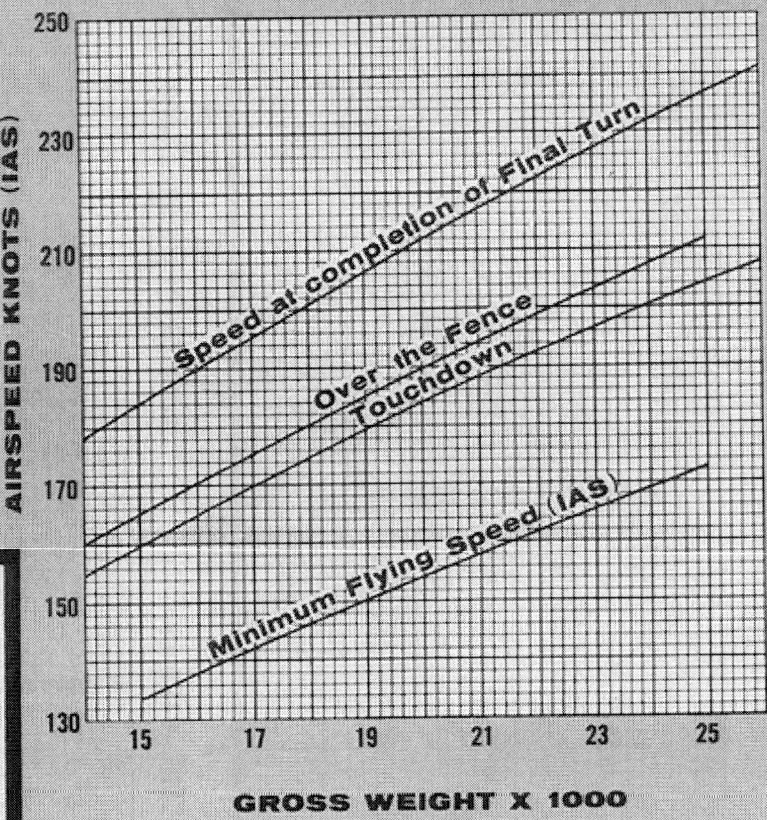

FLAPS DOWN GEAR DOWN

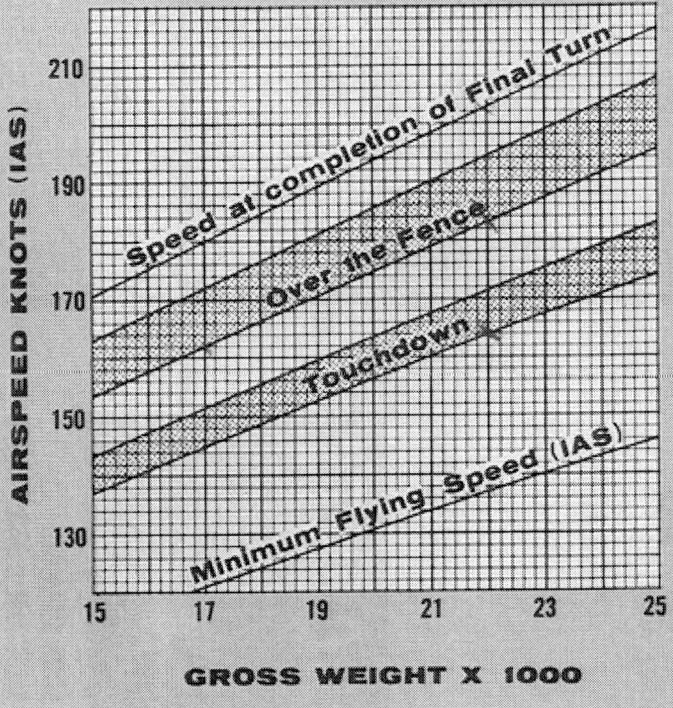

Date: 25 October 1955
Data Basis: Flight Test

Figure A—64

LANDING DISTANCES

WITHOUT DRAG CHUTE
FLAPS DOWN — GEAR DOWN

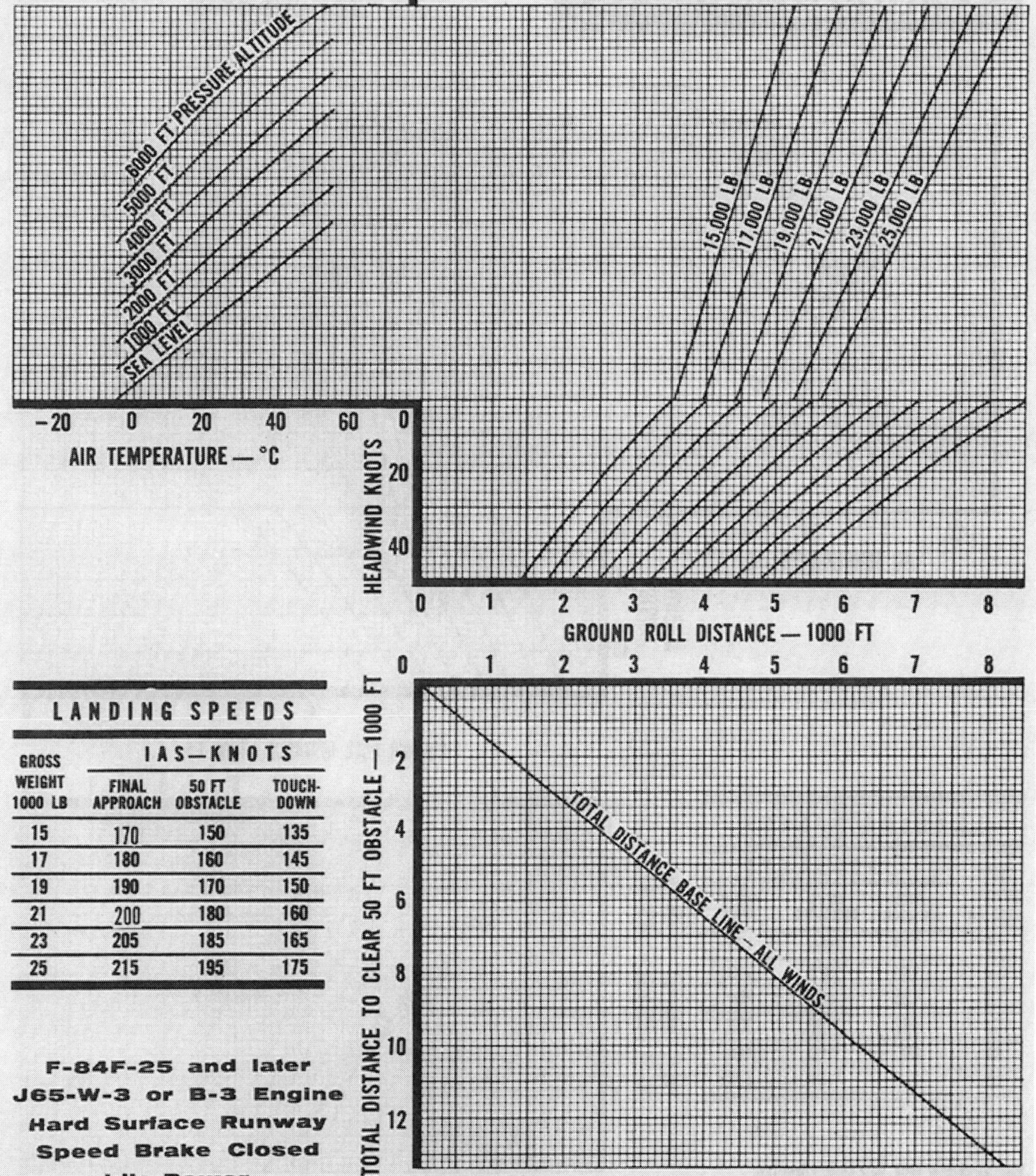

LANDING SPEEDS

GROSS WEIGHT 1000 LB	IAS—KNOTS FINAL APPROACH	50 FT OBSTACLE	TOUCH-DOWN
15	170	150	135
17	180	160	145
19	190	170	150
21	200	180	160
23	205	185	165
25	215	195	175

F-84F-25 and later
J65-W-3 or B-3 Engine
Hard Surface Runway
Speed Brake Closed
Idle Power

Date: 25 October 1955 Data Basis: Flight Test

Figure A-65

IDLE POWER **Hard Surface Runway • F-84F-25 and later**
Speed Brake Closed • J65-W-3 or B-3 Engine

LANDING DISTANCES

DRAG CHUTE INFLATED AT TOUCHDOWN

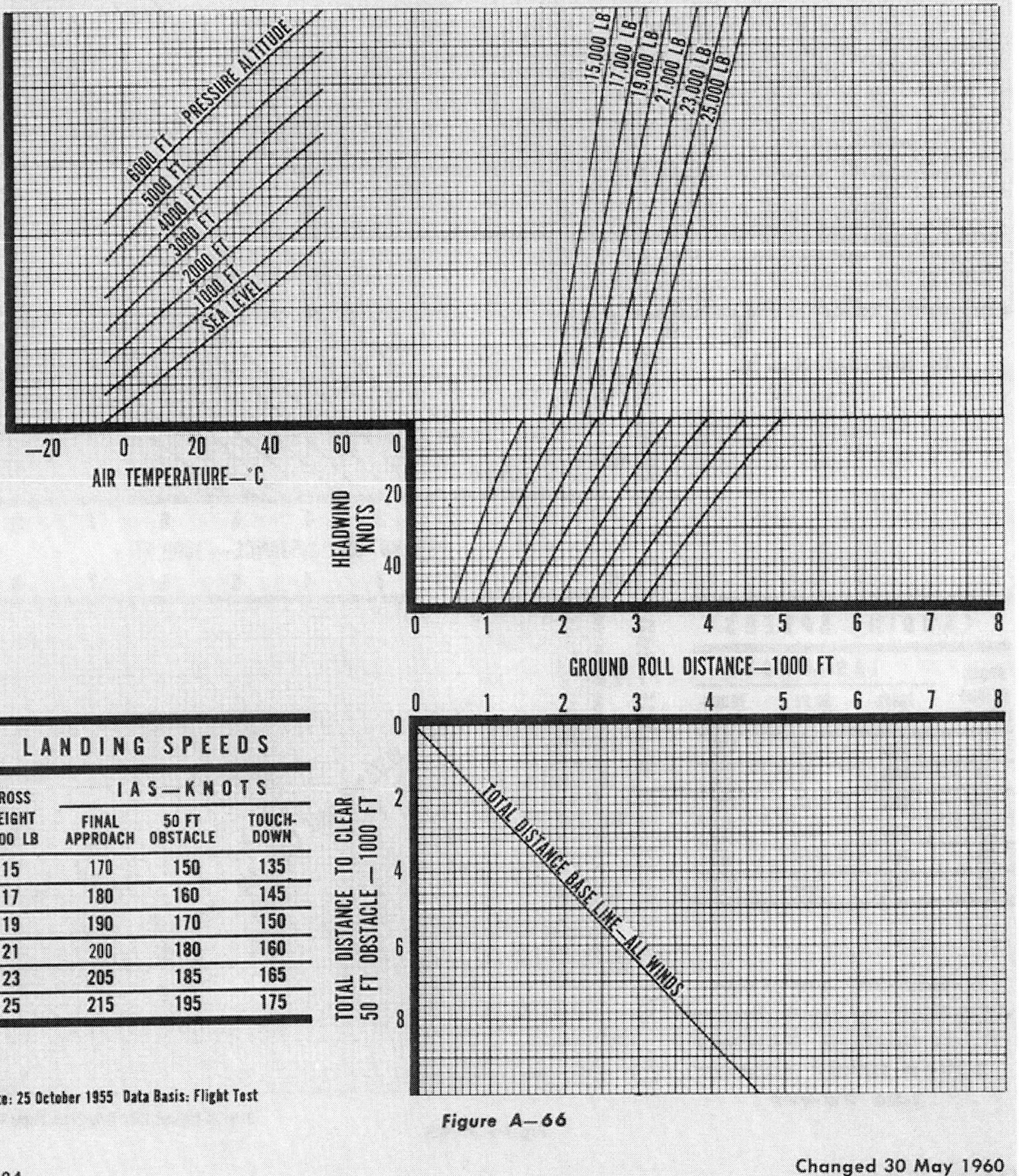

LANDING SPEEDS

GROSS WEIGHT 1000 LB	IAS—KNOTS		
	FINAL APPROACH	50 FT OBSTACLE	TOUCH-DOWN
15	170	150	135
17	180	160	145
19	190	170	150
21	200	180	160
23	205	185	165
25	215	195	175

Date: 25 October 1955 Data Basis: Flight Test

Figure A—66

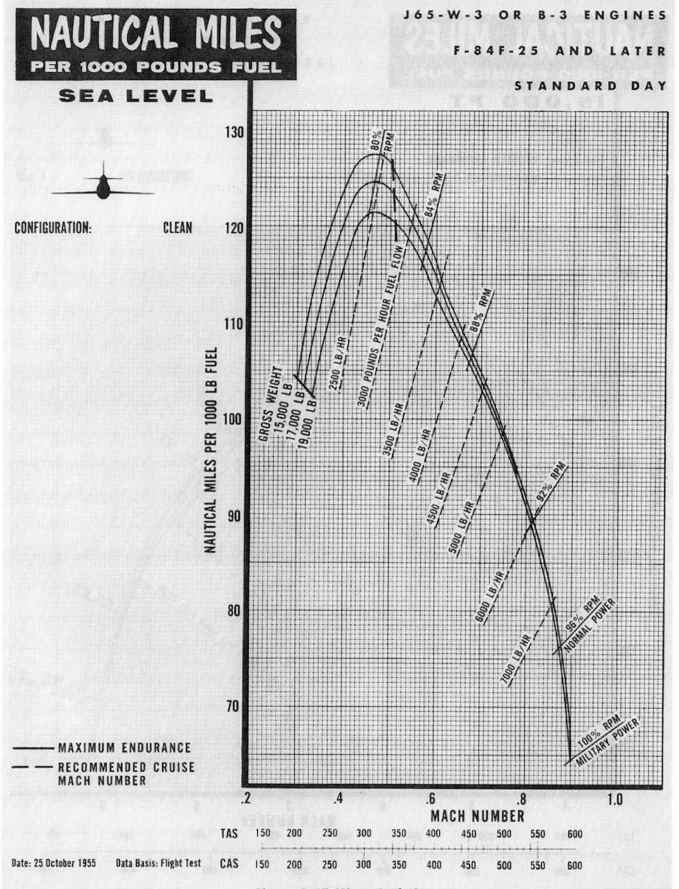

NAUTICAL MILES PER 1000 POUNDS FUEL

SEA LEVEL

J65-W-3 OR B-3 ENGINES
F-84F-25 AND LATER
STANDARD DAY

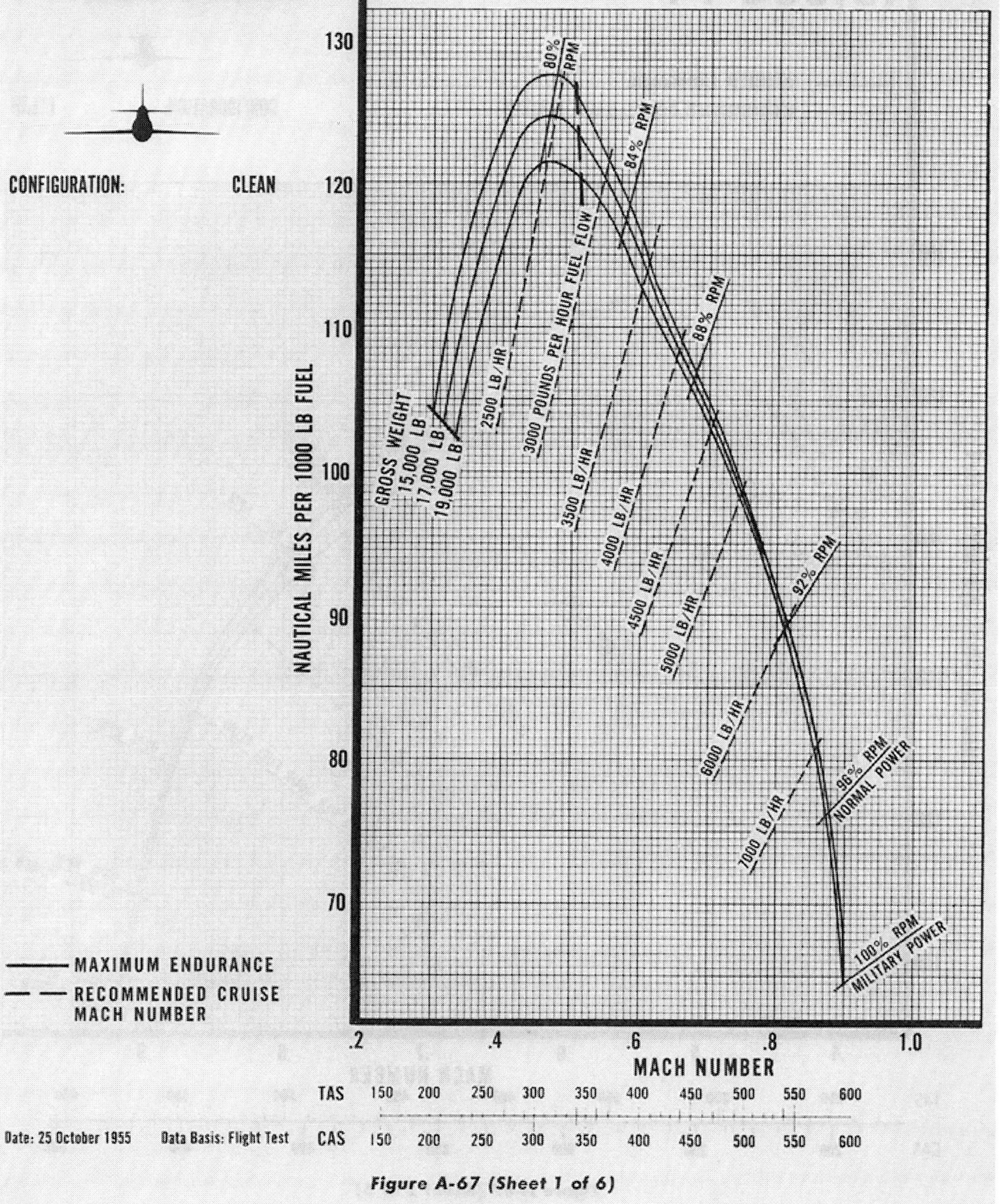

Date: 25 October 1955 Data Basis: Flight Test

Figure A-67 (Sheet 1 of 6)

NAUTICAL MILES
PER 1000 POUNDS FUEL
15,000 FT

F-84F-25 AND LATER
STANDARD DAY
J65-W-3 OR B-3 ENGINES

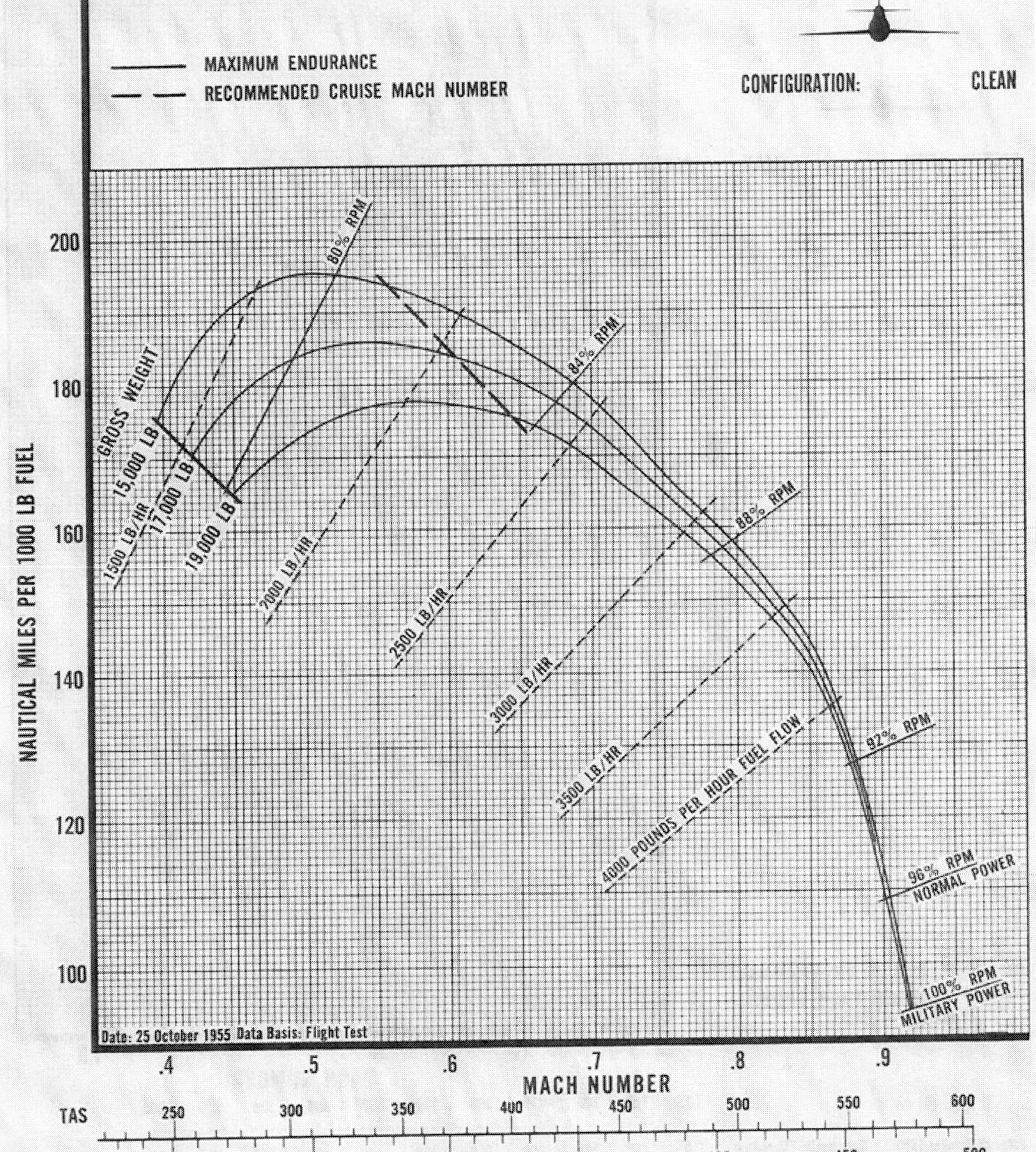

Figure A-67 (Sheet 2 of 6)

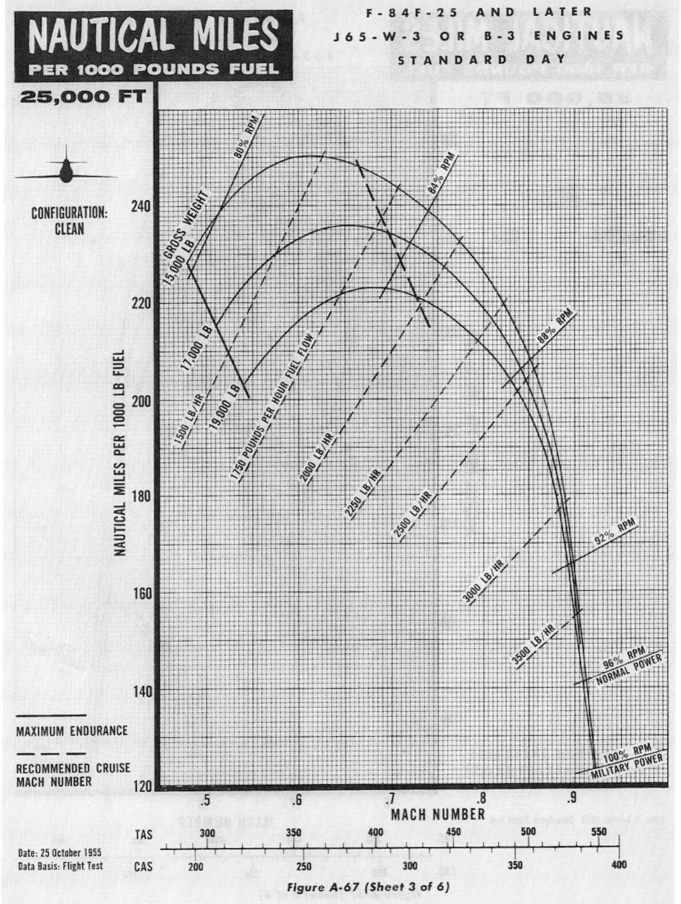

NAUTICAL MILES
PER 1000 POUNDS FUEL

F-84F-25 AND LATER
J65-W-3 OR B-3 ENGINES
STANDARD DAY

25,000 FT

CONFIGURATION:
CLEAN

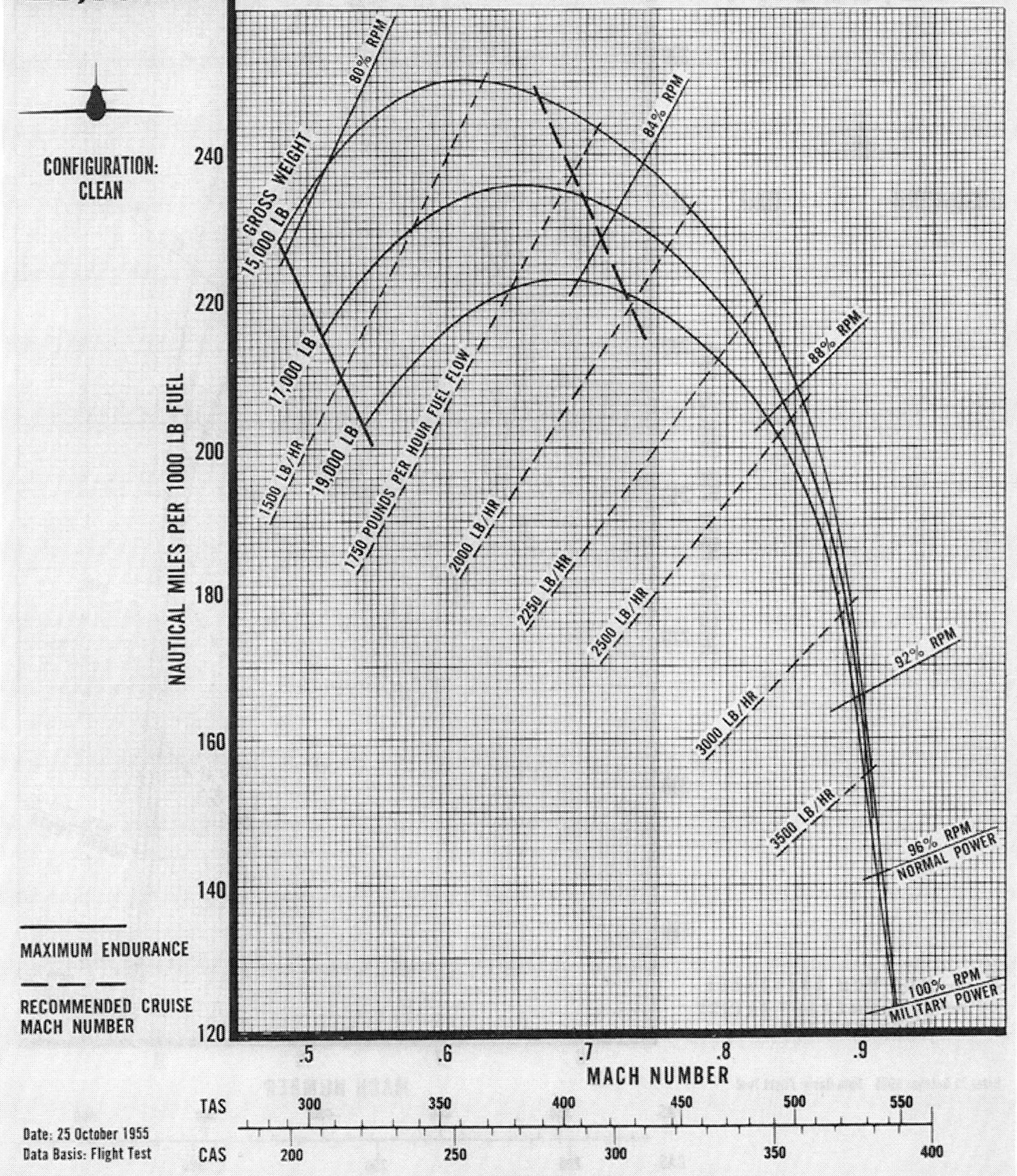

Date: 25 October 1955
Data Basis: Flight Test

Figure A-67 (Sheet 3 of 6)

NAUTICAL MILES PER 1000 POUNDS FUEL

35,000 FT

F-84F-25 AND LATER
STANDARD DAY
J65-W-3 OR B-3 ENGINES

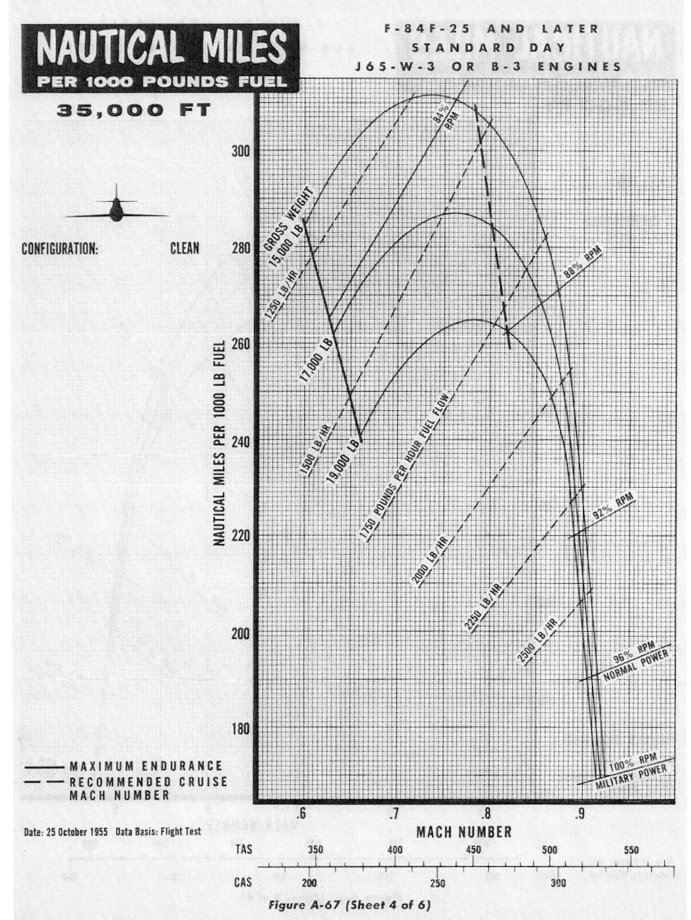

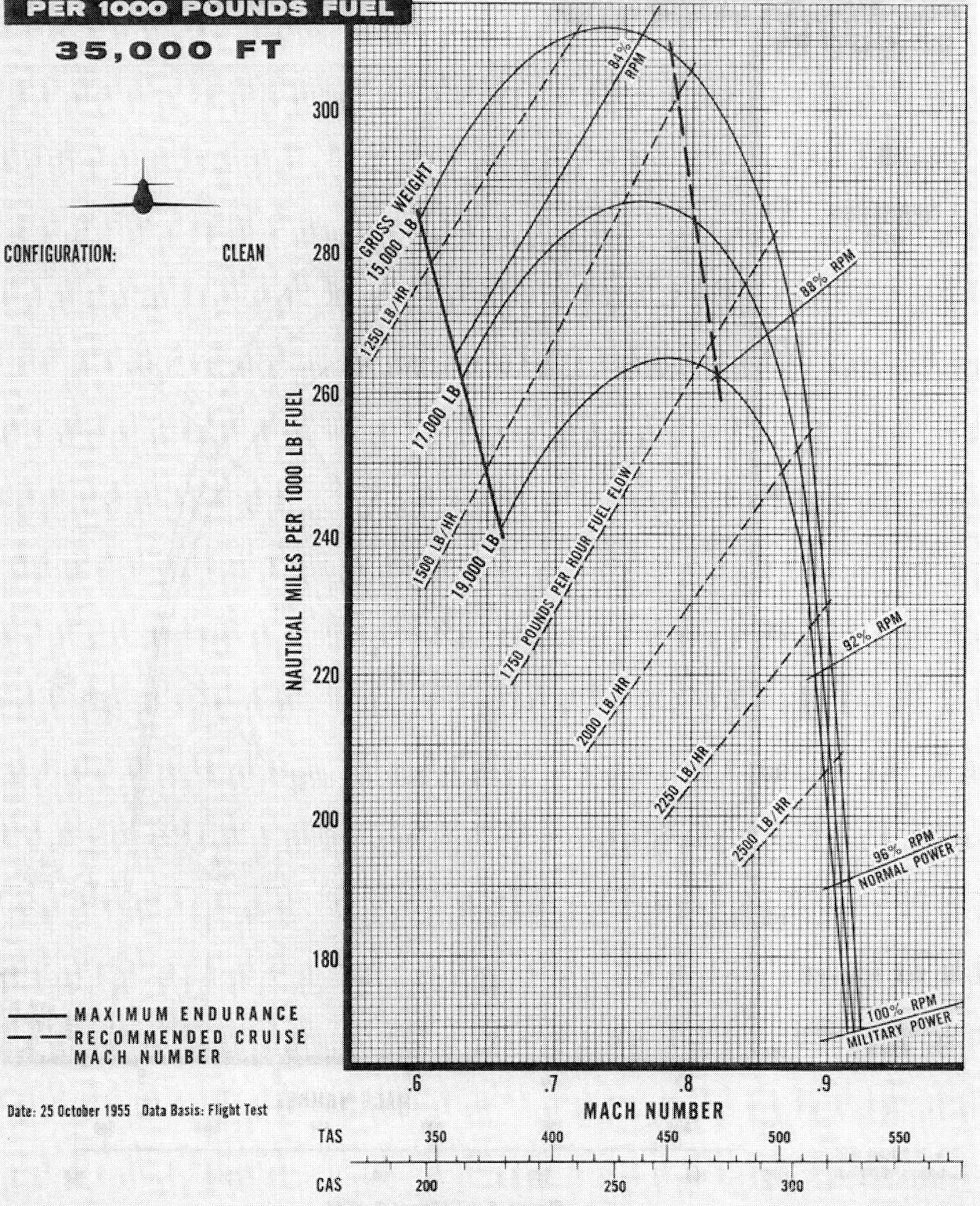

Date: 25 October 1955 Data Basis: Flight Test

Figure A-67 (Sheet 4 of 6)

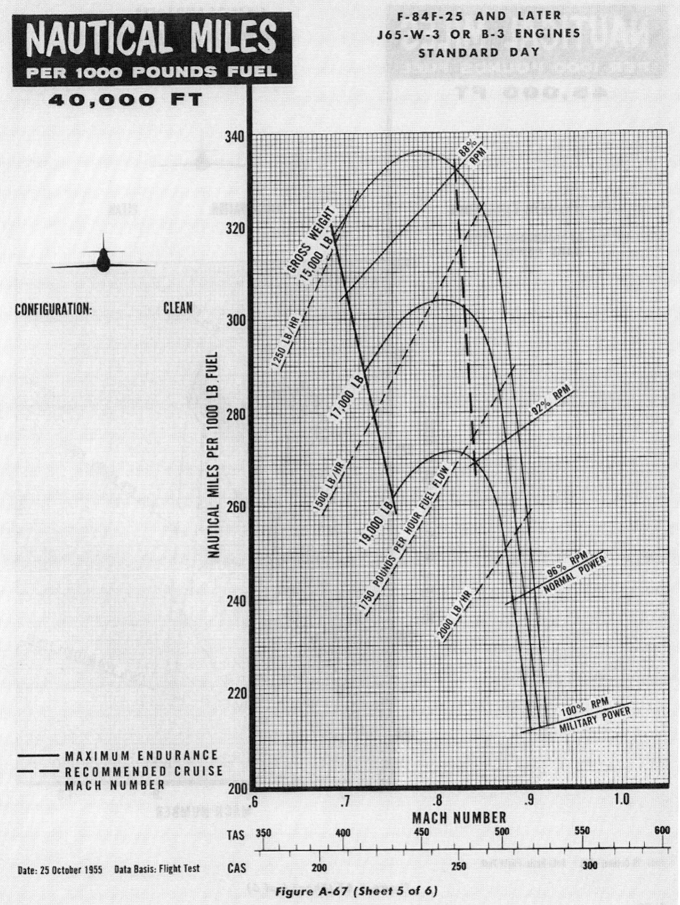

NAUTICAL MILES
PER 1000 POUNDS FUEL
40,000 FT

F-84F-25 AND LATER
J65-W-3 OR B-3 ENGINES
STANDARD DAY

CONFIGURATION: CLEAN

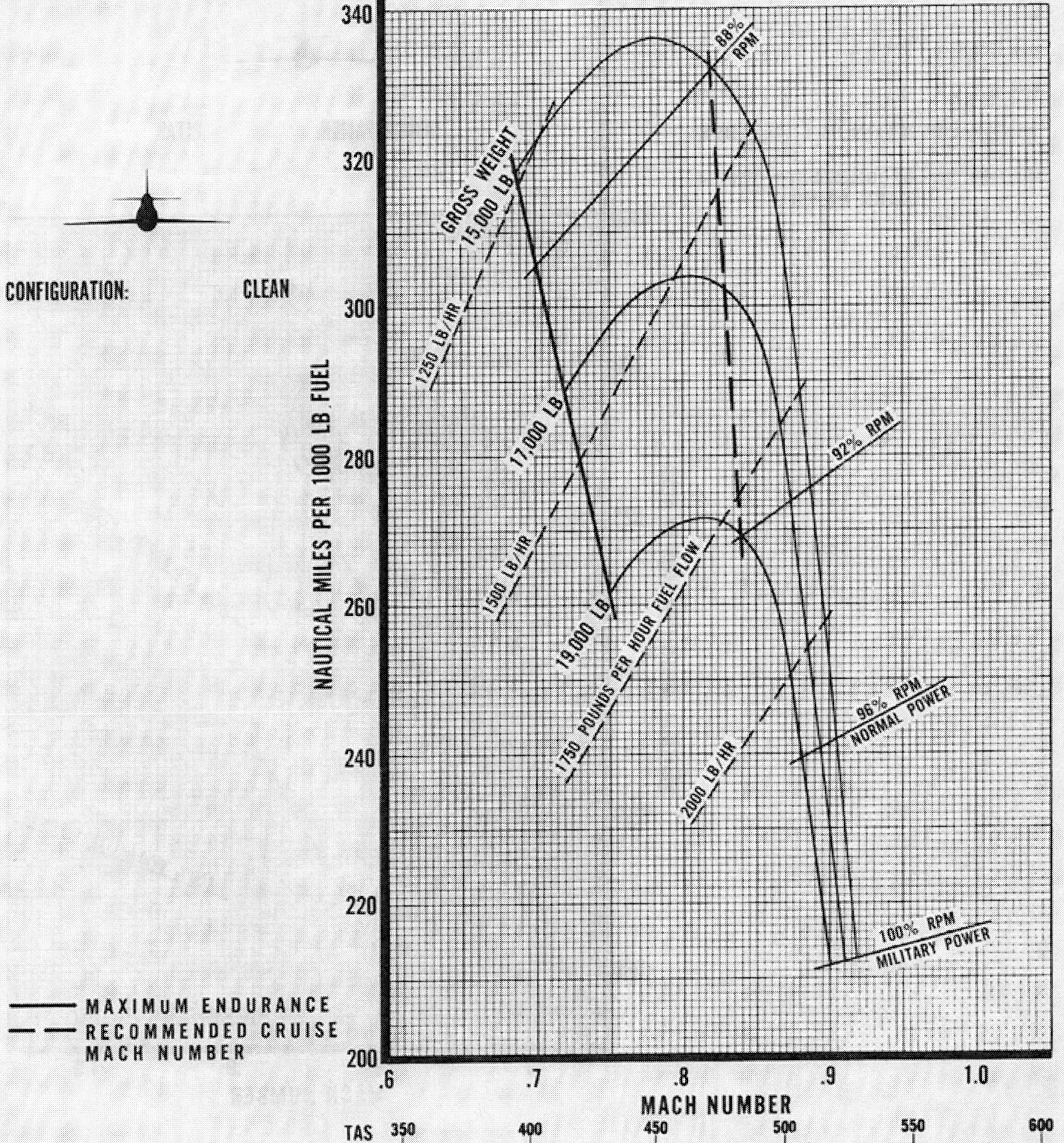

Date: 25 October 1955 Data Basis: Flight Test

Figure A-67 (Sheet 5 of 6)

NAUTICAL MILES
PER 1000 POUNDS FUEL
45,000 FT

F-84F-25 AND LATER
J65-W-3 OR B-3 ENGINES
STANDARD DAY

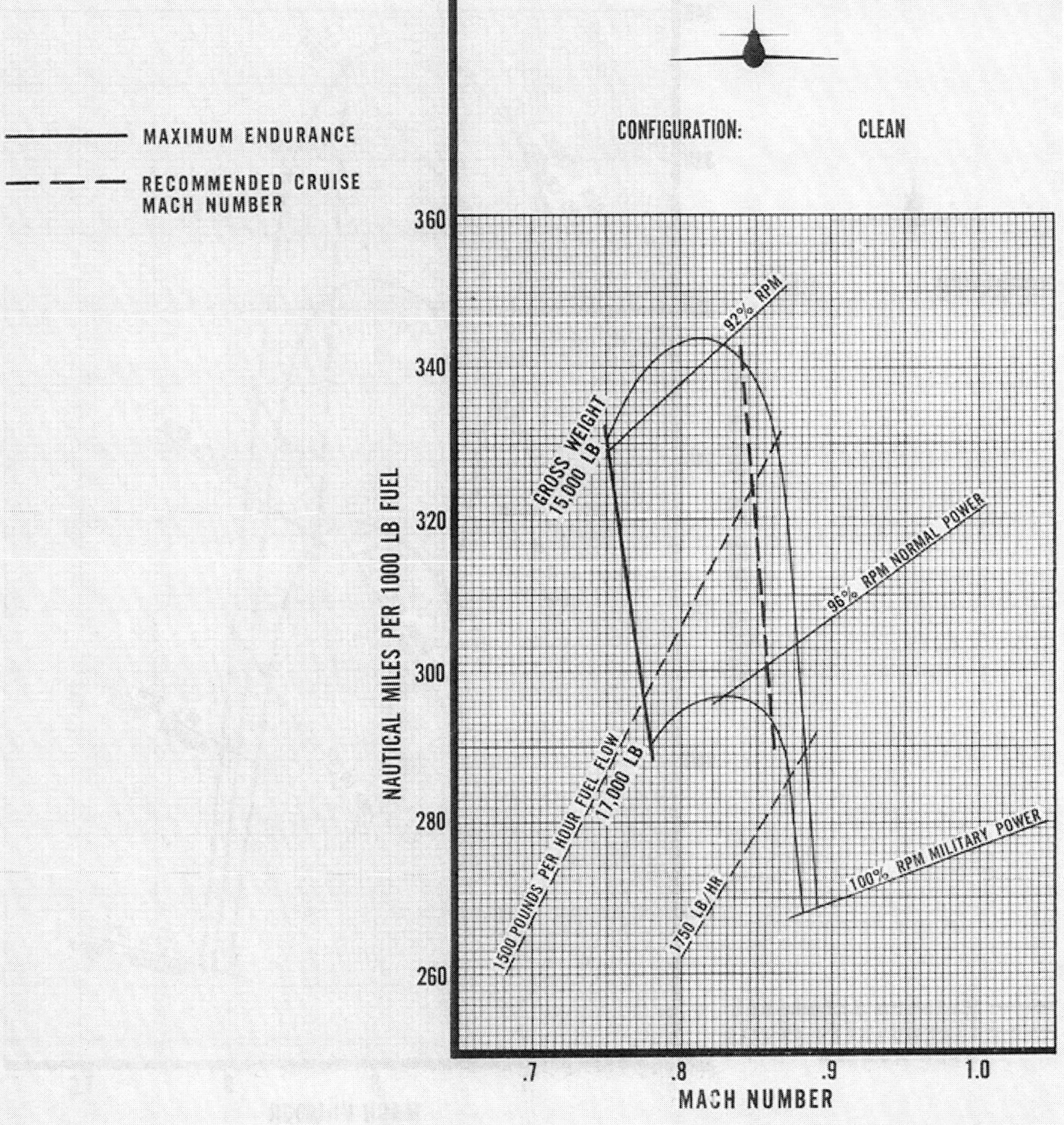

Date: 25 October 1955 Data Basis: Flight Test

Figure A-67 (Sheet 6 of 6)

NAUTICAL MILES PER 1000 POUNDS FUEL

SEA LEVEL

F-84F-25 AND LATER
J65-W-3 OR B-3 ENGINES
STANDARD DAY

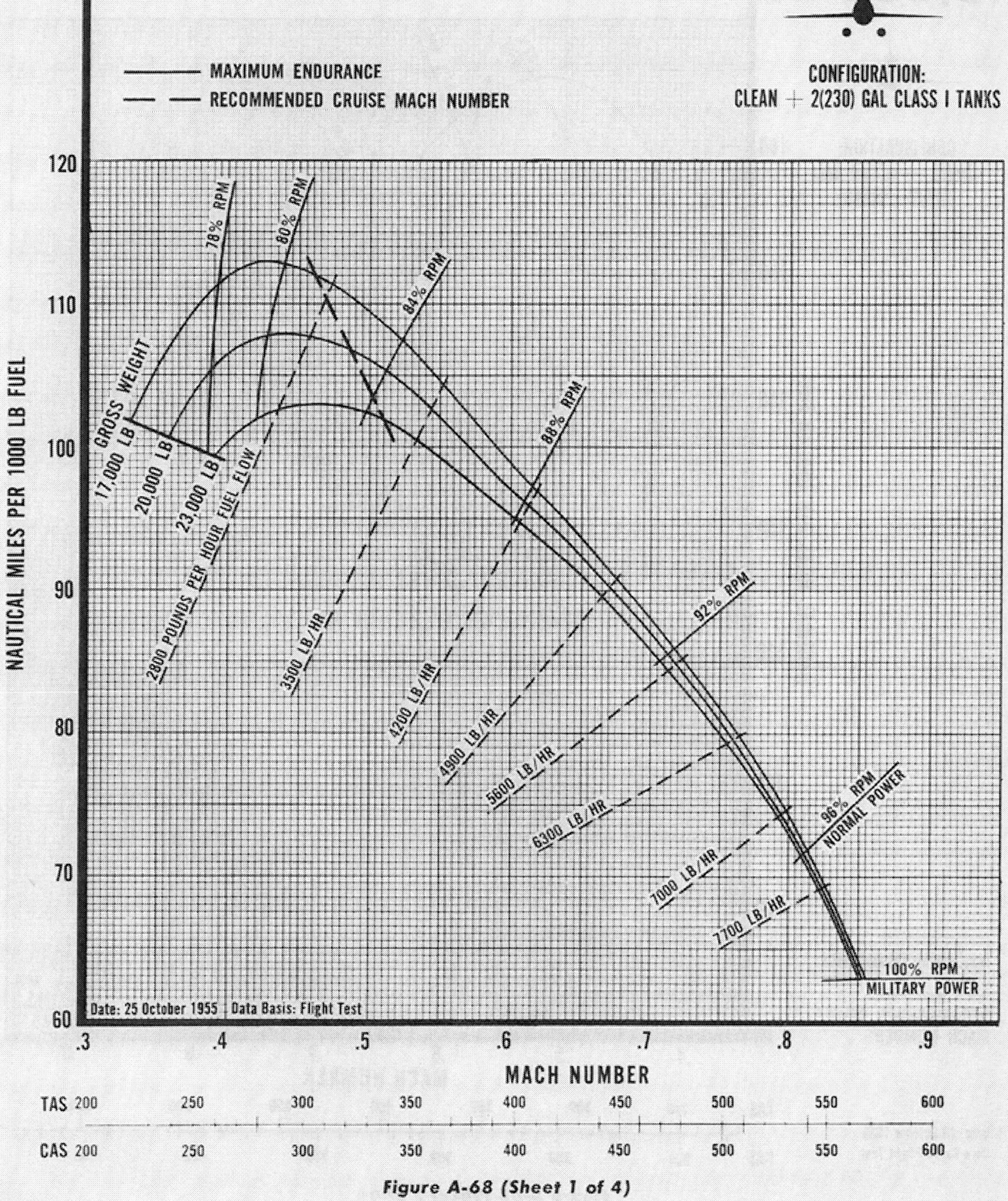

Figure A-68 (Sheet 1 of 4)

NAUTICAL MILES

PER 1000 POUNDS FUEL

15,000 FT

F-84F-25 AND LATER

J65-W-3 OR B-3 ENGINES

STANDARD DAY

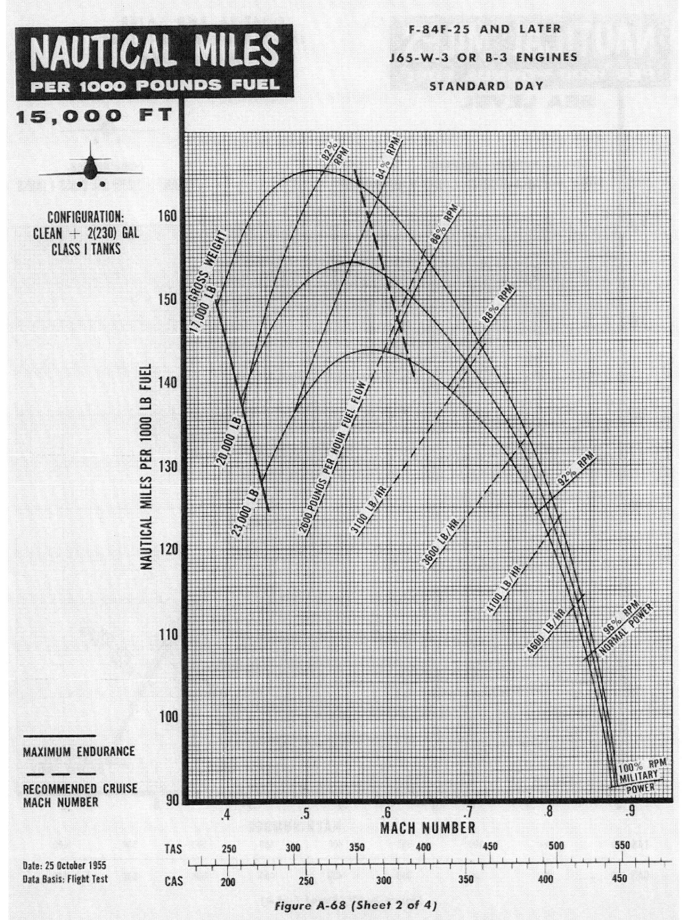

CONFIGURATION:
CLEAN + 2(230) GAL
CLASS I TANKS

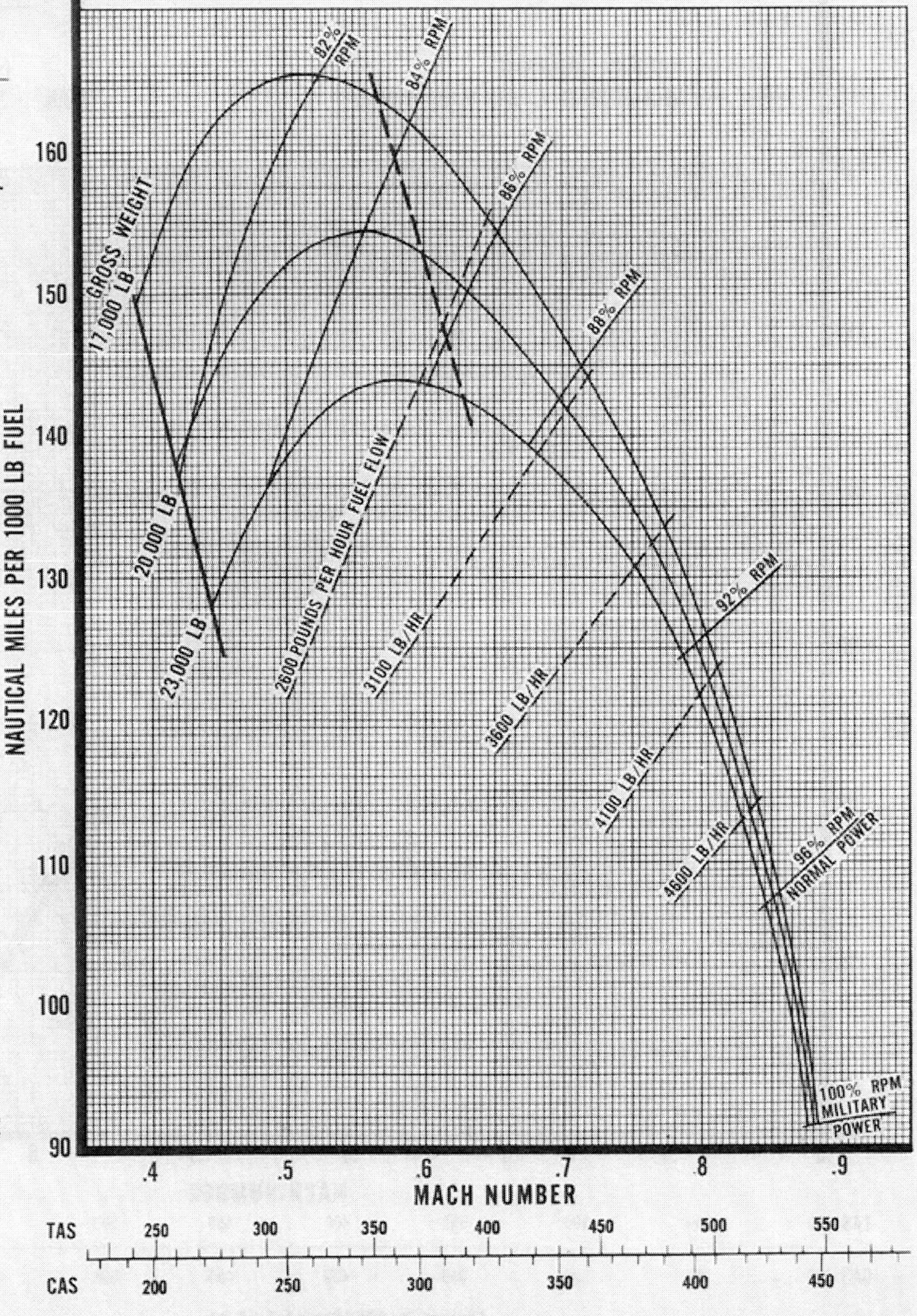

MAXIMUM ENDURANCE

RECOMMENDED CRUISE MACH NUMBER

Date: 25 October 1955
Data Basis: Flight Test

Figure A-68 (Sheet 2 of 4)

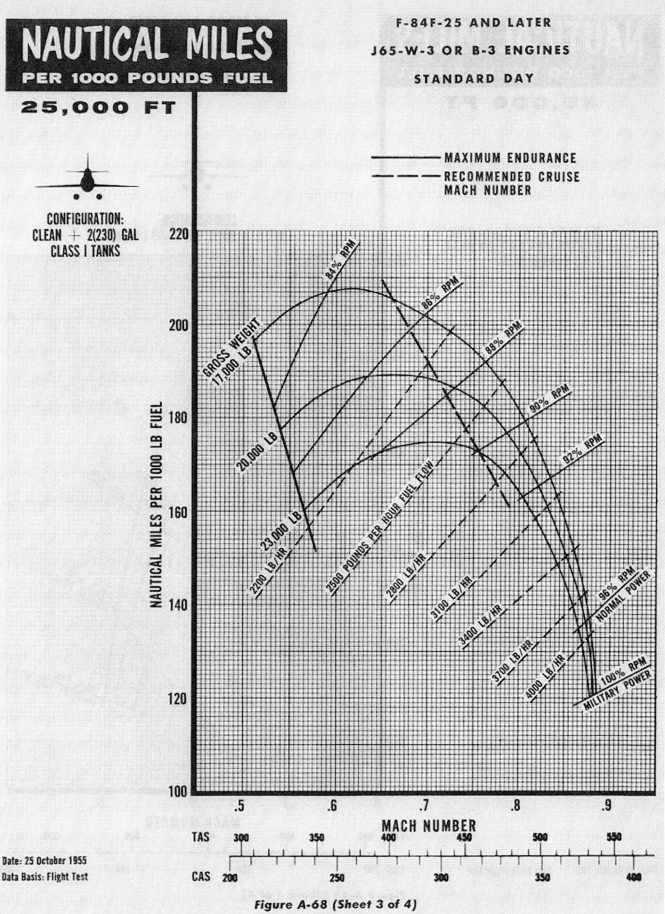

NAUTICAL MILES PER 1000 POUNDS FUEL

25,000 FT

F-84F-25 AND LATER
J65-W-3 OR B-3 ENGINES
STANDARD DAY

CONFIGURATION:
CLEAN + 2(230) GAL
CLASS I TANKS

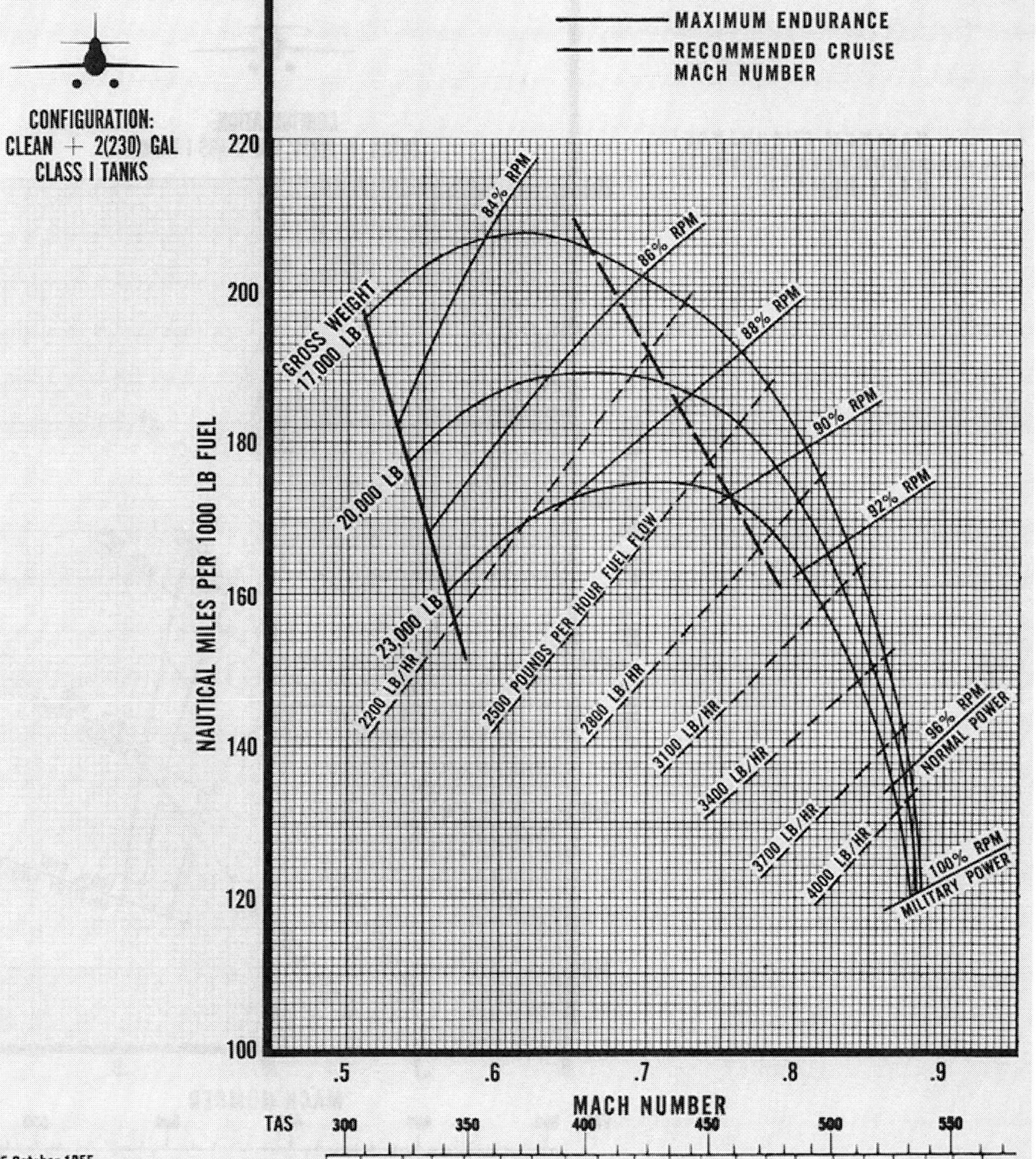

Date: 25 October 1955
Data Basis: Flight Test

Figure A-68 (Sheet 3 of 4)

NAUTICAL MILES
PER 1000 POUNDS FUEL
35,000 FT

F-84F-25 AND LATER
J65-W-3 OR B-3 ENGINES
STANDARD DAY

CONFIGURATION:
CLEAN + 2(230) GAL CLASS I TANKS

——— MAXIMUM ENDURANCE
— — RECOMMENDED CRUISE MACH NUMBER

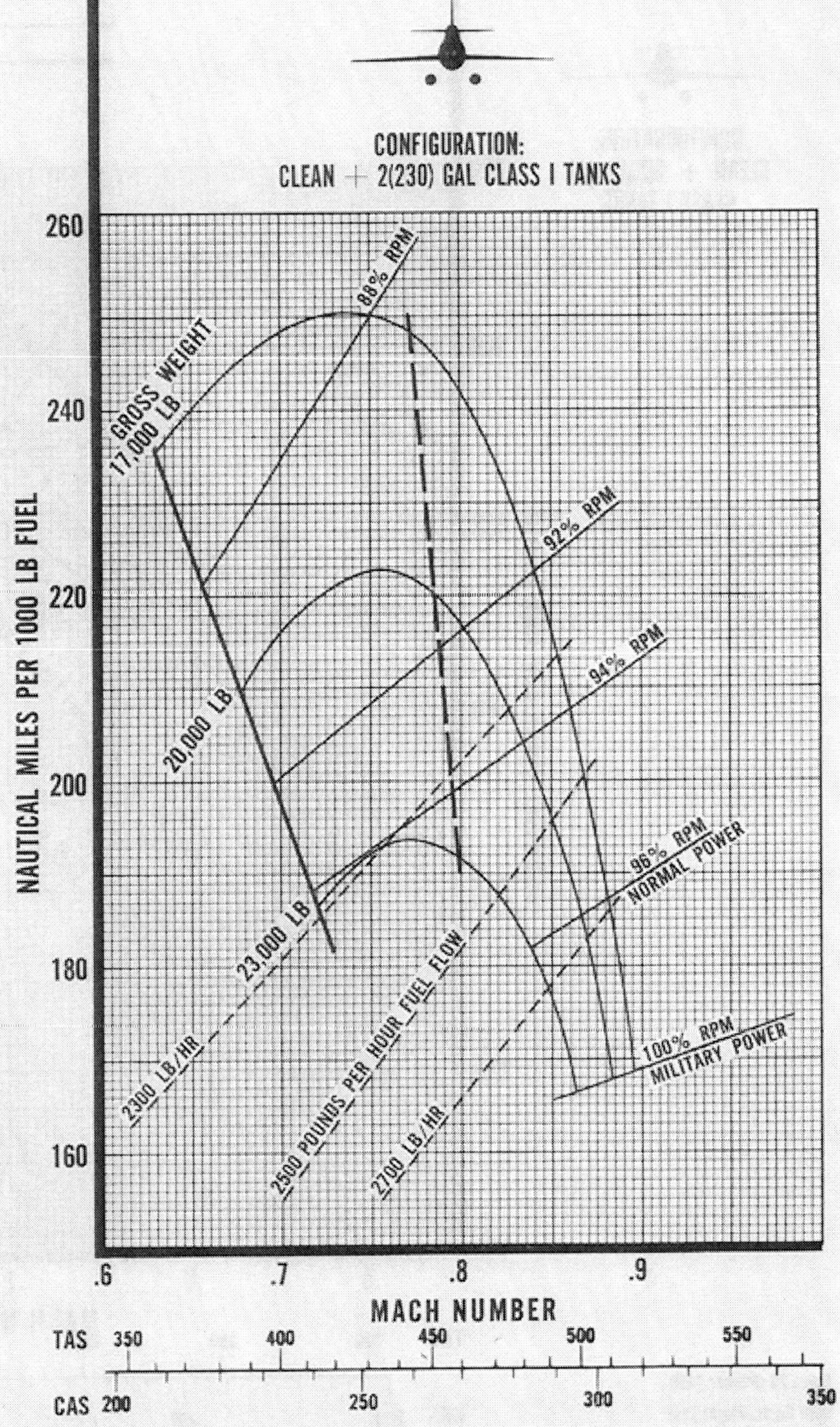

Date: 25 October 1955 Data Basis: Flight Test

Figure A-68 (Sheet 4 of 4)

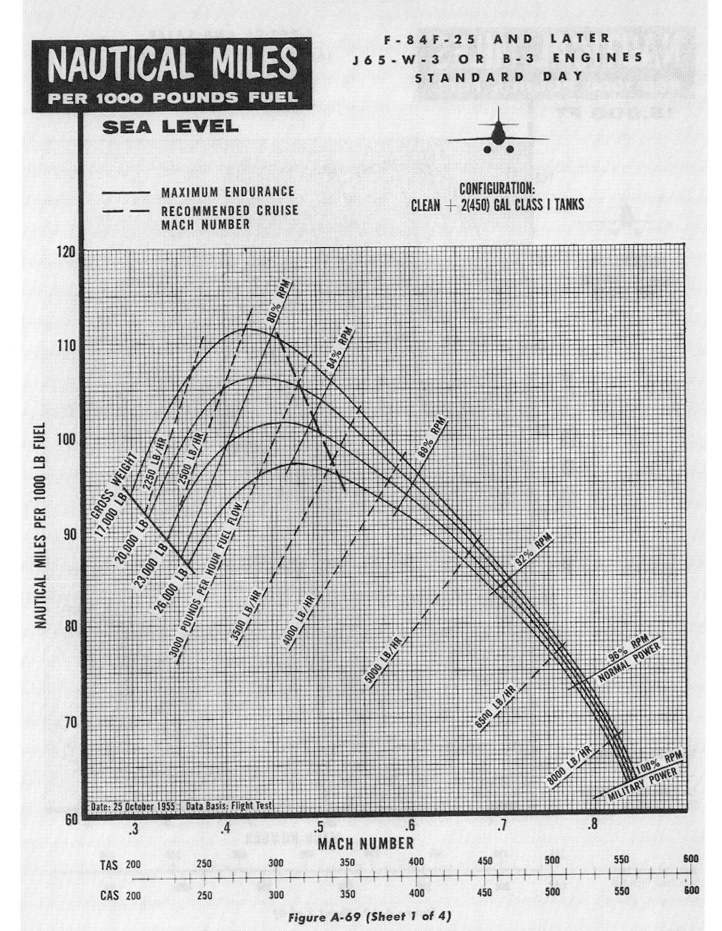

NAUTICAL MILES
PER 1000 POUNDS FUEL
SEA LEVEL

F-84F-25 AND LATER
J65-W-3 OR B-3 ENGINES
STANDARD DAY

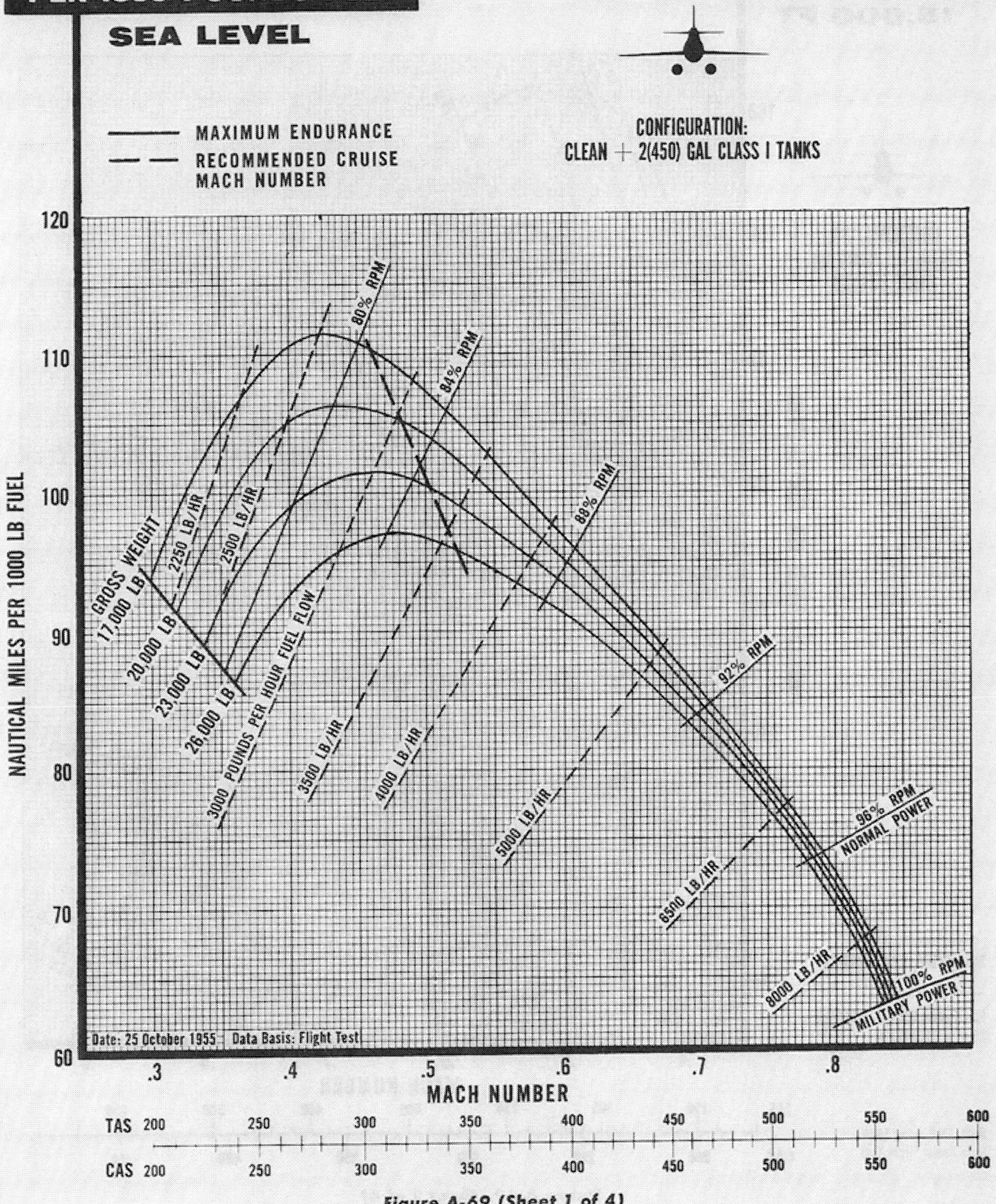

Figure A-69 (Sheet 1 of 4)

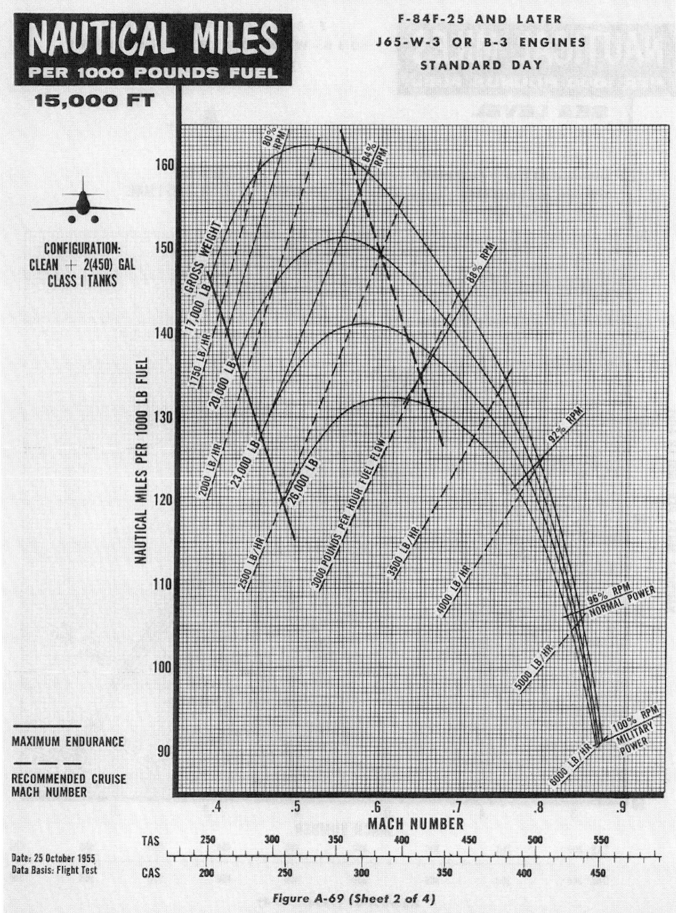

NAUTICAL MILES PER 1000 POUNDS FUEL

15,000 FT

F-84F-25 AND LATER
J65-W-3 OR B-3 ENGINES
STANDARD DAY

CONFIGURATION:
CLEAN + 2(450) GAL
CLASS I TANKS

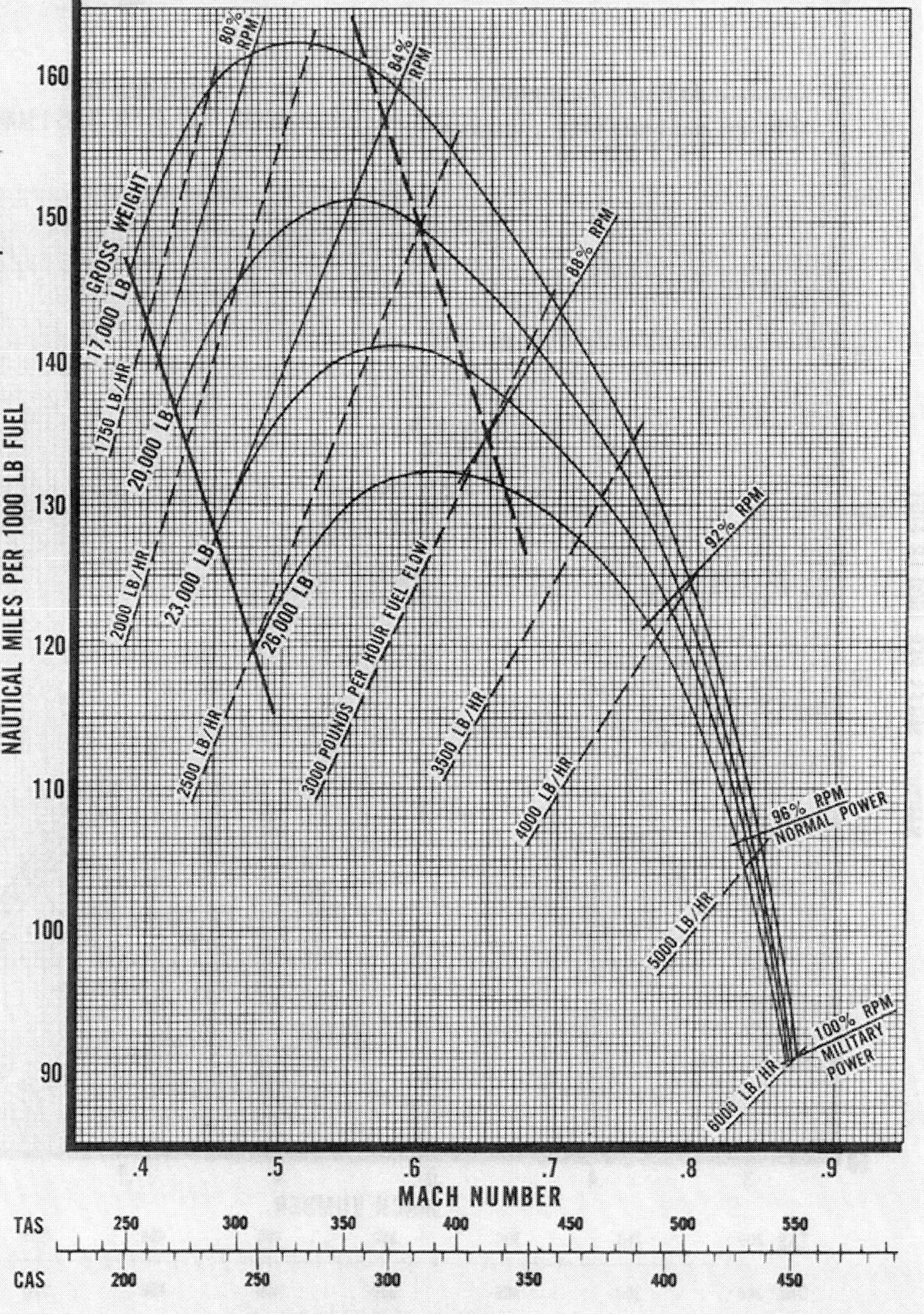

MAXIMUM ENDURANCE

RECOMMENDED CRUISE MACH NUMBER

Date: 25 October 1955
Data Basis: Flight Test

Figure A-69 (Sheet 2 of 4)

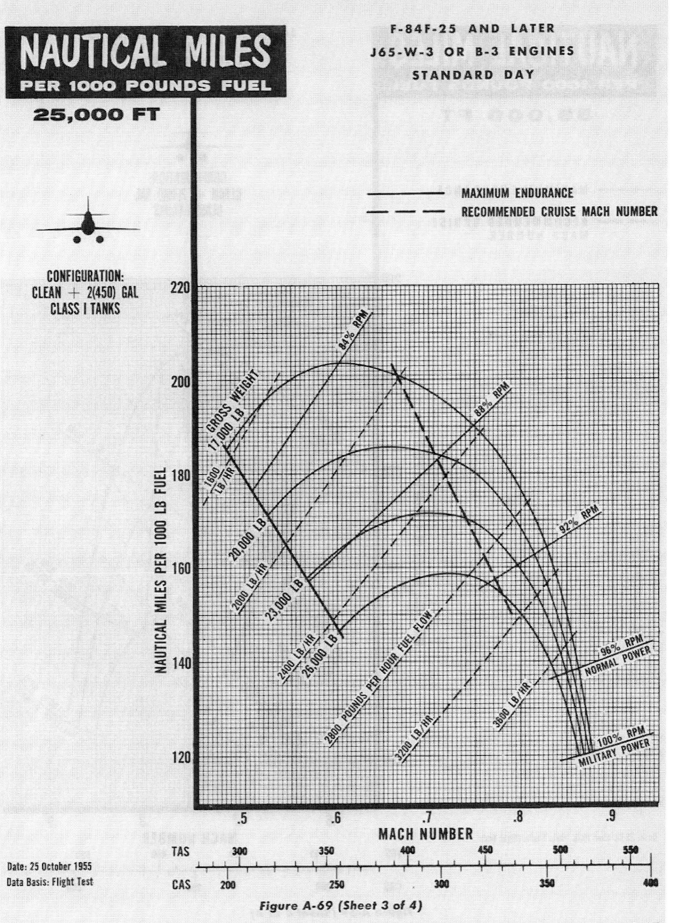

NAUTICAL MILES
PER 1000 POUNDS FUEL
25,000 FT

F-84F-25 AND LATER
J65-W-3 OR B-3 ENGINES
STANDARD DAY

CONFIGURATION:
CLEAN + 2(450) GAL
CLASS I TANKS

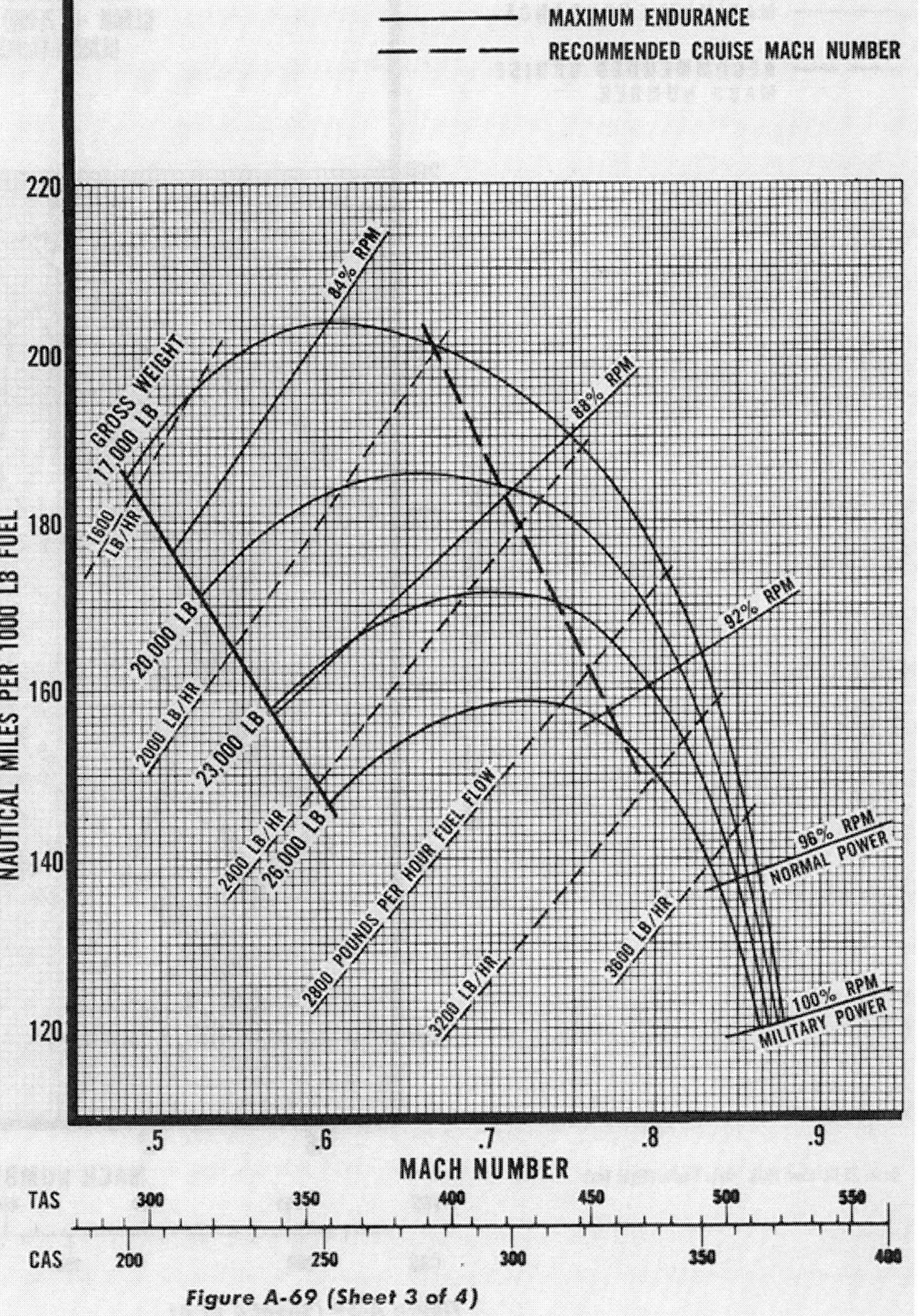

Date: 25 October 1955
Data Basis: Flight Test

Figure A-69 (Sheet 3 of 4)

NAUTICAL MILES
PER 1000 POUNDS FUEL
35,000 FT

J65-W-3 OR B-3 ENGINES
F-84F-25 AND LATER
STANDARD DAY

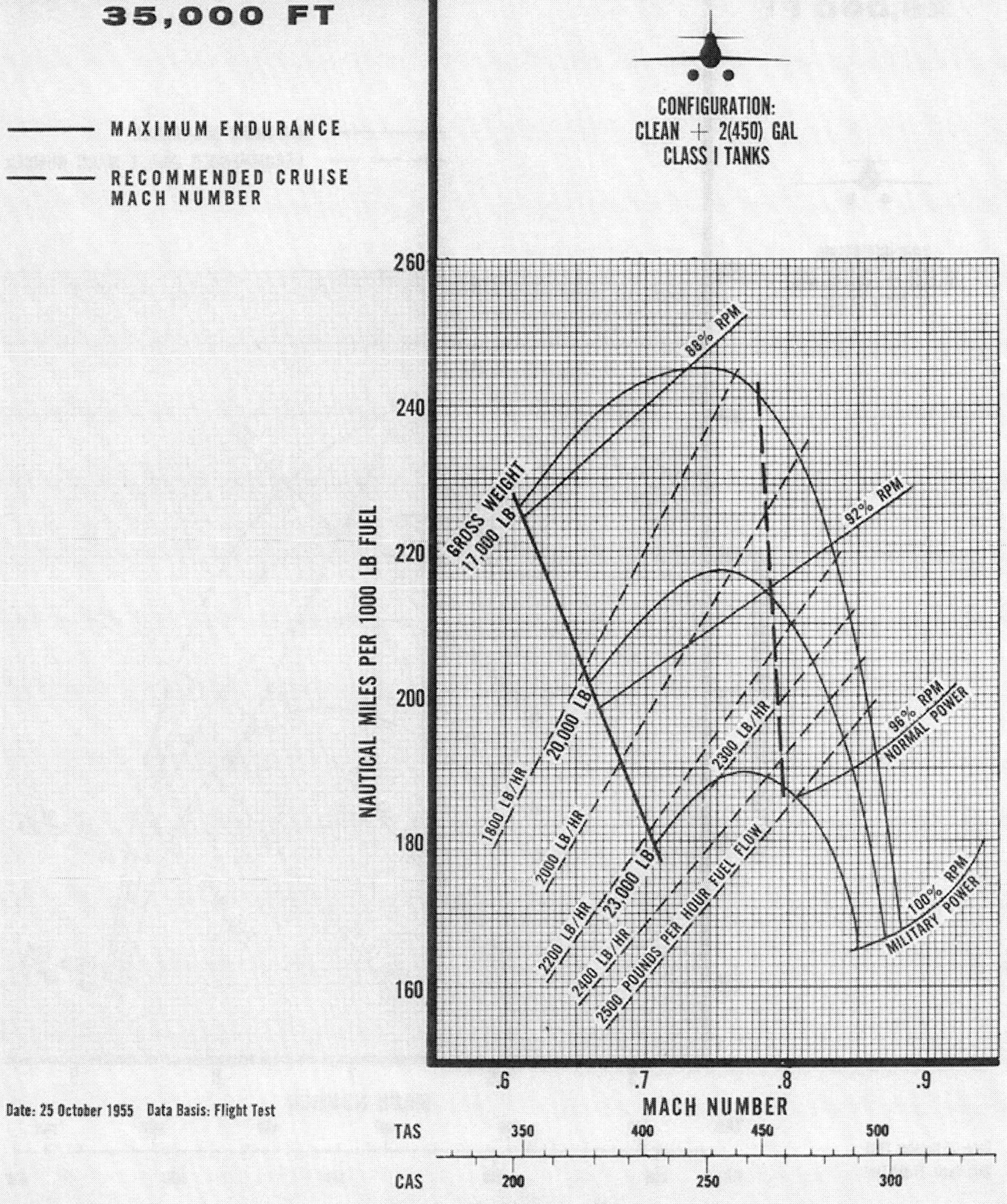

Date: 25 October 1955 Data Basis: Flight Test

Figure A-69 (Sheet 4 of 4)

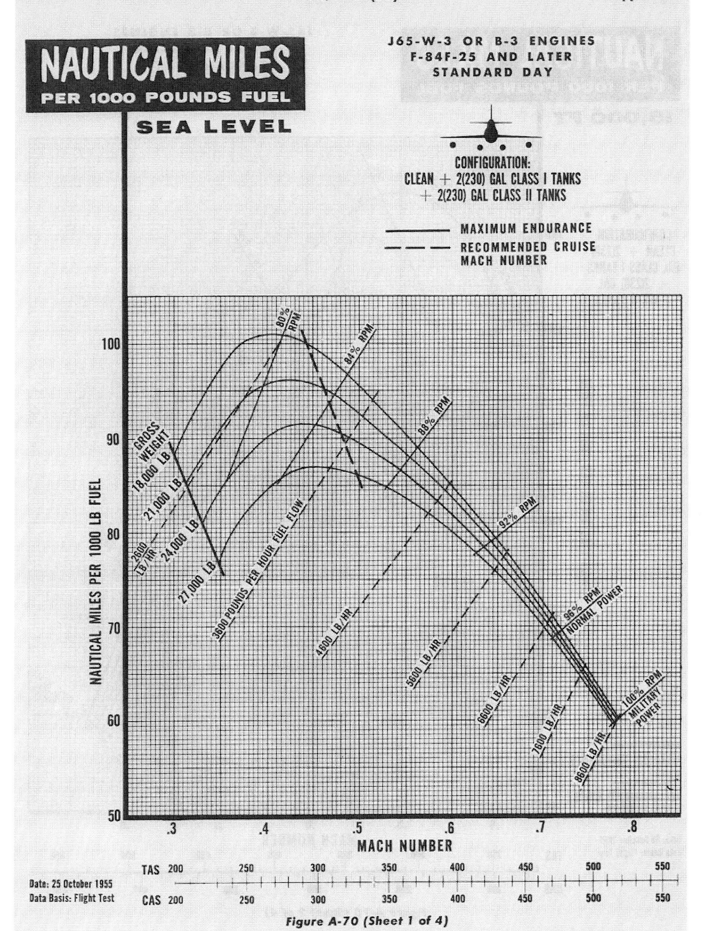

NAUTICAL MILES PER 1000 POUNDS FUEL

SEA LEVEL

J65-W-3 OR B-3 ENGINES
F-84F-25 AND LATER
STANDARD DAY

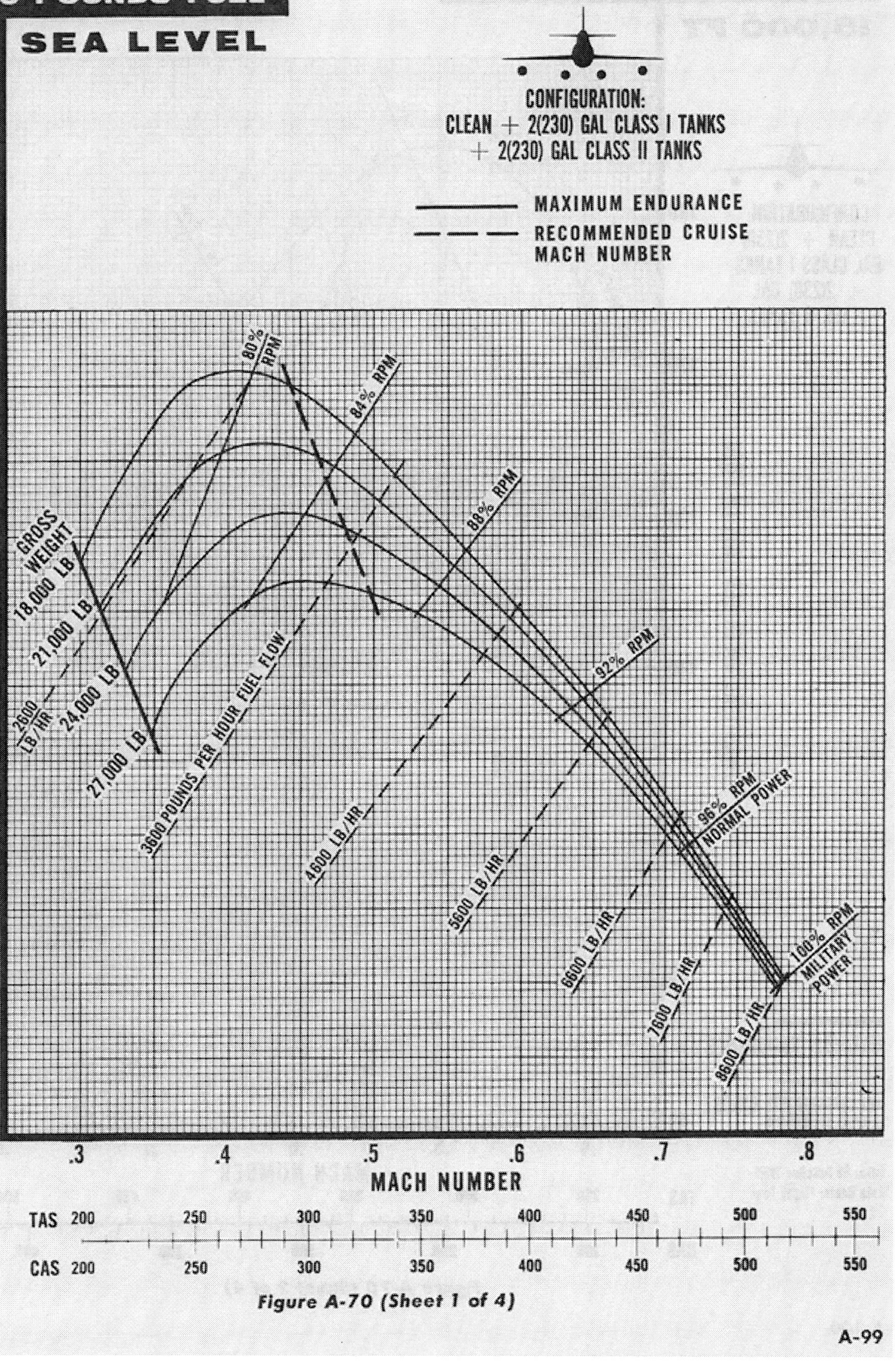

Date: 25 October 1955
Data Basis: Flight Test

Figure A-70 (Sheet 1 of 4)

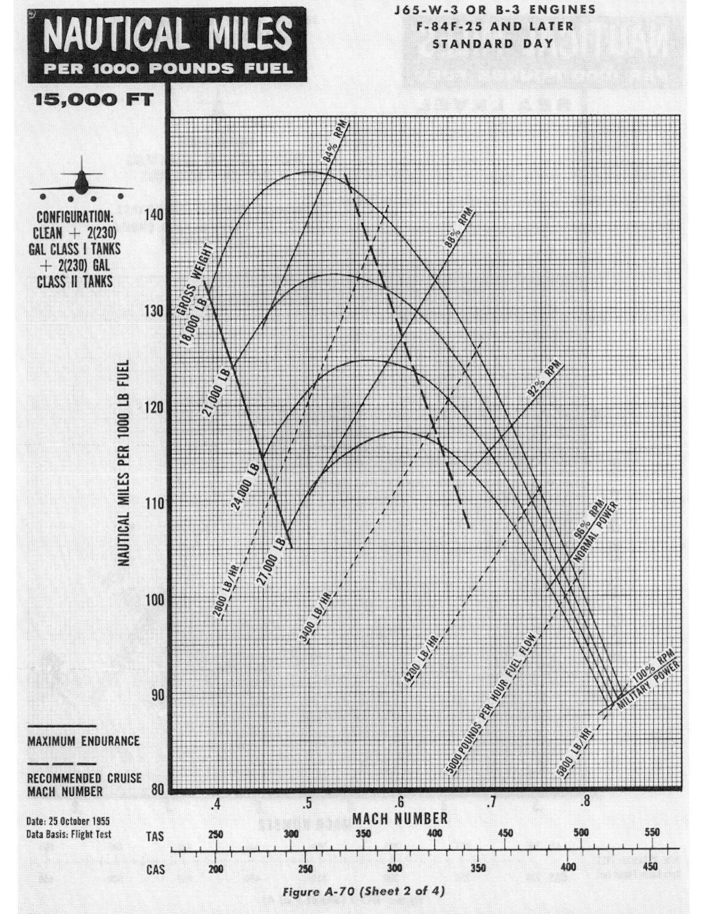

NAUTICAL MILES PER 1000 POUNDS FUEL

15,000 FT

J65-W-3 OR B-3 ENGINES
F-84F-25 AND LATER
STANDARD DAY

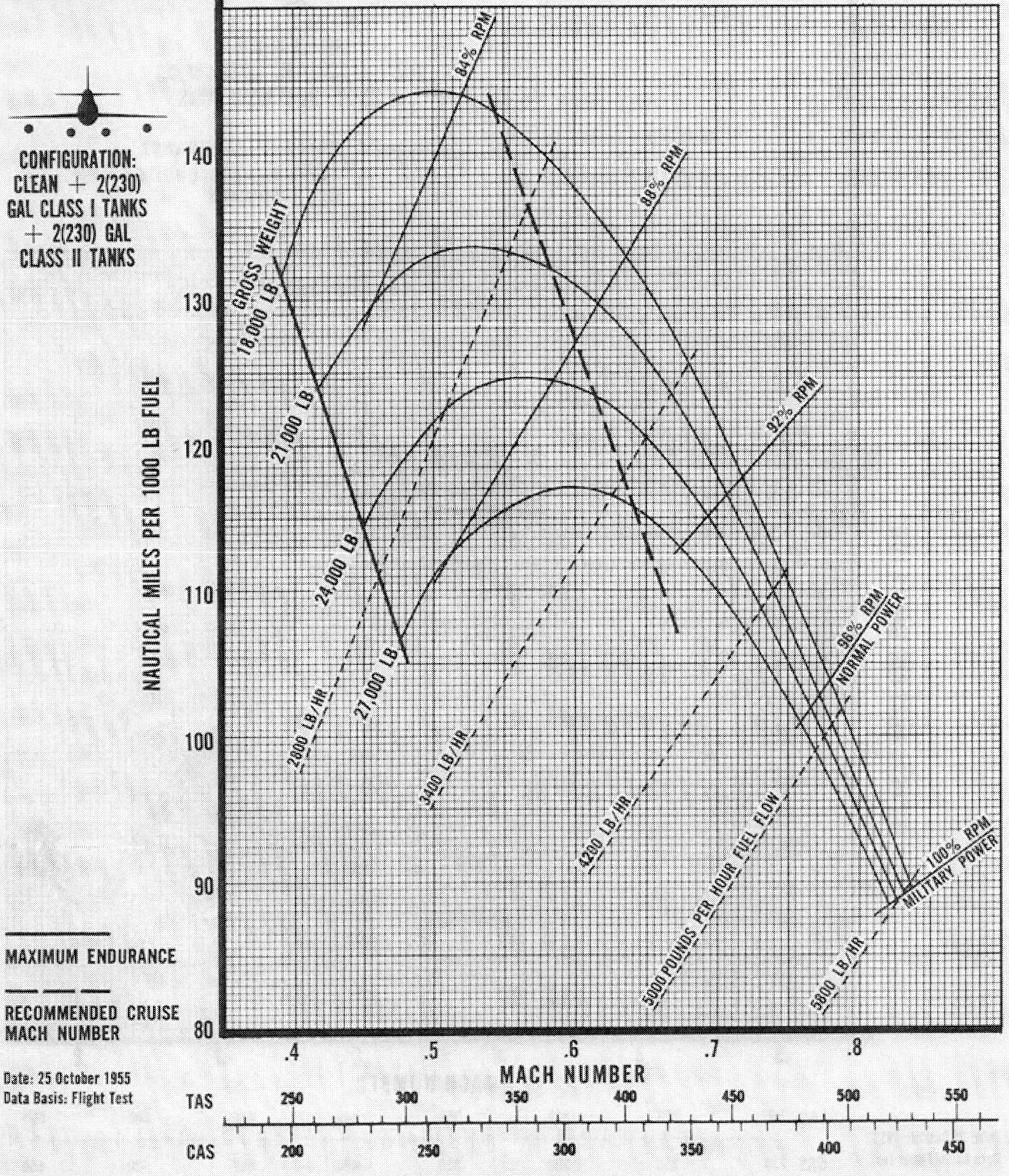

Date: 25 October 1955
Data Basis: Flight Test

Figure A-70 (Sheet 2 of 4)

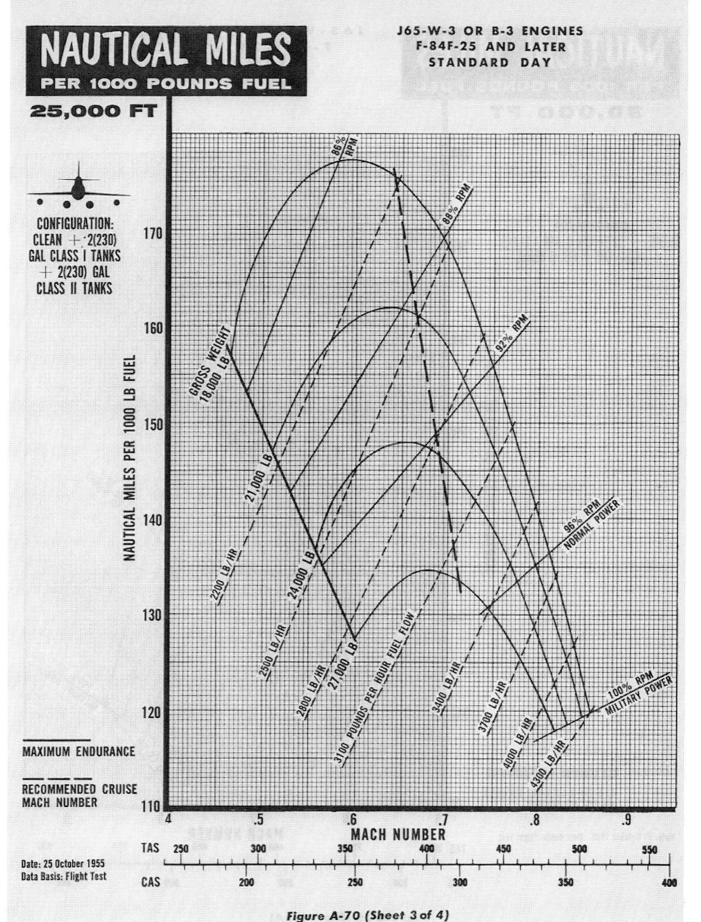

NAUTICAL MILES PER 1000 POUNDS FUEL

25,000 FT

J65-W-3 OR B-3 ENGINES
F-84F-25 AND LATER
STANDARD DAY

CONFIGURATION: CLEAN + 2(230) GAL CLASS I TANKS + 2(230) GAL CLASS II TANKS

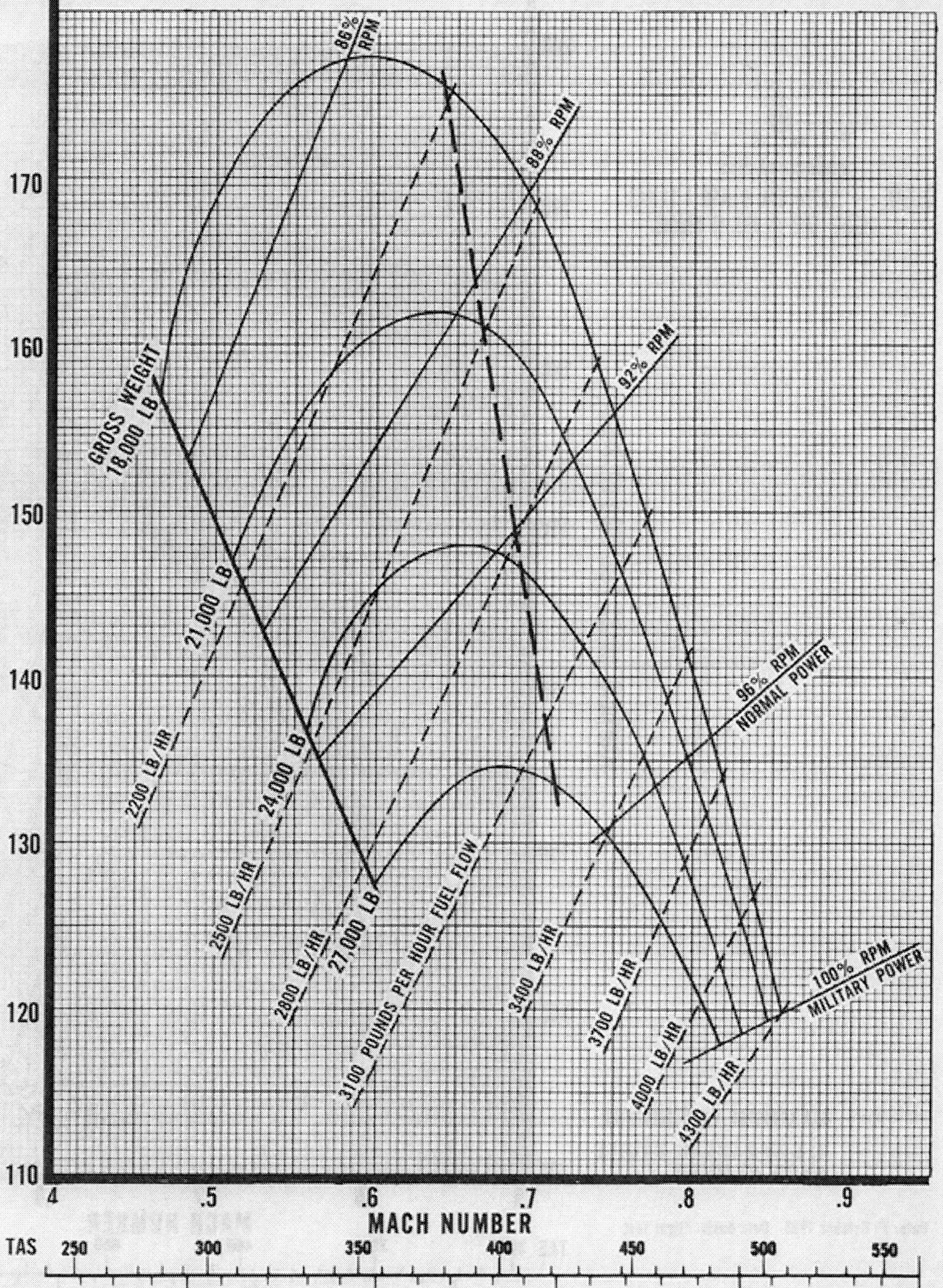

Date: 25 October 1955
Data Basis: Flight Test

Figure A-70 (Sheet 3 of 4)

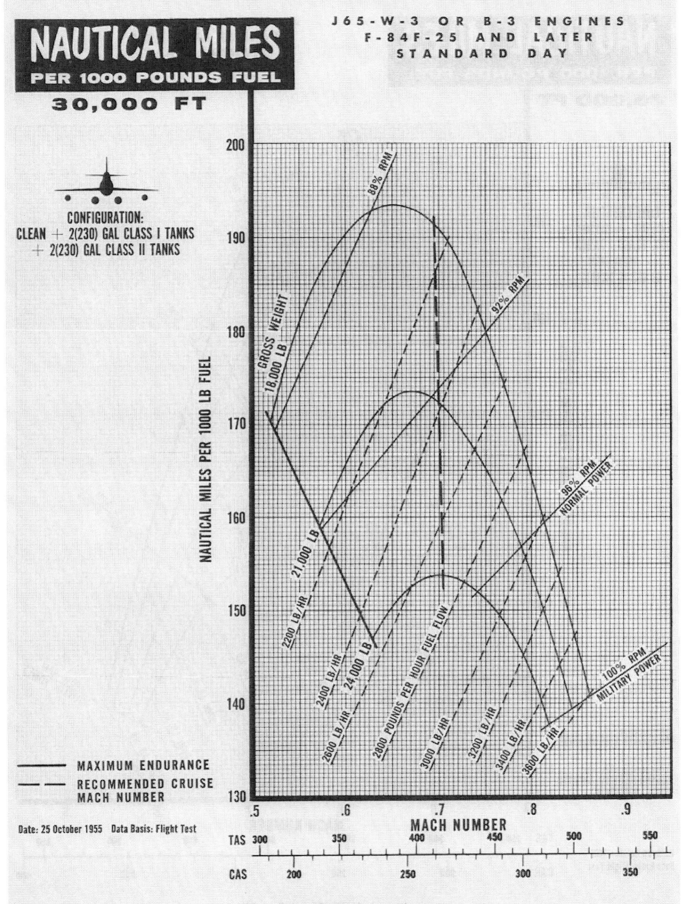

NAUTICAL MILES PER 1000 POUNDS FUEL

30,000 FT

J65-W-3 OR B-3 ENGINES
F-84F-25 AND LATER
STANDARD DAY

CONFIGURATION:
CLEAN + 2(230) GAL CLASS I TANKS
+ 2(230) GAL CLASS II TANKS

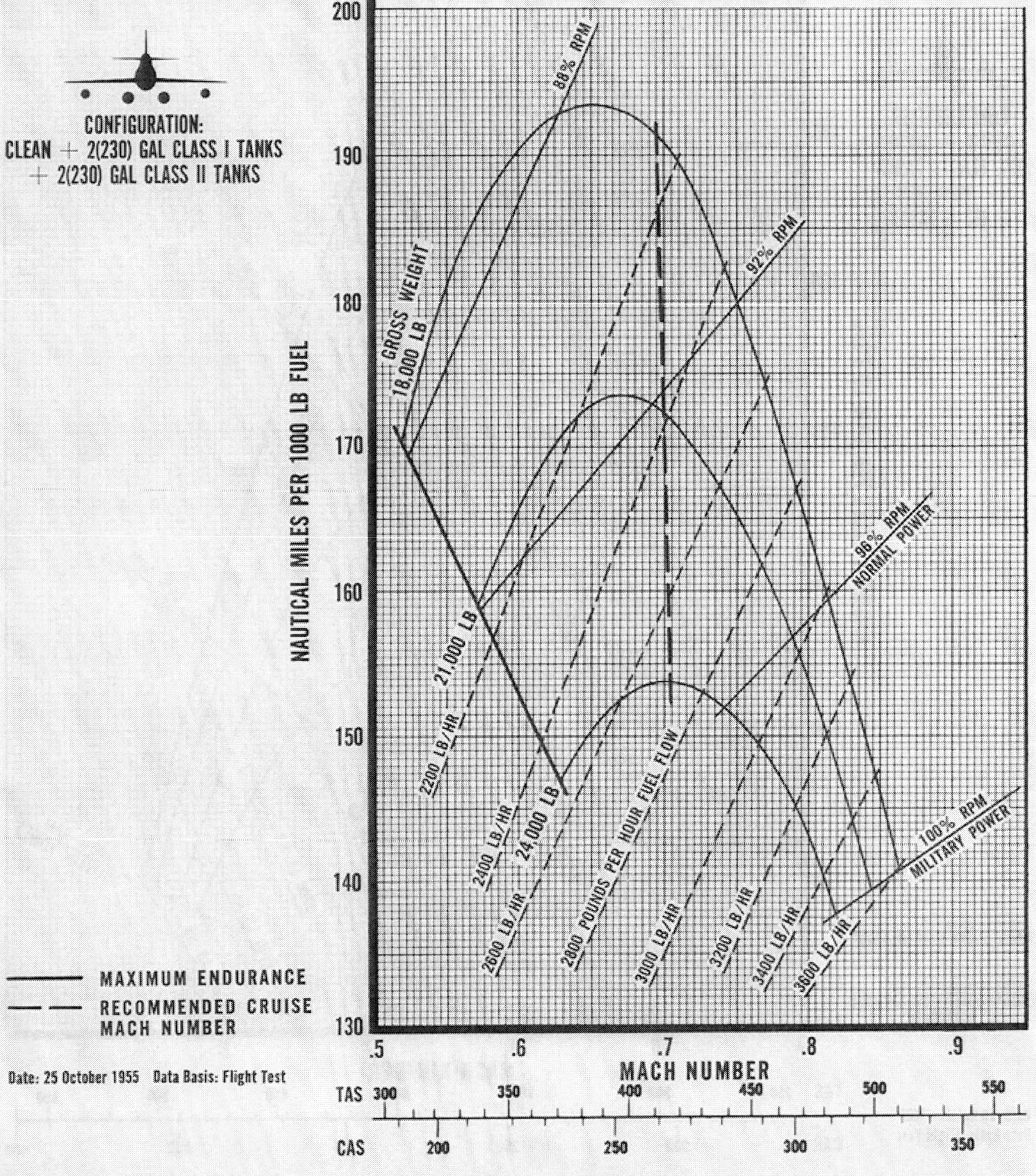

Date: 25 October 1955 Data Basis: Flight Test

Figure A-70 (Sheet 4 of 4)

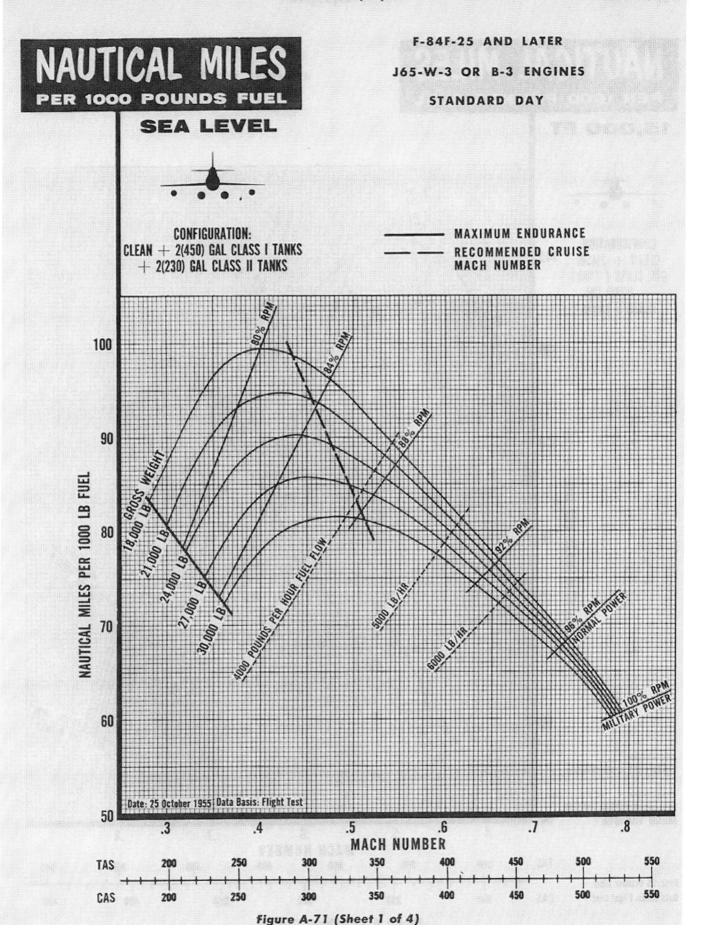

NAUTICAL MILES
PER 1000 POUNDS FUEL

F-84F-25 AND LATER

J65-W-3 OR B-3 ENGINES

STANDARD DAY

SEA LEVEL

CONFIGURATION:
CLEAN + 2(450) GAL CLASS I TANKS
+ 2(230) GAL CLASS II TANKS

MAXIMUM ENDURANCE
RECOMMENDED CRUISE MACH NUMBER

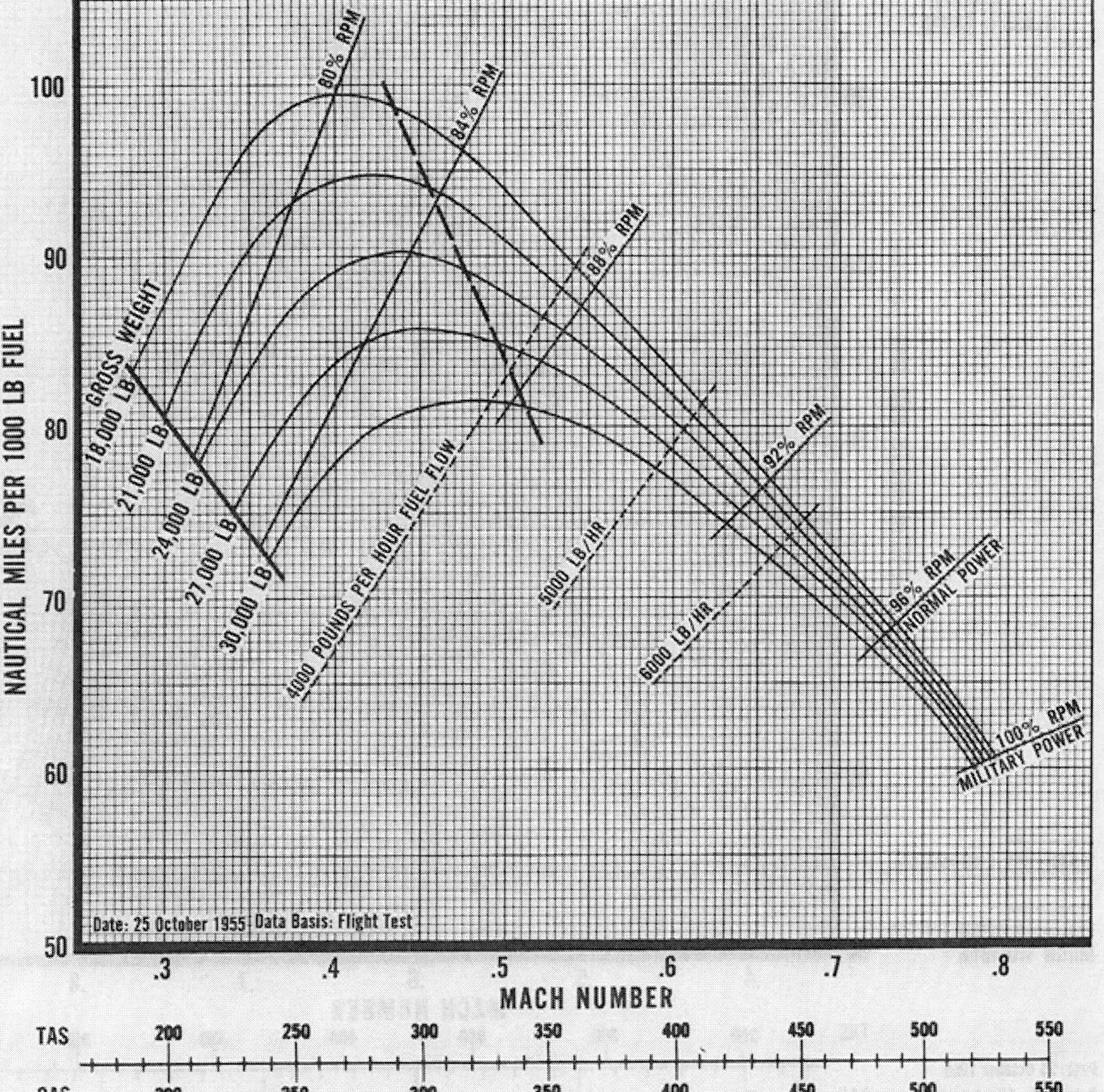

Figure A-71 (Sheet 1 of 4)

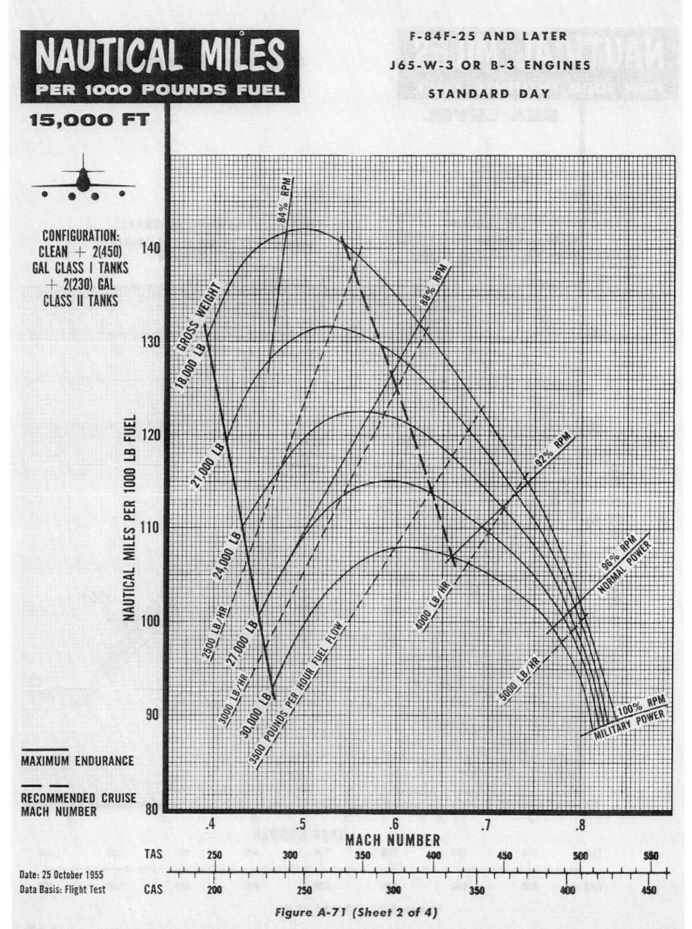

NAUTICAL MILES PER 1000 POUNDS FUEL

15,000 FT

F-84F-25 AND LATER
J65-W-3 OR B-3 ENGINES
STANDARD DAY

CONFIGURATION:
CLEAN + 2(450) GAL CLASS I TANKS + 2(230) GAL CLASS II TANKS

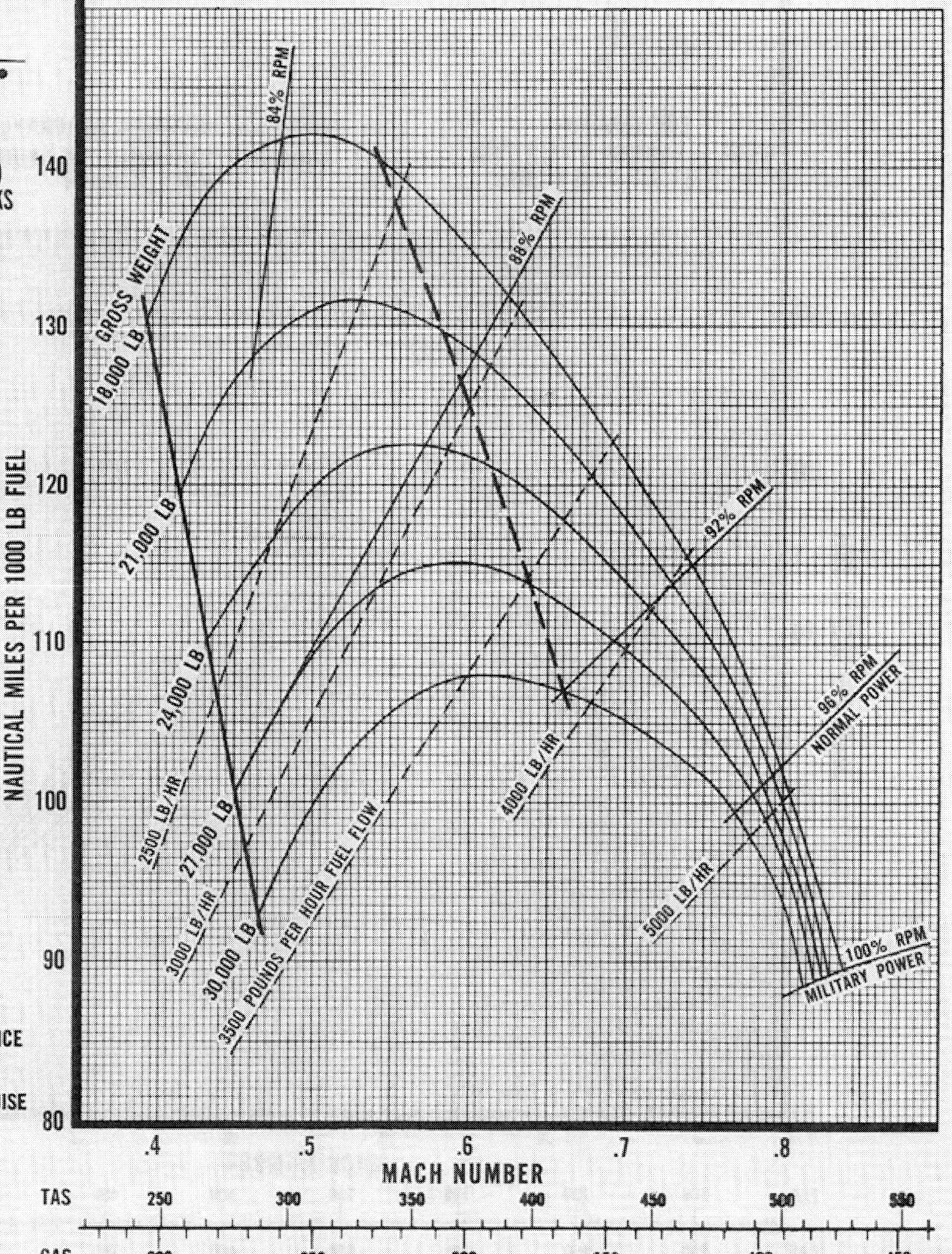

MAXIMUM ENDURANCE

RECOMMENDED CRUISE MACH NUMBER

Date: 25 October 1955
Data Basis: Flight Test

Figure A-71 (Sheet 2 of 4)

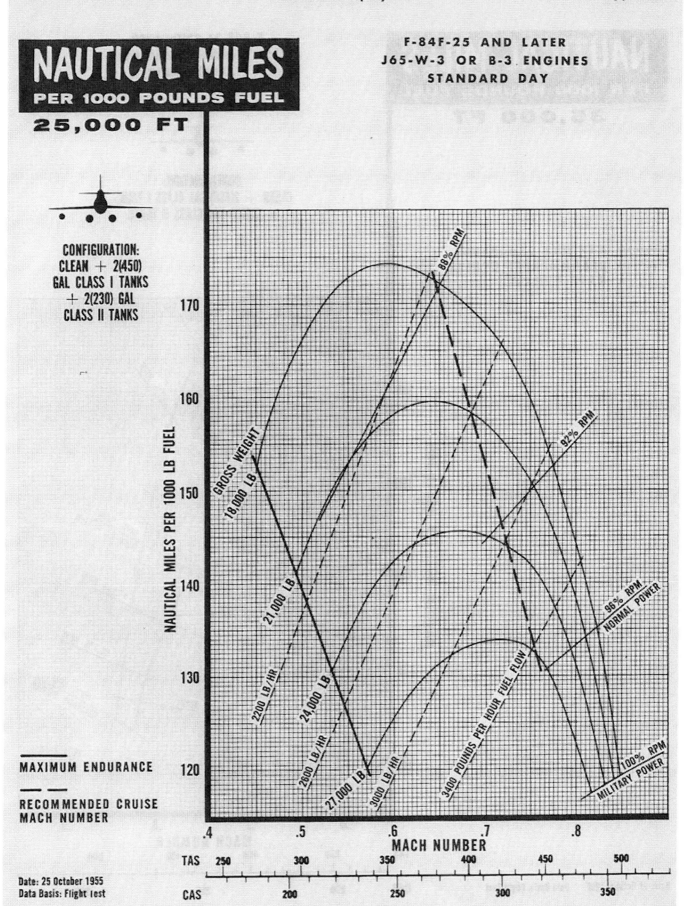

NAUTICAL MILES PER 1000 POUNDS FUEL

25,000 FT

F-84F-25 AND LATER
J65-W-3 OR B-3 ENGINES
STANDARD DAY

CONFIGURATION:
CLEAN + 2(450) GAL CLASS I TANKS + 2(230) GAL CLASS II TANKS

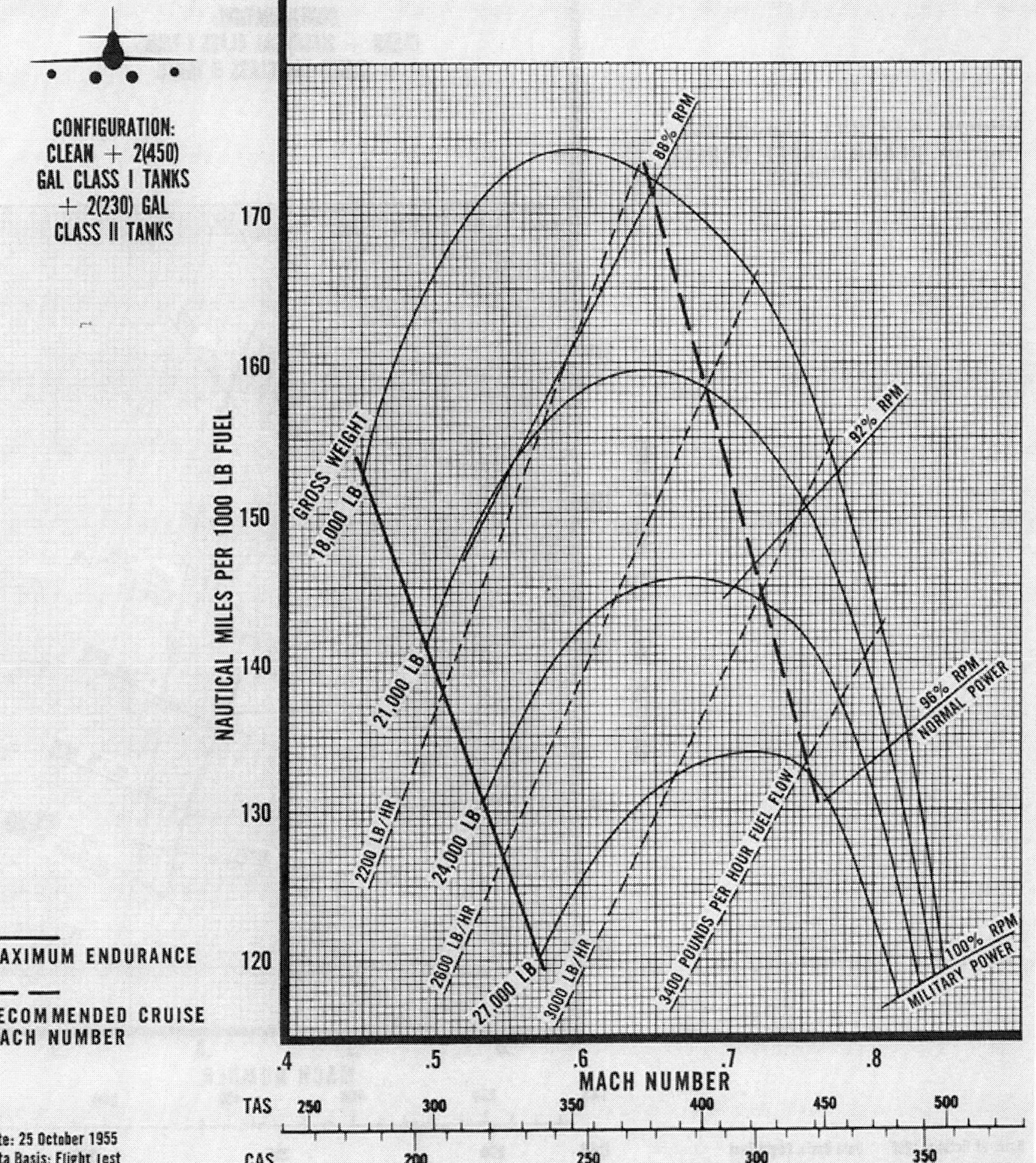

MAXIMUM ENDURANCE

RECOMMENDED CRUISE MACH NUMBER

Date: 25 October 1955
Data Basis: Flight Test

Figure A-71 (Sheet 3 of 4)

NAUTICAL MILES

PER 1000 POUNDS FUEL

35,000 FT

F-84F-25 AND LATER
J65-W-3 OR B-3 ENGINES
STANDARD DAY

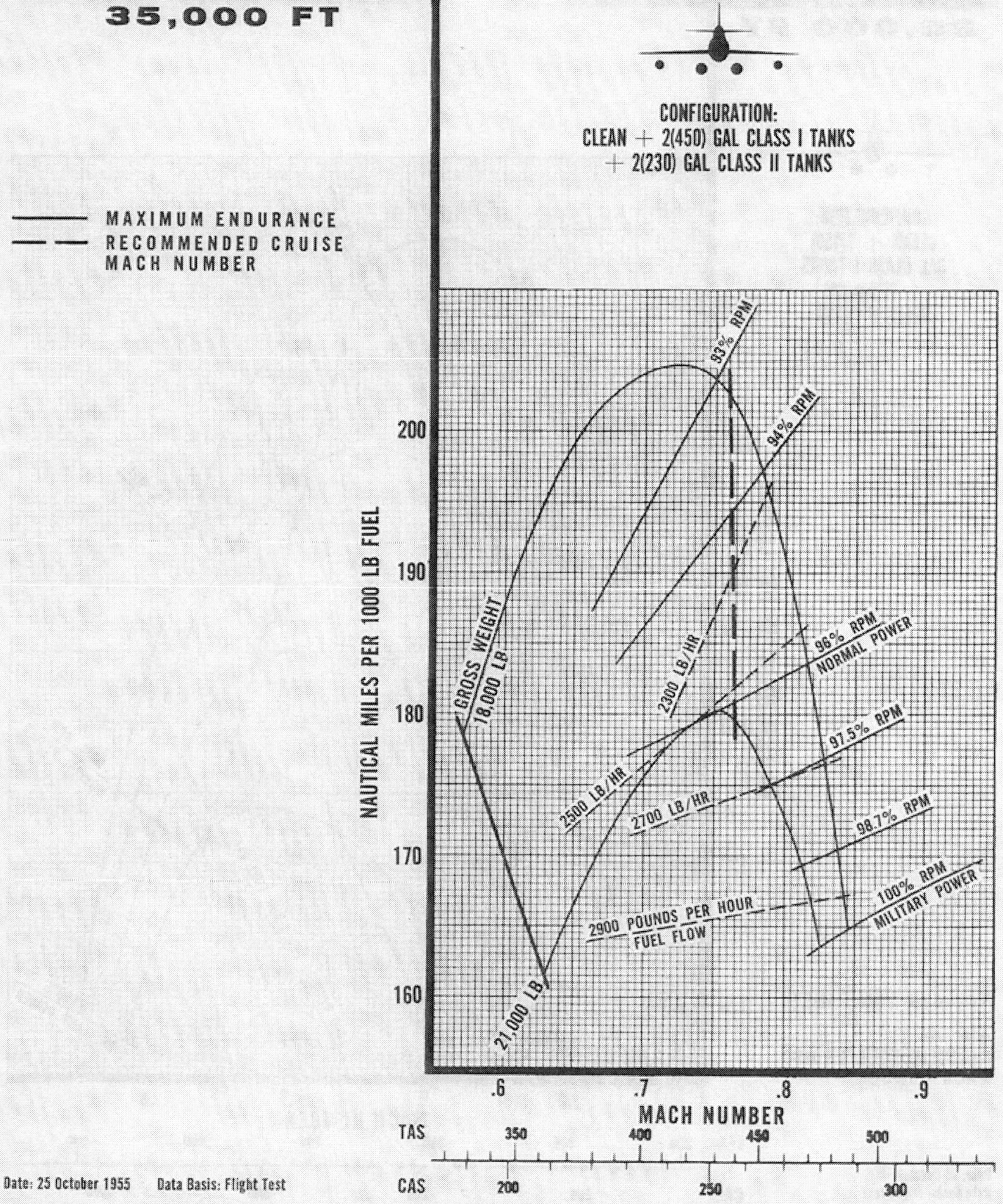

Date: 25 October 1955 Data Basis: Flight Test

Figure A-71 (Sheet 4 of 4)

NAUTICAL MILES
PER 1000 POUNDS FUEL
SEA LEVEL

F-84F-25 AND LATER
J65-W-3 OR B-3 ENGINES
STANDARD DAY

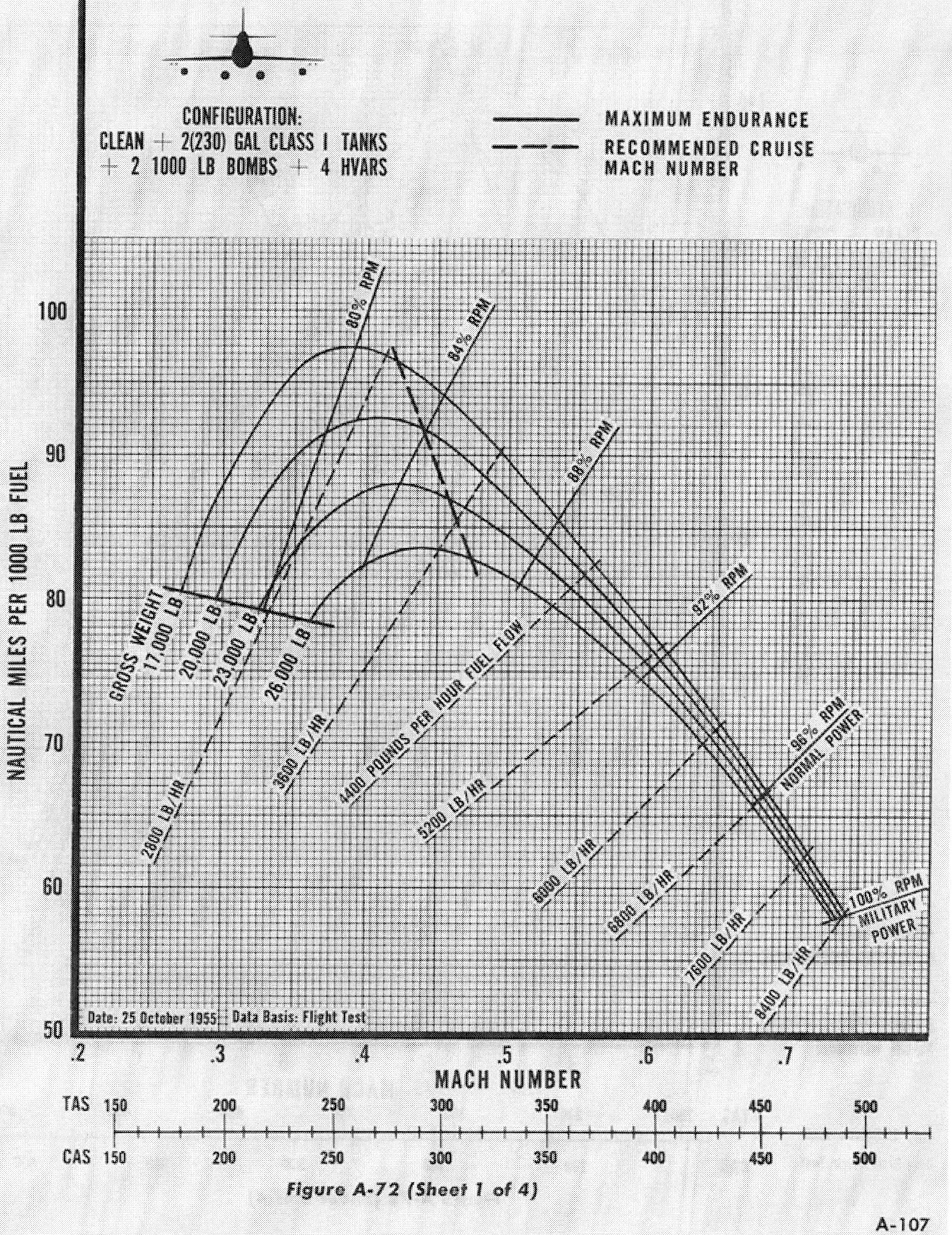

Figure A-72 (Sheet 1 of 4)

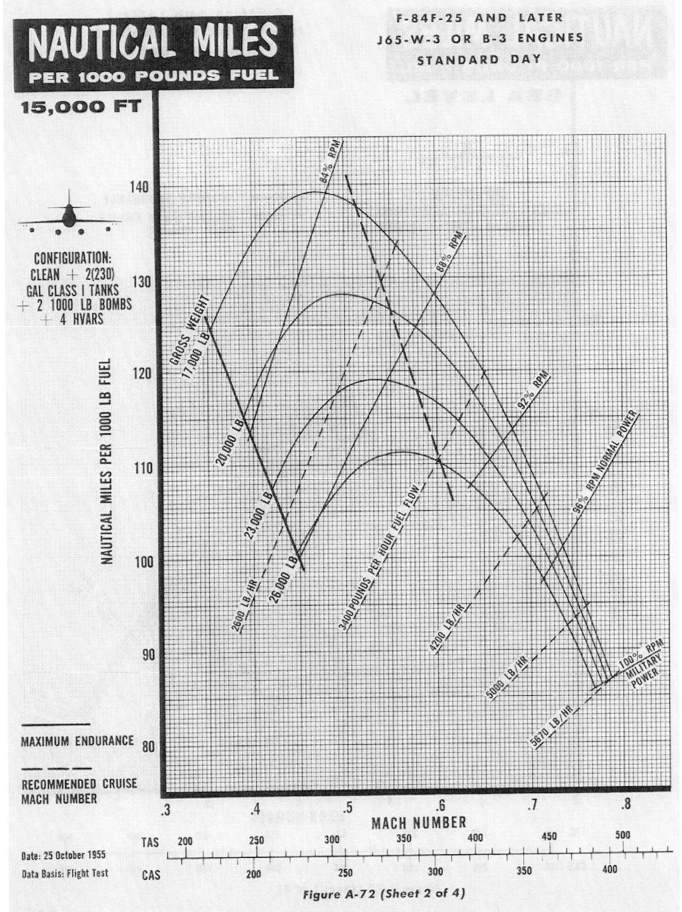

NAUTICAL MILES
PER 1000 POUNDS FUEL

15,000 FT

F-84F-25 AND LATER
J65-W-3 OR B-3 ENGINES
STANDARD DAY

CONFIGURATION:
CLEAN + 2(230) GAL CLASS I TANKS + 2 1000 LB BOMBS + 4 HVARS

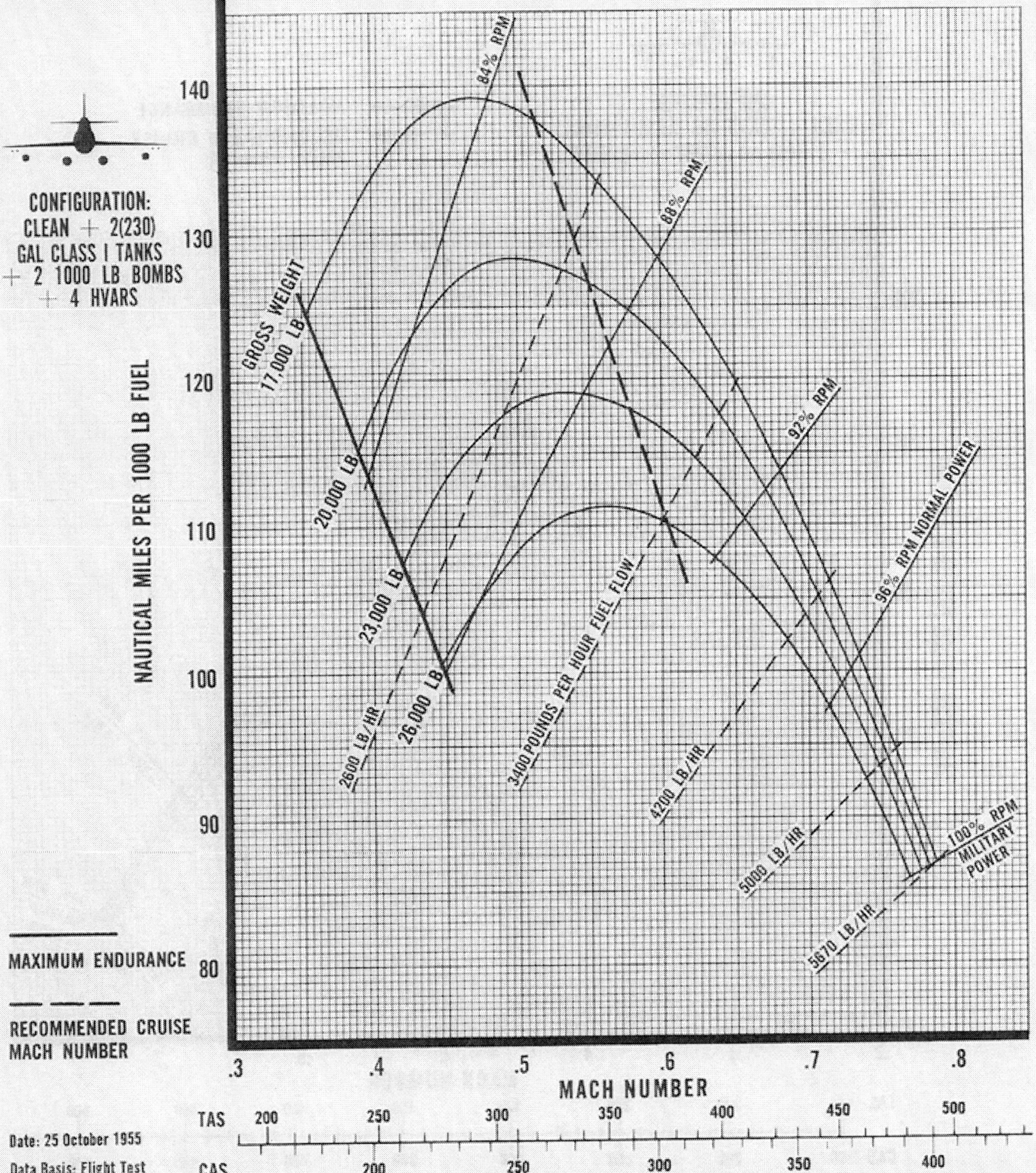

Date: 25 October 1955

Data Basis: Flight Test

Figure A-72 (Sheet 2 of 4)

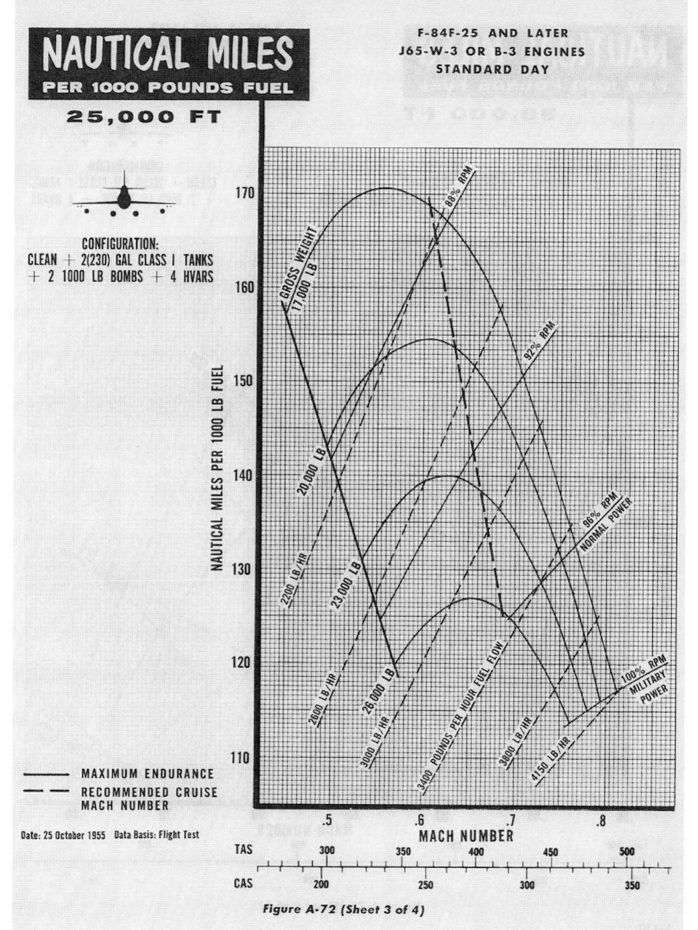

NAUTICAL MILES
PER 1000 POUNDS FUEL
25,000 FT

F-84F-25 AND LATER
J65-W-3 OR B-3 ENGINES
STANDARD DAY

CONFIGURATION:
CLEAN + 2(230) GAL CLASS I TANKS + 2 1000 LB BOMBS + 4 HVARS

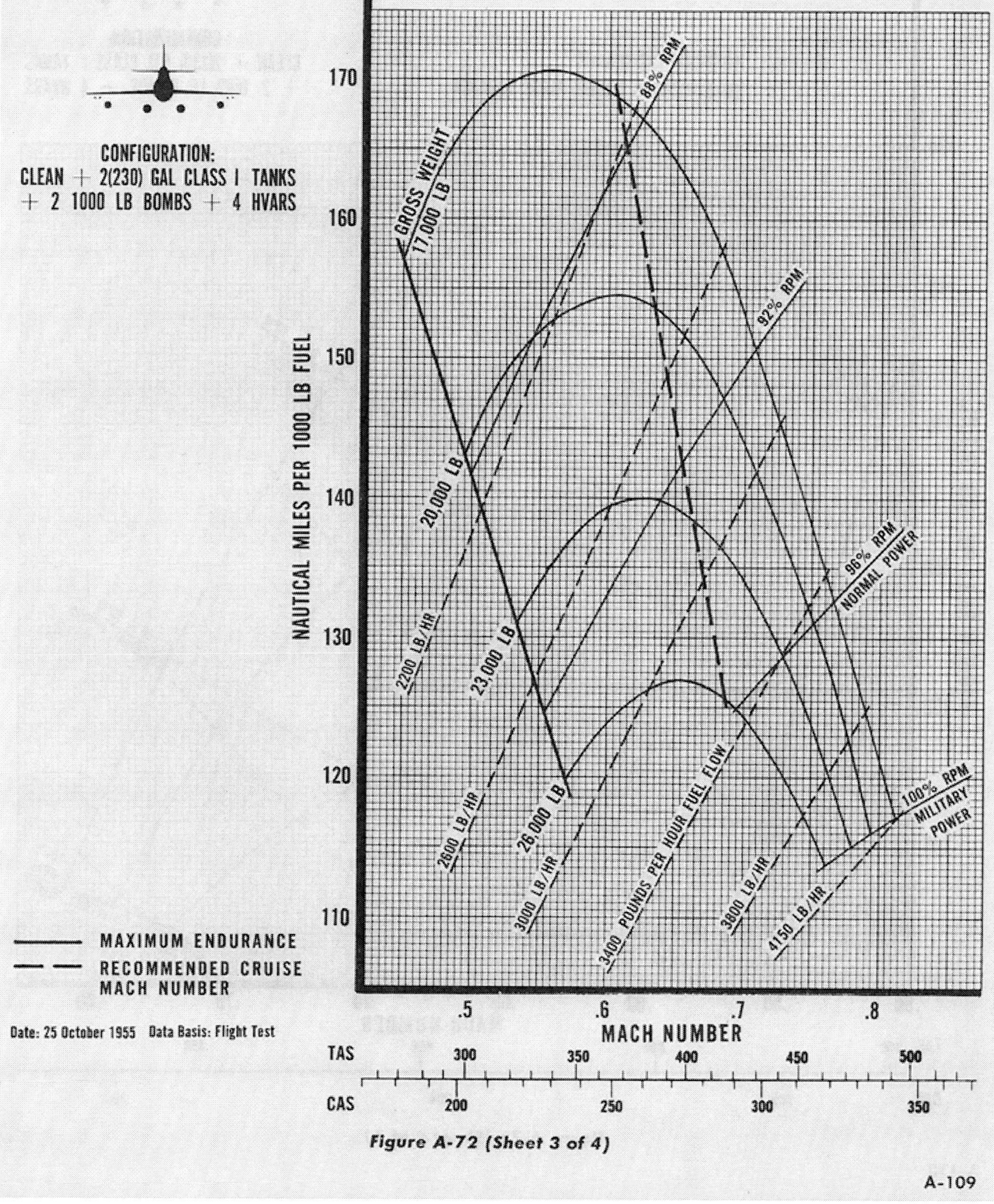

Date: 25 October 1955 Data Basis: Flight Test

Figure A-72 (Sheet 3 of 4)

NAUTICAL MILES PER 1000 POUNDS FUEL

30,000 FT

F-84F-25 AND LATER
J65-W-3 OR B-3 ENGINES
STANDARD DAY

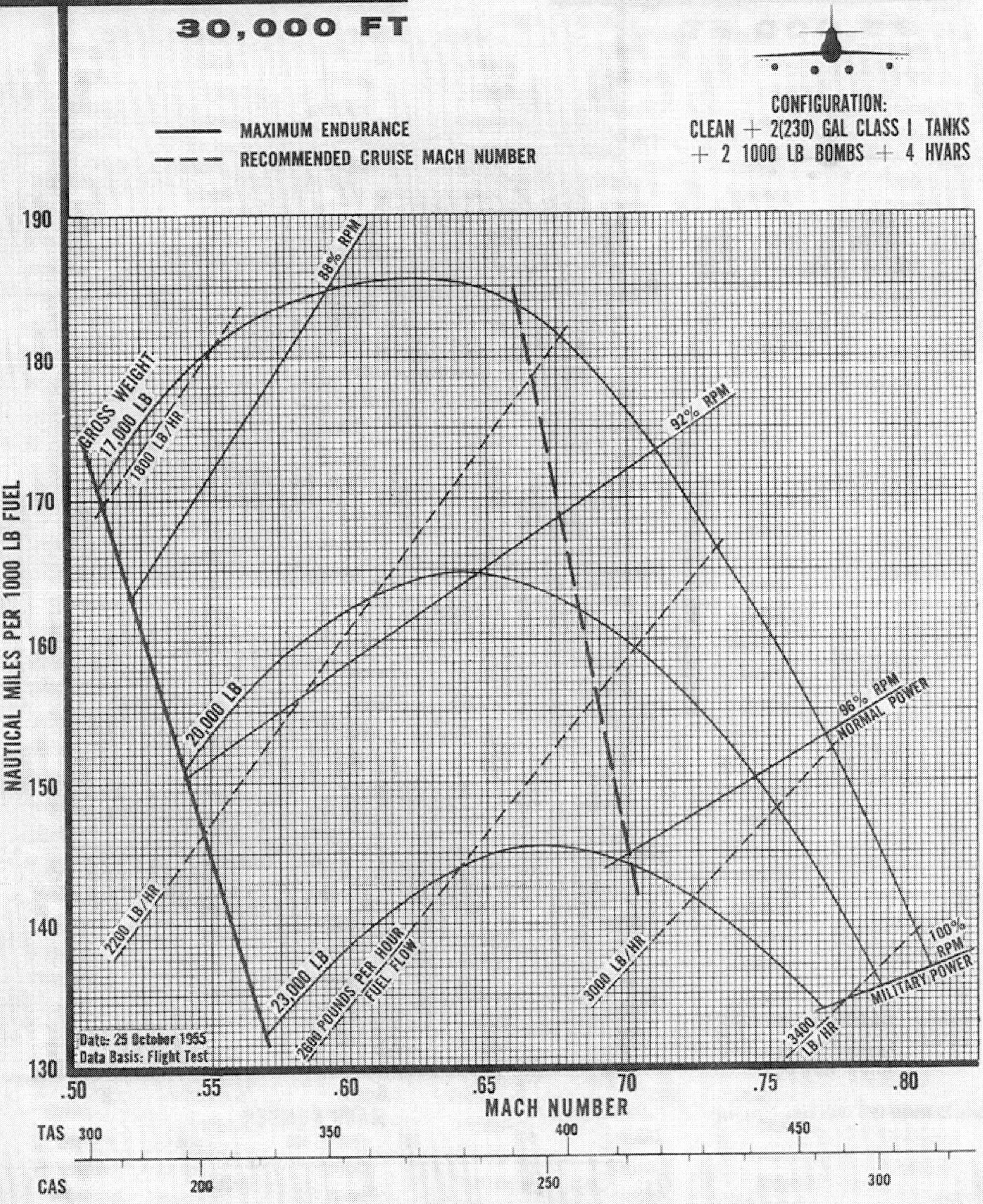

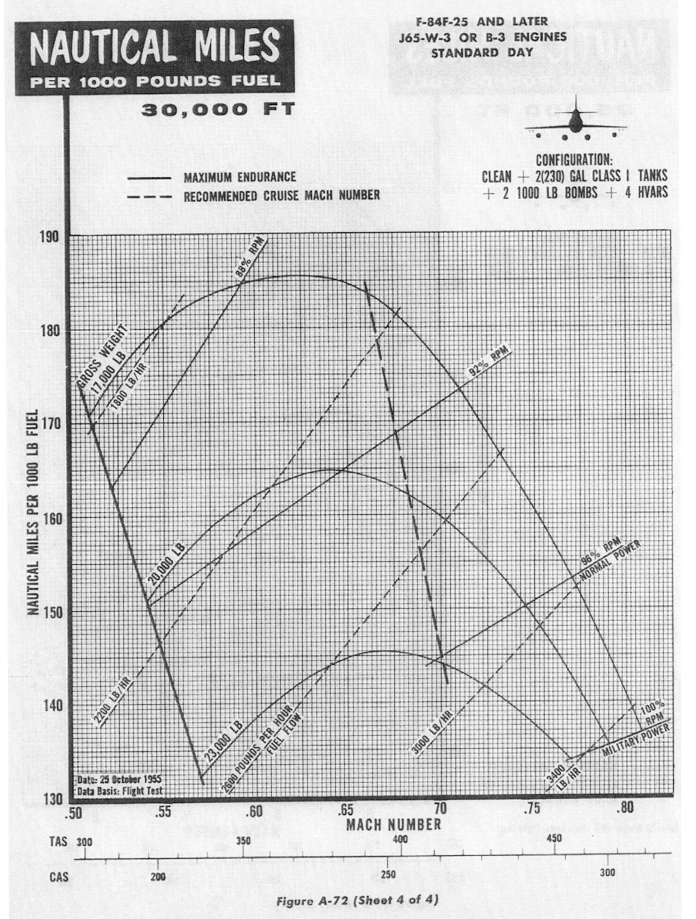

Figure A-72 (Sheet 4 of 4)

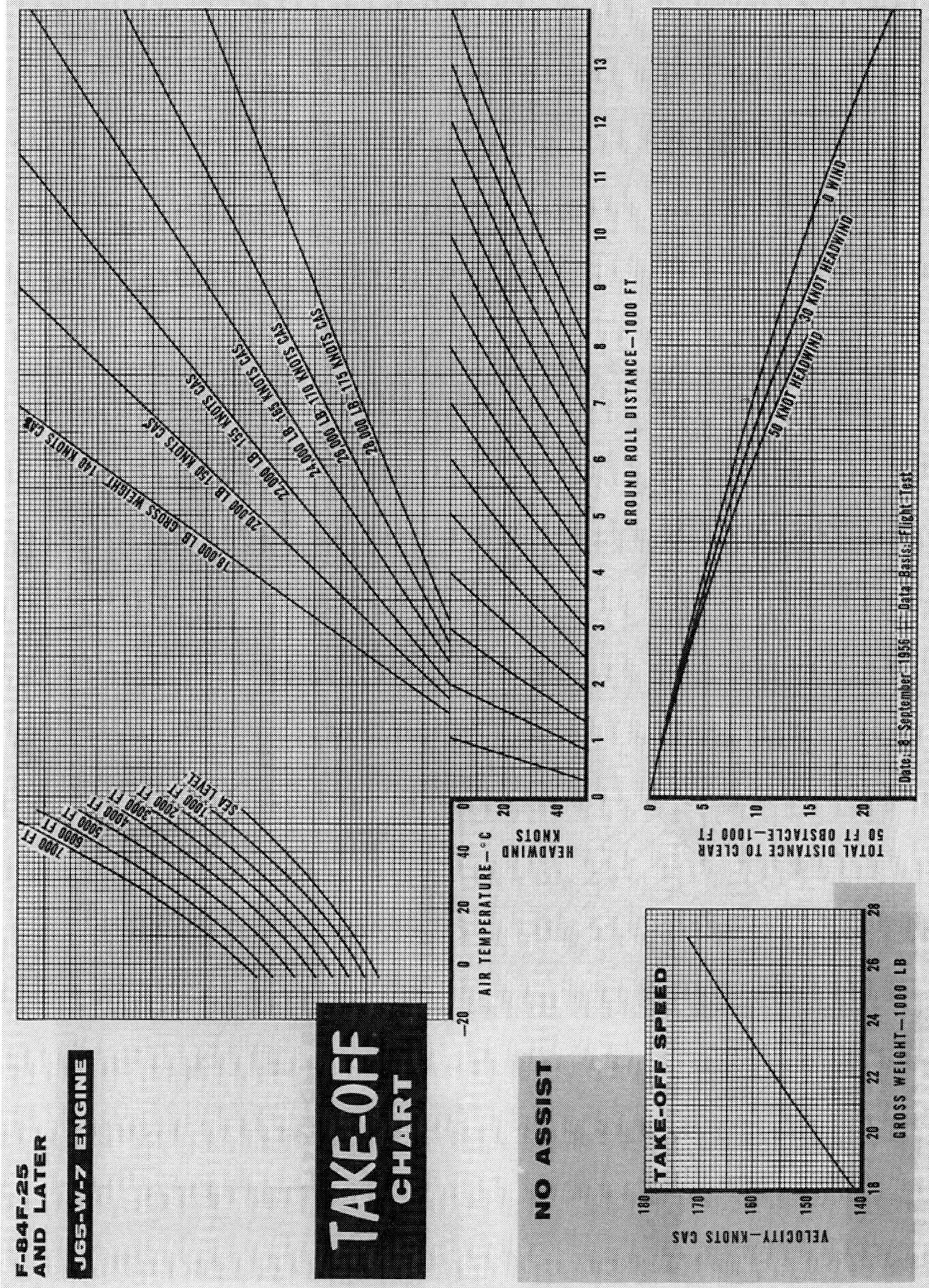
F-84F-25
AND LATER
J65-W-7 ENGINE
TAKE-OFF
CHART
NO ASSIST
TAKE-OFF SPEED
VELOCITY—KNOTS CAS
GROSS WEIGHT—1000 LB
AIR TEMPERATURE—°C
HEADWIND
KNOTS
GROUND ROLL DISTANCE—1000 FT
TOTAL DISTANCE TO CLEAR
50 FT OBSTACLE—1000 FT
SEA LEVEL
1000 FT
2000 FT
3000 FT
4000 FT
5000 FT
6000 FT
7000 FT
18,000 LB GROSS WEIGHT 140 KNOTS CAS
20,000 LB 150 KNOTS CAS
22,000 LB 155 KNOTS CAS
24,000 LB 165 KNOTS CAS
26,000 LB 170 KNOTS CAS
28,000 LB 175 KNOTS CAS
0 WIND
30 KNOT HEADWIND
50 KNOT HEADWIND
Date: 8 September 1956 Data Basis: Flight Test

Figure A—73

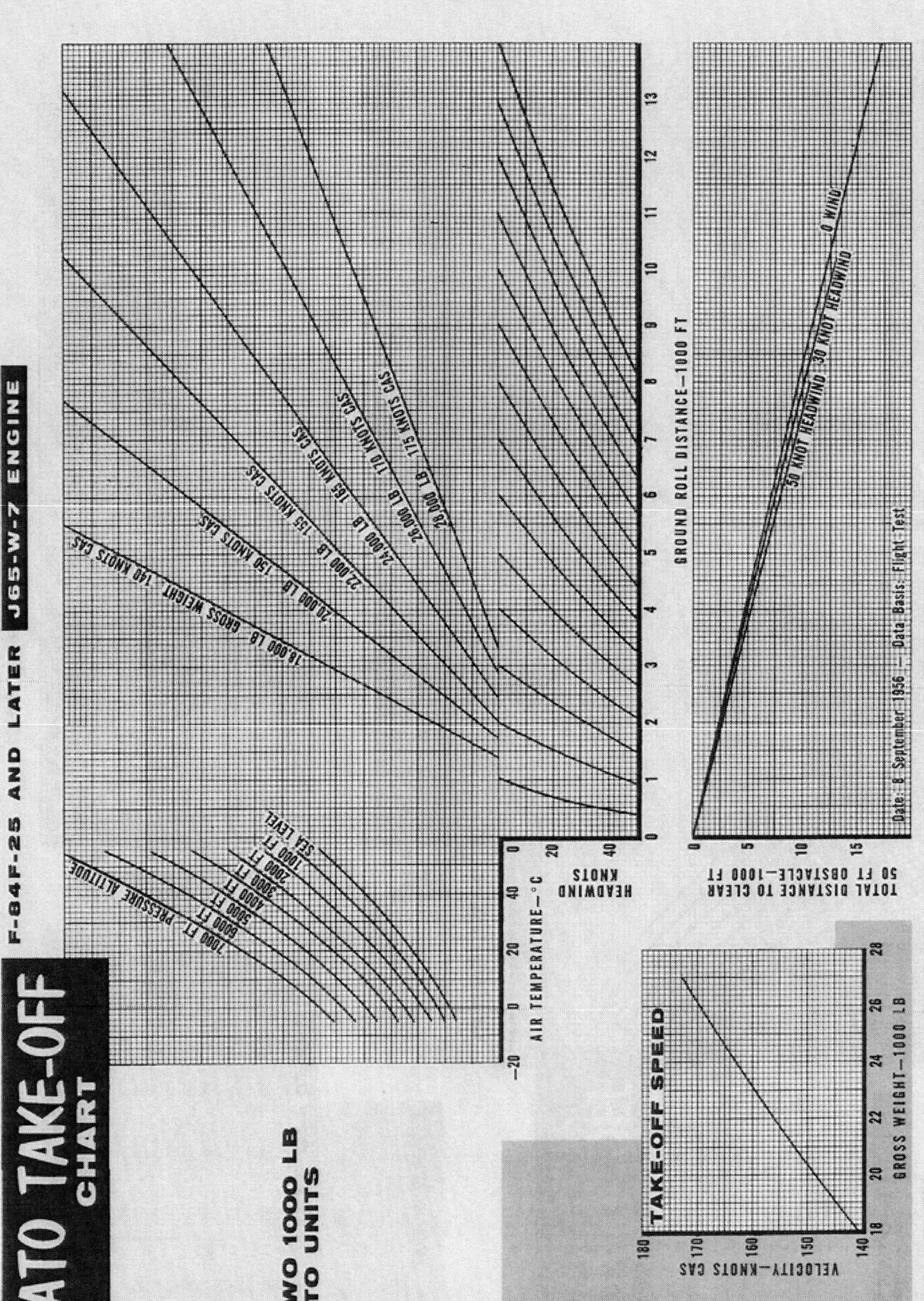

Figure A-74 (sheet 1 of 2)

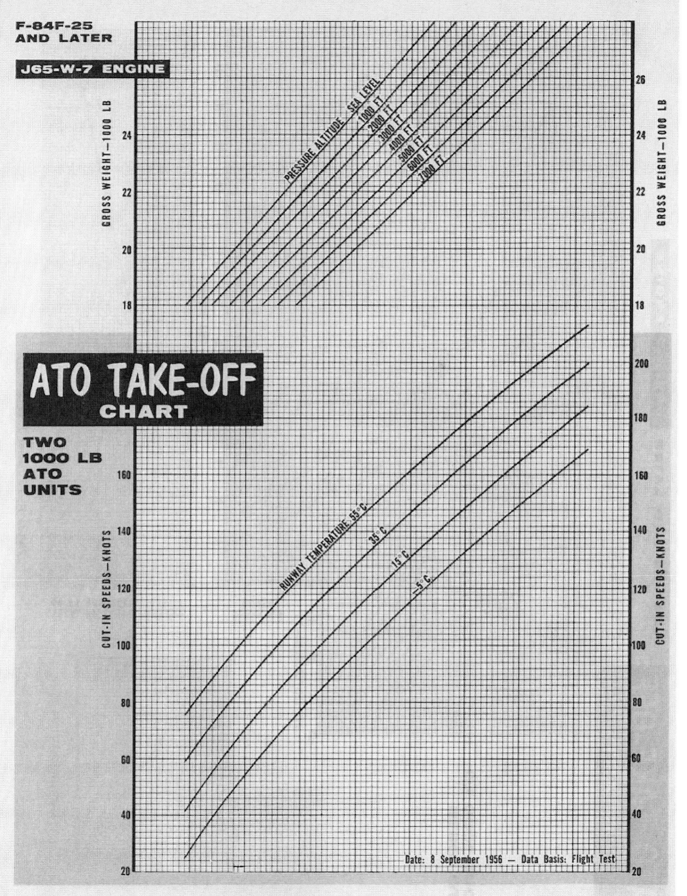

F-84F-25
AND LATER

J65-W-7 ENGINE

ATO TAKE-OFF CHART

TWO
1000 LB
ATO
UNITS

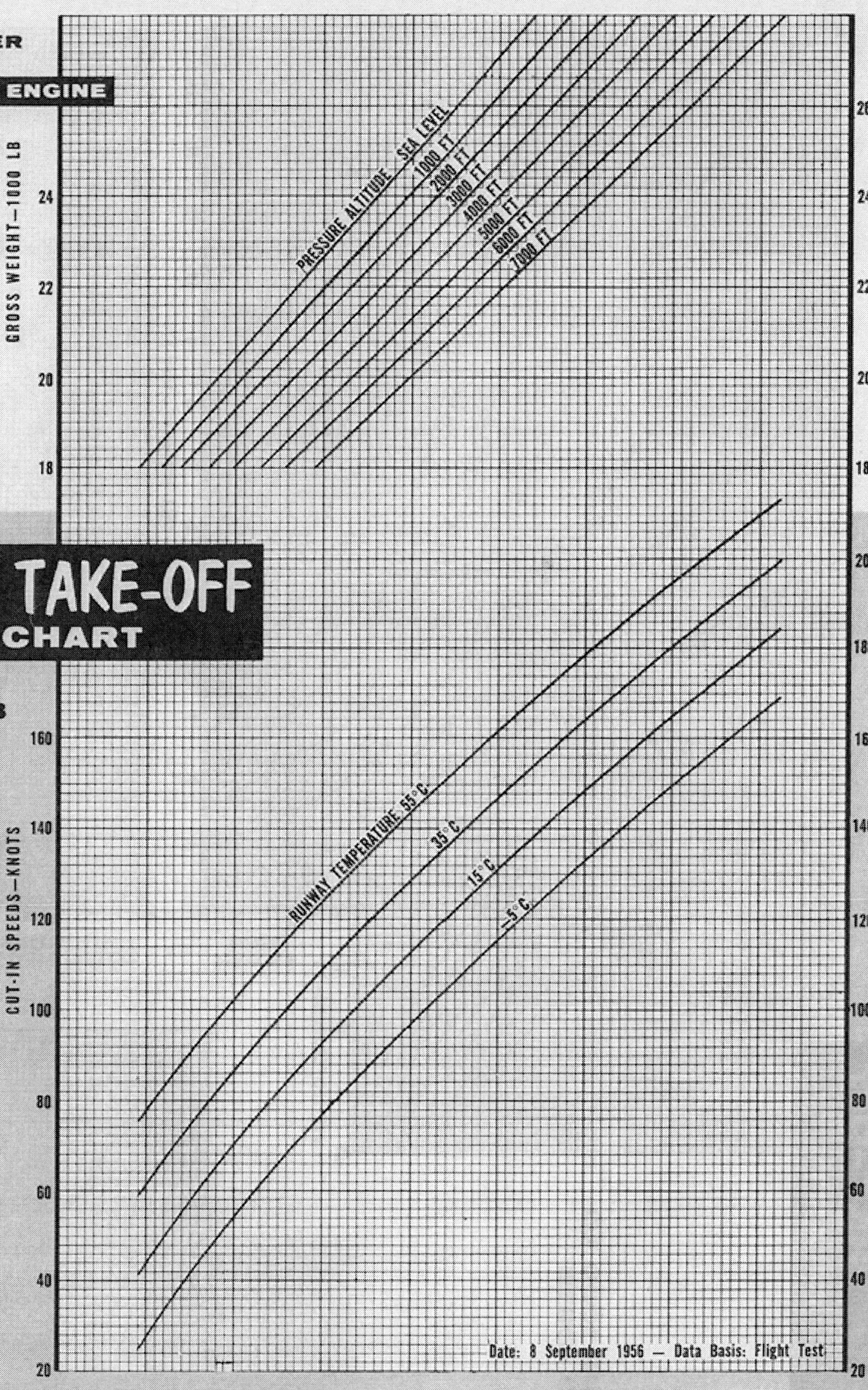

Figure A—74 (sheet 2 of 2)

ATO TAKE-OFF CHART

F-84F-25 AND LATER **J65-W-7 ENGINE**

FOUR 1000 LB ATO UNITS

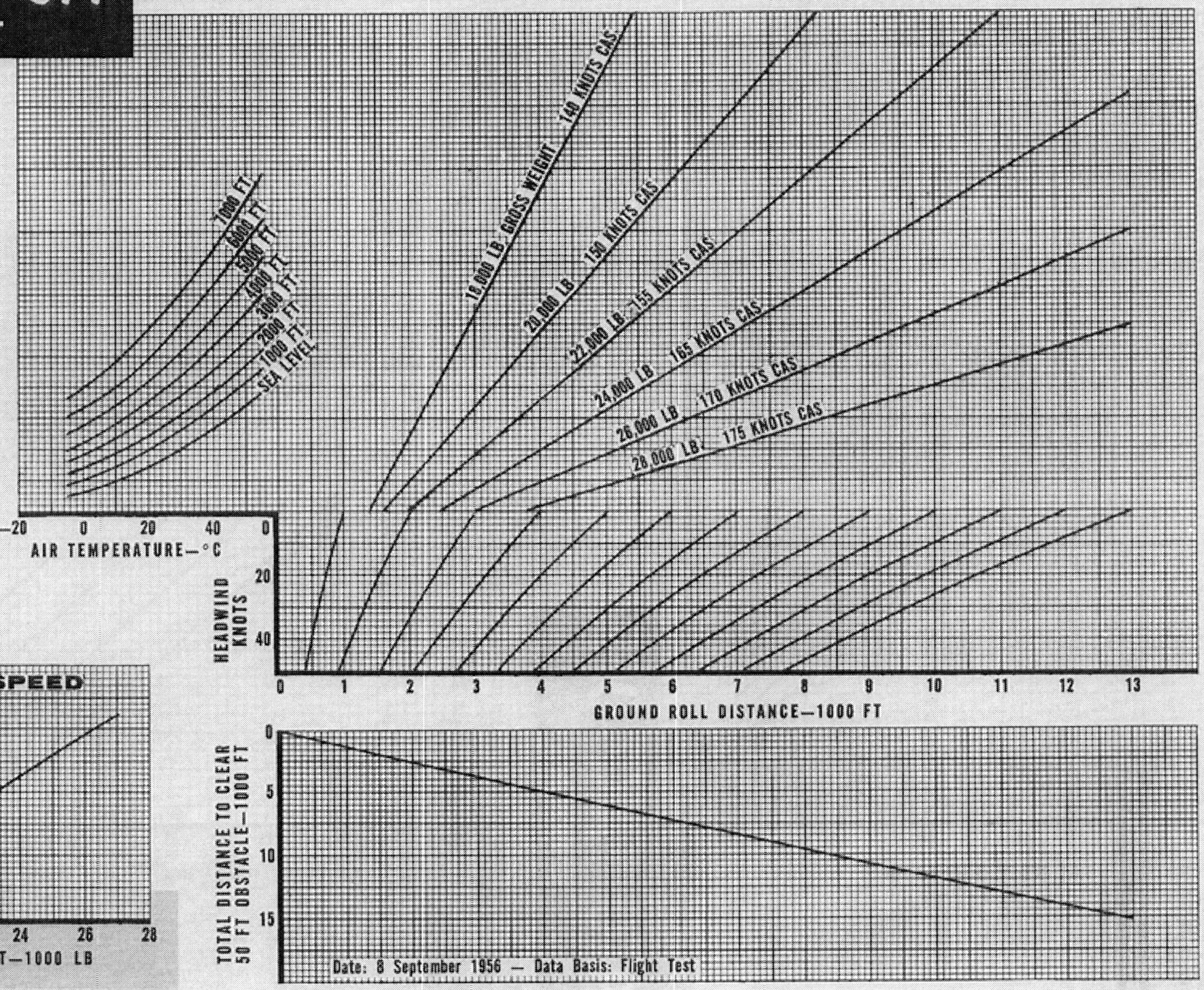

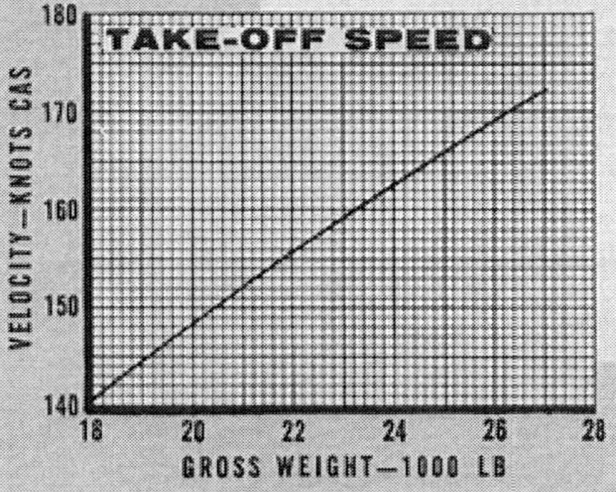

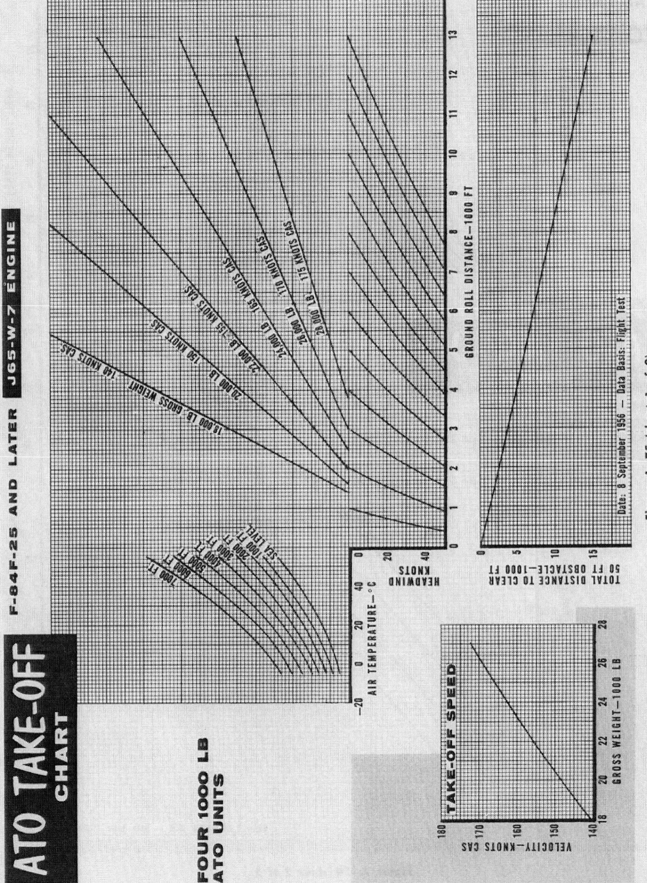

Figure A—75 (sheet 1 of 2)

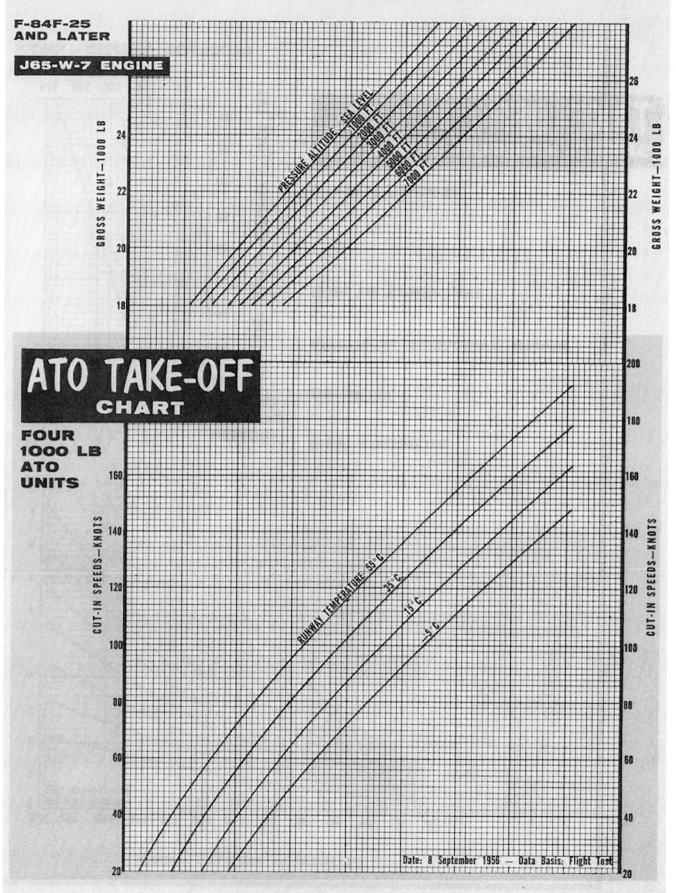

F-84F-25 AND LATER

J65-W-7 ENGINE

ATO TAKE-OFF CHART

FOUR 1000 LB ATO UNITS

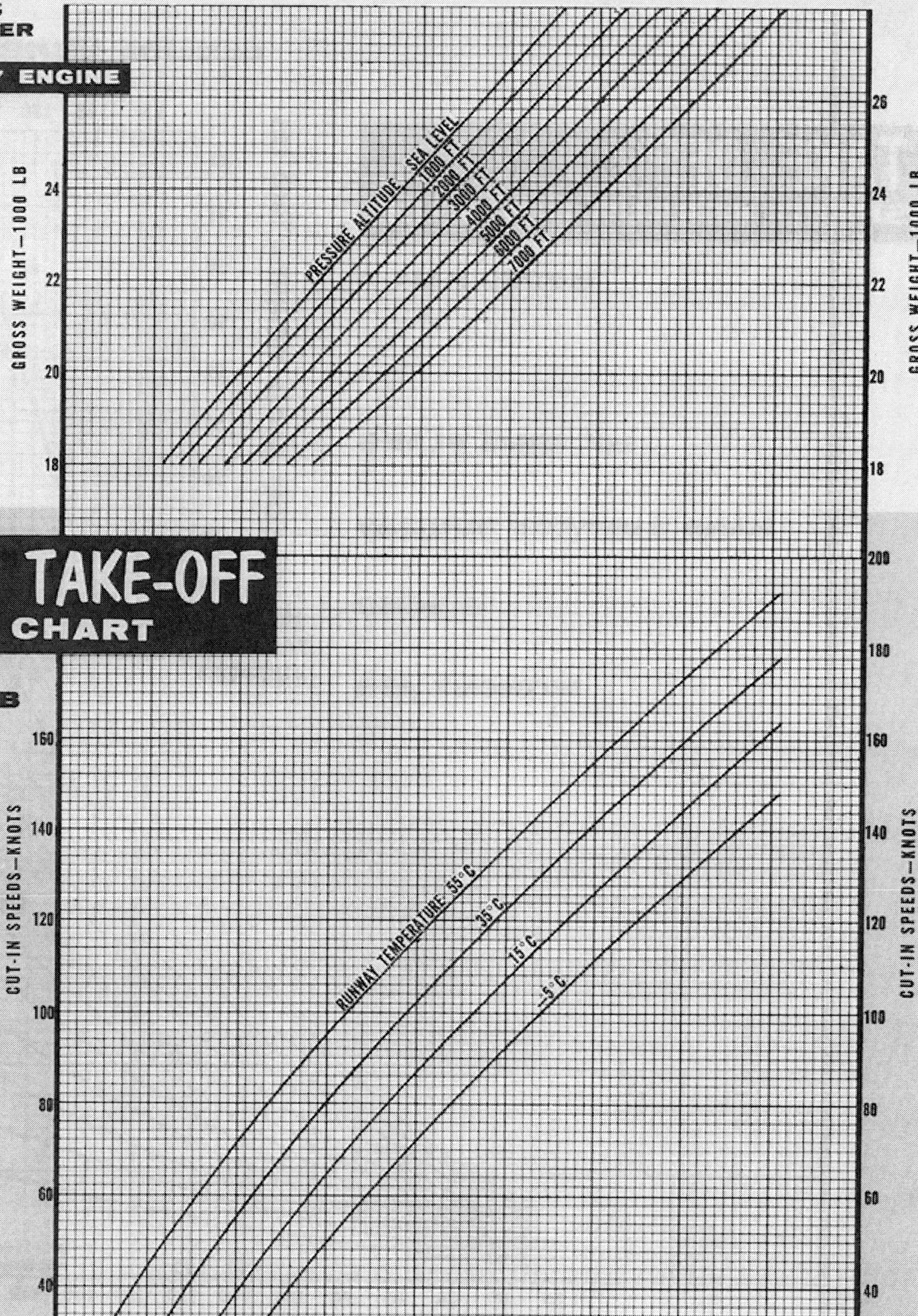

Figure A–75 (sheet 2 of 2)

TAKE-OFF
SPEED VERSUS DISTANCE

MODEL F-84-F

20° WING FLAPS

HARD SURFACE RUNWAY

O WIND

WITHOUT ATO

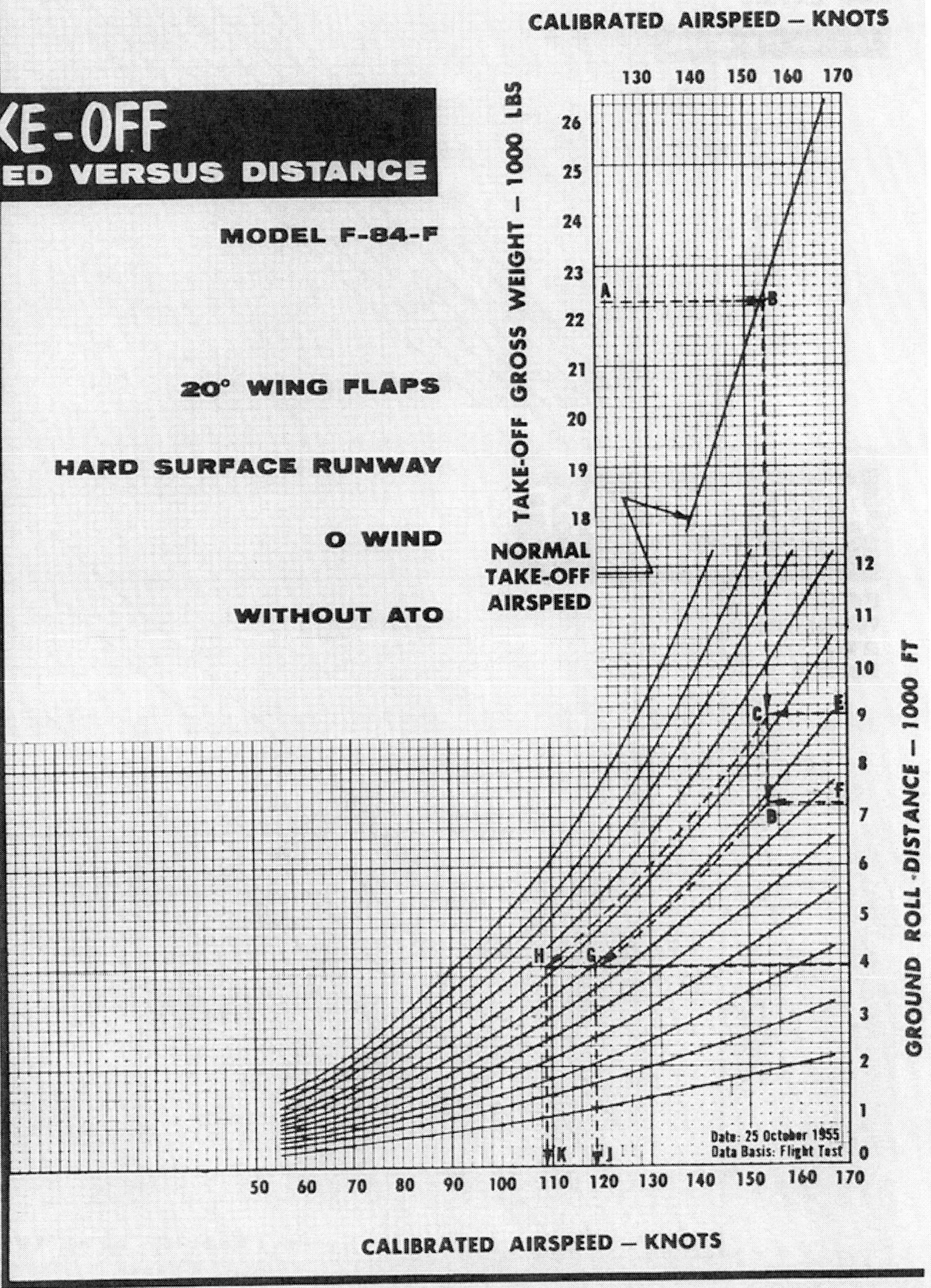

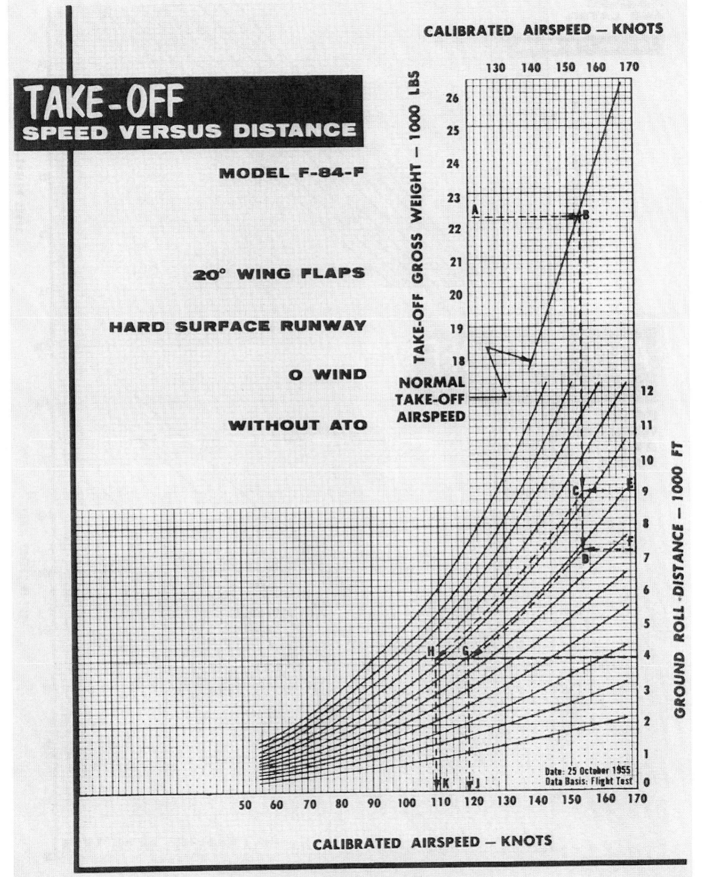

Figure A—76.

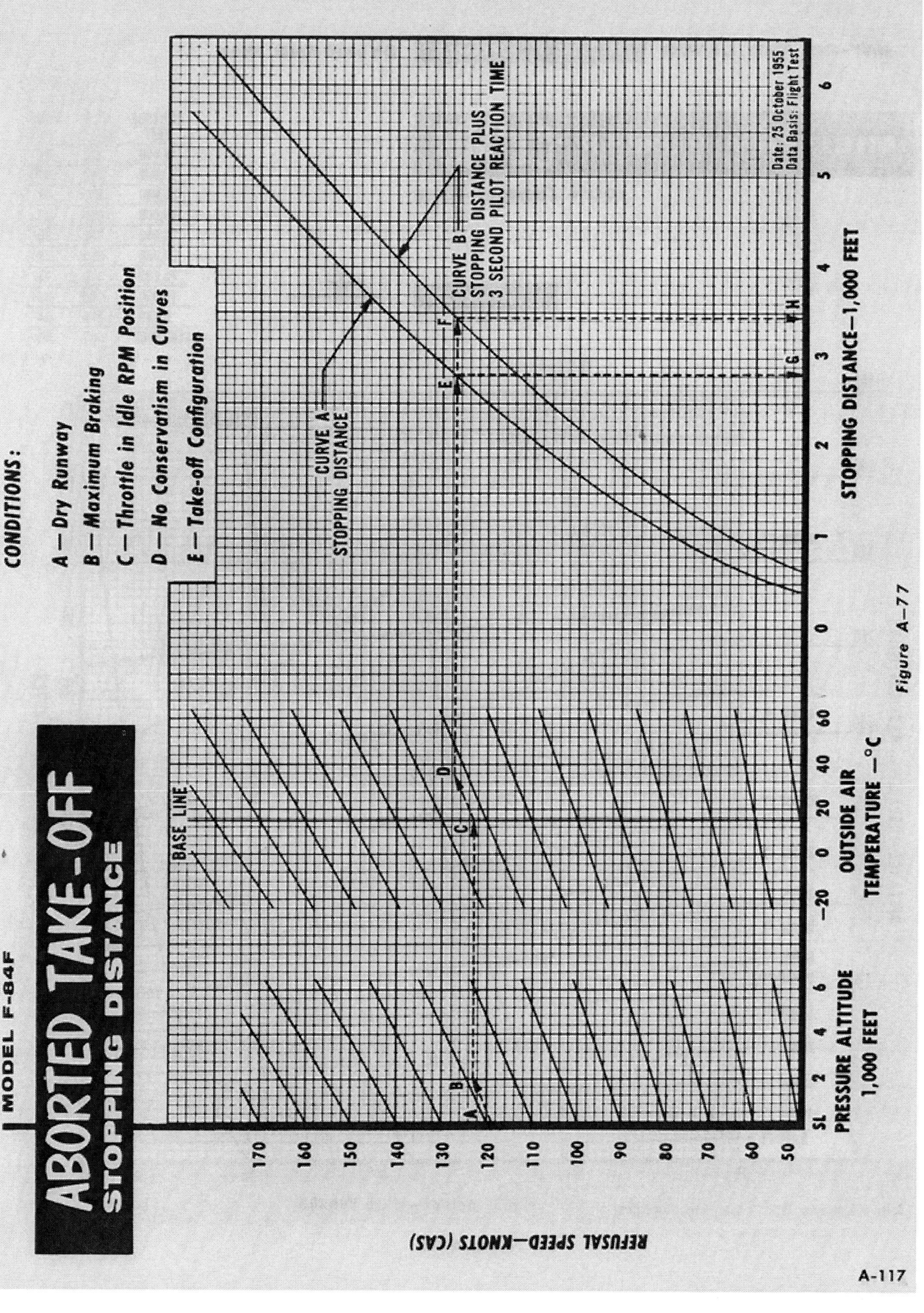
MODEL F-84F
ABORTED TAKE-OFF
STOPPING DISTANCE
CONDITIONS:
A — Dry Runway
B — Maximum Braking
C — Throttle in Idle RPM Position
D — No Conservatism in Curves
E — Take-off Configuration
BASE LINE
CURVE A
STOPPING DISTANCE
CURVE B
STOPPING DISTANCE PLUS
3 SECOND PILOT REACTION TIME
Date: 25 October 1955
Data Basis: Flight Test
REFUSAL SPEED—KNOTS (CAS)
170
160
150
140
130
120
110
100
90
80
70
60
50
PRESSURE ALTITUDE
1,000 FEET
SL 2 4 6
OUTSIDE AIR
TEMPERATURE —°C
−20 0 20 40 60
STOPPING DISTANCE—1,000 FEET
0 1 2 3 4 5 6
A
B
C
D
E
F
G
H

Figure A—77

F-84F-25 AND LATER **J65-W-7 ENGINE** **STANDARD DAY**

MILITARY POWER CLIMB

100% (8300) RPM

CLEAN AIRPLANE

ALTITUDE FEET	CAS KNOTS	MACH NO.
45,000	225	.84
40,000	255	.84
35,000	285	.83
30,000	310	.82
25,000	340	.80
20,000	365	.78
15,000	370	.73
10,000	380	.69
5,000	390	.64
SEA LEVEL	400	.60

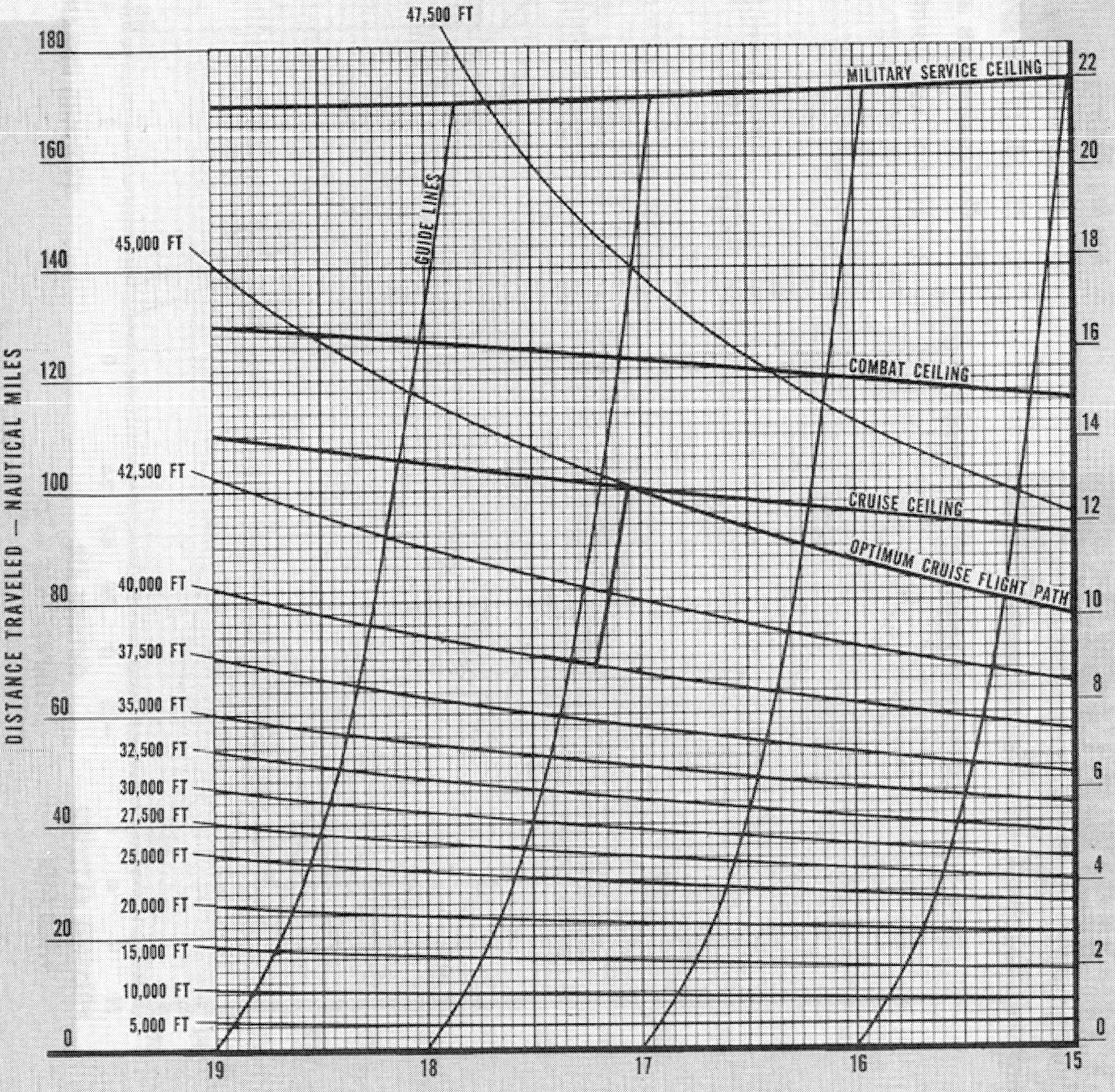

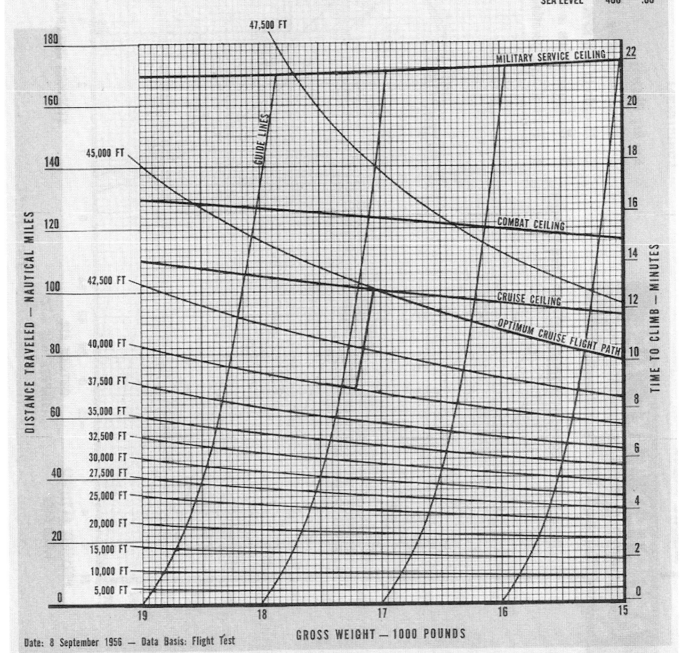

Date: 8 September 1956 — Data Basis: Flight Test

Figure A—78

F-84F-25 AND LATER J65-W-7 ENGINE STANDARD DAY

98% MILITARY RPM CLIMB

(8100 RPM)

ALTITUDE FEET	CAS KNOTS	MACH NO.
45,000	215	.81
40,000	245	.81
35,000	275	.81
30,000	300	.79
25,000	315	.75
20,000	330	.71
15,000	345	.68
10,000	360	.65
5,000	375	.62
SEA LEVEL	390	.59

CLEAN AIRPLANE

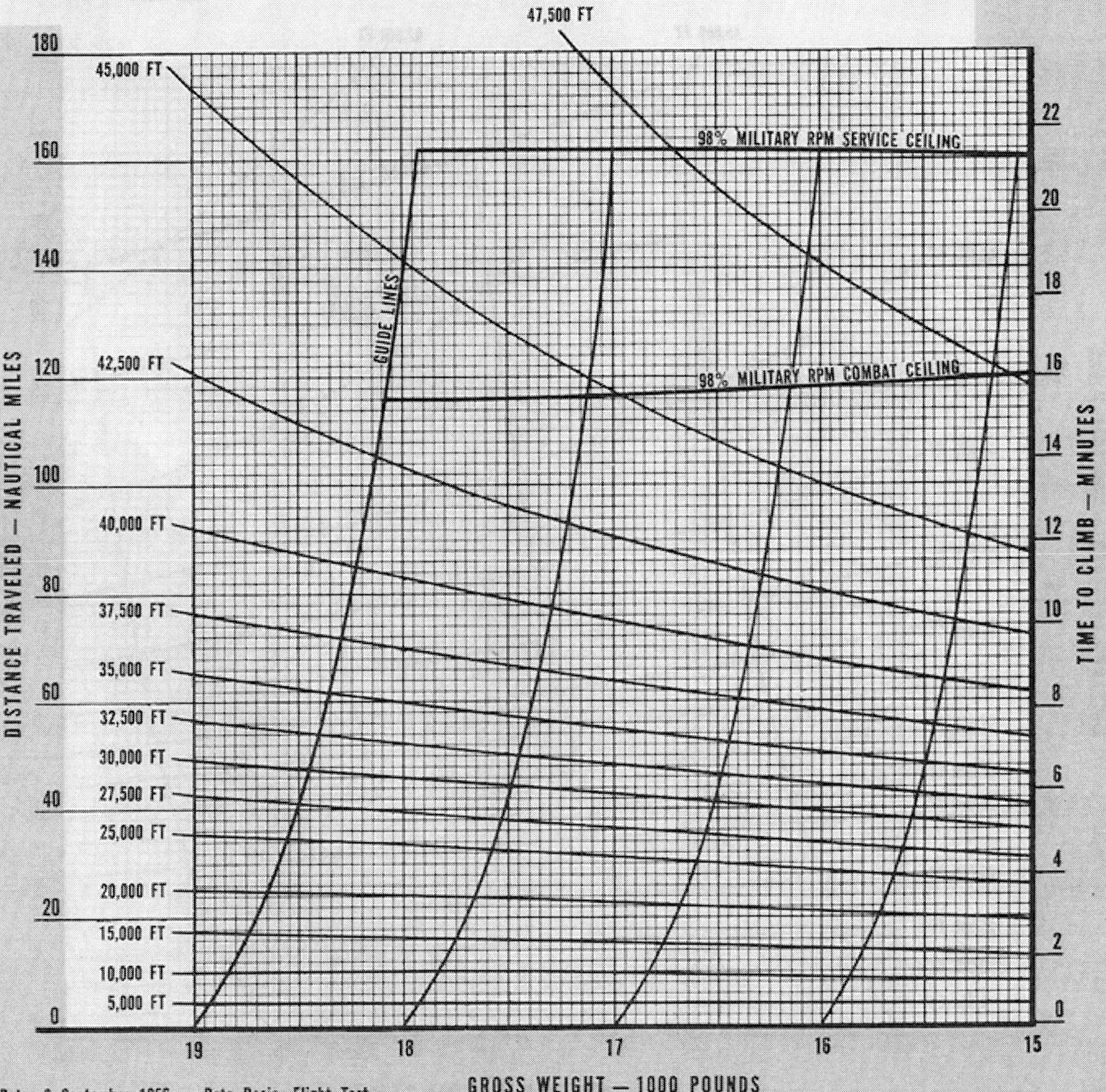

Date: 8 September 1956 — Data Basis: Flight Test

Figure A—79

F-84F-25 AND LATER **J65-W-7 ENGINE** STANDARD DAY

NORMAL POWER CLIMB

96% (8000) RPM

ALTITUDE FEET	CAS KNOTS	MACH NO.
45,000	215	.81
40,000	245	.81
35,000	275	.81
30,000	295	.78
25,000	310	.74
20,000	320	.70
15,000	330	.66
10,000	340	.62
5,000	350	.58
SEA LEVEL	360	.54

CLEAN AIRPLANE

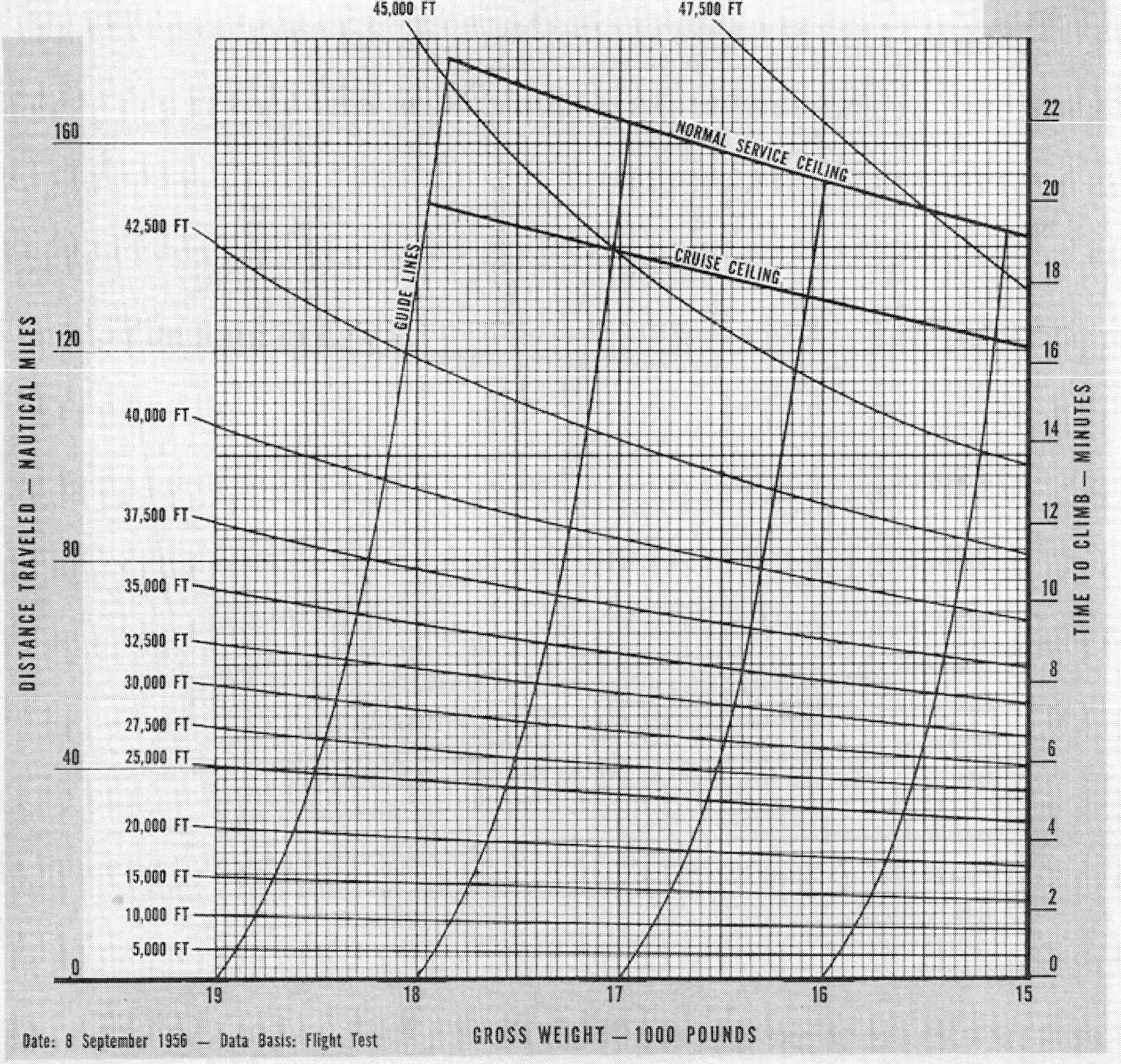

Date: 8 September 1956 — Data Basis: Flight Test

Figure A—80

F-84F-25 AND LATER **J65-W-7 ENGINE** STANDARD DAY

MILITARY POWER CLIMB

100% (8300) RPM

CONFIGURATION: CLEAN +
TWO 230 GAL CLASS I TANKS

ALTITUDE FEET	CAS KNOTS	MACH NO.
45,000	210	.78
40,000	235	.78
35,000	260	.78
30,000	285	.76
25,000	310	.74
20,000	325	.70
15,000	330	.65
10,000	340	.61
5,000	350	.58
SEA LEVEL	360	.54

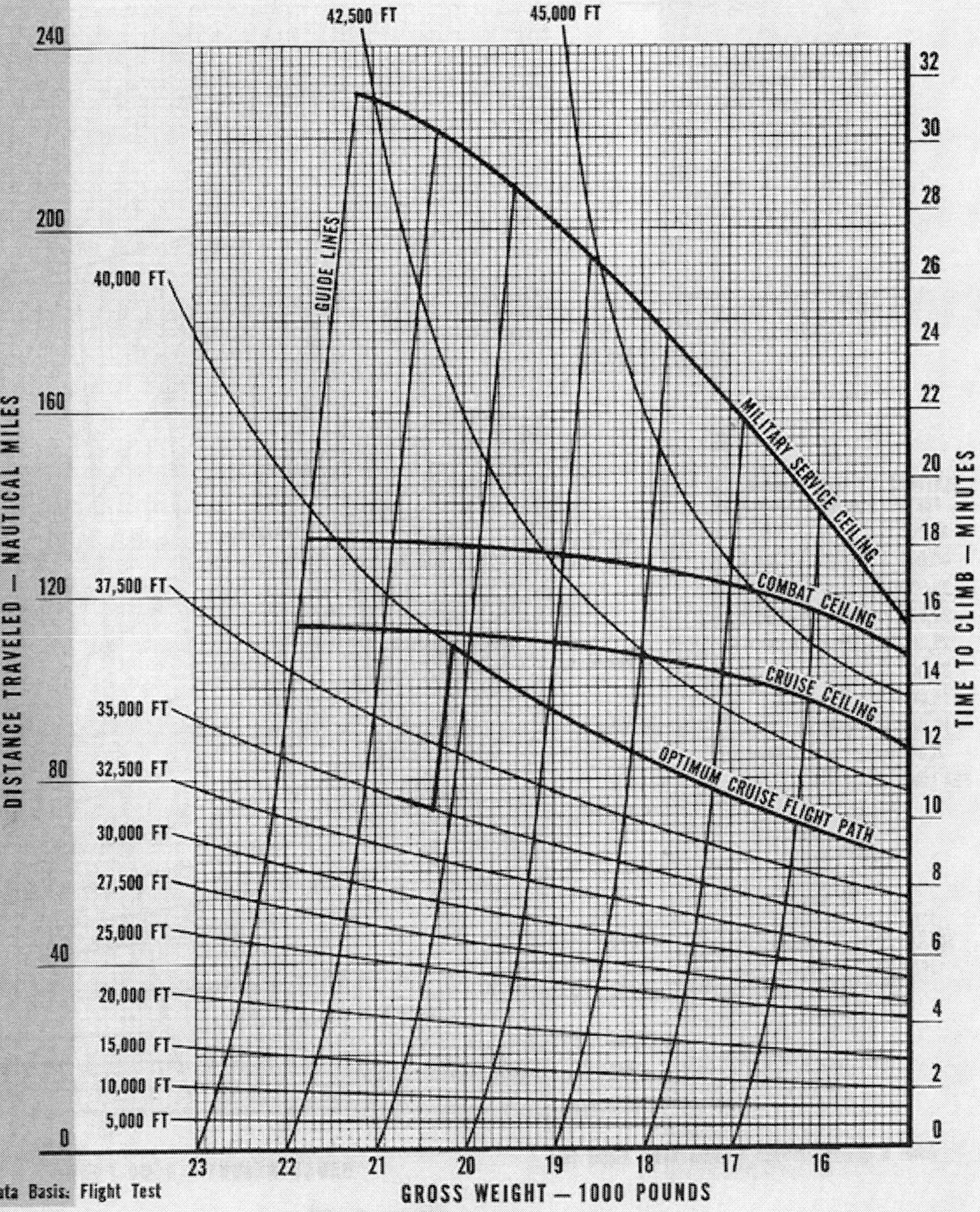

Date: 8 September 1956 — Data Basis: Flight Test

Figure A—81

F-84F-25 AND LATER **J65-W-7 ENGINE** STANDARD DAY

98% MILITARY RPM CLIMB

(8100 RPM)

CONFIGURATION: CLEAN + TWO 230 GAL CLASS I TANKS

ALTITUDE FEET	CAS KNOTS	MACH NO.
45,000	210	.78
40,000	235	.78
35,000	265	.78
30,000	275	.73
25,000	290	.69
20,000	300	.65
15,000	310	.62
10,000	325	.58
5,000	330	.55
SEA LEVEL	335	.51

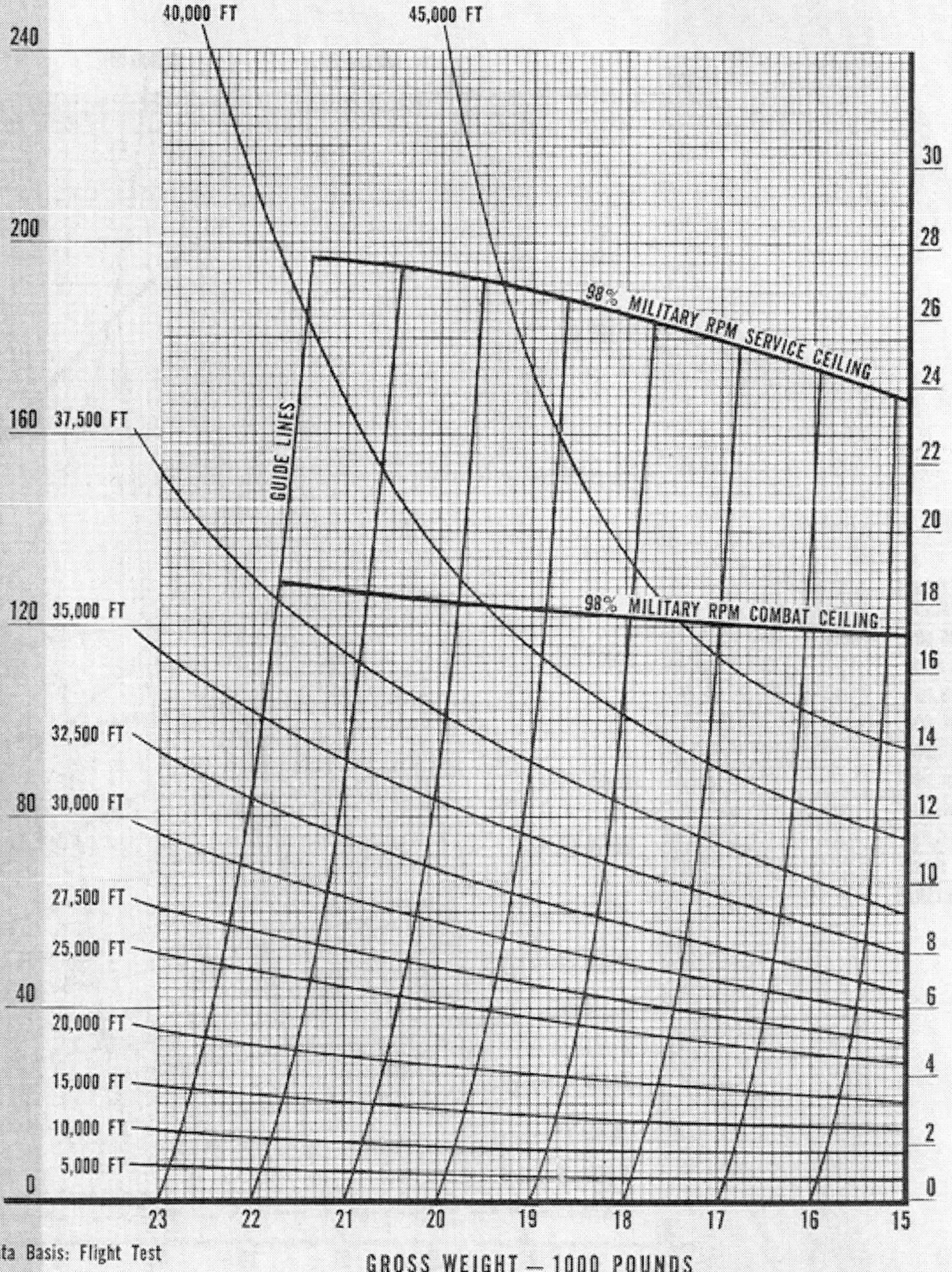

Date: 8 September 1956 — Data Basis: Flight Test

Figure A–82

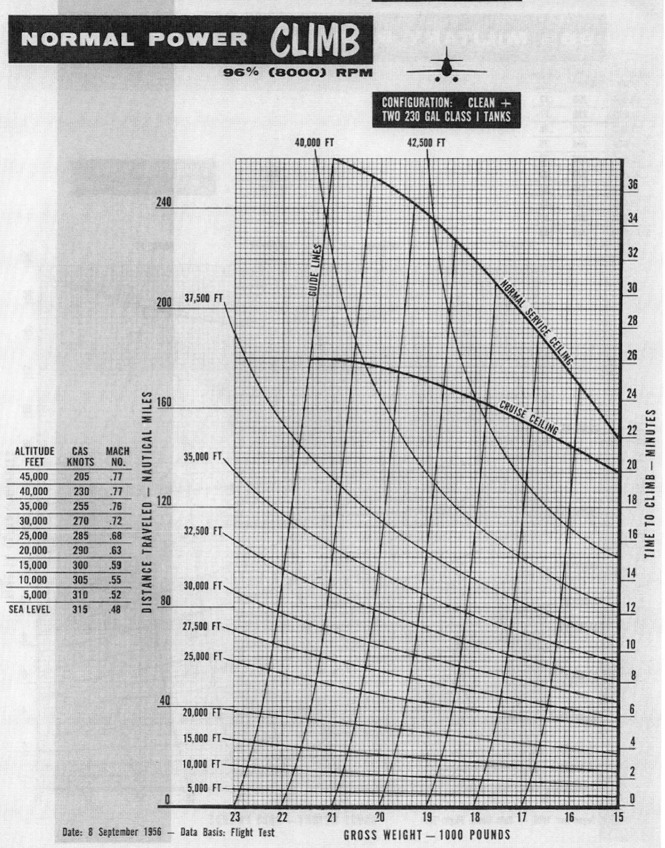

F-84F-25 AND LATER J65-W-7 ENGINE STANDARD DAY

NORMAL POWER CLIMB

96% (8000) RPM

CONFIGURATION: CLEAN + TWO 230 GAL CLASS I TANKS

ALTITUDE FEET	CAS KNOTS	MACH NO.
45,000	205	.77
40,000	230	.77
35,000	255	.76
30,000	270	.72
25,000	285	.68
20,000	290	.63
15,000	300	.59
10,000	305	.55
5,000	310	.52
SEA LEVEL	315	.48

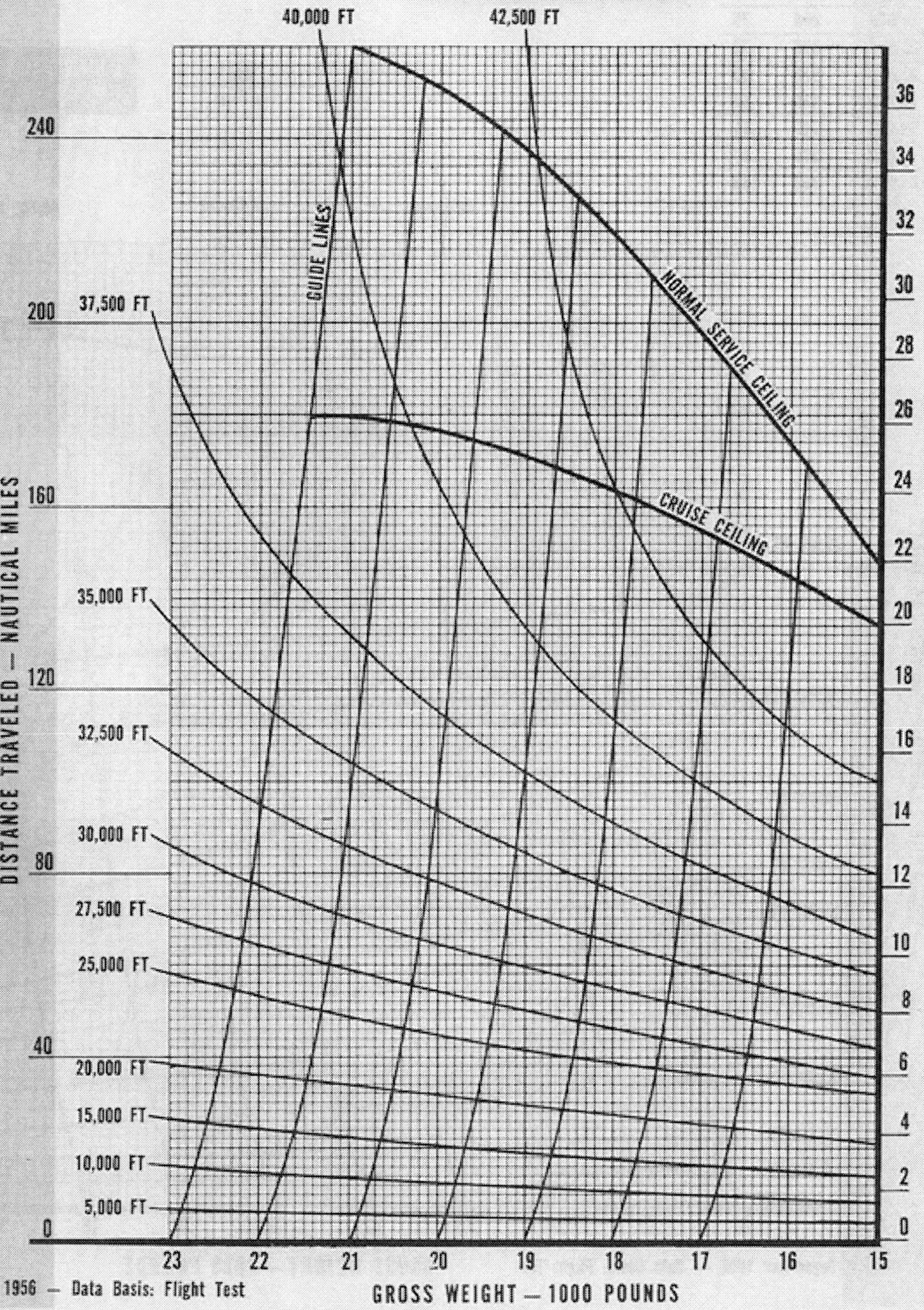

Date: 8 September 1956 — Data Basis: Flight Test

Figure A–83

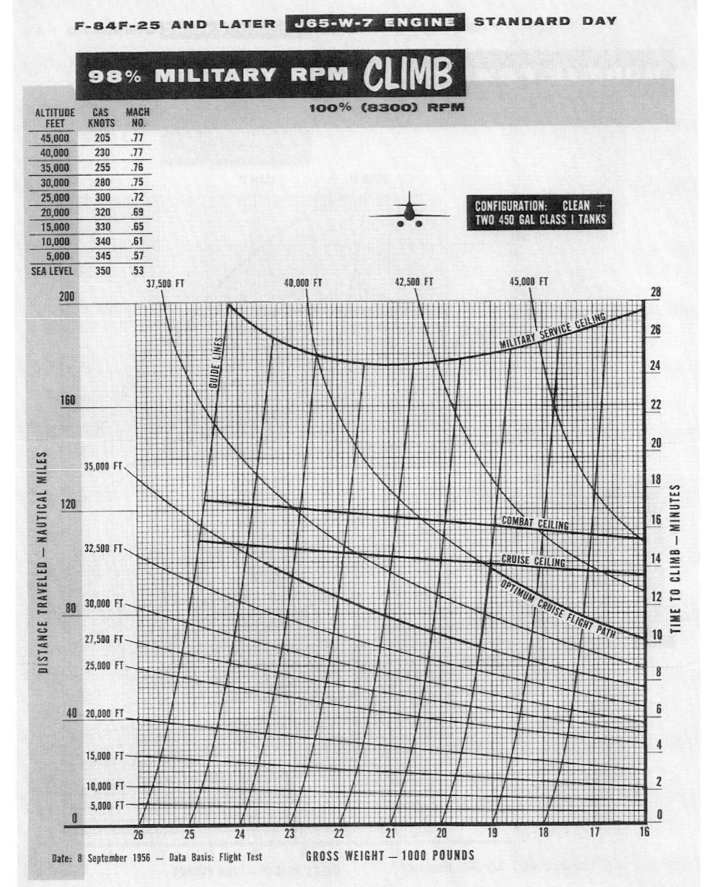

F-84F-25 AND LATER J65-W-7 ENGINE STANDARD DAY

98% MILITARY RPM CLIMB

100% (8300) RPM

ALTITUDE FEET	CAS KNOTS	MACH NO.
45,000	205	.77
40,000	230	.77
35,000	255	.76
30,000	280	.75
25,000	300	.72
20,000	320	.69
15,000	330	.65
10,000	340	.61
5,000	345	.57
SEA LEVEL	350	.53

CONFIGURATION: CLEAN + TWO 450 GAL CLASS I TANKS

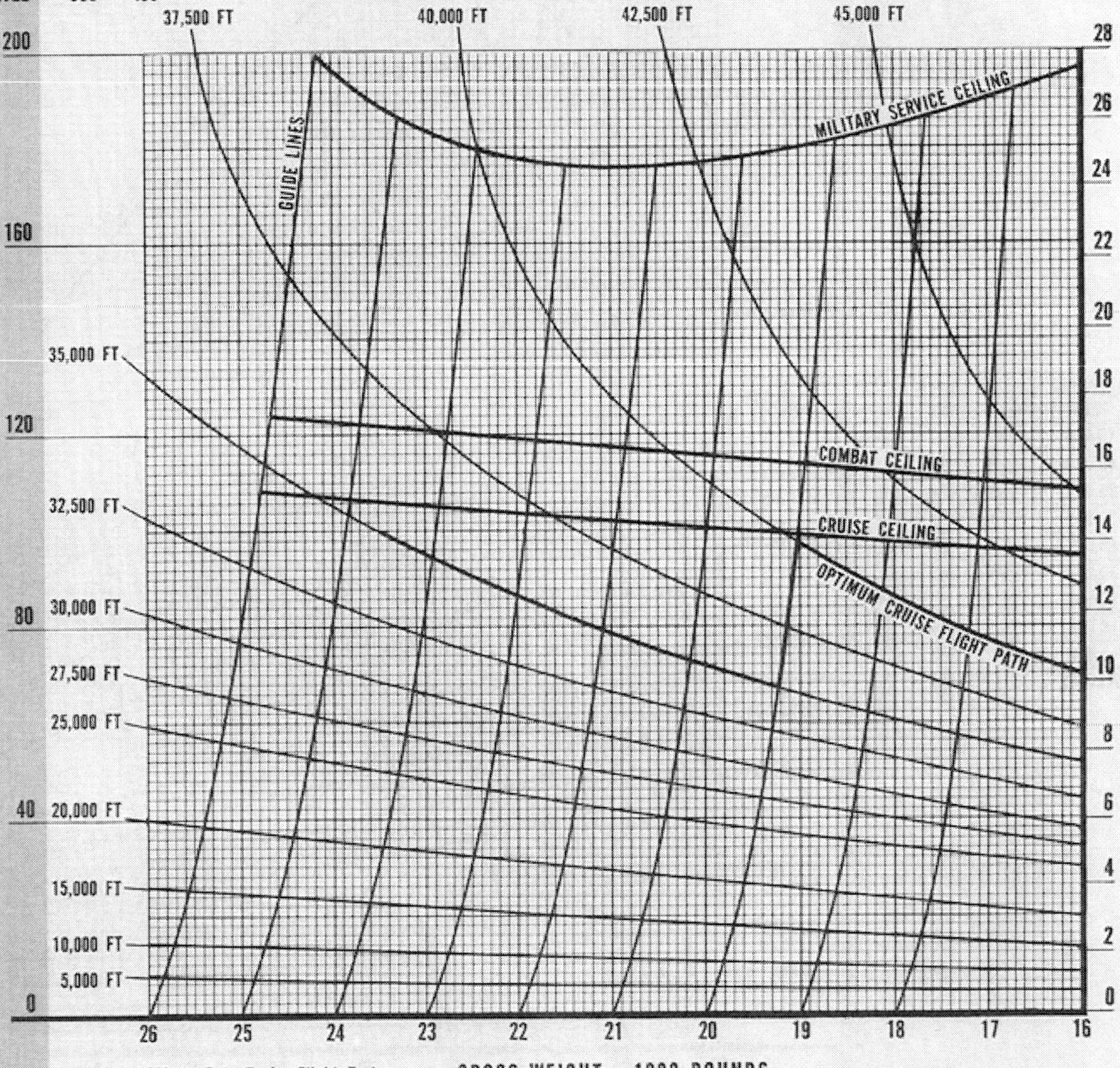

Date: 8 September 1956 — Data Basis: Flight Test

Figure A—84

F-84F-25 AND LATER **J65-W-7 ENGINE** STANDARD DAY

98% MILITARY RPM CLIMB

(8100 RPM)

ALTITUDE FEET	CAS KNOTS	MACH NO.
45,000	210	.78
40,000	235	.78
35,000	265	.78
30,000	275	.73
25,000	290	.69
20,000	300	.65
15,000	310	.62
10,000	320	.58
5,000	330	.55
SEA LEVEL	335	.51

CONFIGURATION: CLEAN — TWO 450 GAL CLASS I TANKS

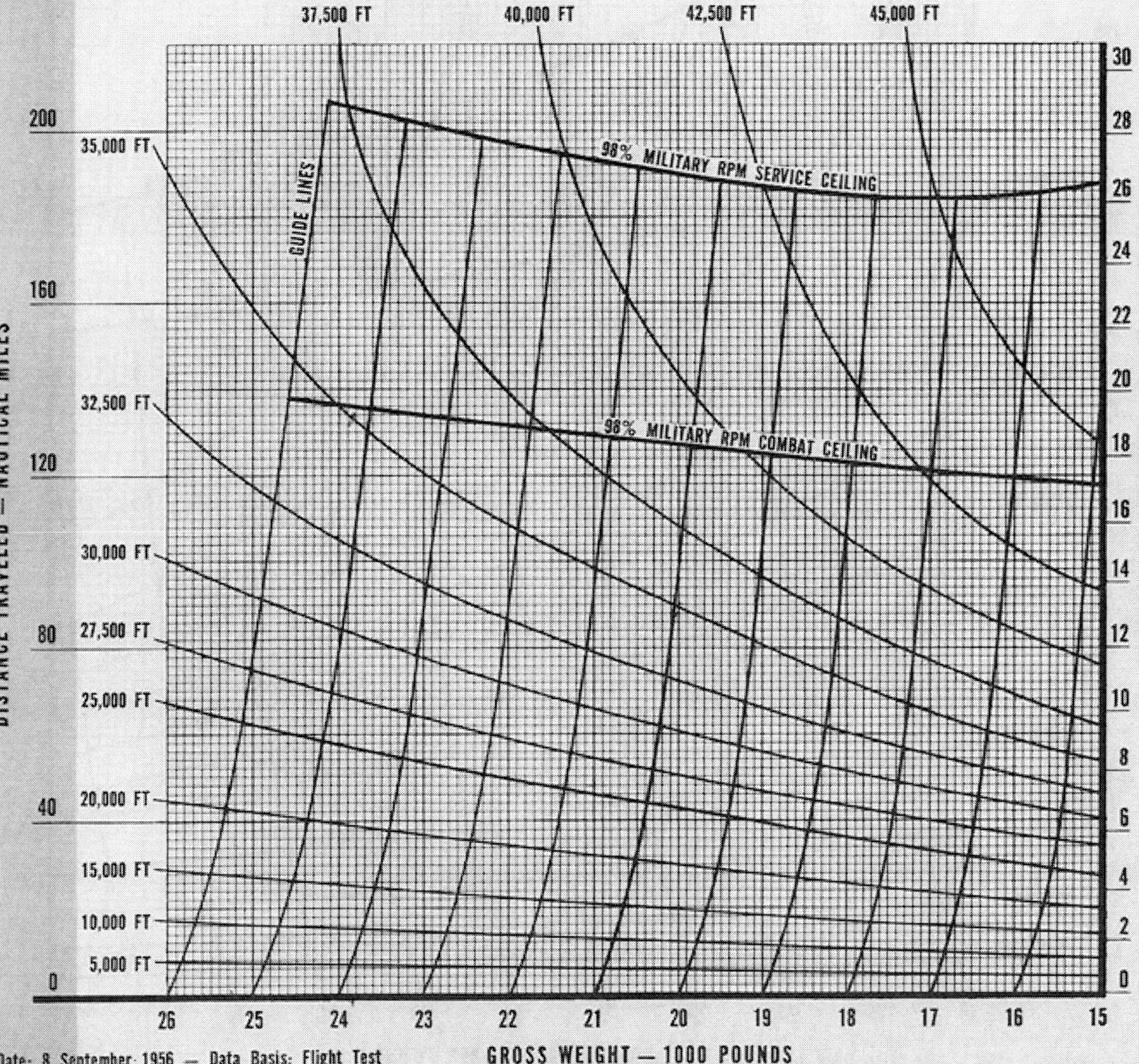

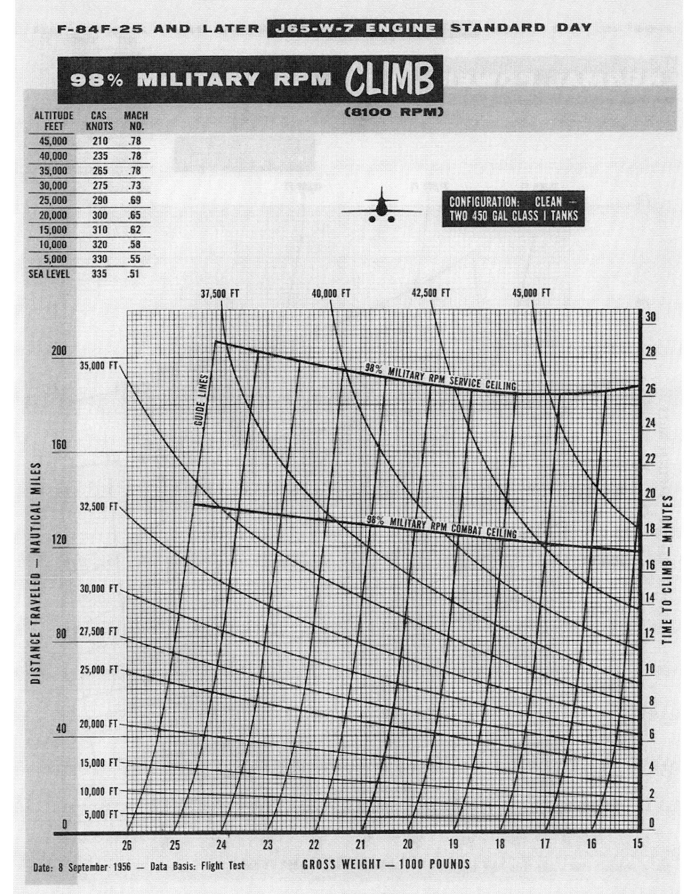

Date: 8 September 1956 — Data Basis: Flight Test

Figure A—85

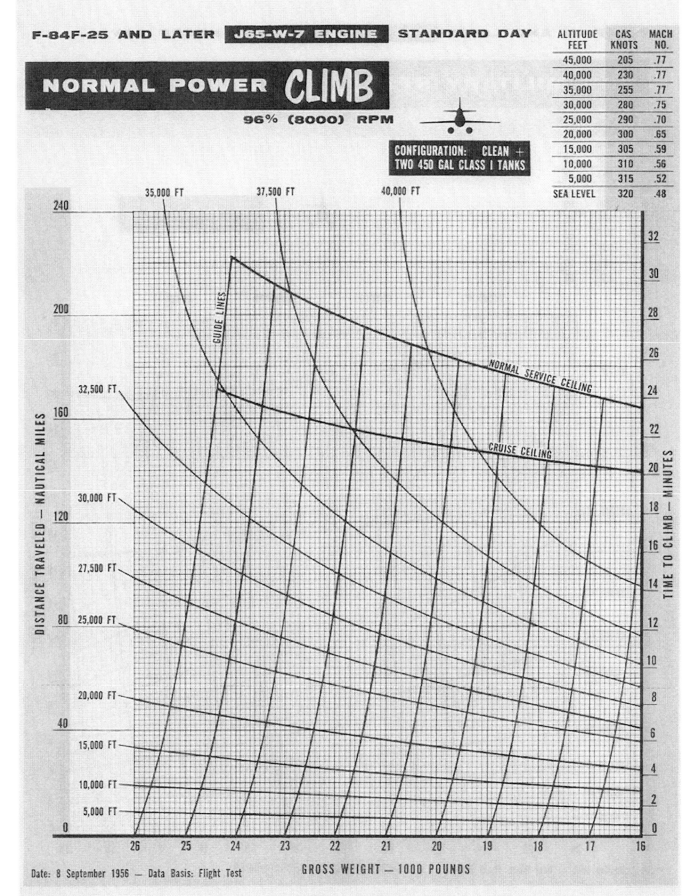

F-84F-25 AND LATER **J65-W-7 ENGINE** STANDARD DAY

NORMAL POWER CLIMB

96% (8000) RPM

CONFIGURATION: CLEAN + TWO 450 GAL CLASS I TANKS

ALTITUDE FEET	CAS KNOTS	MACH NO.
45,000	205	.77
40,000	230	.77
35,000	255	.77
30,000	280	.75
25,000	290	.70
20,000	300	.65
15,000	305	.59
10,000	310	.56
5,000	315	.52
SEA LEVEL	320	.48

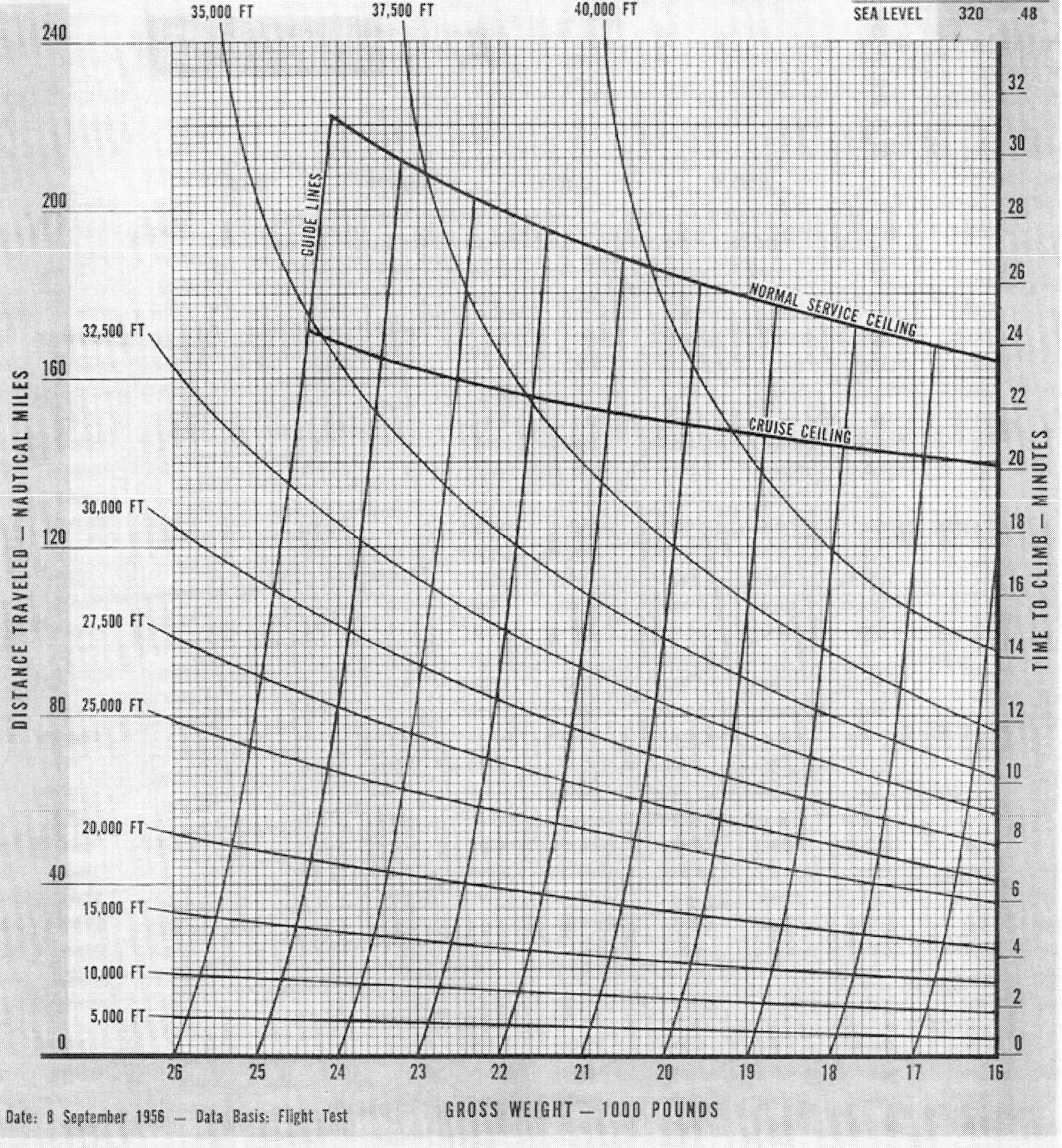

Date: 8 September 1956 — Data Basis: Flight Test

Figure A—86

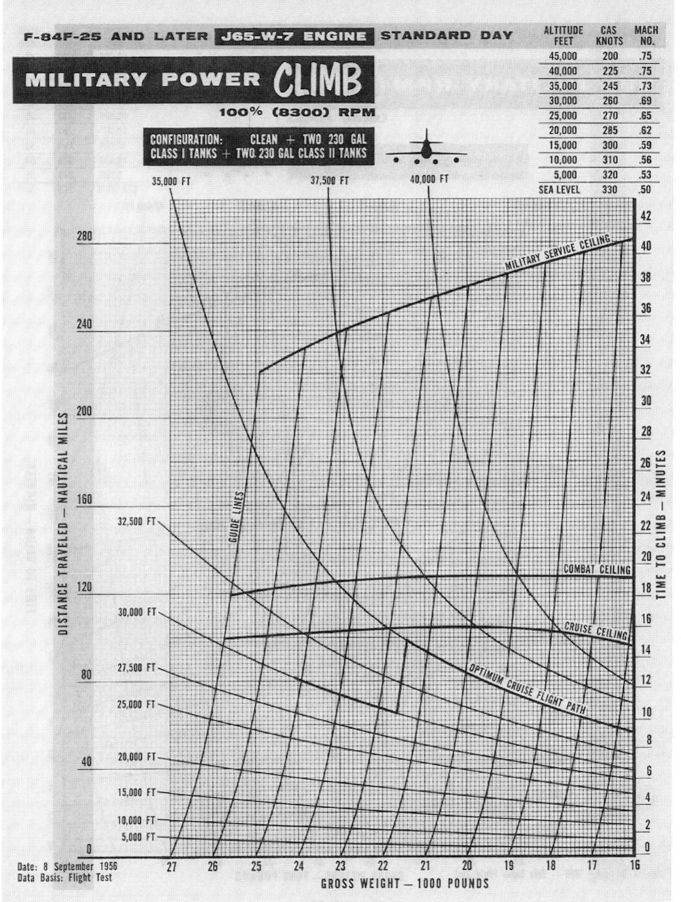

F-84F-25 AND LATER **J65-W-7 ENGINE** STANDARD DAY

MILITARY POWER CLIMB

100% (8300) RPM

CONFIGURATION: CLEAN + TWO 230 GAL CLASS I TANKS + TWO 230 GAL CLASS II TANKS

ALTITUDE FEET	CAS KNOTS	MACH NO.
45,000	200	.75
40,000	225	.75
35,000	245	.73
30,000	260	.69
25,000	270	.65
20,000	285	.62
15,000	300	.59
10,000	310	.56
5,000	320	.53
SEA LEVEL	330	.50

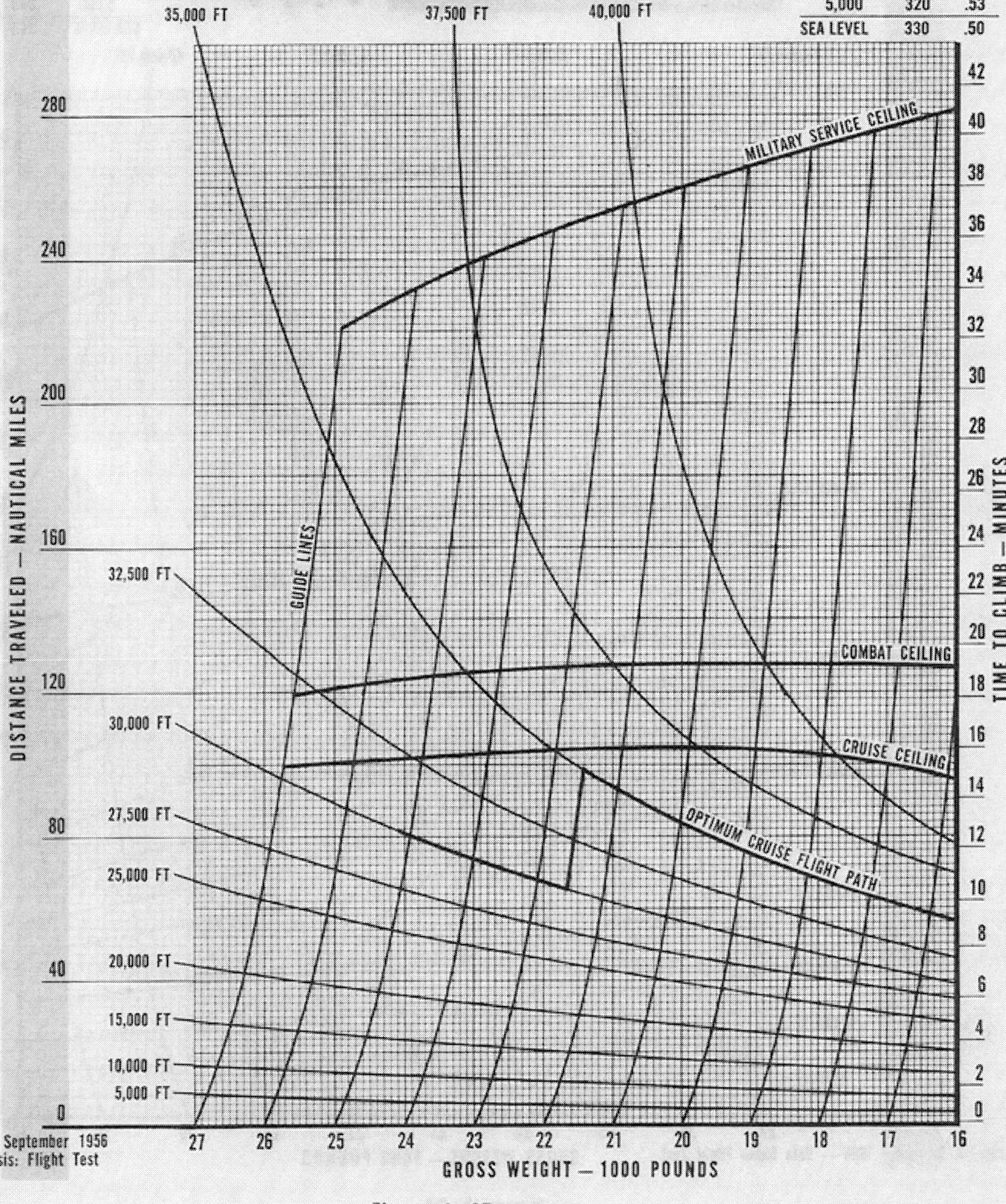

Date: 8 September 1956
Data Basis: Flight Test

Figure A–87

F-84F-25 AND LATER **J65-W-7 ENGINE** STANDARD DAY

98% MILITARY RPM CLIMB

(8100 RPM)

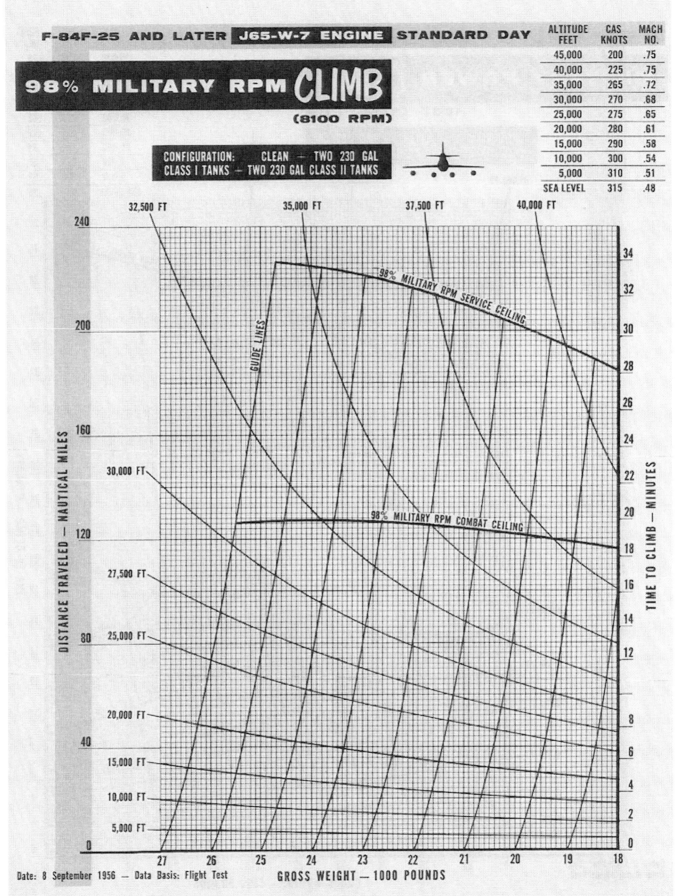

ALTITUDE FEET	CAS KNOTS	MACH NO.
45,000	200	.75
40,000	225	.75
35,000	265	.72
30,000	270	.68
25,000	275	.65
20,000	280	.61
15,000	290	.58
10,000	300	.54
5,000	310	.51
SEA LEVEL	315	.48

CONFIGURATION: CLEAN — TWO 230 GAL CLASS I TANKS — TWO 230 GAL CLASS II TANKS

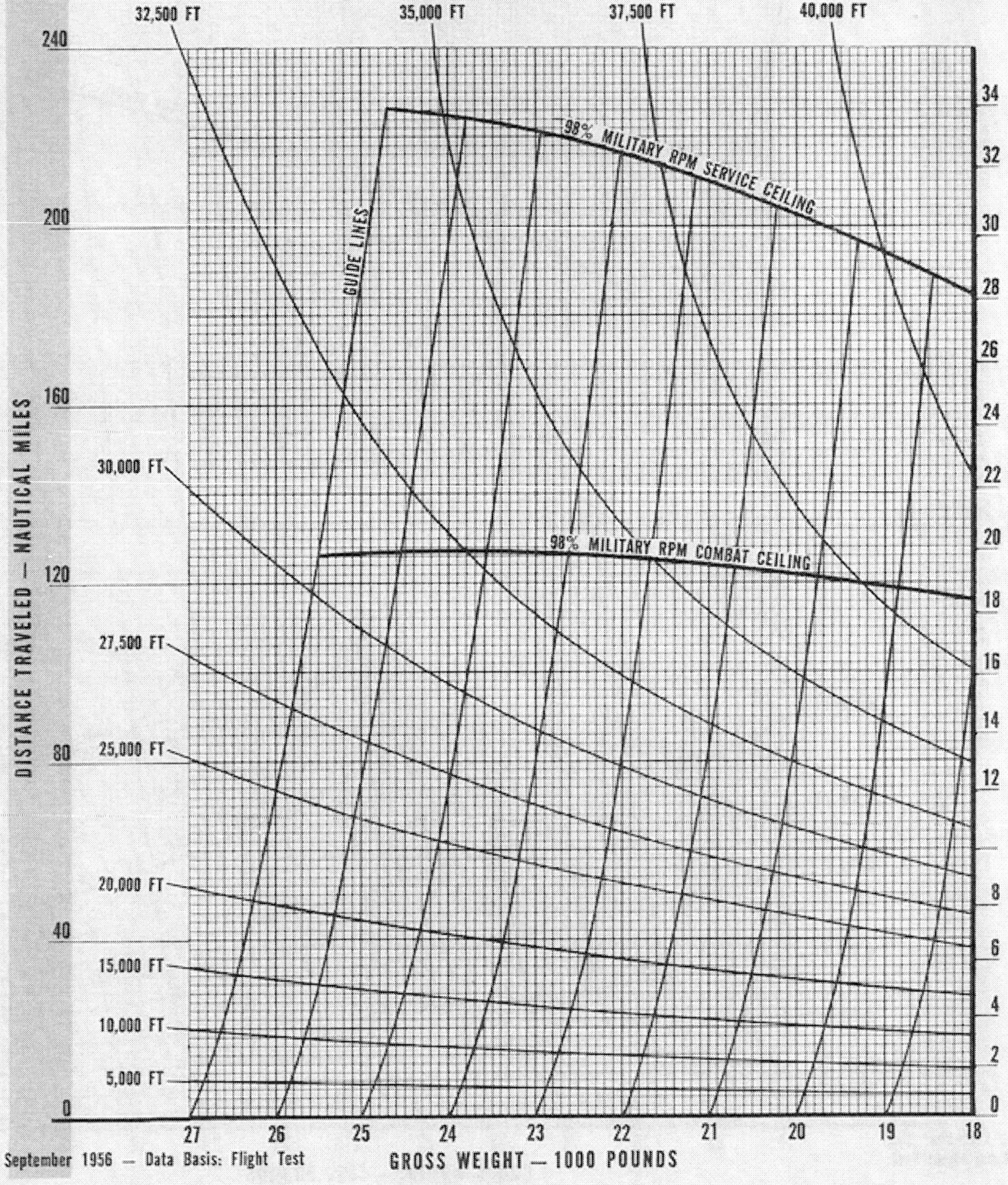

Date: 8 September 1956 — Data Basis: Flight Test

Figure A—88

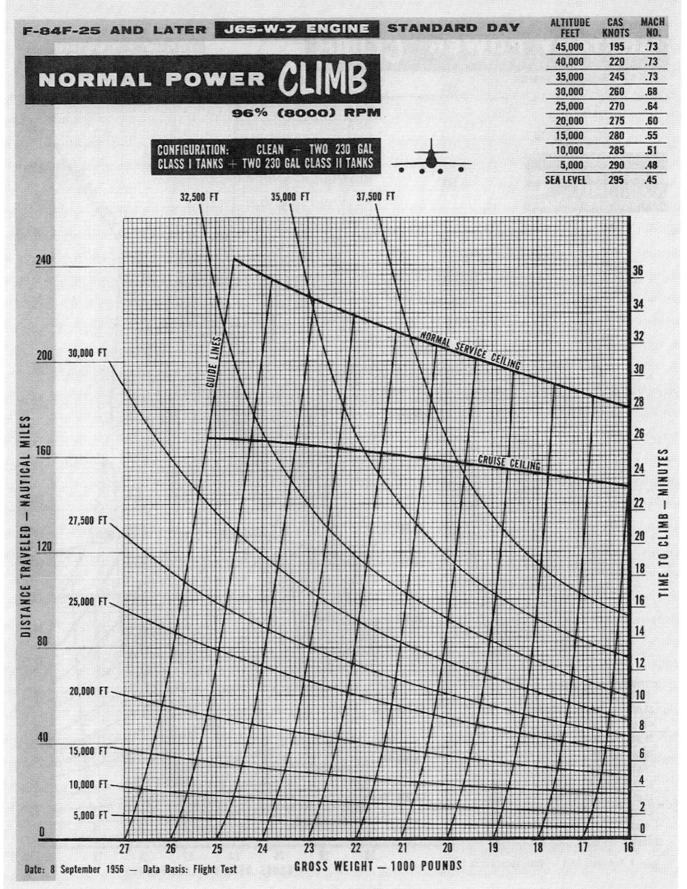

F-84F-25 AND LATER J65-W-7 ENGINE STANDARD DAY

NORMAL POWER CLIMB

96% (8000) RPM

ALTITUDE FEET	CAS KNOTS	MACH NO.
45,000	195	.73
40,000	220	.73
35,000	245	.73
30,000	260	.68
25,000	270	.64
20,000	275	.60
15,000	280	.55
10,000	285	.51
5,000	290	.48
SEA LEVEL	295	.45

CONFIGURATION: CLEAN — TWO 230 GAL CLASS I TANKS + TWO 230 GAL CLASS II TANKS

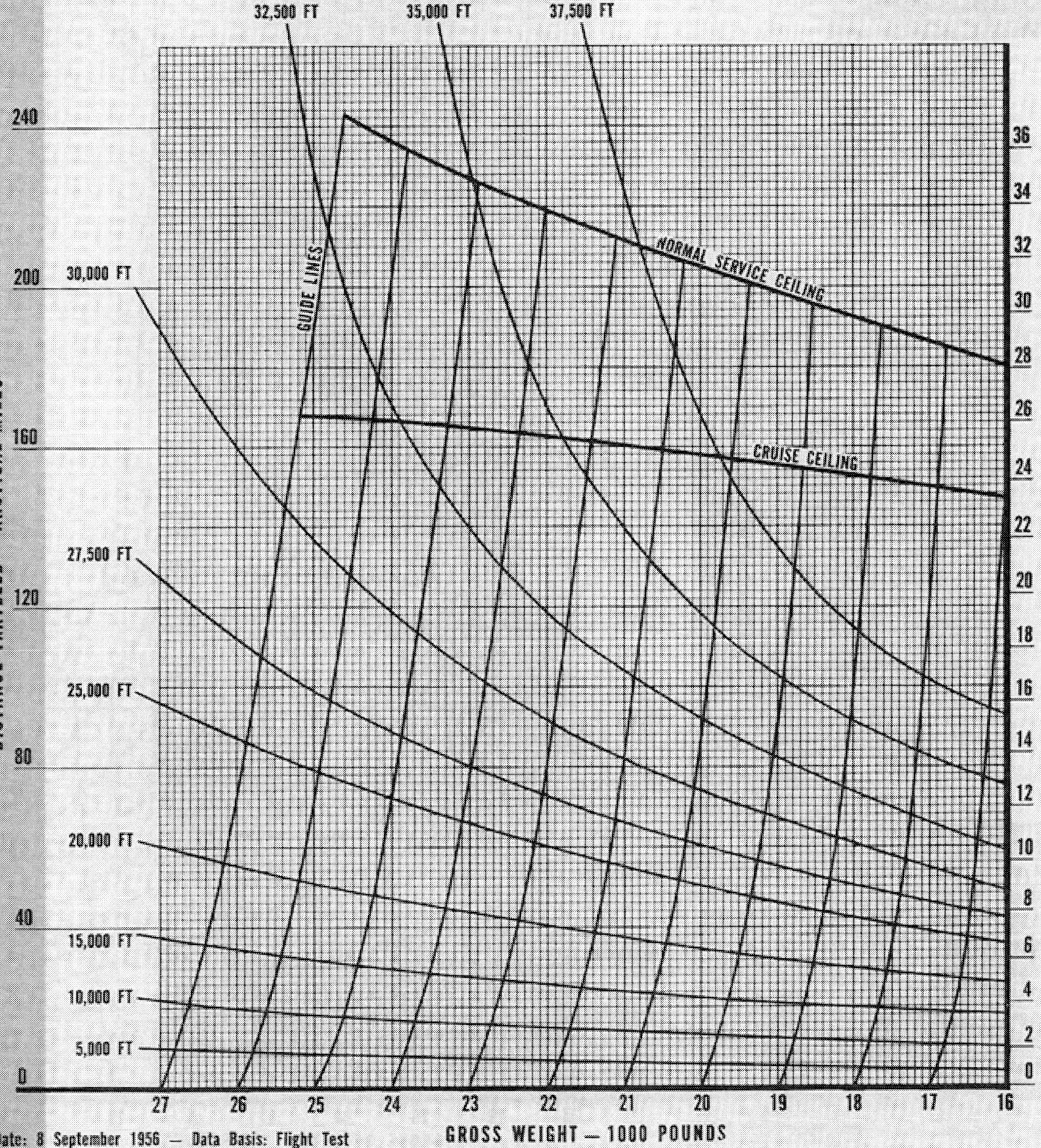

Date: 8 September 1956 — Data Basis: Flight Test

Figure A–89

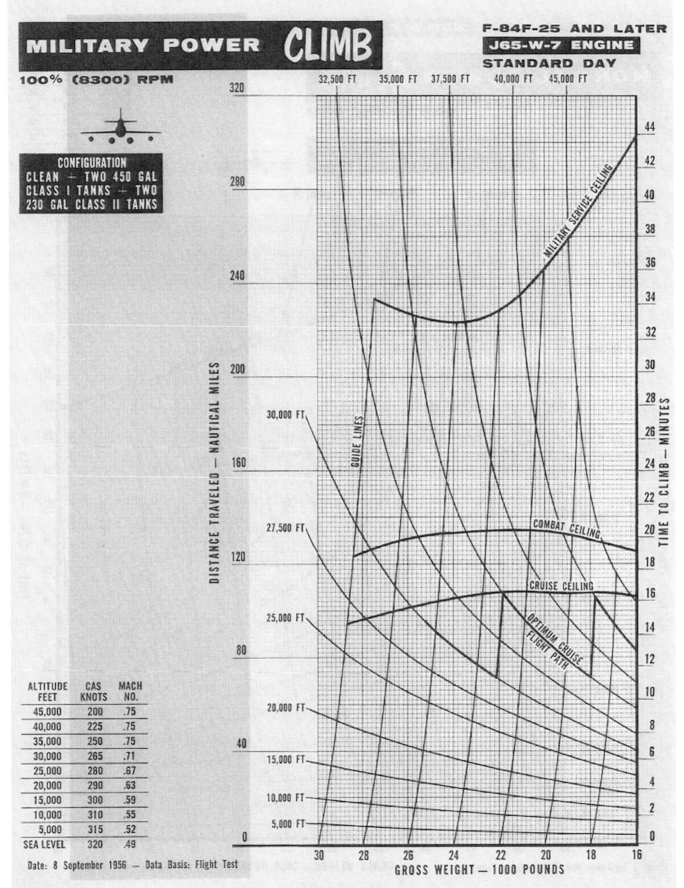

MILITARY POWER CLIMB

F-84F-25 AND LATER
J65-W-7 ENGINE
STANDARD DAY

100% (8300) RPM

CONFIGURATION
CLEAN + TWO 450 GAL CLASS I TANKS + TWO 230 GAL CLASS II TANKS

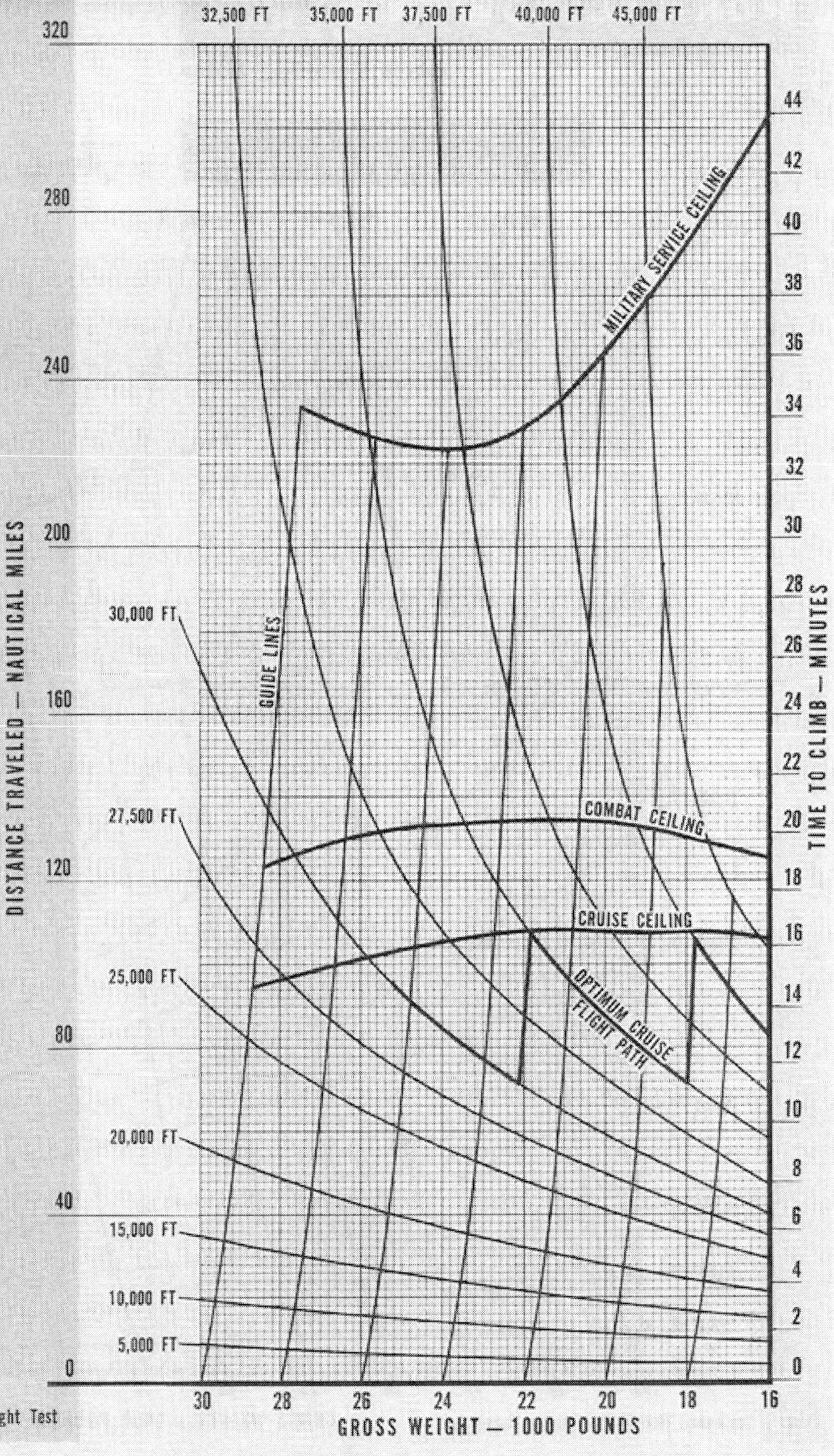

ALTITUDE FEET	CAS KNOTS	MACH NO.
45,000	200	.75
40,000	225	.75
35,000	250	.75
30,000	265	.71
25,000	280	.67
20,000	290	.63
15,000	300	.59
10,000	310	.55
5,000	315	.52
SEA LEVEL	320	.49

Date: 8 September 1956 — Data Basis: Flight Test

Figure A—90

98% MILITARY RPM CLIMB

(8100 RPM)

84F-25 AND LATER
J65-W-7 ENGINE
STANDARD DAY

CONFIGURATION
CLEAN + TWO 450 GAL CLASS I TANKS + TWO 230 GAL CLASS II TANKS

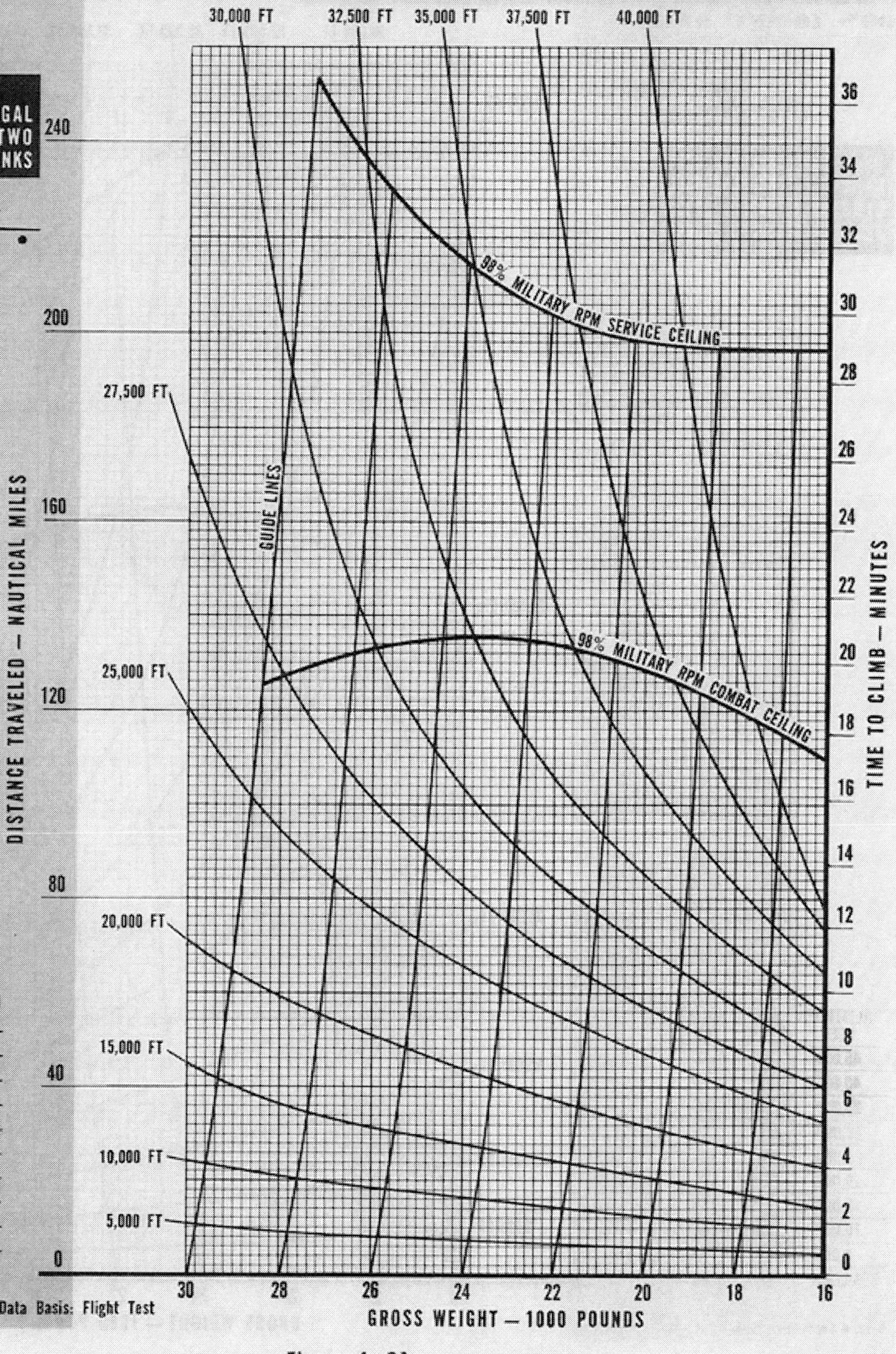

ALTITUDE FEET	CAS KNOTS	MACH NO.
45,000	200	.75
40,000	225	.75
35,000	250	.75
30,000	265	.71
25,000	275	.67
20,000	285	.62
15,000	295	.59
10,000	305	.55
5,000	310	.52
SEA LEVEL	315	.48

Date: 8 September 1956 — Data Basis: Flight Test

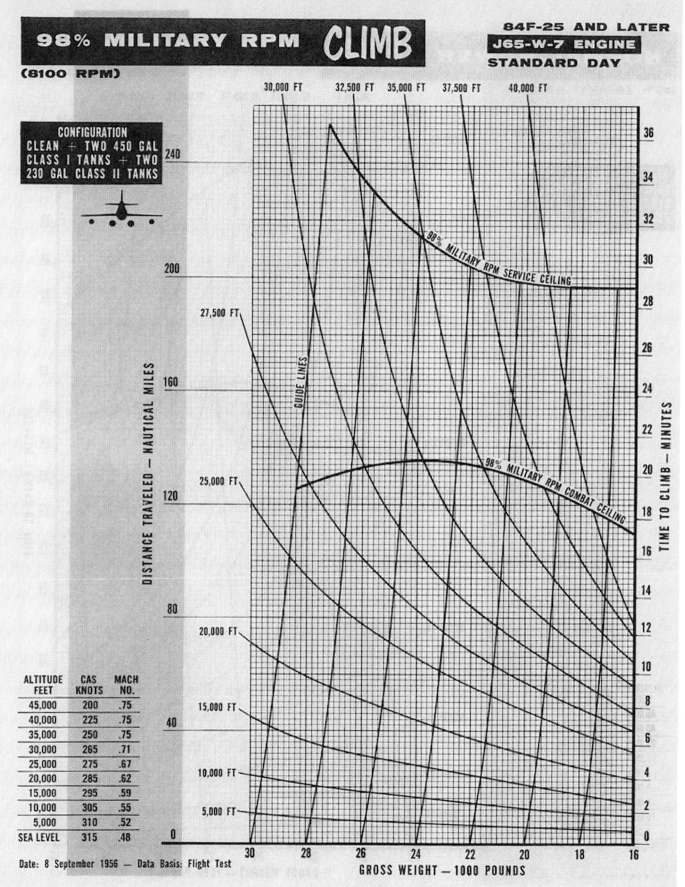

Figure A—91

NORMAL POWER CLIMB

F-84F-25 AND LATER
J65-W-7 ENGINE
STANDARD DAY

96% (8000) RPM

CONFIGURATION
CLEAN + TWO 450 GAL CLASS I TANKS + TWO 230 GAL CLASS II TANKS

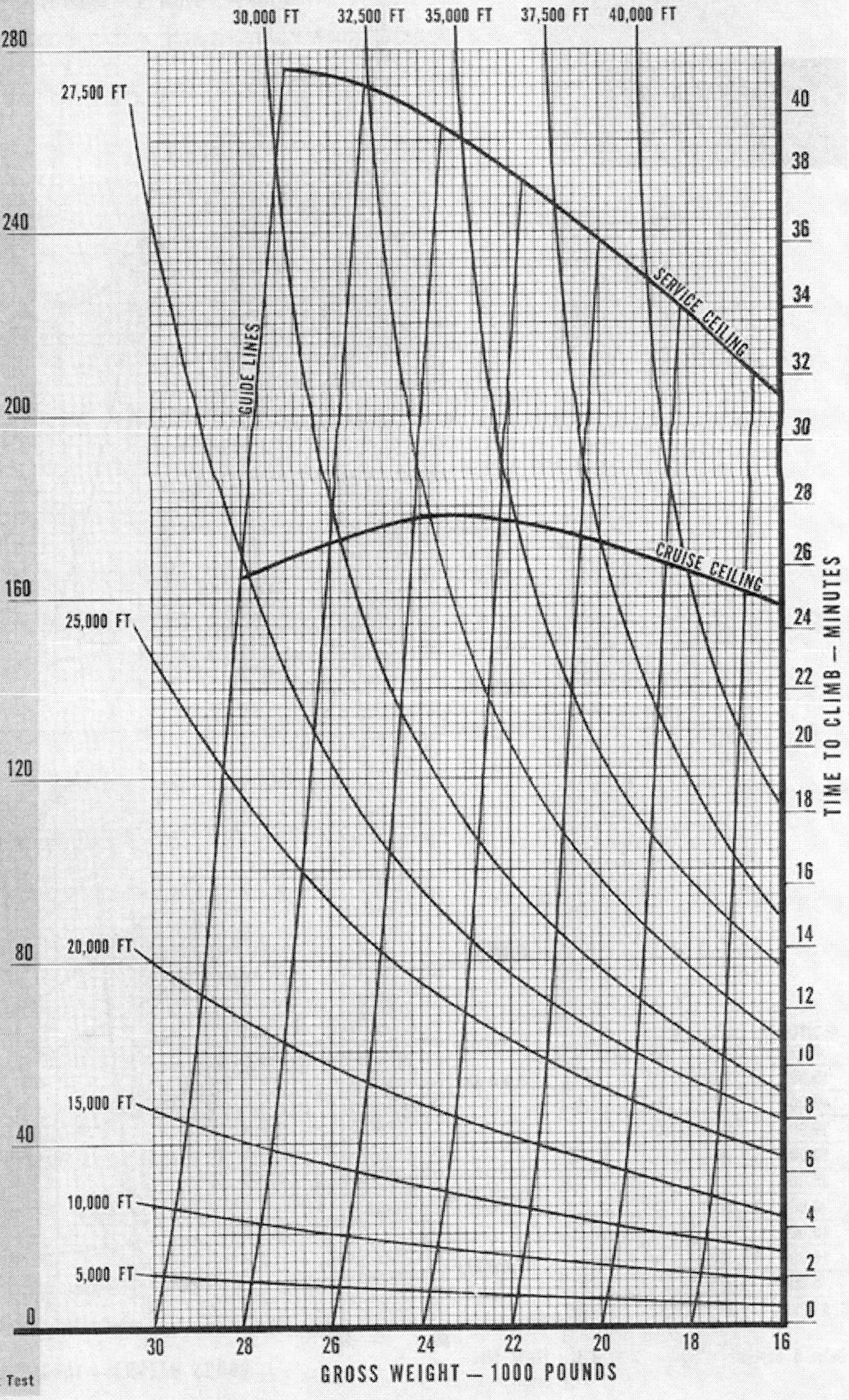

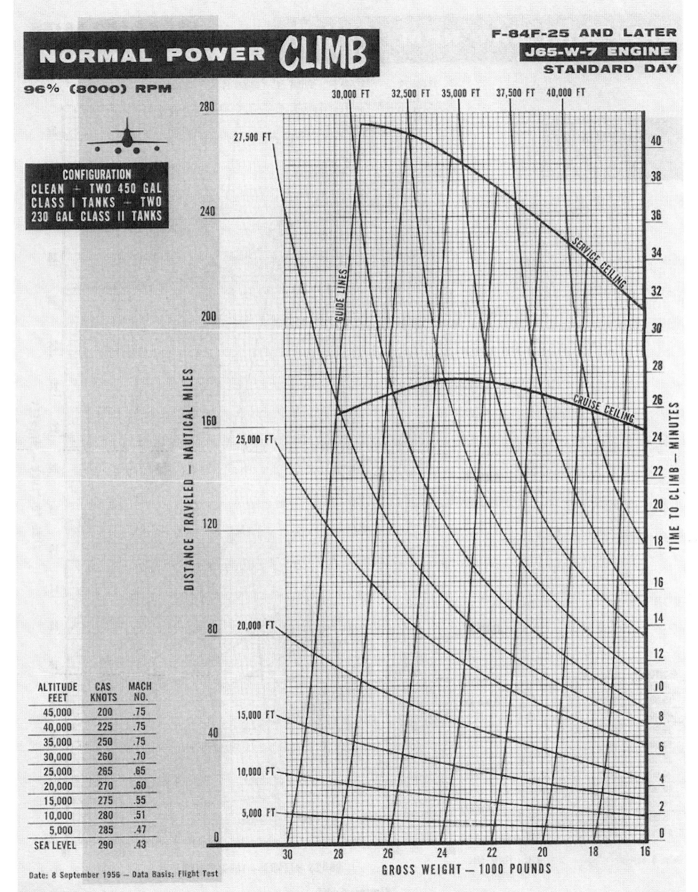

ALTITUDE FEET	CAS KNOTS	MACH NO.
45,000	200	.75
40,000	225	.75
35,000	250	.75
30,000	260	.70
25,000	265	.65
20,000	270	.60
15,000	275	.55
10,000	280	.51
5,000	285	.47
SEA LEVEL	290	.43

Date: 8 September 1956 — Data Basis: Flight Test

Figure A—92

F-84F-25 AND LATER **J65-W-7 ENGINE** STANDARD DAY

MILITARY POWER CLIMB

100% (8300) RPM

ALTITUDE FEET	CAS KNOTS	MACH NO.
40,000	210	.70
35,000	225	.67
30,000	235	.64
25,000	250	.60
20,000	260	.57
15,000	270	.54
10,000	280	.51
5,000	290	.48
SEA LEVEL	300	.45

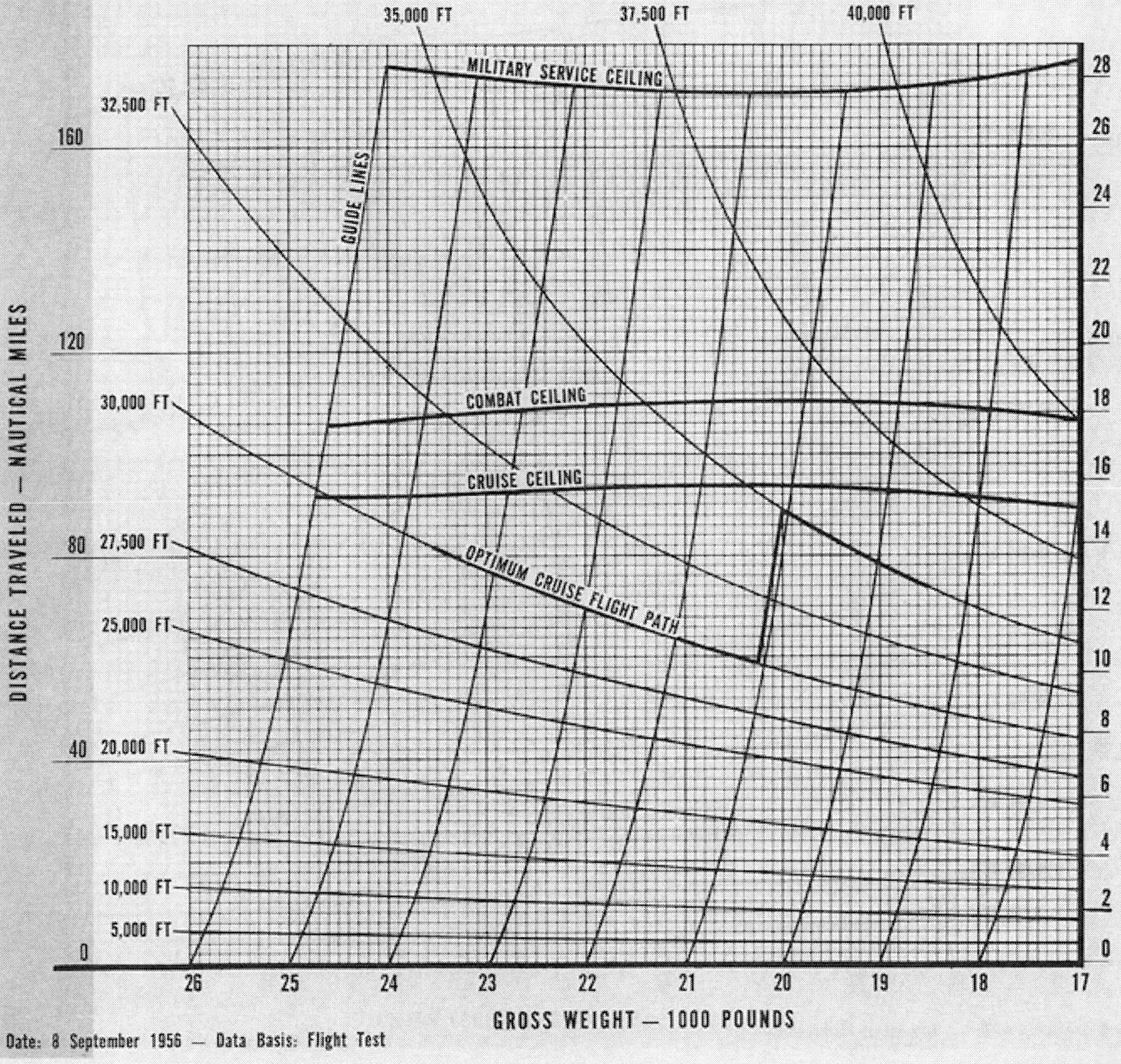

Date: 8 September 1956 — Data Basis: Flight Test

Figure A–93

F-84F-25 AND LATER **J65-W-7 ENGINE** STANDARD DAY

98% MILITARY RPM CLIMB

(8100 RPM)

ALTITUDE FEET	CAS KNOTS	MACH NO.
40,000	215	.72
35,000	230	.69
30,000	245	.65
25,000	255	.62
20,000	265	.58
15,000	275	.55
10,000	285	.51
5,000	290	.48
SEA LEVEL	295	.45

CONFIGURATION: CLEAN + TWO 230 GAL CLASS I TANKS + TWO 1000 LB BOMBS + FOUR HVARS

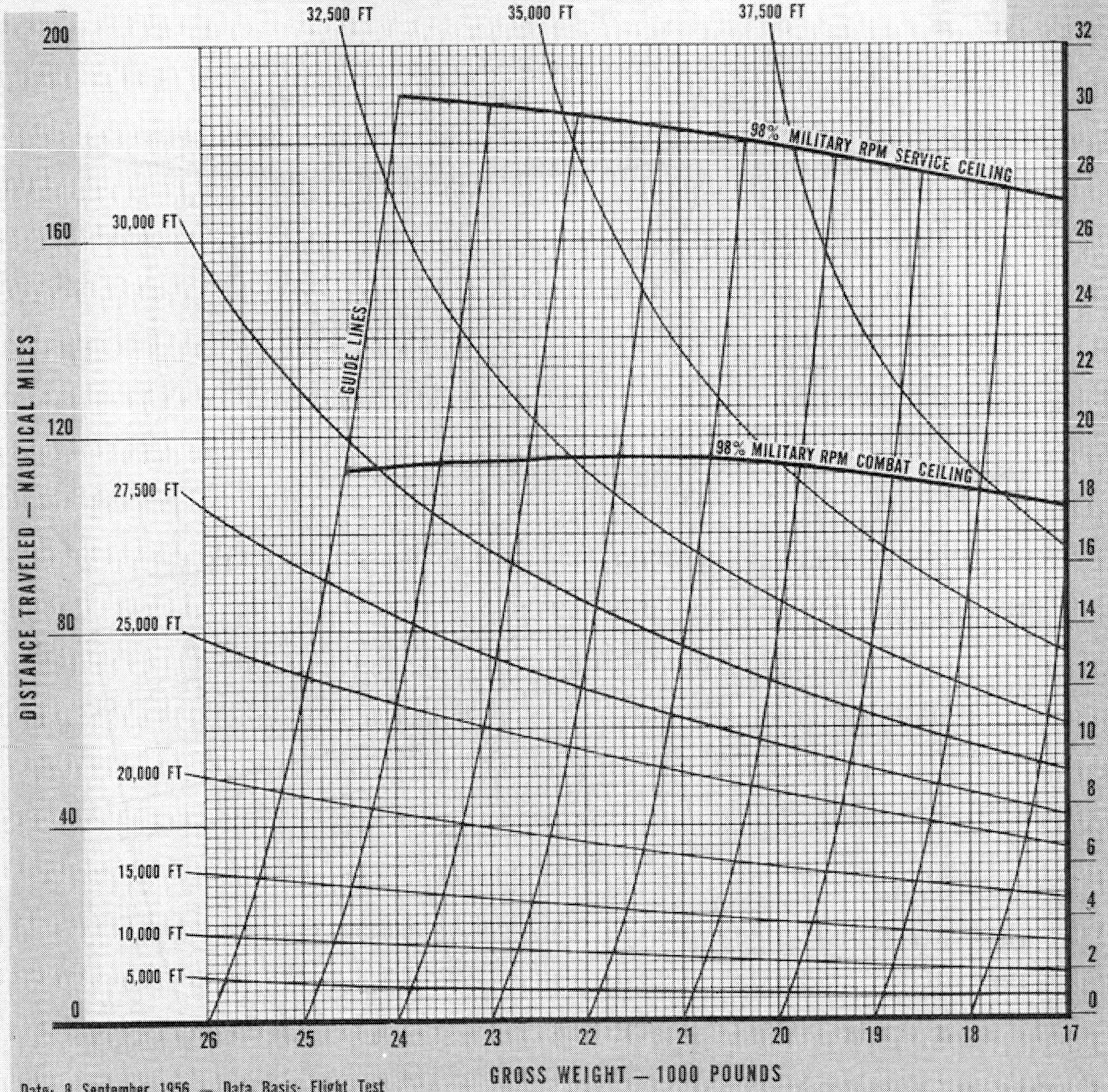

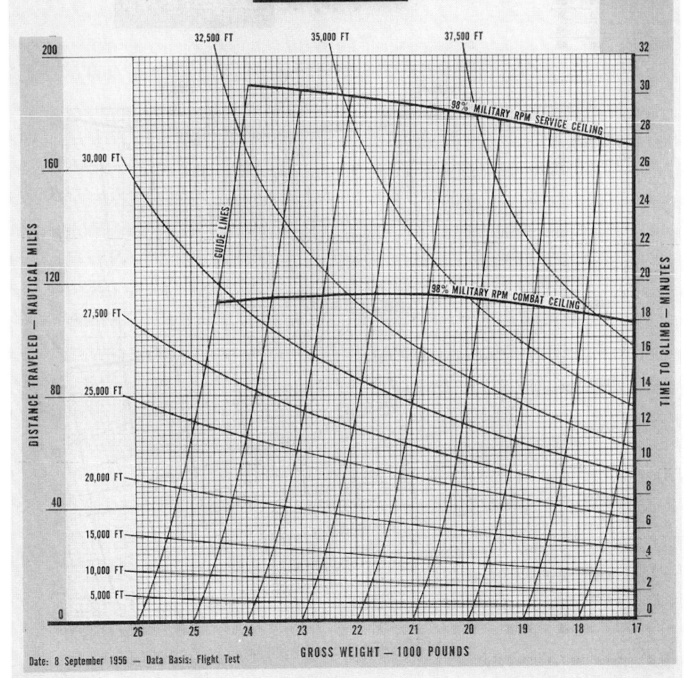

Date: 8 September 1956 — Data Basis: Flight Test

Figure A—94

F-84F-25 AND LATER J65-W-7 ENGINE STANDARD DAY

NORMAL POWER CLIMB

96% (8000) RPM

ALTITUDE FEET	CAS KNOTS	MACH NO.
40,000	200	.69
35,000	220	.66
30,000	230	.62
25,000	240	.58
20,000	245	.54
15,000	255	.51
10,000	265	.48
5,000	270	.45
SEA LEVEL	280	.42

CONFIGURATION: CLEAN + TWO 230 GAL CLASS I TANKS + TWO 1000 LB BOMBS + FOUR HVARS

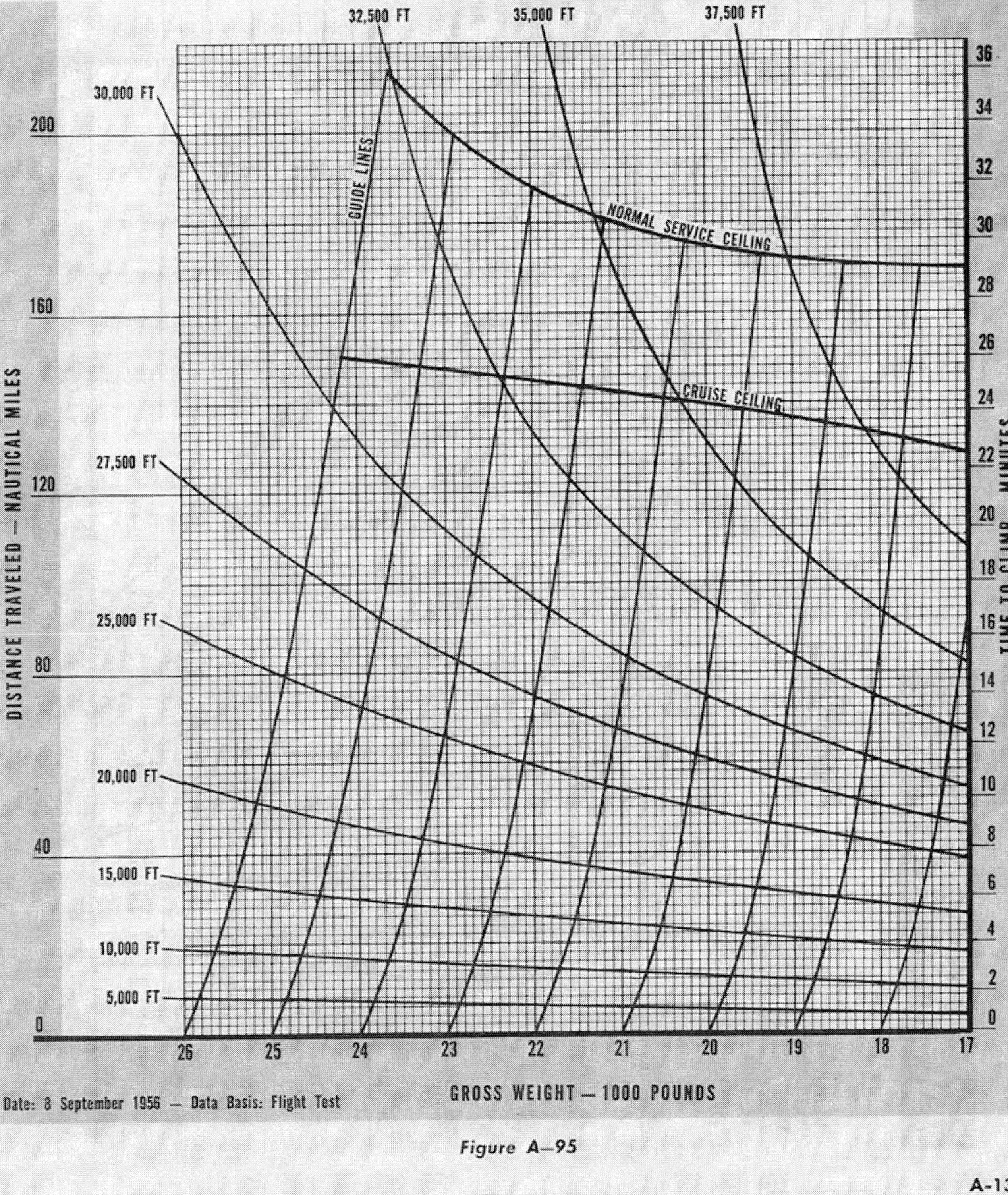

Date: 8 September 1956 — Data Basis: Flight Test

Figure A—95

F-84F-25 AND LATER J65-W-7 ENGINE STANDARD DAY

MISSION PROFILE

CLEAN AIRPLANE
GROSS WEIGHT: 18,700—14,936 LB

ALTITUDE FEET	MACH NO.	APPROXIMATE CAS-KNOTS	TAS-KNOTS	FUEL-LB/HR	% MIL RPM
CRUISE-CLIMB	.85	——	485	1600-1400	91
45,000	.85	230	485	1600	94
40,000	.83	255	480	1550	89
35,000	.81	275	465	1650	86
30,000	.75	280	440	1700	85
25,000	.69	290	415	1800	83
20,000	.65	295	400	1900	83
15,000	.61	310	380	2100	82
10,000	.56	320	370	2300	82
5,000	.55	335	360	2600	81
SEA LEVEL	.53	350	350	2950	81

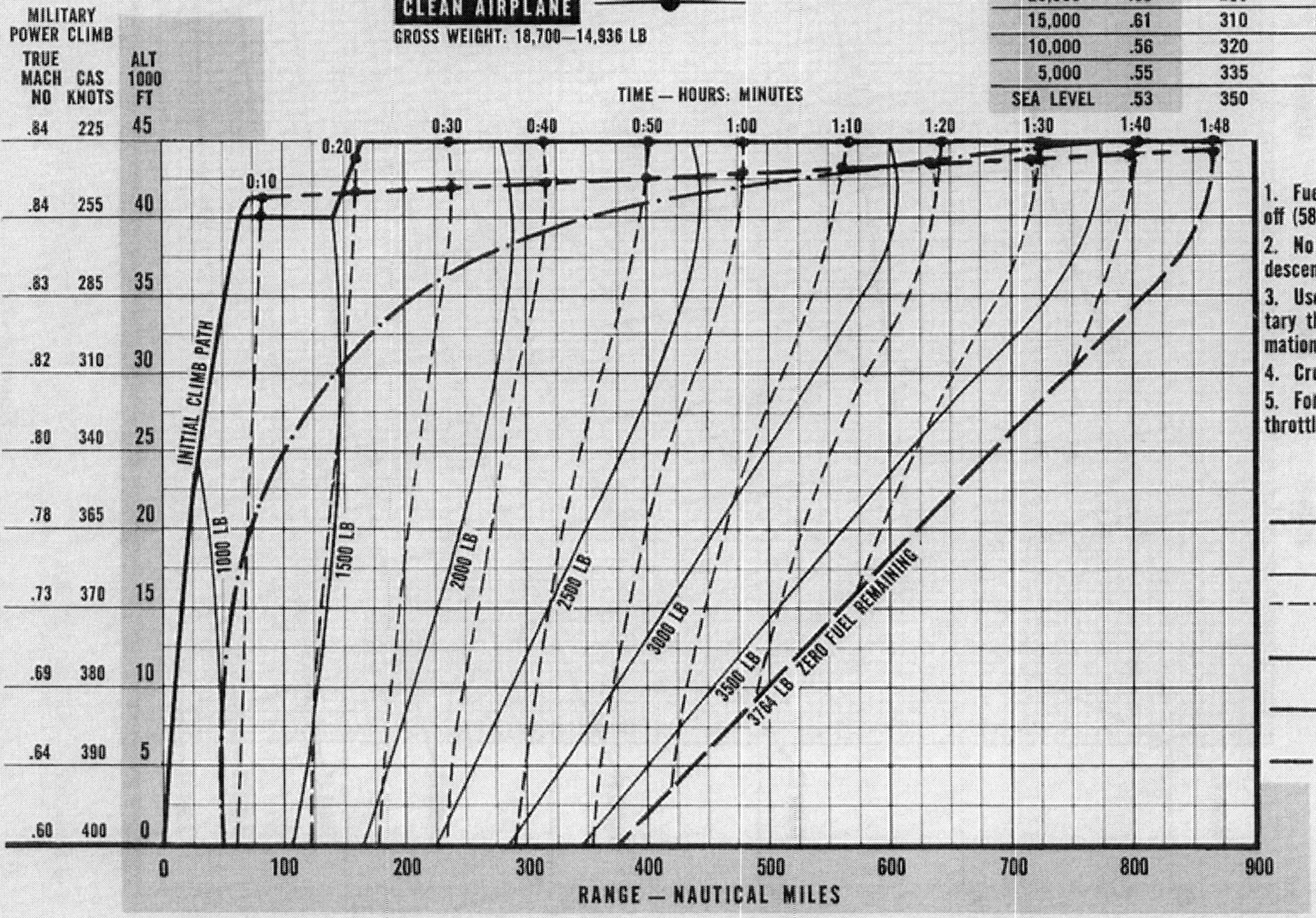

REMARKS

1. Fuel allowance for start, taxi and take-off (587 lb) included.
2. No allowance or reserve for loiter, descent or landing.
3. Use military thrust for climb. (See military thrust climb chart for detailed information.)
4. Cruise at recommended Mach No.
5. For cruise climb, maintain a constant throttle position after initial setting.

LEGEND

- 3764 lb fuel consumed — zero fuel remaining
- Fuel consumed
- Time (start, taxi, and take-off not included)
- Line of best range at constant altitude
- Optimum cruise flight path —step climb
- Optimum cruise flight path —cruise climb

Date: 8 September 1956
Data Basis: Flight Test

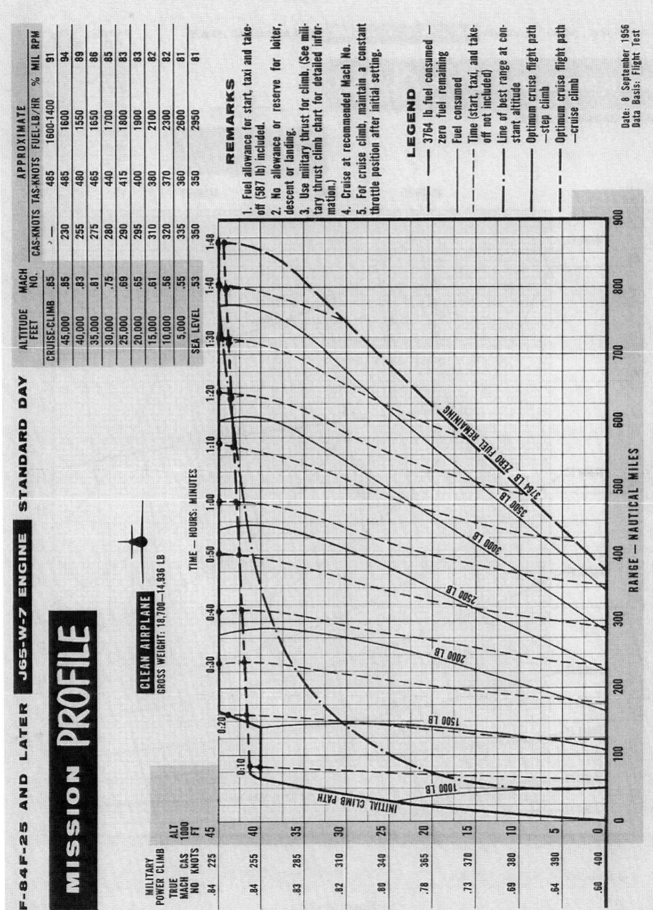

Figure A—96

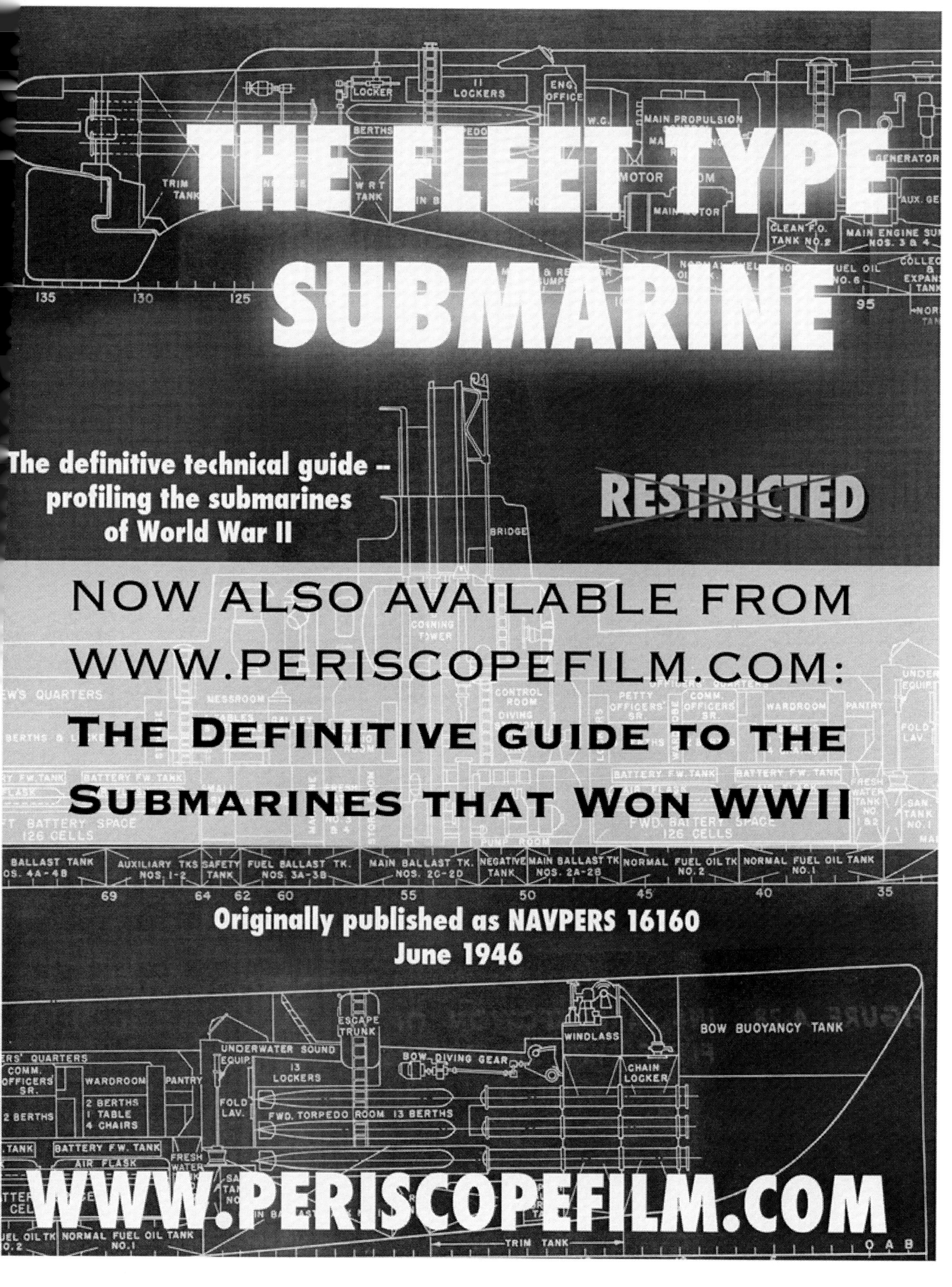
THE FLEET TYPE SUBMARINE
The definitive technical guide -- profiling the submarines of World War II
RESTRICTED
NOW ALSO AVAILABLE FROM WWW.PERISCOPEFILM.COM:
THE DEFINITIVE GUIDE TO THE SUBMARINES THAT WON WWII
Originally published as NAVPERS 16160
June 1946
WWW.PERISCOPEFILM.COM

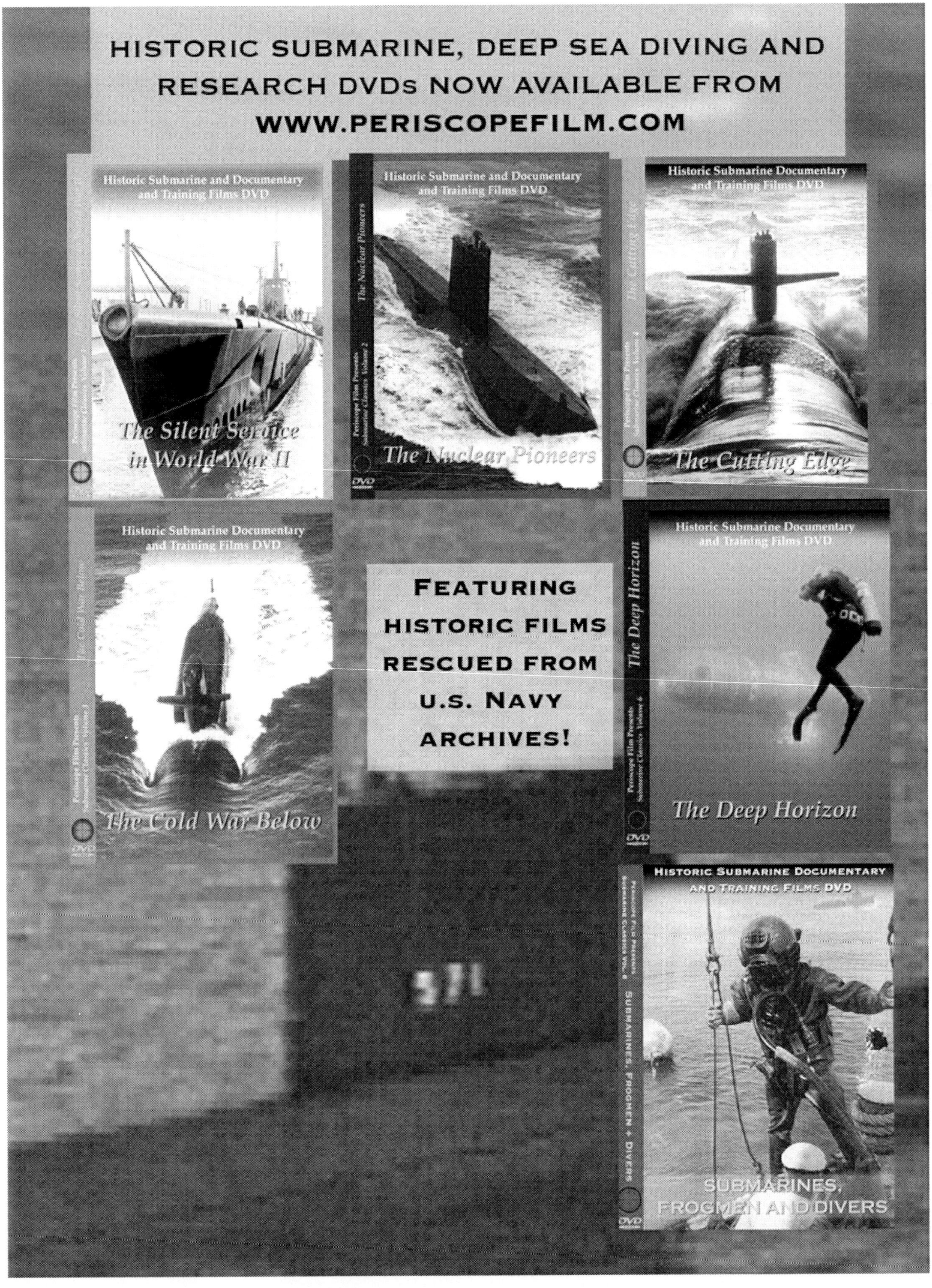

HISTORIC SUBMARINE, DEEP SEA DIVING AND
RESEARCH DVDs NOW AVAILABLE FROM
WWW.PERISCOPEFILM.COM
Historic Submarine and Documentary and Training Films DVD
The Silent Service in World War II
Historic Submarine and Documentary and Training Films DVD
The Nuclear Pioneers
Historic Submarine Documentary and Training Films DVD
The Cutting Edge
Historic Submarine Documentary and Training Films DVD
The Cold War Below
FEATURING
HISTORIC FILMS
RESCUED FROM
U.S. NAVY
ARCHIVES!
Historic Submarine Documentary and Training Films DVD
The Deep Horizon
HISTORIC SUBMARINE DOCUMENTARY AND TRAINING FILMS DVD
SUBMARINES, FROGMEN AND DIVERS

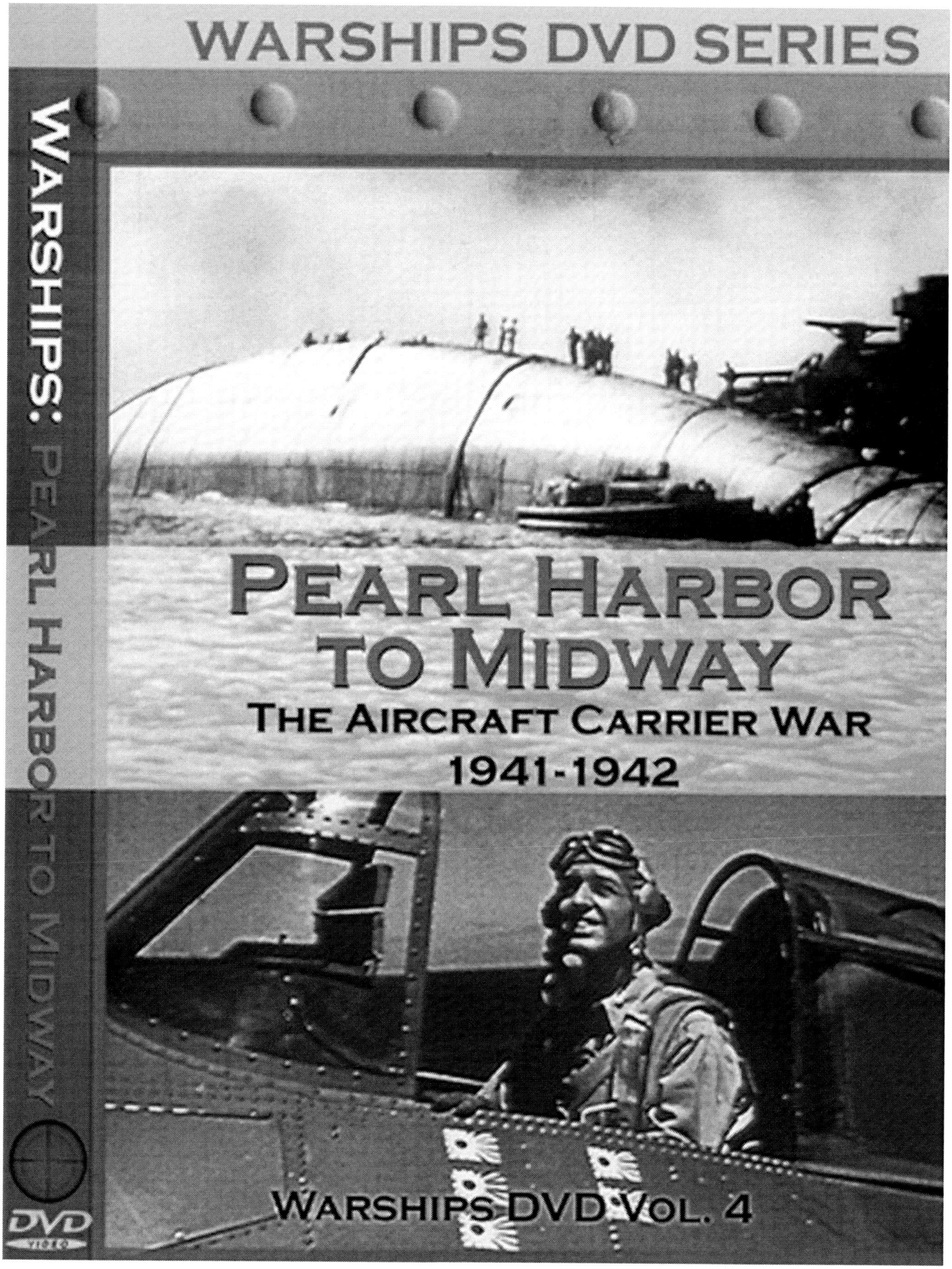

NOW AVAILABLE!

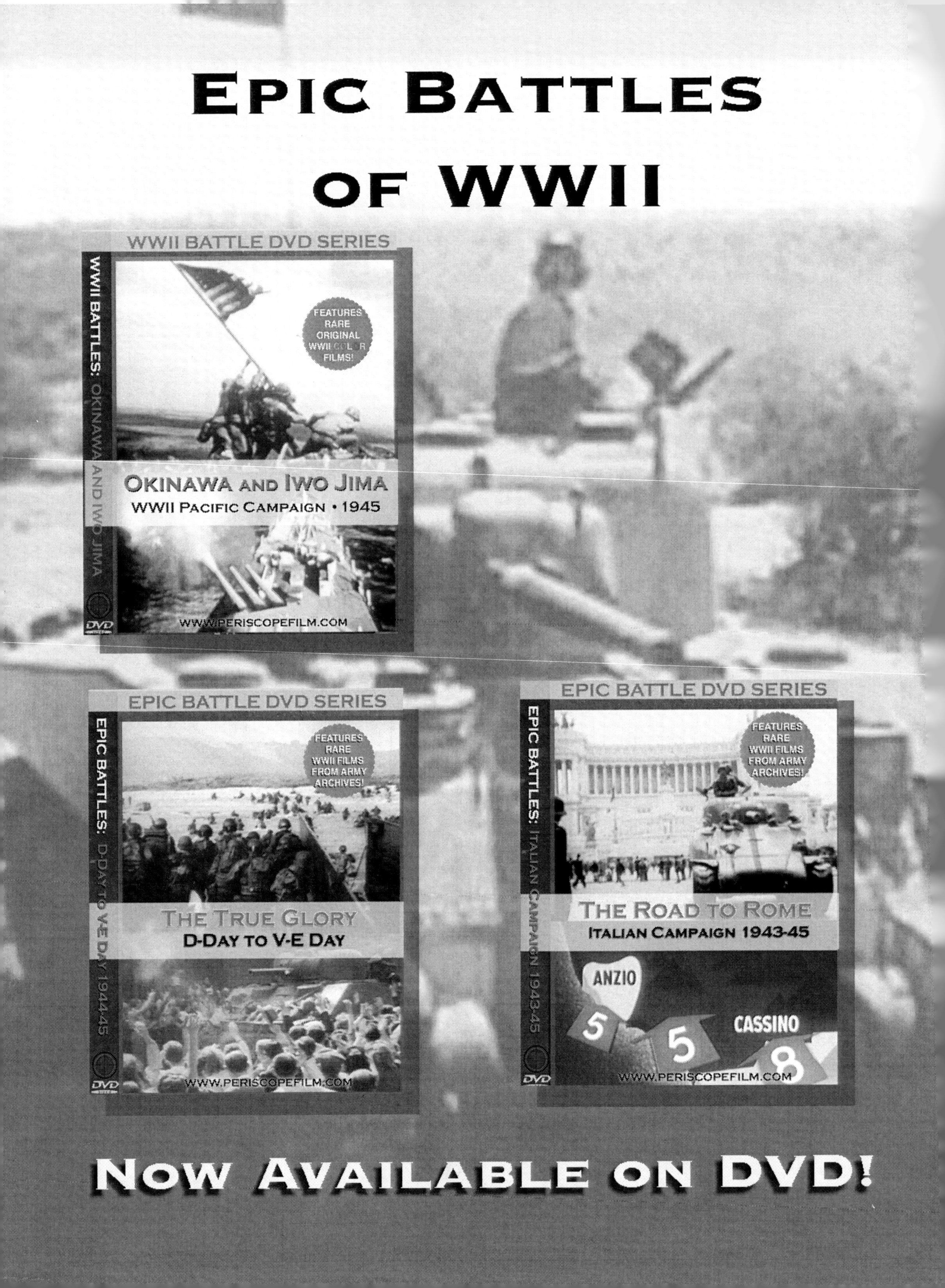
EPIC BATTLES
OF WWII
WWII BATTLE DVD SERIES
WWII BATTLES: OKINAWA AND IWO JIMA
FEATURES RARE ORIGINAL WWII COLOR FILMS!
OKINAWA AND IWO JIMA
WWII PACIFIC CAMPAIGN • 1945
WWW.PERISCOPEFILM.COM
EPIC BATTLE DVD SERIES
EPIC BATTLES: D-DAY TO V-E DAY 1944-45
FEATURES RARE WWII FILMS FROM ARMY ARCHIVES!
THE TRUE GLORY
D-DAY TO V-E DAY
WWW.PERISCOPEFILM.COM
EPIC BATTLE DVD SERIES
EPIC BATTLES: ITALIAN CAMPAIGN 1943-45
FEATURES RARE WWII FILMS FROM ARMY ARCHIVES!
THE ROAD TO ROME
ITALIAN CAMPAIGN 1943-45
ANZIO
CASSINO
WWW.PERISCOPEFILM.COM
NOW AVAILABLE ON DVD!